P9-DTR-606

Virginia Morell

A TOUCHSTONE BOOK

Published by Simon & Schuster

New York London Toronto

Sydney Tokyo Singapore

Ancestral Passions

The Leakey Family and the Quest for Humankind's Beginnings

TOUCHSTONE
Rockefeller Center
1230 Avenue of the Americas
New York, NY 10020

First Touchstone Edition 1996

TOUCHSTONE and colophon are registered trademarks
of Simon & Schuster Inc.

DESIGNED BY BARBARA M. BACHMAN

Manufactured in the United States of America

1 3 5 7 9 10 8 6 4 2

Library of Congress Cataloging-in-Publication Data
Morell, Virginia.
Ancestral passions: the Leakey family and the quest for humankind's
beginnings / Virginia Morell.
p. cm.
Includes index.
1. Leakey, L. S. B. (Louis Seymour Bazett), 1903–1972. 2. Leakey, Mary D.
(Mary Douglas), 1913– . 3. Leakey, Richard E. 4. Physical anthropologists—
Tanzania—Olduvai Gorge—Biography. 5. Physical anthropologists—Great
Britain—Biography. 6. Fossil man—Tanzania—Olduvai Gorge.
7. Excavations (Archaeology)—Tanzania—Olduvai Gorge. 8. Olduvai Gorge
(Tanzania)—Antiquities. I. Title.
GN50.6.L43M67 1995
573'.092'2—dc20
[B] 95-14306

ISBN 0-684-80192-2
0-684-82470-1 (pbk)

Acknowledgments

This is the first full biography of Louis, Mary, and Richard Leakey and, although it is not "authorized," it could not have been written without the family's great generosity, patience, and kind help. They kindly gave me unpublished letters, journals, diaries, and photographs and interrupted their busy lives on numerous occasions to talk to me about themselves, their science, their family, and colleagues.

In particular, I thank Richard Leakey, who was generous beyond measure. He invited me to join his 1984 and 1987 West Turkana expeditions and gave me unlimited access to the Leakey family archives at the National Museums of Kenya, as well as to his personal files. He willingly answered my numerous (and, I'm sure, at times seemingly endless) questions and kindly introduced me to many of his colleagues and friends. I have been a guest in his camp and in his home, and if I have managed to capture some small part of the Leakey family's complexity, it is because of Richard's unwavering and wholehearted support. Meave Leakey was equally kind and helpful, opening her and Richard's home to me, and gently encouraging me to ask the sometimes difficult question.

My deepest thanks go also to Mary Leakey, who granted me numerous interviews and kindly gave me her personal letters. She welcomed my husband and me to her Olduvai Gorge home, and good-naturedly accompanied us on a tour of the gorge's fossil sites as well as the hominid footprint site of Laetoli. She also wrote insightful and pithy replies to my many letters—although I know (because she told me so) that she often groaned when she saw one of my letters in her mail.

I was also warmly welcomed as a guest in the homes of Jonathan and Janet Leakey, and Philip and Valerie Leakey. They shared their memories about growing up in Kenya and thoughtfully answered my many questions.

The British side of the Leakey family was as hospitable and generous as the Kenyan. I owe a deep debt of gratitude to Frida, Louis's first wife, whose delightful memories of Louis and his first East African expeditions provided the core material for my book's opening chapters. Frida was in her eighties when I met her; she picked me up at the Cambridge train station in a sports car with the top down—displaying the same spunk that I'm sure had attracted Louis to her. Sadly, she passed away shortly before I completed my book.

Frida and Louis's two children, Priscilla Davies and Colin, were kind and helpful from the moment we met, warmly welcoming me into their homes and sharing their memories with the same enthusiasm and good humor I had found among the Kenyan Leakeys. Colin even produced a book one evening about the Texan cowboy side of the Leakey family—a branch that I have space only to mention in passing here.

I was extremely fortunate to meet Louis's sole surviving sibling, Julia Barham; his brother's wife, Beryl Leakey; and Frida's sister, Barbara Waterfield. All three women shared their wonderful memories of Louis with me; all three have also since passed away.

Although the Leakey family was cooperative, they exercised no control over the writing of the book and never asked to read the manuscript prior to publication. All interpretations and judgments in these pages are entirely my own.

I can never adequately thank the many people who talked to me about the Leakeys—in person, by telephone, and by mail; who gave me letters, photographs, diaries, and journals; who offered me hospitality, read my manuscript, and kept up my courage. My deepest gratitude to all of these individuals, some of whom have died in the intervening years. For simplicity's sake, the list is alphabetical: Paul Abell, Issa Aggundey, Juliet Ament, Sharon Anderson, Peter Andrews, Margaret Avery, Antonia Bagshawe, Anna K. Behrensmeyer, Frances Bekafigo, Sir Michael Blundell, Bob Brain, Rod Brindamour, Frank Brown, Jean Brown, Fred E. Budinger Jr., Frances Burton, Karl Butzer, Bob and Heather Campbell, Judy Castel, J. Desmond Clark, Ron Clarke, Basil Cooke, Yves Coppens, Garniss Curtis, Glyn Daniel, Michael and Micky Day, Irven DeVore, Gabrielle Dolphin, Jill Donisthorpe, Bob Drewes, Kathy Eldon, Mary Catherine Fagg, Frank Fitch, Dian Fossey, Sir Vivian Fuchs, Biruté Galdikas, Catherine Garnett, Alan Gentry, Diane Gifford-Gonzalez, Jan Gillett, Betty Goerke, Jane Goodall, Nancy Gonzalez, Richard Michael Gramly, Bill Graves, Gilbert Grosvenor, Mary Ann Harrell, Jack Harris, John Harris, Richard Hay, Cathryn Hosea Hilker, Andrew Hill, Ralph Holloway, Sarah Howard, F. Clark and Betty Howell, Elspeth Huxley, Lady Juliette Huxley, Glynn and Barbara Isaac, Toni Kay Jackman, Alan Jacobs, Lou Jacobs, Penelope Jenkin, Peter Jones, Jon Kalb, John and Joan Karmali, Kathryn (Dottie) Kasper, Sir Peter Kent, Kamoya Kimeu, Elisabeth and Barbara Kitson, Maxine Kleindienst, Leo Laporte, Joachim Lentz, Roger Lewin, Dora MacInnes, Wambua Mangao, Anthony Marshall, Graham Massey, Ernst Mayr, Daniel McCarthy, Christine and Ian McRae, Harry Merrick, Elizabeth Meyerhoff, M. E. Morbeck, Amini Mturi, Heselon Mukiri, Joseph Mungai, Ned Munger, Mongela Muoka, Mutevu Musomba, Joseph Mutaba, John and Pru Napier, Charles Nelson, Bernard Newsam, Teresia N'ganga, Bernard Ngeneo, Charles Njonjo, Peter Nzube, Helen O'Brien, Tom Odhiambo, Perez Olindo, Rosalie Osborn, Lita Osmundsen, Bea Patterson, Ethel Payne, David Pilbeam, Tom Plummer and Wahida Muhideen-Plummer, Merrick Posnansky, Richard Potts, Kitty Price, Bill, Debbie, and Dana Richards, Charles and Elizabeth Richards, Rosemary Ritter, Derek Roe, Louise Robbins, John T. Robinson, Alan Root, Michael and Cordelia Rose, Wade Rowland, Walter C. Schuiling, Judith Shackleton, C. Thurstan Shaw, Pat Shipman, Elwyn Simons, Ruth Dee Simpson, Mary Griswold Smith, John D. Solomon, T. Dale Stewart, Chris Stringer, W. E. Swinton, Maurice Taieb, Anne Thurston, Phillip V. Tobias, Joan Travis, Joan Uzzell, John Van Couvering, Hugo van Lawick, Elizabeth Vrba, Alan Walker, Sherwood Washburn, Ron Watkins, Henry West, Sam White, Leighton Wilkie, Lee Williams, Peter Williamson, Milford Wolpoff, Bernard Wood, Marie Wormington, E. Barton Worthington, and Adrienne Zihlman.

It should be noted that Donald Johanson, Tim White, and Vanne Goodall did not wish to be interviewed for this book. Neither Bethwell A. Ogot nor John Onyango-Abuje was available for interviews. Ogot declined, saying only, "Silence is golden," while Onyango-Abuje has been very ill.

I also owe a deep debt of gratitude to Gideon Matwale, the head of the archives of the National Museums of Kenya, who shared his office and staff with me, located Leakey family files and related materials, and spent hours microfilming needed documents. He also introduced me to M. Musembi, the chief archivist at the Kenya National Archives, and his assistants Richard Ambani and Maina Keru, again tireless and resourceful workers. Japhet Otike and his staff at the library of the National Museums of Kenya were equally kind and helpful.

Many of the secondary source materials were collected for me by two excellent and enthusiastic research assistants: John Leedom at the University of California, Berkeley, and Eric Jones, then at Southern Oregon State College, Ashland. In Kenya, Moses Mrabu put in long hours locating documents —and in some cases, because of restrictions, copying them out longhand. In England, T. N. Cooper, Saul Dubow, A. Spanier, and Fiona Stewart ferreted out documents and articles from various libraries and archives for me; while in South Africa, Richard Lunz did the same. I also thank the following archivists, curators, and librarians and their institutions: A. R. Allan at the University of Liverpool; Nancy L. Boothe at the Woodson Research Center, Rice University; E. H. Cornelius and Ian F. Lyle at the Royal College of Surgeons, London; Beverley Emery at the Museum of Mankind Library, London; David

W. Phillipson at the Cambridge University Museum of Archaeology and Anthropology; J. Pingree at the Imperial College of Science and Technology, London; H. Robinson at the Royal Society, London; and Malcolm G. Underwood at St. John's College, Cambridge University. Thanks also to the librarians and staff at the archives of the British Museum of Natural History, London, and the Rhodes House, Oxford University.

For access to the Leakey Foundation archives and for the foundation's great interest in my book, I thank Kay Woods, Barbara Newsom, and Karla Savage. The National Geographic Society's staff was extremely supportive and helpful. In particular, I am grateful to Mary Smith, who provided assistance in a thousand different ways; Ed Snider, who gave me records of the Society's grants to the Leakey family; Dori Chappell, who located photographs from the Leakey expeditions and the photographers; and Niva Folk, who always managed to find answers and/or solutions to the most obscure questions and problems. I also thank John Lampl of British Airways.

This book grew out of a profile of Richard Leakey that I wrote in 1983 for the Canadian magazine *Equinox.* Without the support of my editors there at the time, Frank Edwards, Barry Estabrook, and James Lawrence, who enthusiastically sent me to Kenya, I would never have become involved with the Leakeys.

Had it not been for my agent, Mike Hamilburg, who drove forty miles to the Los Angeles airport to urge me to "think about writing a book" while in Kenya, this volume would not exist. He has been unfailingly supportive, and a wise and trusted adviser.

I am also extremely grateful to Bob Bender, my editor at Simon & Schuster, for his support and encouragement. Bob suggested the book to me, and has stood by throughout the long process of research and writing. His assistant, Johanna Li, was also helpful, especially in the final stages of the book's preparation.

Finally, my deepest thanks of all to my husband, Michael McRae, who understood my love for Africa as soon as he set foot in Kenya, and who was always there as friend and helpful critic during the long years of writing.

Virginia Morell
February 23, 1995
Ashland, Oregon

FOR MY PARENTS, WHO GAVE ME A LOVE OF THE NATURAL WORLD.

AND FOR MICHAEL, WHO HAS SHARED IT ALL WITH ME.

C O N T E N T S

Chapter 1

KABETE

On a rainy April day in 1902, Mary Bazett Leakey stepped off the train at Kikuyu Station, Kenya Colony, clutching her seven-month-old baby, Gladys, in her arms. A small throng of Kikuyu, dressed in skins and beads, had gathered on the wooden platform. Standing among them, conspicuous in his dark clergyman's suit, was Mary's husband, the Reverend Harry Leakey. Nearly five months had passed since he had last seen his family, and he rushed forward to embrace them. A highly strung, emotional man, Harry threw his arms around his wife, baby Gladys, and eldest child, three-year-old Julia. He then turned to Miss Oakes, Mary's companion nurse, and gave her a warm, if formal, handshake. Mary's little party had traveled by steamer from England and by train from the Kenya coast to reach the center of the British East Africa Protectorate (as Kenya and adjacent territories were then called), nearly a monthlong journey. And now, Harry assured them, they were only six miles from their new home, Kabete Mission Station.

It was, nevertheless, a long six miles—mostly uphill and over winding red-earth paths, awash in mud from the rains, and Harry Leakey wanted to waste no time. He had brought with him a party of sturdy young Kikuyu men and women who greeted his family with shrill, celebratory ululations and then bent to the task of carrying luggage, women, and children into the Kenyan highlands. To someone fresh from Victorian England, Harry's helpers must have seemed a formidable group. The men were shirtless and wore only a small leather wrap tied with a beaded belt around their waists; their hair was braided and stained with red ochre; and they all carried either spears or short swords. The women also wore oiled skins, but these were knotted at the shoulder; their heads were shaved, bundles of beaded hoops dangled from their earlobes and necks, and copper bracelets shone on their arms. Both men and women smelled curiously of smoke and rancid butter. Some years later, a similar group came to the station to meet a newly arriving English governess for the Leakey children, and she admitted to being terrified by these "wild men" who she felt certain were about to conduct her to a "cannibal feast."

For their journey to Kabete, Mary and baby Gladys were seated in a curtained hammock, which two Kikuyu men hoisted between them, while Julia and Miss Oakes settled into another. "We all traipsed through the forest then," Julia recalled eighty years later. A small, gray-haired woman, she now lived in a retirement home for missionaries outside of London but had never forgotten her first African safari. "Father was on horseback, we were swinging in our hammocks, and there were two practically naked brown men with oil streaming down their bodies and oil in their hair, singing as they carried us along."

European explorers had ventured into Kikuyuland only twenty years before, in the early 1880s, and the Leakeys were among the first wave of missionaries. The earliest visitors, following in the wake of the great Nile explorers, sought answers to geographical questions. Behind them came traders and missionaries, the one group intent on opening up the African interior, the other on stopping the slave trade and spreading the Gospel. Both groups succeeded at least partially. In 1896 the British began building a railroad from the Kenyan coast to Lake Victoria, 675 miles away. And in 1900 the last Arab slave ship departed from Kenyan waters. Converting the native peoples to Christianity and finding enough trade goods to support the railway were more difficult. But missionary fervor was at a peak in England, fired by David Livingstone's grim tales of human suffering and heightened by the murder of Bishop Hannington—one of the first missionaries to Uganda—in 1885. Missionaries from every possible order—the Church Missionary Society, the Church of Scotland Mission, the African Inland Mission, the White Fathers (Roman Catholics)—were soon riding the new railroad's flatbed cars to the interior.

The Leakeys had been caught up in this humanitarian fever. One of Harry's cousins, the Reverend Richard Herbert Leakey, founded a mission in Uganda in 1892, while a sister of Mary's, Ellen Bazett Gordon, and her husband, the Reverend Cyril Gordon, had traveled by rail and hammock to the same country six years later. Mary herself and two of her other sisters had worked among the Moslem women and children of freed slaves on Mombasa Island in 1892. Mary also started a boys' school there, later called the Buxton High School. It is now defunct. But her time on the mosquito-plagued coast proved costly: she contracted a fever, perhaps malaria, nearly died, and was invalided home. Her doctor told her that she must never return to the steamy tropics. Serving as missionaries together, however, had always been the dream of Harry and Mary Leakey, and in 1900, five years after her return

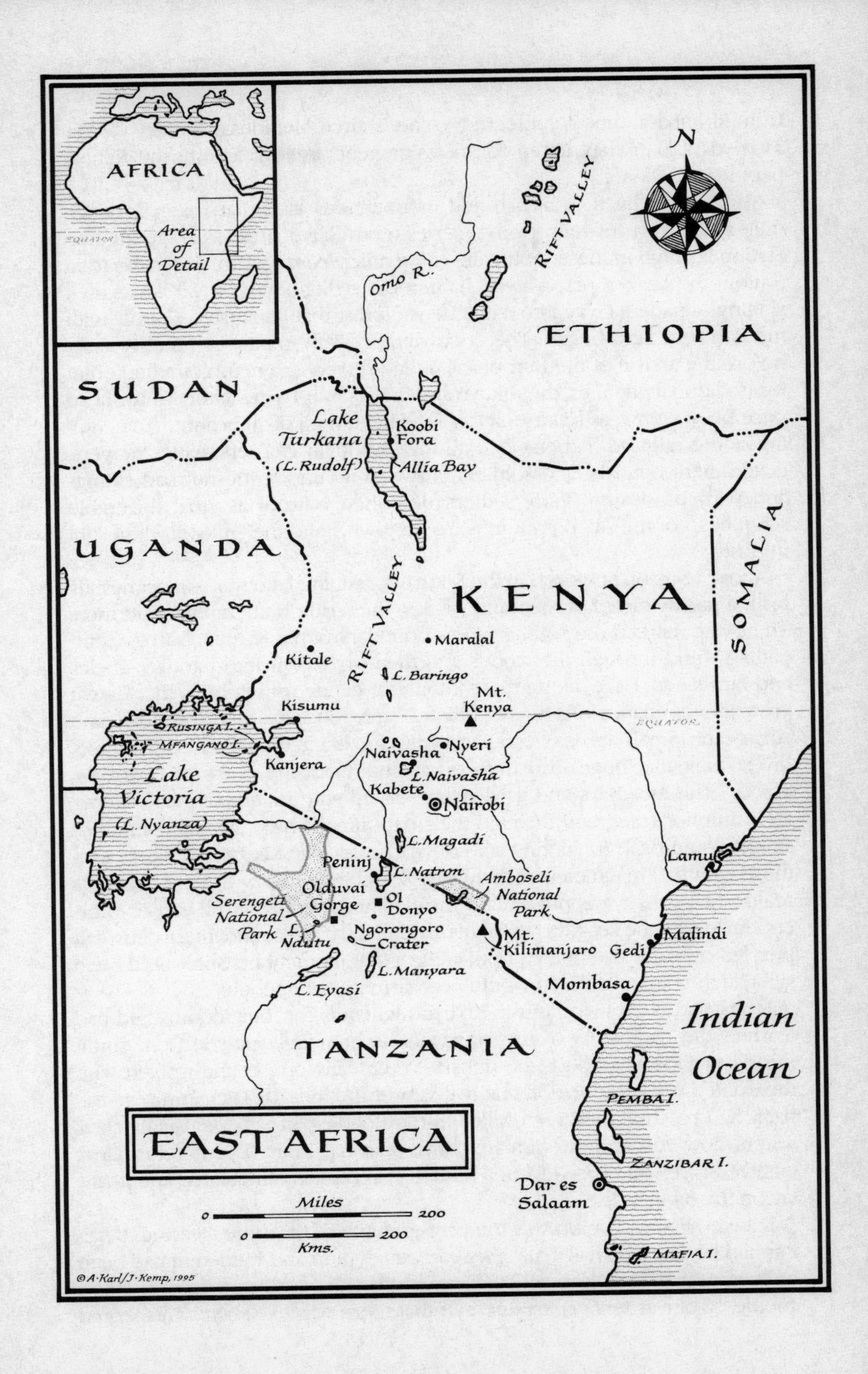

AFRICA
Area of Detail
EQUATOR
N
ETHIOPIA
RIFT VALLEY
Omo R.
SUDAN
Lake Turkana (L. Rudolf)
Koobi Fora
Allia Bay
UGANDA
KENYA
SOMALIA
RIFT VALLEY
Maralal
Kitale
L. Baringo
Mt. Kenya
Kisumu
EQUATOR
RUSINGA I.
MFANGANO I.
Kanjera
Naivasha
Nyeri
L. Naivasha
Lake Victoria (L. Nyanza)
Kabete
Nairobi
L. Magadi
Peninj
L. Natron
Amboseli National Park
Lamu
Olduvai Gorge
Ol Donyo
Serengeti National Park
L. Ndutu
Ngorongoro Crater
Mt. Kilimanjaro
Malindi
Gedi
L. Manyara
Mombasa
L. Eyasi
Indian Ocean
TANZANIA
PEMBA I.
EAST AFRICA
ZANZIBAR I.
Dar es Salaam
Miles
0
200
0
200
Kms.
MAFIA I.
©A·Karl/J·Kemp, 1995

from Mombasa, they volunteered to the Church Missionary Society (CMS). Two years later they accepted an assignment to work among the Kikuyu people.

The Kikuyu lived in a lush and mountainous land that rose above the mud-shack town of Nairobi in a series of knifelike ridges. Kabete Mission Station lay high in these mountains, nine miles from Nairobi. From the train station, the Leakey party—with hammocks swinging and the Kikuyu men singing—made its way into a tall, dense forest that contrasted sharply with the flat, sere land below. There was little reason to stop at Nairobi itself. Before the arrival of the Europeans, it had served as a neutral trading point for traditional enemies, the Kikuyu and the Maasai. To the latter, Nairobi had once been known as Nakusontelon, "the beginning of all beauty." The coming of the railroad had changed all that as Indian laborers, native helpers, government officials, and soldiers crowded in among the railroad sidings and papyrus swamp. Early settlers described Nairobi as "that miserable scrap-heap of tin," a "tin-pot mushroom town," and the "most lawless spot in Africa."

Above Nairobi, however, in the Kikuyu forest, the land was still primevally lush, a patchwork of garden and woods, sheltering both animals and men. The forest marked the southernmost boundary of the Kikuyu. Narrow footpaths wound through the woods and then broke out into parklike glades and farmlands. Here, along the contours of gently rolling hills, the Kikuyu grew an abundance of crops—millet, beans, tobacco, sweet potatoes, and sugar cane. Small herds of cows and goats grazed in the meadows, and beehives hung like huge Christmas ornaments from the limbs of spreading acacias. The meadows and gardens alternated with islands of giant juniper and camphor trees, and through their dark limbs an occasional column of smoke could be seen, hinting at a native's home. The Kikuyus' dealings with the Maasai had made them stealthy and secretive, and few passing strangers realized how many people actually dwelled in these forests. All early explorers (including the Leakeys) to Kikuyuland praised the abundant gardens, but puzzled over the size of the population: at one moment no one would be in sight; at the next, the forest would be swarming with people.

Harry had been living in the Kikuyu highlands for four months and had learned the rudiments of the language. As his family passed from sunlit meadows to shady woods, he translated the calls of greeting, spiked with the excited trills of ululation that rolled over the hills. By late afternoon, the party had reached the green valley of the Bogojee Stream, its usually clear waters now red and swollen from the seasonal rains. The mission land, eighteen acres purchased from a Kikuyu clan for forty-five sheep and goats, lay on the other side.

Kabete Mission Station was the proper name of this little clearing, but it implied far more than actually existed. A small mud-and-wattle hut had been built by Harry's predecessor, the Reverend A. W. McGregor, and it and a couple of canvas tents constituted all that there was of Kabete. This would now be the home of the young Leakey family. It was not the easiest of homes to settle into. Although situated six thousand feet up in the cool African highlands, it lacked a fireplace; heat was provided by small charcoal braziers. Glass was still a scarce commodity in Kenya, and rough wooden shutters kept out the chill night air. The floor was earthen, the thatched roof leaked copiously, and rats, fleas, and chiggers were plentiful. Mary Leakey never wrote about her initial impressions of their new home, but Harry (who once woke to find a rat "making a hearty meal off his moustache") spent a good deal of his first three years trying to persuade the CMS to provide him with funds for a more livable house. Nevertheless, it was their home, and Harry's small family thrived.

Harry Leakey was then thirty-four, a wiry, energetic man with dark sparkling eyes, and a bushy black beard that earned him the nickname "Giteru," or Big Beard, among the Kikuyu. He was anxious to get on with his calling as a missionary, having postponed his dream for six years while he paid off his schooling debts and cared for his ailing mother. She had died in 1899, shortly after Harry and Mary were married, and they had then joined the CMS. On his eighteen acres of mission land he planned to build a church, a boys' school, a girls' school, a dispensary, and workshops. Where the mud hut was, he envisioned a stone house surrounded by vegetable and rose gardens. There would be cows and chickens, orange and lemon trees. He would preach his Sunday sermons in Kikuyu to a neatly dressed (and spearless) congregation of natives. They would follow the Gospel in a Kikuyu Bible that he planned to translate. In the meantime, while he struggled with the nuances of Kikuyu grammar, he gave his sermons in his limited Kiswahili, an up-country variety of the intricate language of the East African coast.

Mary, whom everyone affectionately called "May" since she was so like a "May flower," had been Harry's childhood sweetheart and shared his passion for the missionary calling. Described by family and friends as both quiet and serious, she was also strong-minded and strong-willed. Her father, a retired colonel who had served in the Indian Army, had initially prevented her and her sisters from accepting missionary posts, saying they were too young. They waited until they were all in their mid-twenties, and went anyway. Round-faced and pretty, May was frail from her previous near-fatal bout with the coastal fever, but nevertheless actively joined in her husband's efforts to make the mission a success. Together with Miss Oakes, and an older missionary, Miss Higginbotham, May set up a dispensary in one of the canvas tents to treat the surrounding Kikuyu villagers. Their little clinic attracted many Kikuyu to the mission, some because they genuinely needed medical care, most because they were curious. The mission sat unfenced on a sloping hillside above the Bogojee Stream, and small groups of Kikuyu men, wrapped in calico capes stained with red ochre and holding spears and war clubs, would gather to watch the doings of these "red strangers."

There was, then, something of a ready audience when on August 7, 1903, May felt the first pangs of childbirth. The baby was nearly two months premature. A runner was sent down the narrow paths to the Church of Scotland Mission at Kikuyu Station for Mrs. Watson, the wife of the minister there and a skilled midwife. She had to come "quick, quick, quick," the runner implored. Someone saddled a horse, and Mrs. Watson rode at a fast trot to Kabete, where a party of Kikuyu onlookers had already gathered. Inside, Julia and Gladys were crawling over a barrier of chairs trying to reach their parents' bedroom, where Miss Oakes, Miss Higginbotham, and Harry were doing their best for May. The arrival of the midwife had a quieting effect, and May soon gave birth to a baby boy. They named him Louis Seymour Bazett Leakey after two of his uncles, and then worried about how they would keep him alive. Under such primitive conditions, very little could be done for premature babies, but Harry and Mrs. Watson devised an incubator of sorts by lighting a charcoal brazier, and pulling the bedroom door and shutters tight. They wrapped the baby in layers of cotton and wool, and laid him beside his mother. One Kikuyu boy stayed inside, keeping the brazier going. And Harry prayed.

Louis wrote later, "It was nothing short of a miracle that I survived at all." When he was strong enough, he was placed in a wicker basket on the veranda, where an appreciative Kikuyu crowd could admire him. White children were always a novelty to the local people, and Louis was the first white baby that many of them had ever seen. They wanted to see his skin, to touch him and feel his hair. These displays often ended abruptly because the Kikuyu were also eager to spit on Louis—a customary display indicating that they had not cast the evil eye on him. Julia remembered that her mother had "quite a job guarding him," and so began to keep a sponge close at hand. Later Louis would say that the Kikuyu made him the "best-washed baby" in Kenya.

Louis spent his first two years at Kabete, then was whisked away with his family to England. His father had grown increasingly troubled about their wretched living quarters and, after suffering from insomnia, dizzy spells, and tinnitus (a severe ringing in the ears), had collapsed. One year short of their full four-year missionary term, he and May brought their family home to Reading, England, where they stayed for two years, while he recovered. The cause of his illness, his doctor decided, was neurasthenia, a nervous disorder brought on by overwork. Not that the diagnosis slowed Harry down: he fully intended to return to Kabete, and in order to keep his Kikuyu alive and to continue his work of translating the Bible, he had brought to Reading with him Stefano Kinuthia, one of the Kikuyu boys he had baptized. Still, for the rest of his life, Harry regularly suffered from insomnia and tinnitus.

Just before Christmas, 1906, the Leakey family returned to Kenya, traveling as before by steamer and rail. This time, they disembarked at Nairobi, where May's sister, Sibella, and her husband, the Reverend George Burns, lived.

They had moved from the coast earlier that year to direct the Nairobi mission, and had settled into a stone bungalow on a grassy plain where Nairobi University stands today. Nairobi had grown in the Leakeys' absence. There was a new governor in BEA (British East Africa), Sir Edward Northey, who was eager to continue his predecessor's white settlement schemes. Hundreds of settlers had poured into the protectorate, primarily from South Africa, in response to promotions issued by the colonial office. Land north and west of Nairobi was offered at low prices to any European willing to establish a farm. The fact that most of this area was already settled by the Kikuyu did not trouble the government. Kikuyuland was high and cool, and bore a striking resemblance to England's bountiful Hampshire countryside. Its similarity to northern climes made it "white man's country," according to the colony's first governor, Sir Charles Eliot. He viewed the highlands as a "tabula rasa, an almost untouched and sparsely inhabited country, where we can do as we will." And they did. By 1915 nearly five million acres had been appropriated from the Kikuyu and sold to about one thousand white settlers. They called the area the "White Highlands"; it later became notorious for the decadent lifestyle of some of its titled veranda farmers. The land left to the Kikuyu was designated a "native reserve"—but even it was not secure, and over the following years the Kikuyu were forced to sell more of it to the government.

Much of the land north of Kabete Mission had been set aside as part of the White Highlands. Several British and South African families had already settled on it when the Leakeys returned, and were clearing it for coffee farms. Their homes were all at least a half-day's journey from Kabete, so that the Leakeys' closest neighbors were the Kikuyu villagers they had come to help.[1]

In the Leakeys' absence, the CMS had finally built them a solid stone bungalow, complete with a corrugated iron roof and long, narrow verandas, which soon became one of the children's favorite places to play. But nearly everything else at the mission was in disarray, and Harry and May spent most of their first year home reorganizing it. Not long after their return, May became pregnant with her fourth and last child, Douglas. He was born in 1907, and after his birth, May, who was now forty, found she had little energy to spare for her other children. She entrusted Louis, Gladys, and Julia to the care of a Kikuyu nurse, Mariamu. Photos of Louis, who was then four, show a dark-haired, dark-eyed boy dressed in a navy sailor suit, eyeing the camera quizzically. He was already strongly independent and loved listening to

---

[1] By 1910, however, Kabete Mission was nearly surrounded by European farms, causing Harry to complain to the government that he would soon have no native peoples left to help. The government responded by placing much of the remaining Kabete region in the Kikuyu Reserve. It also established a two-mile boundary between the mission and the closest European farm. But to Harry's dismay, even this promise did not protect his people. A short while later, the government itself annexed most of the two-mile strip to build an experimental agriculture station, forcing the Kikuyu families living there to move.

stories, particularly those that Mariamu told about the clever hare who always outwitted the greedy hyena and proud lion. For much of the next two years, Mariamu was the children's constant companion, bathing them and putting them to bed, and taking them for walks beside the waterfalls of the Bogojee Stream. These excursions and others with his father, whom Julia remembered as being "very keen on natural history," bred in Louis both a deep love and a curiosity about the natural world around him.

Their mission then was still little more than a clearing in the woods, and the children and Mariamu followed narrow, winding paths into dense stands of giant juniper, yellowwood, and wild fig. In the forest lived the birds and animals of Mariamu's fables: casqued hornbills, which wrestled with their ivory beaks; sly vervets and black-and-white colobus monkeys, which the Kikuyu thought of as children of God; duikers (small antelopes) and bushbuck, genets and serval cats. Birds were especially abundant, and Harry built the children an aviary and helped them with an egg and feather collection. They also collected animals: baby gazelles and duikers, wild cats, hyraxes (rabbit-sized creatures reminiscent of guinea pigs), bush babies, and monkeys. "We were very animal-minded," recalled Julia, "and reared the babies and sent them to the Nairobi zoo when they grew up."

Sometimes a spotted hyena stalked boldly into the mission clearing, but the larger cats, the lions and leopards, kept their distance. Lions were commonly seen—and shot—in and around Nairobi, but Louis never saw one in the wild until he was twenty-six and leading his second archeological expedition in Kenya. This seems remarkable, as the family often visited the game reserve on the Athi Plains south of Nairobi, where lion preyed on the herds of zebra, gazelle, and wildebeest.

It may have been the sounds of the African night—the shrill calls of the tree hyraxes and nightjars, the breathy cough of a hyena, the sounds of distant drumming—but whatever the reason, Louis was frightened of the dark as a child and "slept with his head under the sheets." Julia remembered him as being "nervous and highly strung," and anxious, like his father, to fill his day with activity. His parents had hired a tutor for the children, a Miss Laing, and they had lessons every morning on the veranda of their home. After tea, they would go for nature walks. Of the three eldest children, Louis and Julia particularly enjoyed these excursions and despite their age difference—he was five and she was nine—soon became "buddies." Julia said, "Gladys was always into her books. She was as different from us as dots from Ts. And Douglas was a nuisance, so we used to kick him out. He would run to Mother weeping, 'They won't let me play with them.' I was always Louis's friend, or should I say buddy, because we both preferred to look at everything from the natural history point of view and collected every sort of animal and insect and bird. We skinned dead birds, made traps, and collected information." Other times Louis joined the neighboring Kikuyu children in their games, running wooden hoops over the meadows or shooting a toy bow and arrows. He and his sisters now spoke Kikuyu fluently: it was the language of Louis's daily life and soon became the language of his soul. All his life he was to "think and even dream" in its richly rounded sounds.

Harry was gone from home twice a week, traveling by horseback among the Kikuyu villagers—"itinerating," as the CMS missionaries described their work of spreading the Gospel. Since Livingstone's day, Christian converts were typically attracted more by the force of a particular missionary's personality than by the Gospel he was preaching, and Harry, with his enthusiasm for the world and people around him, was magnetic. Now almost fluent in Kikuyu, he began baptizing adult men as well as boys before the first year of his new stay had ended. Young women also began to come to live at the mission, and May offered them her warmth and understanding.

Harry's many projects—itinerating, translating the Bible, building the new church, running his boys' school, preaching on Sundays, and laying out both a vegetable garden and a walkway bordered by orange trees—might have made him a forgotten figure to his children. Instead his dynamic ways delighted them ("We called him the 'Running Clergyman,' " said Julia), and so did his good-natured joking. "He was a *terrific* tease," recalled Julia. "We said we never knew if he was telling the truth or not. He was so full of jokes and funny things, and he teased and teased and teased."

In contrast, May was as reserved and restrained as the English countryside, whose tidy hedgerows she always preferred to the hectic greenery of the African landscape. She rarely joined in the teasing (although she loved her husband's jokes), but instead filled their home and the mission grounds with hymns and ballads, singing to the accompaniment of her harmonium. Her frailty often kept her inside, and in any case she had little interest in the natural world that grew so profusely outside their door. (Julia said, "My mother would not have known how to plant a plant.") Although sickly, May was a motherly figure, who was always calling on the Kikuyu women with kettles of hot soup and spoiling their children. She ran the daily dispensary, too, and started a girls' school on the veranda of their home—the first school for native girls in East Africa.[2]

At the turn of the century, it was widely believed that life under the tropical sun was unhealthy for Europeans and that too much exposure to its harmful rays would cause a person to go mad. Missionaries, settlers, and civil servants were advised to spend only four years at a time in the tropics, and to return to northern climes for a year or more to stabilize their health. The Leakeys now had several trusted workers to look after both their home and mission, and when their second four-year tour of duty ended in 1910 they sailed to England. Again they took a Kikuyu youth with them, this time Ishmael Ithongo, who was now assisting Harry with his translation of the Bible. Stefano, the boy the Leakeys had brought to England with them

---

[2] In recognition of Mary Leakey's pioneering efforts in education, the people of Kabete Station named one of their schools for her. The Mary Leakey Girls' Secondary School stands not far from where the Leakeys built their mission.

in 1904, had left the mission to become the Kenya High Court's first native interpreter.

They expected to stay only a year, and moved in with May's mother in Reading on a Friday afternoon in January 1911. Three days later Louis started school. Now eight, he was accustomed to a life of freedom, and his first formal schooling experience made him feel "like a fish out of water." He disliked having to mix with large groups of strange boys, and was much happier when his father started a preparatory school of his own, where Louis was one of only four boys. But he longed for the African forest and his Kikuyu friends, and was dismayed when their one-year leave was extended to two. The CMS was short of funds and could not afford the Leakeys' passage back to Kenya, and then May's health failed again. She had never fully regained her strength after Douglas's birth, and had begun to suffer severe menopausal hemorrhages. She rested for several months in a nursing home, and was finally able to sail with the family for Kenya in May 1913. Six years —"happy years, and full of incidents"—were to pass before Louis would see England again.

Louis was to have received his secondary education in England, but the Great War intervened. With little warning, the British East African Protectorate was suddenly isolated from Europe, cut off by the threat of mines in the Red Sea. The war itself soon spilled over from Europe into North Africa, and then to the East African colonies, where British and German settlers pursued each other in the tsetse-infested bush. Rumors abounded: the Germans had blown Mombasa to pieces; they had captured the Uganda railway line and were advancing on Nairobi; they were about to attack Nairobi from the air. Settlers in Nairobi were convinced one night that they saw the German Zeppelin flying low over the city on its way to German East Africa; it was, after all, only the planet Venus.

But for Louis and his sisters and brother, life continued at Kabete much as it had before. For a time they studied under a series of governesses, but when their last and favorite, Miss Broome, left to help with the war effort, they were on their own. Harry filled in for Miss Broome when he could, but the children—and Louis in particular—had a great deal of freedom. May had fallen seriously ill again and spent most of her time in bed, with Julia nursing her. Harry was rushing about, Julia said, "trying to keep everything going; Gladys wouldn't stop swotting [studying]; and Douglas played with his trains. I looked after the house and mother, and rolled bandages for the soldiers. That's why Louis had such a lot of freedom then, and most of the time he was out among the Africans."

All Kikuyu boys are organized into groups according to their ages, and shortly after returning to Kabete, Louis had been adopted by his peers, the eleven-to-thirteen-year-olds, who called themselves the Mukanda (meaning the time of the new robes). From them he learned to throw a spear and to handle a war club. He, in turn, taught them to play soccer—barefoot. He organized the team, devised uniforms for them, and was chosen as their captain.

Many of his Kikuyu friends were mission converts, but they were still learning the customs of their tribe from the village elders, and Louis was allowed to participate. When the elders arranged the secret ceremonies (which include circumcision) for the Mukanda passage to manhood, they agreed that Louis, too, would be initiated. They were already calling him their "blood brother," and Louis himself wrote that "in language and in mental outlook I was more Kikuyu than English, and it never occurred to me to act other than as a Kikuyu." All of the initiates were sworn to secrecy, and Louis never divulged either the nature of the ceremonies or whether he had been circumcised. The Kikuyu gave him a new name, Wakuruigi, meaning Son of the Sparrow Hawk, and like his Kikuyu brothers he was treated like an adult.

At home, however, he was not, and, in an effort to gain more independence, he built himself a mud hut at the far end of a grove of black wattle trees his father had planted. His Kikuyu friends had huts of their own where they lived apart from their families, and that was Louis's aim. He had to build three huts—each one a bit larger and better constructed than the last—before his parents approved the idea, but by the time he was 14 he was sleeping and working in his own house. Louis had been an avid collector from an early age, and one room of his hut served as his "museum." In here, he put his collections of bird eggs, bird skins, nests, skulls, animal skins, and stones. Everything in it was both dead and dusty, and Julia recalled it as being "perfectly *awful.*" To Louis, however, it represented "freedom," a place where he could "possess things" of his own, and he was happy.

He had either hunted or trapped most of the animals whose skins decorated his hut, having learned a hunter's skills from a slightly built Kikuyu elder named Joshua Muhia. Louis and Joshua had formed a "great friendship" shortly after the Leakeys had returned to Kabete, and the two spent days together in the forest, hunting and watching the birds and animals around them. They camouflaged themselves with leaves and branches and crept through the woods so silently that once Louis was able to capture a duiker with his bare hands. Later he would credit this training—the "patience . . . and observation"—for his proficiency in searching for fossils. But at the time it was a boy's sport, a "hunting passion, which had nothing to do with reason or logic," and he rose every day before dawn, grabbed his club and spear, and headed off at a quick trot with Joshua to check his traps.

If Louis lived out a kind of Kiplingesque life among the Kikuyu, at home he followed the Victorian dictums of the English. He worked hard at his Latin and mathematics, read his Bible lessons, and spoke French at the dinner table because his father expected him to. His aspirations, too, remained distinctly British, and he hoped that he might one day attend his father's college at Cambridge University to study theology and ornithology

—both of which were his father's interests. For much of his youth, Louis had no other immediate role model, and his father shone in his eyes as something of a paragon. Harry, broad-minded and tolerant, had come to be deeply loved by the native people (Chief Koinange called him the "light of the Kikuyu"), and Louis was strongly affected by the open affection and admiration his father received. Perhaps because of this, and because of his own growing Christian faith, he was, Julia said, "very unselfish," the kind of person who "couldn't do too much for anybody." But he was also "extremely independent" and "dogmatic," and generally refused to believe something unless he could see it or experience it firsthand.

In May, 1914, the first curator of Nairobi's Natural History Museum, Arthur Loveridge, was employed by the East Africa and Uganda Natural History Society. Louis met him soon after Loveridge's arrival and instinctively felt a "kind of hero-worship" for this slim young zoologist who knew the Latin names for all the birds, animals, and flowers. Loveridge was the type of fellow who, as one colleague put it, had been "born with a butterfly net in one hand, a killing bottle in the other." He loved to roam the bush and collect specimens, and despite the "chaotic condition" of the museum, he spent as much time in the forests as he could. Many of these trips took him to the Kabete area, where he stayed with the Leakeys. He had found a kindred spirit in Harry, and saw in Louis an image of his younger self. From Loveridge, Louis learned to classify the birds, to blow the yolk out of eggs, and to prepare specimens for museum collections. And in Loveridge's occupation, Louis began to see a way in which his own life might include the best of both the Kikuyu and English worlds.

For some children, a particular event—the chance meeting of an admired hero, the loss of a parent, a passage in a book—triggers a passion for knowledge. Possessed by a vision, they determine to discover all they can about their chosen subject, and as adults are often single-minded in its pursuit. Their quest becomes their life, and everything they subsequently do appears linked to that initial moment of awakening. Louis was lucky in this way, and by 1916 he knew what his life's work was to be: archeology.

The previous Christmas, an English cousin had sent him as a gift a children's book entitled *Days Before History.* It was an adventure story about the "Stone Age Men" of Britain and featured the exploits of a young boy named Tig. There were pen sketches of "primitive" men—bearded, muscular, and dressed in animal skins—living in caves, making stone tools, and hunting mammoths; and informative chapters—"How Tig Visited Goba the Spear-Maker," and "How Tig Learned to Make Fire." The author described the flint arrowheads and axeheads of these people, and included drawings of each tool. Louis was enchanted. "He lived in that book," said Julia. "It became his Bible, really. I think it made him feel that the place he was living had been full of Stone Age men and that he could find their tools. He began to pick up these pieces of rock that were all over the place, and we teased him about it because we thought they were just stones. But he thought they were tools, and he made a collection. We called them his 'broken bottles.' "

They were, in fact, stone tools, although not made of flint, a rock that is not found in Africa, but of obsidian—a black volcanic glass that takes a very sharp cutting edge when flaked. Louis found his pieces of obsidian—chipped and shaped very much like the drawings in his book—in roadbeds and at the bottom of eroded slopes. Not knowing precisely what an arrowhead or axehead should look like, he decided to keep every piece he found, fearful that he might "throw away some precious piece." His Kikuyu friends knew about these glassy chips and called them "spirits' razors" because many of them appeared after a heavy rainfall; they believed they fell from the sky with the rain. Louis's suggestion that they might be tools of the "very, very oldest people" impressed them and some agreed that it might be possible. The Kikuyu had their own stories of pygmy-sized hunters who had inhabited the forest before them, dwelling in "holes in the ground," and consequently they did not find the idea of an earlier people implausible.

Louis wanted to be certain about his tools, and he shyly displayed his collection the next time that Arthur Loveridge visited. "I'd thought he might laugh at me," Louis wrote, explaining his hesitancy. Instead Loveridge examined the stones with care and assured him that some were "certainly implements." He explained that they were made of obsidian and that there were some good obsidian arrowheads at the museum. The next time Louis was in Nairobi, he would show these to him. Louis was "delighted beyond words." He redoubled his collecting efforts, picking up every piece of obsidian that he saw, and as Loveridge had suggested that he keep a record of his finds, he wrote down the site of each discovery in a catalogue. Stone tools labeled with sticky bits of white paper now lined his own museum's shelves. The shiny black obsidian chips with their sharp cutting edges were more than tools to Louis; they were concrete links to a lost people, and the more pieces he found, the more entranced he became. From the few books about prehistory that Loveridge loaned him, Louis determined that very little was known about these Stone Age men, and nothing was known about those who had lived in East Africa. He decided that he would fill that gap: "I firmly made up my mind that I would go on until we knew all about the Stone Age there." He had just turned thirteen.

Chapter 2

FROM CAMBRIDGE TO OLDUVAI

In the summer of 1919, soon after the Great War ended, the Leakeys made plans for a trip home to England. Six years had passed since their last visit, and Harry and May were eager to see their friends and families again. Louis was ambivalent about leaving. He loved the free life in Africa, and was hesitant about resuming his formal schooling. His studies would take him away from Kenya, and he had no idea when he might return. Yet part of him was "keen to go," eager to begin the training that would lead him to the study of prehistory. Both excited and apprehensive, he packed his trunk with his collections of stone tools and bird skins and eggs, and closed the door on his "museum."

Harry had purchased a small house in the village of Boscombe, near the Dorset coast, during their last leave. Here, along the shelves in a tiny rear room, Louis carefully laid out his treasures and collections. He was now sixteen, tall, lanky, and handsome, with dark hair and eyes, a thin face, and arching cheekbones. Like his father, he had quick, abrupt mannerisms, and rushed about with an intensely preoccupied air. He was already skilled in many areas—he was a good carpenter, hunter, and natural historian; he could cook; and he had enough of a mechanical knack to keep his bicycle running in the African bush—and he wasn't timid about flaunting his talents, a trait that often rubbed the wrong way. "Louis always knew *exactly,*" said Julia. "He was always telling other people the right way to do things. I remember once my uncle told him, 'Now you shut up. I'm older than you and you don't teach your grandmother [meaning elders] how to suck eggs.' "

The Kikuyu had tolerated his impertinence—for Louis was white and of the ruling elite—but his English peers found it irksome.

Shortly after arriving in Boscombe, his father enrolled him in Weymouth College, a boys' public school (which in England meant it was private and charged fees) located in the Dorset hills. Young men often suffered from the cruel hazing of upperclassmen at these schools, and Louis's introduction was no different. His peculiar accent, tinged with the rhythms of Kikuyu, his curious way of walking—he set one foot nearly in front of the other, like an African accustomed to narrow paths—his outlandish tales of his life in Kenya, and his pride and shyness all contributed to his unpopularity and isolation.

In some ways he was more mature than his classmates ("I had, after all, built myself a three-roomed house and lived in it for over two years"), but in others he was as unsophisticated as any country boy. His Kikuyu friends had considered him an adult, but in England he was still a callow schoolboy —and worse, one from the colonies. He had never been to the theater, had never written an essay, played cricket, or learned to swim. Nor had he had to abide by school rules. Weymouth, he soon discovered, had a "fantastic" set of these: he had to obtain a pass to go to town, wear a dark suit and straw hat on Sundays, and "go to bed at a given hour whether I was tired or not." "It was all so very stupid from my point of view," he wrote in his autobiography. "I was being treated like a child of ten when I felt like a man of twenty, and it made me very bitter." Louis felt further affronted when he was chosen as a "fag," a boy who performs menial chores for a student in a higher form, by a prefect only six months his senior. "Oh, he hated it," recalled Julia. "He had to clean all the shoes of this chap. He was always talking about him, complaining about how he'd been a leader of a big clan of African boys and had all that freedom, and now he was cleaning shoes. That was a very harsh spell of his life."

It was an impossible situation. On his first day at school, within an hour of his arrival, a gang of boys locked him in a coal bin. On the cricket field "little boys of thirteen" teased him, and at the swimming pool his classmates left him to stand ankle-deep in the water (he had not learned to swim at Kabete because there was nowhere to do so). "I suppose I could have gone to one of the masters and asked to be taught to swim," Louis wrote in *White African,* the first volume of his autobiography, "but I was too unreasonably proud to go and admit that I could not do what almost every boy could do so well, and I felt rather bitter that no one offered to teach me. It made me feel rather like an animal that had been wounded, and with which the herd would have nothing to do in consequence of its helplessness."

Louis was also behind in some of his classes, and his chances for attending Cambridge University, where his father had gone, seemed desperately slim. He was determined to catch up, though, and amazed the headmaster, R. R. Conway, by requesting permission to work late after the other students had gone to bed. Conway may have been impressed with such drive, but he also considered Louis too poor and too old to attend the university; Louis would be nearly nineteen by the time he finished at Weymouth, a year behind most other boys. When Louis asked Conway's advice about taking the Cambridge entrance exams, Conway "simply shrugged his shoulders" and suggested instead that Louis try for a position in a bank. "I went away utterly miserable," Louis wrote about this encounter, "for I saw all my most cherished dreams falling to the ground; but my despondency did not last long. I was quite determined, and I felt convinced in my own mind that if I tried hard enough I could find a way of achieving what I wanted."

Louis next turned to his English teacher, a Mr. Tunstall, who had attended St. John's College at Cambridge and who openly admired and encouraged his ambitions. With letters from Tunstall, Louis traveled to Cambridge in the spring of 1922 for interviews. Six months later he began his freshman term as an undergraduate at St. John's. He had done so well on his entrance exams that he had won a small scholarship.

"Louis, wherever he went, made an impression, a big splash," said Julia. "People talked about him." At Cambridge, he had gained a reputation as a "wangler" (a wheeler-dealer) even before he arrived. He had managed to convince the authorities to accept Kikuyu for one of his two modern-language requirements, then produced a testimonial signed with the thumbprint of Chief Koinange as proof of his proficiency. When Louis had to train his own teacher in Kikuyu, the legend grew into a quite untrue story that the flamboyant Louis Leakey "had examined himself in Kikuyu." His fellow students now called him the "senior wangler."[1] He was also the first to play tennis in shorts at Cambridge—"Fancy! Tennis in shorts!" the other students exclaimed—and was promptly thrown off the courts for "indecency." Once he accepted a dare to say grace before dinner in Kikuyu rather than Latin, and droned sonorously through it without one of the dons noticing. He joined the Magic Circle, and learned to saw ladies in half. He fell in love and pedaled eighty miles out of town to propose to an unnamed young woman. She turned him down.

He was still zealous about his Christianity and sometimes stood on corner soap boxes to deliver sermons. "My husband, John, was at Cambridge at the same time as Louis, and he said Louis used to come round to the students' rooms and tick them off for not being proper Christians. He said they weren't keen enough," said Julia. He was bright, enthusiastic, but "overcharged," as one female acquaintance put it, and his fellow students tended to be both wary and admiring of him. "Louis's fellow students recognized his eccentricities for what they were—those of an individualist with a big ego," noted one classmate. Yet he made friends—some of them lifelong ones like Gregory Bateson, the future anthropologist, ecologist, and hus-

[1] This was a pun. At Cambridge, the "senior wrangler" is someone who earns top honors in mathematics. Louis, as "senior wangler," was seen as someone who could convince anyone of anything.

band of Margaret Mead, and E. Barton Worthington, who would gain fame for his work on the freshwater fishes of Africa.

Louis's first year passed "very happily and very quickly." He had his own rooms where he could come and go as he pleased, cook, study, and entertain—and he was at Cambridge, preparing himself for the study of prehistory. He returned for the fall term in October 1923, eager to continue his studies and to become a "Rugby football Blue." Every year Cambridge and Oxford teams face off against each other in a number of sports: boat races, cricket, and rugby. Cambridge students chosen to represent their university also earn the right to wear a blazer of light blue; they are the "Blues." Becoming a Blue was one of Louis's "greatest ambitions," and he set out at one of the first autumn games that second year to prove his ability. He was, he wrote, playing the game of his life, when he was kicked in the head and had to be carried off the field. Foolishly, he reentered the game, received a second kick on the head, and was once again carried from the field. That night he suffered from a terrible headache, one that got much worse the next day. His doctor insisted that he take a complete rest for ten days, but the headaches continued. He was unable to study, he was dizzy, and he lost his memory. The blows to Louis's head, his doctor decided, had left him with posttraumatic epilepsy. There was no cure for it, but a prolonged rest, away from Cambridge and in the out-of-doors, might help.

Then, as now, epilepsy was a little understood disorder, and its symptoms —migraine headaches, acute depression, blackouts, muscle spasms, frothing at the mouth—made it frightening as well. Historically, epileptics have been considered marked people, either close to God or the devil. Louis's epilepsy did not set him apart to this extent, but the treatment his doctor prescribed —a change in scenery—contributed to a profound life change.

For Louis the treatment marked the temporary end of his Cambridge studies and the beginning of his fossil-hunting career. Later he would write, "I little thought when I was kicked on the head what a great effect that incident was going to have on my whole career." But Louis also believed in luck, and possibly because of that belief he was lucky. From an old family friend, C. W. Hobley, he learned that the British Museum of Natural History was organizing a dinosaur fossil–collecting expedition to Tendaguru in Tanganyika Territory (now Tanzania). Tendaguru had been discovered in 1914 by German scientists, who had returned with a complete skeleton of *Brachiosaurus,* one of the largest land animals ever to have lived. Now that the British ruled Tanganyika (under a League of Nations mandate issued after the Germans lost the war), they, too, wanted a skeleton of one of these fabulous creatures. The museum had hired a dinosaur expert, William E. Cutler, but they needed someone with African experience to handle the logistics.

Louis landed the job, and on the last day of February 1924 he joined Cutler on board a steamer bound for Dar es Salaam. "My luck," Louis noted, "had certainly turned in a most unexpected manner."

• • •

The headaches Louis had suffered from after his accident disappeared as soon as he stopped his studies—although they would later return—and he threw himself enthusiastically into the work of the expedition. His job was first to locate Tendaguru, as the site was only poorly marked on maps, then to build the camp and hire workers. Cutler stayed behind in Dar es Salaam to read up on the geology of Tanganyika, and to make a collection of shells and mollusks. Louis caught a ride on a Dutch cargo ship south to the tiny port of Lindi, and there, as luck would have it, met up with a village headman from the Tendaguru district named Jumbe Ismaeli. Although Jumbe did not know the fossil locality, he agreed to let Louis accompany him back to Tendaguru. Within three days' time, Louis had organized a safari complete with fifteen porters, a cook, gun-bearer, and personal servant. He and Jumbe took their positions at the head of this long line and struck out through the coastal coconut plantations for the interior. It was a classical African safari, and Louis, dressed in khaki bush shirt, shorts, and pith helmet, felt he had started "on a great adventure."

They marched the fifty-six miles to Tendaguru in a fast-paced three days, making their way through country thick with thornbush and stinging nettles. Around Tendaguru the shape of the land disappeared under a tall covering of elephant grass; in some places it towered fourteen feet overhead. But Louis spotted a small rounded hill that reminded him of photographs he had seen of the site, and a rusted sardine tin and broken beer bottle confirmed his suspicions. Jumbe asked him to fire his rifle in the air to signal that a white man had arrived, then sat down with his signaling drum and beat out a message informing distant villagers that they should come with their knives and axes to help the white man build a house. In a few weeks, they had built grass huts for cooking, sleeping, and working. But two months passed before Cutler was ready to join Louis at the camp.

William Cutler was then forty-two, a man of "little eccentricities and pretensions," who had spent much of the past twelve years collecting dinosaur fossils in the great bone beds of Alberta, Canada. There he had made a name for himself by discovering several new species and excavating complete skeletons. He tended to be jealous of his finds and most often worked alone. He had lived alone in a homestead in the Alberta badlands, and when he decided to move to Winnipeg, he bought a rowboat in Edmonton and rowed north and east on the rough glacial rivers, traveling some six hundred miles between the two cities—alone. C. W. Hobley, the old friend of the Leakeys, noted that Cutler "prided himself on going anywhere with a minimum of kit and the simplest of food." From the outset, he disliked the bulky mass of equipment—tents, camp beds, mosquito nets, tons of stores—that the East African Dinosaur Expedition required. He regarded them as "unnecessary luxuries" and turned the whole business over to Louis. Cutler would have been just as happy to have done without Louis as well. He was used to running his own expeditions and listened coolly to Louis's eager advice.

In the bush, the cocksureness that had merely irritated Louis's schoolmates exasperated Cutler. When the two set out for the Tendaguru camp at the head of a line of thirty-seven porters, Louis's easy command of Swahili and intimate knowledge of the flora and fauna left Cutler feeling awkward and dependent. His Canadian skills and woodcrafts did not translate to Africa, but he was not willing to admit that to this energetic, presumptuous young man. He scoffed at the obsidian flakes Louis showed him ("I think them chipped by the heat of grass fires"), and ignored Louis's warning to leave the velvety seed pods of the Upoopo (buffalo bean) vine alone: their fine hairs, Louis cautioned, would cause painful itching and burning. When Louis looked away for a moment, Cutler wrapped some of the pods in his handkerchief, then in the heat of the afternoon used it to wipe his arms and thighs. In an instant, he leaped up "yelling like a madman and cursing like a trooper." Louis gave him some lotion for the pain but could not suppress a small smile. By the time they reached camp, Cutler was thoroughly weary of Louis.

Yet they worked together from June to the end of November, 1924, excavating dinosaur fossils, making a collection of butterflies and bird skins for the British Museum of Natural History, enduring the heat, and suffering a variety of ailments. Cutler's diary sometimes reads like a handbook of tropical diseases: "Leakey very low with malaria, temperature 104; Leakey still has malarial symptoms . . . but he superintended ditches 4, 5 and 6 all day." He himself was "passing blood and vomiting," and on another occasion the entire camp staff complained of "subnormal temperature, headache, bad stomache and aching bones." They did find a number of large dinosaur bones, but could only get to them by excavating huge trenches, some of them fifty feet long and twenty-five feet wide. The bones themselves were in poor condition and had to be encased immediately in jackets of plaster of Paris. This was Cutler's area of expertise and although he allowed Louis to assist him, he never once let Louis apply the plaster by himself. But Louis was a good observer and years later would write warmly of the "late Mr. C. W. [sic] Cutler, whose knowledge of the technique of excavating and preserving fossil bones of all kinds was unsurpassed. . . . I learned more about the technical side of the search and preservation of fossil bones [from him] than I could have gleaned from a far longer period of theoretical study."

They never did find a complete skeleton of any dinosaur—and apparently never would have, as the site they were excavating lay at the mouth of an ancient river where only random bones had collected. Still Louis was enjoying himself immensely. It was his first taste of field science, and he found the adventure of search-and-discovery exhilarating. He tried to postpone his trip back to Cambridge, but the authorities insisted that he return. Cutler stayed on. He had learned enough Swahili to supervise his workers, but worried in his diary that they were deceitful and cheating him on time. He brooded over Louis, too, fearing that he might be stealing the expedition's glory: "I wonder how the magnificent L. S. B. Leakey is faring? He is sure to be blowing his trumpet hard." (Cutler was not far off the mark. Louis was writing articles with the permission of the British Museum of Natural History, and giving lectures about hunting for the lost world of the dinosaurs in darkest Africa.) When a set of letters and numbers for stenciling disappeared, Cutler noted that it was very likely Louis who had made off with them, then added, "Thankful to say, anyhow, that here at least Leakey is missing."

The isolation, heat, and misunderstandings with his staff took their toll. Cutler alternately cursed and praised his workers, and once fired the cook for making "murderous bread." He suffered from recurring fevers, dosed himself with quinine, then took to his bed on August 21, 1925. "Owing to fever I stay in and only visit the ditches late," he wrote. "Upon returning home, feeling weak and horrible, found my temperature to be 104 so went to bed." Nine days later, nine months after Louis had left, Cutler died, a victim of blackwater fever. "Maybe if he had more fully recognised the dangers of the insidious Anopheles [the mosquito which carries malaria], he would have been spared to carry on his work," noted C. W. Hobley in his obituary in *Nature.*

It was January of 1925 when Louis reached Cambridge again. A light snow covered the stone towers and walkways of the university, and the bells of Great St. Mary's rang out the hours. Louis moved into an attic room of an L-shaped, grey-stone building known as the Third Court. From here he had a view of the narrow, angular streets of Cambridge, and could watch students and dons sweeping along in their black scholars' robes. He was fit and tan from his months in Africa, and as usual, he had a hundred stories to tell. He had done most of the hunting for the camp, rising at four in the morning every Sunday to walk twelve miles to the Mbemkuru River where eland, sable antelope, kudu, and buffalo gathered. At times he had had close calls with elephants and buffalo, and once was nearly bitten by a deadly green mamba snake: when it struck at his neck, he leaped back and "blew it to a thousand pieces" with his shotgun. He had kept a baby baboon as a pet until a leopard padded into his bedroom one night, seized the baboon, and leaped outside with it through an open window. He told about his incredible journey back from Tendaguru to Dar es Salaam, how he had walked the 320 miles in fourteen days, just making the steamer bound for Mombasa. And he told about Jumbe Ismaeli and the signaling drum that he had bought for three shillings and brought back to Cambridge. He often shared his tales in his room with a whole "audience of friends," once gave a formal lecture in borrowed coat and tails to a packed Guildhall in Cambridge, and occasionally made some money by giving talks at various boys' schools.

Louis had returned to Cambridge with a new appreciation for the demands of a scientific career. Before, he had always believed it possible to be both a missionary and part-time scientist; now he was not so sure. He gave the problem "much thought," worrying over his growing knowledge of evolutionary theory and his "more liberal views of native customs." He doubted that a missionary board would accept him. He also doubted that he would "really be content treating science as a 'part time job.'" But with a critical final examination in French looming, he had to postpone his decision. He had failed the exam his first year, but this time, after cramming for an entire term, he received the highest marks possible—"firsts," as they are called in England. His success in French also meant a small scholarship, and any remaining questions about his future career suddenly dissolved. "At last . . . the way was open to train myself in anthropology and archaeology. The dreams I had dreamed as a child after reading Hall's *Days Before History* were coming true."

Alfred Cort Haddon, a tall, lively man with a deep, booming voice, a shock of white hair, and a "real love of teaching," headed the anthropology department at Cambridge. Officially, anthropology came into existence at the university in 1904 when Haddon finally persuaded the authorities to list it in the lecture book. He had first taught the courses sitting on packing crates in an anteroom of the botany department; ever since anthropology had been known at Cambridge as the "Cinderella of the Sciences." Under Haddon's hand, the department blossomed, and by the time Louis began attending his lectures, Cambridge had gained a reputation as a leader in the field. Louis's tutor in archeology was a softer-spoken man, Miles Burkitt, who was known for his field work in South Africa. But it was Haddon, with tales of his collecting adventures in the South Seas and his distaste for convention (he always addressed his secretary as "Colleague" or "Comrade" and often listed her as a coauthor on his scientific papers), who inspired Louis.

Together with Gregory Bateson, Louis spent many Sunday afternoons at Haddon's home, where Haddon held forth "freely, wisely and wittily on any subject under the sun 'except higher mathematics.'" Haddon was eager "to win souls for Anthropology," and his house reflected the exotic nature of his chief enthusiasm: masks, spears, baskets and clubs on the walls, a skull on his desk, and an African drum that he beat to summon his guests to dinner. It was from Haddon that Louis first learned about string figures and the easy entry a well-made cat's cradle could give a visitor to primitive peoples' cultures. The figures and stories that accompanied them often had magical or religious significance, and anyone who was adept at performing them must therefore be a friend. "'You can travel anywhere with a smile and a piece of string,'" Haddon told his students. Later, with probably a touch of exaggeration, Louis would credit his knowledge of string figures with saving his life. But like Haddon, he did become skilled at them and always carried a "suitable piece of string" in his pocket, using it to make friends wherever he traveled in Africa.

Haddon's primary interest was not prehistory but the study of ethnography, collecting handicrafts and artifacts from living cultures. Louis did not readily share his mentor's enthusiasm for material culture, but when Haddon gave a lecture on a German classification system for African bows and arrows, Louis knew parts of it were wrong. He told Haddon about his suspicions, and Haddon helped arrange a small grant enabling him to spend his 1925 Christmas vacation in Hamburg, Brussels, and Paris, where he examined collections in various museums. Ultimately, this research led to Louis's first scientific paper, but even more important it resulted in a meeting with Hans Reck. A prominent German paleontologist, Reck had discovered a complete fossil human skeleton in Olduvai Gorge (then called Oldoway), Tanzania (then German East Africa), in 1913, and had been part of a German expedition to Tendaguru. He and Louis hit it off and developed a "very warm friendship." Louis told him that he was planning to spend the rest of his life "studying the prehistoric problems of East Africa," and Reck shared with him the story of his fossil skeleton. He could not show it to Louis because he had sent it to Munich for study.

Reck's skeleton was very controversial at the time. The remains were clearly those of modern man, *Homo sapiens,* but Reck insisted the skeleton was as old as the extinct Pleistocene animals found nearby. Skeptics argued that it was a recent burial of a Maasai tribesman, dug into the older geological strata. There were no chemical tests then that could decide this issue, and Reck had been planning a second expedition to Olduvai to settle the matter when the First World War broke out. Now, with the gorge in British-held territory (the League of Nations had mandated German East Africa to the British in 1920, and they had renamed it Tanganyika), there was very little chance that he could ever go back. "Half jokingly, half seriously, I said that one day he must come and join me, and that we would visit Oldoway together," Louis wrote. Six years later they would.

Back at Cambridge, Louis again applied himself to studying for his final exams in anthropology and archeology, and again came away with top scores —he had earned, as his fellow students predicted, "double firsts," the equivalent of graduating summa cum laude. He received his baccalaureate in June 1926, and St. John's awarded its brilliant young scholar a research fellowship to head up his first East African Archaeological Expedition. Other organizations, including England's Royal Society (an exclusive society of scientists), the Percy Sladen Memorial Trust, and the Kenyan government also showered Louis with grants, making it possible for him to plan a yearlong expedition.

Only one Cambridge professor seemed to have doubts about Louis's enterprise; he attempted to dissuade the young scholar from wasting his time searching for early man in Africa, since " 'everyone knew he had started in Asia.' " As cocksure as ever, Louis decided to prove him wrong. A month after receiving his degree, Louis and a fellow Cambridge graduate, Bernard

Newsam, sailed to Mombasa and set up a base camp at Kabete Mission Station. Louis was just twenty-three.

For this first of many expeditions he would lead to Africa, Louis decided to excavate one of the sources of his inspiration, a cave he had explored as a boy along the Bogojee Stream. He and his sisters had discovered the cave when playing near a waterfall they had christened "Gibberish." Later they had returned with two "very reluctant" Kikuyu men and a rope ladder, and crawled down into the hole, expecting to find the tools and pots of ancient people lying on the floor. But there was nothing aside from a single obsidian flake, and they climbed out again "very disappointed." "We did not then think of digging," Louis wrote.

When he returned to the cave in the summer of 1926, he, Newsam, and two elderly Kikuyu men dug a trench down the center of the cave, then set to work excavating the floor. Every day a crowd of Kikuyu gathered around the cave's entrance to watch and exclaim over the obisidian "knives and weapons" Louis retrieved from the earth. The elder Kikuyu explained his work as a "search for the Gumba," referring to the semi-pygmy race of people who had lived before them in the forest. Louis, however, was looking for "the oldest culture in East Africa." The Kikuyu were intrigued by this idea and fascinated when Louis showed them how to use an obisidian flake to scrape a skin, but they disapproved of his search for the skeletons of these ancient people. No matter how long ago they had died, their remains were still taboo.

In the 1920s, archeologists believed that the oldest culture was that of the Chellean, a "handaxe culture."[2] These large tear-shaped tools were first identified in France in 1847. Later archeologists found them scattered from India to Palestine, across Europe, and in North and South Africa. Today, the handaxes are known to be the handiwork of *Homo erectus,* who lived between 1.7 million and 400,000 years ago and was the immediate predecessor of *Homo sapiens.* But in 1926 scientists puzzled over which ancestor had crafted the tools. *Homo erectus* was then known only from fossils discovered in Java in the 1890s—although the remains (a jawbone, skullcap, thighbone, and two teeth) were initially classified as *Pithecanthropus erectus,* or the "erect ape man." The designation clearly signaled that scientists (even in the 1920s) did not consider this creature capable of making stone tools. Without a means of dating the handaxes, archeologists somewhat arbitrarily assumed that they belonged to a culture no more than 200,000 years old. The earth itself was believed to be only sixty-five million years old, the Age of Mammals (the Cenozoic) three million years, and the Age of Man (the Pleistocene)

[2] "Chellean" as an archeological term is not used today; this cultural stage is now called the Abbevillean and applies only to Europe.

500,000 years.[3] Within these constraints, a date of 200,000 years for the earliest culture seemed reasonable.

Louis, in preparing for his expedition, noted with great "surprise" that no one had yet found Chellean tools in East Africa. "I argued to myself that if only I could find the right places in which to search, I must be able to find this oldest culture in East Africa too." He failed at Gibberish Cave, and moved his camp to the farm of Major James Alex MacDonald, where he and Newsam excavated a number of burial mounds. This time there was sufficient material—skeletons, stone bowls, beads, and obsidian tools—to describe a prehistoric culture. Louis labeled it the "Gumban" in honor of the small forest folk.

Louis and Newsam then moved to an abandoned farm overlooking Lake Elmenteita. He had heard about the site as a boy of fifteen at a meeting of the Natural History Society in Nairobi. Professor J. W. Gregory, a noted geologist who had been the first to discover stone tools in East Africa, had given a talk, and afterwards Louis went up to greet him. They were joined by one W. S. Bromhead, who produced a fossilized human bone, a shell, and polished stone that he said had been found after rock blasting on a Kenyan farm. Louis had been "struck by the mineralised condition" of the bone, and now, eight years later, he had permission to investigate the site. On his first reconnaissance, Louis noticed a skull jammed into a crack in a cliff face, an "important" discovery. He dug a small trench and found a number of human and animal bones and a bit of pottery all lying undisturbed in stratified sands, and decided that "the site must be examined in detail."

Close to the cliff where Louis had spotted the skull, the Makalia River spilled down from the Mau escarpment. Louis and Newsam set up camp here in a "large and airy" pigsty on a Mr. Keeling's farm, while his Kikuyu workers stayed in tents. They started work at Bromhead's site, as Louis called his new excavation, at the end of January 1927, and within a few weeks had recovered a number of skeletons (eventually he excavated twenty-six), obsidian tools, and bits of pottery. Louis decided that the pottery indicated an earlier cultural phase than the Gumban, and called these people the Elmenteitans. Already, as he wrote in his reports and letters to Burkitt, Louis had begun piecing together the prehistoric cultural sequence of East Africa—something that had never been done, since archeologists then believed that all significant cultural developments had started first in northern climes.[4] He had also found another sixty-four sites worth investigating.

---

[3] In contrast, the earth today is dated to 4.5 billion years, the Cenozoic is given sixty-five million years, and the Pleistocene encompasses the last two million years.

[4] Archeologists then generally agreed that humans had originated in Europe and Asia, and that people first developed stone tools and other cultural skills in these northern regions. Thus the stone tools that were found in the tropics were thought to be of more recent origin, left behind by primitive peoples who had been forced out of the north. The oldest cultures, scientists reasoned (erroneously), would be found in the north, and therefore few paid any attention to Africa.

By now, Louis's excavations were well-known in Kenya (the *East African Standard* carried several stories about his work, and solicited donations on his behalf), and another farmer in the Elmenteita area, Mr. Gamble, asked him if he would like to investigate two caves on his property. They struck Louis as "ideal places for prehistoric man to live in," and he began a trial excavation there in April. He and his team dug fourteen feet down through the cave floor, again uncovering stone tools, bits of pottery, and fragments of human skeletons. Louis was tantalized by the cave deposits, which obviously extended well below the fourteen-foot level. There was not time to excavate the entire cave, but he retrieved enough evidence to show that it had been occupied by at least two distinct cultures, one of which was clearly Elmenteitan.

Earlier in the season, Louis had purchased an old Model T Ford. Each morning he and his crew drove the ten miles from camp to Gamble's Cave, as he called the new site, worked all day, and returned in the evening. One afternoon their usually strict schedule was interrupted by a visit from Mr. Keeling and two young women from Cambridge: Henrietta Wilfrida Avern and Janet Forbes. Frida, as Henrietta liked to be called, and Janet were just completing an eighteen-month tour of East Africa and had stopped by the Keelings' farm on their way back to Nairobi. "We'd gone out in a slap-happy way," Frida recalled fifty-eight years later in the sitting room of her Cambridge home. "I'd got tired of teaching and wanted to see what the world was like. So we'd taken the steamer to Mombasa, then bought a box-body car,[5] a Whippet, in Nairobi, and hired a friend's elephant boy, a Maasai chap we called the 'Long Fellow.' He was stalwart and very good at building bridges if they were down." The three of them had driven through Kenya, Tanganyika, and Uganda, camping and staying with settlers along the way. They had traveled through the great herds in the Maasai Mara game reserve, run a coffee farm for six weeks while the owner recovered from malaria, and spent a week walking with friends in the foothills of the Ruwenzoris, Uganda's Mountains of the Moon.

At twenty-six, Frida was a bright young woman with a "lively mind, keen interest in people and ideas and love of talk." She was not a beauty, but she was vivacious, with sparkling brown eyes and a generous smile. At Cambridge, she had majored in modern and medieval languages, but had also studied some archeology, and Louis found her an astute and receptive audience. He invited the two women to dinner in his "piggery" and two nights later they returned to his camp. "The supper table was covered with stone tools before we had anything to eat," Frida recalled. "There was a python skin drying outside in the sun, and bones and stones covering everything inside. And Louis talked absolutely nonstop about stratified sites and dating. I think he was probably starving for a

[5] Box-body cars were medium-sized trucklike vehicles, with canvas sides that could be rolled up or down, something of a forerunner of the pickup truck.

little reasonably shared and intelligent conversation, and an audience. We talked archeology till dawn."

The next day Louis drove Frida, Janet, and Newsam to the Nakuru site where he discussed his developing ideas on lake levels and dating. He believed that in Pleistocene times several immense lakes covered portions of what is now the Rift Valley, and he thought these old lakebeds might provide a method for dating his archeological sites. "Louis was obsessional about the lake gravels, as he always was when there was something to find out," said Frida. "But it was always something interesting, and he had a way of telling you about the lakes rising and falling, and the ancient people living along their shores that made you see it. There was always an immediacy." That afternoon, they picnicked under the spreading thorn trees that edge the shores of Lake Nakuru. Petal-pink flamingoes rimmed the lake's blue waters, hippos lazily surfaced to snort and bellow, and water buck and buffalo grazed in the distance. It was heady and romantic, and Louis fell in love.

"It was a very daft affair," Frida said. "He came down the following day to see us off to Nairobi—we were on the way to Tanzania—and suggested through the window of the train that we should be married when he came home. I said I wasn't thinking of marriage at that time, that we didn't know each other at all. I thought he was completely mad."

But Frida was also captivated by his flair and dash, the "originality of his thinking and astonishing range of knowledge. He never stopped talking and was never dull." They corresponded ("always about his latest discoveries"), and Louis called on her when he returned to Cambridge in the fall of 1927. As usual, he was full of archeological talk. He had already published a paper ("Stone Age Man in Kenya Colony") about his discoveries in the eminent British scientific journal *Nature,* the first of forty articles and letters he would eventually publish there; he had given a talk to the British Association for the Advancement of Science about his finds; and no less an expert than Sir Arthur Keith, the foremost authority on human origins, was studying the skeletal remains Louis had brought back from his expedition. St. John's, where Louis would spend the next year, had provided its star young scientist with a comfortable set of rooms in which to study and live. But beyond all that, he had a new expedition to plan, and he hoped very much that Frida would accompany him. This time Frida succumbed.

One year later, in July 1928, they were married. Frida's family, upper-middle-class merchants in the cork trade, were "appalled. But they hadn't lived in scientific circles and were much more conventional," she said. Frida, however, liked the prospect of an adventurous life with Louis, and that autumn they set sail together, happy and newly wed, on Louis's Second East African Archaeological Expedition to Kenya.

After a short visit with his parents at Kabete Mission, Louis and Frida set up a base camp at Elmenteita to continue the work at Gamble's Cave. This time

Louis had permission to use the abandoned farmhouse, but it had been left empty for some years and its mud walls were crumbling, full of holes, and rat-infested. Compared to the Elmenteita pigsty, the decrepit house was a step up and Louis proclaimed it "magnificent." Frida gamely agreed, but others who would later join the expedition found the house appalling. "Camp conditions were very rough and bad, though Louis had told my father that they were almost luxurious," said Elisabeth Kitson, who joined the expedition at the end of November. Elisabeth's father must have marveled at Louis's audacity when he received his daughter's first letter home:

> [The camp] consists of a collection of huts made of wattle & daub with thatched roofs, all in a *very* doubtful state of repair & cleanliness. The furniture (such as there is) is made of packing cases & there is a barbed wire entanglement across every window instead of glass. . . . The bedroom part is just like a cowshed—there's no ceiling but just the bare rafter poles & thatch above . . . full of bats & crickets & spiders & doubtless other unknown horrors. The walls are just partitions & don't go up to the top so you can have conversations with your next door neighbour or the person in the bath. The floor is just bare earth.

Louis especially appreciated the house because it saved him the expense of buying tents for all the members of his expedition; then, as later in life, he spent whatever money he had on his work, and where he could cut costs on creature comforts, he did. His Kikuyu workers, many of them friends from his own age group, built their small circular huts behind the house. The camp was well situated for Louis's purposes, six miles from Gamble's Cave and within striking distance of other sites Louis planned to explore in the vicinity.[6]

Elmenteita lies almost in the center of the Great Rift Valley, a massive geological fault that runs from southwestern Turkey, across the Red Sea, south through Ethiopia, Kenya, and Tanzania to the mouth of the Zambesi River in Zimbabwe. Although it drops below sea level at some points, at Elmenteita the Rift is high and broad—almost a mile above sea level, and twenty-five miles across. Three lakes—Naivasha, Elmenteita, and Nakuru—are strung like blue jewels down its center. From Louis's camp the view was of the sheer Mau and Laikipia escarpments that form the edges of the valley, and the waters of both lakes Elmenteita and Nakuru. In the distance rose the imposing cinder cones of two volcanoes, Longonot and Suswa. The air here was exceptionally clear, lending a foreshortened quality to the landscape and fooling the eye. Some days, the distant peaks seemed but a mile or two away, while birds flying overhead appeared tiny one moment, huge the next.

[6] Louis had identified more than sixty archeological sites during his first expedition, and added to this list on his subsequent journeys. Today many of these sites are still being worked by archeologists.

Once Penelope Jenkin, another member of the expedition, spotted what she thought was a flock of starlings spinning over Lake Elmenteita. "But then they began to circle down and down . . . until suddenly, they turned from dark silhouettes to great white birds with enormous golden bills! Sure enough, they were a flock of pelicans." On the plains around their camp, zebra grazed alongside the cattle herds of tall Maasai, and occasionally an ostrich strutted by, turning its long neck and fluffing its feathers.

Louis's first expedition and his reports about Kenya's prehistoric cultures had received such wide acclaim in England that he had been rewarded with sizable grants from all of his previous benefactors, as well as the Rhodes Trust and Royal Geographical Society. With the extra funds, he expanded the size of his scientific staff to eight. He had persuaded his brother, Douglas, to take a temporary leave from Cambridge and come out as his surveyor. He had invited a dark and moody Scot, Donald MacInnes, who was "extraordinarily clever with his hands" and with whom he had struck up a great friendship at St. John's, to make a collection of animal bones at his sites. Tom Powys Cobb, the son of Kenyan settlers, had invited himself along, promising to pay his own expenses. Once camp was established, Louis sent an urgent cable to Professor Haddon asking him to find a geologist, and in December the tall and gentle John D. Solomon came out. Shortly before Solomon's arrival, Elisabeth Kitson and a fellow student, Cecily Creasy, also joined the expedition. Louis had earned extra money at Cambridge by coaching both in their anthropology courses. Kitson, blond, blue-eyed, and twenty-three, was described by a fellow team member as a "lively little person with an immense sense of humour"; Cecily Creasy was a "good-looking and intellectual brunette" of thirty who had lost her husband in the Great War. Louis, ever proper, noted in his autobiography, " . . . as I planned to marry and to take my wife with me, there was no reason why Mrs. Creasy and Miss Kitson should not come too."

Penelope Jenkin was the last arrival. A close friend of Frida's, she had studied the ecology of lakes in Scotland, and while at Cambridge had described her work to Louis. He had responded with an "enthusiastic and vivid account" of the Rift Valley lakes, then invited her to his camp, a kindness that she later decided was "characteristic of Louis." "He [could] see someone else's problem [in Penelope's case, the need for comparative lakes to study] and produce an imaginative response to it, at no small cost to himself in time and energy."

Louis's team—aside from his brother, Frida, and Powys Cobb—were greenhorns in the African bush, and he took somewhat of a "schoolboyish" delight in introducing them to the wilds. For Elisabeth and Cecily, Louis planned a special welcome. Accustomed to the civilities of Cambridge, the two women were deposited, after a long journey, at the Elmenteita train station—"three corrugated iron huts dumped in the middle of a vast plain"—which lay eight miles from camp. It was pitch-dark outside, and they waited anxiously on the platform, surrounded by a group of curious natives, for Louis to arrive. At last he drove up in his Chevrolet box-body car, quickly loaded up their trunks, and headed off into what appeared to Elisabeth as "quite the back of beyond." In her first letter home to her family, she described this "most perilous & exciting drive":

"L. said we might easily meet some rhinos or lions & he had a rifle strapped on to the car by his side in case. The road was a sea of slippery mud covering most awkward undulations & we skidded wildly every few yards. L. said it did not matter as we had the whole of Africa to skid into but I didn't feel convinced by that at all. Further on we got to the bush & L. said we might possibly hear a lion so we stopped the car for a minute & listened & sure enough there was a melancholy roaring going on quite near. I began to wish I had brought a gun."

Later that evening at camp, another "lion" interrupted their dinner:

. . . in the middle of supper . . . we suddenly heard screams & yells & all the natives careened out of the house & down the road as hard as they could go. All the men jumped up & seized their guns & joined in the pandemonium. Several shots were fired & then they came in & said a lion had just gone past. I felt awfully thrilled & wished more than ever that I'd brought a gun.[7]

In the morning Louis shamelessly confessed that both "lions" had been hoaxes, the "roaring" created by someone hiding in the bush. The evening's excitement had been designed "to provide us with some of the thrills of wildest Africa while we were still innocent! . . . but I wish they had never told us," Elisabeth sighed, "as it was so exciting and now seems so tame and disappointing."

Louis himself considered their camp life to be "almost as humdrum as it might be in a suburb of London." And perhaps it was for him. Certainly lions never visited their camp, but hyenas laughed eerily in the night, the hyraxes shrieked, and everyone was ordered to carry flashlights if out after dark to avoid treading on puff adders, cobras, and mambas. By day, the neighboring Maasai with their long spears, glinting brass armbands, and red-ochred braids were regular visitors, and sometimes hunters of the Wanderobo tribe came down from the Mau forest bearing game or honey to sell.

Other provisions—primarily vegetables and maize meal—were bought at the little Indian-run shops in Elmenteita. To supplement their rather

[7] Louis tells much the same story in his autobiography *White African.* In his version, however, the second "lion" is described as a "raid by the Masai," complete with shouts from "young Powys Cobb and MacInnes to 'look after the women.' "

"rough and plain" diet, the team members went hunting at least once a week. "We lived by the gun on that expedition," said Frida. "I had a high velocity, flat trajectory, point five rifle which somebody had given me for a wedding present. And I did some shooting, but I didn't like it. You had to be out by 5 AM, and if you didn't kill at once, you had to go on until you found your wounded buck. I did it as little as possible; I didn't like shooting the animals." Louis viewed the hunting as a matter of practicality, which it was, and nearly always had his gun at the ready, sometimes jumping up to fire at hyraxes and guinea fowl over the windshield of his car.

Like nearly all Europeans in East Africa, the group wore a "bush kit" of khaki clothing and "double terais"—two felt hats stacked on top of each other with a layer of red flannel sewn in between, to protect them from the sun's actinic (ultraviolet) rays, which supposedly damaged the brain and nervous system. "In those days, nobody moved without their terai," said Frida, adding that the wide-brimmed hats made Louis's team look "quite smart."

Life at the camp soon settled into a regular routine, with breakfast at seven-thirty and work until dark. Louis gave each member particular assignments to carry out—assisting on the excavation, sieving the dirt for artifacts and bones, surveying and mapping, or exploring for new sites—and spent much of his own time stretched out flat on his stomach, or edged up on one elbow scratching the earth away from a skull or stone tool with a dental pick. "Louis was the mainspring of all our activity," recalled Solomon, "and as a 'doer' and enthusiast he was wonderful." Although his absolute devotion to his work sometimes irritated others (" . . . he put the work absolutely first regardless of the comfort, safety or happiness of his workers," said Elisabeth Kitson), he did manage to keep his team together for the entire eight months. "If you were part of an expedition of which Louis was the leader," said Frida, "you were totally involved or you went home. But nobody ever did. He did inspire enormous loyalty in a very remarkable range of human beings."

Although Louis had not yet found handaxes similar to the Chellean culture, he was optimistic about their discovery, and believed he was working in the right area. "[W]e were gradually working backwards [in time], and so my hopes were strengthened that one day, sooner or later, I would find traces of the oldest known types of stone tool, and possibly even bones of the men who made them." In 1927 he and Newsam had located two occupation levels in the cave, and now, working from September 1928 to April 1929, he uncovered two more. Obsidian flakes and tools were thick in these lower two levels; they catalogued nearly five hundred tools and thousands of flakes every day, often working into the night by hurricane lamp. "By that time everybody else was exhausted," said Frida. "But Louis would just go on and write the day's report everyday. . . . He had a kind of—dedication is the wrong word, because it sounds pious and it wasn't—it was a sort of monomaniac interest in the thing."

Louis's infectious curiosity and eye for how things worked often gave him insights that more theoretically minded people may have missed. He favored an empirical approach to prehistory, teaching himself how to make stone tools from obsidian and then using them to butcher small antelopes, scrape a skin, or even perform minor surgery. He once lanced a painful boil Elisabeth Kitson had developed, using an obsidian tool from the cave—"a rather typical experiment!" she recalled. "Louis possessed a special antennae," said Frida, "a way of making imaginative jumps" about the objects the team found. The group was especially perplexed by small obsidian fragments they were finding. Louis then unearthed a group of the shiny chips lying side by side *in situ,* "and he made one of these jumps," said Frida. "They were the barbs for making a fish harpoon."

By October 1928 the team had reached the fourth level of the cave, which was littered with obsidian tools. Based on their shapes, Louis decided the tools resembled those made by people of the Aurignacian culture in Europe, who had lived about 20,000 B.C.[8] In his monthly report for October, Louis argued the opposite position. "[I]t would not be surprising," he wrote, "to find that the Aurignacian or Capsian culture originated south of the Sahara, possibly in Kenya itself, and sent off waves north and south." Around the sorting table, Louis was inclined to push his unconventional views even further. "Nobody could possibly fail to notice Louis' fanatical adherence to the belief that the human genus first emerged in tropical Africa," recalled John Solomon. "This belief might be regarded as 'religious' rather than 'scientific,' and he eagerly grasped at any datum which might corroborate it"—a tendency, Solomon noted, that led Louis "to attribute the maximum possible antiquity" to all his discoveries.

Because he had no way of dating his finds accurately, Louis relied on a theoretical dating framework devised by E. J. Wayland, a geologist who, in 1920, had made the first geological survey of Uganda. Wayland, a stockily built, red-haired man with both a strong jaw and constitution, made his survey by foot and canoe, walking nine hundred miles and paddling three hundred. On his travels, he discovered traces of prehistoric lake beaches situated high above the current lake levels, and he decided that they were evidence of past climatic change—an African equivalent of the European Ice Age glaciations. Based on his study of lake stratigraphy, Wayland surmised that the tropics had undergone corresponding periods of rain and drought as the snows on the highest peaks—Mounts Kilimanjaro and Kenya, and the Ruwenzoris—had advanced and receded. By correlating these rainy periods, or pluvials, with the glaciations of Europe, he believed he could develop a dating method for Africa.

---

[8] Carbon 14 tests made of the site in the 1970s showed that Gamble's Cave dates to only 6,000 B.C.

Louis had been following Wayland's pluvial theories, and in 1927, during his first expedition, had gone to Entebbe to meet him. The same year, Wayland visited Louis to inspect his excavations and the ancient lake terraces—resembling giant earthen staircases—above Lakes Elmenteita and Nakuru. In one place Louis showed an excited Wayland a terrace 145 feet thick; in another, a deposit of fish bones three miles from the present-day fishless, alkaline waters of Lake Nakuru. These obvious fluctuations in lake levels indicated to Louis "important changes of climate"; in his view, Lake Nakuru had been 145 feet deep during the Pleistocene and had been dropping ever since. By the end of his first expedition, he believed that he had found data that "prove conclusively that during the Pleistocene there had been at least three pluvial periods, separated from each other by arid periods."

The "Pluvial Hypothesis" was a plausible but altogether fallacious theory; yet it was not until the 1950s that geologists were able to disprove it. "In retrospect," said John Solomon, "I believe that we all overestimated the importance of climatic change in modifying the nature and location of sedimentation." The raised terraces were, in fact, caused by the faulting and uplifts of plate tectonics, not enormous lakes; there was no relationship with the Ice Age. But Louis latched on to the "pluvials," and used them throughout his early writings to explain the climate of prehistoric Kenya and to date his finds. His insistence on giving his sites absolute dates and always pushing for dates of the greatest antiquity, however implausible, became a pattern on these early expeditions, as did his propensity for grandstanding and overstatement.

Yet by any standard, the 1928–29 season at Elmenteita was an overwhelming success. Louis excavated two more *Homo sapiens* skeletons from Gamble's Cave, made an excellent collection of stone artifacts, and, with MacInnes's help, identified the remains of a variety of extinct fauna. He had also organized excavations at sites that other team members found, and turned up more tools and skeletal material. It was not, however, until three short weeks before the season ended that he discovered what he was most hoping for: evidence of a handaxe culture. John Solomon actually made the discovery, but it was Louis who recognized what Solomon had found.

"If I remember rightly," said Solomon,

> it was while returning from Naivasha one afternoon that I took the opportunity to have a cursory look at the gullies running down to the river [at Kariandusi], and picked up a piece of green lava which I thought *might* be an artifact, although I was very doubtful. When I showed it to Louis, he, characteristically, had no doubts whatsoever, and sent Elisabeth and me back there the very next day, with instructions to collect some implements from *in situ* in the deposits exposed in the gullies! I was decidedly sceptical about our chances, but his hunch proved correct,

and we came back with an excellent haul . . . Louis' "hunches" were usually correct.[9]

Louis was ecstatic; the elegant pear-shaped axes were precisely what he had been searching for all these months: relics of one of the oldest known cultures. Based on the shape of the tools and the geology of the site, he estimated them to be from the First Ice Age, or between 40,000 and 50,000 years old. "The discovery was . . . of the very greatest importance," Louis wrote, "as it threw entirely new light upon the age of the great Great Rift Valley [*sic*]." Ironically, Louis was right on all counts except age—today, the Kariandusi tools are estimated to be 500,000 years old. The site is of relatively minor importance now, a little-visited national monument with a humble but well-kept museum and a staff that fairly pounces on the few tourists who do come for a tour. But for Louis, it provided critical evidence of an Acheulean culture in East Africa as sophisticated as those in Europe.[10] Triumphant, he closed the Elmenteita Camp and set out in a half-ton truck with Frida, John Solomon, Elisabeth Kitson, and a Kikuyu assistant for Johannesburg, South Africa. The British Association for the Advancement of Science had scheduled its 1929 annual meeting there, and Louis was eager to present his finds.

In 1929 the Great North Road stretched from Capetown on the southernmost tip of Africa to Nairobi, a distance of three thousand miles. It was part of the British vision of Africa, a road that would link its territories and colonies from Egypt to South Africa. Sections of the road are paved today, but when Louis's party headed south, the road varied from nothing more than a faint track to a pair of dust-filled ruts two feet deep. They were not the first to travel its length; in 1927 a Mr. and Mrs. Court Treat had journeyed by car the entire distance, from the Cape to Cairo. Yet it was a rare enough undertaking that when Louis's group reached Johannesburg six weeks after departing from Nairobi, the local newspaper heralded their arrival with a front-page story: "From Kenya Colony by Truck/Archaeologists' Venture/on £1 a Day Each."

Many of the leading scholars in African prehistory—H. J. Fleure, John Goodwin, C. ("Peter") van Riet Lowe, Raymond Dart, and Gertrude Caton-

[9] In his field notes and in *White African,* Louis attributes Kariandusi's discovery to John Solomon and Elisabeth Kitson. But in later accounts, he takes—or is given—the credit. One photo in the February 1965 issue of *National Geographic* magazine shows Louis bending over a cliff at Kariandusi, pointing to the stone tools below. The caption reads: "A near-fatal step in 1929 led Dr. Leakey to the first known living site of hand-ax man. At Kariandusi . . . he almost stumbled over a cliff and clutched at bushes to save himself. Peering over the edge, he spied an ax embedded in the wall." The *Visitor's Guide to Kariandusi Prehistoric Site,* published by the National Museums of Kenya, reads similarly: "It was Dr. L. S. [*sic*] Leakey, the leader of the expedition, who first noticed stone tools projecting from the cliff. . . ."

[10] In the 1920s, the Acheulean was considered the next step up from the Chellean or Abbevillean culture. Today it is used as a general term for handaxe industries.

Thompson—attended the British Association meeting in Johannesburg. Rumors of Louis's discoveries and the dates he was assigning to his finds had preceded him. Even before leaving Nairobi, he had received two cautionary letters. "I hear you are probably going to the British Association meeting in South Africa," wrote Professor Haddon from Cambridge. "If you do, I entreat you to be careful what you say. Naturally you will tell them of what you have found, but do not go in for wild hypotheses. These won't do your work any good and it's foolish to try to make a splash." E. J. Wayland, the geologist from Uganda, wrote in a similar vein:

> Harm is done to the cause by overemphatic comments with regard to these correlations at this date. (Strictly between you and me and the gatepost, I found it necessary to defend your work . . . for the benefit of some folks at Cambridge who are interested in the bestowal of Fellowships.) Believe me you will serve archaeology, the expedition and yourself best by maintaining a strictly scientific attitude. You have the chance of making yourself, in time, one of the leaders of archaeological thought—don't spoil your chances, for by so doing you will unintentionally let the science down. I must ask you to forgive the sermon, it is delivered in all friendship.

Louis apparently heeded their advice, and presented a straightforward account of his work, "An outline of the Stone Age in Kenya," on the first day of the meeting. In it he described his latest expedition, and the sequence of East African cultures he had uncovered. Only two years before, Peter van Riet Lowe and John Goodwin had worked out the Stone Age sequences in South Africa. Louis's research was a significant addition to theirs and he was soundly applauded.

Yet the meeting was not without controversy. Gertrude Caton-Thompson gave her results on her excavations in what is now Zimbabwe. Contrary to popular belief, which held that the elegant stone towers there had been built by a northern race of people, she asserted that they were the work of Africans. "Raymond Dart didn't agree with that, and he gave a violent diatribe about how disgraceful this work was, that it couldn't possibly have been done by local Africans," said J. Desmond Clark, now professor emeritus of African prehistory at the University of California, Berkeley. Her speech so angered Dart that he stalked out of the meeting. "That really upset Gertrude and I don't think she talked to him again until 1955 when they met [at the Pan African Congress] in Livingstone." Caton-Thompson's conclusions were correct, but the colonial prejudices of the time blinded many people, including scientists.

Ironically, Dart endured the same prejudice over his find, the perfectly fossilized skull of a young, five-to-six-year-old chimplike creature. He had found the skull in 1924 in South Africa and had decided almost at once that it must have balanced on the spinal column of an upright creature—one that walked on two legs, not four. To Dart, the fossil, with its small apelike brain, and small humanlike canines, seemed a transitional creature, neither fully human nor fully ape, and he described it as such in the February 7, 1925, issue of *Nature.* He also judiciously labeled it *Australopithecus africanus,* or Southern Ape of Africa. However, none of the eminent British anatomists and anthropologists supported his views. They had not seen the fossil, but they were certain the creature was really nothing more than a chimpanzee, and a four-footed one at that. Dart believed the 1929 meeting would give him the chance to vindicate his views. Anthropologists and anatomists would see the skull, examine it, and agree with him. But nothing of the sort happened. No one really cared to see his "Taung Baby," as the skull had been nicknamed (after the site where it had been found), and those who did made "noncommittal comments." "It was obvious that few regarded it as anything of real importance in the evolutionary story," Dart wrote forty years later. Like Caton-Thompson, he would eventually be proved right.

For his part, Louis enthusiastically supported Caton-Thompson's theories about Zimbabwe, and they became close friends. He was less sure about Dart's find. "Louis did not seem to have much respect for him," observed John Solomon. The Taung skull did not fit Louis's idea of what early humans should look like; very likely he was one of those making "noncommittal comments."

His own paper had been so well received that nearly sixty scientists decided they would travel to Kenya to inspect his excavations. Louis was naturally pleased by this, but also somewhat dismayed. Instead of lingering in South Africa to visit archeological sites as he and Frida had planned, they were forced to drive back to Nairobi as soon as the conference ended. This time they made the three thousand miles in fifteen days. Groups of scientists began arriving shortly thereafter, and Louis spent the next several days squiring such luminaries as Professor Fleure (then *the* grand old man of prehistory) and Sir Julian Huxley around his digs. "Nobody could seriously fault Louis' excavation techniques or his recording of the provenance of the material obtained," said John Solomon, who had accompanied the Leakeys back to Kenya. The visiting scientists returned to England with words of praise for Louis, and Haddon wrote back congratulating his star pupil and John Solomon on their work.

Louis and Frida sailed for England in November 1929. St. John's College had awarded him a lucrative two-year fellowship, and they settled into a small cottage a short distance from Cambridge. But Louis was already planning a new expedition. "Almost as soon as I was back in England I started making plans for a third season in East Africa, this time with the Oldoway [Olduvai] Gorge as my main objective."

Chapter 3

LAYING CLAIM TO THE EARLIEST MAN

In 1929 only a very few scientists believed that humans had originated in Africa. Instead, attention was focused on China, where W. C. Pei and Davidson Black had just discovered the first skull of Peking Man. Unlike Raymond Dart's Taung Baby from South Africa, Peking Man received great attention and acclaim. Its thick cranial vault and heavy brow ridges marked it as primitive, but it was also clearly a human relative. Newspapers heralded the discovery, and the scientists who had once derided Black's claims for an Asian Garden of Eden now applauded him.

In fact, Peking Man was something of a godsend to scientists studying human origins. Aside from Dart's unrecognized Taung Baby, no fossils of significance had been recovered for fifteen years. There had also been the unfortunate matter of *Hesperopithecus haroldcooki,* a supposedly ancient relative from Nebraska that turned out to be nothing more than an extinct form of pig. Peking Man was unquestionably a hominid (in the human family). But how did it relate to modern people? Where did *Sinanthropus pekinensis* fit in the human family tree?

Most representations of this tree in the 1920s showed a trunk sprouting out of an unknown past, then branching into an array of creatures. At ground level were shrews, lemurs, tarsoids, and monkeys. Above them came orangutans, chimpanzees, and gorillas. And appearing in the upper branches were Negroes, Mongoloids, and Australoids. The zenith position, of course, was held by Caucasians. Curiously, all of the authentic early human fossils known to science—the few Neanderthal specimens, the skullcap (the top of the skull) and leg bone of Java Man, and now the skull of Peking Man—were

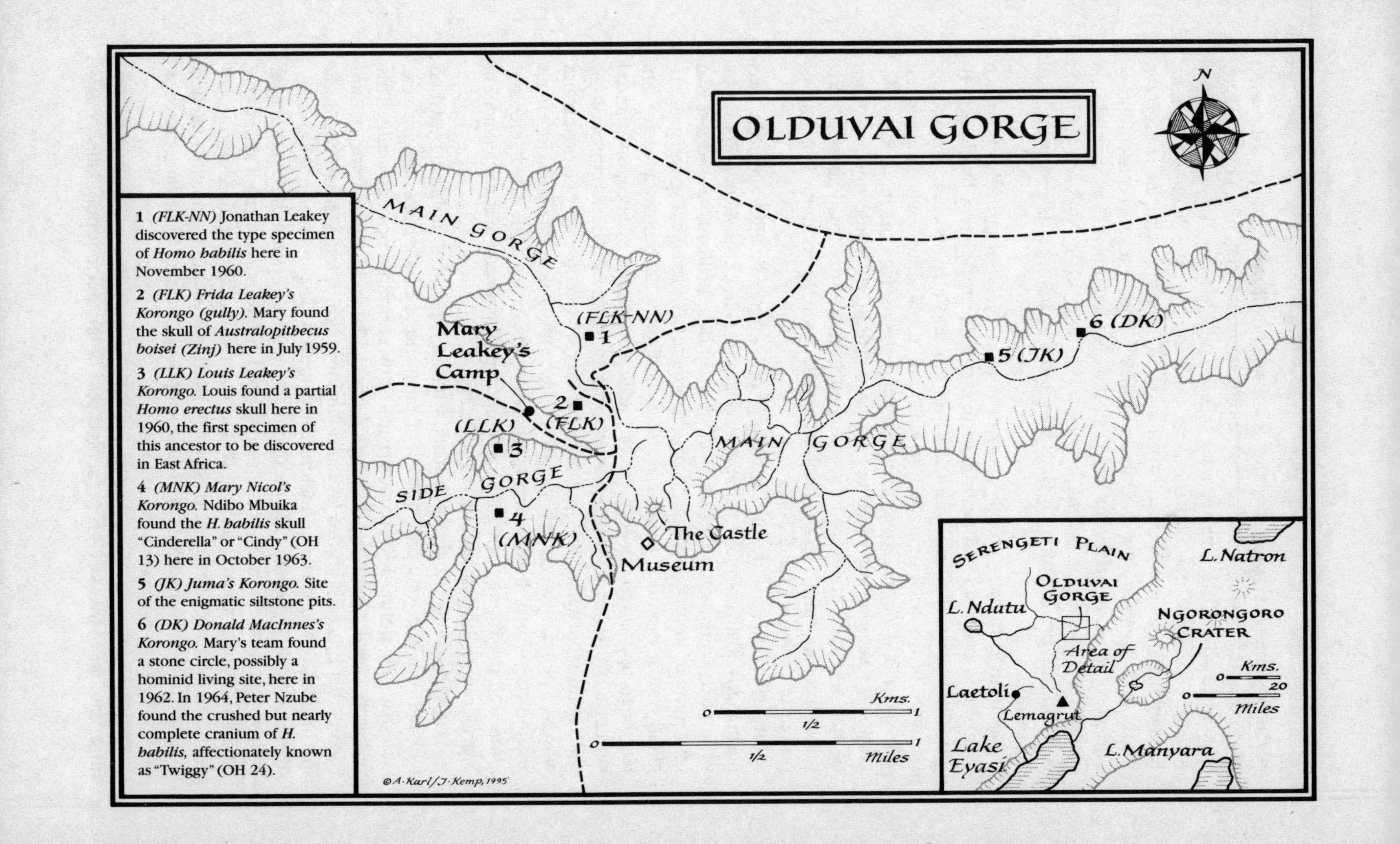
OLDUVAI GORGE
N
MAIN GORGE
(FLK-NN)
1
Mary Leakey's Camp
2
(FLK)
(LLK)
3
SIDE GORGE
4
(MNK)
MAIN GORGE
The Castle
Museum
5 (JK)
6 (DK)
Kms.
0
1/2
1
Miles
SERENGETI PLAIN
OLDUVAI GORGE
L. Ndutu
Area of Detail
Laetoli
Lemagrut
Lake Eyasi
L. Natron
NGORONGORO CRATER
L. Manyara
20
©A·Karl/J·Kemp, 1995
1 (FLK-NN) Jonathan Leakey discovered the type specimen of Homo habilis here in November 1960.
2 (FLK) Frida Leakey's Korongo (gully). Mary found the skull of Australopithecus boisei (Zinj) here in July 1959.
3 (LLK) Louis Leakey's Korongo. Louis found a partial Homo erectus skull here in 1960, the first specimen of this ancestor to be discovered in East Africa.
4 (MNK) Mary Nicol's Korongo. Ndibo Mbuika found the H. habilis skull "Cinderella" or "Cindy" (OH 13) here in October 1963.
5 (JK) Juma's Korongo. Site of the enigmatic siltstone pits.
6 (DK) Donald MacInnes's Korongo. Mary's team found a stone circle, possibly a hominid living site, here in 1962. In 1964, Peter Nzube found the crushed but nearly complete cranium of H. habilis, affectionately known as "Twiggy" (OH 24).

placed by themselves on separate side branches. None of these supported the prevailing notions of what early humans should look like; consequently they were all regarded as separate species, other "types" or "races" of humanlike creatures that had died out. According to the leading theorists, humankind's earliest ancestor would be "a creature with an overgrown brain and ape-like face." Only one fossil met these requirements—*Eoanthropus dawsoni,* popularly known as Piltdown Man. Discovered in a gravel pit in Sussex, England, between 1911 and 1915, Piltdown Man had both a large brain and an apelike jaw, and was immediately accorded full ancestral status.

Not until forty years after its discovery would scientists agree that Piltdown Man was a hoax, an elaborate ruse that brought together a five-hundred-year-old human skullcap with the lower jaw and incisors of an orangutan. Unaware of the trick being played on them, the scientists who announced and described *Eoanthropus* congratulated themselves on their foresight: Piltdown was precisely what they had predicted they would find, a "blend of man and ape." It colored the interpretation of every fossil subsequently uncovered until the 1950s—and influenced several generations of anthropologists, including Louis Leakey.

Sir Arthur Keith, one of the leading proponents of Piltdown Man, was particularly instrumental in shaping Louis's thinking. "Sir Arthur Keith was very much Louis's father in science," noted Frida. Brilliant, yet modest and unassuming, Keith was regarded at the time of Piltdown's discovery as England's most eminent anatomist and an authority on human ancestry. He had been appointed conservator of the Royal College of Surgeons' famed Hunterian Museum and was a regular contributor to the Royal Anthropological Institute journal, *Man*—positions he used to become, in one observer's words, a "one-man 'court of appeal'" for physical anthropologists from around the world. It was largely Keith's opinion that squelched scientific interest in the Taung Baby, and his opinion that assured Piltdown a place on every drawing of humankind's family tree.

A superb scholar, Keith had learned what he called the "alphabet by which we spell out the long-past history of man and ape" after spending a year systematically noting 150 characteristics of two hundred primate skulls. This research, coupled with his dissections of primate and human embryos, and reconstructions of both the Java and Piltdown skulls, led Keith to conclude that modern man was "an ancient form, with a past immeasurably longer than is usually believed." He thought that humans separated from the apes at the beginning of the Miocene (then dated at one million years ago, today dated at twenty-two million years), and soon thereafter acquired both erect posture and a large brain. The Piltdown skull, whose brain capacity matched that of modern humans and which supposedly came from late Pliocene deposits, seemed to confirm his view.

Keith believed that the other major fossil discoveries—Neanderthal, Peking Man, and Java Man—represented part of an "ancient population made up of, not mere varieties of one species of mankind as at present, but of totally different species and genera." He compared humankind's lineage to that of the modern primate world, which is similarly divided into separate but related genera—chimpanzees, gorillas, orangutans, and humans. "Amongst this complex of ancient humanity," he argued, "we have to seek for the ancestors of modern man." These "true" ancestors would be distinguished from the others by the size of their brains. "Man is what he is," wrote Keith, "because of his brain."

In 1915 Keith elaborated on his ideas in a highly influential book, *The Antiquity of Man,* which Louis surely read as an undergraduate at Cambridge. Louis also may have attended some of Keith's popular lectures on human origins. But the two men did not meet until the spring of 1927. Louis then appeared at the Hunterian Museum in London fresh from his first archeological expedition, bearing both "human remains and prehistoric stone implements." His entrance may or may not have been planned for effect, but either way, Keith was impressed. He offered Louis use of a research room in his museum, plus guidance in reconstructing the skulls from Elmenteita, then sent a glowing report to the museum committee: "As to the remarkable ability of this young man—his courage under great difficulties, and the importance of his discoveries, there can be [no] doubt . . ." Already an admirer of Keith's, Louis now became his disciple, learning from this "helpful and kind teacher" the intricacies of skeletal anatomy.

Keith had also assisted Louis in gaining grants for his second expedition. So in November 1929, shortly after Louis and Frida returned from Kenya, Keith once again welcomed the fast-talking, energetic Louis Leakey to his museum. This time Louis arrived with a nearly complete skeleton from the lowest level of Gamble's Cave and an "enormous quantity of artifacts." The skeleton was still encased in its rock matrix, and Louis planned to chip this away, clean the skeleton, then share his latest find with Keith. Louis believed the skeleton was twenty thousand years old, the same date that he assigned to the two small fragments of pottery he had uncovered. Like a growing number of scientists, Keith was intrigued with Louis's archeological discoveries. "[I]f Mr. Leakey's scheme is well founded," noted Keith, ". . . then the men living in the ancient . . . valleys of East Africa were many thousand years ahead—in a cultural sense—of their contemporaries in Europe." Keith was not entirely comfortable with this idea since it went against accepted thought, and he was as curious as Louis to see what one of these supposedly early pottery makers looked like.

As before, Keith arranged a workroom for Louis in the museum's basement. Here, surrounded by the huge white bones of Greenland whales, Louis and the paleontologist Donald MacInnes chiseled away at the skeleton, a painstaking task that required "weeks of work." Frida also worked as part of Louis's little team, helping to sort, clean, and catalogue the artifacts and mend the fragments of animal fossils.

Keith stopped by occasionally to see how they were progressing, and was once photographed by *The Times* with Louis "unwrapping a skeleton." Sometimes the two men sat together discussing the various skulls Louis had unearthed at Nakuru and Elmenteita—Keith, white-haired and elegant in his dress, carefully assessing their similarities and differences; Louis, somewhat rumpled and hasty in manner, eagerly pushing his pet theories. Louis listened to Keith and valued his assessments, but he also had his own opinions. "He was a man who formed his own judgments about things," noted Keith.

One of Louis's more curious opinions—and one not shared by Keith—was his belief that his Elmenteitan people (of which the new skeleton was another example) were "non-Negroid." All of the Elmenteitans had narrow, long-headed skulls, traits that Louis and some other anthropologists considered "Mediterranean" or "Caucasoid." Louis, like his archeology professor, Miles Burkitt, subscribed to the sub-Saharan theory, an idea which held that modern peoples' immediate ancestors had favored the warm regions of sub-Saharan Africa during the Ice Ages. Louis thought his Elmenteitans were representative of these ancestors. He pictured them migrating from Kenya to the north after the glaciers had retreated, taking their supposedly advanced pottery skills and genetic traits with them. Keith was more cautious: "Mr. Leakey believes that the prehistoric people of [Kenya Colony] were more white than black, whereas I am of opinion that the opposite was the case—they were more black than white." Keith, of course, was right: the skulls of Louis's Elmenteitans were not any longer or narrower than those of the Maasai or Tutsi people.

Another opinion Louis formed while working on his new skeleton was, however, correct. To his eyes, the Gamble's Cave skeleton seemed nearly identical to that of Olduvai Man, the skeleton that the German paleontologist Hans Reck had unearthed in Olduvai Gorge in 1913.[1] If so, then Reck's estimate of half a million years for his find was wrong, and Olduvai Man could date only to the middle Stone Age. Just before Christmas, Louis traveled to Berlin to discuss the dating with Reck. More importantly, he wanted to invite Reck to join his upcoming Olduvai expedition.

Their meeting did not resolve the dating issue. Reck obstinately shook his head over Louis's arguments. But he did eagerly agree to join Louis's expedition to Olduvai. He had never expected to see the gorge again and had, said Louis, "resigned himself to his fate." Excited by these new prospects, Reck pulled out his Olduvai rock and fossil collections and generously shared them with Louis. One rock in particular caught Louis's eye. It was pointed at one end, rounded at the other, and decidedly resembled one of the Kariandusi handaxes John Solomon and Elisabeth Kitson had found; but Louis said nothing about this to Reck. Instead, after intently examining the collection,

[1] The Maasai knew the gorge as a source of water in the springtime, and named it *"ol duvai,"* or "the place of the wild sisal."

he announced that "I was now certain in my own mind that the greater part of the Oldoway [Olduvai] . . . deposit was probably of the same age as the beds at the Kariandusi River . . . and I expressed the opinion that we should find implements of this [the Kariandusi] culture at Oldoway." Reck again shook his head. He had already searched the gorge "diligently" for stone tools and found none; Louis would be disappointed if he expected to find more than the "bones of fossil animals."

Louis persisted and then proposed a bet: "I . . . made a small bet that I would find Stone Age implements at Oldoway within 24 hours of arriving there." Reck laughed and offered a hand across his boxes of bones and stones, never catching on to the twinkle in Louis's eye. The two men shook hands, and the bet was on.

While Louis traveled "hither & thither meeting people," Frida worked at home, flakes of obsidian and ink pens spread out around her, carefully sketching the stone tools that would illustrate her husband's first book, *The Stone Age Cultures of Kenya Colony.* Louis and his love of prehistory had quickly become the focus of Frida's life, and she viewed herself as his partner in a "joint adventure."

Louis and Frida had been married a year and a half now, and were very much in love. Frida had a "whole-hearted and lasting devotion to Louis," recalled a mutual friend "and Louis, in turn, always [seemed] ready and willing to discuss every sort of thing with her." They both loved talking and sharing ideas, and if Louis's interests always came first, Frida did not mind. It was the science and Louis's contributions to it that mattered. "I was awfully interested in the science, in the archeology," she said. "And you couldn't possibly not be interested in Louis."

At St. John's, Louis's field research had received great acclaim, with one unnamed reviewer on his fellowship application telling Professor Haddon, "Mr. Leakey's . . . work in Kenya is about the most important work of our generation in Anthropology and . . . it is the duty of Cambridge to see that it is not allowed to get into difficulties." In response, the college once again awarded him a fellowship and provided him rooms and a lab for his research, where he analyzed the stone artifacts he had brought back from Kenya. Louis was also teaching an archeology course, writing two books, and planning his next expedition. He had only a year and a half to arrange this, as he hoped to set out for Olduvai in the summer of 1931.

Louis's third expedition was his most ambitious yet. He enlisted the support of the British Museum of Natural History, and they agreed to send along their paleontologist, Dr. Arthur Tindell Hopwood. Reck, of course, had signed up, as had Donald MacInnes and Vivian Fuchs, who had just graduated from St. John's with a degree in geology. Fuchs, who would later be knighted for his explorations of the Antarctic, had sat in on some of

Louis's rather informal archeology courses, and remembered Louis teaching the students how to make string figures and fire with sticks.[2]

By early summer, 1930, Louis had completed most of the text for *The Stone Age Cultures of Kenya Colony,* then began writing a companion volume, *The Stone Age Races of Kenya.* Almost immediately, he hit a snag: the book required illustrations of the skulls he had uncovered, but for these to be drawn properly, Louis had to spend hours taking "innumerable measurements with calipers." It was boring, tedious work, and when a "mechanically minded" friend, J. Harper, suggested that they design a drawing machine, Louis eagerly turned inventor. Looking something like a Victorian medical contraption, with wire cables, drawing pens, and screws supported on a wooden frame, the thing actually worked. Louis and Harper announced it in *Nature,* and took out a patent on the Leakey-Harper Drawing Machine. One machine was installed in the Royal College of Surgeons, while a second one was sent to Japan—where in Japan is not known.[3] With his usual entrepreneurial flair, Louis demonstrated his machine one evening at a Royal Society *conversazione*—special gatherings of scientists who had new discoveries to show.

The drawing machine did eliminate the tedium of producing the illustrations, but in the meantime Louis had decided to delay his second book. With luck, he might uncover a skull alongside the handaxes he expected to find at Olduvai. Such a discovery would be of immense importance, and Louis took the gamble.

For the time being, Louis's archeological research earned him the most acclaim. Dr. Miles Burkitt, his archeology professor, was "astonished on seeing the finds brought back by Mr. Leakey from Elmenteita." He joined with Professor Haddon, who regarded Louis's research as "first class pioneering work . . . that [has] brought great credit to English Archaeology and Anthropology," in recommending that Louis receive a doctoral degree from Cambridge, and on October 14, 1930, the Board of Research Studies awarded him his Ph.D.[4]

The way to Olduvai seemed smooth now: St. John's again awarded him a grant, as did the Royal Society, the Percy Sladen Trust, and the British Museum. Those monies, combined with Frida's small income from her "Edwardian dress allowance," would pay for another full year's research in Africa. He even had enough, he thought, to visit another possible site in western Kenya, a place called Kanjera that a missionary from the Church Missionary

---

[2] "[W]e didn't write weekly essays," recalled another of Louis's students. "We just listened to LSBL talk and watched him chipping flints. There was no furniture in his college rooms: we sat on packing cases, listened, & sorted flints."

[3] Later Louis had a third one made and shipped it to Nairobi, where it is housed in the National Museums of Kenya.

[4] Louis's doctorate was awarded on the basis of his book *The Stone Age Cultures of Kenya Colony,* as well as research papers on such subjects as the human remains from Elmenteita, the social life of the Maasai, and native peoples' mutilation practices. He also had an oral examination.

Society had recently written him about. The missionary, the Venerable Archdeacon Walter E. Owen of western Kenya, was an avid preacher and fossil collector. He sent Louis both letters and fossilized limb bones from his site, and Louis "made a note that I would visit Kanjera if possible during the 1931–2 season."

Louis and Frida's first child, Priscilla Muthoni—"Muthoni" being a Kikuyu word of affection—was born April 13, 1931, the same day that Frida completed the final drawing for Louis's first book. Priscilla was a round, chubby baby, with the dark hair and eyes of her parents, and both Louis and Frida were enchanted. Still Louis was preoccupied with Olduvai. "It makes me sad to have to admit that until our little daughter, Priscilla, was about eighteen months old I did not see very much of her," he later wrote in the second volume of his autobiography, *By the Evidence.* "I was exceptionally busy with my research work, usually leaving home early in the morning and not returning until after she had gone to bed."

Only two months remained before Louis departed for Kenya, and as usual he was pushing himself to the limit with work and projects. There were expedition details to attend to; galley proofs of his book to read; courses at Cambridge to finish teaching; and papers to publish in *Nature,* the *Geographical Journal,* and the *Journal of the Royal Anthropological Institute.* The long hours and close work took their toll, and Louis again developed severe headaches, which led to fainting spells. These attacks had never actually gone away. Just before completing his undergraduate degree, he had collapsed unconscious on a trip to France, clutching his forehead and crying out, "Oh, my head, my head!" And even on his field expeditions, he had suffered similar "moments of loss of consciousness," as Frida called them. "I was always liable to collapse unconscious if I overworked," Louis himself noted. In fact, his collapses were full epileptic seizures, and frightening for those around him: he passed out, his body stiffened and shook violently, his face flushed red, and he frothed at the mouth.

His attacks worsened in early 1931, and in April he visited a specialist, who wrote his doctor:

> His attacks are typically epileptic. . . . I think they are the result of an old cerebral contusion due to his accident in the rugby match in December, 1923. . . . I would not look upon his case as a severe one and I think there is a good chance of the fits being checked with moderate doses of sedative. All the attacks have occurred in the morning before 11 o'clock, and one of them was when he was doing hard physical work in Africa and two when he was tired. I would suggest that you put him on Luminal gr.I to be taken at bed time and again on awakening. . . . He must quite definitely give up driving a car.

Louis took the Luminal off and on for most of his life, but he never stopped driving. In fact, his first purchase upon arriving in Mombasa at the beginning of his "big expedition to Oldoway" was a box-body Chevrolet. He had left Frida and Priscilla in June 1931 and traveled overland to Marseilles, then sailed by steamer to Kenya, arriving July 13. Frida and Priscilla would join him in September, about the time the full expedition got under way.

As with his previous expeditions, Louis planned to use his parents' home as his base camp. Harry and May had retired as missionaries the previous year and were now living in the cool highlands of Limuru, eighteen miles from Nairobi, where they had purchased a small plot of land. On July 15, two days after disembarking at Mombasa, Louis arrived "completely tired out" in his new car, having driven the 350 miles to Nairobi by himself—but also having spent "considerably less than the price of a second-class train ticket." As usual, his budget was limited and tight.

He now set to work building a three-room mud-and-wattle bungalow for his family on his parents' farm. The hut featured a corrugated iron roof lined with banana bark to lessen the heat, an earthen floor, and "proper doors and windows"—all done at the minimal price of £36. "Some of my less kind friends say, 'It looks like that,' " Louis wrote, "but at least it served its purpose." His parents were especially excited at the prospect of having their daughter-in-law and first grandchild so close by. On the previous expedition they had welcomed Frida "from the bottoms of their hearts," although she acknowledged, "I couldn't have been exactly their cup of tea. Religion was at the center of their lives . . . and they knew that I was much more of a liberal thinker."

Although Louis had abandoned the strict Christianity of his parents (and no longer gave sermons as he used to), he never lost his faith. Instead, he melded his science and religion, preferring to see the biblical story of creation as a metaphor for the actual events. "Louis would always indignantly say that none of his scientific pursuits ever contradicted the Bible," recalled his sister, Julia. "I heard him at one of his lectures. A young man stood up and said, 'Please, sir, how does your theory compare with the Bible?' And Louis said, 'Nothing I've ever found has contradicted the Bible. It's people with their finite minds who misread the Bible.' "

Louis had set his eye on Olduvai Gorge largely because he had come to the conclusion that in East Africa "early Stone Age man" would have inhabited regions below four thousand feet. He based this opinion on his "pluvial" hypothesis, which would have made higher elevations much colder and presumably less "genial" to early humans. Louis also thought that early man was "essentially a hunter who lived where game was most plentiful." This, he reasoned, would most likely be near an ancient lake where herds of animals had once gathered. Finally, any area he investigated had to have fossil exposures. Olduvai seemed to fit all of these criteria—now the only problem was getting there. "[T]he question of finding actual human remains had to depend upon whether Dame Fortune decided to reward us or not."

Olduvai Gorge lies 260 miles south of Nairobi in the trackless savanna of Tanzania's Serengeti Plain. The first European to see it was a German entomologist, Professor Wilhelm Kattwinkel, who in 1911 headed a medical expedition to the distant corners of what was then German East Africa. In lectures and conversations, Louis liked to depict Kattwinkel as a nearsighted butterfly collector who discovered the gorge by nearly tumbling into it while chasing a specimen. But Kattwinkel actually was on a full-blown scientific expedition, researching the natural history of the area and seeking out the insects responsible for sleeping sickness. He may not have lost his footing at the edge of the gorge, but he did explore its eroded slopes and brought back a small collection of fossils. These caused considerable excitement in Berlin, for one turned out to be an unknown species of three-toed horse. Three-toed horses were well known from European deposits of the early Pleistocene, but none had ever been found in East Africa. Plus, it appeared that the small horse had survived in Africa long after it became extinct in Europe. The idea was tantalizing and two years later, in 1913, Hans Reck had set out, with the Kaiser's blessing, to make a complete study of Olduvai Gorge.

Reck had traveled with one hundred porters from the coast at Dar es Salaam, following a route that crossed the Maasai Steppe at the foot of Mount Kilimanjaro, then wound its way up the slopes of Ngorongoro Crater before descending onto the Serengeti Plain. Louis, however, proposed to reach Olduvai by car and truck. "There was a road for the first 110 miles and a rough track for the next 100 miles," noted Louis, "but after that it was a question of finding the best way . . . across the Serengeti plains." For most of the year, the Serengeti is a rippling, golden grassland, dotted with thorn trees, cut by eroded washes, and home to great herds of antelope, zebra, and giraffe. Lions are plentiful as well, and Louis, not wishing to take any chances, hired Captain J. H. Hewlett to guide and protect his party. Hewlett had served in the Indian Army, and had enough of a reputation as a scout and a hunter that he had been employed by the Prince of Wales (later Edward VIII, and still later the Duke of Windsor) on his Kenya safaris. At the beginning of September, Hewlett set off by truck to find a "practicable route to Oldoway."

A little over two weeks later, Hewlett was back, having forged a pioneer track right to the edge of the gorge. The senior Leakeys' small farm was now swept up in activity as Louis's Kikuyu workers sorted and packed food, gasoline, medical supplies, spare truck and automobile parts, camping equipment, and excavating tools. This time Louis could not rely on the little Indian shops for his supplies, for at Olduvai he would be two hundred miles from the nearest town. The piles of equipment were loaded into three one-and-a-half-ton trucks and the box-body car, and a little after twelve noon on September 22, Louis's convoy departed for Olduvai.

His initial party included Hans Reck, Captain Hewlett, and Arthur Hopwood, the British Museum paleontologist. Donald MacInnes was delayed in England by the death of his father, and Vivian Fuchs, who had been exploring Kenya's Northern Frontier District with a Cambridge University team, now lay languishing with malaria in a Nairobi hospital; he and MacInnes would come to Olduvai on the next supply run. Louis also had a "native staff" of eighteen Africans, including "two lorry [truck] drivers, a cook and a general camp boy; Professor Reck's own two boys, Captain Hewlett's two boys, Juma [Gitau] my best bone searcher, a mammal skinner . . . and ten of my excavators."[5]

"[P]rogress was slow," Louis noted, after covering only 120 miles at the end of the first two days. The trucks were heavy, the road "very dusty," and there was nothing to do but inch along. By the end of the third day, after traveling on an "appalling" track that was occasionally used by Indian traders who bartered with the Maasai, they were still seventy miles from the gorge. But now even that track disappeared, and they jounced along in low gear at a painfully slow five miles an hour, with radiators boiling fiercely. Every fifteen miles, they had to stop, uncap the radiators and pour in four gallons of water, "a serious drain upon our water supply." On the morning of September 26, they were only thirty miles from Olduvai and Louis broke camp at the first light. Yet if he was eager to reach his goal, Hans Reck was even more so. "Professor Reck could hardly hide the emotion he was feeling at . . . returning to the scene of his very great scientific discoveries," wrote Louis. Reck was "visibly moved," and Louis drove ahead with him so that the professor would have "the honour of being the first of our party to set foot on the banks of the Gorge."

At ten o'clock in the morning, Louis and Reck stood together at Olduvai, studying its castellated layers of red, buff, and grey earth, and gazing across the wide gash it cut in the savanna to the Ngorongoro Mountains. With its dramatic setting and bounty of fossils, Olduvai was "a scientist's paradise," Louis wrote.

But there was a major problem to overcome: locating an adequate water supply. East Africa was then suffering through one of its cyclical droughts, and the waterhole that Reck had used in 1913 was now completely dry. He had also used a spring on the slopes of Mount Olmoti, a mere thirteen miles away, and, confident that he could relocate it, he set off with Hopwood and one of the Kikuyu workers in Louis's car. Reck's memory was not as good as he thought, and it took him far longer to find the spring. By then night had fallen, and to make matters worse, there was a total eclipse of the moon. Reck's small party did not make it back to camp until after 1:00 A.M.

It had not been an easy night for Louis, either, who had suffered in camp

[5] The "boys," of course, were adult men, yet Louis was not being intentionally racist. British missionaries considered even the adult Africans to be children who needed proper educating, and Louis's use of the word "boy" reflects this thinking. But within the larger colonial world, "boy" often carried with it all of the uglier connotations of racial bigotry and discrimination.

with "awful visions of a sudden and tragic conclusion" to his "long-planned expedition." He had stayed up until Reck and Hopwood's return, tending small "guide fires" on the outskirts of camp, and worrying about the "twenty pairs of eyes, some of which were hyaenas, but many of them lions," that flashed now and again through the firelight. Yet as the sun's rays touched the gorge, Louis was up, unable to sleep late as Reck and the others were doing, and far too excited by the "prospect of great discoveries at Oldoway." The previous evening, one of his workers had found a handaxe near the old waterhole, and "soon after dawn" Louis was there avidly searching for a tool *in situ.* It did not take him long to find one. "I was nearly mad with delight," he wrote, "and I rushed back with it into camp, and rudely awakened the sleepers so that they should share in my joy." Professor Reck lost both his night's sleep and his £10 bet.

In fact, stone tools littered the gorge. By the end of the first four days, Louis's team had recovered seventy-seven handaxes, made from volcanic lavas, chert (an impure rock), and quartz. Reck had missed them on his earlier expedition because he had been looking for tools made from flint—a material that does not exist in Africa.

Louis had come to Olduvai with "four main objects," and in the next four days he and his team achieved them all. He discovered evidence of a "prehistoric culture"; Reck relocated the "exact spot" of his Olduvai skeleton; Hopwood found "numerous fossil remains"; and Louis thought he had established a correlation between the Olduvai beds and those at Elmenteita. Reck's rediscovery was particularly remarkable as he had marked the site only with four wooden pegs, yet there they were, still jammed in the yellowish earth of Bed II. "There it is," said Reck, pointing to the site, "marked with my four markers." Reck then recounted the tale of the skeleton's discovery and persuasively argued that this *Homo sapiens* was not merely a recent burial in Bed II, as Louis and others believed, but was as old as Bed II itself. He dug a trench to show the clear demarcation between Bed II and what he thought was Bed III, then pointed to the excavation site again: the skeleton must have been placed in Bed II when the second layer of the gorge was still young. There was no other way to explain the skeleton's position or the undisturbed deposits above it.

Louis listened, studied the site, perhaps waffled a bit, and then, discounting all his previous suspicions, agreed: "Personally I believe that Professor Reck's man, despite his comparatively modern features, is not only by far the oldest *Homo sapiens* in Africa, but also probably older than any found anywhere else . . . ," Louis wrote in his field report. "To me then it seems that . . . the whole question of the antiquity of the *Homo sapiens* type will need re-examination."

Flushed with enthusiasm, Louis, Reck, and Hopwood drafted a letter to *Nature,* proclaiming that the question of Olduvai Man was solved—the skeleton did come from Bed II; it had been placed there when the bed was still being formed; and its age was thus close to half a million years. Then, with a load of fossils and stone tools, Louis headed back to Nairobi. He had been at Olduvai for one week and he was ready to send a message to the world.

Louis raced back to Nairobi, driving alone, and covering the distance from Olduvai in a day and a half. He traveled in such haste partly because of the excitement of his finds, and partly because one of his Kikuyu workers had become so ill that Louis feared he would die. With the man recovering in the hospital, Louis fired off letters to Sir Arthur Keith and Professors Haddon and Burkitt, and dispatched special reports about his discoveries to the *East African Standard* and *The Times* (London). In Nairobi the headlines ran in four tiers proclaiming: "Earliest Man in East Africa/Mystery of the Oldoway Beds Solved/Dr. Leakey's Expedition Dates Find Before Any Made in Kenya." And *The Times* was only slightly less hysterical, with Louis himself writing that his expediton had established "almost beyond question that the skeleton of a human being found by Professor Reck in 1913 is the oldest known authentic skeleton of *Homo sapiens.*" As usual, Louis had made a splash.

A few days later, on October 7, Louis drove to the Nairobi train station to pick up Frida, Priscilla, Donald MacInnes, and Frances Kenrick, a friend of Frida's. He was bursting with all he had to tell them and after embracing his family launched into a typically rapid-fire and breathless tale about Olduvai and all its wonders. At Limuru, he showed them the handaxes and a sampling of the fossils, and pressed Frida to return with him to the gorge. Frida, however, was still nursing Priscilla, and so was rather amazed at Louis's request; in fact, she thought it completely "batty."

Later in the season, Frida did leave Priscilla with her friend and participated in the fieldwork, going to Olduvai and other sites. Indeed, she found what would in time prove to be one of the gorge's most productive sites, FLK, or Frida Leakey Korongo (gully). But she continued to feel torn between her husband and child. "It was all good fun and tremendously interesting," she once wrote about these field experiences, "yet it was not altogether amusing to be 300 miles either from a husband or from a baby."

Louis was greatly dejected by Frida's refusal to join the expedition immediately. He wanted her at Olduvai where she could share in the excitement of his discoveries. He had also grown accustomed to Frida's attention and praise, and, like many new fathers, felt rather hurt by being shoved off center stage. "When Priscilla arrived, a triangle was created," noted Frida, "although neither of us saw it at the time." Louis spent only three days with his family, and on October 10 headed back to Tanganyika with MacInnes and Vivian Fuchs, who had recovered from malaria.

On Reck's first expedition to Olduvai, he had determined that there were five main layers of deposits in the gorge, and had named them Beds I to V. Bed I, with its black lava flow and sandy wash, marks the bottom bedrock of the gorge, while the other beds correspond to the red and buff-colored layers that so strikingly color the sides of the gorge. Each layer denotes a sequence of time, starting (as we know now) in Bed I at two million years ago, and ending in upper Bed V (now renamed the Naisiusiu Beds) at about 20,000 years ago.[6] Most remarkably from a scientist's viewpoint, the entire sequence from two million to 20,000 years is nearly complete. Standing on the gorge's bedrock, you look up at layer after layer, and millennia after millennia of East Africa's past—an experience that has left several scientists with what they describe as a "mystical feeling" and a "great sense of timelessness."

Olduvai Gorge did not exist two million years ago. Instead, a great alkaline lake spread over the region, fed by streams and rivers spilling down from the volcanic highlands to the south. During the next two million years, the lake slowly accumulated sediments, windblown sand, and volcanic ash. Then a sudden earthquake drained the lake, and later a seasonal river—the Olduvai—cut its way down through the layers of lake sediments and ash, forming the gorge and exposing the past. Fortuitously, the gorge cut mostly along the shoreline of the ancient lake rather than through the lake itself. Animals had once lived beside the lake, and now their fossilized bones, preserved by the volcanic ash, lay open to view. People had lived here, too, evidenced by the numerous handaxes Louis quickly collected. Overall, the gorge and its layers of bones and stone tools was, as Louis noted, a "veritable paradise" for both the paleontologist and the prehistorian.

Olduvai snakes its way across the plains for twenty-five miles in the main gorge, and fifteen miles in a smaller side gorge. Exploring all of it in detail, both horizontally and vertically, in six weeks' time was clearly impossible, and Louis's party concentrated their efforts within a five-mile radius of Reck's skeleton. They set out each day, shortly after dawn when the day was still cool, Reck and Fuchs to work on the geology, Louis, MacInnes, and Hopwood to search for fossils. Although dry and hot, the gorge is thick with scrubby bushes, stands of acacia trees and clumps of spiky sisal, making it uncomfortably easy to walk into the path of a hidden or sleeping lion—and Louis insisted that one person in each party go armed. Giraffe also lingered in the gorge, nibbling at the tops of the acacias, and sometimes small antelopes, the dik-dik and klipspringer, leaped across the rocky slopes.

By the end of Louis's first week back at Olduvai, so many fossils had been recovered that he decided to send a truck every ten days to Nairobi with their finds. Each day as he and his colleagues inched their way up and down the gorge, they made new discoveries: the perfectly preserved tusks of an extinct elephant; the skull of a strange species of hippo whose eye sockets protruded like periscopes; the bones of a small horse; the horn cores of a giant antelope; and the bones of crocodiles, turtles, fish, and flamingoes. Once Louis was exploring exposures along the south side of the gorge with Juma when a flash of white on the north side caught his eye. He scanned

[6] In the 1930s, the gorge was thought to date from 500,000 to 5,000 years ago.

the outcrop with his field glasses and picked out "a large number of white quartzite objects sticking out of the side of the cliff." They crossed the gorge, climbed the cliff and found a site where prehistoric people had sat on a beach working the flat flakes of quartz into handaxes. Within an hour's time, Louis and Juma had uncovered ninety-two such tools.

The next day, Louis and some of his Kikuyu helpers put in a small excavation at the site and unearthed the nearly complete skeleton of an ancient hippo "lying mixed up with the tools." Around the skeleton, they eventually recovered another 470 handaxes, and Louis decided that "a horde of hunters [had] feasted off" the hippo, "and then left their heavy tools . . . preferring to make new ones rather than burden themselves with the ones made on the spot."[7]

MacInnes and Fuchs had each found fragments of teeth from an extinct elephantlike animal known as *Deinotherium*. Unlike the elephants we know today whose tusks curve out of their upper jaw, *Deinotherium*'s tusks grew out of its lower jaw and swept downwards. Deinotheres were shambling thirteen-foot-high creatures, and until this discovery were believed to have died out long before humans appeared. When Louis found another *Deinotherium* tooth together with a stone tool, he knew he had incontrovertible proof of their coexistence with humans, although his report "was received in Europe with frank incredulity." By the season's end, he had collected enough additional evidence to convince even these skeptics.

Every three days a party of Kikuyu and one of the scientists set out by truck to restock the water supply, an undertaking that consumed an entire day—although the water itself, covered with a "green scum," as Fuchs noted in his diary, was not particularly appealing. It was also in such short supply that Louis strictly rationed it in camp. Adding to this discomfort, the wind blew incessantly and with it came dust, "fine black dust that filled every corner of the tents. . . . You breathed dust-laden air, your nostrils were filled with dust, you ate dust, drank dust, slept in the dust-ridden bedding, and in fact everything was dust, dust, dust!" Leakey later wrote.

The team also had to contend with lions, which lay furtively beside the waterhole and at night prowled the edge of their camp with great curiosity. Hewlett and Reck once had a "pitched battle with some lions" while trying to fetch water, and another time Hewlett shot two lions in the gorge at one of the excavations. Louis had to shoot a rhinoceros in "self-defence" when it charged him "point blank," while Hopwood let fly with his pistol one night when a hyena walked into his tent. On another night, Fuchs wrote that he "heard first one rifle shot, then a second so stepping out of the mess tent revolver in hand I was just in time to see Leakey run into the ring of the

[7] Like other archeologists in the 1930s, Louis thought the handaxes had been used for cutting and skinning, although exactly how this was accomplished with such large, cumbersome tools was not clear. Today archeologists believe the handaxes were core tools, and that the hunters simply knocked small razor-sharp flakes off the core as they needed them.

kitchen fire light. Apparently he had shot at a leopard by his torch light, the leopard had charged him, after firing a second shot and then being unable to see the animal he skipped to the fire."

Toward the end of November, one of the springs they had relied on for drinking water began to give out, and Louis decided to pack up camp. By now he had recovered stone tools from each of the five beds in the gorge, giving him a stunning sequence of the "evolutionary stages of the hand-axe culture." In the bottom layer of the gorge, Bed I, he had even found a type of tool older than the Acheulean handaxes of Kariandusi. These were "very simple" tools, made of a "pebble or lump of rock roughly trimmed to a cutting edge along one side." Louis called this the "Oldowan culture," and although he did not then realize it, the crudely flaked pieces of lava represented the oldest stone tools in the world.

Yet despite all the evidence of these early peoples' handiwork, the team had not found any of their bones. The *Homo sapiens* skeleton Reck had uncovered in 1913 remained the gorge's sole human representative. Had Olduvai Man made the gorge's tools? Louis thought it unlikely, and as he packed for Nairobi, he continued to puzzle over this curious juxtaposition of modern humans and primitive stone tools.

On November 23, 1931, Louis's first Olduvai expedition headed back to Nairobi bearing a final load of fossils and handaxes. "[W]e were well content," he wrote, "and really rather glad at the prospect of a change from the unceasing winds and dust that we had endured for the past two months."

In England, the results of Louis's Olduvai expedition had been received with enthusiastic acclaim. Sir Arthur Keith wrote of Louis as "my young friend," and noted that he was making "discoveries of the highest importance." Based on Louis's reports, Keith had even changed his mind about Reck's Olduvai skeleton. "In the light of the discoveries made by Mr. Leakey in the Rift Valley," he wrote, "there can be no longer any doubt as to the antiquity of the Oldoway man. . . . I have had to reconsider my opinion and acknowledge that Dr. Reck was in the right when he claimed Oldoway man as a representative of the pleistocene inhabitants of East Africa."

If Louis and Reck were correct, then *Homo sapiens*—that is, modern humans—had lived half a million years ago (according to the time scale of the 1930s) alongside such strange creatures as the *Deinotherium* and the periscope-eyed hippo. That might be acceptable. But even Louis was concerned about finding modern humans associated with the gorge's very primitive tools, and in his field notes devised a story to explain this. In an entertaining but rather farfetched scenario, he speculated that there may have been other races of people living at the same time who were less evolved than Olduvai Man, and that these people had been responsible for the tools—as well as for Olduvai Man's death. "I have in my mind," he wrote, "the possibility that the famous Oldoway skeleton may just conceivably represent a wandering *Homo sapiens,* who was caught, bound and drowned in the lake by the makers of the Chellean culture at Oldoway."

Similarly, when the excavations in the caves at Chou K'ou Tien in China produced tools and traces of fire that seemed linked to Peking Man, Louis had his own imaginative version of what may have actually transpired. In an article for *The Times,* he wrote that although Peking Man was probably the same age as Olduvai Man, he represented only a cousin and not a true ancestor of *Homo sapiens.* He expected that further excavations at Chou K'ou Tien would reveal that *Homo sapiens* had lived there, leaving behind the tools, the fire, and the remains of Peking Man—"the relics of his meat feast."

Not everyone was convinced about the age of Olduvai Man. The skeleton seemed too modern; the extinct animals associated with it too primitive; the burial method, with knees drawn up in a fetal position, too reminiscent of certain African burial practices. By January 1932, shortly after his Olduvai team disbanded, Louis began to hear rumblings of doubt.

Hans Reck and Arthur Hopwood had stayed in Kenya until the end of December, then sailed for Europe. Hopwood took most of the fossil animal bones with him to the British Museum of Natural History. He also had with him soil samples from the various Olduvai beds, including Bed II, the one that Olduvai Man came from. In England Hopwood heard that John Solomon, Louis's geologist from Elmenteita days, had a new technique for analyzing the mineral content of sedimentary deposits, and he sent some of the Olduvai material to Solomon for testing. "When [the test] proved successful," recalled Solomon, "[Hopwood] sent [to Germany] for a scraping from the inside of the rib of Reck's skeleton . . . and this turned out to contain a very distinctive crystalline volcanic mineral which was completely absent from the older deposits." Specifically, the mineral found with the skeleton did not show up in Bed II. Once again, the age of the skeleton was in doubt.

Louis was shattered and furious. He denounced Solomon's test as "rubbish," argued that Hopwood's sampling was poor, and then announced that he would answer all his critics by finding another such skeleton. "The doubts that you & others have about Oldoway Man must be set at rest," Louis wrote to Sir Arthur Keith, who was now questioning what he had so warmly supported only three months before. "I WANT IF POSSIBLE," Louis continued, "TO FIND ANOTHER OLDOWAY SKELETON." Louis thought he might find one in a cave near Olduvai that might have "the same Oldoway fauna, & cultures, with possibly some skeletal remains of man. Then perhaps I should be able to convince you all!!" He also planned to explore the fossil beds at Kanjera in western Kenya that Archdeacon Owen had written him about the previous year. Louis thought these exposures might be the same age as those at Olduvai, and in his field notes wrote that he had "great hopes that this area would yield information that could be used as a check upon our discoveries at Oldoway."

On March 14, Louis and MacInnes established a field camp at Kanjera, seven miles from the shores of Lake Victoria. The fossil beds here were a colorful mix of pink, orange, and dark brown, and they rolled in small hillocks and deep gullies down to the lakeshore. Almost immediately Louis began picking up fragmented fossils of the same extinct elephants he had found at Olduvai as well as stone tools. Then, on the morning of their second day in the field, MacInnes struck paydirt and stumbled onto "fragments of an intensely mineralised human skull washing out from an exposure." Although not *in situ,* the pieces of skull resembled in color and mineralization the fossils of the extinct mammals, and Louis had "no doubt at all" that they were of the same age. They immediately excavated the site, turning up additional fragments that joined to the first pieces, and also found parts of a second skull. Like Oldoway Man, these were skulls of *Homo sapiens,* and Louis wrote ecstatically to Hopwood:

> I've been working here at Kanjera . . . for 5 days now & we have got a fine piece of skull of something very like *Pelorovis* [a gigantic extinct antelope] with BOTH HORNS & quite complete & with *apparently* most of the skeleton too. Also a nice baboon skull, & some Rhino teeth. But what is *far* more important I have got the greater part of one *Homo sapiens* skull, and a small part of a second. So that will be a bit of a blow to the "Anti-Oldoway-Man" group!!

But Louis also knew that he needed such a skull *in situ,* where there could be no questions at all about which geological stratum it came from. A fossil that merely "looked like" the other fossils was not good enough. And so he continued his search.

Kanjera, set on the southern shore of Lake Victoria's Kavirondo Gulf, has a wet, muggy climate. On many afternoons thunderheads piled up over the lake, then broke into heavy downpours as they passed the summit of Mount Homa, so that the camp was often wet and soggy. "Whereas the constant trouble at Oldoway had been 'not enough water' here our trouble was too much of it," wrote Louis. Even more unpleasant were the hordes of mosquitoes that swarmed up from the lake after dark. "We used to wear long trousers tucked into Wellington boots to protect our legs, and tie towels round our heads leaving only the mouth, nose and eyes exposed, but even so we were terribly bitten, and on one occasion one of us killed over a hundred mosquitoes on his face during one meal," wrote Louis. But in other ways the rain was a blessing as it washed over the fossil beds, bringing new bones into view.

It rained all night March 28, and in the morning Louis decided to drive to a nearby site called Kanam. It was only four miles away, but there was no road, only a footpath, and the ground was "so muddy and slippery" that Louis had to "fix chains to all four wheels of the car." Louis, MacInnes, and Frances Kenrick, Frida's friend, squeezed together on the front seat, while

Juma and four other Kikuyu crowded into the back, finding seats on top of "all the necessary picks and shovels and other excavating implements." After a long hour of "pushing and pulling the car . . . across treacherous ground," they finally arrived at Kanam. They were by now "spattered with mud and water from head to foot," but Louis immediately began picking his way on hands and knees through the eroded gullies. He soon found a fossil deposit with the mandible of an extinct pig poking out and set to work excavating it.

Juma Gitau was at work in another deep gully, digging into an "almost vertical cliff" where he had found part of a *Deinotherium* tooth. Juma had shown this to Louis, who sent him back to search for more. He now "dislodged a large mass of matrix" and when he broke this apart with his pick "noticed these teeth sticking out of a fragment." Juma passed the teeth to MacInnes, who was working at the top of the gully, and MacInnes gave the teeth one glance and shouted for Louis. The teeth were unmistakably human premolars. Back in camp, Louis cleaned away enough matrix to see that he had part of a very badly weathered lower jaw. It was fragmentary material, yet sufficient to convince Louis that "the Kanam mandible . . . represented a man who was a true ancestor of *Homo sapiens.*" Best of all, it seemed to be an *in situ* discovery. Thoroughly delighted, Louis shot several pictures of the discovery site, and then to make certain he could find it again, he marked the fossil's location with four iron pegs set in concrete.

At Kanjera, Louis also found fragments of another human skull *in situ.* Triumphant, he wrote to Hopwood that he had found a *Deinotherium* at Kanam and "definitely *in situ* in this horizon . . . with a pre-chellean tool was a fragment of a human *(sapiens)* mandible. At Kanjera too where at first I only got parts of 3 human skulls washing out of the equivalent of Bed II at Oldoway I know [*sic*] have bits of No 3 actually *in situ.* So Oldoway man has an *in situ* compeer from Kanjera to hold his hand!!"

Meanwhile, in England the death knell was sounding for Olduvai Man. Several independent geological tests had been run on the skeleton and soil samples. These showed that the body had been buried in Bed II in comparatively recent times, when a fault exposed that horizon. Sometime after the burial, Beds III and IV eroded away; then Bed II had been covered over by the deposits of Bed V. Reck had mistaken the soil of Bed V for that of Bed III—an easy enough error to make as both are a deep red in color. As Louis had first guessed, Olduvai Man was not any older than the skeletons he had found at Elmenteita.[8] "[T]he fact remains," Hopwood wrote in reply to Louis's insistent letters, "that no one here now accepts the skeleton as . . . anything else but recent."

The news only momentarily depressed Louis. He was still convinced that the basic premise behind Olduvai Man—that *Homo sapiens* was of great antiquity—was correct, and he felt certain that his Kanam and Kanjera finds

---

[8] Subsequent Carbon 14 tests have dated the skeleton to 17,000 B.C.

carried the proof. "Everyone admits that *Homo sapiens* must go back to the beginning of the Pleistocene at least—somewhere," he wrote to Burkitt about his new fossils. "The question has always been, Where? And the evidence . . . seems to suggest that the answer is 'the region of the great central African lakes.' " But even Louis knew that proving his claim was not easily done. "I can forsee great fights when I get back," he wrote to Hopwood from his last camp at a site called Apis Rock. "Enough. Life is hard but good & somehow we will get results worthwhile & make the world believe."

Chapter 4

LOUIS AND MARY

When Louis and Frida set sail from Mombasa in late November 1932, they had run so low on funds—Louis's ambitious expedition had cost far more than he had planned—that they traveled third-class "without a porthole." Louis had also developed a hacking cough, which soon turned into bronchitis, and Frida divided her time between feeding him oranges and tending Priscilla, now eighteen months old. By the time they reached England in mid-December, Frida had tired of such a rootless life, and decided that they needed a permanent home. "We had to find somewhere to be," she recalled. "We had one baby, a lot of archeology, an enormous amount of work to do, and no money. So we bought this home with the only money we had in the world, which was the capital of my dress allowance." Of her £2,000 dowry, only £200 remained after buying the house.

The Close, as their home was called, was a rambling brick house with mullioned windows and dark-brown shutters. Louis had chosen it for its large garden and the "green fields hedged with may and brambles" that ran from the edge of the backyard to the horizon. The Close lay in the village of Girton, only three miles from Cambridge, and Louis could bicycle the short distance to St. John's where he had a suite of rooms. St. John's had renewed his fellowship, and this, combined with his teaching, provided enough income to pay for his family's daily expenses. They were "awfully penurious," as postgraduate students typically are, but were also comforted by the friendships and community Cambridge provided. Louis's ties to Cambridge at this point in his life ran as deeply as those to Kenya. "Louis loved Cambridge most dearly," said Frida. "St. John's was awfully generous to him, and he was most grateful." The university seemed to be grooming him for a position as a lecturer, and Louis responded with enthusiasm for the academic life.

As he had during his previous tenure at St. John's, Louis lived a somewhat unconventional life, spending far more of his time at the college than he did at home. He brought the "magnificent series" of stone tools from Olduvai and the Kanam jaw and Kanjera skulls to St. John's, and used the two main rooms of his suite as laboratories for sorting, cleaning, and studying his specimens. A smaller side room served as "a combined emergency bedroom, kitchen, and dining room. . . . The great advantage of this arrangement was that when I worked late into the night, as I often did, I could sleep in college and cook breakfast next morning, instead of having to . . . disturb my family in the middle of the night."

Louis was undoubtedly under pressure. His adamant defense of Reck's Olduvai skeleton had somewhat sullied his scientific reputation. In fact, he did not fully withdraw his support for the skeleton's supposed great age until February 1933, after he traveled to Germany to view the specimen again. By then, the general consensus was that Louis was "pigheaded." "He made a bit of a fool of himself by his vehement insistence," observed John Solomon. "It showed that his attitude in those years was not that of a 'scientist,' but of an 'enthusiast.' "

If Louis was disturbed by this mistake, he did not show it. Instead, without any apparent qualms, he simply changed sides, noting that he had, after all, been right in 1929 when he first disagreed with Reck about the skeleton's age. Now he joined his recent critics (notably John Solomon and Professor P. G. H. Boswell) and wrote a letter to *Nature* rejecting a Pleistocene date for Olduvai man. But the matter was not so easily dismissed by others, especially since Louis was claiming a similar great age for his Kanam mandible. Even before he left Kenya, Louis had received due warning from the British Museum's paleontologist Arthur Hopwood about the challenges his fossil would face in England. Louis, Hopwood wrote, would have to somehow "disprove the assertion" that his specimens had either fallen or been washed by rain into the older deposits. "So long as you have plenty of evidence, checked and cross-checked, there will be eventual triumph for you," Hopwood added.

Louis had announced his discovery of both the Kanam and Kanjera fossils while still in Kenya, publishing an article in a May 1932 issue of *Nature,* where he had argued as usual for the "genuineness of the antiquity of *Homo sapiens* in East Africa." He sent the article first to Sir Arthur Keith, who then forwarded it to *Nature* with his approval. Louis had also shown the mandible and skulls to Keith immediately on his return to England, but few other scientists of Keith's stature had viewed the specimens. Keith thus suggested that the human biological committee of the Royal Anthropological Institute hold a special conference to review Louis's evidence for the dates he had assigned to these fossils. The meeting was scheduled for mid-March 1933 with the chief paleontologist from the British Museum of Natural History, Sir Arthur Smith Woodward, presiding over a panel of distinguished scientists. They would decide if, in fact, Louis had the "oldest . . . true ancestor of modern man," as he claimed.

To prepare his defense, Louis took his fossils with him on the train to London every few weeks where he again worked under Keith's supervision. Keith remained a steadfast supporter of Louis's, advising him on the reconstruction of the Kanjera skulls and the cleaning of the Kanam jaw. In fact, he was tremendously excited by Louis's finds, which seemed to confirm many of his own theories about the development of humankind, particularly his belief in a great antiquity for modern humans.

But Keith's idea was definitely "unpalatable" to many other scientists. In particular, Professor P. G. H. Boswell of London's Imperial College, who was then in Louis's view the "most prominent Pleistocene geologist in Great Britain," took a dim view of this theory, believing that it was unlikely that modern humans had existed unchanged for such a great period of time. Boswell had led the geologists in the questioning of Olduvai Man and he now responded with similar skepticism to Louis's claims for the Kanam mandible. A wiry, sharp-featured, and "decidedly dogmatic" man, Boswell placed great value on precision in thought and speech, and having once caught Louis in an error, he suspected there might be others. To his somewhat jaundiced eye, it seemed that Louis had "accepted our views [about the Olduvai skeleton] . . . largely . . . because he now thought he had a better horse to run." He was also astonished that Keith, "indiscreet as ever," had "blessed [Louis's] . . . claims forthwith!" and welcomed the chance to examine Louis's evidence personally.

On March 18, 1933, a chill, blustery day in Cambridge, Louis stood in a small oak-paneled room at St. John's and "lay before the scientific world the evidence which showed that we had found at Kanam . . . the oldest fragment of a real ancestor yet discovered, a real *Homo*. . . ." Twenty-six scientists—anthropologists, anatomists, and geologists—had gathered to review his work, and they listened carefully while he recounted his discoveries and his reasoning for the dates he had assigned his fossils. Sir Arthur Keith was ill and unable to attend, but Louis's professors Haddon and Burkitt, and his colleagues Donald MacInnes and Vivian Fuchs, were among those present. Professor Boswell was also in attendance. After Louis's brief talk, Sir Arthur Smith Woodward divided the scientists into three committees to evaluate the evidence. That evening, the scientists in groups of twos and threes also stopped by Louis's rooms at St. John's, where he displayed his fossils and stone tools from Olduvai, Kanam, and Kanjera. By themselves these were impressive, and Louis's enthusiasm for his finds added to their impact.

When the scientists reconvened the following afternoon, a spokesman for each committee read a short report analyzing Louis's work. Boswell later wrote that he had experienced "some difficulty in ensuring that the geological report was sufficiently cautious." He noted that they had only the photographs Louis provided (which had been taken by others since his camera had failed) to determine the stratigraphy, and thus the age, of the site and fossil. But aside from suggesting that additional geological evidence would be helpful, the committees offered only minor corrections, and Sir Arthur Smith Woodward pronounced the Kanam jaw "a most startling discovery." The conferees then "congratulated Dr. Leakey on the exceptional significance of his discoveries, and expressed the hope that he may be enabled to undertake further researches, seeing that there is no field of archaeological enquiry which offers greater prospects for the future."

For Louis it was a triumphant moment. He had the blessing of his fellow scientists and could rightfully say that he had the oldest "true ancestor" of humankind. His Kanam mandible was placed in the Early Pleistocene and dated at more than 500,000 years. It was thus seen as the contemporary of the Piltdown and Peking men, and confirmed Louis's and Keith's opinion that these latter fossils were merely primitive side branches to the "true stem" of humankind. The meeting and its conclusions were deemed of such importance that *The Times* carried a lengthy article, noting that because of Louis's discoveries "plausibility is . . . lent to the theory, first advanced by Darwin, that Africa is the cradle of the human race." Louis, not yet thirty years old, was already well on his way to becoming famous.

The first warming hints of spring came in April of 1933, Priscilla turned two, and, for the first time, Louis realized that he had a daughter. She had "reached the age when she was running about and beginning to talk, and it was fun to play with her," he recalled. "Instead of having a game of tennis or squash in the afternoon, I would rush back to Girton so as to have time to be with my little daughter. . . . I suddenly realized how charming children —and especially my own child—could be." Partly because of this awakening, his recent success, and own long-standing desire to have a son, Louis and Frida decided to have a second child. But from the very beginning, it was not an easy pregnancy—"I had had a threatened miscarriage & had been 'put on ice,' " Frida said. She was ill much of the time now, and spent long hours in bed, and so "somewhat retreated from archaeology."

In the meantime, Louis's career continued to soar. The Royal Society asked to exhibit the Kanam jaw as did the British Museum of Natural History, and Louis signed a contract with Methuen, the London publisher, to write a popular book about "the latest discoveries concerning the Stone Age." Louis would use this book, *Adam's Ancestors,* as a forum to present his own discovery of "the remains of really ancient true ancestors of modern man," the Kanam/Kanjera fossils. He would also discuss what he considered to be the side branches of the family tree—basically every other fossil that had been found, from Piltdown to Neanderthal Man. He was sought after by the press, offered small fees to lecture before various archeological societies, and in May was invited by the Royal Anthropological Institute to give a talk on early humans in East Africa.

As they often did on such occasions, the RAI Fellows arranged a dinner to honor their guest speaker, and following Louis's talk they met in a restaurant on Bedford Square. Gertrude Caton-Thompson, who had excavated the stone ruins of Zimbabwe and had recently directed a dig in Egypt, joined in this party, appearing with her young, blue-eyed protégée, Mary Nicol. Although lacking academic degrees, Mary was a keen archeologist and talented artist. She had just completed the illustrations for Gertrude's book *The Desert Fayoum,* and Gertrude was immensely pleased with them. Gertrude thought that Louis might be looking for an artist to illustrate *Adam's Ancestors,* as Frida was too ill to help this time, and arranged for Mary to be seated next to Louis at the dinner.

A rather aloof, shy young woman who often felt uncomfortable in social gatherings, Mary had not relished the idea of attending this party. To her, the dinner sounded like "a very stuffy affair . . . I didn't really want to go." Seated beside Louis, however, her perception of the evening quickly changed. As his sister-in-law, Beryl Leakey, noted, "Louis could talk to anyone. . . . Whoever he was talking to, he could do it just right." Others observed that Louis was always "unperturbed by poverty or grandeur," and this combined with his own lack of pretentiousness soon put Mary at ease. He was witty and charming, and they discovered they shared a deep passion for archeology and a love of animals. If he was perhaps a shade too interested in his pretty brunette dinner companion, the twenty-year-old Mary did not notice.

"[Some people] have made a huge romance out of this meeting, but it didn't happen that way at all," Mary recalled fifty years later. "In Sonia Cole's biography of Louis, *Leakey's Luck,* she makes this dramatic meeting at the dinner, and it was nothing of the sort. There was none of this 'swept-off-one's-feet' business. I have no clear recollections of that dinner whatsoever. And I honestly don't know what it meant to Louis. For me, it wasn't anything more than casually liking someone that you meet; or casually disliking them. Nothing special."

Louis himself would only note of this first encounter that "in 1933 . . . I was fortunate enough to meet Mary Nicol." He was, as Gertrude had foreseen, greatly impressed with Mary's illustrations of stone tools. They were "the best representations . . . I had ever seen," he said, and immediately arranged with her to make the drawings for his book.

Very likely, Mary also impressed Louis because, despite her exterior shyness, she was a rather worldly young woman. She smoked, had a sharp wit, spoke French perfectly, and could pilot a glider plane. Perhaps during the dinner, she let Louis catch a glimpse of this side of her. He left the dinner intrigued and that summer began writing her letters while she was on an archeological dig. Since she was working on the illustrations for his book, the letters did not seem improper—although they arrived, Mary noted, "perhaps just a little more frequent than was strictly necessary." At summer's end, they met again, this time in Leicester at the annual meeting of the British Association for the Advancement of Science, where Louis was scheduled to give a talk on Kikuyu customs. The letters may have had more of an effect than Mary realized, for suddenly her attraction to Louis was undeniable. "[F]rom the time we met up soon after our arrivals we became inseparable companions," she recalled, "and it was here that I first felt and instinctively recognized something that was new to me: the mental stimulus and physical thrill of having Louis with me." They managed to maintain a semblance of propriety, returning to their individual rooms each evening, but by the meeting's end "it was clearly understood between us that we would meet again, soon and frequently."

Mary Nicol was not the first romantic liaison Louis had sought since returning to England; in fact he had "several girlfriends" when he met her. His overnight stays at St. John's and in London were not always work-related, although none of this was known to Frida. Nor did she realize that Louis was unhappy in their marriage, or that he had turned to others for "solace" and understanding. Louis also genuinely liked women, a trait that would prove powerfully attractive throughout his life. "He appreciated women," said Betty Howell, wife of Clark Howell, professor emeritus of anthropology at the University of California, Berkeley. "And because of that women responded to him." "Women came to him like moths to a flame," noted another observer. "And he *enjoyed* it; he was a real human that way." But Frida was blind to all of this, although Louis's unfaithfulness was even then "known to many others." "I may have been singularly unperceptive," Frida recalled, "and probably the home and Priscilla took more of my attention than they ought to have done. And maybe I'm just not suspicious by nature. But everyone else knew, and I was the last to know."

Cecilia Nicol, Mary's mother, was shaken "to the core and upset . . . beyond measure" when she learned of the romance. She took an instant dislike to Louis and "set herself to change [Mary's] mind." Others tried as well. Louis managed to find additional tools for Mary to draw for future books, and some of this material was housed in the British Museum in Bloomsbury. There, she worked on her drawings in a room she shared with Sir Thomas Kendrick and Christopher Hawkes, members of the Department of British and Medieval Antiquities. Louis, eager and attentive, would stop by to take her on "surreptitious lunches," and it was not long before "the nature of [their] relationship became clear" to Kendrick and Hawkes. Hawkes ignored their affair, but Kendrick, worried that Mary was "simply dazzled by Louis" and might be hurt, protectively took her to lunch one day. "Genius is akin to madness, Mary," he warned about the mercurial Louis, "you must be careful." "Those were his words, and there was much more along the same lines; but he might as well have saved his breath for all the notice I took," wrote Mary.

They were by now deeply and passionately in love. Whatever other girlfriends Louis might have had fell by the wayside, and he filled his weekends and spare moments with Mary. She found him "very intelligent and very attractive," but above all liked him for treating her "as an equal and a colleague." He, in turn, discovered in Mary a sympathetic confidante, someone willing to listen to both his marital troubles and anthropological musings. He had finally decided that his marriage was a "complete failure," and although he "felt bad over the way he was treating Frida," he saw no way out other than to end their marriage.

Frida, eight months pregnant, was still ignorant of the whole affair. When Louis rather rashly invited Mary to stay at The Close in November 1933, Frida assumed she was simply another of Louis's students. "This was not so unusual," Frida explained. "We'd had many students staying at our house one time or another, and she was just another one. I never suspected anything. I guess I must have been blind—or maybe preoccupied with the [coming] baby. I think most of my thoughts then were centered on the baby, on keeping it alive. So I probably wasn't too aware of what was going on around me."

Mary had come to Cambridge ostensibly to "see the remainder of [Louis's] African collection." She was assisting him with various publications, and he was instructing her in paleontology. She spent a week at The Close, and if she felt any guilt at all about her involvement with Louis, she managed to rationalize it by perceiving the marriage as "not a marriage at all. It was not a marriage in the way I understood marriage to be, such as my parents'," she said. "Louis and Frida were too different; there was a total lack of understanding between them." Seeing them together in their home, where it seemed to Mary that they were "living separate lives," confirmed her opinion.

By the end of the week, Mary's time at The Close was becoming "increasingly awkward." Once Frida's sister, Barbara Waterfield, stopped by in the evening and happened to find Louis and Mary in the sitting room together. "Louis was sitting in an easy chair, holding forth on some topic, and Mary Nicol sat at his feet, a look of adulation on her face," she recalled. The scene struck her as "curious," but she attributed it to nothing more than a student's crush on her teacher. While they were compelled to hide their feelings at The Close, alone in Louis's rooms at St. John's they let their love and desire come rushing out. "Tension mounted," Mary wrote, "until on a day when the atmosphere was almost electric [Louis] told me, quite suddenly, that he now knew the one thing he wanted was to end his marriage to Frida and marry me. I cannot remember exactly what I replied, if indeed I heard myself speak. . . . Eventually I managed to gasp out something that I intended as acquiescence and he clearly understood it as such. After that for some moments neither of us was really in a position to speak."

In England in the 1930s, the only sure way of being awarded a divorce was to commit adultery, and following the week at The Close, Louis and

Mary began spending every weekend together. Mary's mother continued to protest, but Mary disregarded her. On December 13, 1933, Frida gave birth to a boy, Colin Avern Leakey, "the son Louis had so wanted." Louis was home for the birth, and delighted with his baby boy, but not even a new son could persuade him to change his mind. A month later, in January 1934, he confessed to Frida that he had "fallen in love with Mary and was going to take her to Africa."

A few days later, Frida summoned both Louis and Mary to The Close and, as Mary put it, "told us with admirable clarity exactly what she thought of each of us." While Mary remembers Frida more or less labeling Louis "a cad and a traitor," and calling her "a worthless hussy," Frida disagrees. "I don't think those words are in my vocabulary," she said. "What I do remember is coming down the stairs of our home with Colin in my arms, and I was a little wobbly still. And I tried to talk some sense to them. I said, 'Think about what you are doing. Do you really understand what it is you are doing? It's not just me, but it's your children. Think of them.' I was simply so horrified and appalled that Louis could abandon his children like that."

But there was no turning back for either Louis or Mary, and now free of his marriage they proceeded "along [their] chosen path and [took] the consequences."

Mary Douglas Nicol, the young woman at the heart of Louis's grand passion, was, in her own quiet style, as eccentric and rebellious as Louis himself. She was also a woman of determination, quick to judge others, "a tough cookie," as a friend of adolescence phrased it. They were the traits of a survivor, and Mary Nicol, at age twenty, had wounds that ran deep.

Her father, Erskine Nicol, had been an adventurous landscape painter of some success who once spent four years living with Egyptian Bedouins. He had spent much of his adult life traveling abroad, particularly in the south of France, Italy, and Egypt, where he painted watercolors of local scenes in pastel shades. He returned to England only to sell his works. Erskine met Mary's mother, Cecilia Frere, during one of his Egyptian forays. A petite and beautiful brunette of thirty-four, with softly feminine ways, Cecilia had come to Egypt as the companion to a wealthy friend. She had grown up in a villa in Italy and studied painting in Florence—all prior to her family's financial downfall—and Erskine, then forty-four, found her both lovely and companionable. He pursued her to England. Once married, they returned to Egypt, where they spent their first married year living in a houseboat on the Nile. Cecilia soon became pregnant and they went back to England again at the beginning of 1913. They rented a house in Trevor Square, London, and on February 6, 1913, Mary was born.

She was a pretty child, with golden-brown hair and deep blue eyes, and from the beginning was the center of attention. Her mother's three unmarried sisters and their mother lived nearby, and they doted on Mary, as did

Erskine's brother, Percy. Erskine's sister, Elizabeth, was more formidable, often threatening "to drown" Mary's favorite stuffed toy if she did not behave. Yet, as is often the case with "the only child in a world of grown-ups," she was loved by all and generally spoiled. Erskine had planned to take his small family back to the Nile houseboat once Mary reached her first year, but the outbreak of World War I prevented this. He was then forty-six, too old for active duty, but he did join a reserve unit. Still, the war hardly touched their lives. They moved to a cottage in the Huntingdonshire countryside, near the River Ouse, and filled their time with punting expeditions, country walks, and visits to the aunts and grandmother in London.

Then the war was over, and Erskine packed his paints and canvases, and bundled off his wife and daughter to Switzerland, France, and Italy. During the next eight years, from 1918 to 1926, they spent at least half of their time living in small villages on the Continent, particularly in south and southwest France. Erskine was drawn here by the beauty of the region and by its prehistoric past. The Nicols spent two winters living in a small hotel in Les Eyzies, and it was here that Mary "first began to come consciously in contact with Palaeolithic archaeology," as she wrote in her autobiography.[1]

Her father, with his "open nature and great capacity to make friends," befriended Elie Peyrony, one of the first French prehistorians to recognize the antiquity of the cave paintings and engravings. Peyrony was then excavating one of the caves—and rather crudely by today's standards. He would only briefly examine the buckets of earth the workmen brought to him, then dump the rest out on an embankment above the River Vézère. He also did not mind if Mary and her father searched these dumps, and Mary quickly discovered the "sheer instinctive joy of collecting." There were fine points, endscrapers, and elegant flint blades—"all sorts of good things." Touching the tools, feeling their polish and wear, led Mary to wonder about the "world of their makers." And like Louis with his childhood collection of stone tools in Kenya Colony, she set about creating a rudimentary classification system for her finds.

She was now a coltish twelve-year-old, fluent in French and addicted to reading. She had a small group of friends her age, a governess she despised (Mary called this poor young woman "the Uncooked Dumpling"), and a father she adored. "He loved to take walks in the countryside, and he would always take me with him," said Mary. "Anytime that he could take me out to look for stones and bones, he would do so. We were certainly great companions." Mary and her mother, on the other hand, had much less in common. "She had been reared to be an ornamental young lady, to sit in a large house, to socialize and, of course, none of that happened," said Mary. "She lost that life completely."

---

[1] Mary's instinctual love of archeology may be said to have run in her family. John Frere, the first Englishman to recognize a flint artifact as an implement made by humans, was her great-great-great-grandfather on her mother's side.

In the winter of 1925, they moved to Cabrerets, a tiny Dordogne village famous for the prehistoric art in the nearby Pêch Merle Cave. The parish priest, Abbé Lemozi, was a keen amateur archeologist, with a particular interest in Paleolithic art, and he and the Nicols soon became close friends. He often bicycled to their hotel for evening theological discussions, and acted as their guide to the local archeological sites. Once, equipped with small lamps, he guided Mary and her mother through the low, twisting passages of Pêch Merle Cave to a vast chamber where bison and horses seemed to dance on the walls in the flickering lamplight—an experience that affected Mary deeply. And there were walks with her father along the Sagne River and over hills fragrant with lily of the valley. "We so enjoyed doing the same things, and he took a lot of trouble to teach me things, show me things. He knew about the animals and flowers, and he shared all this with me."

All too soon Mary's life of "near perfection" came to an end. Erskine suddenly fell ill, apparently with cancer, and died in the spring of 1926. Mary, who had just turned thirteen, was "shattered . . . [and] no one," she wrote, "could comfort me." Uncle Percy, Erskine's brother, came to stay with her and her mother, and the kind Abbé Lemozi read the service for his friend. But Mary remained inconsolable. To her it was a tragedy of immense proportion, for she had "just lost forever the best person in the world."

"I was extremely close to my father, and my life completely changed after his death," she recalled nearly sixty years later. "Mother and I moved to London, and living in England was such a contrast to my life in France, and not a pleasant one. It was very lonely."

On their return to England, Cecilia arranged the sale of Erskine's remaining paintings and they moved into a boardinghouse in Kensington. Cecilia, a devout Catholic, had spent part of her childhood in convents, and she now decided that Mary, too, would benefit from this more formal education. She sent her to a convent in Kensington Square, but Mary did not last long. She was used to being independent and socializing with adults, and much like Louis when he entered his preparatory school, she "could not find a single kindred spirit among either the pupils or the nuns." In response, she developed a strong rebellious streak. She was expelled from one school for refusing to recite poetry, and from a second for deliberately causing an explosion in her chemistry class. The explosion, she remembers, "was quite loud and quite a lot of nuns came running, which will have been good for some of them." After that, Mary's formal schooling came to an end.

Still, she retained her interest in archeology. Her exploration of the caves in Cabrerets with the Abbé Lemozi had left a lasting impression. "The abbé kindled my interest in prehistory and also gave me a very sound groundwork in excavating," she once recalled. "After that, I don't think I ever really wanted to do anything else."

In the spring of 1930, when she was seventeen, Mary began auditing archeology and geology courses at University College and the London Museum. She also wrote letters to archeologists, asking to join their summer excavations, and was offered a position with Dorothy Liddell on a Neolithic site in Hembury, Devon. Although not a professional archeologist, Dorothy was highly regarded by her colleagues and she was a great inspiration to Mary. "She was an enormous help in training me, showing me how to dig properly and making it quite clear that females could go to the top of the tree," Mary said. The team worked together for three summers, and Mary drew some of the stone tools from the site for Dorothy's publications. Dorothy, in turn, showed these to Gertrude Caton-Thompson, who asked Mary to illustrate *The Desert Fayoum.*

There would be one more summer for Mary at Hembury Hill, but by then Mary's world was considerably different, for Gertrude had introduced her to Louis.

In January 1934, after the last stormy scene with Frida, Louis moved out of The Close and into his rooms at St. John's. He never mentioned this period of his life in his autobiography, but it was unquestionably a time of emotional upheaval. He had given up his family life for Mary, and Frida, crushed by his rejection, refused to let him see the children. Many of Louis and Frida's friends disapproved of Louis's behavior as well, and now treated him with a noticeable coolness. Cambridge University as a whole took a narrow view of such affairs, and had a select panel of heads of colleges and professors of law to investigate gross acts of immorality. Although Louis was not summoned by the Sex Viri,[2] as this body was called, there were few in Cambridge who had "a good word to say" for him. As in any university city, the gossip quickly spread, and neither Louis nor Mary was spared.

Somehow through these tumultuous times—perhaps with the help of quiet visits to the college's wilderness garden he so loved—Louis managed to maintain his typically busy schedule. He continued teaching his archeology course, finished checking the proofs of *Adam's Ancestors,* and wrote the final chapters of *The Stone Age Races of Kenya Colony.* E. Barton Worthington, then also a fellow at St. John's and a specialist in East African fishes, remembers meeting Louis on occasion during this time. "When we met Louis was, as usual, bubbling over with his latest find or hypothesis, and we had good (if somewhat one-sided) talks about Africa, early man, animals, fish & fisheries." Nothing was said about a certain young woman or Louis's pending divorce.

Louis had now also started to plan his fourth expedition, which he hoped to lead in the autumn of 1934. He had greater difficulties raising the funding this time, perhaps because of his affair, but by May had obtained enough

---

[2] "Meaning the Six Men, not the Sex Weary," as the geneticist J. B. S. Haldane wryly noted. Haldane had been called before this panel only eight years before Louis's affair, and deprived of his teaching position because of one adulterous act. He appealed and was reinstated.

money (with grants from the Royal Society, the Sladen Trust, the British Museum of Natural History, and St. John's) for a full year's stay in East Africa. He intended to revisit Olduvai Gorge to determine how his team had made the error in dating the Olduvai skeleton on his previous expedition, and he planned to continue his work at the Lake Victoria sites of Kanam and Kanjera. A team of three students would join him, and Mary would come, too. There would also be a visitor, Professor P. G. H. Boswell, the geologist from Imperial College. Although he had signed the Royal Anthropological Institute's report establishing the dates of the Kanam and Kanjera fossils, Boswell had continued to have misgivings. And he brought the matter up again and again in his "rather incisive voice," until Louis suggested that Boswell come to Kanam and see for himself. "[Louis] was very anxious that I should visit the places while the expedition was at work," Boswell noted in his unpublished autobiography. Sir Arthur Keith and other scientists supported this proposal, and the Royal Society agreed to pay Boswell's way. For Louis, Boswell's visit would afford a chance finally to silence his chief opponent, and prove once and for all that the specimens had been found in Pleistocene deposits and *in situ*.

While Louis pursued his various endeavors, Mary was equally busy with her own archeological interests. She continued to live at her mother's, only going off with Louis on weekend excursions. In April 1934, she returned to the Hembury Hill fort dig, and Louis, as usual, came to "spirit [her] away for weekends." Louis and Mary spent their illicit excursions camping in the countryside of southwest England. They shared a small tent, and took walks to other hill forts, or found quiet spots to sit and watch wild foxes and birds.

By mid-October, Louis's expedition to East Africa was ready to depart, and he spent one final weekend in the countryside with Mary. Partly because of Boswell's impending visit, and out of concern for Louis's parents, they had decided she would wait six months before joining him in Africa. She planned to meet him in April 1935, when he was ready to begin work again at Olduvai Gorge.

Louis visited Cambridge once more, stopping at The Close for a word with Frida and a last glimpse of his children. Frida was still "in shock . . . I kept thinking that somehow the nightmare would end." She drove Louis to the train station, certain that he would reconsider. "I think until he actually left on the train that I kept hoping that somehow everything would be put right." But Louis boarded the train, leaving Frida behind on the platform, and then he was on his way to Africa, and a new life with Mary.

Chapter 5

DISASTER AT KANAM

In late October 1934, Louis sailed for Kenya from Marseilles aboard an "old coal burner," the SS *Compiègne*. He traveled with three student assistants, Peter Kent, G. T. (Peter) Bell, and Stanhope (Sam) White, in second class, where, White noted, "it was de rigueur to appear for breakfast in one's pyjamas and unshaven, as few of the women on board made an appearance!" To pass the time on the three-week voyage, Louis gave his assistants impromptu talks on vertebrate paleontology, taught them the art of string figures, and used them as subjects in thought-transference experiments. Before departing, he had given the three young men the barest outline of what had transpired in his personal life, explaining that Mary would join them in six months' time, and then had sworn them to secrecy. He had not yet told his parents the full story, and did not want any inadvertent leaks.

On this, his Fourth East African Archaeological Expedition, Louis planned first to substantiate his finds at Lake Victoria and search for additional human fossils there, then move south to Olduvai Gorge. As before, he established his base camp at his parents' home in Limuru, using the small mud-and-wattle house he had built for Frida and Priscilla, and hired ten of his Kikuyu friends as workers. Juma Gitau, who had found the Kanam jawbone, also joined the expedition. The Chevrolet box-body car and truck Louis had purchased in 1928, though battered—they had also been used by E. Barton Worthington for his expeditions to Lake Rudolf (Turkana today) in northern Kenya—were still running, and Louis's team soon readied these "old warhorses" for another journey. On November 19, five days after landing at

Mombasa, Louis climbed into his Chevrolet, strapped his rifle above the running board, and set off for Lake Victoria's Kendu Bay. Bell, Kent, and White traveled with him, while a Kikuyu named Ndekei drove the truck, which was heavily packed with gear and the Kikuyu workers.

After a full day's travel, Louis was back at the site of his most famous discoveries. Nearly three years had passed since he had explored Kanam/Kanjera, and in the interim the fossils he had uncovered at the two sites had brought him the attention and acclaim of England's most distinguished anatomists and anthropologists. A year before leaving England, he had added a touch more glory to his Kanam mandible by pronouncing it a species of humankind separate from *Homo sapiens* and giving it a name, *Homo kanamensis.* This was the first human ancestral species that Louis would name in his career, and it secured for him a kind of immortality—whenever the specimen's full name was used, it would be written *Homo kanamensis* Leakey. Louis based his new species on "the general massiveness of the whole mandible" as well as the shape of the roots of the molars and premolars. But unlike the Piltdown, Java, or Peking men, which scientists then thought represented three genera[1] of humanlike creatures, Louis believed his Kanam Man was a true *Homo,* a representative of "a human stage very close to *Homo sapiens.*"[2] He thus placed the fossil in the genus *Homo,* but acknowledged its differences from modern humans by assigning it a distinct species name.

Louis had announced his new species in October 1933 at a meeting of the Royal Anthropological Institute with such luminaries as Professor Grafton Elliot Smith and Sir Arthur Smith Woodward in attendance. Both men publicly accepted *Homo kanamensis,* while Dr. W. L. H. Duckworth, an anatomist at Cambridge, "suggested that Dr. Leakey was over-modest in classifying Kanam man only as a new species," and should call it a new genus as well. Woodward then noted that "geologists and palaeontologists were convinced that Dr. Leakey had proved the antiquity of these fossils" and "congratulated him on his most valuable results."

Now once again at Kanam, Louis could show his students the actual site of *Homo kanamensis,* and "bursting with information" about his discoveries, he led the way down into the western Kanam gullies. But almost immediately he hit a snag. To his dismay and "horror" the four iron stakes he had cemented into the ground to mark the fossil's site were gone, apparently taken by the local Luo fishermen to hammer into fishing harpoons and spears. Further complicating matters, he had only a single photograph of the site, taken by Frances Kenrick, Frida's friend, in 1932. Louis had also taken pictures of the site shortly after Juma found the mandible, but his

---

[1] All living things are classified into a hierarchy of categories: kingdom, phylum, class, order, family, genus, species. Humans are in the class Mammalia; the order Primates; the family Hominidae; the genus *Homo;* and the species *sapiens.* Related species are grouped together into one genus.

[2] Peking Man and Java Man were later recognized as direct ancestors of *Homo sapiens,* and in 1950 were grouped together as *Homo erectus,* the immediate predecessor of *H. sapiens.*

camera had failed and all his film had turned out blank. When Kenrick gave Louis her photograph, she supposedly remarked, "I am not sure that this is the exact spot," and Louis is said to have replied, "Near enough." Louis had used this photograph "to show the general nature of the [fossil's] site" at the Cambridge conference, and had also exhibited it at the Royal College of Surgeons together with the jaw. But now, walking through the maze of little gullies, he could not locate the area shown in the picture. The photo did not look at all like the site where he and Juma remembered finding the jawbone. Louis finally decided that during his three-year absence the landscape had been considerably altered by heavy rains, and by villagers cutting down trees. "[We] tried to identify this photograph exactly and failed," Louis wrote in his third monthly field report, which went to a mix of sixty scientists and supporters in England, "but, as many trees had been cut down at Kanam West main gullies, and as a village had been moved, I was not entirely surprised that we could not get the position exactly."

The missing iron stakes made it equally "impossible to relocate the exact spot where the . . . jaw had been found," and Louis knew that there would be at least "raised eyebrows" now about the validity of his claims. Still, he believed he had managed to "fix the position" of the fossil "within reasonable limits."

Louis and his party then drove the three miles to the Kanjera fossil beds, where Louis had discovered the fragmented skullcaps of three "early examples" of *Homo sapiens.* At the Royal Anthropological Institute meeting, and in his book *The Stone Age Races of Kenya,* Louis had described his Kanjera skulls as coming from "very primitive and generalized men of *Homo sapiens* type." Very likely, they were the descendants of his *Homo kanamensis.* Despite their "primitive" appearance, these Kanjera men had, according to Louis's reconstructions, a "brain capacity greater than the average of human beings today." Most importantly, said Louis, they had been found alongside Chellean handaxes, and so showed "for the first time what the makers of this great and well known [handaxe] culture were like."[3] Taken together, the Kanam mandible and the Kanjera skulls seemed to suggest an evolutionary progression of early humans, and Louis concluded that "it appeared not improbable that Africa was the home both of the *coup-de-poing* [the French term for handaxe] culture group and of the *Homo sapiens* branch of the human family." This was precisely what Louis had set out to prove beginning with his first archeological expedition in 1926—that humankind, as Darwin had suggested, had originated in Africa. Only six years later, it appeared that he had the evidence.[4]

---

[3] Prior to Louis's claim, scientists had argued over whether Neanderthal Man or some other type of *Homo sapiens* had made the handaxes, which are found across Africa, Asia, and Europe. They are now known to be the work of *Homo erectus.*

[4] At this time, few Europeans were willing to acknowledge a kinship with the peoples of the "Dark Continent." The idea of "parallel evolution," which suggested that the various human races had evolved independently of each other, was still very much in vogue.

But now at the Kanjera discovery site, Louis was faced with the same situation he had found at Kanam: the iron stakes missing, inconclusive photographs, and only his memory "to carry him through." Again, Louis thought he had relocated the site "within ten yards or so . . ."

Louis was visibly shaken by this turn of events; White remembers him looking "completely flabbergasted" and "dismayed." Professor Boswell, the eminent Pleistocene geologist and emissary of the Royal Society, was due to arrive in one week to verify the age of Louis's fossils, and Boswell, with his reputation for meticulous precision, would surely be unhappy with such statements as "within reasonable limits," or "within ten yards or so." White, the expedition's surveyor, already foresaw the professor's complaints: why had not someone made a "rough sketch map, . . . or marked the trees in some way, [or] at least paced off the distance as best they could?" Louis later explained that he did not have enough money to pay a surveyor to map the area in 1932—"I was saving all the money I could for a second visit to Oldoway"—but, as White pointed out, a simple sketch map would have sufficed. There was, of course, nothing Louis could do now to correct his mistake. He had to rely on his memory of the sites and hope that Boswell either accepted his word, or that his team uncovered additional human fossils *in situ*.[5] He also had an expedition to lead, and putting aside whatever misgivings he had about Boswell's impending visit, he turned his attention to it.

Sam White was Louis's surveyor and Peter Kent his geologist. Both men were recent university graduates, and together they provided the expertise Louis had so badly needed on his previous expedition. "Louis would tell us what he wanted doing," recalled White. "For me [it was] to continue mapping this bit or that, Kent [was] to investigate the geology of such and such a place, and so we would work on our own jobs." Meanwhile, Louis, Juma, and Heselon Mukiri, who would later become Louis's most famous Kikuyu assistant, continued exploring for fossils. Whenever anything noteworthy was found—an antelope's skullcap and horns, a partial skeleton of a pygmy hippopotamus, the massive remains of a giant tortoise—Louis put his Kikuyu assistants to work on the excavation. Allen Turner from the Nairobi Natural History Museum had also come to help, and he spent part of his time hunting for fossils and the rest assisting Peter Bell in collecting birds and mammals for the British Museum of Natural History.

Camp conditions were extremely spartan, for "as always, Louis was short of funds. . . . We lived almost exclusively on fish . . . with very little in the way of vegetables or fresh fruit," said White. "We had no camp beds but slept on mattresses on the ground; we had mosquito nets but did not take prophylac-

[5] David Pilbeam of Harvard University has pointed out that in other situations, the scientist's word was accepted for the discovery site. For example, the details of Charles Dawson's discovery of the first pieces of Piltdown Man were utterly imprecise; not even the exact date of the find was known. Further, Dawson was an admitted amateur fossil hunter; yet the location of his find was never questioned.

tic quinine—as we should have done. . . . [White was the only one who did not contract malaria.] Cold drinks were unknown, and we had no alcohol. [At night] we were too dirty and tired to keep us long from our beds." Rising at five-thirty every morning, they worked until noon, took a break until three o'clock during the hottest part of the afternoon, and returned to camp at six in the evening. Dinners were served by kerosene lantern light, "with swarms of insects—a real cloud round the lamp and falling into our soup. One soon ceased to be particular, for it was a case of no insects, no food." It rained nearly every evening, but the dawns were beautiful and the sunsets glorious.

While Louis organized his camp, Professor Boswell set out for East Africa. Among scientists in England, Boswell's trip was viewed as a matter of some importance, and had even been announced in *Nature*. Louis's African Kanam Man had, after all, replaced the British Piltdown Man as the oldest ancestor of humankind—a demotion that not all scientists were ready to accept. "Notwithstanding the close and expert scrutiny to which Dr. L. S. B. Leakey's evidence for the early occurrence of man in Kenya has been subjected," wrote *Nature,* "the far-reaching effect of the conclusions to which it leads make it eminently desirable that no means of verifying and substantiating the data should be neglected." The editors therefore "welcome[d]" this opportunity to bring "final and decisive verdicts" to what had come to be known as the "Kanam controversy."

Commercial flights from London to Capetown, South Africa, had been inaugurated in 1932, and Boswell boarded an Imperial Airways "flying boat," a passenger-and-mail seaplane, on November 21. Five days later, after stops in Paris, Athens, Alexandria, Aswan, Khartoum, and Entebbe, he landed at Kisumu, a major port on the Kenyan side of Lake Victoria. Louis had taken a ferry from Kendu Bay two days before, and now rushed forward to greet his "Grand Inquisitor," as Kent billed the professor. Boswell was in good spirits, enjoying this break from his usual academic and administrative tasks, and initially, noted Kent, he and Louis "got on very well." But they were very different. White described Boswell as "a smallish, dried up sort of man." To Kent he was "dogmatic," "occasionally irascible," fastidious in his habits, parsimonious in movement and speech. Louis, however, was garrulous and expansive, "a man of untold energy and vision," somewhat zealous, and "prone to jump to conclusions," said White. Louis lacked the cautious thoughtfulness Boswell deemed important in a man of science, and so Boswell watched him warily.

In his typical enthusiastic way, Louis wanted to show Boswell as many of his sites in East Africa as possible. He had also arranged for Boswell to meet Sir Joseph Byrne, the governor of Kenya, and to give a talk in Nairobi. After his long flight to Kisumu, Boswell was ferried back to Kendu Bay, given a hurried half-day tour of the fossil beds at Kanam, then bundled into the car and driven to Nairobi. From there, Louis planned to make a "hasty visit" to Olduvai. In 1932 Boswell had played a key role in demolishing Louis's report asserting that Hans Reck's very modern-looking Olduvai skeleton dated to Pleistocene times. Louis now wanted Boswell to examine the Olduvai deposits himself, and then explain how Louis's team had misinterpreted the geology, and thus the age of the skeleton. Whether Louis was also stalling for time is unclear; certainly the meetings in Nairobi had been scheduled with Boswell's approval, and the professor relished the idea of seeing Olduvai. Louis, too, had been honest with Boswell about the problems at the Kanam and Kanjera sites. Shortly after Boswell's arrival, Louis told him about the missing stakes; but, Louis reassured Boswell, he had virtually relocated the original sites. As evidence, Louis displayed "one large fragment of human skull and several smaller pieces from the surface" that Juma had uncovered that morning close to the site of the Kanjera skulls. It was a "promising" find, and Louis hoped more pieces might turn up *in situ* while they were gone.[6]

With long, hard driving, Louis made the round-trip journey from Lake Victoria to Olduvai in an amazingly short twelve days. En route to the gorge, Louis's small party "bumped over the plains" at a tedious ten miles an hour, driving through "roving herds" of giraffe and antelope, and encountering an occasional group of wandering Maasai. There were lions and hyenas at night, and once, after the car broke down, a lonely two-day wait in "waterless country" while Louis traveled 150 miles back to Nairobi for spare parts. Along the way, they feasted on whatever meat Louis could provide, including buzzard and warthog—"the toughest meat ever!" in Boswell's words; and at the gorge, they came to know intimately Olduvai's notorious "red and black volcanic dust." Boswell noted that on return his first bath had turned "the water into soup."

By the time they returned to Kendu Bay, Boswell was eager to begin his study of the Kanam and Kanjera fossil beds. He particularly wanted to analyze the sites where the human fossils had been uncovered. At the Cambridge conference, Louis had stated that his discoveries could be "dated geologically, palaeontologically, and archaeologically"—geologically, because they had been found *in situ;* paleontologically, because they had been discovered with the remains of extinct animals that were known from the Pleistocene; and archeologically, because stone tools of distinct cultural phases had also been found in the same horizons as the fossils. Based on this evidence, he had assigned the Kanam mandible to the Early Pleistocene, and the Kanjera skulls to the Middle Pleistocene. According to the timescale of the 1930s, this meant that early humans had lived in Africa nearly one million years ago.

Boswell and other scientists questioned that date partly because of the shape of the Kanam Man's jaw.[7] The fossil appeared to have a "very pro-

---

[6] When Louis wrote about this journey in his autobiography *By the Evidence,* he transposed the dates, saying that the Olduvai trip took place after Boswell had finished his work at Kanam. This seems to have been an unintentional error.

[7] In Germany and France, Louis's claims were dismissed "as meaningless and without foundation," according to Dr. Phillip Tobias, an anatomist at the University of Witwatersrand (South Africa).

nounced" chin and lacked a "simian shelf"—a ledge of bone that joins the two halves of the lower jaw in apes.[8] Both of these features were then considered to be indicative of modern humans, and Boswell, who thought our primitive ancestors should look more like Piltdown Man, "had a foreboding that all might not be well" with the provenance of *Homo kanamensis.*

In heavily eroded areas with clay soils, such as those at Kanam/Kanjera, fossils can sometimes be washed out of one geological horizon, then redeposited in another. Boswell suspected that this had happened with Louis's discoveries. If the fossil sites had this slumping or "reworked" look, then, he believed, it would be impossible to assign them to any geological level, and thus to date them.

Nothing of significance had been uncovered while Louis and Boswell were at Olduvai, but two days after their return, Louis found a skull and palate of *Hipparion,* the three-toed horse, while Juma discovered a skull of an unknown species of giraffe at a new site nearby called Rawi. Louis put his excavators to work here, and the next day led Boswell and Kent through the Kanjera exposures. They found a number of "good fossil remains," but then Louis was off again to see what was transpiring at Rawi. When he heard from a settler that there were fossils near a town called Chemelil, fifty miles away, he jumped in his car and drove to that site. "I hoped that his fossils might prove to be Miocene," wrote Louis. "However, I was disappointed, as the 'fossils' were all bones of very recent age." He salvaged this trip by stopping at another set of Miocene beds he had discovered in 1931, called Songhor, where he found "a half of the mandible of a small anthropoid ape."

Other men may have been impressed by Louis's energy and enthusiasm, but Boswell was not. "The country was so rich in archaeological sites and fossil bone that [Louis] was for ever darting, like a bee, from one new site to another," Boswell noted with some irritation in his unpublished autobiography. He wanted Louis to settle down and excavate the Kanam and Kanjera sites. Louis, of course, could not locate them precisely, and continued to turn his attention elsewhere.

Christmas Day came, and—after asking his Kikuyu friends to waken the camp with "Noel" sung to a five-note scale—Louis headed off to the Kanam gullies to remove "the most complete skull of *Hippopotamus gorgops*" he had ever seen. Finally, on December 28, a thoroughly exasperated Boswell "insist[ed] that Leakey . . . get on with excavations at the sites where he found the jaw-bone and skull, the age of which I had come out to confirm or question." In his field report, Louis writes that he began this work on December 25 "on a big scale," but the next day he was once again at Rawi.

---

[8] In 1960 Tobias reexamined the chin of Kanam Man and showed that it "was in reality largely a bony reaction" to a tumor that had grown to the left of the jaw's midline. The jaw was thus virtually chinless.

As White had foreseen, Boswell now began to ask "plaintively" why the sites had not been more carefully marked, and dismissed as "unlikely" Louis's argument that the area had been altered by heavy erosion. He grew "increasingly irritated when Louis prevaricated over the site of the Kanam man discovery," said Kent, "prevarication because he could not identify it." Boswell himself noted with annoyance that "all attempts to find the particular sites of the photographs or the precise places where Leakey made his finds had failed. And as he had not marked the sites on the ground, we could only excavate [when] he was certain he was at the right spot."

Boswell was somewhat willing to accept the site that Louis pointed to as that of the Kanjera discoveries because Louis and Donald MacInnes had found these fossils themselves. But the Kanam mandible had been discovered by Juma, one of the Kikuyu workers, and despite Louis's faith in him and his belief that Juma was "a born scientist," Boswell never regarded him as anything more than "a native boy." The fossil had not been seen *in situ* by a scientist, and Boswell was not about to accept Juma's word for its original location.

On January 13, 1935, E. J. Wayland, the director of Uganda's Geological Survey and the geologist who developed the pluvial hypothesis, came over from Entebbe to assist Louis. He, too, was uneasy about Louis's reliance on heavy erosion to explain the discrepancies between the photographs and the sites, but he joined in the group's efforts to solve the problem. The next day, Louis, Boswell, Wayland, Kent, and White all worked at Kanjera, seeking to match a photograph of the locale with the site where Louis had found what he called "Kanjera [skull] No. 3."[9] "But I could not place the site . . . exactly after the lapse of time, though I placed it within ten yards or so to my satisfaction," Louis wrote in his field diary. White then turned up a few pieces of fossilized skull, and Louis "spent the rest of the day riddling the surface and getting quite good material (small fragments which fitted)."

They visited Kanjera again on January 17, and this time "Juma found a bit of human skull very heavily mineralized within thirty feet of where I claim the Kanjera skull to have been before it was eroded away . . . and I found a second bit." Still, none of the fossils was found *in situ.* Then, to Louis's chagrin, the team discovered that the supposed photograph of the Kanjera No. 3 site was actually of a cliff of volcanic ash some distance away. Boswell was openly annoyed by the discrepancy. Deciding that he had now seen enough of Kanjera, he announced that he would spend the next day at the Kanam site, the area Louis referred to as "Kanam West main gullies."

Meanwhile, Sam White began to map the eastern gullies of Kanam. On the afternoon of January 18, he was out surveying when, as he recalled,

---

[9] All of Louis's pictures of the Kanjera site, like those he shot at Kanam, had turned out blank. He relied on a photograph taken by Sir Alfred Kitson, a visitor to the site in 1932, to identify the locality for Boswell.

> I realised I was on the site where [Miss Kenrick's] photo had been taken. I left Gilbert [his Kikuyu assistant] with the plane table [a portable drafting table for mapmaking] and went back to camp to get my quarter plate camera; I found Kent and he came with me; I set up the camera with a ground glass view plate in position, and holding the copy we had in the camp of the famous photo, I moved my plane table until the view on photo and ground glass was the same.

Unfortunately for Louis, the photograph was again of an entirely different site and "some distance away" from where he said the Kanam jawbone had been found and where his workers, Boswell, and Wayland were excavating.

Kent confirmed White's identification, then brought Boswell and Wayland to take a look. Louis was working that day at Kanjera, and did not hear the news until late in the afternoon when he returned to camp. "He was utterly dismayed," recalled White, ". . . and Boswell's attitude to Louis became frosty to say the least." Louis "admitted the error," but this did little to placate the professor, who considered it a "bad mistake."

White and Kent both pointed out that these were simply mistakes—admittedly "mistakes that should never have happened," according to White—but neither believed that Louis was being intentionally deceptive, although other scientists in England would soon spread a rumor to this effect.[10] Louis's first reaction, in fact, was to attempt to correct the error before it went any further. Oxford University Press was then readying his *Stone Age Races of Kenya* for publication, and the now notorious photograph was set to run on the book's opening page. Louis knew he had to stop this, and immediately drove to Kisumu to cable his publisher.

He returned to camp late that night, but sat up, weary and shattered, to record the day's downward spiral of events:

> Professor Boswell, Wayland and Kent spent the day at Kanam and they say—and I suppose they are right—that they have located the exact position of Miss Kendrick's *[sic]* photo and that it was not Kanam West main gullies at all, but Kanam West Fish Cliff gullies. I've cabled to the Press to hold up distribution of my book pending insertion of an erratum notice. The Professor is in a bad humour over it. Apart from the absence of trees and of details in the gullies it is terribly like Kanam West and the mistake is not surprising though very regrettable.

To White's surveyor's eye, however, the sites were considerably different. He noted that if the photo was examined closely one could see that the gullies in it rose in the east and sloped to the west; but at the Kanam mandible site the gullies were cut in the opposite direction, and sloped west

---

[10] Fifty years later, White wrote, "I am quite certain that there was never any question of duplicity on the part of Leakey—indeed I would stake my head on it."

to east. "If this site was the epoch making site it was then held to be, I would have thought it would have been so indelibly stamped on the memory of those involved that they could have drawn it in their sleep. . . . The famous photo . . . to anyone who had left the place less than three years before could NOT be mistaken for the Kanam West gullies."

The next morning, Boswell packed his bags in a huff. "[He] did not in the least appreciate wasting two months on a wild goose chase," said Kent. Despite Louis's insistence that he had relocated the sites of the fossils "within a few feet," Boswell was unconvinced. In his opinion, the sites remained "uncertain," and so it was impossible for him to determine "whether the deposits . . . were in place or not." "Thus, effectively, my journey had been useless," Boswell noted. He bade Louis a cool good-bye and left with Wayland for Nairobi. Louis, shaken and "very down," collapsed in his tent with a high malarial fever.

Professor Boswell returned to England at the end of January 1935, just as Louis's book *The Stone Age Races of Kenya*—which described the Kanam and Kanjera discoveries—was being warmly reviewed by several scientific journals. In *Nature,* Sir Arthur Keith hailed Louis's fossils as being "of the utmost importance to anthropologists. . . . The great enigma, the origin of the modern races of mankind, is brought a big step towards solution. No one can now discuss the rise of humanity and the evolution of early 'cultures' without taking Dr. Leakey's discoveries in East Africa into consideration." Boswell thought otherwise, and he set about making his opinions known.

At Kendu Bay, Louis suffered through a two-day malaria attack, and then, although weakened by the fever, once again pressed on with his search for human fossils. Far from England, he was unaware of the impact Boswell was about to have on his scientific reputation.

In his January field report, Louis had written only a cursory description of his "unfortunate mistake," and noted that also "unfortunately certain of the [geological] problems still remain unresolved." He was hesitant to elaborate partly because, as Kent said, Louis was "making the best of a poor show." But Louis also believed that he and Boswell had agreed to say nothing about the problems at Kanam and Kanjera until Louis returned to England and could defend himself. Boswell had a decidedly different impression. "I had been compelled, of course, to tell [Louis] exactly what I proposed to report on my return to England . . . ," he wrote in his autobiography.

On February 1, Boswell attended a council meeting of the British Association for the Advancement of Science, where he very likely discussed his recent trip to Kenya Colony, although he does not specifically say so in his autobiography. Two days later, he gave a full report about the expedition to Professor W. W. Watts, his predecessor at Imperial College and the president of the British Association. It is fair to assume that at these meetings Boswell presented an unfavorable picture of the events at Kanam/Kanjera, for within a week's time, "the Leakey business," as Boswell called it, "had become a *cause célèbre* in the scientific world." He had yet to write a word, and had declined giving a talk to the Royal Society "until Leakey could be present," but the scientific community was nevertheless buzzing with the scandal. Colleagues of Boswell's at the Royal Society and the Anthenaeum clustered around him to ask "interminable questions," while at a meeting of the Royal Anthropological Institute, archeologists Dorothy Garrod and J. P. T. Burchell pulled him to one side so they could hear "the Leakey story 'from the horse's mouth.' "

Louis's fall clearly delighted many of his peers. Some were jealous of his youthful success, while others strongly disapproved of his pending divorce ("divorce was non-U in Cambridge then," noted one friend). Louis's old nemesis, his cocksureness, had also galled and alienated a number of scientists. "He had crowed over his British contemporaries so blatantly that he had few friends in the business!" recalled Peter Kent. Louis had particularly irritated "some of the older archeologists," said Vivian Fuchs, with his "new ideas about early man. He was not all that careful when he trod on other people's toes by saying they were wrong. And therefore, a lot of people who wanted to get their own back started saying that Louis had been deceptive."

But if Louis's intent was to deceive, he had done a remarkably incompetent job of it, beginning with his invitation to Boswell to visit the sites. As Boswell noted, there was a certain slapdash quality to Louis's work—more than once, he jotted "important notes" on the back of a cigarette pack or envelope, then later absentmindedly tossed the notes away. And in his eagerness to make discoveries, Louis did not always use his limited funds wisely: thus, in 1932, he had chosen to spend his remaining money on another expedition to Olduvai, rather than hire a geologist to map Kanam. Louis was also professionally ambitious and longed to be considered a peer by the British scientific aristocracy. His colonial roots were a strike against him, but Louis thought he might gain admittance to the club simply on the strength of his discoveries. Unfortunately, he lacked the art of politicking, and never outgrew his irritating, "know-it-all" style. Many influential people in British science considered him arrogant and cavalier, and instead of reaching the inner sanctum, Louis became an outcast.

Not given to reflection, Louis never seemed aware of these failings—or at least he never mentioned them in his diaries and journals. The only lesson he seemed to draw from the Kanam controversy was that he should carry two cameras. "Ever since then I have carried with me no fewer than two cameras, and I have always photographed anything of importance with both of them," he wrote in his autobiography. After Kanam, Louis's fossil discoveries were always well documented. But he did not change his way of relating to his fellow scientists, nor did he stop pushing the most sensational interpretation of his fossil finds.

February passed, and Louis, oblivious to the distant gossip, dispatched his fourth monthly field report from Kanam. He had put Boswell's unpleasant visit out of his mind, and was busy pursuing several new lines of research. He and his team had started searching Miocene deposits on Lake Victoria's Rusinga Island and found these to be "rich in remains of anthropoid apes." By the month's end, he had "parts of at least 16 Miocene apes" from Rusinga and Songhor, and "over 30 species" of Miocene animals. He had also started interviewing two "very old Kikuyu men" for a book he proposed to write about the tribe, and had continued to work at Kanam, although "no finds of any great interest" had been made. Then, on March 16, he opened a letter from Dr. Haddon, his old anthropology professor.

"I have been shown Boswell's report to the Royal Society and also your field report, December 24–January 24, and I must confess that I am disappointed at the casual way in which you deal with the matter," Haddon wrote sternly.

> So far as I can gather it is not merely a question of a mistaken photograph, but a criticism of all your geological evidence at Kanam, Kanjera and Oldoway. The conference at Cambridge had to rely implicitly on your statements, and from what I hear there is much annoyance in view of recent developments. It seems to me that your future career depends largely upon the manner in which you face the criticisms. I am not in a position to know to what extent they can be rebutted by you with scientific evidence, but if you want to secure the confidence of scientific men you must act bravely and not shuffle. You may remember that more than once I have warned you not to be in too much of a hurry in your scientific work as I feared that your zeal might overrun your discretion and I can only hope that it has not done so in this case.

The letter stunned Louis; he had no idea that Boswell had reported on his expedition to the Royal Society, still believing that the matter would wait until his own return to England. Deeply troubled, he wrote in his diary: "A letter from Dr. Haddon which is very disturbing. In it he suggests that Boswell's findings may ruin my career. Apparently he has reported most unfavourably to the Royal Society and is writing to *Nature.*"

In fact, Boswell's letter to *Nature* had already been published; it appeared on March 9, 1935, and although Boswell felt that he had "let [Louis] down (as everyone realized) as lightly as possible, while being very firm about the scientific side of the business," others considered the letter to be "scathing." Boswell never mentioned any of the extenuating circumstances at the site: that Louis's camera had failed, or that the iron stakes marking the fossils' locations had been stolen. Instead, without mincing words, Boswell addressed Louis's mistakes one by one.

Louis, Boswell noted, had been unable "to find the exact site of either discovery, since the earlier expedition . . . neither marked the localities on the ground nor recorded the sites on a map." Boswell then questioned

Louis's understanding of basic geology, suggesting that he was unable to distinguish between clay and volcanic deposits. And beyond that, dates at the two sites would always be problematic, because the deposits were prone to slumping. Finally, Boswell concluded,

> [I]n view of the uncertain location of the Kanam and Kanjera sites, and in view also of the doubt as to the stratigraphical horizons from which the remains were obtained and the possibility of disturbance of the beds, I hold the opinion that the geological age of the mandible and skull fragments is uncertain.... It is disappointing after the failure to establish any considerable geological age for Oldoway man [Hans Reck's skeleton, which Louis had also erroneously dated]... that uncertain conditions of discovery should also force me to place Kanam and Kanjera man in a "suspense account."[11]

Boswell's letter to *Nature* drew immediate response. "[It] gave rise to almost a fan mail and to requests for interviews by Pressmen [journalists] and others," Boswell wrote. "In popular parlance, the earliest human being had been debunked!"

Newspapers around the world picked up the story. "Oldest Fragment of Man Disputed/British Professor Reports He Found in Africa No Support for Leakey Discovery Claim," pronounced the *New York Times;* "Modern Man Not So Old/Piltdown Skull Still Holds Its Place" said the *Morning Post*—unaware of the irony of replacing Louis's doubtfully dated, but genuine, fossil with what would later be revealed as a forgery.

Louis, camped far away by the shores of Lake Victoria, did not receive his copy of Boswell's letter to *Nature* until March 28, 1935—over two weeks after it was published. Flushed with anger, he wrote hotly in his diary, "Got a copy of Boswell's attack in *Nature,* to which I must reply when I calm down. At present I'm so angry that I'd probably say things which I'd regret afterwards."

Louis, however, did not calm down. Instead, he wrote bitterly in his March field report, "Professor Boswell explicitly told me and my colleagues that he proposed to publish *nothing* until the Expedition returned to England." Boswell, a highly esteemed scientist and a senior fellow of the Royal Society, was not to be lightly accused of breaking promises. He wrote to Louis,

---

[11] Today, paleoanthropologists generally accept the authenticity of Louis's Kanam mandible. The Kanam beds are now known to range in age from about 200,000 to 6 million years. "There is no reason to believe that the mandible did not come from where Louis said it did," noted Harvard anthropologist David Pilbeam in 1987, "and no reason that it may not be two or more million years old." Pilbeam and others believe the fossil may represent *Homo habilis,* a species dating to two million years, which the Leakeys discovered in 1960 at Olduvai Gorge. The Kanjera skulls, however, date to less than 10,000 years. "They were probably burials into the older sediments," says Tom Plummer, a paleoanthropologist at the University of California, Los Angeles, who has done extensive research at the site and on Louis's original fossils. "The situation was exactly analogous to Reck's [Oldoway Man] skeleton."

apparently telling him that he would be wise to retract his earlier statements.[12] Louis may have received other letters as well, for in his final field report, written in June 1935, he was contrite.

> I most certainly never intended making an unjustified accusation of bad faith against Professor Boswell, and what I wrote I wrote believing it to be absolutely true. In his letter *to me* Professor Boswell implies—but does not actually state—that he had told me he was going to publish a criticism in *Nature* or elsewhere, but from what he says in a letter to one of my colleagues I understand that he says that he told me on January 18th. I can still say quite frankly that I have no recollection whatever of his saying so. . . . I am, of course, fully prepared to accept Professor Boswell's word that he did tell me . . . and I sincerely apologise for having mistakenly made an unfair accusation of bad faith.

But the damage had been done. The cumulation of events—Louis's separation from Frida, his "colossal blunder" at Kanam/Kanjera, and his accusations against Boswell—were not soon forgotten or forgiven. Louis did not then realize it, but he would never again sail from England at the head of an East African Archaeological Expedition.

---

[12] Louis refers to this letter in his combined Ninth and Tenth Monthly Field Reports, but the letter is not extant.

Chapter 6

OLDUVAI'S BOUNTY

Before Louis departed for Kenya and Kanam, Mary decided that she would not "be left kicking her heels" at home while he was out exploring, and so arranged a trip to South Africa with her mother, Cecilia. In early January, 1935—while Louis was arguing with Boswell over fossil locations—mother and daughter set sail from England on a Union Castle liner bound for Cape Town. They planned to visit various prehistoric sites in South Africa, then take the train north to Rhodesia (now Zimbabwe). From there, Mary expected to fly to Tanganyika and Louis's arms, and filled with excitement about the trip, she happily and lightheartedly left England behind. Cecilia, of course, secretly hoped that the journey would somehow do what she could not: get Louis out of Mary's mind and out of her life.

But for Mary this was the start of a great adventure. "Louis wanted me with him," she wrote, ". . . and I could hardly wait to see Olduvai." There was only one hitch: Louis was not at the airport at Moshi when Mary arrived. He had not left any messages either, and disappointed and cross, Mary checked into a "rather sleazy" German-run hotel.

Meanwhile, Louis was doing his utmost to get to Moshi. A month before, he had closed his camp at Kanam/Kanjera and headed back to his parents' home in Limuru to prepare the second half of his expedition, an exploration of Olduvai and "reported sites" south of the gorge. This was the trip he had promised Mary the year before in his rooms at St. John's. But at Limuru he learned that the expedition's funding was in doubt; a final decision about the grant would be mailed to him at Arusha, Tanganyika, a small colonial town on the Great North Road of Africa, 192 miles south of Nairobi, and only forty miles from Moshi.[1] Louis had made the trip to Arusha several times before and expected to cover the distance in a day and a half. It seemed that he had ample time to pick up the letter in Arusha and still reach Moshi before Mary arrived. But then it began to rain, the red earth roads turned slick as grease, and Louis's imagined easy jaunt turned into a four-day "nightmare journey." The expedition's car and truck sank axle-deep time and again in the muddy ruts; on one day, Louis covered all of two hundred yards in five and one-half hours. He finally reached Arusha's post office on the morning after Mary's arrival at Moshi. The promised letter was waiting for him and it brought good news: there would be money for Olduvai.

With that settled, he drove to Moshi and found Mary eating breakfast on the veranda of her hotel and marveling at the sight of snowy Mount Kilimanjaro. They did not linger, Mary said. Louis was, as usual, "in a hurry" and so he simply tied her suitcase on the car, asked Mary to squeeze in beside him, and then set off for Olduvai. "It was what you expected with Louis," said Mary. "The thing was to get on with the job."

Mary had arrived in Tanganyika in the middle of the "long rains," a six-week period of heavy monsoons. To Louis, the rain meant the possibility of a good water supply at Olduvai, and when they left Moshi, he was pleased to see clouds gathering again on the horizon. But the clouds also promised another epic mud journey, though neither Louis nor Mary gave this much thought. They were at last together in Africa, and now they were "going out into the blue," as Louis put it.

Louis had decided to try to reach Olduvai via a new road that led up the eastern slope of 7,500-foot Ngorongoro Crater. From the crater's rim, he thought he could find his way down onto the Serengeti Plain and there pick up a track he had made in 1931 leading to the gorge. This was actually the most direct route to Olduvai from Arusha, and the district commissioner gave Louis permission to try the new road. After another brief stop in Arusha to collect Sam Howard, a college friend of Louis's who was joining the expedition, Louis and Mary set off for Ngorongoro. Sam White and the Kikuyu staff had started for the crater the previous day in the truck; the two parties planned to meet at the crater's rim in three days' time.

White had little trouble in reaching the rim, but heavy rains once again delayed Louis. It took his small party two days simply to reach the foot of Ngorongoro, and when they started up the volcano's steep incline they immediately ran into a quagmire of sticky black mud. Louis had expected an easy journey and had sent most of the food and supplies ahead with White. Now stuck in the mud, Louis's party faced the wet and cold on a diet consisting only of "several bunches of bananas and a few tins of sardines."

---

[1] These letters are not extant, so it is unclear which granting agency was involved.

Yet they pushed on, manhandling the car toward the summit. "On some occasions, [we] practically carried the car and the equipment," wrote Louis. "Sometimes we unloaded the car and carried the luggage ahead for half a mile or so, and then returned to push and carry the car. Sometimes we took the car ahead and went back to get the luggage." Mary, who did not want to be thought soft, joined in this struggle, pushing the car and carrying loads alongside the men.

They finally reached the crater's rim on the afternoon of April 24, six days after leaving Arusha. It had yet to stop raining. "That night," recalled White, who had helped with this final effort, "the only cover we had was the tarpaulin off the lorry so this was laid on the ground, we lay on it, and then pulled it over us—me, L[ouis] and Mary and Sam Howard!!!"

For Mary, the six-day journey had been a rigorous initiation into African travel. But if all the hardships caused her any regrets or doubts about life with Louis, these were soon dispelled in the morning. "The clouds had lifted," Louis wrote, "and we had a magnificent view down into Ngorongoro Crater. Through our field glasses, 2,000 feet below us we could see countless wildebeest, hundreds of zebra, and an occasional rhinoceros. I had never seen the crater from this angle before, and the others, of course, had never seen it at all. It was a most exciting moment." They camped at the crater's edge for two days, then drove around the rim to begin the descent down to the Serengeti. "In a few more miles," recalled Mary, "I was looking spell-bound for the first time at a view that has since come to mean more to me than any other in the world." Standing beside her, Louis eagerly pointed out the landmarks he already knew so well: to the west, the rugged slope of the extinct volcano Mount Lemagrut; to the east, the flanks of the crater Olmoti; and stretching to the horizon, the vastness of the Serengeti, its grasses lushly green from the rains. Several small hills rose from the plains—Naibor Soit, Engelosen, and Kelogi. And there, close to the hills, were the "two narrow dark converging lines" of Olduvai Gorge.

Almost fifty years later, after she had made her home at Olduvai, Mary would write of the landscape Louis had first shown her: "I shall never tire of that view, whether in the rains or the dry season, in the heat of the day or in the evening. . . . It is always the same; and always different. Now, nearly half a century later, that view means to me that I am nearly home."

But this time it was simply a new and exciting land, and Louis and Mary stood together at its edge, watching the great dark herds of animals move across the plains, and the thunderheads drift across the sky. They then started down the crater's slope, making their way past lava boulders and blossoming thorn trees to the grasslands of the Serengeti. Louis picked up the track he had made in 1931 and followed it to the gorge.

Louis's expedition had arrived at the height of the great game migration, when thousands of grazing animals mass together to follow the rains and the greening of the grass. Bumping along the track, it seemed that his party had stumbled into Eden. All around them, for as far as the eye could see, stretched lines of loping, bleating wildebeests and sleek zebras. There were giraffes, too, peering over the tops of the acacias, and agile, lyre-horned impala, gazelles with their dark painted eyes, and the gentle, dewlapped eland. Sometimes a zebra or gazelle would stop to stare at the rattling vehicles, then turn with a snort and gallop away. Louis kept up a nonstop monologue of natural history lore, describing the habits and behaviors of the animals they passed, while Mary sat quietly beside him. On that journey, Africa "cast its spell" on her, she would later write, and Mary Nicol "would never be the same again."

They reached Olduvai late that afternoon, six hours after leaving the rim of Ngorongoro. The trip from Arusha had taken far longer than Louis had anticipated, but the road he made has since come to be used as the standard route to Olduvai—although it can now be covered in a single day.[2] "From the tourist point of view, the change is obviously advantageous," Louis observed almost forty years later, "but I still look back with pleasure at the hardships of that pioneer trip and would gladly endure them again for the mere satisfaction of achievement in the face of so many difficulties."

Louis managed to drive the vehicles partway down into the gorge, away from the constant winds of the plains. He set up his camp on a gentle slope facing Ngorongoro, near the junction of the main and side gorges. The Kikuyu slept in the truck; Sam Howard and Sam White shared a tent; and Louis and Mary slept in another, on an air mattress on the rocky ground. Late that night, a pride of lions prowled the edges of camp, filling the air with their deep, shivering roars. "[M]ost of the party had never heard lions at close quarters before," Louis wrote, "[and] they experienced a great thrill."

On this visit to Olduvai, Louis planned to explore the beds of the side gorge, an area that he had not previously prospected thoroughly. His main intent was to pinpoint "the places where fossils or stone tools were being exposed by erosion"; at a later date, he would return to excavate them. Of course, Louis was also keeping an eye open for human fossils that might vindicate his work at Kanam/Kanjera and prove to Boswell that human ancestors had lived in Pleistocene Africa.

As on his 1931 expedition, the abundance of fossil and archeological sites at Olduvai staggered Louis's teammates. Mary found the gorge to be "incredibly beautiful and . . . nearly every exposure produced some archaeological or geological excitement." Olduvai's bounty so stunned White that he found himself thinking, " 'What the hell were we doing messing around in the pittling little gullies of Kanam-Kanjera when this wonderland existed?' And it was a wonderland," he continued, "everywhere the sides of the gorge . . . were fronted by pinnacles of soft rock; they had been protected from erosion by the presence of a fossil or hand axe which in many cases still sat on the top." For a time, they simply "pluck[ed] these plums . . . the 'pinnacle

[2] Sam White notes that "a few years ago" he made the journey to the top of Ngorongoro Crater in seven minutes "in a Leakey Landrover [*sic*]."

caps,' " collecting in this manner an elephant's fossilized tusk, the complete skull of a *Hippopotamus gorgops,* and numerous stone tools.

Louis was again working with greatly reduced funds, and so all exploring in the gorge—and all retrieving of fossils—was done on foot. This sometimes meant hoisting a weighty skull, like that of a hippopotamus, onto a stretcher and carrying it miles back to camp, but it also saved precious fuel. "If we found fossils," Mary noted, "we had to carry them back to camp ourselves."

A week after Louis set up his camp, Peter Kent and Peter Bell arrived, also via the Ngorongoro road, and the team soon settled into a regular routine. Kent and White surveyed and mapped the gorge; Bell collected birds; while Louis, Mary, Sam Howard, and several of the Kikuyu workers searched for fossils.

"Every day we used to get up before dawn, have breakfast, and leave on foot, taking with us a little food and bottles of drinking water, together with notebooks, dental picks, [and] brushes," Louis wrote. They fanned out over the slopes and gullies, where "every step [was] challenged by bayonets of sisal and a hundred other needle points," and often stooped to crawl on hands and knees across the exposures. Their care and diligence paid off: in their first six weeks at the gorge, Louis's team found twenty new sites that were rich in either fossils or stone tools. "Louis liked to name [such] areas after their discoverers," Mary recalled, "and soon everyone's initials had been given to at least one or another site . . . which they had first identified." After their initials, as a further identification mark, Louis added either a 'K' for *korongo,* the Swahili word for gully, or a 'C' for cliff. Every archeological map of the gorge now bears these abbreviations: SHK, for Sam Howard's Korongo, an area littered with artifacts and fossils; BK, for a gully Peter Bell found in Bed II that had hundreds of artifacts; SWK, for a site where Sam White discovered the skull of a giant antelope with a horn span of "something over 6 feet." Louis would later name this previously unknown creature *Pelorovis oldowayensis.*

But it was at MNK, Mary Nicol's Korongo, that Louis felt his hopes soar. Crawling across the sediments one day in early May, Mary picked up two "rather thick" pieces of human skull from the lower level of Bed IV, one of the gorge's four main geological horizons. Handaxes and the fossilized bones of antelopes and pigs lay nearby, and it seemed that here they might have found another Olduvai Man. Louis put in a small excavation, but the rest of the skull eluded him.[3] Still, the find left Louis and Mary "very excited," and in his monthly field report he wrote with renewed confidence, ". . . I still am convinced—that somewhere at Oldoway we shall sooner or later find the fossilised remains of the men who made the Chellean and Acheulean tools which are so plentiful."

---

[3] Although nothing more was ever found of this skull, the two pieces, known as OH2, for Olduvai Hominid 2, have been identified as belonging to *Homo erectus.*

Mary's find improved her standing with the members of the expedition, none of whom had entirely approved of her presence. Although Louis's British colleagues were "all polite" to her, Peter Kent remembered them initially feeling "strongly adverse" to her coming. "Our joint objections . . . were partly 'puritanical,' since unmarried bliss was still frowned on in the 1930's, and partly what may be called the 'men's club syndrome,' " said Kent. Yet in camp he found her "pleasant and easy to get on with," as did White, who noted that "she was quiet and confident," and "fitted in at once." The Kikuyu, who had known and liked Frida, were somewhat more cautious in forming an opinion. At first, White recalled, they were "completely incredulous" that Mary was coming at all. On their journey to the top of Ngorongoro, the Kikuyu asked White again and again whether it was true that "Bwana Louis" was taking a new wife. "[They] couldn't believe Bwana Louis, son of missionaries, would break the vows and social norms that the Mission was laying down for them," said White. White did not think they disapproved of either Louis or Mary, but they did feel "shocked wonderment" at Louis's behavior. "Probably at the back of their minds was the belief that this was a leg pull and all would be as before."

But Mary's quiet ways and her archeological skills slowly won over Louis's Kikuyu friends as well. At another site in the gorge, she found a very fragile fossilized pig's skull that needed to be encased in plaster. Louis had taught her the technique on one of their digs in England, and now with some finesse she successfully plastered and lifted the skull—all the while being scrutinized by Louis's chief Kikuyu assistant and "virtual scientific colleague," Heselon Mukiri. He had been "somewhat suspicious of me up to this point," Mary noted, "but he now accepted me as a worthwhile member of the team and as someone who might even be worthy of Louis."

Camp conditions again were extremely spartan. There were, of course, no shops close by where they could replenish supplies, so they had to rely on the canned food they had brought from Nairobi, occasionally supplementing this with a ram purchased from the local Maasai, or with a gazelle Louis shot on the plains. The hunting bothered Mary, who had a deep affection for animals, but she made herself eat this meat, which she "genuinely disliked . . . without complaining." Their water supply was far worse. On his previous expedition, Louis had used the water from springs twenty miles away. This time, his limited fuel supply made such water runs a luxury, and he decided to draw on several pools in the gorge itself. When they first arrived, these were deep and relatively fresh from the recent rains, but as time passed and the pools dwindled in the sun, the water became increasingly unpleasant. Already alkaline and muddy, the pools turned into filthy wallows as local rhinos rolled and urinated in them. "An attempt to filter this water through charcoal met with no success," Mary wrote in *Olduvai Gorge: My Search for Early Man,* a popular account of her discoveries, "so that our soup, tea or coffee all tasted of rhino urine, which we never quite got used to."

Then one day a sudden shower fell on the camp, leaving clear pools of fresh water "lying in hollows in our tent canvases." Not realizing that the canvas had been treated with a copper-based insecticide, Louis's team rushed about "eagerly collecting the water from the tents in every possible container," then poured it into tall glasses. "We sat round gloating at our glasses of clear water," said White, "then Louis drank his in one gulp; I did the same with half of mine, then both of us were violently sick"—as were all the others. When they later had the water analyzed in Nairobi, White learned that "we had had a more than fatal dose of copper arsenates . . . it was so strong that our stomachs rebelled!!"

Other times, as they suffered with their bad water, they watched in frustration as rain poured down all around them on the plains and the distant Ngorongoro range. At last the clouds stopped overhead, and for two days it rained continuously. "[A]fter which, for five memorable days, the river at the bottom of the Gorge flowed strongly and with a roar of rushing water that could be heard half a mile away," Louis wrote in some amazement. A small flock of ducks showed up to enjoy the water, and Louis himself delightedly took a swim in the Oldoway River. But within a week, the water had subsided and they were once again forced to drink from the rhinoceros wallows.

To supplement the poor water, Louis arranged for the local Maasai people to bring fresh milk to his camp every day. He had befriended these Maasai on his 1931 expedition by running a small medical clinic, and this time he brought the tribal elders gifts of tobacco. He also set up the clinic again, and in the late afternoon, just before sundown, treated his patients for everything from tapeworms to spear wounds.

One day in mid-May, a half-Maasai, half-Kikuyu visitor named Sanimu showed up at the camp and stayed to join in the fossil hunting. He watched Louis and his team work with "great interest," and finally asked if Louis "would like to know about another site" where they could find more "bones like stone." "Naturally, my reply was yes," Louis wrote in his autobiography. The fossils, explained Sanimu, were near his flocks and village, on the far side of Lemagrut Mountain in a river valley he called Laetoli. For further proof, Sanimu walked the long distance to his home, then returned ten days later with a small bag containing fossilized pig and antelope teeth.

"That decided me," Louis wrote. "I immediately made plans to close down the camp at Olduvai . . . and leave with Sanimu to investigate the new area." A few days later, Louis and his team (except for Sam Howard, who left for Nairobi) set out in the car and truck for Laetoli. There was, of course, no road and the journey took them two days (now it can be made in about an hour). At Laetoli, Sanimu led Louis upstream two miles through undulating grasslands to a heavily eroded area of grey-colored sediments banded with layers of volcanic ash. And as Sanimu had promised, fossilized bones and teeth "in excellent condition" lay everywhere. Although Louis had also hoped to find "stone spears"—the term he used to describe handaxes to Sanimu—he only turned up a single "crude pebble tool" and a "surface scatter" of tools from the Middle Stone Age. But they did find the remains of a rhinoceros, numerous tiny rodent bones, antelope horn cores, two species of extinct elephant, several complete tortoise shells, and a clutch of perfectly preserved birds' eggs.

Many of these fossilized animals were slightly different from those that Louis knew at Olduvai. For example, the Laetoli elephants appeared to be somewhat more primitive than those in the gorge, and Louis rightly concluded that Laetoli's fossil beds represented an older geological period. Still, evidence of early human activity was not abundant, and after two weeks Louis drove back to Olduvai, never realizing that among his Laetoli fossils lay a precious hominid canine tooth.[4]

Louis's expedition returned to Olduvai in the middle of June. The team had now run seriously low on both food and gasoline, and the remaining water in the gorge had turned into a foul swamp. Sam White and Peter Bell were also scheduled to return to England, so Louis sent them off in the truck to Nairobi. They traveled with two of the Kikuyu, including the mechanic Ndekei, and sufficient food and gasoline to see them to Loliondo, a town sixty-five miles to the north. From there, it would be a simple matter to reach Nairobi, where Ndekei could purchase fresh supplies before returning to Olduvai. "I was hopeful they [his Kikuyu staff] could be back in ten days," wrote Louis.

In the meantime, Louis, Mary, Peter Kent, and the remaining Kikuyu workers continued to explore Olduvai. Almost immediately, Louis turned up the "greater part of a very fine elephant skeleton [in] the lower part of Bed I," the lowest geological horizon in the gorge, and put in an excavation to remove it. This was hard, hot work under very trying conditions. With the end of the rains, the formerly green plains had turned dry and golden, nearly all the game had moved north, and Louis had given up any attempt at hunting. Instead, they concocted what they could from their remaining stores: rice, canned sardines, a little maize meal, and apricot jam—a combination that "finally became quite revolting," wrote Mary. They also had a little tea left to dilute their "muddy, ammonia-tainted water." Worst of all, they had run out of cigarettes. Both Louis and Mary and several of the Kikuyu were heavy smokers, and in the evenings they searched their campsite for all their old cigarette stubs. They then sat about rerolling the tobacco "in toilet paper, which," Mary noted, "was . . . of a substantial variety known as Bronco."

---

[4] Based on its primitive characteristics, Louis identified the tooth as that of a cercopithecoid (a monkey) and sent it along with the other Laetoli fossils to the British Museum of Natural History. In fact, it was the canine of an australopithecine—the first such hominid fossil to be found after Dart's Taung Baby in South Africa, and the first australopithecine discovered in East Africa. But the tooth remained misidentified until 1979 when Tim White, now a paleoanthropologist at the University of California, Berkeley, recognized it for what it was and reclassified it as *Australopithecus afarensis*.

After a long three weeks and a series of calamities, including a broken rear axle on the truck and a local Maasai uprising, the expedition truck finally returned to Olduvai laden with canned food. The fresh supply gave Louis an extra week in the field, and he redoubled his efforts to find an early human fossil.

Back in England, at least one colleague was betting that Louis would succeed. Arthur Hopwood, Louis's paleontological friend at the British Museum of Natural History, had read Boswell's report in *Nature* and sent Louis a sympathetic note (the only one he received): "I was very sorry to hear the results of Boswell's visit to the Kanam-Kanjera sites, and sympathise deeply on your disappointment. Doubtless your infernal luck will not desert you and you will confound your critics yet."

But as his three-month exploration of Olduvai drew to a close, Louis had nothing new or dramatic to show. His team had, of course, found quantities of wonderful fossils and scores of handaxes, but aside from Mary's two small pieces of human skull, they had not discovered any bones of early humans. Still, handaxes lay strewn about the gorge, offering tantalizing evidence of a people long forgotten, and Louis remained certain that someday at Olduvai he would find their fossilized bones. "I am more and more convinced," he wrote in his monthly report, "that we have yet a very great deal to learn and to discover in the Oldoway Gorge, and I want to lay bare more of the secrets hidden there."

On July 6, with water "very scarce" in the gorge, Louis closed his camp at Olduvai and drove back over Ngorongoro Crater to Arusha. He had sufficient funds to spend another six weeks in the field with Mary, and they decided to investigate more recent archeological sites in Tanganyika. Louis also had a promise to keep to Mary. In England he had told her of the rock paintings he had seen in 1929 during his overland trip to South Africa. They would visit these sites together after the Olduvai expedition, he had assured her, and they now took the Great North Road south from Arusha to an area on the edge of Tanganyika's Maasai Steppe known as Kisese. They traveled in the car, while Ndekei drove the truck, accompanied by several other Kikuyu workers and Sanimu, the Maasai who had shown them Laetoli; Peter Kent returned to England.

Kisese is a hilly, wooded area, with blocky granite outcroppings forming steep walls and caves. These were apparently inhabited in the Late Stone Age, and the people covered the smooth rock faces with paintings of antelope, rhino, elephants, and people. Louis and Mary pitched their camp about a mile from one of the main rock shelters, then spent the next ten days finding and recording as many of the paintings as they could. Many Maasai inhabited the region, and through Sanimu, Louis and Mary learned about several sites "we never would have found by ourselves." The Maasai here were "unsophisticated" and fascinated by Mary's red watercolors (which she had brought to make copies of the paintings) and red lipstick. "Once we were offered a fat ram in exchange for one small tube of red paint and even a young sheep for my lipstick!" she recalled.

There was plenty of springwater at this camp, enough even for hot baths, and the local Warangi people sold them honey, bananas, huge juicy tomatoes, and fresh green vegetables. "It was like a luxurious holiday," Mary wrote, "after the rigours of the past few months." Perched on a ladder, she spent long hours recording the paintings on sheets of greaseproof paper. Louis found the art to be "extraordinarily naturalistic," and was impressed with their "immense size. . . . At Kisese No. 3 shelter," he wrote in his monthly report, "there is an elephant which is nearly 9 feet long and 5 feet high." But their time at Kisese was limited, and they packed up once again, this time for Nairobi, on August 16. The paintings had "absolutely entranc[ed]" Mary and she vowed that she would "return one day to make a proper study of this beautiful rock art."

They arrived in Nairobi six days after leaving Kisese. Louis went to his parents' home in Limuru, while Mary quietly checked into the Salisbury Hotel. "Louis was nervous about upsetting his parents by telling them he was planning a divorce to marry me," she said. But they had already heard rumors that a young woman was traveling with the expedition, and Louis now did his best to ease their fears.

Earlier, when Sam White and Peter Bell returned to Nairobi, they had also stayed with Harry and Mary Leakey. "We had the embarrassing experience of having to sit through a prayer by Canon Leakey asking for help for Louis, Kent and the Kikuyu left in Oldoway without being able to add 'and Mary is also there,' " said White. The mechanic, Ndekei, had also gone to the Leakeys', and White surmises that he may have been the one who "let the cat out of the bag." At any rate, by the time Peter Kent returned to Nairobi, Louis's parents were well aware that something was amiss. They were thinking of visiting Louis in Arusha, they told Kent, and he immediately sent word to Louis. "I have told them to let you know in advance, because you might be out on safari. They asked me (twice at least) whether I had left you alone and I said yes—with a face (almost) like an angel. But it seemed rather rotten to be like that and then afterwards join in family prayers. If they eventually hear all about it, please convey my apologies for such conduct."

Perhaps other members of the Kikuyu staff expressed their own confusion and concern about the situation to the Leakeys, because a week after Kent's departure, Harry sent Louis a letter filled with dismay and worry:

> Since you told us a year ago about your feelings with regard to Frida, Mother and I have hardly ceased to think about the matter, and it has been like a millstone around our necks. Never has a day passed that we have not prayed God to show you your error and so change your heart that you must feel that, cost what it may, you simply cannot carry on with such a dreadful plan. For your sake, and because we love you so much,

> we have accepted Frida absolutely as a daughter and love her almost exactly as we do Julia and Gladys. And as for our dear little grandchildren, Priscilla and Colin, whom you have given to us, we love them more dearly than we can describe. Cannot you understand what it means to us to think of the life before all these three if you are divorced?

Whatever discussions Louis had with his parents at the end of his expedition are not recorded. But Mary was not "produced for their inspection or even mentioned," she wrote later. Louis kept her protectively out of sight. He dismantled his expedition, paid off his staff, and readied his fossils and artifacts for shipping. Then he and Mary took the train to Mombasa. Early in September, they boarded a steamer for England, and although unmarried, they traveled as husband and wife.

Chapter 7

CONSEQUENCES

Louis and Mary arrived in London at the end of September, 1935. Both were almost penniless, so Mary returned to her mother's flat in Kensington, while Louis searched for inexpensive housing. As on his previous stays in England, he planned to carry out research in Cambridge and London. But this time, he would be without a Cambridge base. The repercussions of the Kanam affair and the scandal of his pending divorce had effectively ended his academic career at St. John's. His research fellowship was not renewed and nothing—not even a part-time teaching position—was offered in its place. "This meant that I had to move all my accumulated collections of specimens, papers, and books elsewhere," Louis noted. It also left him without a salary, so when he found a cottage that was renting for only twenty-five shillings a quarter he took it on the spot. Mary and her two dogs, a spaniel named Fussy and a Dalmatian named Bungey, soon joined him.

Steen Cottage, their new home, was in the village of Great Munden, a rural community of farmers and blacksmiths, situated midway between London and Cambridge. The cottage had been built in the sixteenth century and had neither plumbing nor electricity, so Louis and Mary hauled water from the well, bathed in a tin tub set before the fireplace, and worked at night by the soft light of oil lamps. After the privations of Olduvai, Louis deemed the primitive facilities luxurious. No one in Great Munden knew that Louis and Mary were "living in sin," and, Mary believes, no one would have cared if they had known. In the larger nearby town of Ware, however, Louis was thrown out of the tennis club when his affair was discovered. Although

Mary's mother and aunts thoroughly disapproved of her actions, they visited regularly as did her uncle Percy, who offered only one comment. "Mary, you haven't tied up your dahlias properly."

Louis intended to wed Mary as soon as his divorce from Frida became final, but Frida did not file for a divorce until 1936. The delay may actually have suited Louis, for his rush to marry Frida, he now realized, had been a mistake. He had become a supporter of "trial marriage," and had even written a play about the subject, although this is not extant.

Louis and Mary lived together at Steen Cottage for the next eighteen months, a time that Mary remembers as being "blissfully happy." But they were also nearly destitute. Two months after his return to England, Louis was down to his last eleven pounds, and bills were piling up—most urgently, £42 for his life insurance, which he had set aside as an education policy for his son, Colin. "If I can't get a grant somehow," he wrote plaintively to the Royal Society, "I shall have to stop my research work for a bit and write my 'life' which a publisher wants and is prepared to pay for, but I don't want to do that yet."

Louis had not completely fallen out of favor with the scientific community, and the Society awarded him a grant of £100 "to relieve [your] situation . . . [and] to enable you to continue your work on the collections already made." This money, together with royalties from *Adam's Ancestors* and article fees, provided him with just enough to live on. Methuen also gave him a small advance for a new book, *Kenya: Contrasts and Problems,* fifty thousand words of history, social commentary, and travelogue that he had dashed off on the three-week voyage home. And Edinburgh University invited him to give its Munro Lectures in the spring of 1937 for the respectable sum of £200. These lectures became the basis for yet another book, *Stone Age Africa.*

Although he did not have an academic position—and chances of securing one seemed poor—Louis continued to pursue his study of human origins, setting himself an intensely rigorous schedule. He rented an attic in London, not far from the British Museum of Natural History, and spent three days a week there analyzing his Olduvai stone tools for a future book. The next three days found him back at Steen Cottage working on his lectures; and on Sundays, he studied a small collection of ape fossils he had found during the last Kanam/Kanjera expedition. Still, he was no longer a fellow at St. John's, and he felt its loss keenly. "Losing [his connections with] Cambridge was a great blow to Louis," Mary recalled, who divided her time between gardening and illustrating their Olduvai stone tool collection. "I think he was hurt, because the cause for it—the divorce—seemed so extraneous. What did that have to do with his ability as an archeologist? He thought that was unreasonable. As it was. But he was very resilient. I think he believed he would bounce back."

Louis probably had been treated rather harshly. At the same time that the academic community was snubbing him, they were eagerly supporting the archeological ambitions of Mortimer Wheeler, who seduced many of the young women working on his excavations. Wheeler was married, and according to his biographer, often carried out his seductions in full view of his wife: "Whether they were working at [home] or in the field, his amours could be conspicuous and Tessa [his wife] had to witness all the evidence: scenes of jealousy when one mistress was being supplanted by the next; girls turning crimson at the sight of her husband, and in one case jealous hostility toward herself." Although his peccadilloes were common knowledge, Wheeler's career soared and he was knighted for his work.

The Munro Lectures at Edinburgh University seemed to offer Louis the best chance of getting his career back on track. They were a prestigious series, and at thirty-two, Louis was one of the youngest scholars invited to present them. His ten lectures, he decided, would not merely review his own work in East Africa, but would look at prehistoric archeology on the entire African continent. No one had surveyed the subject in this manner in English before, and Louis's lectures and subsequent book, *Stone Age Africa,* were regarded as "a milestone." "That book probably did more than any other . . . to focus the interest of prehistorians on the possibilities of research in Africa," recalled J. Desmond Clark, who read Louis's book as an undergraduate at Cambridge University, and went on to an illustrious career of his own in African prehistory.

In his book and lectures, Louis noted that all too often archeologists working in Africa had simply collected the stone artifacts, unaware of what information could be gleaned from their surrounding context, such as the geology and paleontology. He thought, for example, that scientists should pay particular attention to the animal fossils at a site since they provided the "surest way of determining" its age. Louis's book thus alerted his colleagues to important data they had been missing and, Clark believes, "established the basis for subsequent archaeological research in the [African] continent."

But neither the lectures or book landed him a job. He applied for professorships in anthropology at Oxford and Cambridge. He tried to interest the colonial governments of Kenya and Tanganyika in establishing an Archaeological Survey Bureau as South Africa had in 1935. He contacted museums in the colony of Rhodesia about a position. Nothing turned up, and after returning to Steen Cottage from Edinburgh, Louis spent the next few weeks writing the one book he had tried to avoid—an autobiography covering the first thirty years of his life. The London publishing firm Hodder & Stoughton had sought this book for some time, and early in 1936 gave him a generous advance. Louis was eager for the money, but dismissed the book itself as "simply and solely . . . a 'potboiler.' "

Yet his autobiography, entitled *White African,* was neither a potboiler nor an exercise in self-promotion, but an engrossing narrative of his childhood and the adventures that led to his career. *The Times* called it "unique and enthralling."

In the preface to this book, Louis wrote that he had recently completed his fourth East African Archaeological Expedition, then added, "I hope that there will be a fifth and sixth in due course." He had expressed the same sentiments in his Munro Lectures, where he also championed the cause of his Kanam and Kanjera fossils. Despite Boswell's report to *Nature,* which effectively removed these specimens from scientific consideration, Louis remained convinced that they supported the idea of a long ancestry for *Homo sapiens.* All he needed was a grant for another expedition and he would find further proof. He also hinted that the entire subject would be aired in a public debate. "As the whole evidence will probably be rediscussed soon at a conference," he announced loftily to his audience, "I do not propose to say more about a matter that is *sub judice.*"

But such a conference was not to be. The Royal Society and Royal Anthropological Institute concluded that there was not sufficient new evidence to warrant such a meeting, and Louis had to content himself with sending a letter to *Nature.* He had written one in 1935 shortly after Boswell's appeared, but the editors rejected it as being too long and detailed. Louis finally had his say in the October 10, 1936, issue, eighteen months after Boswell's report had been published. By then, the subject was stale, and Louis's letter had the desperate sound of a man falsely judged by a jury that was no longer interested. "In spite of everything," he pleaded, "I maintain that I showed Prof. Boswell the actual stratum from which the Kanam mandible was obtained. . . ." Boswell did not write a reply.[1]

By midsummer 1936, Louis was still unemployed and worried about how to continue his explorations in East Africa. New discoveries of early human fossils in England, South Africa, and Tanganyika only increased his frustration. In England's Thames Valley, near a town called Swanscombe, A. T. Marston, a dentist and amateur paleontologist, uncovered fragments of a human skull together with handaxes and fossilized elephant teeth. Though scientists did not realize it at the time, Swanscombe Man, as the specimen was billed, was England's only genuine early human fossil. Instead, arguments erupted over whether Swanscombe Man was allied to modern humans, possibly a descendant of Piltdown Man.[2]

Farther afield, Dr. Robert Broom, a physician and paleontologist, announced the discovery of "a new ancestral link between man and ape" from South Africa. In fact, it was not a new link but a partial skull of an adult *Australopithecus africanus,* or southern ape-man—the same creature that

---

[1] Boswell did, however, keep a close—and lengthy—watch on any attempts Louis made to resurrect *Homo kanamensis.* In a book review Boswell wrote in 1950 about Robert Broom's *Finding the Missing Link,* he took Broom to task for referring to the Kanam/Kanjera fossils as evidence that "man of some sort lived in the Upper Pliocene." And in his unpublished autobiography, written in 1960, Boswell noted that "during the past 25 years, Leakey . . . has frequently referred to these discoveries, often in misleading and incorrect terms."

[2] Swanscombe Man has since been classified as a Neanderthal—and as a female.

Raymond Dart had first shown the world in 1925.[3] Broom triumphantly dashed off a report to *Nature,* asserting that his fossil showed that australopithecines were "on or near the line whereby man has risen." Debates immediately erupted over the find, and by September, Broom's "missing link" was making international headlines.

Like many anthropologists in England, Louis thought that australopithecines were more like chimpanzees than humans, and Broom's discovery did not change his mind—but Broom at least was in the field in Africa, while Louis continued to cast about for a steady means of support. More disheartening news came from Germany, where a friend of Professor Hans Reck's, Dr. Ludwig Kohl-Larsen, announced that he had recovered parts of three fossilized human skulls from Lake Eyasie, Tanganyika. The site was only fifty miles from Laetoli, where Louis and his team had explored in 1935. Even worse, Kohl-Larsen found some of the remains just three months after Louis had returned to England. Reck invited Louis to Berlin in July to help reconstruct the specimens. Louis stayed for a week, then wrote a description of the skulls for *Nature,* noting that the most complete skull resembled Peking Man. But Louis did not merely want to write descriptions of fossilized human skulls; he wanted to discover them himself. And to do that, he had to find a way to continue his field work in East Africa, "to which," Mary wrote, "he was longing to return."

Louis's plight was not unknown to other scientists, and in August 1936 Professor R. Coupland, an anthropologist at Oxford University, suggested that he contact the Rhodes Trust for a grant to write an anthropological study of the Kikuyu people. Louis himself had often contemplated making such a study, and had started collecting notes about the tribe's customs during his last expedition. He immediately sent the Trust a letter of application.

"I was born and bred among the Kikuyu tribe," he wrote, "and speak their language better than I do English, I am a recognised member of one of the age groups (Mukanda) and also an initiated first grade elder. . . . I feel that it is a duty which I owe to science and also to the Kikuyu people, to make use of my rather unusual position as regards the Kikuyu and publish a detailed account of them, and from January 1937 I could for a time put my

[3] In 1925 Dart claimed that his fossil, which was popularly called the Taung Baby, represented a human ancestor, but few scientists supported him, choosing instead to label it a chimpanzee. What Dart needed were more fossils, particularly specimens of adults, but as Broom noted, "Dart was not much of a fighter," and he quietly gave up both the debate and fossil search. Broom had been one of the few scientists to side with Dart. He had even made a special trip to Dart's laboratory two weeks after the announcement of the Taung Baby's discovery, and according to Dart, "strode over to the bench on which the skull reposed and dropped on his knees 'in adoration of our ancestor' as he put it. . . . Having satisfied himself that my claims were correct, he never wavered." Eleven years afer that visit, in May of 1936, Broom decided to fight Dart's battle for him. He would "return to the charge" and look for an "adult Taung ape." Only three months after setting out on his search, Broom recovered an adult australopithecine's skull and several teeth from the limestone caves at Sterkfontein, South Africa.

Prehistoric work on one side and devote a year or two to completing my study of the tribe. . . ."

At the same time, Sir Julian Huxley suggested that Louis apply for the Darwin Research Studentship from the Eugenics Society, which wanted a scholar to "elucidate the problems of race-crossing in Africa." Either study would take him back to Kenya, and Louis seriously pursued both. He also asked the Rhodes Trust for a grant to help with the race-crossing research, noting that "a critical survey of racial crossing is an urgent Empire problem." While some authorities considered racial mixing to be "biologically unsound," Louis planned to "approach the problem, in an entirely unbiassed way as I do not know in my own mind what I think about it as a scientist."

Louis's applications to the Trust were sent out for review, together with a note from the Trust's secretary, Lord Lothian. "Leakey is a first class man, I think," Lothian wrote. "I should certainly be inclined to help him." Yet even in fields not directly related to the study of human origins, Louis's reputation was not free from the stigma of the Kanam affair. "I think that some further enquiry into Leakey's credentials would be useful," responded H. A. L. Fisher, an Oxford professor of archeology, to Lothian's letter, "for I seem to have heard that they have recently been assailed by a writer in *Nature*." Coupland, the Oxford don who first suggested that Louis apply to the Trust, wrote a more supportive letter, but also acknowledged Louis's shortcomings as a scientist. "He [Louis] is an impulsive chap, & perhaps lacks the patience to be a first-rate scholar or scientist: & a good deal has been made recently of a blunder he is said to have made in prehistoric archaeology. But that doesn't, to my mind, make much difference. Even if he makes a slip or two (and who doesn't?) the publication & lifelong knowledge of his Kikuyu is of first-rate importance. . . ."

Lord Lothian and the Rhodes Trustees chose to listen to Coupland, and on October 5, 1936, awarded Louis a grant of £500 for a year's study of the Kikuyu people. Without any other prospects in view, and although he had no way of knowing if the Kikuyu would cooperate, Louis accepted the grant. "It was that or starve," recalled Mary. "We were in Steen Cottage and very hard up, and we couldn't see any future. So when that offer came, he took it." He also dropped his proposal for studying racial crossing, noting that he would be far too busy researching the Kikuyu to pursue this topic as well. But once in Kenya, he expected to find time for some archeological work, and he could also pursue the possibilities of establishing an Archaeological Bureau for the government. Wisely, Louis and Mary did not look too far beyond the one-year grant. "[We] had already developed a relationship of mutual trust in which our philosophy was not to fret over difficulties but to accept opportunities . . . using whatever resources were to hand," Mary wrote in her autobiography.

Other matters also fell into place. Frida had filed for divorce in January 1936, "on the ground of [Dr. Leakey's] adultery at Great Munden, Hertfordshire, . . . with a woman named Mary Nicol," and a settlement was reached in mid-October. Frida was given custody of the two children and claimed support for them, but none for herself "until Dr. Leakey's financial position improved." He had stopped by The Close once in the fall of 1935 to see his children, and a photograph taken then shows the baby, Colin, smiling at his father, and Priscilla reaching out to him. For the next twenty years, "we had no further sight, or letter from him," Frida said.

Louis and Mary were now free to marry. On Christmas Eve, 1936, they walked into the registry office in the town of Ware and exchanged their vows in a short civil ceremony. Louis wore his best and only overcoat, its hemline frayed where Mary's dog Bungey had chewed a hole; Mary says that she "cannot possibly remember" what she wore. Mary's mother and Aunt Mollie came along as "reluctant witnesses," and Louis's childhood Kikuyu friend, Peter Koinange, the son of Senior Chief Koinange, served as a somewhat baffled best man. Peter had just completed an undergraduate degree at Columbia University in New York, and with Louis's help was beginning additional studies at St. John's. He had been in England only a few weeks and had been staying with Louis and Mary in their cottage, but had never realized that they were not married. He was therefore understandably surprised to discover that there was to be a wedding.

Mary then gave her two dogs to her mother to look after, and three short weeks later, on January 21, 1937, she and Louis sailed for Kenya and "whatever might await."

Of all the tribes in Kenya, the Kikuyu were the most affected by the arrival of the Europeans. The Kikuyu farmed the cool central highlands that extended from the outskirts of Nairobi north and west to the Aberdare Mountains, and east to the slopes of Mount Kenya. It was and is rich agricultural land, and its gentle, misty climate reminded many of the British settlers of home. Much of the region seemed to be unfarmed bush—"good land lying uncultivated," as one settler phrased it—and in 1904 a government surveyor simply drew a line through it, dividing the inhabited from the supposedly uninhabited land.[4] Four thousand square miles were opened up to white settlers in this first survey. Those Kikuyu who lost their property had to find new homes, either as "squatters" on European farms or in one of the three Kikuyu Reserves. By the time Louis and Mary returned to Kenya in February 1937, a total of twelve thousand square miles of Kikuyuland had been appropriated by the British and sold to white settlers.

Further changes had come to Kikuyuland in 1919 when the government introduced a hut tax, which had to be paid in shillings. To earn their tax money, the Kikuyu men left the reserves for jobs as agricultural laborers,

---

[4] In fact, all of the land, even that lying uncultivated, was owned by individual Kikuyu families, although the early colonial officials did not realize this. The Kikuyu used the virgin land for grazing their livestock, much as ranchers in the American West used the open range, and parceled out sections of it to their children when they came of age.

houseboys, or grooms on estates; others traveled to Nairobi to work as delivery boys, janitors, cooks, or laborers. The Kikuyu, of course, were not the only native peoples required to pay this tax. But their proximity to Nairobi and the European farms brought them increasingly in contact with an alien way of life—one that, in combination with the tax and the loss of their land, undermined their independence and destroyed their traditional way of life.

"From that time," a Kikuyu elder named Kabetu recalled about the arrival of the Europeans, "the state of things began to change more and more rapidly, and ceased to be at all like it was in the olden days. The country became like a new country that was unknown to us." What Louis hoped to capture in his study was the daily life, customs, and traditions of the Kikuyu before these changes—in the days before 1900, the year when the first Europeans settled in their country.

As soon as Louis and Mary arrived in Nairobi, Louis bought a secondhand Chevrolet and drove to the village of Kiambaa to obtain the support of Senior Chief Koinange Mbiyu. Kiambaa straddled a forested ridge of the Kiambu Reserve, only eight miles from Nairobi, and within walking distance of Kabete, where Louis had grown up. Like many of the Kikuyu living so near Nairobi, Koinange had adopted European ways, and was considered by Louis to be "an exceedingly progressive man." He had traded his cloak of monkey and hyrax furs for trousers, shirt, and tie. Instead of living in a "dirty windowless hut," as he had when Louis was a boy, Koinange now lived in a "well-built stone" house, grew roses and carnations in his garden, and acted as chairman of the Kikuyu Association, the first native Kenyan political organization.[5] Several of his children had attended the Leakeys' missionary schools, then gone on to study overseas; he himself had once traveled to England to lobby the Colonial Office about Kikuyu land rights. A wise, courteous, and gentle man, he believed what Canon Leakey told him about the justice of the British legal system, and was seen by the local district commissioner as "a sober and moderating influence over the Kikuyu in times of potential trouble. . . . [He] generally backs Government in any scheme which he is satisfied is for the good of his tribe."[6]

---

[5] The Kikuyu Association was founded under the guidance of Canon Leakey, Louis's father, after he assisted one of his mission families in a land dispute with the government. The family of Stephen Kinuthia had lost most of their land in 1908; subsequently, in 1919, the government marked off their remaining sixty-seven acres. Kinuthia then "asked Canon Leakey to write a letter from us [the Kinuthia clan] to the Government saying that 240 acres of our land had been taken in 1908 and now the Government was taking some more and we had nowhere to go. The letter was written, . . . and it was sent to the Government. They listened and the land was not taken and we all thought the power of this letter a most wonderful thing." Based on this success, Canon Leakey suggested that the chiefs band together to protest other land appropriations. In 1920 the Kikuyu Association was born with Chief Koinange serving as Chairman.

[6] Later, Chief Koinange's attitude changed and he became bitterly hostile toward the government and white settlers. In 1937, after spending twenty years working through the British legal system to regain his land, he was awarded only one-tenth of his previous holdings. The white owner had planted the land in coffee bushes, an export crop that the colonial government forbade the Kikuyu to grow. Koinange protested the law, lost, and was forced to uproot the coffee plants.

The Leakeys were great friends of Chief Koinange, and he warmly welcomed Louis and Mary to his home, where Louis explained the purpose of his study. Despite his close association with the tribe and his status as a first-grade elder, Louis was a white man, suspect and untrustworthy to most other Kikuyu. After a long consultation, in which Louis pointed out that after the Roman invasion of Britain, the ancient customs of the Britons had been lost because no one had the foresight to write them down, Koinange agreed to help. He provided Louis and Mary with a guest hut behind his cottage, then called a meeting of nearly one hundred elders from the Kiambu Reserve. They gathered in the shade of a sacred fig tree, and Louis presented his case, asking for their help in writing a book "so that the young people of future generations would really know and understand how the Kikuyu had lived." He also gave them his "solemn promise" that he would "act only in the tribe's interest." There was more discussion, a delay of a week, and finally the elders granted his request. They also appointed a panel of nine of their senior members to act as his advisers.

With his usual enthusiasm, Louis decided that his study would comprise "the most complete record of a tribe that has ever been written." He planned an ambitious three-volume work, covering every topic from tribal legends to the methods for building a home, Kikuyu botanical lore, witchcraft, family life, warfare, and religion. To obtain this material, he met daily with two of the Kikuyu elders in Koinange's guest hut, then interviewed other elders to corroborate his data. Louis and the visiting Kikuyu would sit together at a rough-hewn table; he would ask questions, listen intently to their replies, then write down everything in longhand. "Louis put a tremendous amount of energy into his study," Mary recalled. "It was something that he had always wanted to do, and once he got engrossed in it, he was very content to do that." By the end of May he had already written five hundred pages, but he estimated that this would make up only half of the first volume. "[I]t's going to be a *gigantic* work," he wrote in a progress report to the Rhodes Trust. "Vol. 1 will be about 1000 pages so far as I can see!"

While Louis pursued his Kikuyu study, Mary began a trial dig in nearby Waterfall Cave, where there was evidence of a Middle Stone Age culture. "[W]hat I wanted most, was a site to excavate," she wrote in her autobiography, "it did not much matter what period, because I was interested in everything prehistoric." She hired several of Louis's Kikuyu staff, including Heselon and Thairu, who had worked with them at Olduvai, to assist on this excavation.

Then quite suddenly and shortly after starting her project, Mary was stricken with a high fever, which the local doctor diagnosed as malaria. He gave her a quinine injection, but instead of subsiding, her temperature soared, and that night Louis rushed her to Nairobi Hospital. There, the diagnosis was double pneumonia, almost always fatal in the days before penicillin. So certain was the doctor that Mary would die, he advised Louis to cable her mother telling her to catch the next plane to Kenya.

It was still a four-day flight from England to Kenya, and the journey was surely an anxious one for Cecilia, given the grim prognosis. Without antibiotics, the only cure available for pneumonia was intensive nursing and constant use of an oxygen mask. The hospital lacked the staff for such attentive care, so Louis offered his assistance, and stayed long hours by Mary's bedside. He was determined that "I should not give up: somehow he managed to convey this to me," Mary recalled. She was dangerously ill for a long time, but finally, "by the grace of God," as Louis thankfully phrased it, Mary did recover.

By the end of August, 1937, Mary had made a full recovery and was eager to tackle a more ambitious dig than Waterfall Cave offered. Louis had also completed most of the interviews for his three-volume study, and they decided to move from Kiambaa to an area near Lake Nakuru and the Rift Valley sites Louis had excavated on his first and second expeditions. In the hills above the lake, Louis knew of a Neolithic site, called Hyrax Hill. Mary remembers being "delighted with this plan . . . [as] I could not match Louis's enthusiasm for the Kikuyu study. The archaeology of the Nakuru region would certainly offer me the kind of opportunities for which I was longing."

Hyrax Hill, a low ridge of black lava rising one hundred feet above the Nakuru basin, was named for the colonies of buff-colored hyraxes that live in its rocky fissures. Louis and Mary set up their camp in a grassy area at the base of the hill. With the aid of their Kikuyu staff, they built a comfortable grass *banda,* or thatched hut. In one room, they placed tables for Mary to sort and catalogue her archeological finds; in another, Louis had his small writing table; a third room served as their private living area. They and their staff slept in tents behind this hut. It was inexpensive housing, but with a view that would be hard to match. Below them stretched the golden plains of the Rift Valley and the bright waters of Lake Nakuru, often fringed with over a million rose-tinted flamingoes; behind them loomed the dark, blocky wall of Menengai Crater.

Two of the Kikuyu elders had accompanied Louis to Hyrax Hill to help with his remaining questions, and while he sat inside, writing his voluminous study, Mary began to excavate a low mound of massive stones along the southeastern side of the hill. Below the stones, she uncovered nine pits containing nineteen skeletons, "lying in heaps with the skulls detached from the bodies." From pottery shards, a few iron objects, and glass beads, Mary dated these people to some three hundred years ago. Because the skeletons were primarily those of young men, she speculated that they may have been warriors killed in battle. On the opposite side of the hill she excavated a small village of pit dwellings, and another cemetery—all considerably older than the Iron Age site. Several of these burials held carved platters, bowls, and pestles of stone, while dark grey and bottle green obsidian tools and spouted clay vessels were found in the village. Taken together, the material showed that Hyrax Hill had been occupied intermittently from Neolithic times (two thousand years ago) to between the seventeenth and nineteenth centuries.

Mary again hired Heselon and Thairu to help her with this excavation, which was very much her own project—Louis never disagreed with her interpretations or how she ran the dig. From the beginning of their relationship, Mary had asserted her independence, although in at least one instance, her willful determination to do as she pleased nearly brought disaster. One day on their 1935 expedition, while out exploring the badlands at Laetoli, Louis spotted fresh lion tracks. "I felt the lion might be quite close," he wrote in his autobiography, "[and] called out to Mary not to wander too far." Mary ignored his warning. "[B]eing still new to Africa, I felt entitled to walk where I wished," she recalled. She continued walking, disappeared behind a small hummock, and nearly stepped on a sleeping lioness. "The next minute I heard two sounds—a sharp cry from Mary and a low feline growl!" Louis wrote. "In a flash I saw two figures running fast in opposite directions —Mary towards me and the lioness up a grassy slope. The lioness was every bit as frightened . . . as Mary was." Although Louis could not refrain from an "I told you so," Mary did pay more attention afterwards to his advice about the dangers of the African bush.

While Mary excavated Hyrax Hill, Louis began to research the botanical lore of the Kikuyu people. He decided to "locate each and every one" of the plants they used for medicinal or economic purposes, and make a collection of pressed flowering specimens. He was also helping his elders build a traditional Kikuyu hut, one that did not use nails, wire, or other European building materials. Even before deciding to take on such time-consuming projects, Louis realized that he would never complete his Kikuyu study in the one year covered by his Rhodes Trust grant. He checked on other sources of funding, discovered he was too late to apply, then sent an appeal to the Trust. "I don't feel it is fair to ask the Rhodes Trust again but I must . . . as I have *NO* private income. . . . Also I *must* write this stuff here among the people if it is to be first class (as it must be)." The trustees sympathetically extended him another £500 one-year grant. "They felt that having launched you on the task they could not desert you," responded Lord Lothian to Louis's plea.

The additional sum meant that Louis and Mary could now stay in East Africa at least until the end of 1938. Louis had also received a £60 grant from the Royal Society to make a quick reconnaissance of the Lake Eyasie region in Tanganyika, where Dr. Kohl-Larsen had discovered his fossilized human skulls. Partly to celebrate the extension of his grant, and partly to take a break from their projects (writing the first draft was "exciting, but . . . also exceedingly tiring," said Louis), they set out for Eyasie in November 1937.

Louis followed what was now becoming his customary route to Tanganyika: from Nairobi he drove south to the Tanganyikan town of Arusha, then west to the little village of Oldeani, at the base of Ngorongoro Crater. As Lake Eyasie lies fifty miles due south of Olduvai Gorge, Louis continued west and south this time, instead of driving over the crater. Today a good dirt road leads to the lake, where both European and native Iraqw farmers raise melons and onions; but in 1937, the two-wheel track grew fainter and fainter until it nearly vanished in the tall grass. Louis and Mary finally reached a native settlement beside a cool spring, where everyone assured them it was impossible actually to drive to the lakeshore. Kohl-Larsen had apparently hired porters here and walked the last thirty miles to the lake. Louis was naturally "determined to try and get our vehicle through," and after spending three days exploring possible routes, succeeded. But although Louis and Mary found fossilized bones and stone tools, they could also see that the material was younger in age than Olduvai Gorge, which, they decided, remained the prime spot to search for early human fossils.

After this brief holiday, Louis and Mary resumed their projects at Nakuru. By now, Mary's excavation at Hyrax Hill had come to be regarded as of great importance by the local white settlers. "So many people . . . thought there wasn't any history to the country," explained Mary Catherine Fagg (then Davidson), a local teacher who worked as a volunteer on Mary's dig. "The excavation was a surprise to them." Africa was still deemed scientifically unimportant by many historians and prehistorians; few thought that the huge continent had any history at all, prior to the arrival of Europeans. An excavation such as Mary's, which clearly showed otherwise, was thus novel and exciting—although Mary did not realize just how exciting it was to the local populace until she opened her site for three days to visitors: four hundred showed up for a tour on the first day. Her visitors also generously contributed to her work, donating more than £100, and then the *Kenya Weekly News* inaugurated a fund to support her dig through 1938. One settler, Mrs. Arthur Selfe, whose farm was adjacent to Hyrax Hill, was so impressed with Mary's work that she bought the hill to preserve it, and later donated it to the Kenyan government. Five years later the government proclaimed the site a national monument, and in 1964 a small museum containing artifacts from the dig was opened to the public.

Among the many visitors to Mary's excavation was Mrs. Nellie Grant, the mother of the writer Elspeth Huxley. Elspeth later immortalized her parents by turning them into the characters of Tilly and Robin, the aristocratic pioneer settlers in her novels *The Flame Trees of Thika* and *The Mottled Lizard*. When Nellie came to visit Mary's dig, she and her husband, Colonel Joscelyn Grant, were running a thousand-acre farm at Njoro, high on the Mau Escarpment and south of Hyrax Hill and Lake Nakuru. She was an extremely observant woman, and after visiting Mary's excavation, she discovered an archeological site on her own land. She collected a flat stone bead, some obsidian flakes, and pieces of pottery and brought these to Louis and Mary, then asked if they might like to explore the "exciting cave" she had found in the forest behind her farm. Both Louis and Mary were intrigued, and they drove across the plains of the Nakuru basin to Nellie's farm in December 1937.

The cave lay above the Njoro River in a narrow, steep-sided valley, which was thickly forested with olive and cypress trees. The river had carved the cave's greyish-black walls into one main room, and two narrower side caves. On this first visit, Louis and Mary dug only a small trial trench in the main shelter, but quickly hit paydirt. "[I]mmediately we found the remains of a human cremated burial with stone artifacts and more beads," Louis wrote. Like Nellie, they were excited by the cave's potential and agreed that it should be properly excavated. Nellie had a grass *banda* built for them, and on April 18, 1938, Louis and Mary began the excavation of what they now called Njoro River Cave. Louis welcomed the work as a holiday from his Kikuyu study. "I had got so tired writing about 8000 words a day on [the] Kikuyu," he said in a letter to a colleague.

Njoro River Cave proved to be even more rewarding than Hyrax Hill. The cave had been used as a crematorium by Late Stone Age people, and Louis and Mary removed the skeletons of about eighty individuals.[7] The cremation process had hardened some of the bones, but reduced others to a flaky, powdery state. Louis preserved many of these with shellac; the remainder simply disintegrated. Buried with the bodies were animal skins, baskets, handwoven fabrics, gourds, necklaces of agate, quartz, and chalcedony (a milky blue quartz with an opal-like luster), and one elaborately carved wooden vessel.

Like Louis, Mary had an enormous capacity for work. After excavating all day, she spent the evenings sorting and cataloging every obsidian flake and bead they had discovered. "There were hundreds and hundreds of stone tools, and we would go on sorting tools by lantern until we were too tired to do any more," recalled Mary Catherine Fagg. "I had a little of that, but Mary had the brunt of it. She did it because it was her life."

The excavation fell naturally into two categories: the archeological and the skeletal, or, more simply, the stones and the bones. It was a natural division of labor for Louis and Mary as well, and in all their future work, as on this dig, Louis devoted his attention to the bones, while Mary worked on the stones. They followed this same division when they wrote a monograph describing the cave's excavation, with Louis contributing chapters about the skeletons and Mary covering the archeology. But they both considered the excavation to be primarily Mary's, and when *Excavations at the Njoro River Cave* was published in 1950, Mary's name was listed first.[8]

By mid-May 1938, Louis was eager to return to his Kikuyu study, so they closed the Njoro dig and moved to a small stone bungalow in Nairobi. He had also been captivated by the stone beads they had found in the cave and sent some to an archeologist in England, who reported back that they

[7] Later the cave was dated by radiocarbon to 960 B.C.
[8] The book's publication was delayed due to the outbreak of World War II.

resembled "Predynastic Egyptian stuff." Louis himself mistakenly believed that the beads were carved from opal and jade, instead of the more common but similarly colored chalcedony and quartz, and concocted a fanciful story for the press about prehistoric opal mines and Mesopotamian traders. Louis suggested that the people who had frequented the cave—whom he also decided were "non-Negroid"—had mined the opal, then traded their opal beads for jade ones from the Mediterranean regions. It was a wildly speculative tale, and found its way into several British newspapers, including *The Times.* There it ran under the headline "Prehistoric Man in East Africa/Opal-Mining in 4,000 B.C.," and described the "mausoleum and extensive opal-mining" discovered by Dr. and Mrs. L. S. B. Leakey. This was the first story *The Times* had carried about Louis's work since the stories proclaiming the Kanam disaster. But as with Kanam, Louis's geology was faulty. After specialists examined the beads, they were determined to be made of only semiprecious stones. Louis made no further references to opal mines, and the story vanished from the papers. Yet the incident showed once again that Louis was easily misled by his hyperactive imagination and was too quick to hypothesize in public. He had not yet learned scientific caution.

By the end of June, he sent off all 250,000 words of volume one of his Kikuyu study to Oxford University Press, by late August dispatched volume two, and by February 1939 completed volume three.[9] Throughout this period, he continued searching for an academic position. He sent letters of inquiry to England, Rhodesia, and South Africa, exploring the possibilities of both teaching and museum curatorial positions, and he talked up the idea of an East African Archaeology Department with the Kenyan government. Partly because of the Leakeys' excavations at Hyrax Hill and Njoro River Cave, there was a stirring of support for such a bureau.

"Plans for a Government Dept. of Archaeology or Antiquities are coming before Legislative Council soon," Louis wrote to Professor Clarence van Riet Lowe, who headed a similar department in South Africa. "H.E. The Governor is keen, but he says he won't push it if the elected members representing the settlers don't support it strongly." Louis then asked van Riet Lowe if he thought General Jan Christiaan Smuts, who held a powerful cabinet post in the South African government and was deeply interested in African prehistory, would send a letter of support. Louis and van Riet Lowe had struck up a friendship after Louis's first visit to South Africa in 1929, and van Riet Lowe now did all he could to help. He appealed to the general, and Smuts wrote a long letter to Lord Francis Scott, head of Kenya's Legislative Council, praising the "remarkable discoveries already made in East Africa by Leakey" and supporting the idea of a Kenyan Department of Archaeology. "I hope that the missing link [in the archeological departments] in Central Africa could be supplied so that we may have an unbroken chain from North to South,"

---

[9] These were finally published in 1977 by the Academic Press with assistance from the L. S. B. Leakey Foundation.

he wrote. Van Riet Lowe kindly sent Louis a copy of this letter with the note "I hope this does the trick."

Unfortunately, it did not. In December the Legislative Council voted against channeling any money to a Bureau of Archaeology, on the grounds that funds were needed for Kenya's defense. This was at least partly true. World affairs were unsettled and war seemed imminent: Italy had invaded Ethiopia, then formed a disturbing alliance with Germany, which had annexed Austria, occupied the Rhineland, and dismembered Czechoslovakia. But Louis was also unpopular with many in the white settler community. In his book *Kenya: Contrasts and Problems,* he had bluntly reviewed the problems between the white and black races in Kenyan society, and listed all of "the reasons the Africans hate and distrust the settler community." He had even dismissed the idea that Kenya might be a white man's country. "A vital question in connexion with colonization in Kenya is whether the country can ever really be a 'white man's country' in the fullest sense of those words. Personally I have grave doubts." Although the issues he discussed in his book were valid, and would lead to the native peoples' struggle for independence in the 1950s, many of the settlers viewed Louis as a traitor to his race.

Louis's divorce, too, may have been a factor in the Legislative Council's negative decision. He apparently suggested this in a letter to van Riet Lowe, who agreed and responded that he thought Louis would be wise to leave Kenya for the time being.[10]

The Council's decision left Louis and Mary in an extremely precarious position. When they were camped at Njoro, Nellie Grant noted that they seemed to be living only "on the smell of an oil-rag," but they at least had money from the Rhodes Trust. Now that support had ended as well, and Louis, who did not want to leave Kenya, wrote in some desperation to van Riet Lowe, "[W]e are stranded for 1939 with no funds at all. Maybe I can earn a living by devious ways—teaching the Kikuyu language, lecturing on Native Law and customs and selling beads!! Also I think buying beeswax in the [Kikuyu] Reserve and exporting it."

Ironically, salvation came from the Kenyan government, which hired Louis as a civilian intelligence officer. "Certain subversive elements" were spreading anti-British rumors among the Kikuyu, and Louis, with his command of the Kikuyu tongue and his friendships with the people, seemed the perfect spy. Although Louis sympathized with the Kikuyu over the losses and wrongs they had suffered, he was against an outright revolt. He accepted the job, and traveled among the natives as a wholesale trader, selling soap, canvas shoes, tea, sugar, and coffee to the small Kikuyu shops, while keeping his

---

[10] Despite the rather loose morals of many of the aristocratic settlers (Karen and Bror Blixen, Denys Finch-Hatton, Beryl Markham, and their set)—whose actions gave rise to the phrase "Are you married or do you live in Kenya?"—divorced couples were viewed with opprobrium by most of the European community in Kenya. In 1925, following a Royal Command from Queen Mary, the governor's wife, Lady Grigg, issued an order that divorced couples were not to cross the threshold of Government House.

ear tuned for rumors. The money he earned paid "our rent for awhile," Mary recalled, but it was scarcely the type of position Louis had envisioned for himself when he was at Cambridge.

Ostracized by the academic community, it seemed that he had finally hit rock bottom, though he left no records indicating that he ever despaired over what had happened to his career. And by September 1939, he probably did not care. Britain had declared war on Germany, and Louis was drafted into the African Intelligence Department, Special Branch, Section 6, and was running guns to Ethiopia.

Chapter 8

CLOAK-AND-DAGGER

Ethiopia, Kenya's northern neighbor, had been invaded in 1936 by the Italians and annexed in 1937. The Italians also occupied Eritrea and a portion of Somalia on the Red Sea, where they had established strong military bases. The British feared that the Italians would use these countries as a staging ground to attack the Suez Canal and break Britain's hold on the Middle East with its vital oil supplies. By the summer of 1940, it seemed that the Italians might succeed. Britain suffered two major setbacks in the region, quickly losing British Somaliland and the northern Kenya garrison of Moyale to Italy. But the Italians did not press their advantage, and the autumn of 1940 found Kenya bristling with soldiers and arms waiting to move north.

The main road to Ethiopia from Nairobi passed through the fertile highlands bordering Mount Kenya, then descended across the Northern Frontier District—a forbidding expanse of thorn trees, volcanic desert, and eroded badlands. Historically, the NFD was home to pastoralists, nomadic tribes who wandered with their herds of camels and goats. Now the region also harbored small bands of Ethiopian guerrillas. It was to these soldiers that Louis and others in the Special Branch funneled weapons and supplies.

It is not clear from his records how often Louis made this trip, or what routes he followed, but he had sufficient time to explore for archeological sites and collect some stone tools. The region intrigued him, and in several letters later in life he would refer somewhat wistfully to the potentially fossil-rich badlands of the Northern Frontier. Louis also helped organize an intelligence network in the deserts, composed partly of his Kikuyu age-mates, whom he had trained in paleontology. They, too, kept their eyes open for fossils and artifacts, and after unloading the guns they ferried north, filling up their trucks with prehistoric treasure. Once Louis boasted to his friend, Archdeacon Owen, an Anglican cleric in western Kenya, "My boys from the OMO [a river that runs south from Ethiopia into Kenya's Lake Turkana] are back with 14 cases of fossils!" Similarly, the paleontologist Donald MacInnes, who had been sent as a mosquito controller to army camps in British Somaliland, managed to ship Louis several crates of handaxes from that country's deserts, with the tantalizing note "I spent as much time as possible on the exposures where there is some pretty startling material . . . and I had to leave behind vast quantities of things which I would have liked to bring."

With his knowledge of native languages, Louis was often called on to interrogate suspicious individuals, sometimes receiving a summons or cryptic note in the middle of the night. Other times he would vanish for days or weeks, then surface again at home in Nairobi, explaining his absence with a theatrical air of mystery and danger. "Louis loved that 'cloak-and-dagger' stuff," recalled Mary. "It was fun for him, but he would never say exactly where he had been."

Sometimes Louis assumed the role of an eccentric bird collector. While hunting and then stuffing the birds he shot, he casually quizzed curious villagers who had come to watch. What rumors had they heard about the war? Why were the young men in their village refusing to join the King's African Rifles? Apparently his ruse worked well enough on one occasion to uncover a "prophetess" among the Kamba people who was warning the local youth not to enlist; they would never return from this journey overseas, she warned.[1] Other times, he simply called meetings with the village headmen and made propaganda speeches about the need for volunteers. Among his other responsibilities, Louis also wrote weekly radio broadcasts in Swahili and Kikuyu, and served as the colonial government's primary handwriting expert.

Louis had become interested in handwriting as an undergraduate at St. John's. While studying medieval French texts, he had been told by his tutor that certain passages had been added by someone other than the original author. "I disliked having to rely on my professor's opinion in such matters," Louis recalled, "and was determined to find how I could decide the truth myself."

Handwriting analysis was perhaps the most arcane of Louis's many talents, but it served him well as an intelligence officer. The colonial government was particularly worried about the fifth-column leanings of some native political organizations, and in May 1940 the Criminal Investigation Department raided the offices of the three most outspoken groups—the Kikuyu

---

[1] In the First World War, native Kenyans had been barred from carrying weapons and had served only in East Africa as porters in the Carrier Corps. This time around the British wanted them as soldiers, and by 1942 several thousand black Kenyans were fighting in North Africa, the Middle East, Ethiopia, and Burma.

Central Association, the Taita Hills Association, and the Ukamba Members Association—and confiscated thousands of documents. To make certain that political agitation would not disrupt the colony during the war, the government then banned these associations and detained their leaders. Many of the seized documents were handwritten, and Louis was given the unenviable task of sorting through the KCA's papers and identifying their authors. Eventually, he was called on to give evidence before a government tribunal, where his analyses were used to help prove that the KCA's leaders had intended to aid the enemy. Louis was uneasy in this role, and the whole affair left him saddened. Although he did not approve of the KCA's more radical approach to native problems—he supported the middle-of-the road Kikuyu Association—he considered the government's banning decision unwise.

While Louis worked at the Special Branch, Mary continued to pursue archeology. The government had not called her for service as it had some other women, and this left her free to put in a dig at Naivasha, about fifty miles northwest of Nairobi. In late 1939 Louis had learned that the realignment of the railway at Naivasha would destroy a rock shelter he had discovered ten years earlier. The railway authorities gave Mary permission to excavate the site and even provided her £400 for the task, although they allowed her only two months for its completion. To make the most of her time, Mary moved to a friend's unoccupied cottage at Lake Naivasha. Louis came to help as often as he could, primarily on the weekends, and sometimes brought two young women who had volunteered their help. For companionship during the week, Mary had the four Dalmatians that she and Louis had acquired after thieves had ransacked their camp at Hyrax Hill in 1938. From that time to the present, Mary has always kept several Dalmatians as pets.

Naivasha Railway Shelter, as Mary called the site, proved to be extremely prolific—and extremely educational about the potential pitfalls of doing archeological research in Africa. From the single trench she had time to dig, Mary recovered, sorted, and labeled more than 75,000 finished stone tools and two million waste chips, all dating from the Late Stone Age. But before she completed her study of this material, termites destroyed the tools' cardboard containers, and all the implements collapsed into a single confused pile. Mary never found the time or desire to sort them out again.

In the spring of 1940, Mary closed her dig and returned to Nairobi and the Coryndon Memorial Museum.[2] She was pregnant now with their first child and decided to postpone further excavations, though she was not, she wrote, "one of those for whom pregnancy brings sickness and frailty." Instead she spent most of her days at the museum, where she and Louis had

[2] The museum was built in 1929 and named for Kenya Colony's late governor Sir Robert Coryndon, a keen naturalist. At independence, in 1964, the museum's name was changed to the National Museum of Kenya.

been given laboratory space for sorting and storing their finds. Many of their friends also worked here either as volunteers or in poorly paid staff positions. In January 1941 Louis was appointed honorary curator of the museum, and it quickly became the focal point of their lives, both socially and intellectually.

Although Louis's new position was unsalaried, it did include the use of a furnished wood bungalow on the museum grounds. For Louis and Mary, the cottage was a godsend. Only two months before, on November 4, 1940, Mary had given birth to their first son, Jonathan Harry Erskine. Louis was away on an intelligence mission at the time, but as soon as he heard the news, he rushed back to Nairobi, where he found Mary and the baby "in fine shape." At the time, the Leakeys were living in a tiny rented house, full of dogs and artifacts, on the outskirts of Nairobi. They left these cramped quarters shortly after Louis was appointed curator, and with Jonathan and their four Dalmatians, moved into the rambling curator's cottage under the pepper trees on Museum Hill.

"I quite liked having a baby," Mary wrote about her first son, ". . . but I had no intention of allowing motherhood to disrupt my work as an archaeologist." Although she was nursing Jonathan, she continued to spend several hours each day at the museum, analyzing and writing up her finds from Hyrax Hill. When Peter Bally, the museum's botanist, announced that he was planning a collecting expedition to Ngorongoro Crater and suggested that Mary come along to excavate some burial mounds, she jumped at the chance.

The burial cairns in Ngorongoro had been discovered about thirty years earlier, when a German farmer, Adolph Siedentopf, was collecting stones to build his home. Subsequently, in 1913, the geologist Hans Reck had excavated several of the cairns, uncovering Stone Age skeletons, bowls, and tools. No other scientist had investigated the site since that date. From a brief report Reck wrote about the burials, it seemed to Louis and Mary that the mounds had been left by people similar to those who once lived at Njoro River Cave. If so, the crater would provide them with another and more distant site to relate to their East African Neolithic cultures. "[I]t seemed too good a chance for her to miss," said Louis, who encouraged Mary to go.

Bally's two-week expedition was set for April 1941, and a few weeks beforehand Mary began weaning Jonathan. Louis, who was busy with intelligence activities, had agreed to care for their baby while Mary was gone. "[S]he had spent many months carrying her unborn baby and it was now her turn for a change and a rest," he said. Perhaps because he had missed much of the early months of his other children, Louis happily looked after his son, mixing the baby formula, changing Jonathan's diapers, and "patting his little back" when putting him to bed. He also had the help of a daytime nurse.

In addition to Peter Bally and Mary, the expedition included Bally's wife, Joy (who would later marry George Adamson and gain fame as the author of *Born Free*), and Jack Trevor, an anthropologist from Cambridge University and a friend of Louis's. Trevor was actually supplying the funds for the excavation, having persuaded Cambridge to donate £50. Trevor's little team traveled together in one car to the rim of Ngorongoro. There were no roads leading to the crater's floor, so they hired sixteen porters from the Wambulu tribe to carry their stores, then hiked down the steep escarpment into the caldera and crossed the crater to the ruined Siedentopf farm where they set up camp. They made the hike in one long, exhausting day, which was momentarily fraught with tension when the Ballys' terrier decided to challenge a rhinoceros. The little dog advanced with shrill barks; the irritated rhino snorted and pawed the ground, and then, to everyone's relief, moved off.

During their stay, Peter and Joy collected specimens in the crater's forest, while Mary and Jack Trevor excavated two of the burial mounds. The first cairn they examined was only of recent origin, but in the second they uncovered several skeletons, stone bowls and pestles, obsidian tools, ornamental beads, and ivory lip plugs (used by natives to decorate their lips). As Louis and Mary had surmised, much of this material resembled their finds from Njoro River Cave. The mounds were eventually dated by Carbon 14 to 700 B.C.[3]

Throughout the war, Louis and Mary made as many of these short archeological expeditions as they could, using their own funds from Louis's annual £600 salary from the Criminal Investigation Department and their carefully saved gasoline ration coupons. Whenever Louis had a few days to spare, they would pack up Jonathan, the Dalmatians, and any friends who longed, as they did, for "freedom from all thoughts of war" and head off into the bush. Once everyone piled into two old Chevrolets and bounced down the two-wheel track to Olduvai Gorge for five days of collecting. Other times, they drove to Lake Victoria, took a boat to Rusinga Island, and explored its fossil deposits. But they made their most stupendous wartime archeological find at a site called Olorgesailie, only forty miles south of Nairobi in the Rift Valley.

Since 1929 Louis had been searching Olorgesailie for a handaxe site that the geologist J. W. Gregory had discovered in 1918. Unfortunately, Gregory's directions were imprecise, and the handaxes he reported were never found again—until Easter morning of 1942. On that day, Louis drove to the bushy flats that marked the edge of an ancient Pleistocene lake—now completely dry—and parked under an acacia tree. The Leakeys' party this time included two of Louis's Kikuyu staff; an Italian prisoner of war named Ferucio Men-

[3] Mary did not publish her report about the artifacts from the burial mounds until 1966. She delayed publication, hoping that Jack Trevor would complete his description of the human remains; unfortunately, he never did.

engetti, who had been paroled as a museum technician under Louis's supervision; and Mary Catherine Fagg, a regular member of Leakey expeditions. Carrying their food and water, they fanned out across the sediments, heads down, eyes on the ground. Suddenly and almost simultaneously, Louis and Mary stumbled independently onto enormous quantities of handaxes the size of dinner plates. In the next minute, Davidson and Menengetti discovered similar sites as well, and the air filled with everyone's shouts to come and see what they had found.

Louis tied a handkerchief to a bush to mark his find, then rushed to look first at Mary's. "When I saw her site I could scarcely believe my eyes," he wrote in *By the Evidence,* the second volume of his autobiography. "In an area of about fifty by sixty feet there were literally hundreds upon hundreds of perfect, very large handaxes and cleavers, as well as a few flakes and some bolas stones (round stone balls)." The tools lay as close together as paving stones and looked "as if they had only just been abandoned by their makers," said Mary. Louis's site and those of Menengetti's and Davidson's held the same mix of tools, but not in such quantity. Mary's area was so impressive that everyone in the party agreed that the implements should be left undisturbed and permanently in place. This was done, and five years later, in 1947, Olorgesailie was opened to the public as an open-air museum. Today visitors can see the artifacts just as the Leakeys discovered them.

Louis and Mary longed to stay and begin a complete survey and excavation, but they had planned only a day's outing. Jonathan had been left this time with a baby-sitter, who expected them back by five o'clock. Reluctantly, they returned to Nairobi, discussing all the way home "just how and when [they] were going to be able to work [Olorgesailie] satisfactorily." They did manage a brief trip in December 1942, but Mary was heavy with her second child, and the bulk of the work had to wait.

In the meantime, tragedy struck. Mary's new baby, a daughter they named Deborah, died in April from dysentery when only three months old.

In her autobiography, Mary does not reveal how traumatic this loss was, and buries the incident in a paragraph mainly about the deaths of three of her Dalmatians. But she suffered greatly at the time, as did Louis, who, Mary recalls, "was openly upset." Donald MacInnes's wife, Dora, lost her infant son that same year and vividly remembers Louis and Mary visiting her in the hospital. "The one thing that comes clearly to my mind [from that visit]," said Dora, "was Mary saying with great feeling to me, 'It's absolute hell, isn't it?,' & I knew she had been deeply hurt by her loss."

"How does one know how long it takes to get over [a death like that]?" Mary said nearly fifty years later. "I think at the time we both just carried on." The wound was deep, and when Archdeacon Owen wrote to ask Louis about some archeological material he had sent, Louis uncharacteristically waited several weeks to reply, then wrote, "I am afraid I have been a very long time in answering, but . . . I have had my children very ill and one has died, and I have not been touching any of my work."

. . .

The spring rains that had started in March 1943 drew to a close by May. They left Nairobi green and washed, with masses of purple and pink bougainvillea cascading from rooftops and piled atop garden fences. Big puffy clouds drifted overhead, and the Nairobi River, just below the Leakeys' bungalow, ran red and swollen from the rain. For Louis and Mary, the mornings were the best time there. Their home was sheltered by jacaranda, pepper, and eucalyptus trees, and each dawn a chorus of birdcalls greeted the day, making it, Louis wrote, "a real joy to wake up."

For a time following their baby's death, Louis stayed off his schedule, but he gradually resumed his long hours, heading off to the intelligence department by eight o'clock each morning, returning by four-thirty, then striding across the lawn to the museum, where he worked until midnight. He attended to the museum's affairs, answered letters from people with natural history questions, arranged exhibits, and still found the time to write a number of scientific papers. At Naivasha Railway Shelter he had helped Mary uncover a skeleton, and he wrote a report describing this find. In quick succession he then turned out two lengthy papers on the extinct fossil pigs from Olduvai and the Omo region of Ethiopia, a short report about rare Verreaux's eagles he observed in Kenya's Kamba country, a description of stone tools from Ethiopia, and a paper on extinct baboons. At the same time, he agreed to assist Archdeacon Owen with the reconstruction of a skull Owen had discovered—and broken—in a rock shelter near Kavirondo in western Kenya, though in the end, Mary completed this task. Mary, too, put in long hours at the museum, assisting with exhibits, and writing and illustrating her study of Njoro River Cave.

Louis had a three-week leave due in August 1943. Putting aside their sorrow, he and Mary decided to spend this time beginning an excavation at Olorgesailie. As the expedition drew nearer, Louis's old enthusiasm returned. In letters to Archdeacon Owen, he declared Olorgesailie a "perfectly magnificent site," a "second Oldoway," and urged Owen to visit if he possibly could. On their previous brief surveys, Louis and Mary had located several places where stone tools protruded from the hard clay beds in such quantities that they thought they might discover the "actual camp sites and living floors" of early man. Louis hoped they might uncover his skeletons as well.

With a party of their friends, Louis and Mary set up a tented camp under a spreading thorn tree near a dry river. They built a rather large encampment because Mary intended to stay on for another six weeks after Louis returned to Nairobi. Jonathan and his *ayah,* or native nurse, would remain with her as well, as would a houseboy, several Kikuyu assistants, and the Dalmatians. Since there was no water at Olorgesailie, Louis arranged with the foreman of the Magadi Soda Company, a firm fifty miles farther south, to drop off several forty-gallon drums of drinking water each week.

Sometimes after a sudden downpour, water briefly filled the dry river, though not long enough to retain sufficient supplies for the camp. But the rains did turn the desert a soft green, and in their wake came antelope, rhinos, and tall Maasai warriors, herding their cattle. For a while a pride of five lions lingered nearby, but aside from raising the curiosity of the Dalmatians they caused no harm. The Maasai thought the Leakeys' digging a ludicrous activity, and only grudgingly accepted the idea that the fossils might be very old when Mary showed them the remains of a hippopotamus. It was very unlikely, they agreed, that any hippo had recently lived in the desert.

While Louis was still in camp, the team of excavators began to dig three trenches in the white sediments. Almost immediately, they uncovered what Louis and Mary interpreted as a living floor, a site where the Acheuleans, or handaxe people, had camped while hunting. By the season's end, they had discovered *"not one, but a number of ancient camp sites,* . . . of the hand axe culture," as Louis emphatically phrased it in an article he published after the war. He believed that these were the first such living sites of Acheulean man that "had . . . been found anywhere," and Mary excavated them as if they were layered deposits such as one would find in a cave.

Louis thought that the campsites would enable them to re-create the "way of life of the makers of the [handaxe] culture"—something archeologists had not yet been able to do. At Olorgesailie, the Acheuleans had lived along the shoreline of a large lake for many centuries. Sometimes after heavy rains, the lake had flooded, and the people had abandoned their camps and heavy stone tools. The rising waters buried the implements under layers of silt. In time, the lake receded, the silts dried out, and the people returned, making their new camps on top of their previous sites. In this way, the occupation layers had built up over time.

Or so it seemed to Louis and Mary. Archeologists who later worked at Olorgesailie in the 1980s and who had the benefit of more sophisticated geological techniques showed that these sites with the many handaxes were not living floors at all, but ancient stream channels where the tools had accumulated. But in 1943 it seemed entirely plausible that they were Acheulean campsites, and Louis created a picture of the Acheulean way of life based on the artifacts and fossils Mary uncovered.

In Louis's eyes, the Acheulean people had been great hunters who skillfully herded animals into shallow lake water where they stoned and clubbed them to death. He based this on the numerous fossils of extinct giant animals —pigs the size of rhinos, baboons as hefty as chimpanzees, horses as large as draft animals, huge giraffe with mooselike antlers—that Mary found interspersed among the tools. All of the animals' skulls and bones had been smashed, and Louis assumed the Acheuleans had done this to extract the brains and marrow. He also concluded that they had eaten this meat raw, as none of the bones appeared to have been burned. Based on the extinct animals, many of which were the same as those he had found at Olduvai, Louis dated Olorgesailie to the Middle Pleistocene. In 1943 this era was dated at 125,000 years ago; today, geologists using paleomagnetic techniques have dated the lowest deposits at Olorgesailie to 900,000 years before the present.

As the excavations proceeded, Mary uncovered more of the round stone spheres they had discovered on their first trip to Olorgesailie. These puzzled everyone: why would the Acheulean people have taken such care to make round stone balls? With his usual imaginative flair, Louis decided that the balls were most often found in groups of three and must have been used for making hunting bolas, much like those that South American gauchos throw at fleeing animals. This image captivated Louis; he even argued that the bolas had been Acheulean man's primary weapon. "It has always been a matter of speculation what weapon Acheulean man used for hunting, since his common tools—the hand-axe and cleaver—seem to have been domestic implements," he noted in an article he wrote about Olorgesailie for *The Times.* "Now we have found a number of round stone balls . . . strongly indicating the use of the bolas . . . as a hunting weapon." He imagined his Acheulean hunters hurling their prehistoric bolas "to capture and kill the giant baboon."

But Louis's hypothesis did not win many supporters. It was impossible to say if the stones had actually been used in sets of three. The stones were also twice the size of those the gauchos employed, making the contraption heavy and unwieldy. The resulting weapon, as Louis discovered, was perhaps more dangerous to the hunter than to the hunted. "I personally tried an experiment with a bolas on one occasion," he wrote in his autobiography, "but was forced to give up after I nearly killed myself."[4] Today archeologists know that the round stone balls—which have since been found at many prehistoric sites, dating from 1.8 million to 40,000 years ago—were never designed as special tools or weapons. Instead, the spheres are the result of using a chunk of quartzite as a hammerstone for making other tools; the repeated bashing, by itself, turns the chunk of rock into a round ball.

Mary worked at Olorgesailie through mid-October 1943, and was happy enough with her first season's excavation to call it "a great success. We have material *in situ* from 3 different levels with associated fauna and a most primitive flake culture," she confided to Archdeacon Owen. "It is most amusing stuff." Yet ultimately Olorgesailie did not yield all that Louis and Mary hoped for. Mary continued to direct excavations here over the next ten years, but even with the assistance of several geologists, she found it nearly impossible to correlate the trenches she uncovered. This meant that she could not connect the various campsites through time, and so had difficulty extracting "any useful archaeological information" from her dig. She never published a report about Olorgesailie. Nor did any skeletons of Acheulean men turn up, though Louis often speculated that at some point at least one of

[4] Louis and Mary also found these round stone balls at Kariandusi and Olduvai, although Louis did not think they were used as bolas at these sites because they were never found in sets of three.

the hunters must have drowned. Eventually, other sites claimed the Leakeys' attention, and they turned Olorgesailie over to younger archeologists. "I fear the best moments [at Olorgesailie] were probably the very first ones, when we discovered the dramatic spreads of hand axes," Mary observed.

Nevertheless, archeologists today regard Olorgesailie as the most famous Middle Pleistocene site in Africa. It was treated with particular importance throughout the late 1940s and 1950s, and numerous archeologists toured the excavations. Olorgesailie also caught the attention and, in at least one case, the imagination of the press. All of the major British newspapers printed articles about the site, and in 1946 the *London Illustrated News* carried a two-page spread of a charcoal sketch showing hairy Acheulean hunters hurling bolas at fierce baboons in a "Prehistorian's Paradise."

On another brief leave from his wartime responsibilites, Louis visited Rusinga Island in Lake Victoria to explore its Miocene fossil deposits. Louis had first noticed Rusinga's potential in 1926 when taking a steamer across Lake Victoria; normally the steamer made the crossing at night, but this time it sailed during the day, enabling Louis "to examine the . . . island with field glasses and make a note that [it] looked a very promising place to search for fossils." Subsequently, the Ugandan geologist E. J. Wayland explored the island in 1927, and Louis himself spent time surveying the fossil beds on his third and fourth archeological expeditions in the 1930s. Based on the animal fossils he collected, Louis knew that Rusinga's deposits dated as far back as twenty million years ago (an estimate that has been subsequently verified), a substantial increase over the Middle Pleistocene deposits of Olorgesailie. But he also believed that even in strata this ancient he would recover important clues to the development of humankind.

Louis had first accepted the possibility of a Miocene ancestor in the 1920s after reading Sir Arthur Keith's *The Antiquity of Man.* In the closing paragraph of his book, Keith suggested that some stone tools might be dated to that era. Further, he asserted, "there is not a single fact known to me which makes the existence of a human form in the Miocene period an impossibility." Ever the student of Keith, Louis had quickly incorporated this idea into his own theory of human evolution. And in 1931, when Arthur Hopwood had turned up the fossils of three Miocene apes near the shores of Lake Victoria during Louis's third archeological expedition, his enthusiasm for the era was firmly established.

One of the apes Hopwood discovered looked remarkably like a chimpanzee—at least its teeth did—and he named it *Proconsul africanus,* after a famous captive chimp in England called Consul.[5] When Hopwood described

[5] The first *Proconsul* fossil was found at another Lake Victoria site, Koru, in 1927 by H. L. Gordon, the medical officer there. Louis guided Hopwood to Koru in July 1931 with the hope that additional specimens would be found, and they were.

his new ape, he had only a partial lower jaw with one tooth, a broken upper jaw, and three molars. This may seem a scanty amount of material from which to re-create a complete animal, let alone decide who its relatives were. Yet since the time of Georges Cuvier, the early-nineteenth-century French zoologist and father of paleontology, paleontologists have done exactly that. Cuvier compared the process to a mathematical equation: "The shape and structure of the teeth regulate the forms of the condyle [the jaw joint], of the shoulder-blade, and of the claws, in the same manner as the equation of a curve regulates all its other properties. . . . Thus, commencing our investigation by a careful survey of any one bone by itself, a person who is sufficiently master of the laws of organic structure, may, as it were, reconstruct the whole animal to which that bone belonged."

Hopwood found the similarities between the dentition of the chimpanzee and his Miocene ape so striking that he labeled *Proconsul* "ancestral to the Chimpanzee." Other anatomists of the day agreed with Hopwood, and Louis eagerly extrapolated from their conclusions. "It means that the animal which we call a chimpanzee was already almost fully differentiated at that very remote period, and that it has changed comparatively little since," he wrote in *Adam's Ancestors* in 1934. Then, echoing Keith, Louis added, "If I interpret this aright, it means that we must not be surprised if one day we find that at that same period—the Miocene—a true man-like common ancestor of the various human genera was already in existence."[6]

Several writers have suggested that Louis's belief in a long antiquity for humankind somehow stemmed from his missionary upbringing. But mystical sentiments never appear in any of his books or articles about human evolution. Instead, Louis's theory sprang from an acceptance of evolutionary change as the long, gradual process Darwin proposed. Louis also accepted Keith's opinion that a great length of time was required to explain the various races of humans and their wide distribution.[7] "The subdivision of a species into a number of distinct races is a slow and gradual evolutionary process, for which ample time must be allowed," Louis wrote in 1934, borrowing heavily from Keith. Since archeologists had already found what they believed to be a variety of human races (Piltdown, Java, and Peking men) that dated to the last Ice Age, it seemed logical to conclude that the "real ancestors of the *Homo sapiens* type would eventually be found in

[6] Physical anthropologists today agree with Hopwood that *Proconsul*'s teeth are very similar to a chimpanzee's. But several skeletons of the entire animal have now been discovered, revealing that it was more a mosaic of ape and monkey throughout. Molecular evidence also suggests that *Proconsul* was far too old to have any direct bearing on human origins. Our close genetic relationship with chimpanzees and gorillas indicates that humans separated from the apes between seven and five million years ago. Richard and Meave Leakey have also recently discovered a variety of apes from Kenya's Miocene deposits, revealing that an explosive radiation of apes occurred during this era. One or more of these animals eventually led to the modern apes and early humans.

[7] In *Adam's Ancestors,* Louis, following prevailing ideas, divided humankind into four races: Australoid, Negroid, Mongoloid, and European. Today, anthropologists no longer use those categories. They may speak of Caucasians, Asians, and Africans, but in terms of geography, migration, and genetics.

deposits at least as old as the beginning of the Pleistocene," and that "their common ancestor must be looked for in deposits at least as old as the Miocene," as Louis wrote.[8]

Following Hopwood's fossil discoveries, Louis visited the Lake Victoria region several times in the 1930s, exploring the Miocene deposits on Rusinga Island and the beds at Songhor on the mainland. At both sites he turned up the broken jaws and loose teeth of apes like those Hopwood had described. The deposits on Rusinga seemed to hold the most promise, but Louis lacked time and money for further explorations.

Once in 1940 he managed a short two-day excursion to the island, and then returned in September 1942 for five full days while on leave from the Special Branch. Mary and Jonathan joined him, as did their friend Mary Catherine Fagg and Louis's Kikuyu assistant Heselon Mukiri. Louis planned only to locate fossil beds that might prove worthy of study at a later date. But this time luck was with him. "By the greatest good fortune I found a jaw of *Proconsul* projecting from the face of a small cliff of the Lower Miocene beds, and 15 feet away we found a jaw of *Xenopithecus* [another Miocene ape][9] a few minutes later," Louis wrote about his discovery.

Louis's *Proconsul* jaw was in excellent condition with the chin region intact and a full set of teeth on one side. Soon after his return from the island he circulated photographs of the jaw and letters describing the find to several leading anatomists, including Sir Arthur Keith, Oxford University professor Wilfrid Le Gros Clark, and Dr. William K. Gregory at the American Museum of Natural History in New York. His fossil was the "most complete Miocene anthropoid jaw ever discovered," Louis announced in one letter, while in another he noted, "As you will see . . . this specimen has many characters which seem to link it more with man than with the great apes." Although Keith did not agree with this last assertion—"I see no suggestion of humanity"—he did think that it might be the "common ancestor of chimp & gorilla." He also agreed with Louis that the jaw represented a different species from that of Hopwood's, and urged Louis to name it *Proconsul leakeyei.* Louis declined the honor. In a paper published in 1950 with Le Gros Clark, he grouped the specimen with similar ones previously discovered in the region, and named the ape *Proconsul nyanzae,* after the native name, Nyanza, for Lake Victoria.

Louis also published a brief description of the fossil in *Nature* in 1943, in which he stressed the similarities between *Proconsul*'s jaw and that of man's. As in humans, the mandible lacked a simian shelf, the bony ridge that connects the two sides of the lower jaw in modern apes; further, *Proconsul*'s molars and dental arcade resembled that of humans. Dr. Gregory agreed with Louis's analysis, and Louis quoted his opinion that *Proconsul* stands

---

[8] Biologists now know that races in any species can form quickly; they believe that the modern-day human races probably began separating between 200,000 and 500,000 years ago.

[9] It is now classified as *Proconsul heseloni.*

"near to the common ape-man stem." "[I]n *Proconsul,*" Louis wrote, "we have a near approach to a form of ape-like creature from which the human stem eventually was evolved."

Congratulations for his discovery poured in, and it seemed unlikely that Louis would ever again have to sell beads or beeswax to support his family.

By 1943 Kenya Colony no longer feared an Italian invasion. The King's African Rifles had routed Italy's armies in Ethiopia, Eritrea, and Somalia, bringing an end in 1942 to what had been Italian East Africa. Yet almost at once a new threat to Kenya appeared in the east. Japan had entered the war in late 1942, and in 1943 its submarines were patrolling Kenya's coast, contacting spies about the movements of the King's African Rifles and taking on food supplies from sympathetic villagers. Louis was part of the Special Branch team called on to investigate these rumors. Throughout 1943 and 1944, he made many trips to the coast, sending back "MOST SECRET" reports. The spy business had lost none of its appeal for Louis, and his reports had the melodramatic flourishes of a pulp detective novel. "Bausi is the alias of Yunis Bin Famau," he wrote in one report, "a Bajun Arab who has been strongly suspect[ed] of pro-enemy sympathies before, and who was found in possession of a framed picture of Hitler some time ago. He has no regular employment but seems to have plenty of money and goes about the country *taking photographs.*"

Despite the drama of his tales, Louis did help uncover a ring of Somali and Arab prostitutes, train ticket collectors, and dhow captains who had been gathering information about troop movements in Central and East Africa, then passing it to agents across the Red Sea in Yemen. Louis also pinpointed one of the key coastal chiefs responsible for supplying the Japanese submarines with food and water. Nevertheless Kenya was not attacked, nor was the colony involved in any military actions in the last years of the war, although many native Kenyans served overseas with the King's African Rifles, and the colony's farmers devoted extra effort to raising crops and cattle for the Allied forces.

As the war drew to a close, Louis's espionage activities ceased to be a major source of income, and once again he had to find a way to support his family and his archeology. Since Mary was due to have another baby in December 1944, Louis hoped to land a position with good pay, one that would enable him to set aside funds for his children's education. In Kenya his most immediate prospects were with the Coryndon Museum—providing he could persuade the trustees to change his honorary curatorship to a full-time, paid post.

Louis and Mary had both generously given time and energy to the museum's cause throughout the war, putting in long evening and weekend hours. "I do C.I.D. 8–4:30 daily & Museum 6 pm to late in the night also every day & Mary does Museum all day & most of the rest of the time as well!" Louis wrote in a hurried note to Kenneth Oakley at the British Museum of Natural History. But Louis did not always feel that the museum's trustees appreciated his efforts, and in late 1943 he considered resigning. The trustees, he complained in a memo, had never thanked him for all he and Mary had done. Furthermore, when he had asked for some payment for his work, they had told him that he would need to "state a case. This practically amounts to a request that I should agree to make myself dependent upon the charity of the Trustees, and also to an indication by the Trustees that I am a spendthrift, which I strongly deny," Louis wrote angrily. They had further antagonized him by "not even allowing me to use the curator's house rent free." Apparently, the trustees had charged Louis a deposit on the furniture after one member, a Mr. H. William Gardener, visited the Leakey home and found the Dalmatians sprawled inside. "We certainly ought to have a complete list of the furniture which has been bought," he bristled. "I didn't like the way his four large dogs were allowed to occupy our chairs especially in the rains!"

The curator's house itself was another sore point, though the Leakeys had been happy enough with it at the outset. Built in the early days of the colony, it was a sagging wooden structure, faced with corrugated iron on the outside and honeycombed with holes on the inside. Bees, termites, and rodents infested the dwelling, and once, when Jonathan was still a baby, fierce army ants, called *siafu,* climbed up his cot by the thousands and attacked him. "[T]hey would undoubtedly have killed him if we had not heard his screams immediately and rescued him, badly frightened and still covered with swarms of biting ants," Mary wrote in her autobiography. Louis and Mary fought back with kerosene, spraying it down the holes and covering them with rush matting; they also set the legs of Jonathan's cot in shallow pans of water and kerosene. Louis thought the trustees might try to find a more suitable dwelling, but not until late 1944 when Nairobi's Medical and Municipal authorities declared the house "unfit for human habitation" was anything done. The trustees then agreed to build a new home, and the Leakeys moved temporarily to a house in Karen, a Nairobi suburb named after the writer Karen Blixen (Isak Dinesen).

Louis's complaints finally brought an honorarium of £150 and a carefully worded message from the secretary: "I am directed to inform you that at a meeting of the Museum Trustees . . . it was decided to offer you an honorarium . . . as an appreciation of the very valuable services rendered by you to the Museum. . . . I am directed to . . . make it clear that such will in no way impose any obligation on the Trustees to make similar, or other payment in future years."

The trustees' reports do not say why they were so reluctant to give Louis more money. They had paid his predecessor an annual salary of £900 and, despite the drain the war created on the national treasury, the museum still had the funds to pay that salary. And in Louis the trustees were getting not only an administrator but an enthusiastic natural historian, anthropologist, archeologist, and effective promoter. He had a flair for arranging special exhibits, speakers, and films, for inventing competitions that amused the public, and then presenting lively talks about these events over the radio. Nor did he regard anyone's request for information too insignificant or obscure. Parcels bearing dead spiders, snakes, birds, fossils, and stone tools arrived every week, and Louis dashed off letters identifying each. Once a soldier stationed at Mombasa wrote to say he was fascinated by the variety of butterflies in the area and asked the curator to tell him how to make a collection. Louis responded in detail, explaining how to use a killing bottle, make a collecting net, and construct mounting cases from cigarette tins. Another letter writer from the Adventurers Club in Chicago told how he had killed "American Bears" with the throw of the spear; what types of spears did the local natives use against lions and rhinos? Louis sent back a two-page letter describing the spears and throwing sticks of the Kipsigis and Maasai.

Most importantly, soon after assuming the curatorship, Louis opened the museum for the first time to all races; previously only Europeans had been able to visit. "[A]t first there was a considerable outcry from the white population of Nairobi," Louis recalled about his decision. "The museum was the first public institution to be opened to nonwhites, and the Europeans disliked the idea of viewing the exhibits side by side with Africans, who, they claimed were 'smelly,' or Asians, who were 'overscented.' " For a while the number of white patrons dropped off, but Louis ignored this, choosing instead to welcome the Asians and Africans whose visits "increased by leaps and bounds." Within a few months' time, Europeans accepted the change, and the museum soon became one of the few public places in Nairobi where everyone could mingle freely and equally.[10]

Unsure of what the trustees might offer him after the war, Louis applied for a number of positions outside of Kenya, including lectureships in anthropology at both Cambridge University and Gordon College in Khartoum, and the secretary's post at the Royal Geographical Society in London. But he and Mary decided that "neither of us could stand the life in Cambridge," so he withdrew his application there. Louis was also torn between leaving Kenya for a traditional career and staying on to continue his archeological explorations. London and the secretaryship began to seem far too distant and staid, so Louis withdrew this application as well. But in 1945 when Gordon College offered him a full appointment worth £1,500 a year and free housing, Louis hesitated, nearly said yes, then declined, but with the request to be reconsidered in two years' time. "The trouble is that I have been asked to arrange and organise a Pan African Congress on Prehistory . . . here in Nairobi. . . . And I must get some real work done on my new Olorgesailie site [which] is one of the most important in the world," Louis explained in his letter to the college.

[10] Racial segregation had been a de facto policy since the colony's founding, and restaurants, hotels, theaters, and swimming pools all had white/nonwhite designations.

Then the museum trustees came forth with their own postwar offer for Louis: they would appoint him full-time curator, give him a house, and pay him an annual salary of £750—precisely half the sum he would have received at Gordon College. That position was still open, and Louis eyed it hungrily. He now had two children in Kenya to support, for Mary had given birth to a second son, Richard Erskine, on December 19, 1944. But Louis stuck to his decision to remain. "It is perhaps hard on my children not to accept much better pay but scientific work must come first in my opinion," Louis wrote in his letter to Gordon College. Mary concurred.

On November 16, 1945, Louis sent the museum's board his formal acceptance, but he let the trustees know he regarded them as a miserly lot. "While I have agreed to accept the appointment . . . I have only done so for the reasons set out in this letter," he wrote, "and not because I consider that the terms are satisfactory or that they are commensurate with my qualifications." He noted that he was being paid less than his predecessor, and was not being granted funds for leaves to England as was customary for overseas positions. Nevertheless, he would take the post: "I feel it is my duty to science to remain in Kenya for the present."

In truth, Louis's heart was also in Kenya, and in a few years he would never have to consider leaving the country again.

Chapter 9

RACE FOR THE MIOCENE

With the war's end in May 1945, Louis and Mary's thoughts turned to England. They had not set foot in the motherland in eight years and longed to reestablish their ties. Mary's mother was ailing as well. In a steady stream of letters, penciled in a fragile scrawl, she begged Mary to come home, and to bring with her Jonathan and Richard. "[My years] are I think fairly limited," she wrote in June, "& I do so long to see the little pets so much not to speak of how I long to see *you.*" Mary fretted over her answers: Richard was an infant, Jonathan screamed whenever separated from Louis, and a passage on a troop ship home was nearly impossible to come by. But her mother's condition continued to worsen, and in the autumn Mary placed hers and the children's names on a government list of those seeking a way home. Louis would stay behind awhile in Kenya. He was busy organizing his Pan-African Congress for prehistorians and thought he might be able to use the conference to wangle a free passage to England at a later date.

In early March of 1946, Mary, Jonathan, and Richard boarded a special transport ship taking women and children to England from the East African colonies. Thousands of people had been stranded by the war and were desperate to return home, and Mary found herself squeezed into a tiny windowless cabin with two other mothers and their two children. It was a long, unpleasant journey, hot and stifling through the Red Sea, chilly and windy in the Mediterranean. Richard developed a painful earache, and Mary paced the ship's corridors at night, rocking him to soothe his crying, and wondering—as all the passengers did—what they would do if the ship struck a mine.

Mary was thirty-three now, trim and pretty, with her hair combed smooth across the crown, then fluffed softly around her face. Though dressed in plain, unstylish tweeds, she retained the cool, confident air of her youth. She was also a concerned and attentive mother, and although years later she would say that "babies are boring," she had been almost bewitched by Richard. He was a particularly winsome baby, with full, round cheeks, large brown eyes, and an impish smile. "I don't wonder you have fallen for your Baby," Mary's Aunt Kathleen wrote to her after receiving photographs of a four-month-old Richard. "[H]e is perfectly sweet . . . [and] most adorable." He was also fearless, unafraid of wrestling with his older brother, or taking flying leaps from his mother's arms. She called him her "tough guy," and watched his every move with delight.

Despite the hazards of the voyage, Mary and the children reached London safely in April. The city had changed drastically in her absence. Rather than feeling like home, London, with its gutted, blackened buildings and weary citizens, seemed a "strange city" to Mary. A heavy greyness hung in the air, and she checked into the boardinghouse room her mother had booked for her with a growing sense of unease. Her mother's flat was only a short walk away.

Mary and the children were soon making this walk every day, then climbing up the stairs to visit Cecilia, who was far more ill and frail than Mary had realized. In fact, Cecilia was dying. She passed away one night only two weeks after Mary's arrival, and Mary was overcome with guilt and sorrow. "I can't get over this tragic business," she wrote in anguish to Louis. "It's so cruel that she should have had only these two weeks with us after the years & years she waited, just living for the time to come. It's really heartrending. It keeps me awake at night thinking about it and also that I was not with her although so near."

If London felt like a strange city before to Mary, it now seemed a prison. There was little chance that she would be able to return soon to Kenya—passages continued to be extremely difficult to come by—so she spent her days settling her mother's affairs, visiting with her aunts, and writing Louis lengthy letters. Louis had not yet confirmed his own trip to England, and Mary, stranded in the dingy boardinghouse where the dinners seemed to consist solely of boiled cod, began a persistent campaign. "I can't decide from your last letter whether you intend to come here or not," she wrote on May 7. "I feel more strongly all the time that you *should*. You are so out of contact with everyone & it is a bad thing." Two days later, she wrote again. "Really it is *so* awful being separated like this & I feel so very strongly that you should come over, not only for [planning] the Congress but for *yourself*." One day she took the children to the zoo, where the grunting and roaring of the lions nearly made her weep, "it made me so homesick for

Olgasalic [Olorgesailie] & Oldoway [Olduvai] & the best parts of Kenya—safaris." But it was Louis she truly yearned for. "Life is so flat and empty without you," she wrote at the end of one letter, "& there is nobody I can talk to as I do you about everything."

Louis was not being intentionally difficult. Besides the problem of getting to England, Louis found that his Pan-African Congress of Prehistory was shaping up to be a major event. Planning it consumed most of his time. He had first conceived of this conference in 1944, and had discussed it in letters with colleagues in South Africa and England, all of whom responded enthusiastically. "I was delighted with . . . your plans for an archaeological congress," wrote van Riet Lowe. He offered to enlist the support of South Africa's prime minister, Jan Smuts, and suggested inviting delegates from every African country, "from the north to the south," plus European and American scientists. They would discuss the geology, archeology, and paleontology of Africa's prehistory, a field that, as Louis noted, had grown dramatically in the twenty years since Raymond Dart's discovery of the Taung Baby's skull. More skulls and bones of the australopithecines (as Dart's find was called) had been uncovered in South Africa, Louis had established the importance of the Miocene apes in East Africa, and both he and Mary, as well as other scientists, had documented the presence of people of the handaxe culture across the continent. "[S]cientists were at last beginning to believe Charles Darwin's prophecy that the birthplace of both man and the great apes would be discovered in Africa," Louis later wrote in *By the Evidence.* "I felt strongly that the moment was most opportune to inaugurate a Pan-African Congress of Prehistory."

Louis also thought that the conference might rehabilitate his career, for his standing among many of his fellow scientists was still tainted by the Kanam/Kanjera affair. It was common knowledge that Louis languished in Nairobi—an academic backwater—because of the errors he had made fifteen years earlier. He was now forty-three, an age when most accomplished academicians are settled in comfortable teaching posts, elected to honorary societies, and awarded substantial grants. There was none of this for Louis. Instead, he continued to have the "reputation of being a bit of a maverick," recalled Desmond Clark, then a young curator at the Rhodes-Livingstone Museum in Livingstone, Northern Rhodesia. Louis's constant use of newspapers to discuss his discoveries had further eroded his position. Then, as now, scientists were suspicious of colleagues who caught the public's eye and were comfortable in the limelight, as Louis was. Uneasy with what Frida had called his "entrepreneurial, showbiz manner," many loftily dismissed Louis's publicity seeking. For example, in 1936, when the *East African Standard* carried an enthusiastic account of archeological discoveries made by the Ugandan geologist E. J. Wayland and Archdeacon Owen, Wayland actually demanded an apology from the newspaper, and then told Owen, "This is exactly what ought not to have been published. That sort of thing is best left to Leakey."[1] The Pan-African Congress gave Louis a chance to restore some of the luster he had lost among his colleagues, and he was determined to make it a memorable meeting.

Armed with letters from many leading scientists—van Riet Lowe, the Abbé Henri Breuil, William Gregory, Wilfrid Le Gros Clark—Louis had little trouble selling the conference to the Kenyan government. Sir Philip Mitchell, Kenya's governor, approved a grant, as did the governors of Tanganyika and Uganda. By the middle of May, Louis had also persuaded the Coryndon Museum's trustees to fly him to London to talk up the congress with his British colleagues. No flights were available until the end of June, but he cabled Mary anyway. He would board a seaplane on June 29 at Lake Victoria and be reunited with his family three days later.[2]

To Mary, Louis's cable was more than welcome news. "Jonathan & I are in such a fever of excitement," she wrote to Louis the moment she had the telegram in hand. "I'm already counting the days." It had been the longest separation since their marriage, and she had had time to reflect on how much Louis meant to her. "I get so excited when I think [of your coming] for I really can't do without you. When I'm with you, I'm often so horrid—I must be a dreadful person, not to realise when I'm well off. Now that I'm away from you, all I want is to be with you again, so *terribly* badly."

A little more than a month later, in early July of 1946, Louis's plane landed at the London airport. Mary and the children rushed forward to greet him, and they then made their way to a small, furnished flat she had found for them in the city.

Louis made the most of his time in England. Over the next two months, he renewed his ties with colleagues at the British Museum of Natural History, paid visits to the anatomist Wilfrid Le Gros Clark at Oxford and his old archeology professor, Miles Burkitt, at Cambridge, and made a quick trip to Paris to discuss his Pan-African Conference with the French scientists Camille Arambourg and Abbé Henri Breuil, who was then regarded as the world's leading prehistorian. By September, Louis had signed up all of these scientists and had learned that many others, including the South Africans Raymond Dart and Robert Broom, would be attending as well.

Louis also used his stay in England to promote his Miocene research. He had brought his *Proconsul* jaw from Rusinga Island with him, carrying it in

[1] Louis himself was sensitive to the charges that he was a publicity seeker. The same year that Wayland complained to the *East African Standard,* Louis contemplated suing the *London Daily Mirror* for writing an article about him without his knowledge. Titled "Thirty-three—that's all: but —He Discovered Adam's Ancestor," the story drew on Louis's writings and public lectures to portray him as an athletic, brilliant young scientist, "wise to the ways of the jungle," but equally at home in "the gloom of a Cambridge study." The *Daily Mirror* denied that the article had in any way defamed Louis or his career, and lacking funds to pursue the matter, Louis dropped his suit.

[2] In his autobiography *By the Evidence,* Louis mistakenly writes that he made this trip in December 1945.

a biscuit tin lined with cotton and producing it, with a bit of a flourish, to everyone he visited. He even managed to show it on British television. Not only was this Louis's first appearance on TV; it also marked one of the first television appearances of a fossil relating to human origins. Louis also attempted to convince the Royal Society about the importance of his find, hoping that they would fund an expedition to Rusinga Island, but they gave him "very little hope."

By early October, Louis, Mary, and their sons were back in Nairobi, where Louis quickly pulled together a small volunteer staff to organize the upcoming conference. "[W]e only had a single secretary . . . working in a small office in the Museum and using a portable typewriter," recalled Mary, who also joined in the flurry of preparations. On the fourteenth of January, 1947, sixty scientists representing twenty-six countries were due to descend on Nairobi, where they would spend a week giving papers and holding discussions. Louis then proposed to take all sixty on a grand tour of his sites: the handaxe sites of Kariandusi and Olorgesailie; his excavation at Gamble's Cave and Mary's at Hyrax Hill; the layered, fossil-rich beds of Olduvai Gorge and the rock paintings at Kisese. The delegates would be able to see Louis's work firsthand and, Louis hoped, would acknowledge his pioneering contributions.

With bright flags and bunting snapping in the wind at Nairobi's Town Hall, Kenya's governor, Sir Philip Mitchell, welcomed the delegates to the two-week conference. For the next seven days, the scientists delivered papers, argued over ancient climatic conditions, debated the terminology of Africa's Early Stone Age cultures, heard reports about discoveries in the Sudan, Morocco, and Rhodesia, and listened attentively to new assessments of the australopithecines and Louis's Miocene apes. There were parties and dinners to attend—Mary had purchased her first set of dinnerware just before the conference opened—and at the end of the week, daily jaunts to see the Leakeys' sites that were closest to Nairobi, followed by a weeklong journey to Olduvai, Ngorongoro, and Kisese.

The conference marked the first time that such an eclectic group of scientists—anatomists, paleontologists, archeologists, and geologists—had gathered together to discuss Africa's prehistory. Those who attended, particularly scientists working in far-off corners of the continent, remember it as a splendid, almost euphoric affair. "When I arrived in Africa [in 1938]," recalled Desmond Clark in a personal memoir, "there were only two or three professional archaeologists in the whole of the continent south of the Sahara. [We] were separated by hundreds of miles and met only on rare occasions." Clark and others felt as if they had been working in "watertight compartments" until they joined in the conference. "We had all been working in isolated enclaves," noted Basil Cooke, then just starting out as a geologist in South Africa. "So one of the highlights of the congress was bringing all these people together who had never before exchanged ideas." The congress was so successful that at its end the delegates voted to hold similar meetings every four years.[3] There was also generous praise for Louis and for his "energy and enthusiasm," as an anonymous commentator wrote in *Nature* two weeks after the congress closed.

As Louis had hoped, the congress marked a turning point in his career. None of the delegates could fail to notice that he had almost single-handedly traced East Africa's prehistory from the Miocene to the Early Stone Age. They left the conference realizing that he was less a maverick and more an inspired, dedicated scientist. Mary's excavations at Hyrax Hill, the Njoro River Cave, and Olorgesailie had also created a stir. She had presented a paper about her excavations, although when the scientists visited the sites she took a backseat to Louis, letting him hold forth about their significance.[4] Nevertheless, Desmond Clark and the other visiting archeologists were well aware of her accomplishments. "Hyrax Hill was beautifully excavated," Clark recalled nearly forty years after viewing her dig with other members of the congress.

> Everything was left exactly in position so that you could see the relationships of stones to one another, and bones and mounds. She applied these same techniques at Olorgesailie so that all the handaxes were lying along their [geological] horizons. As far as I know she was the first one to excavate a Paleolithic site this way, with reference to their horizons. Now it's become the standard for people excavating these sites; I applied it myself in 1953 when working at Kalambo Falls [a Paleolithic site in Zambia]. Hers was pioneer research.

Perhaps most important for Louis, his Miocene discoveries intrigued the delegates, in particular the eminent anatomist Wilfrid Le Gros Clark, known to his friends simply as Le Gros. Le Gros was then forty-nine, a tall, handsome, studious man and a cautious scientist. A specialist in the comparative anatomy of primates, he had written a book on primate evolution, and following Sir Arthur Keith's retirement in 1933 had replaced Keith as the arbiter of disputes in this field. Le Gros had come to the congress straight from South Africa, where he had spent a month studying Dart's and Broom's australopithecines. Like most of the British scientific establishment, he had long argued against their place on the human family tree. But now after a careful assessment of all the australopithecine fossils, he announced himself a convert. "I am afraid there is no escape from the fact that these specimens are very closely related to man and are survivors of the group that gave

---

[3] Over the years, the Pan-African Congresses became, according to Desmond Clark, "the major instrument for exchange and rapport in Quaternary studies in Africa." The most recent was held in 1995.

[4] Mary was so quiet about her accomplishments that even several of her closest friends in Kenya did not realize that she had directed her own digs in the 1940s until they read her 1984 autobiography. "It was my impression that she was very much Louis's disciple," said Joan Karmali, who met the Leakeys in the mid-1940s. "It seemed that he was really the knowledgeable archeologist and she tagged along behind. So reading her autobiography was a great surprise."

origin to man," Le Gros told the assembled scientists. The australopithecines were, he said, "man in the making." Le Gros's words stunned his colleagues and delighted both Broom and especially Dart, who had made much the same pronouncement when he had discovered the Taung skull twenty-two years before.[5] His speech also had an immediate impact in England, where his stature was such that Sir Arthur Keith, whose writings had left the australopithecines among the apes, felt compelled to send a letter to *Nature.* "I am now convinced . . . that Prof. Dart was right and that I was wrong; the Australopithecinae are in or near the line which culminated in the human form."

It is not clear when Le Gros and Louis first met, but very likely their paths crossed initally in the 1930s, during one of Louis's visits to England. On Louis's 1946 trip, the two men had met again, and Louis had shown Le Gros his *Proconsul* jaw. The jaw, hinting as it did at a possible ancestor for both apes and hominids, fascinated Le Gros. He was further tantalized at the congress by Louis's exhibit of the teeth, jaws, and limb bones of the three different Miocene apes—*Proconsul, Limnopithecus,* and *Xenopithecus*—along with the many other Miocene mammal, reptile, and fish fossils Louis had collected. Mary recalled that this exhibit, and Louis's and Donald MacInnes's papers about the Miocene, left a "tremendous" impact on the scientific gathering.

Privately, Louis laid the fossils out for Le Gros's inspection, then proceeded to tell him, in his hurried, breathy way, about the richness of the fossil beds at Rusinga and Songhor. All the teeth, jaws, and limb bones of the apes had been collected in hasty visits of only a few days' duration, Louis said. Imagine what could be found if a proper expedition were mounted. Pressing his point, Louis suggested that it might even be possible to find complete skulls of fossil apes. Perhaps the fossils and the puzzle they presented, along with Louis's enthusiasm, hit a treasure hunter's nerve in Le Gros, for by the end of the congress he had agreed to help Louis raise the money for such an expedition. Very likely Broom, who had established a firm friendship with Le Gros, also put in a good word for Louis. Broom greatly admired Louis, and found him daring and courageous.[6] Le Gros, too, was captivated by Louis's confidence, and only two weeks after the conference closed he began discussing the possibility of a grant with the Royal Society. "I have become fairly 'bitten' with these Miocene deposits," he

---

[5] Other scientists, notably William K. Gregory of New York's American Museum of Natural History, had accepted the hominid status of australopithecines as early as 1930. But until Le Gros's pronouncement, the leading British scientists had continued to regard them as apes, ancestral only to the chimpanzee or gorilla.

[6] Broom also liked to support any scientist who had been attacked, as he had been, by the "English doubters." Thus he had never doubted Louis's story about the Kanam/Kanjera discoveries. In 1945, while writing his book *Finding the Missing Link,* he told Louis: "I say [in my book] that I fully accept all you say about Kanam & Kanjera & pay no attention to the geological critics. I put Boswell in the same group as Sir John Evans, Sir Boyd Dawkins, & Sir J Prestwich as rather light weights. . . . [W]hat Boswell has done re Kanam & Kanjera will not I think enhance his scientific reputation."

confided to Louis from Oxford, "& would like to take part in their investigation." He hoped that Louis could return to Rusinga that summer.

There was another reason behind Louis and Le Gros's urgent interest in Kenya's Miocene deposits. A brash young American named Wendell Phillips had surfaced at the Pan-African Congress, boldly proposing to lead an expedition to Kenya and South Africa to solve the riddle of humankind's origins. No one seemed to know much about Phillips, but he addressed the conference with such authority that Robert Broom was heard to whisper, "What a lad, he talks as though he were Uncle Sam." Phillips subsequently wrote Louis a letter loosely outlining his plans for Kenya—the expedition would "undertake extensive scientific research in the fields of Palaeontology and Archaeology"—and signed it "Leader—UNIVERSITY OF CALIFORNIA AFRICAN EXPEDITION." Louis sent back a cautious but straightforward reply. "I am in favour of co-operation in the scientific field and . . . I welcome suggestions of American research work here, for the area is so big that there is ample room for several expeditions to work." But Louis also made it clear that he regarded Rusinga Island and its Miocene deposits as his territory. "I told Wendell several times," he explained in a letter to a colleague of Phillips, "[that] Rusinga [is] very much 'our pigeon.' "

At the same time that he was laying down the ground rules for Phillips, Louis was making inquiries. Phillips was then twenty-five, a slightly built, mustachioed man whose glib manner put Louis and Mary on guard. "Louis couldn't stand him, and neither could I," Mary recalled. "He was obnoxious; a pushy American." Theodore McCown, then head of anthropology at the University of California, Berkeley, apparently agreed. "Phillips is not a scientist, but a promoter," McCown wrote in answer to Louis's letter. Nevertheless, he had the full backing of the university, and of Charles Camp, the director of the university's museum of paleontology. McCown explained that the expedition had come about "principally because of the persistent promotion on the part of Phillips, and he has quite extraordinary gifts in terms of getting to important people and getting them to promise help and assistance." Yet no one seemed to know precisely what Phillips expected to achieve. "At one time or another we have heard that work will be done in musicology, ethnology, motion picture photographs of African ceremonials and dances, and excavation for more South African fossil anthropoids." Having met Louis and heard about his Miocene discoveries, it seems that Phillips decided to add one more item to his list.

Louis soon found out just how good a promoter Phillips was. Three months after the congress closed, he received a long letter from Phillips tallying the booty he had rounded up for his expedition: $150,000, ten cars and trucks, an airplane, tents and camping gear, ten thousand gallons of gasoline, and round-trip airline tickets for twenty scientists. The U.S. Army and Navy were helping as well, for Admiral Chester Nimitz and General

James Doolittle had become Phillips's close friends. "We are today the largest scientific expedition of its kind ever to leave America," Phillips declared. Louis could expect them in July.

Phillips's letter staggered Louis. The Royal Society had not yet approved Le Gros's proposal, and Louis knew that Kenya's governor would be only too happy to welcome the Americans and their dollars. "If we have no definite plans by the end of July I feel fairly certain that the Governor will insist on giving Wendell Phillips a licence for [exploring] Rusinga and the adjacent Miocene areas," Louis wrote to Le Gros at the end of April. Determined to lead his own expedition to the island, Louis then approached the Aga Khan, the wealthy leader of the Ismaili sect of Muslims, for funds—after all, if Phillips could appeal to the high and mighty, so could Louis. He explained that while he was not opposed to other expeditions, "and, in fact, welcome them as there is room for work by all, I am, nevertheless, very greatly worried that I, who discovered the sites and the finds shall not be able to be continuing my work . . . and also making fresh discoveries through lack of funds." Louis's appeal worked: the Aga Khan, who made his home in Kenya and was interested in prehistory, gave him £250 for the expedition.

Louis also took another lesson from Phillips. After the Royal Society awarded a £1,500 grant—for Le Gros had brought all his influence to bear following Louis's last letter—Louis dashed off letters to various companies asking for donations. "I have approached the Shell Company for 1,000 gallons of free petrol [gasoline], the Motor Mart for free lorry transport, Ahmed Brothers [a Nairobi safari outfitter] for free tents and BOAC for a free air passage for a Geological Consultant." Louis did not get all he asked for, but Shell did sell him the gasoline at a discount, Ahmed Brothers supplied two tents, and BOAC gave him one ticket.

Meanwhile, Phillips's expedition continued to expand. By summer he had raised half a million dollars, added seven more vehicles plus a motor boat, and increased his scientists to thirty-five. Admiral Nimitz was also supplying a "special Navy ship," which would carry the first wave of explorers to Egypt to search for Oligocene-era primates (dated at thirty million years ago). They would then proceed to Kenya and its Miocene deposits, while another group went to South Africa. But the size of the operation had so mushroomed—Phillips was now touting it as a Cairo-to-Cape archeological extravaganza—that Phillips was making little headway. There had been political squabbling as well and Phillips's old professor, Charles Camp, had quit the expedition to lead his own to South Africa. As a result, Phillips would not be coming to Kenya in July but in September; he then pushed the date back to October, and finally settled on the spring of 1948.

With each delay Louis breathed a little easier, although he never altered his own plans. The Royal Society's grant was, after all, the first Louis had received since 1934 and he was determined to "achieve maximum results," as he told Le Gros Clark. Le Gros, too, was excited. He himself would not join the expedition, since he had no experience as a field collector, but he wished Louis "the very best of luck." A few weeks later, just after twelve midnight on July 8, 1947, Louis and Mary, along with six-and-a-half-year-old Jonathan and two-and-a-half-year-old Richard, left Nairobi at the head of the first British-Kenya Miocene Research Expedition. With the grants and donated supplies, this was the best-equipped expedition Louis had yet led. In addition to Louis and Mary in their car, the expedition consisted of the Kikuyu Thairu driving a rented truck with all the gear, and two other native assistants. Heselon Mukiri, Donald MacInnes, and two British geologists, Ian Higginbottom and Robert Shackleton, would join them on Rusinga Island later.

Rusinga is a small island in Lake Victoria, only nine miles long and five miles wide, and is shaped something like a long-necked gourd, with a thin peninsula jutting to the north. By motorboat, the island is about a five-hour journey from Kisumu on the mainland. On his preceding trips to the island, Louis had had to rely on local transportation—canoes or dhows—and the trips lasted anywhere from ten to twelve hours. This time, he had sufficient funds to rent a launch, the *Maji Moto* (Swahili for "hot water"), and two days after leaving Nairobi had pitched his camp in the shade of a fig tree near the island's highest point, Hiwegi Hill, and close to one patch of fossiliferous gullies. The camp also lay a few hundred feet above the lakeshore, where there were fewer mosquitoes and lake flies, and had a good landing. From here Louis could take the *Maji Moto* quickly around the island, stopping to inspect other fossil beds he spotted from the water.

Almost every square foot of Rusinga harbors fossils, though there are three main sites: the pinkish-white, clayey silts at Hiwegi Hill, the light yellow hardpan below Kiahera Hill, and the buff-colored gullies of Kathwanga Point. These latter two localities lie on the opposite side of the island from Hiwegi. Much of the rest of Rusinga is given over to the maize and millet fields of the Luo people. All the fossils date to the Miocene, eighteen to twenty million years ago, when a volcano called Kisingiri on the mainland erupted with heavy ashfalls. The ash buried and preserved many of the plants and animals then living on Rusinga, in much the same way that Mount Vesuvius preserved Pompeii. Rain and wind had now brought the fossils to the surface, and in the eroded gullies bits of hardened bone lay everywhere.

On their third day at Rusinga, Louis and Mary took the launch to Kiahera, where Louis had found the *Proconsul* jaw in 1942. Within minutes, Louis picked up a canine from the same animal, "a good augury for the success of the expedition," he noted in his journal. Just fifty yards away, a Luo worker spotted a rhino jaw lying at the bottom of a gully. Louis found the animal's skull weathering out along one wall, and he put a team of excavators here. Over the next few weeks, they uncovered most of the rhino's skeleton as well—a new species of *Dicerorhinus,* a genus related to today's Sumatran rhino.

The expedition stayed in the field until the end of November, though Louis and Mary made trips back and forth to Nairobi. They spent most of the time on Rusinga, but also explored the mainland site of Songhor, where in 1932 Louis had found a fragment of a large canine tooth similar to that of a gorilla. Now, fifteen years later, Louis picked up a lower jaw fragment containing part of that same tooth; he and Le Gros would later name this large Miocene ape *Proconsul major*.[7]

Altogether, Louis and Mary's team collected more than 1,300 fossils by the expedition's end, including sixty-four specimens of Miocene apes (teeth, upper and lower jaws, and palates), and an astonishing array of mammals new to science. There was a giant, horse-sized hyrax; several primitive carnivores, some as small as a weasel, others as massive as a grizzly bear; ancestral elephants, rhinos, and pigs; an odd hippolike animal called an anthracothere; and primitive ruminants related to giraffes. New rodents, fish, insects, reptiles, gastropods, and plant fossils rounded out the picture. It was as if the island were a fossilized ark, bearing with it all the forms of life that had once roamed its shores. With such a bounty, scientists would be able to recreate the complete environment of the Miocene apes, something that was seldom possible for any animal let alone a possible human ancestor.

Le Gros, who pronounced himself "very pleased" with the expedition's results, was especially delighted by the discovery of a partial upper jaw belonging to the same *Proconsul* individual Louis had found in 1942. "It does look as though there *must* be a skull or two just round the corner," he wrote to Louis at the end of September.

Meanwhile, the Boy Wonder, as Phillips had been dubbed by the American press, was readying his forces. To maximize his efforts, he had split his team in half. Basil Cooke, the South African geologist, would lead one group to Egypt's fossil beds and then to Kenya; Phillips himself would lead the other contingent to South Africa and join Cooke later in Kenya. The question now was where in Kenya they would be allowed to explore, a decision that was entirely up to Louis.

In many of his letters to Phillips and Cooke, Louis talked generously of offering his support and advice, of directing them to islands near Rusinga, even of letting them collect at Olduvai—but undoubtedly he was making the best of a bad situation. Nearly every site under consideration had been explored by Louis using his own private funds, and though he tried to view the matter disinterestedly and consider only what was best for science, he had difficulty stifling his possessiveness. "I would have personally rather that [Phillips] had not worked in Kenya at all," Louis confided to Le Gros, "but I did not want to be selfish, and I do think it is right to let his Expedition come to try, provided they play fair." Le Gros, too, disliked the idea of handing over any of the Miocene deposits at Lake Victoria to the Americans, and considered Phillips's request to explore this region as highly unethical. "[No] reputable scientific expedition, I think, would even suggest working

[7] Astonishingly, in 1962 a schoolboy found another piece of this same jaw, which Louis fitted to the fragment he had discovered in 1947.

on sites at which work is already in progress by another expedition," he wrote to Louis.

Still, Louis did not think it "a wise policy to exclude other scientists"; besides, he had also already promised to help Phillips. Louis and Le Gros then bandied about several possible sites the Americans could work.

From the sites Louis offered, Phillips chose the west side of Lake Rudolf, a vast region that Louis assured him was "virtually unexplored." It was also agreed that any hominids Phillips's team discovered would be given to Le Gros to include in his study, though Louis thought their chances of making such a find slim. "[T]heir lack of local knowledge will be a great hindrance to them," as would be the expedition's "large and unwieldy size—they are bringing eleven team members."

By now the story of the two expeditions—one British and one American—and their search for human ancestors had reached the press. Le Gros had asked Louis to send him a brief report about his work on Rusinga to publish in *Nature,* and Louis mailed a similar one to the London *Times.* A reporter there added a single provocative sentence to this article: "A well equipped and richly endowed expedition from the University of California is now in Africa to explore these [Miocene] sites discovered by British scientists." Immediately there was a general harrumphing and clacking of tongues in England; Louis even had a letter from his old nemesis, Percy Boswell, who offered to help him secure another grant from the Royal Society! But the most promising letter came in January 1948 from a wealthy London businessman, Charles Boise. Writing on a small pad of hotel stationery—he was on holiday in the Swiss Alps—Boise told Louis that he thought the Miocene research should continue "under British auspices," and that he might be of assistance.

Though Louis had never heard of Boise before, he took him at his word and asked for £1,000. Funds were "urgently needed," Louis explained, for the "wonderfully equipped American Expedition" was soon due to arrive. "[They] will certainly try to beat us to our goal, which is to obtain a skull or skulls of these fossil apes." One month later, in February 1948, Boise mailed Louis the check. The Kenyan government, spurred into action by the American interest in Louis's discoveries, also provided a £1,500 grant. He would use the funds to tackle once again the exposures at Rusinga.

In the meantime, Louis flew to England to discuss his new Miocene ape fossils with Le Gros and to attend a Geological Congress. But his trip was cut short when a cable arrived from his sister Gladys, saying that their mother was dying. May had been living alone, with Gladys and her husband nearby, since the death of Louis's father, Canon Harry Leakey, in 1940. She was now terminally ill and died not long after Louis's hasty return. "I am sorry to say . . . it was rather a sad homecoming," Louis later wrote to a friend.

• • •

In March 1948, only a few weeks after the death of Louis's mother, Basil Cooke arrived in Kenya at the head of the American Expedition. They had driven in convoy south along the Nile, through the Sudan to the border town of Juba, and from there to Nairobi and the Coryndon Museum. Louis greeted them there politely, and Cooke recalled nearly forty years later that Louis "was very cooperative, providing us with some of his African preparators and collectors." But he was being cagey too, assessing the team and ultimately dismissing them: "[T]he party doesn't impress me greatly, and I don't think they will do much serious work," he wrote to Boswell. By now he had come to view the Americans as rivals, and Cooke sensed that Louis's help was given reluctantly. Cooke would not be the last scientist to run up against what he called Louis's "proprietary air."[8]

Despite Louis's assessment of the party's skills, they discovered a partial lower jaw and a single canine tooth of a Miocene ape, and made some lesser finds ranging from Late Stone Age to Miocene.[9] All passed without incident —Louis graciously praised their finds; Cooke thanked him for his assistance —until the arrival of Wendell Phillips and a film crew in mid-April. Apparently in a report Phillips wrote about the fossils' discovery, he included Louis's name as a member of his expedition, implying that Louis was working under him. This ignited a quarrel of such intensity that rumors eventually reached Phillips's old professor, Charles Camp, that Wendell had "threatened [Louis] in some way with fire arms." Louis denied the gossip, and the Americans continued to explore Rudolf, remaining in the field through October.

In the meantime, Louis and Mary readied their own new expedition. They used some of Boise's donation to buy and outfit a Commer truck with bunks and cupboards. They could now travel in comfort, without having to stop and pitch their tents, on their drives to the field. Worried that the Americans might find additional ape fossils—or even a complete skull—at Rudolf, Louis decided to first explore some deposits he had not told them about near Loperot. He and Mary made a reconnaissance trip there in June, but Louis quickly realized that the area was so barren and water so scarce that he could not "possibly [explore] the area . . . as I had hoped." He lacked the vehicles and field equipment the region demanded.[10]

Louis decided to stick to his original scheme and explore Rusinga again, though his plans to be in the field by July were delayed by an onset of gall bladder trouble. He had been suffering with these pains—brought on by

---

[8] Desmond Clark notes that Louis was not alone in such feelings of possessiveness. He himself and almost every other prehistorian working in Africa in the 1940s and 1950s felt much the same. "In fact," he has written, "it is still a trait among prehistorians today as can be seen when one hears such expressions as 'my Neanderthals', 'my Acheulian' or, indeed, 'my site'!"

[9] In 1985, Richard and Meave Leakey's West Turkana Expedition found additional fossils of this type of ape. They plan to name it for Kamoya Kimeu, head of their fossil-hunting team. The fossils are now known to date to the Oligocene, and are about twenty-eight million years old.

[10] Most of this region remained unexplored until 1983, when Richard Leakey launched his West Turkana expeditions.

his love of greasy food, particularly cold mutton fat—for some time, and was temporarily hospitalized. His recovery was swift, and by the end of September he and Mary were on their way to Rusinga with their two boys and all the Dalmatians.

On this second British-Kenya Miocene Research Expedition, Louis planned to spend more time exploring the exposures at the sites of Kathwanga and Kiahera on Rusinga. The Leakeys set up camp under a spreading fig tree near Kathwanga on September 30, and the next day were out early to search for fossils in the nearby gullies and cliffs. "For some time we had both had a hunch—if that is the word for it—that something very important was near at hand," Louis wrote in his autobiography. "The problem was to locate it." Louis soon found a crocodile's skull and settled in to excavate it. The skull was large and nearly complete, and Louis continued his excavation the next day. Mary, however, had no interest in Louis's project. "I have never cared in the least for crocodiles, living or fossil—so I continued to explore the eroded surfaces not far away," she recalled in her 1984 autobiography.

Walking slowly through the crumbly beige-colored gullies, Mary noticed some interesting bone fragments lying along one slope and followed them up the hill with her eye, where she caught the glint of a tooth. "Could it be . . . ?" she wondered. Within minutes, "I was shouting for Louis as loud as I could and he was coming, running." Together, they gently brushed away the earth, revealing not only a tooth, but a jaw as well. It was unmistakably that of a *Proconsul.* Using a dental pick, Mary carefully scratched at the hillside, and now "the greater part of a skull became visible," Louis later wrote. Although it was fragmented, they could see that much of the facial region was also present, and that both jaws had a complete set of teeth in perfect condition. "This was a wildly exciting find," Mary wrote, "which would delight human palaeontologists all over the world, for the size and shape of a hominid skull of this age, so vital to evolutionary studies, could hitherto only be guessed at. Ours were the first eyes ever to see a *Proconsul* face. . . ."

Here, at long last, was the skull Louis had promised Le Gros. Heady with excitement, he and Mary decided to celebrate their find that night at camp. "We were exhilarated and also utterly content with each other," Mary recalled, "and we thought that quite the best celebration would be to have another baby. . . . [So] that night we cast aside care. . . ." Their third son, Philip, was born nine months later, on June 21, 1949.

Louis had not rented a boat on this expedition, and they could not send out word of their discovery until a launch from the Fisheries Department called at the island. Waiting for the boat was a period of "intense frustration" for Louis, but he consoled himself by hoping that the *Proconsul* skeleton would turn up as well. Heselon Mukiri took over the sieving at the site, searching for each minute bit, while Mary painstakingly glued the fragments —many of them, Louis noted, "the size of a match head"—back together at camp. "Once I dropped a tiny piece into the dust on the tent floor," she wrote, "only a crumb of bone, but a vital link in joining two larger pieces. It took ages to find, but we got it."

By October 4, Mary had much of the skull together, and Louis made a sketch of it in his journal, then scribbled it out. But next to the drawing, he wrote, "The ape skull now begins to look very nice with jaw complete & face & frontal coming into position. The frontal [to] glabella region is remarkably human in contour much more so than even in Broom's ape men." In other words, Louis thought the brow region of *Proconsul* more humanlike than that of the australopithecines from South Africa. The frontal bone was "almost infantile as I fully expected it would be, & as I have argued so often." With this skull, Louis thought he had proof that humankind's earliest ancestors were present in the Miocene, nearly twenty million years ago. He had also skunked Wendell Phillips and the American Expedition.[11]

When the Fisheries boat finally docked at Rusinga on October 8, Mary had finished her reconstruction. Louis left Heselon in charge of the excavation, and sailed with Mary to Kisumu. That afternoon he sent Le Gros both a cable and a letter. "We [have] got the best primate find of our lifetime," he wrote. It was so "remarkable," so "outstanding," that the only way to deliver it to Le Gros was by air. Louis surely would have loved to have made this trip himself, to stand in the limelight showing the world the skull he had predicted he would find. But Mary had found it, and Mary, he decided, would carry the skull to England. "I feel that Mary, as the finder, should be the one to bring it to you," he told Le Gros.

---

[11] Phillips had better luck in South Africa, where he teamed up with Robert Broom to investigate the caves at Swartkrans. They recovered additional skulls of *Australopithecus robustus,* a somewhat larger and heavier-jawed form of *A. africanus.* Phillips was back in America in 1949 organizing an American Foundation for the Study of Man, and promoting a new expedition—this one to search for the hidden city of the Queen of Sheba in Yemen. Again he raised great sums of money, but the expedition came to an abrupt end when Phillips ran afoul of some Yemeni soldiers. He fled to Oman, where according to a 1966 *Time* magazine story about him appropriately entitled "The Great Iam," he became a "good friend of Sultan Said bin Taimur." With this friendship, Phillips apparently came into his own. The sultan showered him with oil wells. Almost twenty years after his grand entrance at the Pan-African Congress, Phillips was telling reporters, "I am the largest private oil concessionaire in the world." His interest in finding the "missing link" was a thing of the past.

Chapter 10

A LIFE IN THE SEDIMENTS

"Your beast is grand," proclaimed Robert Broom to Louis, a few weeks after Mary's discovery of the *Proconsul africanus* skull.[1] "As a collector you are one of the world's wonders." Louis had sent Broom and other colleagues a photograph and letter describing the find, and congratulatory letters and cables poured in. "Well done, indeed!" enthused Henry Field of Harvard University's Peabody Museum, while Sir Arthur Keith said, "My dear Leakey, I do congratulate you on a discovery of first importance to all students of Primate Evolution. You are opening up a New World of life." Le Gros, too, was beaming. He thought the skull a "marvelous find," and only one week after receiving Louis's telegram advised Louis that "it is already exciting the greatest interest over here."

Louis was extremely pleased to hear this, for he intended to capitalize on the skull. "I do very much feel that we have got to make the fullest use of publicity in connection with this skull, for the raising of funds for further work in the next few years," he wrote to Le Gros, who concurred. Together Le Gros and Louis planned a minor media blitz: press announcements, a report to *Nature,* a press conference at Heathrow Airport upon Mary's arrival

[1] In 1993 the anatomist Alan Walker and his colleagues reclassified the *Proconsul* fossils from western Kenya, and in the process identified a new species, *P. heseloni.* They included Mary's *Proconsul* skull in this group, so that it is now known as *Proconsul heseloni,* after Louis and Mary's longtime field assistant, Heselon Mukiri. Louis and Le Gros Clark had actually suggested this name in 1948 for another ape specimen from Rusinga, but they later decided that all of the fossils belonged to *P. africanus.* It was their correspondence, in which they first used the name *P. heseloni,* that led Walker to name the new species for Heselon. The term *P. africanus* is now used to describe certain Miocene ape fossils from Koru and Songhor.

with the skull, and a larger one at the Colonial Office in London after Le Gros had studied it. Louis's efforts at public relations had already begun to pay off. Shortly after returning from Rusinga, he had approached BOAC for free airfare for Mary, and the airline's officials had happily complied.

The reconstructed *Proconsul* skull was fragile as eggshell and small enough to fit in the palm of Mary's hand. She carefully placed it in a biscuit tin lined with cotton, and early on the morning of October 29, 1948, boarded the flying boat at Lake Naivasha. It was still a tiring flight to London, with many stops en route, and throughout Mary held the tin safely on her lap. In the meantime, Louis dispatched his press releases. When Mary's plane touched down at Heathrow, *Proconsul* was already front-page news. She stepped from the plane, box in hand, into a mob of newspaper, radio, and TV reporters. Photographers called to her to please climb the airplane's stairs and come down one more time. After several such ascents and descents, she was escorted to a VIP lounge to meet the press formally. As the reporters gathered round, Mary, dressed in a stylish tweed suit, her head bent shyly, knelt on the carpet and held the skull up for display. The crush of the media was daunting, "something to which I was completely unaccustomed," she later recalled; but her nerves eased as she realized that "everyone's attention [was] fixed on the skull." She answered a few questions, promising more details after Le Gros completed his study, and then was whisked away under police escort to Paddington Station and, after a train ride, to the quiet of Oxford. Le Gros was waiting for her there, and with great relief she handed over her "precious burden," as Le Gros always called the fragile skull. Mary then cabled Louis: "BOTH ARRIVED SAFELY OVERWHELMING RECEPTION AIRPORT."

At first, the press was somewhat taken aback that Mary had made the discovery. Until now she had remained in Louis's shadow. It was he, after all, who had so eagerly courted the press. Given the scarcity of professional women at the time, her sudden emergence as a first-class scientist was also unexpected. "She is an archaeologist in her own right," one paper reported gratuitously, "and this is not her first important find." After the flurry of attention, Mary retired to Le Gros's lab for a few days. While he puzzled over the fossil's relationship to apes and humans, she tried to join a few skull fragments to it that Louis had found near the *Proconsul* site in 1947. These did not fit, but more than thirty years later, another paleontologist working in the British Museum of Natural History came across bits of bone that had also been collected in 1947 and did belong to this *Proconsul* skull.

The media was eager for further news about the fossil. Louis's first press releases had tantalizingly suggested that *Proconsul* might show "resemblances to the human condition," and this had touched a common nerve. All of Britain, it seemed, was speculating on what the skull might mean.

For two weeks, Le Gros sat sequestered in his lab, analyzing the fossil in detail, measuring the angles of the jaws, the cusps of the teeth, the slope of the forehead and length of the face. He then compared *Proconsul* to fragments of other Miocene apes as well as the skulls of the South African australopithecines, modern monkeys, apes, and humans. He and Louis were also corresponding about their interpretation of the fossil. Finally, Le Gros issued a press release of his own. "Professor Le Gros Clark . . . has found this discovery to be so important," reported the *London Daily Herald,* "that . . . he is calling a Press conference . . . to explain its significance to the world. When this skull . . . was first found, it was thought it might alter all our views on human evolution. Is man descended from the ape, or did they grow up side by side? . . . Have we at last found the 'missing link'? We wait for the Professor to tell us."

Louis, of course, also wanted to tell the press what the skull meant, and he did his best to influence Le Gros via the mail. Mary played no part in this, explaining in her 1984 autobiography that "the Miocene [never] was one of my major interests." She was comfortable with the division of labor she and Louis had evolved: he studied the bones, and she the stones.

In his first letter to Le Gros, written soon after the fossil's discovery, Louis had described at length what he had only hinted at in his press release: the humanlike characteristics of the skull. To his eye, the shape of the frontal bone (forehead), eye sockets, and jaw joints appeared "more human in form . . . than apelike." He was also struck by the absence of thick brow ridges and a simian shelf, found only in modern apes. He did not believe that any ancestor of "true man"—those on the direct line to *Homo sapiens*—had ever possessed either feature. *Proconsul* also seemed proof of another pet theory, "that it was about in Miocene times—say 25,000,000 years ago—that the stock which gave rise to humanity broke away from the stock which gave rise to the great apes." Here, he thought, was the very creature that had taken that first tenuous step.

Initially, Le Gros was impressed by *Proconsul*'s human similarities, too. The day after seeing the skull for the first time, he wrote enthusiastically to Louis of the "human-like contour of the supra-orbital [brow ridge] region . . . and refined forehead. . . . Clearly an astonishingly *generalised* ape, and a good deal more easy to conceive as playing a part in human evolution than anything like the modern apes." But as he studied the skull more closely, Le Gros backed away from this position. What he saw now was a creature linking monkeys and apes. *Proconsul*'s pointed muzzle and narrow nose opening seemed features left over from its monkey ancestry, while its teeth and jaws pointed to the apes. It seemed to have little to do with human ancestry. But these views of Le Gros's did not reach Louis until after the press conference.

Le Gros had slated the event for November 19, and nearly every paper in England as well as the BBC turned up. In a room packed with reporters and cameramen, jostling for space and shouting questions, Le Gros and Mary did their best to explain *Proconsul*'s importance. Undoubtedly, the reporters hoped the professor would utter the magic words "missing link." Instead, as he reported to Louis, "I [took] the line that [the skull's] main interest is the light which it throws on the origin of the apes—that its importance in relation to human origins is *indirect* rather than *direct*—as I think must be the case since, of course, it is too early to be otherwise." Some of the reporters apparently expected more, and one at least was "nasty" to Mary, said Le Gros. Overall the conference was not the dignified, decorous meeting he had envisioned. "[It was] a harrowing experience," he wrote to Louis. "[W]e did not feel ourselves again until Sonia [Cole, who would later write Louis's biography] had given us each a double whiskey at her club! However —it has certainly led to publicity... & you & Mary have become quite famous!"

Even demoted, *Proconsul* made news—papers across England carried reports of the press conference—and, when placed on exhibit at the British Museum of Natural History, it drew large crowds.

Louis was delighted with the publicity, though he privately protested to Le Gros that "*Proconsul*... throws a lot of light, direct light, on what was happening in the ape-human stock which gave rise to the *Hominidi* [*sic;* the human family] on the one hand and the great apes on the other...." Le Gros, however, continued to insist on a more restrained interpretation, and in their joint reports, at least, Louis followed suit. Their 1951 monograph on the Miocene hominoids (a grouping which includes apes and humans) presented *Proconsul africanus* as a creature that "approximate[s] rather closely to" the common ancestral stock of humans and apes; they did not suggest that it led to humans alone. Today paleoanthropologists consider *Proconsul* a key fossil in understanding the split between monkeys and apes.

Mary did not wait to bask in the glory of her newfound fame and left England the day after the conference. She was home with her family a few days later, and planning a new expedition with Louis to Rusinga Island for the spring of 1949. The *Proconsul* skull remained behind in London on long-term loan to the British Museum of Natural History.[2]

As Louis had hoped, the publicity surrounding *Proconsul* led to a steadier flow of research funds. He tucked a bundle of press clippings into his 1949 grant application to the Kenyan government, which responded with £1,500. "It would seem that this sort of scientific inquiry gathers impetus," noted the chief secretary, "and work can be found as long as money is available." Over the next ten years, the government would continue to provide this annual stipend for Louis's Miocene research. A second generous check

[2] Although the chief secretary of the Kenya government instructed the British Museum of Natural History that the fossil was on temporary loan, the museum regarded it as part of its collection. Throughout the 1970s, Richard Leakey tried to have the fossil returned to Kenya but did not succeed until Mary's secretary discovered a copy of the chief secretary's letter in the files of Kenya's National Archives. The fossil was finally brought home to Kenya in 1982. It is now housed in the National Museums of Kenya.

came, too, from Charles Boise, the wealthy London businessman who had partially underwritten the Leakeys' 1948 Rusinga expedition. Mary had called on him during her stay in England to give him a private viewing of the skull. Immensely pleased with the results of his first grant, he promised Mary additional support. When Louis suggested that funds for a motorboat would be most useful, Boise immediately cabled him £1,600.

After a brief search, Louis found a secondhand forty-two-foot twin-engined cabin cruiser on the coast. He had it shipped by rail to Lake Victoria, where he christened it the *Miocene Lady* and hired a skipper, Hassan Salimu. The Leakeys were now free to travel on the lake as they pleased. Louis looked forward to exploring islands he had not yet visited, as well as continuing his search for Miocene primate fossils on Rusinga Island. The yearly grants from the Kenyan government were also liberating, and throughout the early 1950s the Leakeys made regular, monthlong excursions to the lake. "We always took the children with us when we went," Mary recalled in her autobiography, ". . . so the whole thing was quite a family affair." For their three young sons, the journeys were what all children crave: "real adventures."

Weeks before setting out, lists were made, food and equipment ordered, and messages dispatched to Hassan to ready the *Miocene Lady.* The boys' excitement grew as piles of potatoes and onions, hefty bags of posho (maize meal), bottles of Bear's Honeydew (a soft drink), and flats of cigarettes were stacked in and around their little house. They joined in the work where they could, or stood by and watched as tents and sleeping cots were inspected, the kerosene lamps refurbished, and mosquito netting mended. A special box would arrive, too, from England, bearing a season's supply of Louis's own "Leakey patent" picks, lightweight picks he had designed for excavating. On the eve of departure, everything was sorted and packed into rough wooden boxes, and loaded into the Commer truck and Louis's Dodge box-body car (Louis had replaced his old Chevrolet). Kikuyu and Swahili commands filled the air, then died away as night fell. In their beds, the boys slept fitfully, waiting for their father's early morning call, the signal that the adventure had at last begun.

Mary laid out bedding for the children on top of the crates in the back of the Dodge. Here they could lie and watch the passing landscape through the car's wire cage. Just before dawn, while the African sky was still a deep luminous blue, Louis pulled away from Nairobi and started up the steep incline of the Kikuyu escarpment, heading northwest for Naivasha and the Rift Valley. "In those days there were few cars on the road," Richard wrote thirty years later in his autobiography, *One Life,* "and it was exhilarating to drive slowly through the countryside, breathing the early morning fragrances: the cedar-wood smoke from the homesteads, the smell of cattle, sheep and goats as we passed by villages, and the dank but wonderful smell of a forest, dripping with mist in the light of a Kenya dawn."

By nightfall, they arrived at Kisumu on the edge of Lake Victoria, four hundred miles from Nairobi. Louis preferred to travel on the lake at night, when the winds and seas were light. He and Mary roused the boys and with the help of Hassan and their Kikuyu staff stowed all the gear on board the *Miocene Lady.* While Louis and his workers moved to and fro, Richard and his brothers lay in their bunks, watching the bustle, listening to the jumble of languages, and reveling in this "atmosphere of urgency and expectation." When the last box was put away, Louis called to Hassan, who nudged the boat into gear. Then they were out on the lake, with only the steady rhythm of the boat's engine breaking the quiet. Some eight hours later, they would drop anchor at Rusinga.

On these expeditions, Louis and Mary's work came first, and the children were expected to entertain themselves. The family was up at dawn, and after a quick breakfast of fresh pan-fried fish, they marched off to whatever fossil site Louis and Mary were excavating. Sometimes these were near cool, shady fig trees, or close to a village where the boys could find other children to play with. But other times—and all too often in young Richard's opinion—the fossils were in hot, open terrain far from the cooling breezes of the lake. He grew to dislike especially the site of Mary's *Proconsul* skull, which was marked by one scrawny, nearly leafless tree. He was four at the time of his mother's discovery and longed to be near her while she worked. But there was no shade, the sun was intense, the flies pesky, and he ended up crouching sullenly against the wall of a gully. The experience was so unpleasant that he decided then that he would have "nothing to do with a life in the sediments."[3]

Richard's feeling was reinforced two years later, when, at the age of six, he found his first fossil bone. This time his parents were excavating at Kanjera, and he had been "firmly instructed" not to get in their way. He could also see that they were deeply involved in their work, and not likely to stop soon. Once again there were flies everywhere and no shade, and Richard began to whine, "I am hot, I am thirsty, I am bored . . ." Finally, in exasperation, his father looked up and said, "Go and find your own bone!" "I was partly pleased," Richard recalled in his autobiography, ". . . [s]o I moved off to look for my bone—at least I had something to do!" He had his own set of dental picks and a small brush, and after spying a small bone a short distance from his parents, he settled down to dig it out. "My bone quickly showed signs of being large . . . and, more important, the shiny enamel surfaces of some teeth quickly appeared after my initial scraping and brushing." He was soon entranced and no longer complaining. But Richard was too quiet for Louis, who came to take a look, and "was instantly

---

[3] In February 1983, Richard returned to Rusinga for a quick walking tour of its fossil sites. A staff member of the National Museums of Kenya was surveying them and Richard had stopped by to check on the work. Although he had not been to the island in twenty-three years, he had no trouble finding the *Proconsul* site; the scrawny little tree still grows at the top of the hill, and Richard pointed it out at once. "I remember that awful tree quite clearly, and I remember quite clearly why I never wanted to be a paleoanthropologist," he said, laughing. "You're always hot, sticky, wishing for shade and swatting at lake flies!"

alert." Richard had discovered the first complete jaw of an extinct species of giant pig. Although his parents heaped praise on his discovery, they also took over his excavation, leaving Richard "furious and deeply upset"—and wondering thirty-three years later if the incident had "contributed to my original firm decision to avoid at all costs a profession that involved excavation and the search for fossils!"

Fortunately from the children's point of view, Louis and Mary did not spend the entire day excavating on these expeditions but usually returned to camp for lunch and a welcome swim. Louis always fired his shotgun into the water to frighten away any nearby crocodiles, then stood guard while the boys splashed in the lake. There was time, too, for dangling fishing lines from the *Miocene Lady,* and in the cool of the afternoon there were long nature walks with their father. Louis took these walks to explore for new fossil sites, but along the way would pause to point out birds and butterflies, to identify plants that were good for making twine, a tree that could be used for a bow and another that would provide arrows. Or he would stop and with a few deft moves bend branches into a snare to trap a francolin (an African partridge) or guinea fowl. He showed them how to chip stones into tools, make fire by rubbing sticks, and creep up on animals without being detected. "Father didn't teach from a set plan," Richard explained, "but showed us things as we were growing up so that we could track an antelope or trap a bird without any modern technology." The lessons turned the boys into "ardent naturalists," and living animals rather than fossils became their passion. Still, surrounded as they were by fossils, it was perhaps inevitable that they would learn to identify the bones of the extinct animals as well. One friend of the Leakeys' has a vivid memory of five-year-old Philip looking at a tray of fossils and saying confidently, "This is a *Deinotherium,* that is a snake vertebra . . ."

Louis continued to hope that skulls and limb bones of the other Miocene apes—*Proconsul major, Proconsul nyanzae,* and *Limnopithecus*—would turn up on Rusinga or one of the other islands. Covering 26,828 square miles, Lake Victoria is the second largest freshwater lake in the world and is vast enough to create its own weather system. Its islands are regularly swept by heavy rains and strong winds, which bring a new sampling of fossils to the earth's surface each season. On every expedition, Louis spent time at old sites and searched out new ones. Between 1947 and 1954, when he paused long enough to write an interim report for *Nature,* his team collected more than fifteen thousand fossils, representing mammals, reptiles, fish, insects, mollusks, and plants. He and Le Gros recruited several specialists to describe these fossils. Believing that the mammals were of particular importance, they arranged for these reports to appear in a new series published by the British Museum of Natural History and entitled "Fossil Mammals of Africa." The first volume in this series was Louis's and Le Gros's joint monograph, "The Miocene Hominoidea of East Africa," the publication that first described *Proconsul africanus* in detail. Over the years, twenty-one more volumes would appear, an invaluable contribution to paleontology.[4]

Among the mammals were bones of 450 individual apes—evidence that there must have been a large ape population in East Africa in Miocene times. Louis and Mary never did find another complete ape's skull, but they did uncover the jaws and limb bones of *Limnopithecus,* the gibbonlike ape, as well as fragmentary skulls and limb bones of different species of *Proconsul.* Louis's Miocene research and his collaboration with Le Gros had by now greatly enhanced his scientific reputation, and Le Gros kindly gave it another boost by arranging for Oxford University to award Louis an honorary doctorate of science in July 1950 for his contributions to physical anthropology.[5]

One of the most spectacular discoveries on Rusinga Island was made by Thomas Whitworth, a geologist from the University of Durham, whom Louis had engaged in 1951 to map the island. (After his disaster at Kanam, Louis always sought a geologist's help on his expeditions.) Louis and Mary had gone to Tanganyika to study the rock paintings at Kondoa Irangi and nearby Kisese for part of that season and had entrusted the island's deposits to Whitworth. While surveying the island's peninsula, Whitworth discovered a six-foot-wide pothole packed with fossils—pigs, antelopes, rodents, lizards, and most exciting, a partial skull, jaws, arm, hand, and foot of what is now called *P. heseloni.*[6] These bones, together with Mary's *Proconsul* skull, gave scientists the most complete skeleton of any fossil ape then known.

There were other exciting, unexpected finds as well. Once while exploring Mfangano Island in 1950, Louis and Mary paused for a cigarette break in a patch of fossilized wood. While idly talking, they suddenly noticed that all around them were "what seemed to be small pebbles, many of identical shape and size. . . . There could be no doubt that they were seeds," Mary recalled. This chance discovery led to a vast collection of eighty different kinds of nuts, fruits, and seeds, and enabled paleobotanists to re-create *Proconsul*'s forest habitat.

---

[4] The British Museum of Natural History discontinued publication of this series in 1967 due to lack of funds.

[5] In some ways, Le Gros Clark acted almost as a one-man public relations firm for Louis. For example, at one point Le Gros was annoyed that *Nature* had not written a word about their "Fossil Mammals of Africa" series. "I do think," he told Louis, "that some publicity should be given to your work on the Miocene beds in Kenya." To get the journal to take notice, he suggested that Louis write the interim report about the results of the British-Kenya Miocene expeditions; Louis would then send it to Le Gros, and Le Gros would forward it to the editor. *Nature* published this report in February 1955.

[6] New excavations by Alan Walker and staff from the National Museums of Kenya in 1984 showed the "pothole" to be the fossilized remains of a hollow tree. In Miocene times, many animals—bats, pythons, and monitor lizards—made their home in the hollow. A small carnivore occasionally resided there as well, and brought the carcasses of various species, including the *Proconsul* Whitworth found, to its lair. Walker's excavation turned up additional bones of this individual as well as a bounty of other animal fossils. That same season, his team also discovered ten fragmentary *P. heseloni* skeletons—with partial skulls, but hands, feet, and limb bones intact—in a corn patch on the island.

Another time on the same island, Louis picked up a fossilized beetle, while another team member found a fossilized caterpillar. These were beautifully preserved, looking as if a sculptor had fashioned them from stone, and everyone immediately began to look for more. In a few days, Louis, Mary, and the children had amassed a collection of fossilized grasshoppers, worms, snails, cockroaches, spiders, and ants. Most surprising, as Louis wrote to a colleague, all of the insects were "wholly uncompressed and undeformed . . . [and] looking as though they were alive." Over the years, they found similarly preserved insects on Rusinga Island, and once Mary discovered a fossilized tree ants' nest, complete with eggs, grubs, soldiers, and workers. "It was unbelievable to work with insects in the round, especially the ants complete with eyes and every leg perfect," said a friend who joined the Leakeys on this expedition.

Prior to these discoveries, fossilized insects had only been found in amber, shales, and sedimentary rocks. Such soft-bodied creatures usually rot before they can be preserved. Scientists do not yet fully understand what processes were at work on Rusinga and Mfangano islands, but apparently the insects were so quickly covered by a thick mud that decay could not set in.

A similar process may explain two other discoveries—one a bird with part of its flesh and muscles preserved, the other the head of a large lizard with its scales, eyes, and ears intact and its tongue hanging out of its mouth—that sent Louis's hopes soaring. "[O]ne day we might even find a head or a foot of a primate similarly preserved," he wrote to Le Gros, "which would indeed be exciting." Though Louis did not live to see it, another scientist working on Rusinga in 1984 did discover fossilized nerve endings in the ear cavity of a fragment of a *Proconsul* skull.[7]

Not all of Louis's expeditions panned out. In February 1951, he traveled to the southern end of Lake Rudolf in an attempt to explore the region around Losodok, where Wendell Phillips's team had turned up two Miocene ape fossils. But he found the deposits to be "very limited" and the wear and tear on his vehicles expensive. After only a week in the field, Louis cabled MacInnes at the museum: "CLOSING DOWN RESULTS DONT JUSTIFY COSTS." Ironically, thirty-five years later, Richard and Meave Leakey's fossil-hunting team would turn up a wealth of Miocene ape fossils only ten miles from where Richard's father had explored.

Louis and Mary's journey to study the rock paintings of Kondoa Irangi and Kisese in central Tanganyika was far more productive. They had promised

---

[7] Louis's hopes for a primate fossil with flesh and muscles intact were fueled by a tantalizing letter he had received in 1943 from a Major G. Grundy, who wrote: "About 1932, I was prospecting and the locals on Mfanganu [as the island's name was then spelled] brought us some fossils. The one in question was about the size of a hand, was obviously part of a skull, and was remarkable for the fact that parts of the brain had been preserved. On our return to Kisumu . . . a number of people saw the skull (portion) and we had intended to hand it over to Archdeacon Owen. Then someone unknown lifted it and that's all there is to it."

each other on their first expedition in 1935 when they viewed the paintings together that they would one day return to study them in detail. Van Riet Lowe, the South African archeologist, had encouraged this ambition. He and others had already recorded similar art work in North and South Africa, and he was anxious to "link the art" of the two regions. "The important 'link' needed to fill the gap undoubtedly lies in Tanganyika, and . . . I wish you well!" he wrote after Louis told him of their plans. Early in 1951 Louis secured a £2,000 grant from the Wenner-Gren Foundation for Anthropological Research in New York, and in July he and Mary were on their way with their sons and Dalmatians.[8]

"Those three months at Kondoa Irangi . . . will always count as one of the highlights of my life and work in East Africa," Mary wrote in her autobiography. She was enchanted by the countryside, a hilly, rocky landscape forested with slender Brachystegia trees, whose spring-green foliage reminded Mary of English oaks. The paintings, too, intrigued her, as did the task of deciphering them. "It was the combination—the beauty of the country, the beauty of the paintings, the fascination of disentangling the art, the feeling that we were achieving something by putting these paintings on record—all of that, and having the family there. It was great fun," she recalled. Perhaps most important to Mary, this was a project that she and Louis equally appreciated. "During our lives together," she wrote thirty years later, "we shared many archaeological projects, but probably never one that was so mutually satisfying."

In addition to their food and camp equipment, the Leakeys brought with them ladders, scaffolding, reams of cellophane paper, paints, and pastels, for Mary planned to make accurate copies of as many of the paintings as possible. They then hired sixty porters to transport all their gear to a campsite close to the blocky granite outcroppings where the prehistoric artists had once gathered to paint. This region was home to agriculturalists of the Warangi tribe. They were naturally curious about the Leakeys' work and advised Louis that he should sacrifice a goat "to propitiate the spirits of the painted site, which [were] regarded as very powerful."

The goat was duly sacrificed the next day, and with the spirits appeased, Mary's work could begin in earnest—for on this expedition, she was the leader and Louis the "bottle washer," as he called himself. Louis set up the ladders and scaffolding, mixed her paints, and attached the cellophane sheets to the rock walls. Then Mary worked contentedly, tracing and untangling individual human and animal forms from what "at first sight seemed a completely bewildering mass" of figures. Around her drifted the sounds of the Warangi people, their herds of cattle, sheep, and goats browsing under the trees, and the herd boys piping on their flutes. As she traced, distinct pictures emerged: there were zebras, lions, ostriches, a kudu antelope with an exaggerated eleven twists to its horns, a beehive with bees; there were

[8] The Wenner-Gren Foundation was then called the Viking Fund.

undulating snakes and prancing giraffes, a pair of rhinos engaged in courtship, and images of the Late Stone Age people themselves. These particularly appealed to Mary because they revealed "details like clothing, hair styles, and the fragile objects that hardly ever survive for the archaeologist—musical instruments, bows and arrows, and body ornaments. . . . No amounts of stone and bone could yield the kinds of information that the paintings gave so freely," Mary wrote.

While she worked, Louis and the boys searched for other paintings, a hunt that Jonathan especially enjoyed, but Philip found wearying. He was only two, and would tag along calling out, "Poor me, poor me!" Altogether, the Leakeys recorded 186 rock art sites, and Mary copied forty-three of their paintings. They also dug trenches at some of the sites, turning up odd bits of Late Stone Age material, but found nothing that would help them date the paintings themselves. Based on its appearance, Mary believed the artwork to be only a few thousand years old.[9] Many of the original paintings have been defaced since Mary made her copies, and it seemed for a while that Mary's copies, too, might be forgotten and destroyed. The Leakeys' original grant did not include funds for publishing, so Mary stored her paintings in a tin trunk for thirty years. They were rediscovered by Richard's wife, Meave, in 1980 and published in a handsome volume, *Africa's Vanishing Art: The Rock Paintings of Tanzania,* in 1983.

Life for the family while on these expeditions was not without its hazards. Once after discovering a rock shelter with especially "magnificent paintings," Louis started back to camp with Philip perched on his shoulders. Seeing what he thought was a fallen log lying in the tall grass, Louis set his foot firmly on top of it, only to have it rear up in front of him: it was the biggest python he had ever seen. Louis jumped back, and the snake slithered into the forest. On Rusinga Island, Richard, who could not swim, once fell overboard and was rescued by an alert Hassan. One time, the entire family contracted malaria, and on another occasion after walking through a swamp on Maboko Island, everyone—except Philip—came down with bilharzia, a particularly debilitating and long-lasting parasitic disease that affects the intestines, liver, and spleen. Philip was again riding on his father's shoulders, and so escaped the parasite-infested waters. In 1951 Donald MacInnes, the Leakeys' paleontologist, nearly died from an appendicitis attack on Rusinga, and then at the end of the field season in February 1952, Mary's appendix also burst. Just as the *Miocene Lady* docked at Kisumu, Mary collapsed and had to be rushed to the hospital. "I know it was 1952," Mary recalled in her autobiography, "because Louis was away acting as interpreter at the trial of Jomo Kenyatta."

---

[9] In 1956 another archeologist, Ray Inskeep, made further excavations at a site called Kisese II and found coloring materials in deposits dating to 29,000 years ago. Mary thinks that it is doubtful, though, that any of the surviving paintings are that old.

Chapter 11

LOUIS AND KENYATTA

In 1964 Jomo Kenyatta would be elected the first president of an independent Kenya, and hailed by blacks and whites alike as a brilliant statesman and a hero of African nationalism. But in October 1952 he had been charged by the British with being the leader of Mau Mau, a secret Kikuyu society allegedly bent on the violent destruction of Kenya's colonial government. The government hoped that Louis, with his extensive knowledge of Kikuyu ways, would help them convict Kenyatta and put an end to Mau Mau.

Kenya had entered the 1950s a prosperous but troubled colony. Its agricultural industry was thriving, and its European settlers busy expanding their acreage of tea, coffee, and dairy farms in the White Highlands, north and west of Nairobi. The settlers had used the war years to consolidate their political and economic position, enacting laws that prevented Africans and Asians (most of them from India) from owning property in the highlands, the country's most fertile region. They had also managed to stave off a British Colonial Office suggestion for a multiracial government composed of equal numbers of Africans, Asians, and Europeans. Instead, the British settlers remained firmly in control of the Legislative Council, tolerating the four appointed Africans and six elected Asians but ensuring that the colony remained segregated and the Africans disenfranchised. It was not a policy that could endure. Beneath the prosperity of the colony's 32,000 whites lay the poverty and resentment of five million native people.

A host of other laws, many similar to the apartheid laws then being established in South Africa, compounded the Africans' bitterness. They were forced to carry identification papers (the hated *kipande*), forbidden to grow coffee and other export crops (they were not permitted to compete with the white farmers), and required to pay a high hut tax, which compelled them to work for the settlers. Other rules enforced a strict color bar, dictating where an African might travel, make his home, eat, and sleep, and what wages he would receive—always less than those of a European or Asian, regardless of qualifications.

Shortly after World War II, the native peoples' frustration peaked and Kenya began to feel the first stirrings of African nationalism. With it came a time of social and political upheaval, and no one in the colony, including the Leakeys, escaped the fear and uncertainty of these turbulent years.

The Kikuyu people were at the heart of the turmoil. Of all the tribes in Kenya, they had lost the most land to the settlers. More than a million Kikuyu were confined to three overcrowded reserves, while another quarter million were either homeless or living as temporary squatters on white farms. But few had given up hope of regaining the land they considered rightfully theirs. "When someone steals your ox," observed Senior Chief Koinange, "it is killed and roasted and eaten. One can forget. When someone steals your land, especially if nearby, one can never forget. It is always there, its trees which were dear friends, its little streams. It is a bitter presence."

For Louis the plight of the Kikuyu presented a dilemma. He had grown up among them, spoke their language, was an initiated first-grade elder, and considered them his people. "I am in so many ways a Kikuyu myself," he often said. Like his father, who had helped Senior Chief Koinange and others try to regain their property, Louis sympathized with the Kikuyus' land grievances. As early as 1929, he had served as their interpreter on a government committee inquiring into the Kikuyu system of land tenure, and had done his best to persuade the colonial government to grant Kikuyu property owners the title deeds to their land.[1] In his 1936 book, *Kenya: Contrasts and Problems,* Louis had further aligned himself with the tribe by asserting that Kenya would never "really be a 'white man's country,'" and advising the Europeans "to work towards co-operation [with the Africans] instead of domination." These sentiments had not endeared him to the settlers. They considered him pronative, which was tantamount to being a traitor in their eyes.

In spite of his Kikuyu sympathies, Louis had allegiances to both the British and colonial governments. He had never supported the tribe's politically radical members, believing that the problems were best resolved by peaceful means. Nor had he ever cut his ties to the Special Branch, the colony's equivalent of the Federal Bureau of Investigation. After World War II, he had continued to assist with handwriting cases, and occasionally investigated

---

[1] Title deeds had been issued to the white settlers, but the colonial government had postponed giving them to Africans on the grounds that they did not have enough surveyors available for the job. By 1950, the government had still not granted the Africans title deeds, another factor contributing to the rise of Mau Mau.

robberies and murders—partly to supplement his income, and partly because he loved detective work.[2] But he also acted as an adviser ("a sort of unofficial *éminence grise,*" according to Cambridge historian David Throup) on subversive activities among the Kikuyu.

When Louis heard in 1949 about secret oathing ceremonies among the Kikuyu—oaths designed to unite the tribe—he immediately reported them to the Special Branch and sought a meeting with the colony's governor, Sir Philip Mitchell. Louis knew that the ceremonies broke too many Kikuyu taboos: they were held at night, the oaths were often given forcibly, and to women and children, as well as to men. But Mitchell brushed aside Louis's concerns that the oaths held a "grim significance," and that there were urgent problems on the reserves. "[O]f course, I was not popular," Louis wrote twenty years later, attempting to explain his inability to influence Mitchell. "Had it been possible to make the government open its eyes to the realities of the situation, I believe that the whole miserable episode of what is frequently spoken of as 'the Mau Mau rebellion' need never have taken place."[3] As it was, within two years of Louis's warning, the oathing ceremonies had spread from one to all three Kikuyu Reserves, and with them came murder and terror.

Mau Mau was a secret society that began among Nairobi's radical Kikuyu youth, but it became almost impossible for any Kikuyu—man, woman, or child—not to join.[4] The basic oath united the tribe, but others required participants to kill Europeans and their African collaborators. Those who refused the oaths were murdered, as was anyone suspected of betraying the movement. Kikuyu policemen and staunch Christians were found dismembered; bodies turned up in rivers, their hands tied with wire; and chiefs and headmen loyal to the colony were assassinated. Settlers' farms also came under attack. Their homes and granaries were torched, livestock maimed and poisoned, and sometimes the owners themselves murdered, often slashed with *pangas* (African machetes) by their servants. Wrote one settler,

---

[2] In *By the Evidence,* the second volume of his autobiography, Louis devotes more than thirty pages to the cases he helped investigate. Probably the most famous was the 1941 murder of the handsome playboy Lord Erroll. Erroll's death created a great scandal in Nairobi, and was later the subject of James Fox's book *White Mischief.* Louis happened to be in Superintendent Poppy's office when the call came saying that Erroll's body was in the morgue. He accompanied Poppy to view the body, and subsequently played a minor role in the investigation—though his own retelling of the case makes him sound like a central figure. Louis very likely enjoyed such detective work for many of the same reasons he enjoyed searching for early human fossils: both tasks required careful observation, a love of ferreting out clues, deductive reasoning, and imagination.

[3] Louis was not the only one to try to warn Mitchell, but Mitchell dismissed everyone else's apprehensions as well. He considered Mau Mau essentially "just another . . . fanatical religious movement." When he retired from the governorship in June 1952, he still had not mentioned Mau Mau to the Colonial Office and assured his successor, Sir Evelyn Baring, that all was well in the colony.

[4] To this day, no one has been able to determine what "Mau Mau" means. The secret society that started in Nairobi was actually nameless, a tactic devised to keep it hidden from the government. The first official report of the organization was made by a district commissioner in Nakuru in 1948, who recorded its name as one word, "maumau." In Rosberg and Nottingham's 1966 book *The Myth of "Mau Mau": Nationalism in Kenya,* which analyzed the development of African politics in Kenya, the authors suggest that this may have "been a corruption of the Kikuyu word for an oath—*muma.*"

"Men and women live night and day with arms ready to grasp, watching second by second for treacherous attack. The strain is terrific and cannot be maintained."

The terrorist acts affected Louis, too, as he tried to straddle the fence between African and European—and found himself not fully trusted by either side. Mau Mau brought a "constant mental worry," he wrote, "of never knowing whether you and your family will be the target for a sudden raid, of never knowing whether when you come back . . . home, you may not find your family dead, [and] your house in flames."

His activities on behalf of the Special Branch had increased as Mau Mau spread. Besides keeping in touch with his Kikuyu informants, he interrogated Mau Mau suspects, translated their oaths and special hymns, broadcast propaganda to the tribe, and initiated a counter-oathing ceremony. All of this made him a prime target for the rebels.[5] To them, Louis was "the supreme example of a prominent European working against Mau Mau in spite of his own Kikuyu connections," Mary observed. The rebels did not keep their intentions secret, but sent him death threats in the mail and let it be known that they had placed a £500 reward on his head.

The Leakeys did what they could to protect themselves. They started carrying handguns, a fashion that came to be known as "European National Dress." Mary kept a .22 pistol on her belt during the day and tucked it under her pillow at night, while Louis wore a revolver at all times. The government also assigned two armed and brawny Luo tribesmen as his personal bodyguards; there were guards to watch the children as well. "There were police everywhere we went," recalled Richard. "We all had police guards. We weren't taught not to trust Africans because of this. We were told that there were some bad ones who were trying to kill us, kill our daddy. But I have no recollection of being afraid; I think I was too young to figure that out."

The Leakeys' home, too, was heavily guarded, night and day. When Louis decided to build a new house on five acres in the suburb of Langata in 1951, he chose a fortresslike design: square, with a central courtyard, small exterior windows, and only one door leading inside. As further precautions, Louis screened their Kikuyu servants and museum employees, trying to determine if they had taken the oath, and he and Mary varied their daily routines, and debated sending the children to England for safety. The tension left them jumpy. Jean Brown, a newly arrived staff member at the museum, remembers once joining Louis on an outing to Gamble's Cave and watching with some surprise as he edged up the trail "with pistol out at the ready in case of a Mau Mau ambush." He seemed to her "very overcharged and strung up at that period [but] with good reason." Mary was equally nervous and once nearly shot their gardener by mistake. Even when she was

[5] After independence, the onetime "rebels" were immediately hailed as "Freedom Fighters"—a term even Louis adopted in his 1974 autobiography, *By the Evidence*.

on leave in England in 1954, a friend noticed how "difficult" it was for her to relax, "so used was she to peering round the corner of every wall before she turned it." "It was a very worrying time for all of us," Mary wrote in her autobiography.

Then in October 1952, after a spree of Mau Mau murders left twenty-four Kikuyu headmen (all government appointees) dead as well as thirty-six witnesses to the crimes, the government arrested the tribe's most popular leader, Jomo Kenyatta, and eighty-two other nationalists, and proclaimed a state of emergency. Several weeks later, Louis was summoned to Kapenguria, a tiny outpost three hundred miles west of Nairobi near the Uganda border, where Kenyatta and five of his closest associates were to stand trial for masterminding Mau Mau.

Kenyatta's trial opened on November 24, 1952, under heavy guard. Apparently the government had selected Kapenguria—a simple administrative post, consisting of two prisons, the district commmissioner's house, and a small agricultural school—because of its isolation. Located at the end of a dusty dirt road, Kapenguria was also off the railway line, and lacked both telephones and hotels. Still, rumors circulated that a Mau Mau attack was imminent, and the town bristled with soldiers, armored cars, and barbed wire, while a helicopter circled overhead. During the four months the trial was in session, it became a popular outing for the nearby white farmers. Their wives came, too, to picnic in their summer frocks and wide-brimmed hats under the town's cedar trees. Africans could attend the trial only if they had passes—which were seldom given—so the only natives in view were the local Suk people, tall nomadic pastoralists. They sometimes wandered through town, with their bows and arrows, but little clothing, shyly covering their eyes with their hands.

At the time of his arrest, Jomo Kenyatta was fifty-nine years old, a stocky man with intense light-brown eyes, a trim beard, and a theatrical manner. He had spent most of his life fighting for an independent Kenya led by Africans. From 1931 to 1946, he had lived in the U.S.S.R. and England, working as a representative for the Kikuyu Central Association, a society the colonial government had banned at the beginning of World War II. Kenyatta had returned to Kenya in 1946 to enthusiastic acclaim among his people, and had immediately been elected president of the Kenya African Union (KAU), a nationalist organization seeking a peaceful path to independence. At his rallies, he dressed flashily, sometimes draping a cape of monkey skins around his shoulders (a gift, he said, from Ethiopia's emperor, Haile Selassie), wore a large carnelian signet ring on his left hand, and carried a massive, elephant-headed ebony walking stick in the other. Natives turned out by the thousands to hear him speak, and hailed him as the "Hero of Our Race," "Saviour," and "Great Elder." But to the Europeans in Kenya, he was "a bogey-man . . . an agitator, a Revolutionary, a Red, fully equipped with horns and a tail," as Elspeth Huxley wrote in *Out in the Midday Sun,* her memoir about life in Kenya Colony.

The government had asked Kenyatta several times to denounce Mau Mau at his rallies, and on a few occasions he had appeared to comply. But some Kikuyu witnesses averred that Kenyatta's verbal denunciations meant nothing, that he had nullified them by visibly twisting his ring, or giving other signs to the crowd. In the government's eyes, Kenyatta's Kenya African Union was merely the banned Kikuyu Central Association in disguise, and both organizations were a cover for Mau Mau. This was the charge the court brought as well: that Mau Mau was actually part of the KAU, and that Kenyatta and the other defendants, all high officials in KAU, therefore managed Mau Mau.

The prosecution had little hard evidence to back this claim, and Kenyatta's British lawyer, Denis Nowell Pritt—as well as historians reviewing the case today—regarded the whole trial as a legal farce, a political frame-up. The judge, Ramsley Thacker, and the prosecutor, Anthony Somerhough, appeared to be in collusion: the prosecution's room was connected by a door to the judge's chambers, and the two men often traveled together to Nairobi, apparently to consult with government officials. Though Pritt did not know it at the time, the prosecution's key witness—who testified that Kenyatta had given him the Mau Mau oath—had been bribed and would later admit to perjury.

Louis's role as the court interpreter was also suspect. On the surface, he appeared to be simply a neutral translator, admittedly one of the best in the colony, and a known expert on Kikuyu customs. In reality, he was "up to his neck in government preparation of evidence for the trial," asserts historian David Throup. "He was an integral part of the prosecution's team." All of his work for the Special Branch—his translations of oaths and hymns, his interrogation of suspects—was used by the prosecution. His role was further compromised by a short book, *Mau Mau and the Kikuyu,* which he published shortly after the trial began. In it, Louis sketched the Kikuyu people's customs and history, their land grievances and distrust of the government. Writing as an anthropologist, Louis contended that Mau Mau was not so much a political movement as an "evil campaign" that had its origins in the breakdown of Kikuyu traditions. He thought that "probably . . . the speed of progress has been too rapid . . . [and] has made a part of the population unbalanced in their outlook and thus paved the way for movements like Mau Mau, in the hands of the unscrupulous few." He then linked Kenyatta to Mau Mau, charging, as the court did, that the KAU was merely a cover for Mau Mau.

Though he did not know about the book when the trial began, Kenyatta could not have been happy to see Louis as the interpreter. The two men's lives had been curiously entwined since childhood. As a child, Kenyatta had lived at a Scottish mission not far from Kabete, where Louis had grown up, and had taken as his first wife Grace Wahu, a young woman educated by Louis's mother. When Kenyatta was preparing to leave Kenya in 1929 for his first visit to England, Louis's father did his best to dissuade him from going. He disliked Kenyatta's political leanings and considered him a troublemaker. Louis shared this opinion, and since he was in England at the time, he kept an eye on Kenyatta there. When he heard that Kenyatta was giving a seminar on the Kikuyu custom of female circumcision at the London School of Economics, Louis made a point of attending. He had also written a paper on the subject, and thought that Kenyatta was distorting the nature of the custom for political reasons. In the end, the two men shouted at each other in Kikuyu, a discussion lost on the rest of the class. Both Louis and Kenyatta had also written anthropological studies of the tribe, though at the time of the trial, Louis's had not been published—it would not appear until 1977—while Kenyatta's, *Facing Mount Kenya,* was actually a political manifesto, containing the seeds of Kenyan nationalism.

Kenyatta may have told Pritt about some of this history, for at the trial Pritt wasted no time in attacking Louis's interpretations. "I keep getting notes from my clients that translations are wrong and that other things are interpolated," Pritt protested at the end of the first week of deliberations. "I must insist that only what I ask be translated to the witness and that only the witness's answer be translated. There ought not to be a lot of conversation between the translator and witness in Kikuyu."

When Louis's book on Mau Mau appeared shortly after this exchange, Pritt lodged another angry complaint. "It is impossible to read Dr. Leakey's book . . . without being convinced that he was making as plain as possible an innuendo that my clients, or some of them, are the people who are dominating and leading Mau Mau." Somerhough, the prosecuting attorney, rejoined, "It is certainly a book against Mau Mau. If [your] clients like to wear the hat, it will fit."

Pritt went after Louis on three more occasions, arguing that he was misconstruing what the witnesses had said. Then, on January 7, 1953, a white-robed Muslim woman answered a question with the Kikuyu phrase for "seven months." "She probably means seven Islamic months," Louis added. Pritt stalked to the front of the room. "It is your job to say exactly what the witness says!" he shouted at Louis. Louis protested that he was only trying to be helpful. But Pritt had already turned to the judge, insisting that another interpreter be called. "[T]his is intolerable. . . . I have been pressed for five weeks now by all my clients. . . . They do not trust this interpreter." Judge Thacker attempted to smooth things over, assuring Pritt that Louis was the best interpreter in the country. Pritt only redoubled his attack. "He is my clients' enemy. . . . [He] is biased and has written a book against my clients. In the course of the trial he publishes it. I hope it will be the subject of criminal proceedings."

Pritt's words left Louis little choice. He sat quietly for a moment, then asked the judge's permission to withdraw, which Thacker reluctantly granted. Louis stood up, gathered his papers together, and walked out of the courtroom, "his face drawn and ashen-gray," one paper reported. He left for Nairobi that afternoon, and the court eventually located another interpreter—though two more were dismissed before one was found who was acceptable to both the prosecution and defense. Some of Louis's translation troubles probably stemmed from the Kikuyu language itself, which is extremely allusive, filled with metaphors and double meanings. Historian David Throup believes that Louis was not "mistranslating, but was applying certain subtle nuances to what was being said, so that perhaps it became more coherent and intelligible to the European mind."

Pritt's attack on Louis was not well received by the European community. "This attack on a scholar of Dr. Leakey's eminence has naturally caused considerable indignation and disgust," Sir Evelyn Baring cabled Britain's Secretary of State. Somerhough was equally offended, and apparently had words with Pritt about his accusations, for he soon sent Louis a letter explaining what had happened since his departure, and asking him to return as a witness on document translations.

Deeply interested in the outcome of the trial, Louis welcomed the chance to return. He arrived back in Kapenguria on January 26, 1953, and stayed for the trial's duration, completing the translation of the many KAU documents, political songbooks, and letters seized at Kenyatta's home after his arrest. Somerhough hoped that somewhere in these Louis would find positive proof of Kenyatta's ties to Mau Mau, but Louis succeeded only in turning up vague references. Nevertheless, Judge Thacker ruled that they were sufficient evidence—together with the testimony of one witness—to convict Kenyatta and his codefendants.[6] On April 8 the six men were sentenced to seven years' hard labor in the deserts of Kenya's Northern Frontier District. To ensure that Kenyatta never returned to Kikuyuland, a restriction order was added: even after he completed his sentence, he would not be allowed beyond the NFD. The government then tore down his home, destroyed a small college he had started, and turned his farm into an agricultural school. Kenyatta had become a nonperson.

Louis left no record of how he regarded Kenyatta's conviction, though from his close association with the prosecution it seems likely that he approved. After the trial, the attorney general sent him a £250 honorarium and a letter thanking him for his

> splendid work on behalf of the Crown in the case of *R* [Regina] v. *Kenyatta*. . . . I know, and probably no one knows better, how much the Crown owes to you for your invaluable advice and help in the preparation and presentation of that case and although I know you put in countless hours

[6] The colony's Supreme Court later acquitted one of the defendants. Pritt carried his appeal on the others' behalf all the way to London and the Judicial Committee of the Privy Council. After a two-day hearing, the Council's lordships decided against reopening the case.

of work and study with no other thought than that of assisting the cause in which you so sincerely believed, nevertheless the Government of Kenya thinks it right that it should make some acknowledgement of its indebtedness to you by paying you this honorarium.

The Mau Mau rebellion did not end with Kenyatta's removal. An estimated 15,000 Kikuyu militants fled to the forest enclaves of Mount Kenya and the Aberdares, where they continued their battle against the government, Kikuyu loyalists, and white settlers. British troops arrived to reinforce the colony's security forces, and also enlisted 20,000 Kikuyu "Home Guards"—an act that further divided the tribe. For the next three years, Kenya Colony was under siege.

Throughout the Emergency, Louis continued to work as an officer with the Special Branch.[7] From his network of informants, he attempted to locate where the Mau Mau gangs were hiding, and who might be bringing them food and supplies. Once he solicited information from his sister Julia, who was working as a missionary among the Kikuyu near Mount Kenya. "Louis said to me, 'Do you think you can do a little scouting for us to find out who are the leaders?' After some killings, the African women told me that one certain man was the instigator up where I was, and I told Louis. And then I felt awful. . . . How could I be among them and be their friend and do that?" Though she was horrified to see several Kikuyu loyalists who had been "terribly beaten up" for refusing to take the Mau Mau oath, Julia could not bring herself to pass on any more news about the rebels to Louis. "It stopped me being honest and being their friend."

Louis perceived his own friendship with the Kikuyu in a different light, and when the government invited him in mid-1953 to serve on a committee investigating the "sociological causes underlying Mau Mau," and to suggest ways to defeat it, he gladly accepted. In their final report, released in January 1954, the committee wrote that Mau Mau was not a political cause but "a dangerous obsession based not on intellect, but on feeling and emotion. . . . To overcome this obsession . . . an attack must [also] be made on feelings and emotions." The committee regarded Mau Mau adherents as psychologically ill, and urged that they be detained in government-run camps where they could make "confessions" and be "rehabilitated." Eighty thousand Kikuyu, approximately one-third of the tribe's adult males, were sent to these camps.

[7] "As for Mau Mau, and [Louis's] position [in] the CID [Criminal Investigation Department], no one should ever doubt Leakey's deep and abiding love for the Kikuyu," observed Greet Kershaw, an anthropologist who studied the Kikuyu in the mid-1950s and was well acquainted with Louis. Kershaw disagreed with Louis's interpretation of the movement and his role with the Special Branch, but never doubted that he had taken the job "from a deep desire to serve the Kikuyu and their culture."

While on the committee, Louis wrote a second book about the movement, *Defeating Mau Mau,* and published a series of articles in the *Observer,* the *Manchester Guardian,* and the *New York Times Magazine.* These were well received in Britain and the United States, and established him as "the leading literary authority on Mau Mau," a reviewer wrote in *The Times* of London. In *Defeating Mau Mau,* Louis carried the committee's idea of Mau Mau as obsession one step further. By this time, in late 1953, seventy percent of the Kikuyu had taken the Mau Mau oath—a figure that surprised and puzzled Louis. As before, he argued that Mau Mau's political goals could not account for its attraction. Instead, the movement "was in fact a religion," steeped in black magic, ritual, and mystical acts. How else could it turn the "normally peace-loving Kikuyu into . . . fanatical, murdering maniacs," Louis wondered.

In detailing some of the crueler murders and bizarre rituals, his book and newspaper articles popularized the view of Mau Mau as savage, antiwhite, and anti-Christian—precisely the image the white colonial government wished to convey to the outside world. *Defeating Mau Mau,* however, was not consciously written as propaganda. Louis sincerely believed that Mau Mau was evil, and hoped that his books and articles might "help to defeat" it. He also hoped that they would help the Kikuyu overcome their "mental unstability." In so doing, Louis became both self-appointed interpreter and protector of "his" tribe, a not uncommon role for anthropologists as recently as thirty years ago. While his book should be faulted for its melodramatic portrayal of Mau Mau, it also dealt squarely with the Kikuyu people's troubles. All of his writings appealed to the government and settlers to open the White Highlands to Africans, abolish the squatter system, raise Africans' wages, and work toward a multiracial government. He advocated substantial changes in native peoples' education, housing, and agriculture, and called for independent Christian churches that would accept the Kikuyu customs of polygamy and female circumcision, and optimistically envisioned a time when Kenya would exist as "a state of inter-racial harmony and co-operation." Louis's views were far more liberal than those of most Europeans in Kenya, yet many of these reforms came about in his lifetime.

Louis was not alone in his erroneous interpretation of Mau Mau. Today historians regard the movement as a political organization with nationalistic aims that, for a variety of reasons, collapsed into a civil war among the Kikuyu. Ultimately, the all-out assault against Europeans that Louis and the other government advisors perceived never did materialize, though Mau Mau did hope to terrorize the settlers into leaving the country. Of the approximately 13,000 people who died during the uprising, only thirty-two were white civilians. Two thousand Kikuyu loyalists died at the hands of Mau Mau terrorists; another 11,000 Kikuyu, presumed to be rebels, were killed by government forces.

Among those Europeans murdered by Mau Mau rebels were Louis's elderly cousin, Gray Leakey, and Gray's wife, Mary, who had a farm at the base of Mount Kenya. (Gray had originally come to Kenya in 1906 to help Louis's father build the church at Kabete.) Their murders, in October of 1954, were particularly brutal. Mary was strangled in front of her husband; Gray was then dragged away to the forest, where he was apparently buried head downward while still alive. His death may have been part of a black magic ritual, as a sheep's sacrificed remains were found nearby. Louis always believed that because Mau Mau terrorists could not get to him, they killed Gray instead. And the day after Gray's death, Louis did receive another death threat. Whether or not the murder and threat were connected, they underscored Louis's belief that Mau Mau was essentially atavistic, a movement demanding "acts of incredible beastliness and depravity" from its adherents. Still, no harm came to Louis or his immediate family during the Emergency, though once when en route to Kisumu, Louis discovered that a crucial bolt on their car had been deliberately loosened. "If the bolt had come adrift at speed, the results would certainly have been disastrous," said Mary, who had been driving.

It seemed likely that a member of the museum staff was behind the attempted sabotage, though this was never proved. There were then six Europeans employed at the museum, and all of them were armed in the event of a surprise Mau Mau attack on Louis. The sight of Louis's secretary with a "large revolver strapped around her middle" startled more than one newcomer to the country.

Throughout the Emergency, and in spite of his intelligence duties, Louis's fossil research and position as museum director remained at the center of his life. This curious mixture of careers was not lost on one London reporter, who described Louis in terms that would seem to fit the much-later fictional hero Indiana Jones: "A lean and boyish figure, active, provocative, and intensely alive, Leakey is not only something of a genius: he seems to belong peculiarly to Africa. Where else would one find a serious-minded, middle-aged archaeologist engaged in dangerous secret intelligence work in between visits to inaccessible caves in search of the Missing Link?" At the time, July 1954, Louis was in London writing a report on the Pleistocene pig fossils he had discovered in Kenya and Tanganyika. Nothing, it seemed—not even death threats from terrorists—could slow him down. In the three preceding years, he had overseen a major expansion of the museum; participated in the second Pan-African Congress of Prehistory—the gathering of scientists that Louis had first organized in 1947—in Algiers, where he presented three papers; published the first descriptive volume on the handaxes of Olduvai Gorge; revised his 1934 book, *Adam's Ancestors;* written the text for *Animals in Africa,* a popular book about African wildlife; and prepared several short articles on various archeological and paleontological subjects.

He and Mary had also started once again to explore Olduvai Gorge. Their search had been given a fresh boost in 1951 when their benefactor, Charles Boise, who had generously assisted them with funds for the British-Kenya Miocene expeditions, offered to support their explorations in the Gorge for the next seven years. "This particular promise . . . marked something of a turning point in our work at Olduvai," Mary wrote in her autobiography, "because hitherto we had gone there whenever an opportunity arose and concerned ourselves mainly with exploring the Gorge in general. . . . [Now] we could begin a more planned and concentrated kind of research." For Louis, the money brought great hope that they would at last find their "Olduvai Man."

Chapter 12

"OUR MAN"

When Louis and Mary renewed their explorations of Olduvai in 1951, Louis was "absolutely certain" that they would discover the fossil of an early human, Mary recalled. "We talked about it a good deal, both theoretically and practically—when it might turn up. We were both equally interested in the subject, and we were both convinced that Olduvai was the place we were going to make finds of considerable importance."

No other site could compare. Not only was Olduvai bristling with artifacts and fossils, it was also a layer cake of evolution. In its distinct geological horizons—Beds I through V—one could easily see the emergence over time of new animal species and the disappearance of others. What the Leakeys hoped to find was a similar clearly delineated sequence of human ancestors.

But where in the gorge should they begin their explorations? Together, Olduvai's main and side gorges stretched for thirty-five miles, while the fossil deposits rose three hundred feet above the sandy floor. A careful search of each bed would require years of work, and even Louis and Mary underestimated the size of the task. "All one realized [in the early 1950s] was that there were five levels of deposits with stone tools and bones coming out from every separate level, and that it had to cover a very long period; that there was a tremendous story to be uncovered," said Mary. "But in those days, one didn't quite realize how big a story it was."

At the time, Louis and Mary believed that Olduvai's lowest deposits (Bed I) dated to the Middle Pleistocene, or 400,000 years ago, while the highest deposits were 15,000 years old. (Today Olduvai's Bed I is known to date to

Early Pleistocene times, or two million years ago.) From Bed I, Louis and Mary had unearthed flaked tools and crude choppers—palm-sized rocks with only one edge roughly hammered off—but no handaxes. In his 1951 book, *Olduvai Gorge: A Report on the Evolution of the Hand-axe Culture in Beds I–IV,* Louis labeled these primitive stone tools the Oldowan Industry. They represented some of the earliest artifacts then known; only the tools of Uganda's Kafuan Industry (which have since been shown to be the product of natural forces and not tools) were regarded as older. Using the Oldowan artifacts of Bed I as a baseline, Louis traced a cultural sequence progressing from these primitive tools to the rough handaxes of Bed II, and the more skillfully rendered versions of Beds III and IV.[1] Theoretically, the creators of these artifacts would show a similar development, with the more primitive species *in situ* in the lower beds.

Yet Louis and Mary did not begin their search in Bed I, but instead set about excavating two promising sites in Bed II. Louis particularly wished to find the more advanced inventors of the handaxe culture, since the question of who had made these tools remained, as a colleague of Louis's noted, "one of the more abiding mysteries of prehistory." Louis planned to resolve it at Olduvai. He reasoned that because he and Mary had not discovered any handaxes in Bed I, they would not find their creators there either. Sticking to his belief in the antiquity of modern humans—and to the validity of his Kanam and Kanjera fossils, which he believed were early forms of *Homo*—Louis argued that when the fossils of "Handaxe man" were discovered at Olduvai, they would prove to be "of *Homo sapiens* type, but primitive."

The Leakeys thought that the best place to find their "Man," as they often referred to the as yet undiscovered ancestor, would be in the deposits of an ancient living floor. Their excavations during the mid-1940s at the handaxe site of Olorgesailie had suggested to Louis and Mary that these people had lived for some time in one area, perhaps in brushy shelters. The two sites they selected as possible living floors—BK and SHK (Peter Bell's and Sam Howard's Korongos)—had been discovered in Olduvai's side gorge on Louis's 1935 expedition and were, Louis noted, "extraordinarily rich in fossils and artifacts."

BK proved to be especially worthwhile. After digging a trial trench here in 1951, Louis and Mary launched a full-scale excavation in 1952, and were soon rewarded with hundreds of stone tools, thousands of flakes, and masses of fossilized mammals, particularly giant pigs, buffalo, and antelopes. Many of these huge creatures would have towered over their descendants today: one skull of the extinct buffalolike *Pelorovis oldowayensis* had horns with an almost eight-foot span, while the hippo-sized pig *Afrochoerus nicoli* (which Louis playfully named after Mary) had yard-long tusks and was, understandably, first mistaken for an elephant. There were complete skulls,

[1] Although Louis then thought that handaxes were not part of the Oldowan Industry, Mary has since shown that it did include primitive handaxes.

limb bones, and, in some instances, entire skeletons, and surrounding these lay the tools of early humans, perhaps dropped after a surfeit of killing and butchering. "We have not yet got the man," Louis reported to a colleague at the British Museum of Natural History, "but I think we have found a most likely place to get him. . . . We found a 'slaughter house' of 'Chellean Stage I' man [the first stage in Louis's sequence of handaxes] at the very base of Bed II, where he drove hundreds of animals into a sticky bog of clay and then, when they were bogged down, killed and cut them up."

Close to the "Slaughter House," as the Leakeys named this swampy site, was a dry land surface, also heavily littered with tools and bones. Here they uncovered what Louis had expected to find: a "living site of early Chellean Man." Although later disputed, Louis and Mary believed that they had unearthed an undisturbed living floor—a place our early ancestors may have thought of as home, or at least used regularly.[2] There were stone implements, animal bones that had been broken to get at the marrow, and lumps of red ochre, which the inhabitants may have used to decorate their bodies, much as the Maasai do today.

Louis and Mary continued this excavation in 1953, took a break in 1954, and then planned another major Olduvai expedition for 1955, just after the New Year. Jean Brown, the archeologist who had arrived at the Coryndon Museum during the Mau Mau troubles, would accompany them, along with a staff of ten Kikuyu workers. However, Mary was suddenly called to London to tend one of her ailing aunts, and so Jean, Louis, and the workers set off for Olduvai.

"We were a very rattletrap shoe-string expedition," Jean recalled. "The ancient vehicles [two trucks and a Land Rover] boiled and held us up at regular intervals." After climbing to the top of Ngorongoro Crater, they descended to the Balbal Depression where they "traveled on a track a foot or more deep in choking white powder, then cut over trackless country to the gorge." Men, tents, digging equipment, and crates of live chickens (Olduvai was now part of the Serengeti National Park and hunting was no longer permitted) were piled on one truck, while behind the other they towed a two-hundred-gallon water tank. Because of the difficulties of traveling to Olduvai during the rainy season, Louis now ventured there only in the driest months. But this meant that water was always a problem, and the water tank only partly solved it. Twice a week, some of the men were sent to a spring thirty-five miles away on the rim of Ngorongoro to fill the tank, a journey that Louis estimated was costing him three Kenyan shillings a gallon—more than the price of gasoline at that time. Much of the hauled water went for plastering the fragile fossil bones they unearthed, and what remained was

[2] Most archeologists now think that the accumulation of stones and bones on the living floors at Olduvai were caused by geological processes—perhaps the action of streams or the ancient Olduvai lake's waves—as were those at Olorgesailie. Yet the hominids also left large numbers of their handaxes at these sites, perhaps because they visited them regularly to butcher scavenged animals; or perhaps after finishing a butchery job with the heavy implements, they simply abandoned them.

strictly rationed. "We had literally only a teacupful [of water] each day to wash in," Jean said, which was "quite a hardship when we were covered from head to foot in white dust by the end of the day." Nevertheless, she found it a "wonderful, wonderful experience to live at Olduvai."

Jean had been working for Louis at the museum for a year, and had discovered that the "august director" preferred being called "Louis"—an informality that surprised her after the staid atmosphere of the British Museum of Natural History—and that he tended to rush about in a haphazard and unpredictable manner. "I would work for six weeks or so and then, as I was about to go home in the evening Louis would precipitate himself into the room to say, 'Don't come to work tomorrow; I have no more money to pay you.' Two or three weeks later he would suddenly ring up to say, 'Come, come quickly, come tomorrow. I have got some more money.' " He always seemed to have a thousand projects going at once, and "scurried around the museum so fast that he almost fell over his own feet." At times, he also acted as comically helpless as any absent-minded professor. Jean remembered that he "frequently lost his spectacles and had all the museum staff from curators to messengers searching high and low for them only to find after some lapse of time, that they were hanging round his own neck!"

At Olduvai, Jean found Louis similarly headstrong, impetuous, and impossible to control. On his 1953 Olduvai expedition, he had suffered such a severe case of sunstroke that his hair, Mary recalled, "turned from brown to white, literally overnight." Worried that he would not stay out of the sun, Mary had taken Jean aside before leaving for London and pressed her "to nag him more than a wife" about wearing his hat and not working in the midday sun—a task Jean found about as easy as trying to keep a child's mittens on in winter. She and Louis arrived at Olduvai in the late afternoon after an exhausting two-day drive, but Louis was so excited about being back that he "charged straight off down the gorge with me in tow. . . . We had gone about a mile or more down the gorge when Louis suddenly collapsed completely." It may have been sunstroke again, or a return of his epilepsy, but gradually he recovered and was able to lean on Jean and walk slowly back to camp. As she had promised Mary, Jean pestered Louis about wearing his hat, but even when he remembered to bring it with him, it was frequently not on his head, but rather holding "some precious find." He dressed without care in the field, wearing baggy old khaki coveralls that flapped about unbuttoned "most of the way down for lack of buttons" and a pair of old white tennis shoes "often with holes in the toes."

Just as others who had been on safari with Louis had discovered, Jean found him "a wonderful companion in the bush." His years of hunting and exploring had left him extremely aware of his surroundings, and even while on his knees searching for fossils, he was alert, heeding the soft cluck of a guinea fowl or the click of dry pebbles that might mean a rhino up ahead. In the afternoons, as they sorted and labeled the fossils, Louis shared his past adventures, taught her to speak Kikuyu and to make string figures, told her about the ways of the animals and the cultures of the African peoples. Jean was, however, less impressed with the excavation he was directing. She had been trained by Sir Mortimer Wheeler, Britain's eminent archeologist, and was "horrified by Louis's haphazard excavation methods." While admitting that "he got results," she remembered watching in amazement when he strung a "couple of Kenya cents" with holes in their centers on a string for a "make-shift plumb bob. So much was makeshift. Louis wrote his notes and observations on the back of his cigarette packets and then forgot and threw them away. The site was so untidy that it was often difficult to see what was going on."[3]

Louis's sloppiness took its toll. One day, as his Kikuyu workers dug hastily at the BK site, a hint of tooth enamel glinted briefly in the sun, but the tooth was "crushed into fragments by the unskilled excavator before we knew what was happening." Louis called for his workers to stop; then he and Jean swept up the pieces. She spent the next two days gluing the fragments back together, and then handed to Louis what he had long hoped to find: an early human fossil. It was a milk molar, a child's tooth, Louis thought, but extremely large and oddly shaped, not at all like that of the primitive *Homo sapiens* Louis had expected to discover. This child's skull would not have resembled those from Kanjera. But the dissimilarity did not deter him. Jean remembered that Louis "became tremendously excited and quite carried away with his enthusiasm when [the tooth] was found."

In an article for the *London Illustrated News,* Louis had once speculated that the people who had slaughtered the giant animals in the swamp might be giants as well, and he now eagerly pursued this possibility. "Recently a really gigantic human milk tooth has been found at Olduvai Gorge," he reported to readers of the *News.* For emphasis, he included scaled drawings of the single tooth "so the huge dimensions of this giant baby can be better appreciated."

The illustrations were startling, for Louis believed that the tooth came from a child only three years old, yet the tooth was nearly double the size of a present-day child's tooth. Clearly, as Louis observed, even if the child was not a giant, its jaw must have been "very, very massive."

Just before leaving Olduvai, Louis's team uncovered a second large hominid tooth at the same site—a lower canine, which Louis also believed to be a child's.[4] But in spite of the direct association of teeth and handaxes, Louis was not ready to credit these people—whoever they were, giants or not—with the invention of the handaxe. Perhaps they belonged to "some other human race" and had been a "victim of Chellean man's hunting activities,"

[3] Desmond Clark, professor emeritus of African prehistory at the University of California, Berkeley, had the same impression. He praised Mary as a "meticulous excavator," but found that Louis was "not very good. . . . He was too excited to get things done and find out all that was to be found out . . . and go on to the next one. But you don't find out everything by hastening the excavation; you usually miss a lot."

[4] Today both the canine and molar are classified as australopithecine.

he suggested. The only point Louis was certain about was that the tooth had not come from an australopithecine, one of the fossil "near men" from South Africa. Instead, he argued that he was "dealing with a human . . . possibly some form of man not hitherto described. We must hope that future work will yield more complete remains." A little more than a year later, Louis's wish was granted.

After one more season at Olduvai, the Leakeys decided to explore again the deposits at Laetoli, a site thirty miles south of the gorge. They pitched their camp along the Laetoli River on June 13, 1959, but stayed only ten days. The deposits were disappointing, yielding few worthwhile fossils and no associated stone tools. On June 23 the Leakeys, their crew of Kikuyu and Kamba workers, and Dr. Peter Davis, a British anatomist, were back at Olduvai.

"We decided then to change the scene of our operations to Bed I," Mary said in 1984. "It was just a chance to look somewhere else." This time luck was with them. The day after arriving at the gorge, Louis's longtime field assistant Heselon Mukiri spotted a hominid tooth protruding from a block of limestone in Bed I. *"13th day of the safari!"* Louis wrote joyfully in his journal. "We *got* a human molar . . . of huge size & quite comparable to the 2 human teeth from BK2, at the Base of Bed I." Scattered around the tooth were tools and fossilized bones, and the Leakeys immediately decided to excavate the site, believing that they might uncover more of this hominid as well as another living floor. But they had also exhausted most of their season's research funds, so Louis made a quick trip to Nairobi to arrange an overdraft on their research account. He also invited their friends and neighbors, Armand and Michaela Denis, who produced a popular British television show called "On Safari," to come to Olduvai to film the excavation. Since the Denises wanted to photograph the dig beginning with the first trowelful of earth, Louis promised to delay the work a few weeks until their cameraman, Des Bartlett, could arrive—a rendezvous they scheduled for July 17. Then Louis headed back to Olduvai.

The promise to the Denises kept the Leakeys from working on their excavation—a delay that would actually prove to be a stroke of luck, one of many that benefited the Leakeys on their explorations over the years.

While waiting for Bartlett, Louis and Mary went out together every morning with their Dalmatians to explore other sites in Beds I and II. These excursions turned up the usual bounty of fossils: a skull of the bug-eyed hippopotamus gorgops, two hyrax skulls, and the complete skeleton of an extinct giraffe. But on the morning of July 17, Louis awoke with a slight fever, and Mary insisted that he stay in camp. "I'm sorry," he remembered her saying, "but you just cannot go out this morning. . . . You're not fit for it, and you'd only get worse." While he dozed fitfully in bed, Mary headed off alone in the Land Rover, taking with her their two dogs, Sally and Victoria.

Louis and Mary had intended to explore an area some miles down the gorge, but she decided instead to search a site in Bed I that was closer to camp. These exposures were called FLK (for Frida Leakey's Korongo), and stone tools were fairly common there. In fact, Louis had found the very first Olduvai artifacts at this site on his 1931 expedition. Since then, FLK had been searched only cursorily.

The morning was still cool when Mary carefully began exploring the slope's buff-colored sediments. Fragmented fossils and broken stone tools littered the surface, and Mary, a wide-brimmed straw hat shading her face, bent close to the earth. She stooped and crawled slowly up a rocky incline, and where the fossils seemed most promising, let her eyes linger on each bit of bone. By eleven o'clock she had found very little to reward her efforts. She would stop at noon, she decided, when the sun's overhead glare would make it difficult to distinguish fossils from mere stones. Just then she noticed a scrap of bone that "was not lying loose on the surface but projecting from beneath. It seemed to be part of a skull. . . . It had a hominid look, but the bones seemed enormously thick—too thick, surely," she recalled in her autobiography. Gently brushing away the earth, Mary then saw two large teeth set in the curve of a jaw, and her doubts vanished. "They *were* hominid. It was a hominid skull, apparently *in situ,* and there was a lot of it." Rounding up the dogs, she leaped into the Land Rover and drove madly back to camp.

"I've got him! I've got him! I've got him!" she cried to Louis as she wheeled up to their encampment. But Louis was groggy with the flu and could only manage a confused "Got what? Are you hurt?" "Him, the man! *Our* man," she replied. "The one we've been looking for. Come quick. I've found his teeth!"

"I became magically well in a matter of moments," Louis later recalled in a National Geographic Society film about their discovery, "and I went down with her and sure enough, there were these teeth sticking out."

In another account written for *National Geographic* magazine, Louis said that he then turned "to look at Mary, and we almost cried with sheer joy, each seized by that terrific emotion that comes rarely in life. After all our hoping and hardship and sacrifice, at last we had reached our goal—we had discovered the world's earliest known human."

In reality, Louis's enthusiasm was somewhat tempered. He immediately saw that the teeth were not human, but very much like those of the australopithecines from South Africa, creatures that he had long asserted had nothing to do with the evolution of "true man." "Louis wanted a *Homo,*" Mary explained, "and he could see that the teeth were those of an australopithecine. So he was excited, but he was also disappointed."[5] But his disappoint-

---

[5] Louis did not, however, swear, "Why it's nothing but a goddamned robust australopithecine," and stomp angrily back to bed as recorded by Donald Johanson and Maitland Edey in *Lucy: The Beginnings of Humankind.* With his missionary upbringing, Louis was unaccustomed to swearing; Mary asserts that he "never swore." It is highly doubtful that he would have ever used the very American expression "goddamned"—particularly at this time, since he had not yet visited the United States.

ment was only momentary. After quickly surveying the site, he could see, as Mary had, that most of the skull was there and *in situ.* Whether *Homo* or australopithecine, it was a remarkable and significant discovery.

Still haunted by the Kanam/Kanjera affair, Louis vowed not to touch any of the bones until Des Bartlett arrived with his cameras—a wait that Mary remembers as excruciating, so eager were they to see what the skull actually looked like. Bartlett finally appeared at camp late in the afternoon with his wife and daughter, and Richard Leakey, then fourteen years old, in tow. Louis immediately hauled them off to the site, talking breathlessly all the way, and Bartlett took the first photographs of the Leakeys' new find.

As Bartlett snapped pictures, Louis sat and studied the teeth, puzzling about how something that looked so much like an australopithecine could be found with stone tools. Few scientists then believed that the australopithecines, or "near men," discovered in South Africa had actually made such tools.[6] They may have used tools, but their brains seemed too small to give them the capability of "systematic tool-making," a skill that required "foresight based on memory," as Kenneth Oakley of the British Museum of Natural History wrote in 1956. "[T]he most satisfactory definition of man from the scientific point of view is probably Man the Tool-maker," Oakley believed, and Louis shared his opinion. In their eyes, toolmaking was the final dividing line between human and prehuman. Now, faced with a skull that seemed prehuman but was directly associated with stone tools, Louis considered various explanations.

In the past, when confronted with unexpected hominid remains—that is, hominid fossils that did not conform to Louis's expectations—in the presence of tools, Louis had always invented a cannibalistic feast to explain the discrepancy. He had first imagined this scenario in 1931, when stone artifacts and charred animal bones had been discovered with the Peking Man fossils —fossils that Louis believed were "aberrant branches" of the true *Homo* line. He used this same explanation when stone tools were found with australopithecine fossils in a South African cave in 1956. It seemed to Louis "that these 'near men' were contemporary with a type of early man who made these stone tools and that the australopithecines were probably among the victims which he killed and ate."[7]

But this time Louis rejected such a feast. The FLK discovery marked the third time the Leakeys had turned up a large-toothed hominid in direct association with stone tools at Olduvai. If toolmaking defined the threshold

---

[6] Anthropologists did not fully accept the idea that the australopithecines made stone tools until 1988. Recent discoveries in a cave at Swartkrans, South Africa, have turned up several hand bones, including the thumb bones of a robust australopithecine. Anatomists who have studied the thumb believe that it would have been as capable of a precision grip as that of modern humans. Thus these creatures might have been capable of making stone tools.

[7] The suggestion of a more advanced hominid who was contemporary with the australopithecines was actually made by John T. Robinson, an anthropologist then at the Transvaal Museum in Pretoria, South Africa. Robinson also theorized that this unknown creature had made the tools, an interpretation Louis supported and embellished with his claims of a cannibalistic feast.

of humankind, then perhaps these creatures were not australopithecines, but some new, unknown hominid, one capable of hammering out tools to a set pattern, and therefore, in Louis's eyes, one directly ancestral to humans.

By nightfall, Louis was already pulling back from his initial interpretation. *"Mary got it!"* he wrote elatedly in his record of the day's events. She had discovered a hominid skull, he noted, and though it was "quite lovely," the "premolars *recall Australopithecus.*" He emphasized the "recall," as if weighing just how similar the teeth were to those of the South African fossils. After penning a few more lines about the arrival of Bartlett and Richard, he ended the day's entry: "I wonder what tomorrow will reveal. Is it a very primitive man or skull of an *Australopithecus*?"

Louis did not take long to reach a verdict. At the end of the first day's excavation—all filmed by Bartlett with a motion picture camera as well as five still cameras (Louis was not about to repeat the Kanam/Kanjera photo disaster)—they had removed the upper jaw and palate, and part of the base of the skull. Studying these pieces, Louis soon convinced himself that their creature was not, after all, an australopithecine. "My fresh feeling [after] getting more of the specimen into my hand and putting it together is of a man with strong *Australopithecus* affinities and presumably derived from the same stem with *Australopithecus* and *Paranthropus* [Robert Broom's name for the robust australopithecines] as offshoots to [the] side, while this is [the] true *Human* line," Louis wrote that evening. After removing more of the skull the next day, Louis drew rough sketches of their find and scribbled his thoughts alongside: "The zygomatic arches [cheekbones] are huge. The occipital condyles [the neck joint at the base of the skull] *are very small.*" He was now thoroughly enchanted with their hominid and unabashedly proclaimed, "He is a fabulous creation. *'Titanohomo mirabilis'* [Wonderful Titan-like Man] would be a good name. People will say he is *NOT* human but he is." Here at last, he thought, was the *Homo* he had so wanted to find.

Mary, too, was delighted by her discovery, and she and Louis began to refer affectionately to the massive skull as "Dear Boy." Just as she had pieced together the skull of *Proconsul,* Mary began reconstructing "Dear Boy." She spent the mornings at the dig, but retired to the shade of the camp's thorn trees in the afternoon to work at what Louis called their "three-dimensional jigsaw puzzle." The excavation itself was a tedious process lasting nineteen days. After brushing away the top layer of soil, they found most of the skull embedded in the earth, but broken into some four hundred fragments. Pieces of it lay scattered in the scree, and other bits were found after sieving and washing the soil. By the end of the dig, they had recovered the upper jaw and all its teeth, the facial bones, and almost all of the top and back of the skull; only the lower jaw was missing. Mary's careful reconstruction revealed a creature with a flat face, wide, arching cheekbones, enormous teeth, hardly any forehead, and a bony crest running the length of its skull, much like the ridge on a gorilla's skull. Based on the teeth, Louis determined that "Dear Boy" had been about eighteen years old when he died, probably of an illness. Around the skull lay a scatter of Oldowan tools and broken animal bones, certain evidence, Louis and Mary thought, that the site was also a living floor—the "man" had been found in his home, just as Louis had predicted.

By August 7, Louis and Mary knew that they had recovered as much of the fossil as they could, though they planned to continue excavating the living floor the next season. Since they were also heavily overdrawn at the bank, they closed their camp, packed up the skull in a biscuit tin, and headed back to Nairobi, eager to announce their find. They knew "Dear Boy" was important, but they had no idea of the fame and fortune—and changes—this fossil would soon bring to their lives.

Chapter 13

FAME, FORTUNE, AND *ZINJ*

As soon as the Leakeys arrived back in Nairobi, Louis sat down to draft a report about their splendid new find for *Nature.* The three-page article, simply entitled "A New Fossil Skull from Olduvai," was scheduled to appear in the journal's August 15, 1959, issue, less than a month after Mary made her discovery. Louis had planned a press announcement as well, but a printers' strike in England delayed the publication of that issue of *Nature* for several weeks, forcing him to hold back news of the discovery.[1]

In his report, Louis further developed the thoughts he had scribbled in his field diary about the relationship of "Dear Boy" to the australopithecines. He first acknowledged that the skull did indeed bear certain resemblances to these creatures, enough to warrant including the fossil in the subfamily Australopithecinae, a grouping William K. Gregory had created in 1939. However, Louis also noted twenty "major differences" between his fossil and the other genera within this subfamily, a sufficient number, he argued, to justify placing "Dear Boy" in a genus of its own. "I am not in favour of creating too many generic names among the Hominidae [the human family]," he wrote—apparently forgetting that he had given names to nearly every hominid fossil he had discovered, at both the species and family level —"but I believe that it is desirable to place the new find in a separate and distinct genus. I therefore propose to name the new skull *Zinjanthropus*

[1] Scientific protocol requires major discoveries to be announced in a recognized science journal or forum before being released to the public. Though not printed until early September, the delayed issue of *Nature* still carried the August 15, 1959, dateline.

boisei." Though the skull might still be *Titanohomo mirabilis* in Louis's heart, the new name was far more suitable. *Zinj* came from an ancient Arabic word for the coast of East Africa, while the Greek *anthropos,* meaning man, denoted the fossil's humanlike qualities, and *boisei* honored Charles Boise, the Leakeys' longtime patron. Translated, *Zinjanthropus* became the "Man from East Africa."

But would *Zinjanthropus boisei* be accepted by other anthropologists as a new and distinct genus? In spite of the printers' strike, it would not be long before Louis found out: the fourth Pan-African Congress on Prehistory was due to open in Leopoldville in the Belgian Congo (now Kinshasha, Zaire) at the end of August, and all of the leading scientists in the field were expected. Louis had been elected president of this congress at its 1955 meeting, and he planned to unveil *Zinj,* as the fossil soon came to be called, during his presidential address on the opening day.

In the meantime, Louis could not help dropping hints to his colleagues. He sent tantalizing cables to Le Gros Clark and Camille Arambourg, France's leading African prehistorian, and teased Peter Davis, the anatomist who had joined the Leakeys for part of their expedition. "We . . . have got our 'Bed I man,' " he wrote to Davis. "Do you know where we got him? . . . [W]e went to FLK where I sent you and Thairu those two mornings. You had missed him!" Louis worried about these conspiratorial disclosures, and ended the letter by begging Davis (as he did all his confidants), "Please don't talk about him just yet."[2]

But Louis himself was bursting to show off his prize, and when a young American anthropologist, Clark Howell, came to the Leakeys' home for dinner one evening, Louis promised him a special dessert. "Louis said, 'I have a little surprise for you,' " recalled Howell, who is now professor emeritus of anthropology at the University of California at Berkeley. "He went away and came back with a cookie tin, and with a little smile he opened it up and there inside was *Zinj,* a magnificent skull. I was the first scientist to see it." At the time, Howell was only an assistant professor at the University of Chicago—a minor rank in academia—and so was stunned by Louis's act. "Most people of Louis's stature would never have shown such an important find to someone as young and inexperienced as I was. But it was characteristic of him to be generous like that, especially with younger colleagues."[3]

A few days later, on August 17, Louis restaged this scene in a small hotel room in Johannesburg. He and Mary made a brief detour to South Africa en route to Leopoldville, and were met at the airport by Phillip Tobias, newly

---

[2] Apparently, however, Louis never actually sent this letter. He wrote it at Olduvai, stuck it in the back of his field diary, and there it is to this day.

[3] Louis could also be as canny as a poker player about showing off his fossils, if it suited him. In her biography of Louis, Sonia Cole tells the story that soon after *Zinj*'s discovery, the distinguished British archeologist Sir Mortimer Wheeler called on Louis at the Coryndon Museum. He had heard rumors about the Leakeys' find and asked to see the specimen. But Louis, who considered Sir Mortimer a rival, told him the fossil was locked away in a bank vault. "[I]n fact," wrote Cole, "at that very moment Sir Mortimer was leaning against the safe in which *Zinj* was lodged!"

appointed professor of anatomy at the University of Witwatersrand and a protégé of Raymond Dart's. Though it was well after midnight by the time the Leakeys settled into their hotel, they invited Tobias to their room. They had brought *Zinj* with them in a specially made wooden box fitted with a small padlock. Smiling mischievously, Louis placed their "glory box," as Tobias called it, on the dressing table, turned the key in the lock, and pulled out three plastic bags. Each one contained part of the skull wrapped in green tissue. Louis carefully fitted the pieces together, then turned the skull's massive face toward Tobias. "I knew they had found something of importance, but I never expected them to show it to me like that. It absolutely sent shivers down my spine," Tobias said.

Louis had arranged the stopover in Johannesburg in order to compare *Zinj* directly with the South African australopithecine specimens—an examination that was somewhat after the fact since he had already completed his *Nature* paper explaining how the fossils differed. In his paper, he had even written that he had "personally" made a "very close examination [of] and direct comparisons" with these fossils. Certainly he was familiar with many of the specimens, having spent three weeks in South Africa in 1949 to study them. He also had a collection of plaster casts of the majority of the fossils, had corresponded with Robert Broom and Le Gros Clark in detail about them, and had discussed the australopithecines in several publications, most extensively in his 1955 edition of *Adam's Ancestors.* Nevertheless, he had not made the direct comparison with *Zinj* that he described in *Nature,* nor did he arrive in South Africa with an open mind about what he was going to see.

In Johannesburg Louis and Mary spent a day at the university with Tobias and Raymond Dart looking at their fossil collection. Since 1945, when Tobias had rediscovered the fossil-rich caves of the Makapansgat Valley, the two men had added steadily to this collection, particularly with bones from a creature that Dart originally had called *Australopithecus prometheus*—"Southern Fire-Making Ape." Dart derived the name from what appeared to be the fire-blackened animal bones that were found at one site—Makapansgat Limeworks—together with the remains of several australopithecines.[4] Based on this evidence and the crushed lower jaw of one adolescent australopithecine, Dart depicted these ancestors as "murderous and apparently cannibalistic." While most of his colleagues were skeptical of his claims and doubted that *Australopithecus prometheus* had made fire, no one disputed the importance of the fossils themselves. These included partial skulls, upper and lower jaws, as well as hip and pelvis bones—the latter proving that *Australopithecus prometheus* was fully capable of standing upright.[5]

Over the years, even with his on-again, off-again approach to fossil hunting, Dart had amassed a greater collection of hominid material than the

[4] The black coloration was actually caused by manganese that had leached out of the surrounding dolomite.

[5] *Australopithecus prometheus* is now classified as *Australopithecus africanus,* the original australopithecine genus and species that Dart named for his Taung Baby in 1925.

Leakeys, but he never gloated about his achievement. On the contrary, his eyes filled with tears when Louis presented him with *Zinj.* Holding the skull in his hands, Dart looked up and whispered, "I'm so glad that this has happened to you of all people." Dart then opened his cabinets and laid out the bones of his australopithecines, including a cranium that he had just finished extracting from its hard limestone matrix.

None of the skulls were as complete as *Zinj,* though it was clear that they all shared the general australopithecine look: a small, flattened braincase and large teeth and jaws. The *Australopithecus prometheus* specimens were, however, smaller and more delicately formed than hefty *Zinj.* They also lacked his bony sagittal crest and had smaller eyebrow ridges, and their faces were slightly less protruding. The number of differences could certainly add up to a new species, possibly even a new genus, and neither Dart nor Tobias pressed Louis on the issue of his skull's name.

More problematical for Louis were the robust australopithecines that Robert Broom and a colleague named John T. Robinson had collected. (The energetic and ebullient Broom, who had promised that he would "wear out, not rust out," had died in 1951, moments after writing the last lines of his final monograph on the australopithecines. "Now that's finished," he had dramatically whispered to his nephew, ". . . and so am I." He was eighty-five.) Like Dart and Tobias, Broom and Robinson had unearthed a stunning collection of fossils—a dozen nearly complete but badly crushed skulls, limb bones, complete pelvises, and vertebrae. Both the slightly built and more robust forms of australopithecines were represented. Broom, who like Louis was drawn to the differences rather than the similarities in the fossils, had given new genus and species names to many of his finds, while placing them all under the subfamily Australopithecinae. One group of the larger australopithecines, which Broom called *Paranthropus crassidens*—Large-Toothed Robust Equal of Man—was of particular concern to Louis.[6] Like *Zinj, Paranthropus* was broad-faced, heavy-jawed, with huge molars and a sagittal crest.

During his 1949 visit to South Africa, Louis had examined many of these same specimens. That study had led him to agree with Le Gros Clark that the creatures were hominids, although he was certain that they were not direct ancestors of "true man." "I am convinced," he had written at the end of his stay to Ralph von Koenigswald, who discovered in the late 1930s several fine specimens of Java Man (now *Homo erectus*) in Indonesia,

> that they [the australopithecines] were (a) very close to the stock from which man came, (b) much too late in geological time (they are Pleisto-

[6] Today Broom's and Dart's more gracile australopithecines are classified as *Australopithecus africanus.* Broom's *Paranthropus* genus is, however, still the subject of debate. While most scientists agree that Broom's two species, *crassidens* and *robustus,* are actually one—*robustus*—they disagree on what genus they belong to. Some prefer the designation *Australopithecus robustus,* while others insist that *Paranthropus robustus* remains a valid genus.

> cene) to be ancestral to man, (c) are far too specialised in certain unexpected ways . . . to be regarded as representing even a Pleistocene survival of the type of creature from which man evolved. In my opinion they represent an offshoot from the stock from which man arose.[7]

The intervening years had not changed Louis's mind.

Although he disagreed with some of Louis's ideas, Robinson offered to pick up Louis and Mary in Johannesburg and drive them to the Transvaal Museum in Pretoria, where all of Broom's and Robinson's fossils were housed. For the whole hour-long trip Louis "talked volubly, telling me what he then believed about *Zinj* and its relationship to the South African australopithecines," Robinson recalled. Once they arrived at the museum, Robinson turned over his fossils, desk, and office to the Leakeys, where they worked steadily for more than an hour. Robinson then rejoined them and was amused to find that Louis "had not modified his opinion at all. . . . After he had examined my material he explained what he now believed about '*Zinj*' and he said just what he told me on the way over from Johannesburg." When Robinson attempted to point out similarities between *Zinj* and the *Paranthropus* fossils, Louis waved him away. "He had the attitude that I simply did not understand the material well enough to recognise the difference between them."[8]

In these discussions, Mary simply sat and listened; if she had any opinions, she did not offer them. Her silence could be taken in any number of ways, and Robinson chose to believe that she "did not appear to be strongly supporting Louis's views." Undoubtedly, Mary's silence meant exactly the opposite to Louis. Either way, her quiet demeanor was becoming a powerful tool. She might be the faithful spouse, the questioning ally, or astute skeptic —one never knew. Her silence gave her the freedom to assess the fossils, the scientists, and Louis on her own, and to learn from them. In turn, she gained a reputation as an independent thinker, and as the careful and cautious scientist that Louis was not.

The fourth Pan-African Congress of prehistorians gathered at the Jesuit University of Louvanium on the banks of the Congo River. Twelve years had passed since Louis had initiated the first Pan-African Congress, and in that time the conference had grown into a major event. This time sixty delegates representing fifteen countries were expected. From Paris came Camille Arambourg, who at seventy-four was still directing excavations in Northern

---

[7] Ralph von Koenigswald's summary of the australopithecines was much pithier. He made his own pilgrimage to view the South African fossils in 1951 and sent Louis a brief note: "Brooms material is phantastic [*sic*]. Human, but off the line leading to Alexander the Great."

[8] Clark Howell had much the same experience. When he peered into the cookie tin at the Leakeys' dinner party, he remarked on how much *Zinj* resembled the robust australopithecines. But Louis shook his head emphatically, "No, no, no, no," and insisted that his fossil was something new.

Africa; Desmond Clark flew in from the small natural history museum he headed in Livingstone, Northern Rhodesia; Raymond Dart arrived with Phillip Tobias in tow; and Clark Howell traveled from Nairobi to present a paper on his archeological dig in Tanganyika.

Louis and Mary were among the last to arrive. After landing at the Leopoldville airport, they took a shuttle van to the university and there happened to spy Desmond Clark standing in the main lobby. "They were sitting together on the bus, and Mary had a little square box on her knees. They were looking very sort of mum," Clark recalled, imitating their arched eyebrows and pursed lips. "And Louis whispered, 'Ahhhh Clark . . . ahhh we have something new. . . . Really, I'll show you.' " After revealing his prize, Louis then pressed Clark to please keep *Zinj* a secret and Clark agreed. But it wasn't long before someone else came up and whispered to Clark, "Louis has got something interesting, hasn't he? He's shown it to me. But don't tell anyone." Soon there were small groups of people scattered around the university grounds all murmuring about Louis's secret. The air hummed with intrigue and drama, and Louis walked from group to group enjoying it all immensely. "He loved that kind of thing," said Clark, "the cloak-and-dagger business." But among the whispers were also criticisms, for, like Robinson and Howell, most people thought the new skull not distinctive enough to warrant its own genus.

It was even the subject of debate in England. Although Louis had asked Le Gros Clark not to discuss *Zinj,* Le Gros had immediately written a letter about the "remarkable discovery" to Kenneth Oakley, head of paleontology at the British Museum of Natural History. And as soon as Louis and Mary left South Africa, John Robinson wrote to Le Gros, telling him all about the skull and Louis's decision to give it a new name. "I am a bit disappointed in [Louis]," said Robinson, "because he insists on making [the skull] a new genus. To me it has all the major characteristics of *Paranthropus* so clearly developed. . . . it seems highly improbable to me that the Olduvai form could be a different genus."

This news was troubling to Le Gros. An impetuously assigned name would not only harm Louis's reputation (which Le Gros had helped salvage only ten years before) but damage the science of paleoanthropology as well—which was already regarded as a less than rigorous field. After receiving Robinson's note, Le Gros wrote to Oakley, "I do hope Louis will not be so silly as to create a new genus simply because the dimensions of the post canine teeth are a little larger [than those of *Paranthropus*]—but I feel he will do so—& thereby complicate the issue quite massively."

"The issue" was the cluttered taxonomy that surrounded the early human fossils. Nearly every hominid fossil that had been discovered seemed to have been given a distinctive label, so that by the mid-1950s there were twenty-nine generic names and more than one hundred specific names for the ancestral specimens. In part this was due to the paucity of hominid remains. They were—and are—rare, their discovery always a momentous event. Naming them was a way for Louis and many of the older fossil hunters—Robert Broom, Raymond Dart, Ralph von Koenigswald, Franz Weidenreich, Camille Arambourg—to celebrate what they had found. But in strict biological terms, the plethora of names meant nothing. They did not illuminate relationships and lineages, as taxonomical classifications should do. Instead they announced the individuality of each fossil—and the fossil's finder. It was not a rigorous approach to doing science, and in 1950 the zoologist Ernst Mayr took the field to task, challenging the paleoanthropologists "to harmonize [their] categories . . . with those of the rest of zoölogy." In Mayr's eyes, zoological names were tools, the building blocks that enabled one to construct evolutionary theory; they could be given only when a scientist was familiar with an animal's overall population and understood the variations that age, sex, and disease could cause. Assigning a generic and specific name to a single fossil, as Louis planned to do with *Zinj,* "simulates a precision that often does not exist."[9]

By 1959 many paleoanthropologists, particularly the younger scientists, had accepted Mayr's basic premise—that lumping the hominid fossils into larger groups would throw more light on the evolutionary development of humankind.[10] In this new approach, there would be only two genera: *Homo* or *Australopithecus.* All of Broom's names for his australopithecines would be swept aside, and there would be no new introductions.

Mayr's criticisms reflected another feeling as well (one shared by Le Gros Clark)—that paleoanthropology as a science needed to mature, that it could not simply be run by individuals with wildly speculative ideas. An additional impetus came in 1953 when Le Gros Clark, Kenneth Oakley, and Joseph Weiner proved that Piltdown Man, *Eoanthropus dawsoni,* was a fraud. Instead of representing some Pleistocene ancestor, the skull was actually that of an early form of *Homo sapiens,* while the lower jaw and teeth were those of a modern orangutan, stained and filed to resemble a fossil. (Ironically, Piltdown was the exact opposite of the australopithecines. It matched a large skull—and brain—with an apelike jaw, while the genuine fossils had small, apelike skulls and humanlike jaws and teeth.) Forty years had passed since Charles Dawson had "unearthed" the Piltdown remains, and in that time nearly every anthropologist, including Louis, had studied and written about

---

[9] Mayr also noted that six hundred species of the fruit fly, *Drosophila,* were contained in a single genus, while there were twenty-nine genera of human ancestors. "If individuals of these [*Drosophila*] species were enlarged to the size of man or of a gorilla, it would be apparent even to a lay person that they are probably more different from each other than are the various primates," he wrote.

[10] In 1950 the anthropologist Sherwood Washburn explained the difference between "lumpers" like Mayr and "splitters" like Louis this way: "The number of names is a function of the kind of interest of the investigator. If one is primarily interested in classification, in type specimens and priority, then the less is known about fossil primates the more names there will be. If one is interested in the mechanics of evolution, in the understanding of process, a cumbersome and constantly changing classification is a great liability and the tendency will be to lump, to leave fragmentary [fossil] bits unnamed, and to create new groups only when absolutely necessary."

the fossil's place in the human family tree.[11] Piltdown's influence on human paleontology had been enormous; to discover now that it was nothing but a forgery was acutely embarrassing. For Louis's old mentor, Sir Arthur Keith, the revelation was particularly humiliating. He had built much of his reputation on his analysis of Piltdown Man, had written considerably on the subject, and used it as the primary evidence for his thesis that humankind's Pleistocene ancestors would closely resemble modern humans.[12] When told of the forgery, Keith, who was then eighty-six, said, "I think you are probably right, but it will take me some time to adjust myself to the new view."

Without *Eoanthropus dawsoni,* the human family tree was far more straightforward—for all of the fossils, from the australopithecines to *Homo erectus,* combined in some degree a small brain with a humanlike jaw. Piltdown's removal strengthened the stance of the "lumpers," those scientists who emphasized the similarities rather than the differences among the fossils. For the first time in his career, Louis found himself behind the times.

When the governor general of the Belgian Congo officially opened the Pan-African Congress, there were few among the delegates who had not at least heard rumors about Louis and Mary's new skull. Nevertheless, when Louis took his place at the podium and held *Zinjanthropus* aloft for the audience to see, pandemonium broke out. "It was put over to us as a most fascinating and important find," recalled Phillip Tobias, who was seated in the audience.

> For thirty years, Louis and Mary had been pulling out stone tools from Olduvai Gorge, but the maker had never been found. And now, after thirty years, this staggering discovery had been made. There was tremendous applause, and not just a murmur of conversation, but quite hysterically excited conversation from everyone present. It was really something very, very thrilling—the excitement heightened by the fact that he had the actual specimen with him. He didn't just do it by [showing] slides.

Louis had not yet spoken the wonderful new name, *Zinjanthropus boisei,* knowing that it would likely be printed in the popular press and so violate the rules of scientific nomenclature. Instead, he saved it and his detailed

[11] As early as 1947, however, Louis had expressed doubts about the Piltdown remains, writing to Le Gros that he did not believe that the skullcap and lower jaw belonged together, but rather represented two different individuals, one a human, the other an ape. But he never guessed that the fossil was a hoax, perhaps because, like other anthropologists who studied the specimen, he was given only a plaster cast and not the original bones to work with.

[12] Keith wrote in his 1950 autobiography, "To such a degree did the Piltdown problem take hold of me that in 1913 I resolved to make it the subject of a book." This decision led to his systematic inquiry into all of the human fossils that had been discovered at that time, and to his book *The Antiquity of Man*—which was the primary text for students of human evolution when Louis was a student at Cambridge.

description of the fossil for the special section on human paleontology, from which the media was excluded. Louis was listed fourth on this program behind Clark Howell, and sat in the front row squirming impatiently through the other scientists' talks. So eager was Louis to announce his new genus that "about halfway through my talk," Howell recalled, "Louis suddenly said in a very loud whisper—loud enough so that everyone could hear—'Clark, Clark? Can't you hurry it up, Clark? I've got a very important thing here.' " Clark brushed off Louis's plea, but when he was finished, Louis fairly leaped onto the stage.

With *Zinj* balanced next to him on the podium, Louis read much the same report that he had sent to *Nature*. And as everyone in the audience expected, he did exactly what Le Gros was praying he would not do—he pronounced the skull Mary had discovered at Olduvai a new genus and species, *Zinjanthropus boisei.* It was markedly different, he insisted, from the other australopithecines and represented a bridge between them and humankind. Those other forms, *Australopithecus, Plesianthropus,* and *Paranthropus,* had all been dead ends, he argued, while *Zinjanthropus*'s characteristics "do quite clearly . . . indicate something more in the direction of Man." Once again there was applause and praise, and then Louis and Mary invited the physical anthropologists to examine the fossil with them.

A table had been set up under the palm trees in the university courtyard, and Louis took a seat at the head, while Mary sat at the opposite end. Seated between them were Phillip Tobias, Raymond Dart, Camille Arambourg, Clark Howell, William Howells from Harvard University, and Kenneth Oakley from the British Museum of Natural History.

"Each one of us had the chance of looking at and handling the specimen under the blistering Equatorial sun on the banks of the Congo River," recalled Tobias.

> Louis then said, "Well, gentlemen, I'd like to hear your reactions to it." And round the table it went, Camille talking in French, the others in English. And one by one the comments went something like this: "What a magnificent specimen. But is it really a new genus? Isn't it very similar to those robust australopithecines from South Africa?" The next one saying, "I congratulate Dr. and Mrs. Leakey on this magnificent specimen. But I must confess, I have some misgivings about the name that's been given. Is it a new genus and species?" And so it went. And Louis's hackles were beginning to rise. He was getting a little pink, a little perspiration of anger was beginning to appear on his face. Then it came to Dart's turn, and they all looked at Dart to see what his comment would be, as it was obvious that the tension had mounted around the table. And Dart defused the whole situation with a timely and beautiful wisecrack. He said, "Louis, I can't help wondering what would have happened if Mrs. Ples [a nickname for one of Broom's *Plesianthropus* skulls] had met Dear Boy one dark night."

The tense little gathering exploded with laughter. Louis even joined in, laughing in his curious, breathy way at Dart's wit.[13] For a successful mating—one that produced fertile offspring—Mrs. Ples and Dear Boy would have to belong to the same genus, *Australopithecus,* although they might represent different species. This would necessarily make them members of the same species of humanlike creatures.[14]

Although *Zinjanthropus* had a larger skull and larger teeth ("an enormous set of 'nutcrackers,' " as Tobias called them) than the robust australopithecines, it still shared their basic features, and it was these similarities rather than the minor differences that Tobias and the others noticed. They did not press the issue further with Louis; there would be time for that in the science journals. But the line between his vision of the new Olduvai skull and theirs was clearly drawn.

By September 3, Louis and Mary were back in Nairobi. They immediately held a joint press conference at the Coryndon Museum, with Louis describing *Zinj* as "the connecting link between the South African near-men . . . and true man as we know him," and the "oldest well-established stone toolmaker ever found anywhere." He estimated the fossil's age at between 600,000 and one million years old. As Louis anticipated, their announcement made headlines around the world. The *New York Times, The Times* (London), and the *Illustrated London News* carried the story on their front pages, the latter with dramatic photographs of *Zinj*'s massive jaw and the Leakeys under the headline, "A Stupendous Discovery: The Fossil Skull from Olduvai."

As Louis had also anticipated, scientists and the general public alike wanted to see and hear more about *Zinjanthropus.* Le Gros arranged for Louis to come to London to address the British Academy, a prestigious body of scientists that had elected Louis a fellow the year before. From there, he would travel to the University of Chicago to present a lecture at the Darwin centennial celebrations. The Wenner-Gren Foundation for Anthropological Research offered to pay his way, and Louis decided to make the most of this free trip—his first to the United States—by giving talks at various universities and museums. He thought he might be able to raise additional funds to continue the work at Olduvai—just as he had made money for his school fees when a student at Cambridge by lecturing about the dinosaur collecting expedition he had helped lead in Tanganyika.

Although Mary had discovered *Zinj,* she had no desire to face the hordes of reporters certain to descend on Louis. She also disliked speaking at public

---

[13] Louis would later recite Dart's quote to audiences, adding "I have no doubt at all she [Mrs. Ples] would have run away."

[14] Different species from the same genus are sometimes able to produce young, such as when a horse is mated with a donkey. But the offspring, in this case the mule, will be infertile. Only members of the same species can give birth to fertile young.

functions, and felt that her time would be better spent organizing the next expedition to Olduvai, scheduled to depart in February 1960.

With *Zinj* packed carefully in its traveling box, Louis flew off alone to London the first of October. He met with Le Gros shortly after his arrival, but whatever discussions the two men may have had about the fossil and Louis's decision to place it in a new genus were not recorded. Regardless of Le Gros's opinion, Louis was determined to continue calling the skull *Zinjanthropus boisei*—as he did when he presented the fossil to the British Academy. By now Louis and his fossil were both celebrities. The Piccadilly lecture hall, where the Academy met, was designed to seat three hundred people, but more than five hundred turned out for Louis's talk, wedging themselves in the aisles and along the walls. It was so crowded that even "television stars like Sir Mortimer Wheeler" were forced to stand, reported one newspaper. "It was the night of Dr. Leakey's life," wrote another reporter, who apparently sensed Louis's great pleasure at being recognized, at long last, by his colleagues. Ever since the disaster with Boswell and the Kanam/Kanjera fossils, Louis had struggled for this moment of glory, and now he reveled in it. He pulled *Zinj* from its wooden box, set it in a glass aquarium, and next to it displayed the gigantic remains of an extinct ostrich, hippopotamus, and buffalo—all from Olduvai. It was an impressive exhibit, which left "the experts . . . no words to describe the doctor's remarkable discoveries."

From London, Louis flew to the United States, where in one month he delivered sixty-six lectures at seventeen universities and was met with an outpouring of warmth and curiosity. It was not only the fossil skull that Louis's American audiences responded to; they were enchanted by Louis himself. He was fiery, enthusiastic, and inspiring. The story he told of searching Olduvai for thirty years, of never losing hope, of persevering against great odds with no money—all to find *Zinjanthropus*—was a story his audiences understood, one that embodied all the elements of the quintessential American success tale. While his British audiences might cast a skeptical eye on Louis's optimistic determination, the American public loved it. They found Louis so inspiring that many gladly gave donations for the Leakeys' research at Olduvai. These funds along with Louis's speaking fees produced enough money to start the next season's work.

There was also the chance of obtaining a grant from the National Geographic Society. Louis's photographer friend Armand Denis had given him an introduction to Dr. Melville Bell Grosvenor, then president of the Society. Louis met with Grosvenor in early November and the two formed an immediate and lasting friendship. "My father and Louis were two peas in a pod," said Gilbert Grosvenor, who succeeded his father as the Society's president. "Louis spoke MBG's language, and MBG instantly recognized the popularity of early-man studies and the importance of it." Melville Grosvenor arranged for Louis to meet with the National Geographic's research and exploration committee. Ever the showman, Louis produced a cast of *Zinj*'s jaw with a characteristically dramatic flourish, then sparked the committee members' imaginations with visions of all that remained to be found at Olduvai. Though Louis did not know it, and would not until early 1960, the committee voted that day to award him a $20,200 grant. It was by far the largest sum of money the Leakeys had ever seen, and it would make possible their next series of major discoveries.

Grosvenor also arranged for Louis to write an article for the magazine *National Geographic* about finding *Zinjanthropus.* This appeared in the September 1960 issue under the title "Finding the World's Earliest Man." In it, Louis estimated, as he had in all his articles and speeches about the skull, that *Zinj* "lived more than 600,000 years ago." But shortly after the article appeared, Louis was to learn that he had been wrong.

Two young geophysicists from the University of California at Berkeley, Jack Evernden and Garniss Curtis, had employed a new dating technique on volcanic ash samples found in the bottom of Olduvai Bed I, where *Zinj* had been discovered. Their analysis showed that the skull was not 600,000 or even one million years old, but rather 1.75 million years old. Thus *Zinj* instantly tripled the time that scientists believed humankind's ancestors had existed on earth. "One thing is certain," Curtis wrote to Louis in May 1961, after obtaining the first test results, "Olduvai man is old, old, old!" He was so old that at first, few scientists believed the date. "It was considered with a lot of skepticism everywhere," explained Yves Coppens, then a young paleontologist working in Chad, who was later to become the director of anthropology at Paris's Museum of Mankind.

> Everyone said the dates were good for the lava, but not for the skull. But as paleontologists are very clever as you know, they thought there must be many australopithecines in East Africa. So there was a sort of "bone rush" like the gold rush in the far West. Officially, the paleontologists were saying, "No, no, it's too old," but behind their backs they were packing their suitcases to go to East Africa. They came like locusts.

The combined interest of the National Geographic Society, the American public, and the eager paleontologists brought an end to the struggling life that Louis and Mary had known. After *Zinj,* the days of shoestring budgets and quiet times exploring Olduvai together were finished.

Chapter 14

MARY'S DIG

With their funds from the National Geographic Society and the Wenner-Gren and Wilkie Brothers foundations, the Leakeys planned a full-scale excavation of the *Zinj* site beginning in February 1960. It was an enormous undertaking, by far the most extensive dig they had made at Olduvai. They would eventually excavate an area forty feet wide, sixty feet long, and forty feet deep at this single site, a task that they estimated would require seven months of hard work. As much as Louis loved expeditionary life, it was impossible for him to get away for such an extended stay. His position as curator of the Coryndon Museum limited him to weekend and vacation visits to the gorge, and so he "most generously" made Mary the director of the excavations, as Mary wrote in *Olduvai Gorge,* Volume 3, her 1971 monograph about this research. Mary was not at all daunted by the size of the dig, having directed several others in England and Kenya, and lost no time in making this one her own.

Of utmost concern to her was hiring her own staff. In the past, the Leakeys had always employed Kikuyu workers—men who were often childhood friends of Louis's. But Mary was not fond of the Kikuyu (although she did retain Heselon and Thairu), and she decided instead to select her crew from the Kamba tribe, a farming people who live east of Nairobi in a hilly region called Machakos. "The Kikuyu couldn't really do anything wrong in Louis's eyes," Mary explained. "And that was a bit irritating for me . . . [because] they took advantage right, left and center. So I thought, if I'm going to have my own dig, I'll have Kamba."[1] She

[1] Mary was not the only one who felt this way. As early as 1931, on Louis's first Olduvai expedition, Vivian Fuchs had written in his journal, "[T]hese boys are some of the laziest I have seen since I have been out here, taking advantage of Louis Leakey their blood-brother as much as they can."

had first become acquainted with the Kamba people in the mid-1950s through Mutevu Musomba, a quiet, dignified man the Leakeys employed to care for their children's ponies. Mutevu was a traditional Kamba, his teeth ceremonially chipped and filed to sharp points. Although he had never been on a Leakey expedition, Mutevu had a gentle way with animals and so had won Mary's trust.

Mary also knew that the Kamba made beautiful wood carvings and intricate metal chains, and thought that their skills would translate well to the delicate task of excavating. But it was not easy explaining the precise nature of this work to people completely unfamiliar with it. "What we were told is that we were going to be digging for bones," remembered Kamoya Kimeu, who was a young man of twenty when his uncle Mutevu recruited him for the Leakeys. In time, Kamoya would become the preeminent fossil hunter in East Africa, finding more hominid bones than any one else. But when Mutevu asked him to come to Olduvai, Kamoya had no idea that fossils even existed. "We didn't know then about hominid bones, that there were such things. I thought we were coming to dig some graves of dead people. I didn't like that very much." None of the Kamba did. Their culture, like that of most African tribes, proscribes touching the bodies of the dead. "People think a dead person must be very much different," explained Kamoya. "You must never touch it because you may die like it. That's why we were worried."

Mutevu drove his recruits—who signed up because they needed the money—from the Kamba Reserve to Nairobi, where they camped on the museum's grounds. Mary had already departed for Olduvai, so the job of organizing the new workers fell to Louis. He came to greet them in the morning, a meeting that Kamoya and several of the other Kamba workers remember distinctly. "He spoke to us in Kikuyu [a language very similar to Kamba]," recalled Kamoya. "He said we were going on safari to Tanzania [then Tanganyika] in three days, and that he wanted us to work hard doing many things. He would give us food, blankets, everything, and he would pay us seventy shillings a month [about ten dollars]. I could see that this *mzungu* [white man] was not like the others; he was talking to us like a person. So I knew he must be a very good man, and it would be very good for me to stay with him."

At the gorge, Louis offered a further explanation of what he expected the men to do by giving them a lecture on evolution. But most of the Kamba recruits had attended a Catholic Mission School, and so, in spite of Louis's persuasiveness and obvious importance, they listened to him skeptically. "You see we believed in religion," said Joseph Mutaba, who is now the chief fossil preparator at the National Museums of Kenya. "We had been told that the first parents were Adam and Eve, and we had been convinced of these things. Then Louis told us about the ages of things, how one day in the Bible was actually a couple centuries, and that was very interesting. I liked that very much. But people coming from the apes—I didn't believe that." On other occasions, Louis demonstrated how to make stone tools and once used a sharp-edged chert flake to skin a ram for them—an impressive show, but not convincing enough to make the Kamba change their minds about Adam and Eve.[2]

For the next seven months, Mary and her crew would be living at Olduvai. They built a permanent camp, called Camp 5, on the north side of the main gorge, nearly opposite the *Zinj* site. Consisting of tents, a cooking hut, and simple grass *banda* furnished with wooden cases and shelves, tables, and canvas-backed camp chairs, Camp 5 would in time become Mary's primary residence. Some distance away from Mary's *banda,* the crew set up their own tents and cooking area—for the camp was to be run on strict British lines: management (Mary; nineteen-year-old Jonathan Leakey, who acted as camp manager; and Louis when he was in camp) would have its own living and eating quarters, while the help lived apart.

Unlike Louis, Mary was not fluent in African languages (though she could speak "kitchen Swahili," a patois used by most of the colonials) and did not have a natural curiosity about the different tribal customs. She was at Olduvai to work on what most keenly interested her: the stones and bones of the past. So at night she never joined the Kamba at their campfire to share their songs and laughter, but instead sat bent over her worktable, sorting and sketching the fossils and artifacts by lantern light. She also wrote long letters to Louis, and started her days by calling him on the radio telephone to report on the excavation's progress. Olduvai's remoteness and the camp's protocol were tremendously isolating, and the sound of Louis's voice on these calls was a comfort to her. Though she would never confess to loneliness, Mary did admit that she missed Louis "very much."

The *Zinj* discovery site, which Mary planned to excavate, lay on a rocky slope about ten feet above the gorge's sandy bottom. Above it, the earth rose up in a steep cliff of hard, fawn-colored sediments, barren for the most part, but occasionally spiked with a stand of sisal, a small thorn tree, and tufts of dry grass. On this expedition, Mary and her crew would cut back into the cliff face some thirty feet, then dig down through twenty feet of cement-like deposits (composed of layers of volcanic ash, limestone, clays, and sandstone) to the level she and Louis had identified as the *Zinj* living floor.

From their previous season's work, they knew the living floor was rich in fossils and artifacts. Just before closing their 1959 camp, the Leakeys had uncovered a hominid leg bone, a tibia, about eleven feet away from the *Zinj*

---

[2] Even today, after working for the Leakeys for more than thirty years and discovering many hominid fossils, Joseph, Kamoya, and Peter Nzube all dismiss the idea of a link between humans and apes.

skull.[3] Louis initially thought that because the two bones lay so close together, the tibia very likely belonged to *Zinj*. If so, *Zinj* would have been a creature of small stature, "probably like a pygmy," as Louis suggested.[4] Most important, the bone held the promise of additional discoveries, leading the Leakeys to hope that further excavations would reveal more of the skeleton along with *Zinj*'s lower jaw—a bone that Louis thought might show "whether our man . . . had the power of speech." But as Mary observed during a return visit to the gorge in 1984, "You dig for one thing and you find something quite different. . . . There are so many stories to be discovered and told here. And every one you find is different from the one you expected. It's always a surprise." The 1960 season's surprises proved to be many and varied, and they began almost with the first shovelful of dirt.

The excavation itself was a tedious process, especially with an untrained crew—and Mary was extremely particular about how the work would be done. While the overburden of loose soil and rocks could be removed with heavy picks and shovels, only ice picks, dental probes, and small, soft-bristled paint brushes could be used on the archeological levels.[5] In this manner, Mary and her crew eventually would remove several hundred tons of earth, excavate twenty-one levels that showed some sign of hominid activity, and expose 3,384 square feet of the *Zinj* living floor. But at the outset, the Kamba lacked any archeological skills, and it fell to Mary, with Heselon's assistance, to teach them the finer points of excavating. They also remained uncertain about the purpose of the dig—in spite of Louis's lectures—and watched Mary's initial preparations with some apprehension.

Particularly worrisome were the wooden stakes tipped with white paint that they watched Mary hammer into the ground. "That frightened me very much because I thought those must mark the graves," recalled Peter Nzube. Much to his relief, the stakes only indicated the first trench she wanted the men to dig. "We made a line with our shovels and picks," said Kamoya, "and we started digging. Then, to make it go fast, we sang a Kamba song we always sing when we start to make a new *shamba* [garden]. But Mary said,

---

[3] Jonathan actually found this bone. While helping at the site, he noticed one of the bones Mary's crew was unearthing nearby. "Does any animal have a long thin bone like this?" he asked his mother, tracing in the air with his finger the shape of the fossil he had spotted. Mary said she couldn't think of one. "Oh, then I think it must be hominid," Jonathan replied. Mary dropped her drawing, and the two rushed to see it.

[4] In 1964 the tibia would be assigned to a new species the Leakeys discovered, *Homo habilis*. Based on a partial female skeleton Richard Leakey's team discovered in 1972 at Koobi Fora, Kenya, scientists estimate that *Australopithecus boisei (Zinj)* females were three feet seven inches tall. Males were probably taller, about four and one-half feet, and heavier. Overall, the australopithecines had somewhat odd body proportions compared to modern humans, with large molars and premolars set in hefty jaws, and diminutive bodies. "Even early *Homo* specimens have disproportionately large teeth and small bodies," notes Henry McHenry, a physical anthropologist at the University of California at Davis, who has studied the sizes of the early hominids.

[5] A few months into the excavation, the Kamba crew devised their own version of an ice pick for digging through the sediments. They made the chisel point from a sixteen-penny nail and hafted this into a wooden handle, which they carved to fit comfortably in the palm of the hand. The resulting "Olduvai pick" is now the standard tool for East African excavations.

'What is this? This is not a potato *shamba!* Stop that singing!' " (It would not have mattered to Mary if Mozart had struck up a tune at her dig—she cannot abide background music of any sort. Consequently, there was never any music at her camp nor in the Leakey household.) The line of men instantly fell silent. In the days to come, they learned that talking was equally forbidden—"This is not a school! I want quiet on my dig"—and mistakes not easily forgiven. "Mary," said Kamoya, "didn't want somebody to do wrong and then say 'Sorry' to her. She didn't like that. If you did that, she'd say, 'I don't want your *sorry!* You go away.' She was a very *kali* [hot-tempered] teacher."[6]

Coming from a patriarchal society, where the women submissively wait on the men, the Kamba workers were stunned by Mary's stern manner. But they were also amused, and at night sat by their campfire shaking their heads and laughing at this *kali mzungu*—prickly white—woman and imitating her high-pitched commands. "Mary was another woman we hadn't seen before," Kamoya said. "If a Kamba lady makes a mistake, the husband can beat her. But Mary was doing the opposite. If Louis makes a mistake, she almost beats Louis. So when we talk about Mary, we laugh very much." In the daytime, however, they silently obeyed her edicts—or, as some did, quit angrily and returned to the Kamba Reserve.

Mary expected her workers to be as cautious, keen-eyed, and observant as she was herself, for they had to distinguish between the fossils and stones, pick out the tiniest fossilized rodent's bone on a sieving screen, or note the chipped edge of an artifact. "She knows that if you are talking you do not see well," said Kamoya. "To see those very small bones you want to be careful and quiet. So she says, '*Yamaza!*' [Swahili for 'Silence!'] and everybody is quiet."

Sometimes Mary explained the nature of her work in Swahili, but more often she taught by example, showing how she wanted each archeological level divided into a grid of four-foot squares, and then how to remove the earth carefully within each square in three-inch layers. Each fossil and artifact was to be left in place for her to number, map, and draw; fragile fossils were to be coated with a preservative before being gently lifted with a dental pick; then all the loose soil had to be swept up and washed and sieved through a mesh one-sixteenth of an inch wide, and every bone fragment found among the sieving debris set to one side. "This work was not easy. It was hard to know and a very big job," said Kamoya. "If Mary was not *kali,* I'm sure we would not learn. So everybody was working hard, running hard, studying hard."

Such diligence paid off. At the beginning of March, two hominid teeth and some skull fragments turned up in the scree, the eroded material lying in the gully at the base of the cliff. The fragments, no bigger than a half-dollar,

[6] In spite of this treatment, Kamoya and Nzube say they "like Mary very much. Because she taught us all very much. We have good work today . . . because of Mary."

included the season's first surprise: some pieces belonged to *Zinj,* others clearly did not. They were too thin to be part of his massive skull. Likewise, the teeth, an incisor and a premolar, were too small. Louis drove down from Nairobi to view them and initially thought they must have come from a female *Zinjanthropus.* "[W]e have some bits already of Dear Boy's wife, mistress, mother or sister!" he wrote delightedly to Phillip Tobias, adding his usual conspiratorial refrain, *"(but keep it quiet!)".*

Nothing more of their mystery woman turned up until June, when Mary started the work of excavating the actual *Zinj* floor. Finding the floor—and determining its dimensions—had been a task in and of itself. Mary and her crew had first excavated downwards in a series of six-foot-wide steps until they hit the basalt at the bottom of Bed I, a depth of forty feet. This "Trial Trench," as Mary termed it, enabled them to see clearly the different geological strata and to mark the precise level of what she interpreted as the living floor. The floor lay midway down, twenty feet below the surface. They then cut back into the hillside, dividing it into eight twelve-foot-wide strips, and excavating each to the *Zinj* level.

Mary could tell when they were about to reach the *Zinj* floor, as they first encountered a nine-inch layer of hard volcanic ash. The ash had the unfortunate habit of sometimes sticking to the clay and fossils below, forcing the workers to pare it off an inch at a time. Water proved even more effective at loosening the ash, and despite the tremendous shortage, gallons of it were poured over the site. In contrast, the men often were allotted only a cupful for washing at night. But after the water had soaked in, the ash came up in slabs, revealing a wealth of remarkably well-preserved fossils and artifacts underneath.

Scattered across the clay surface lay 3,150 fairly large fossils, several thousand bone fragments (some mere slivers), 2,470 large implements, and 2,275 stone flakes and chips. Mary recorded and mapped every item except those that were only millimeters in size, a painstaking task that produced what Desmond Clark termed a "new dimension in Palaeolithic research." Prior to her work, archeologists had assumed that any evidence of early human living or activity sites had vanished long ago—particularly after 1.75 million years. The idea of a place that the hominids regularly returned to, for eating and sleeping, also seemed implausible, although the Leakeys had been certain for some time that exactly such sites would be found. Mary's excavations not only seemed to substantiate their theory, but also gave scientists a first glimpse of early human behavior.

As she recorded the material, Mary was struck by one particular pattern. In an oval area measuring twenty-one by fifteen feet lay a dense accumulation of smashed animal bones, small tools, and stone flakes. Some distance away, there was another concentration of material, but here the artifacts were heavy and the fossils large and mostly unbroken. A narrow strip of ground, nearly devoid of fossils and tools, separated the two collections. The artifacts and bones were unweathered, indicating that they had been buried quickly, perhaps by rising lake silts, and had not been disturbed since the time of *Zinj*. What could explain the difference between the two groups of stones and bones?

After closely studying the types of fossils and artifacts found in each area, Mary suggested that the oval marked a place where hominids had gathered to eat the meat from animals they had either hunted or scavenged. Apparently they had come to this same spot for many years, for the "food debris" accumulation was very thick. She pictured them seated together, striking razor-sharp flakes from blocks of quartzite, using these to carve the meat from the bones, then smashing the bones with hammerstones to extract the marrow. While many of the bone fragments in the oval area came from the limbs of antelopes, those in the adjacent collection were largely ribs, jawbones, hipbones, and vertebrae—bones lacking any marrow. Perhaps after stripping off the meat, the hominids had cast these bones aside, tossing them over a brushy fence they had built to protect themselves from the wind. Such a shelter would have fit neatly into the strip of barren ground separating the two collections.[7]

The hominids had apparently availed themselves of whatever food was at hand, for the oval was thick with the tiny bones of rodents, chameleons, lizards, birds, fish, and snail and mollusk shells. There was no evidence of fire, so they very likely ate their food raw. (They may have consumed some of the smaller animals whole as well. At another site Mary excavated, they uncovered what she interpreted as "the residue of human faeces"—little piles of crushed bones from small animals like mice and lizards.)[8]

Mary's massive excavation revealed no further sign of *Zinj,* although some distance away from the oval, her crew uncovered a second hominid leg bone, a fibula (the outer bone in the lower leg) this time. Like the tibia they had found the previous season, the fibula was a slender, short bone, and Louis surmised that the leg did not belong to *Zinj* after all, but to his consort.

Louis again drove down to the gorge to view the bone, bringing with him Philip and Richard, free on their school vacation. They would stay for the summer, while Louis, rushed as usual, had to return to Nairobi in a few days.

---

[7] In volume three of *Olduvai Gorge, Excavations in Beds I and II, 1960–1963,* Mary writes that Desmond Clark first made the suggestion that this empty stretch of ground might have been caused by a shelter. Today, however, archeologists think it more likely that a low shrub or row of tall grasses left the blank space. The idea of shelters or home sites at this early stage—nearly 2 million years ago—has fallen out of favor, although the idea of hominids gathering together at certain sites for some activity has not. "That's the main thing that Mary's research revealed," said Rick Potts, an archeologist at the Smithsonian Institution. "She pointed out the association between the stones and bones. And that is as exciting now as it was then, because it means that the hominids were engaged in cooperative behavior of some sort. They also had sites they returned to on a regular basis, although we don't know what the magnet was."

[8] Louis tried to persuade several young men, including his sons and the anthropologist Irven DeVore, to eat a raw rat so that Louis could compare the final product with this coprolite. "Louis put his arm around me," recalled DeVore, "and said, 'My dear boy, let me make you famous,' " and then explained the experiment. When DeVore asked Louis why he didn't eat the rat himself, Louis said, "It wouldn't do any good. I'm nearly toothless; it would be a poor experiment. But you're a strapping young man—just what I need." DeVore declined the honor, as did Louis's sons.

During his visit, he discussed the various finds with Mary, planned a future visit for one of the members of the National Geographic research committee, and did some digging and exploring of his own. He also examined an unusual jawbone Jonathan had discovered lying on the surface about one hundred yards away from the *Zinj* site. The fossil was that of a saber-toothed tiger, an animal that had never been found before at Olduvai, or anywhere in East Africa, and Louis urged Jonathan to sieve the site; perhaps he would find more of the animal's skull or skeleton.[9]

A few days later, back in Nairobi, Louis's radio telephone crackled with the report of an important new find. No additional bones of the saber-toothed tiger had turned up in Jonathan's sieving. Instead, he had unexpectedly discovered a hominid tooth and toe bone, and, with his mother's guidance, was now starting an excavation at site FLK-NN, or, as the Leakeys came to call it, "Jonny's site."

Although photographs of Mary at Olduvai typically show her standing above her crew, giving instructions, she was more often in the excavation pit, crouched on hands and knees, and wielding an Olduvai pick and brush. She swung a heavy pick alongside the men, too, when trenches were being dug, and joined them in the chill of the morning to wash the dirt in the sieving pile—and always while she worked, a cigarette dangled from her lower lip "like a French person," as one friend remarked. The tough work called for plain clothes, and she dressed simply in sleeveless blouses, khaki drill trousers, and a battered wide-brimmed cloth hat. Trailing at her heels, wherever she went, were five black-and-white Dalmatians.

In mid-August, the Dalmatians were at her side when Mary carefully extracted the first of several hominid foot bones at Jonny's site. Within an area of one square foot, she uncovered fourteen foot bones in all, almost a complete specimen. It was a stunning discovery, the first foot of an early human, and Mary was on the radiophone early the next morning, summoning Louis.

Louis arrived at the gorge the following day, having driven the 357 miles of dusty dirt roads with scarcely a break.[10] Mary spread the bones out for Louis to examine and he quickly confirmed what she had suspected: they had discovered the world's oldest hominid foot. In the next few days, more

---

[9] Three years before, when he was sixteen, Jonathan had made his first important discovery at Olduvai. On that expedition, he had picked up the jawbone of an extinct giant baboon, an animal as large as today's gorilla. Louis was particularly delighted by the find, calling it in a *National Geographic* article "the most fantastic and dreadful creature [at Olduvai]," and had named it *Simopithecus jonathani* after his son.

[10] Louis was notorious for his driving—which bordered on the reckless—and his stamina. In 1957 he once drove almost nonstop from Nairobi to the town of Iringa, Tanganyika—a distance of nine hundred miles—to inspect Clark Howell's nearby archeological dig. Betty Howell, Clark's wife, had waited at a small hotel to greet him, and when Louis, covered with dust, jumped out of his Land Rover, she suggested he might like a brief rest and a cup of tea before heading out to the site. "Tea, girl!" responded Louis. "I didn't drive nine hundred miles to drink tea! Let's get on with it." They went immediately to the dig.

hominid fossils turned up: several finger bones, a collarbone, and tiny bits of skull. "[I]t looks as though we are gradually moving towards what might be a body," Louis wrote elatedly to Le Gros.

The skull fragments were particularly tantalizing. They were fragile and thin, like those from the unidentified hominid at the *Zinj* site, and Louis began to wonder if this creature was something other than *Zinj*. All of these bones had been unearthed in deposits a full foot lower than the *Zinj* living floor—strong evidence that whatever hominid it was, it had camped beside the ancient Olduvai Lake several thousand years before *Zinj*.

Only one month later, in mid-September, Louis's suspicions were confirmed. Mary had gone to Arusha to meet a former archeology student of Clark Howell's, Maxine Kleindienst, whom Mary had invited to begin excavations in Beds III and IV. Jonny had stayed behind in camp and was digging at his site when the two women arrived. "He sort of had this Cheshire cat grin on his face," Maxine recalled. "But it wasn't until we sat down to lunch that he brought out his handkerchief and handed it to his mother with [a] couple of skull fragments." Mary studied the fragments for a moment, said nothing, then passed them to Maxine, who instantly saw why Mary had been silent. "They were obviously the parietals [the bones that form the top of the skull], but they did not have a sagittal crest, and I said to Mary, 'No sagittal crest.' Mary said, 'Yes,' and then with a great sigh of resignation, 'Now, we, too, have *Telanthropus*.' " Her sigh acknowledged the inevitable: Louis, though he did not yet know it, was about to become embroiled in another debate.

Telanthropus was the name Robert Broom and John Robinson had given in 1949 to what they believed was a new genus of hominids. They based their genus on a single specimen: a hominid's lower jawbone, which Robinson had extracted from South Africa's Swartkrans Cave two years earlier. Prior to this discovery, they had unearthed several skulls, upper and lower jawbones, and other skeletal remains from the robust australopithecine Broom called *Paranthropus crassidens*—the creature that bore the most resemblance to *Zinj*. But Robinson's little jaw, with its delicate shape and small molars, did not look at all like *Paranthropus*. The teeth, in particular, were much more like those of modern humans, and Broom began referring to *Telanthropus* as a relative of "true man." The name and the paucity of specimens did not upset other anthropologists nearly as much as the idea that *Paranthropus* and *Telanthropus* must have been contemporaneous.[11] Many scientists in the field, including Le Gros Clark, argued that because of the peculiar adaptations of early humans—the upright stance, the free hands, the large brain—only one type of hominid could have been alive at any point in time, a hypothesis that came to be known as the "single species" theory. But now it seemed—as Mary knew the minute she glanced at the

---

[11] Robinson reclassified the *Telanthropus* specimens as *Homo erectus* in 1961.

skull fragments Jonathan handed her—that there had been two species of hominids living side by side at Olduvai as well.[12]

If she and Louis had any remaining doubts, these were dispelled on November 2, when Jonathan unearthed a lower jaw with thirteen opalescent teeth. The fossil was somewhat distorted, fractured and incomplete, but nevertheless sufficiently intact to verify their hunch: this hominid was not another *Zinj,* not even a female *Zinj.* Well aware that the suggestion of two humanlike creatures, each standing upright, and living together on the African savanna was certain to provoke criticism, Louis broached the subject with a mix of caution and defiance in his letters to Le Gros.

"Let me add," he had written in September, after telling Le Gros about the skull and hand fossils, and his growing suspicions that they came from some creature other than *Zinj,* "that I am not in the least concerned about what outside observers say, or do not say, the evidence is such that nobody can gainsay it if they are honest but some of it is a little disturbing."

By November, with the hominid's mandible in hand, Louis discussed his idea more openly. Enclosing photographs of the hand, foot, and skull bones, he told Le Gros that he did not think the fossils "fit into the australopithecine pattern. . . . I have an increasingly strong feeling that, in point of fact, we have two entirely distinct hominids living side by side in Bed I at Olduvai, just as we have some eight different genera of pigs side by side and so on."

Le Gros replied immediately. Despite Louis's previous dismissal of other people's opinions, he was unprepared for what he read. Le Gros first admitted that the skull bones did appear larger than he had imagined, but then confessed to being "very puzzled about the jaw. It is really not unlike that of a smallish female gorilla . . . but it may be that the canine and the first premolar are quite different. I suppose by the way this is a genuine fossil! [Here he added a handwritten asterisk and the note, "Shades of Piltdown arise before me!!"] Perhaps it is ridiculous of me to make such a suggestion but I do find it difficult to suppose from the photograph that it can possibly be a hominid jaw. . . ."

Louis was stunned—first by the suggestion that the fossil might be fraudulent, and then that he was incapable of distinguishing a hominid's jawbone from that of an ape's. As soon as he finished reading the letter, he dispatched a telegram: "STAGGERED YOUR DOUBTS INTELIGENCE [*sic*] AND . . . INTEGRITY/SUGGEST YOU COME SOONEST, IF NECESSARY MY EXPENSE SEE SPECIMENS AND SITE EVIDENCE."

Le Gros's reply arrived within a week and was all apology. "How badly I must have worded my letter to give you this false impression of my meaning! I am afraid I was trying to be a bit funny (not very successfully!)—& you know me well enough, I do hope, to realize that I have never for one

---

[12] Although the Leakeys did not realize it, they had actually unearthed the first fossils—a molar and a fragment of the jaw—of this new hominid in 1959, shortly before the discovery of *Zinj.* Phillip Tobias later classified these fossils as *Homo habilis.*

moment doubted your intelligence—& least of all your integrity. This new jaw does puzzle me immensely, however, & I still wonder whether it can be a hominid jaw. . . ." He could not, however, get away and would wait for Louis to bring the specimens to England.

Perhaps Le Gros was only joking, but the suggestion of fraudulence must have wounded Louis deeply. Similar innuendos about his Kanam/Kanjera hominid fossils had set back his early career. Yet he graciously forgave Le Gros and thanked him for his apology—though he also added somewhat testily, "I think both Mary and I do know a genuine fossil when we see one and can distinguish a pongid [ape] from a hominid without any difficulty." He also described the fossils in further detail, explaining why he believed they differed from other hominids, and emphasizing that they now had at Olduvai a "quite distinct and more manlike, though very primitive, hominid." He suggested that it might be the ancestor of Robinson's *Telanthropus.*

But Louis and Le Gros would never agree about the small hominid from Jonny's site. Indeed, this fossil would eventually upset all of paleoanthropology, and make Louis's previous battles over the Kanam jaw and the naming of *Zinjanthropus* look as civil and polite as an English tea party.

The Leakeys made one more spectacular discovery at the end of the 1960 season. Mary had extended her stay at the gorge for six more months after the National Geographic awarded the Leakeys another $14,560 grant. By November she had three digs in progress: the *Zinj* living floor; Jonny's site; and another one midway between these two called FLK-N, where they had discovered a hominid toe bone and quantities of artifacts surrounding the complete skeleton of the elephantlike creature *Deinotherium giganteum.* They had also added two new members to their collection of extinct giants —a giant species of swamp antelope and a giant porcupine.

Louis was naturally delighted by the great variety of discoveries, and kept up a massive "confidential" correspondence with his colleagues, announcing everything that Mary and her crew found. He kept in check any frustration he might have felt at having to watch the work from the sidelines. As often as he could, he traveled to Olduvai, pitching in with the digging and exploring the gorge for new productive sites.

On the first of December, Louis joined Ray Pickering, a geologist attached to the Tanganyikan Geological Survey who was helping to map the gorge, for such an exploration. The two men walked a mile away from camp, climbed a knoll, and paused a moment to look back at the *Zinj* excavation site. It was then that Louis noticed an area of deposits in Bed II that he had not yet surveyed, and he decided to investigate it the following day.

The next morning, he set off with young Philip and Pickering, and the trio were soon scrambling over rocks and picking their way through dense, spiky vegetation. As they reached the slope Louis had spotted the day before, he had a sudden "premonition that something important was about to turn up," and said "half-laughingly" to Pickering, "This is the sort of place where we'll find a skull." Moments later he did just that, spying a small, rounded pile of bones poking up through the soil. Louis was stunned, and at first thought it might actually be only a tortoise's shell, which can look deceptively like a human skull. But another glance told him his first hunch was right. "I fell on my knees beside the spot, and . . . for a moment I could not speak coherently," he would later write in an article for the *National Geographic.* "At last we had found what countless prehistorians had sought for more than a century . . . Chellean Man."

Although the tear-shaped handaxes of the Chellean culture littered the gorge and were distributed across Africa, Asia, and Europe, very little was known about the people who had made them. The only fossils that seemed to be directly associated with handaxes were the Swanscombe skull from England, and three jaws Camille Arambourg had discovered in Algeria in 1954. Louis's new skull was obviously more primitive than either of these, and also had the advantage of being datable. In time, the geophysicists at the University of California, Berkeley, would assign a date of 1.4 million years to this fossil. It was thus younger than either *Zinj* or the new hominid from Jonny's site and, Louis thought, might be a descendant of one of these.

Ecstatic over his discovery, Louis left Pickering and Philip to guard the bones and drove madly back to Jonny's site to tell Mary. She met his announcement that he had found Chellean Man with the suspicious reply, "What do you mean, you have Chellean man?" "A skull, a skull!" he answered almost breathlessly. "I've found a Chellean *skull.*" In the next instant, Mary dropped her work and was off to the new site with Louis. Together they searched the slope for additional pieces of the skull, and eventually found nearly all of its cranium plus the brow ridges, which Mary described as "unbelievably massive." Although the skull's face was missing, the shape of the braincase and the thick brow ridges indicated that this hominid was a close relative of Java Man and Peking Man—fossils that would soon be grouped together as *Homo erectus.* Louis had always maintained that such creatures were only side branches on the true tree of humankind. But now, with this thick-browed skull in hand, he glossed over its resemblances to other fossils and claimed that he had uncovered "a direct Chellean ancestor of *Homo.*"

Six days after making this discovery, Louis was back in Nairobi, sending letters to everyone from Le Gros Clark to his old and new patrons, Charles Boise and the National Geographic Society. Almost simultaneously, a report Louis had written describing the Olduvai fossil discoveries through September appeared in *Nature,* together with photographs of the hominid foot and leg bones. It was a stunning array of finds—to his colleagues it seemed like a yearlong run of incredible good luck—and they responded with a mixture of admiration and amazement. "I have just seen the photograph of the foot in *Nature,*" wrote Peter Davis, an anatomist at London's Royal Free School of Medicine, "and am trying to be objective about it in spite of growing excitement. Surely it is a relatively short, broad, but *clearly bipedal* foot?" Le Gros had similar words of praise for Louis's Chellean Man, as did Desmond Clark, who had toured the excavations in September.

Louis and Mary were finally making the kinds of discoveries that Louis had always known were possible. If it seemed surprising to others, it was not to him. He enjoyed the acclaim he and Mary were receiving at long last, but he was also a little piqued by his colleagues' astonishment that they were finding so many fossils. "You may be wondering," he told Le Gros, "as others have done, why we are finding so much hominid material now. . . . The answer is simple. We have been working *continuously now since February with adequate funds* and a huge labour staff, and have already put in some 72,000 man hours this year. Had we been able to do this sort of thing before, we would have had the results before."

The Leakeys' success did not end soon. Throughout the 1960s they continued to unearth important fossils at such a rate that their colleagues dubbed their good fortune "Leakeys' luck."

Chapter 15

MURDER AND MAYHEM

Four months before finding the Chellean skull, Louis had celebrated his fifty-seventh birthday, toasting it with a glass of Scotch with Mary at Olduvai. Despite his active and vigorous life, Louis's hair and mustache were snowy white, his skin weathered, his teeth badly decayed, and a decided paunch had settled at his waistline. He had arthritic pains in his hip joints as well, but chose to ignore these as he had every other ailment. (Over the years, he had suffered numerous attacks of malaria and three bouts of schistosomiasis, as well as kidney stones, gallbladder attacks, glandular fever, and an occasional epileptic fit, but seldom had bothered to see a doctor.) But if he had aged physically, Louis was still young in spirit. His eyes shone mischievously beneath his arched brows, giving his face a pixieish look, and, in spite of his extra girth, he was solidly built, "a bear of a man," as one friend phrased it. He walked with a youthful spring to his step, too, his neck craned slightly in front of him, as if his head and thoughts were rushing on ahead and his body were merely something to drag along behind.

It was this well-worn yet boyish figure, dressed in a dark suit and tie, who strode to the podium at Constitution Hall, Washington, D.C., in late February 1961 to deliver the first of what would be many lectures for National Geographic Society members. Louis had titled his talk "Finding the Earliest Man," but it was more a naturalist-anthropologist's travelogue of East Africa. Narrating a film Des Bartlett had shot of wildlife on the Serengeti, of the Maasai people, and of the Leakeys working at Olduvai, Louis charmed the audience with his eccentricity and unabashed corniness. Not only did he know the habits of the animals, he spoke for them as well: "Here you see just immediately after birth the baby [zebra] struggling to stand up . . . 'Ah, I am free. Now, I have got to get steady. I have got to learn to balance. . . .' " He rhapsodized over the images of snowy Mount Kilimanjaro ("What a beautiful sight, indeed . . .") or of wildflowers at Olduvai ("They are *wonderful* flowers"); joked about using dental picks for excavating ("My wife and I are working with those horrible tools a dentist uses in your mouth. We use them for a better purpose"); and cracked one-liners about him cleaning *Zinjanthropus*'s teeth ("This was the first time, of course, he had ever had his teeth cleaned with a tooth brush.") When the lights came up, the audience rose for a standing ovation, and the National Geographic officials knew that in Louis they had found a star. "He was, and is, an *extraordinary* find," wrote Melville Grosvenor to Armand Denis, who had introduced Louis to him.

Louis had come to Washington mainly to lecture—at the National Geographic headquarters, the Smithsonian Institution, and the National Academy of Sciences—but also to hold a press conference and to appeal for a new grant from the Geographic. As usual, Mary had elected to stay behind. She wanted nothing to do with either lectures or the media, though she readily acknowledged that "an exciting report [would] persuade our financial backers to continue their support." In fact, the National Geographic's Research and Exploration Committee took little convincing, awarding the Leakeys another $28,000 to continue the work at Olduvai. They also made a separate and unusual arrangement with Louis. To free him from his duties as curator of the Coryndon Museum, the National Geographic would pay his salary (now a modest $7,800) for the next three years. He could thus pursue fossils full-time.

"It was an exceptional arrangement," said Mary Griswold Smith, now a retired National Geographic Society editor, who worked closely with Louis for twelve years. "Normally the Geographic doesn't pay the salaries of people receiving grants. But in Louis's case—part of it was that they [the Leakeys] had no money and part of it was his persuasiveness. He could charm the birds out of the trees."

Yet Louis seemed unable to duplicate this success in Great Britain. With the 1960 season's bounty of fossils in hand—two types of hominid skulls, as well as hominid leg, hand, and foot bones—Louis was invited to deliver several speeches, including the prestigious Herbert Spencer lecture at Oxford University and the Thomas Huxley lecture at the University of Birmingham. But the crowning glory for any British scientist—to be elected FRS, a Fellow of the Royal Society, Britain's most elite scientific organization—continued to elude Louis.

For Louis, the initials "FRS" after his name would, he felt, finally erase any lingering suspicions that Boswell and the Kanam/Kanjera affair had raised about his scientific credibility.[1] He would unquestionably be part of the British scientific establishment, no longer the slightly disreputable colonial hungering after recognition from home.

Although the Society's election process is supposedly secret—even the nominees are not told that they have been suggested for membership—Louis was well aware in early 1960 that Le Gros Clark was laying the groundwork for his nomination.[2] Le Gros himself had been elected a Fellow in 1935 for his research on primate brains (he had also been knighted in 1955, and so was properly known as Sir Wilfrid). The key criterion for selection was "original research," but it was common knowledge that politics and personalities played a part as they do in any club. This element and Louis's love of speculation worried Le Gros the most, and he urged Louis again and again "to be moderate" in his claims for his new discoveries, and to focus his energies on the Olduvai excavations. "In particular," Le Gros advised, ". . . it would be highly advisable to avoid *controversial* issues or *other* matters, for this *might* possibly give a handle to critics who might refer to them in a disparaging way."

Doing his best to heed Le Gros' advice, Louis was uncharacteristically cautious during this period. He did not give a hint to the press about his Cheliean skull, or the jaw and skull bones of the slightly built creature from Jonny's site. Nor did he create a special zoological name for this hominid as he had for *Zinj,* although he was already convinced that it was a new species. Instead, he drafted a straightforward descriptive report for *Nature* entitled simply "New Finds at Olduvai Gorge." The article appeared on February 25, 1961, the day after he held his first National Geographic press conference.

A crush of reporters gathered at the Geographic's headquarters that morning to hear Louis announce his new discoveries. Displaying casts of the fossils, Louis cautiously referred to the hominid from Jonny's site as "earliest man." The skull, he said, was that of a child about eleven years old, who

---

[1] Nearly thirty years after Professor Boswell assigned Louis's Kanam and Kanjera fossils to a "suspense account," the incident still had not been forgotten in England, and many scientists there continued to consider Louis a charlatan. "Louis was not highly regarded in England," recalled Ernst Mayr, Harvard's distinguished zoologist who spent several years at Oxford University. "His reputation at Cambridge was very poor because of his frauds."

[2] In fact, various people had made efforts on Louis's behalf since the mid-1950s. Le Gros Clark, Kenneth Oakley, and Errol White, chairman of the paleontology department at the British Museum of Natural History, had exchanged letters about Louis's chances for an FRS in 1958, and concluded that they were slim. "White thinks that the chances of Louis Leakey getting into the Royal as a palaeontologist are small at any rate for a very considerable time," Oakley had written to Le Gros. "He had himself made tentative moves for support for L.S.B.'s nomination, but so far with no success." The three men thought this was partly because of Louis's reputation, but also because very few archeologists or anthropologists were elected to the Royal Society. They decided that Louis would have better luck trying to become a Fellow of the British Academy, though Louis was at first hesitant about pursuing this honor. "If I accept the suggestion [to be elected to the British Academy]," he wrote to Kenneth Oakley in January 1958, "will it remove or lessen my chances for FRS? I feel that my Miocene discoveries since 1940, my Pleistocene work at Olduvai and elsewhere, and my work on pluvials is more appropriate to FRS than FBA, and that is what I would like to aim at. If accepting your proposal would ruin my chances for FRS later on, then I'd rather wait." On Oakley's and Le Gros's advice, Louis went ahead with the FBA and was elected in 1958.

had lived many thousands of years before *Zinjanthropus.* He pointed out differences between *Zinj* and the new fossil, suggesting, as he had in his *Nature* paper, that two different hominids may have lived together at Olduvai. With a string of "might have beens" and "possiblys," he described the child as a massive creature with a large chest and short legs. "Beyond that," he joked, "your guess is as good as mine. They [the hominids] might have been as green as a jersey dress." Perhaps it was then that a reporter asked Louis about the visible hole in the child's skullcap; or perhaps Louis pointed to it himself, although he always insisted that he had only answered a reporter's leading question.[3] Either way, the hole and fracture lines radiating from it were noticed, and Louis readily supplied a story. He thought the child had died as a result of the injury, but doubted if it had been caused by a fall. Instead he told the reporters, "I think we can take it for granted that the child was hit on the head by a blunt instrument. It was murder most foul."

Instantly, there were headlines worldwide: "World's First Murder?" queried one paper, while another stated "Earliest Human Killed by Blunt Instrument," and a third, echoing Louis's words, shouted, "Murder Most Foul." But in England, Louis's tale of a "blunt instrument" met with a decidedly cool and skeptical reception. The *Times* (London) hung quotation marks around its headline: " 'Murder Verdict' on Hominid," followed by a single, cutting subhead, "Anthropologists Puzzled." Noting that Dr. Leakey's report to *Nature* had not mentioned a blow to the head, the *Times* correspondent wrote, "British anthropologists were left wondering during the weekend what new consideration of evidence had led Dr. L. S. B. Leakey . . . to bring in a verdict of murder, hundreds of thousands of years after the event. . . ." Louis's claims created such a stir in the papers and among his British colleagues that even the humor magazine *Punch* lampooned him. Entitled "More Secrets from the Past: Oboyoboi Gorge," the satire starred the well-known anthropologist Dr. C. J. M. Crikey. More damaging were the serious questions raised in *The New Scientist.* "It would be interesting to know what Dr. Leakey actually said at his press conference," the magazine editorialized.

> It is entirely possible that he did nothing but answer a leading question on this point. But whatever happened, it is extremely unfortunate that wild speculations—and they can be nothing else—about the way in which beings 600,000 years old met their death should be given the wide currency which modern methods of communication make possible.[4] It is not merely that some ordinary people are misled. . . . [But] because tales like

---

[3] Mary apparently accepted Louis's version of what transpired. In her autobiography, she wrote about this incident: "[The press] pounced on his description of a certain fracture in the centre of one of the parietal bones and cornered him into agreeing that it could have been made with a blunt instrument. That was enough: here was evidence of 'the first murder.' "

[4] When this article appeared, the dates for Olduvai Gorge, establishing that the hominid fossils were nearly two million years old, had not yet been published.

this are all too obviously unsubstantiated, the public image of this important field of science becomes more than a little ridiculous.

Louis always insisted that he never uttered the word "murder." Perhaps he did not; records of the press conference are not extant, so it is impossible to verify his wording. But in a *National Geographic* article written by him that appeared later in 1961, he elaborated on the child's death: "It would seem . . . that our child died by violence rather than disease. . . . The child's left parietal shows clear signs of having received a blow. There is an obvious point of impact, a break in the skull that reaches even to the inner wall, and fractures that radiate from the break." Discounting the possibility of a fall, Louis reiterated what he had told the press: "I think it is reasonable to say that the child received—and probably died from—what in modern police parlance is known as a 'blow from a blunt instrument.'" Although he did not use the word "murder," he nevertheless left his readers with the impression that this is what had occurred.[5] Scientists who study the process of fossilization say that the hole and fracture lines in the skull were most likely caused by the surrounding clay soil expanding and contracting over the millennia.[6]

By the time Louis returned to Kenya in early March, his nomination for FRS was dead. Several years would pass before he and Le Gros discussed the subject again.

Mary had not waited idly while Louis traveled abroad. She kept the Olduvai camp running through the end of February, then packed up the crates of stone tools, fossils, and all the camp gear and headed back to Nairobi. When Louis arrived, she was already ensconced in the museum's laboratory, sorting through the previous season's haul of artifacts. She and Louis planned a series of monographs on the Olduvai excavations, with hers focusing on the archeology and his on the animal fossils. Not a trained anatomist, Louis turned over the job of writing the detailed analysis of *Zinjanthropus* and the other hominid skulls to Phillip Tobias, Raymond Dart's protégé.

Tobias apparently had won Louis's approval by making a fresh study of the Kanam jaw, and attempting to reinstate it as a valid find.[7] "I concluded that it [the jaw] was a very interesting, very important and primitive speci-

---

[5] It was not the first time that Louis had speculated about how the hominids he found had died. In 1960 he told a reporter from the *Sunday Post* that *Zinj* "probably was gored by [a] rhino or a giant pig and crawled back home to die. That's why we were able to find him. . . . If he had died normally he would have been eaten by [a] hyena as no one ever thought of burying bodies in those days."

[6] In *Olduvai Gorge,* Volume 3, Mary Leakey suggested both possibilities: "The fact that the cracks radiate from a central point may be due entirely to natural causes, resulting from earth pressure being greatest on the most convex part of the bone, but it is also possible that they may have been caused by a blow."

[7] It was Tobias who realized that the bony growth on the Kanam jaw that looked like a chin was in fact an abnormal growth, resolving a puzzling anatomical mystery.

men," said Tobias, "and that Boswell's case had been overstated." He presented his results at the 1959 Pan-African Congress—the same meeting where Louis and Mary first revealed *Zinj*.

> I think they liked my approach to the Kanam study . . . because [a few days later] Louis and Mary said, "We'd like to talk to you, Phillip." And they took me on one side, and they had the little glory box [containing *Zinj*] under their arm, and we wandered off a little among the trees, and they said they would like me to undertake the study of this marvelous specimen. Well, my knees went absolutely gelatinous on me. I was quite overwhelmed. With all the great scientists and anatomists in Great Britain to whom Leakey had ties—they had chosen me, working all the way down at the other end of the continent.

It was the beginning of a lasting association and friendship.

Tobias was not the only expert the Leakeys sought. Louis engaged specialists in Britain, France, and America to study everything from the mollusk and rodent fossils to those of the giraffe and hippos. He asked the British anatomists John Napier and Peter Davis, who analyzed the *Proconsul* limb bones, to turn their attention to the hominid hand and foot bones. The Leakeys needed a more highly qualified geologist, too, and through colleagues at the University of California at Berkeley found one in Richard Hay. He would join them at Olduvai in the summer of 1962.

Many of these people, as well as other visiting scientists, came to Nairobi to examine the collections at the museum, and nearly all of them were invited to the Leakeys' home either for dinner or an extended stay.[8] It was not a quiet household they stepped into: five Dalmatians bounded up in greeting, little duikers stepped daintily from room to room, hyraxes cowered behind the toilet, an African eagle owl peered from his nest on top of a cupboard, and Simon, the mischievous Sykes monkey, chattered bright-eyed from his perch on the couch. Eight tanks of tropical fish—primarily fancy, prizewinning guppies that Louis bred for a hobby—lined the walls, while a cage of American rattlesnakes lived next to the telephone. ("[They] invariably rattled in alarm whenever the telephone rang," Mary noted in her autobiography.) There were more snakes in two large brick-lined pits in front of the house—one to hold Jonathan's pet python, the other his collection of venomous vipers. Alarmed by this writhing mass, Betty Howell, who had come for a three-week stay while Clark was in the field, asked Louis, "But don't they get out?" "Oh, sometimes," he replied nonchalantly. "But we try to keep them in"—an answer that did not settle Betty's nerves at all. She was further dismayed that evening when washing up. "I suddenly felt

[8] So many visitors trooped through the Leakeys' home that young Richard once confided to a visitor, "This isn't like a home at all. . . . It's like a hotel. My mother and father spend all of their time with colleagues, and none with us. And as soon as I'm old enough I'll go out and build my own house and make it a home."

an uncomfortable feeling at the nape of my neck, and I looked up and what do I see in a window above me but a civet cat [a small carnivore]! I thought, 'Oh, my God! Where's the nearest hotel? I don't have to put up with this.' " She changed her mind only after Louis explained that if she showed no sign of fear, the animals would not bother her. Nevertheless, she was bitten by Simon the monkey and one of the hyraxes.

Stories about the Leakeys and their menagerie quickly circulated and visiting scientists arrived ready to take the animals in stride. "I knew that one always had to be particularly careful with animals in [Mary's] presence," recalled Alan Gentry, a specialist on fossilized antelopes, who first dined at the Leakeys' in 1961. "I had been told, although it didn't happen to me, that if a hyrax came walking along the dinner table and wanted to help itself to what was on your plate, you weren't allowed to smack it out of the way or cast it down or reprimand it in any way." The hyraxes were particularly fond of cigarette tobacco and gin, and more than one guest found the furry creatures nibbling at their pockets or tippling their drinks—a habit that Simon was also allowed to indulge.

The Leakeys' backyard was equally chaotic. Their African staff lived here in small cinderblock buildings with their wives, children, and visiting relatives. "It was like a little village out there, like something you'd see in the old South," said Betty Howell. "I saw fifteen or twenty people sometimes, with fires going and people cooking and babies crying. I asked Louis and Mary many times, 'Who *are* all these people? There are women and babies and old men and old women. What are they all doing here?' And they didn't really know. Nobody seemed to know who they were, and they didn't care either."

None of this pandemonium appeared to bother Louis and Mary in the least. Mary had early on adopted a laissez-faire attitude toward household affairs, letting the servants tend to what needed to be done. Yet every Saturday she baked a cake, and once a year she stirred up a potful of orange marmalade. "I made that with her once on their porch," said Betty. "She always had a little cigar in her mouth, and the ashes would fall into the marmalade. But that didn't make one whit of difference to her."

If the general tone of the house seemed casual, dinnertime was not (although there were the five Dalmatians crowded underfoot and the odd hyrax sauntering from plate to plate). "A real Victorian atmosphere" prevailed at the table, one visitor recalled. Louis sat at the head of the table, with the plates stacked in front of him. Mary sat at the other end, and the three boys and any guests were seated in between. Once everyone had settled in, Louis rose to carve the meat (for dinner was always a traditional English meal of roast beef, pork, lamb, or chicken), and then passed bowls heaped with potatoes and green beans, and a platter of fresh bread, which Louis had often made himself. (Louis loved to cook, but aside from his bread, which everyone enjoyed, his cuisine received mixed ratings. Betty Howell watched Louis cooking on several occasions, and recalled that he "wouldn't cook anything that took more than fifteen minutes. He used to play a game with himself. He would try the most complicated dishes and test himself . . . to see if he could prepare them in fifteen minutes. Once he made ravioli. Well, they were a disaster, but they were finished in fifteen minutes and he thought they were wonderful.") The Leakeys liked their meat on the rare side, and Jean Brown, who worked with Louis at the museum, remembered the boys, when small, crying, "More blood Daddy, more blood!" as Louis spooned the juices over their plates.

Dinners were lively affairs, the conversation spirited, bantering, and diverse. Archeology and anthropology (and archeologists and anthropologists) were the main topics, but Louis was liable to discuss everything from guppy breeding to the potential medical uses of zebra fat. "Louis thought something about every subject in the world," said Mary Smith of the *Geographic,* who joined the Leakeys for dinner on several occasions. "You could bring up needlepoint and somehow or other in five minutes he'd be telling you some anecdote in his life that had to do with needlepoint." He could also engage anyone, whether an eight-year-old child or an eighty-year-old grandmother, in conversation—and then mesmerize them with his tales of animal encounters, ESP experiences, and speculations about the past. Although not one to initiate conversations, Mary Leakey did listen attentively to all that was said, and often surprised people with her wry, irreverent, and sometimes caustic remarks. "She is a woman of choice words," remarked Betty Howell.

Mary's reticence was a given, but the Leakeys' sons were also curiously silent at the table, not speaking unless spoken to. This was partly a result of what Richard termed their "semi-Victorian, missionary upbringing," but the brothers also genuinely despised one another. They behaved with such icy hostility among themselves that visitors were alarmed. "Jonathan and I hated each other so much that there was often no discussion [at the dinner table]," said Richard. "We couldn't bring ourselves to talk to each other, *ever,*" not even for simple requests. If, for example, Richard wanted the salt shaker, he would turn to Louis and say, "Father, would you please ask Jonathan to pass me the salt?" And Louis would oblige, apparently as oblivious to this behavior as he was to the chaos in his home. "What could we do?" Mary asked years later. "It was something they [the children] had to sort out for themselves."

Disorder and confusion marked Louis's life at the Coryndon Museum as well. "There was no peace for him at home, and none at his office," remarked Clark Howell. "He just had so many things on his plate, ten or twenty things going at once. I don't think I ever saw him in Africa when he wasn't harrassed and irascible." Up before six in the morning, Louis would first make a list (which sometimes ran for two pages) of everything he intended to do that day. If it was his turn to drop the boys and neighborhood children at school, he would pile them in the car and be off by seven. It could be a harrowing ride. Not only was Louis an erratic driver at best, but he had the unnerving habit of reading or writing notes while driving. "He would frighten me," recalled Christine McRae, one of the children, "because he would have a newspaper beside him on the seat, and he'd be sitting with one hand on the steering wheel, the other with a finger over his lip, and he'd be reading the newspaper, instead of looking at the road! But I think it was probably the only few minutes he ever got to look at the newspaper—that was it, on the way to school." With the children deposited, Louis would travel on to the museum and dash up its stairs carrying bundles of papers and bunches of fruit, all stuffed in the wide sisal market baskets the Africans call *kikapus*.

There were always piles of correspondence for him to attend to, and the telephones rang incessantly. In addition to a steady flow of fan mail (which grew as his articles appeared in the *National Geographic*), there might be requests to write books (which he turned down), take on handwriting analyses for court cases (which he did for the money), look into problems with the museum's water supply, help with a Kikuyu employee's familial troubles, explain the tribal practices of a particular East African people to a new district officer, arrange an exhibit on swordfish, assist a young American student who longed to come work with the Leakeys, and collect thirty zebra skulls and jaws for a colleague in England. Louis handled all of these affairs, regardless of how petty. He also managed to see the many visitors who hovered outside his office door. Often these were elderly men from Kabete, the village where Louis had grown up. Still very much a member of the tribe, he was expected as a "big man" (someone of importance) to help his childhood friends and their families with advice and money. He never turned them away. "Louis was a pity man," said museum technician Joseph Mutaba, who saw many of these exchanges between Louis and the Africans. "He was someone who feels pity for someone who is all-suffering. I saw the old Kikuyu men come to his office. . . . They would put their head in, wanting to greet him and also to get help."

Theoretically, the salary arrangement that Louis had made with the National Geographic Society should have freed him from many of these responsibilities. Perhaps, initially, it did. On June 1, 1961, Louis turned over his administrative duties to the museum's entomologist, Robert Carcasson, who was given the title of acting director. Louis kept a seat on the museum's board of trustees, and was also named Honorary Keeper of Palaeontology, thus assuring that he maintained control of the museum's fossil collection. Yet whatever newfound freedom he had was soon filled with a hundred fresh projects and plans. He wanted to start an independent center for African prehistory and paleontology; to study and save the wild African primates; to devise a new taxonomic classification for the modern antelopes; to live alone on the Serengeti without any tools as early man had done (Mary strictly forbade this); and to explore a new site, one that dated to Miocene times, in western Kenya. Le Gros's pleas that Louis devote all of his energies and attentions to Olduvai went unheeded. Instead, once again, Mary shouldered that load.

"Most of our time together was delightful," Mary said about her relationship with Louis. "[But] he was irritating at times because he would rush off and do something else when I would just like to stick with one subject and see it through. It was a little exasperating." Did she feel that she was being left behind to look after the details? "Oh, I was, yes. I was."

During the summer, Louis and Mary discussed the agenda for the upcoming season at Olduvai. There were two new promising archeological sites (one found in January 1961 by their youngest son, Philip, who was then nearly twelve) to excavate in the lower beds, and Mary would put part of her crew on these. Maxine Kleindienst, Clark Howell's former student, would tackle the archeology of Bed III, while John Waechter, an archeologist from London's Institute of Archaeology, would start work on Bed IV. Most important, Mary would continue the dig at Jonny's site. With luck, she might uncover additional bones of the small, gracile hominid the Leakeys had come to regard as "really the most important find from Olduvai to date," as Mary wrote Phillip Tobias.

Louis was by now absolutely convinced that this specimen (affectionately called "Jonny's Child" by the Leakeys) was not an australopithecine. But he hesitated about giving the fossil a new name. "I doubt if I shall name the pre-*Zinjanthropus* child until we have done more work in the field," he told a colleague, adding confidently, "and I am sure we shall get additional material." This was, of course, the "wiser" route to follow in naming new species, as Louis said to Tobias, but Louis was seldom restrained by prudence. He delighted in provoking people with his unconventional ideas, relished being the underdog in scientific disputes, and most of all loved being right and proving everyone else wrong. Naming the fossil would upset all of his colleagues, including Le Gros Clark. Le Gros had seen "Jonny's Child" in England and listened to Louis's arguments, but still considered the fossil "inseparable from the Australopithecines." "I do hope that you will not commit yourself too strongly [to a new species] at this stage," he implored Louis in July.

Le Gros's entreaties may have had some influence, since Louis did hold back. But like a reckless poker player, he could not resist giving his colleagues a little tip of his hand. After making a comparative study of the child's lower jaw and teeth with the australopithecines, he decided to write a further description of the fossil for *Nature*. In this report, published in July 1962, Louis stressed the differences between the teeth of the South African fossils and those of the child (which had relatively large front teeth and small cheek teeth, the opposite of the australopithecines), and included precise measurements of each tooth to back up his claims. He also made it very clear that he thought the specimen represented a new hominid. But he did not name it. Instead, he hedged his bet: "If I am right in believing that the juvenile from FLK.NN.I is not an Australopithecine, but a very remote and truly primitive ancestor of *Homo,* then it is possible . . . that it was this branch of the Hominidae [the human family] that also made the Oldowan tools at the site . . . where *Zinjanthropus boisei* was found. We do not yet know the final truth of this problem." The truth could only come with more fossils, and it was Mary's task to find them.

Mary reopened her Olduvai camp at the beginning of October 1961. She and her Kamba crew planned to spend another full year in the field, and soon had the camp buildings that had been constructed the previous year rethatched and refurbished. The men also pitched a row of tents for visitors and built a separate grass sleeping hut for Mary and her Dalmatians. Louis stayed behind in Nairobi, finishing his volume on the fossil fauna of the gorge, planning his new prehistory center, and arranging his now-annual lecture tour in the United States. He would visit Olduvai as before, on weekends and holidays.

Within a few weeks, the camp was in full swing, with an abundance of people at work. Since the Leakeys' eldest son, Jonathan, had decided to specialize in snakes rather than fossils, Mary asked Richard to serve as the camp manager. Now seventeen, he had driven down in mid-October with Glynn Isaac, a young archeologist who had just graduated from Cambridge.[9] Maxine Kleindienst also turned up with her field assistant and a crew of fifteen Kamba workers. Other scientists, visitors, and friends drifted in and out throughout the year, so that Olduvai felt at times more like a small village than a simple research camp.

Many of these people were new to Africa, unprepared for the vastness of the land and the colonial relations among natives and Europeans. "I particularly remember crossing the border into Tanzania," said Alan Gentry, whom Louis took to Olduvai for a brief familiarization tour before beginning his study of fossilized gazelle bones at the Coryndon Museum. "In those days it was just Tanganyika Territory, and so the border was marked simply by a sentry box with a Union Jack. Nobody was there. And Leakey stopped the Land Rover, walked in, signed his name in the book as crossing the frontier and drove on. That was all." At Olduvai, Gentry found himself similarly surprised by unexpected customs.

> Coming out from England, I hadn't realized that the whites in Kenya were well-served by the blacks. Of course, this came up first at Nairobi in the hotel, but it wasn't really noticeable because you expected the hotel staff

[9] Isaac had actually been hired by Louis to reopen the excavations at the handaxe site of Olorgesailie, but the Leakeys felt that he first needed to be trained at Olduvai. "Louis said, 'You've dug in South Africa [Isaac's birthplace], and Europe, but you should go down to Olduvai and dig with us, and we'll show you how we do it in East Africa,'" Isaac recalled. "So I had only a fleeting visit at Olorgesailie before I was sent off to dig with Mary."

to serve the food and make the beds. But [at] Olduvai I realized for the first time that field work wasn't quite as I had imagined it would be. The Leakeys and their visitors had all the domestic side of the camp run for them by black servants.

This meant a cup of tea served to everyone in their tents at sunrise (6:00 A.M.), followed by breakfast in the main *banda* at six-thirty. The Kamba staff cooked and served lunch and dinner as well, did the dishes, washed and folded everyone's laundry, swept out the tents, made the beds, filled the washbasins with water, and in general kept the camp trim and tidy. "It was not the daily routine I had suspected," said Gentry. "Living at Olduvai was a lot easier and far more enjoyable than I had imagined."

Work in the gorge began at 7:00 A.M. and continued until 1:00 P.M. Then everyone, scientists and crew, returned to camp for lunch and a rest break. During the hot afternoons, fossils and tools were sorted and catalogued in the shade of the camp until 5:30 P.M. Following the British system, people dressed for dinner after quick showers in the shower tent when water was plentiful, or a cool splash in the washbasin when it was not. The scientists met up again for cocktails in the main *banda* at 6:00 P.M., "the narcotic hour," as Mary liked to call it. "Why, they had a regular bar up there," recalled the National Geographic's science adviser, T. Dale Stewart, marveling at the selection of Scotch, gin, and beer. After pouring herself a Scotch and lighting a cigarette or small cigar, Mary would lean back in her camp chair and watch the evening light wash distant Mount Lemagrut ("a heavenly mountain") in blues and purples.

The twilight hour was quiet, but after downing the three-course dinner, conversations turned lively and gossipy. "Every professional visitor to Olduvai knows that after dinner you get out the brandy and you tend to discuss people. You analyze them and their weaknesses," said Glynn Isaac. "People tended to be classified either black or white." Though Isaac thought that "none of this was in any way nasty," others disagreed. The geologist Richard Hay, who arrived in camp in the summer of 1962, remembered Mary and another English archeologist sitting up at night, "bad-mouthing every other archeologist in the world, except one, a Berkeley man who somehow came off clean. But it was very irritating. All I saw was two defensive, or envious or insecure people criticizing everyone else, and it really put me off." (Hay excused himself from these sessions, and perhaps because of this and the excellence of his work, was always held in the highest regard by Mary.) Often the group sat together, Scotches and brandies in hand, gossiping and sorting micromammalian fossils by lantern light until midnight. But there were also nights when long after the others had gone to bed, Mary would sit up drinking, alone with her dogs.

For as long and hard as the days were at Olduvai, there was never a guarantee of an undisturbed night's sleep. The wildlife of the Serengeti was still abundant then, and rhinos, giraffes, lions, leopards, antelopes, and zebras all found their way to the gorge. Lions were particularly noticeable during the migration season, when thousands of gazelles, wildebeests, and zebras moved across the plains. More than once, an entire pride leaped over the camp's thorn fence in the middle of the night, then paced restlessly around the tents. "Their roaring would shake the ground," Mary said. She made it a practice to keep the keys in the Land Rovers' ignitions and advised everyone to keep their kerosene lanterns burning all night outside their tents—though the lights did little to lessen the animals' curiosity.

Another time, Mary woke in the night to the sound of eight leopards coughing and snorting outside her grass hut. They could smell the Dalmatians inside and padded back and forth, trying to find a way in. Once a leopard did succeed in breaking into the hut of Mary's assistant, Michael Tippet, and carried off his pet baby wildebeest. Other nights, hyenas broke into the camp, making off with most of the meat supply. "The stupid hyaenas *[sic]* opened the fridge, night before last," Mary wrote to Louis about one such incident. "I didn't hear that, but heard them fighting over the proceeds outside our *banda* at 3 A.M. I threw a thunderflash [firecracker] within a few yards of them & all they did was to retreat behind the nearest bush. I was so furious, that I got into [the Land Rover] and pursued 1 hyaena & 1 jackal all down the slope to the gorge. . . . They left no dog meat and very little for us, either. . . ."

Mary tossed firecrackers at the lions, too, on nights when they were particularly annoying, or stood outside her hut banging a pot to frighten off the leopards. "The leopards always sounded like big pussy cats running off, going bumpity-bump, bumpity-bump just like my cats at home, only louder," said Maxine Kleindienst. But for the most part, the Leakeys accepted their nocturnal visitors as part of the experience of living at Olduvai, and advised their visitors to do the same. "Louis told us, 'If you hear lap, lap, lap [outside your tent], don't worry, that's just the lion and that's the way he gets water,' " one visitor recalled. "And he [the lion] did. We woke up and heard him drink every single bit of water out of the wash bowl."[10]

Regardless of the night's events, Mary and her crew were down in the gorge at 7:00 A.M. every day except Sundays, working on the excavations. This season, she had her crew first expand the digs at Jonny's site and FLK North. The previous year, they had recovered a *Deinotherium* skeleton

---

[10] Louis always enjoyed teasing newcomers to Africa, often advising them to leave their tent flaps open so the lions could walk through. For the French paleontologist Yves Coppens, Louis devised a special warning. "Louis took me, my wife and an African assistant to a site . . . to excavate an elephant's jaw," said Coppens. "He left us there very early in the morning without any car and said he would be back at 5 PM to pick us up. Then he drove away, but suddenly came back and called me and gave me a *panga* [an African machete], saying, 'I am sorry, I forgot to give you this. Be careful. There are rhinos around.' And he left. Which means that between 6 in the morning and 5 at night, I was thinking what I could do with my *panga* in front of a rhino. [Every time] there was a strange noise in the bush, we were looking around. Could this be the moment when we would have to use the *panga* against the rhino?" Coppens did see fresh rhino tracks, but the animals themselves kept their distance.

surrounded by stone tools at this latter site in the lower deposits of Bed II; Mary now wanted trenches dug down to the bottom of Bed I, the oldest part of the gorge. "I had a plan," Mary explained in 1984. "I wanted to get all the successive Stone Age industries, and any possible hominid fossils and fauna that were with them." Once again every stone tool and fossil was carefully mapped, recorded, and collected. As always, it was slow, paintstaking work, not helped at all by a sudden change in Olduvai's weather pattern.

"It rained," recalled Maxine Kleindienst. "It rained and rained continuously—probably the worst rains in fifty years. Often the river [in the gorge] was so high you couldn't cross to the south bank, where we were working." Periodically, high, roiling flash floods swept through the gorge, sometimes stranding a group of workers on the south side. The men stretched ropes above the raging river as a makeshift bridge, but these were wet and slick, making it all too easy to lose one's footing. "If you slipped and lost your hold on the rope, you were just going to go," said Maxine. "I remember at least two occasions when Richard [Leakey] and my field assistant had to jump into the river to save one of the crew from drowning." The rain also turned the excavations into muddy pits, making work virtually impossible. The scientists and field workers used sand bags, tarps, and plastic sheeting to protect the sites from sudden deluges, and in spite of the dramatic weather, they made discoveries.

"Confidentially," Louis wrote to Phillip Tobias in early December 1961, "we have what I am nearly sure is part of the right temple of the 'child.' " At the end of the month, a hominid's fossilized toe bone turned up in one of the trenches at FLK North. As luck would have it, the bone was the terminal phalanx of the big toe, one of the few bones missing from the foot Mary had found in 1960. By June 1962, a hominid's upper jaw bone had been found on the surface at a new site in Bed II, while the pieces of a crushed skull of Louis's Chellean Man *(Homo erectus)* were discovered in Bed IV. (When Mary reconstructed the skull, it had such a small cranial capacity that the Leakeys jokingly referred to it as "Pinhead.")

The crew also discovered another butchery site, where early hominids had cut up an elephant's skeleton. And at the site that Philip Leakey had found in 1961, they unearthed what Mary interpreted as the world's oldest hominid habitation. This lay in the gully DK (Donald's Korongo) at the base of Bed I, just a few feet above the black basalt that marked the very bottom of the gorge, making the structure nearly two million years old. Philip had first been attracted to the site by the dense accumulation of stones and bones, and under his mother's watchful eye had dug a trial trench, exposing part of a hominid living floor. As the excavation progressed in 1962, Mary's team uncovered several large lava rocks, which Mary plotted on her map and then had the men remove. But more of the stones were found, some of them stacked in what seemed to be loose piles. Struck by the pattern, Heselon Mukiri, who was overseeing the dig, called Mary to one side. "He had dug with me on Neolithic sites where we had found stone shelters and he said to me that this was just like those. And he was right," Mary recalled.[11]

The stones had been stacked in a rough circle, about fourteen feet in diameter, with each pile placed about two feet apart. "They were probably a foundation of sorts that supported branches or poles, which would have been covered with animal skins or grass, much like the huts the Turkana [a desert people in northern Kenya] build today," Mary said. Very few tools or bones were found inside the circle, but outside, these were found in abundance—just as the area beyond the wind shelter at the *Zinj* site was thick with artifacts and fossils. From the great number of fossilized rhizomes (the roots of reeds, much like papyrus), as well as crocodile and flamingo bones, Mary determined that the hut had been built beside the ancient Olduvai lakeshore.

Hoping that other such structures might turn up, she extended the dig another two thousand square feet. But the volcanic tuff here was nearly three feet deep and so hard that picks and shovels proved useless against it. "I then got a fellow from the Mines Department in Nairobi," Mary recalled, "and he dynamited the tuff. It was a bit hairy as the living floor was exposed, but he did a fine job." Although there was no evidence of another rock circle, there were more tools and fossils buried beneath the hard ash. These came from the very lowest level of the gorge, making them the oldest stones and bones recovered from Olduvai—a position they still hold today.[12]

But although Mary's crew dug three new lengthy trenches at Jonny's site and removed several more tons of earth, nothing more of the child turned up. Mary closed this excavation at the beginning of May, 1962. In spite of the lack of additional evidence, Louis decided he would introduce the child as a new species after all. Unlike the naming of *Zinj,* however, Louis decided he would not stand alone. Instead, he would enlist the support of the anatomists Phillip Tobias and John Napier. Together, they would form a solid triumvirate, making it easier for Louis to fend off any attackers. All he had to do was persuade Tobias and Napier that the child was something new.

---

[11] Some archeologists today question whether the stone circle is actually anything more than a geological formation. "I trust that what Mary excavated did indeed look like a shelter," said Rick Potts, an archeologist at the Smithsonian Institution who has restudied the Olduvai living floors. "But if you ask, 'Are there other agencies—such as tree roots—that can cause a similar circle of stones?,' the answer is 'Yes.'" Mary, however, stands by her initial interpretation, noting that when first uncovered, the circle was much more clearly defined. "It was a shelter of some sort, of that I'm sure," she said. Still, it is the only known example of a shelter at that time period, and so archeologists seldom cite it as an example of early hominid behavior.

[12] In the late 1960s, a shelter was built over the rock circle to protect it and to enable visitors to see our ancestors' earliest home. One year, a group of young Maasai warriors broke into the shelter and roasted a leg of beef in the stone enclosure. "So then a group of tourists came rushing up to tell Louis and me how wrong we were about this circle," Mary recalled in 1984. "There was a sign of fire!"

Unfortunately, this outdoor museum has not been maintained. The site and stone circle are now badly damaged, and many of the artifacts and fossilized animal bones Mary left in place have been removed.

Chapter 16

THE HUMAN WITH ABILITY

Unlike Louis, the two anatomists he invited to name "Jonny's Child" were, in Napier's words, "lab men." Their science was ruled by calipers. Through precise measurements and comparative studies of hominid and ape bones, they sought an objective description of each fossil. Louis, too, could wield a pair of calipers to prove a point, and did so when it suited him, but he worked less by measurement than by intuition. Eyeing the six fragments of the child's skull, which had been discovered in 1960 near the *Zinj* site and which barely covered the palm of his hand, Louis decided the child was *Homo*—"a very primitive *Homo,*" as he wrote Tobias in the summer of 1962, but *Homo* nonetheless. He urged Tobias to join him in recognizing it as such.

At the time, despite the nearly one-hundred-year search for human origins, the genus *Homo* was only vaguely defined. Most authorities considered brain size the distinguishing feature, although they had never agreed on the boundaries of this "mental Rubicon," as Sir Arthur Keith called the dividing line between humans and apes. Keith himself thought the transition occurred when a hominid's brain measured 750 cubic centimeters, while Franz Weidenreich, an anthropologist who specialized in *Homo erectus,* thought a volume of only 700 cc would do, and Henri-Victor Vallois, who studied the Neanderthal specimens, boosted the figure to 800 cc. Keith based his estimate on studies that gave 650 cc as the highest known brain volume in gorillas, and 855 cc as the lowest known brain volume in what he termed a "primitive race of mankind, the aborigines of Australia." Although the numbers were somewhat arbitrary, most anatomists accepted the idea that a large brain was a "distinctive human trait," as Le Gros Clark noted, and that humankind probably began when the brain attained a capacity between seven hundred and eight hundred cubic centimeters.[1]

To Tobias, at this early stage in his study, the child could be considered *Homo* only if its brain fell within this range. But determining the child's brain volume was not an easy matter, since only the parietals (the bones that form the top of the skull) remained. The vault of the child's skull would have to be reconstructed, a cast made of its interior, calculations performed to estimate the brain's capacity, and the results compared to other hominid and ape brains. Although he had not yet completed his detailed study of *Zinjanthropus,* Tobias willingly took on this new task in the spring of 1962. "It was extremely exciting," he recalled. "I had to devise a method to find out what brain size could have lain within such a skull."

He also quickly discovered that he had to be careful about how he presented his ideas and results to Louis, who already knew how they should turn out. At thirty-six, Tobias was regarded as a talented anatomist. Three years earlier, he had been appointed head of the department of anatomy at the University of Witwatersrand, replacing Raymond Dart. Describing the Leakeys' fossils—by this time, some of the most famous hominid fossils in the world—would greatly enhance his stature. Nevertheless, Tobias was just beginning his career while Louis was nearing his peak. In Kenya Louis was known as the "Father of Kenyan Archaeology," and at the Coryndon Museum the staff deferentially referred to him as "Father." It was not the relationship between equals that it might have been if, for example, Louis had asked Le Gros Clark to describe the fossils. Nor did Louis have any qualms about playing the part of the magnanimous bully with the younger man.[2]

Tobias had discovered just how overbearing Louis could be in October 1961, when he opened a letter that began, "I have just received your letter . . . , and Mary and I are frankly very disappointed." Louis's letter went on for three devastating pages, chastising Tobias for foot-dragging on the *Zinj* study, for suggesting that *Zinj* resembled a hominid specimen from Chad, and for requesting that he (Tobias) be allowed to write an article about *Zinj*'s discovery. The depth of Louis's anger was made clearest, though,

---

[1] After Charles Darwin suggested in 1871 that humans were related to the apes, scientists attempted to tease out the central difference between human and ape brains. Thomas Huxley first suggested that the human brain was simply a richer and larger version of an ape's. Subsequently, many anatomists, including Sir Grafton Elliot Smith, carried out comparative studies on the shape of each brain. "No structure found in the brain of an ape is lacking in the human brain," Smith wrote in 1928, "and, on the other hand, the human brain reveals no formation of any sort that is not present in the brain of the gorilla or chimpanzee. . . . The only distinctive feature of the human brain is a quantitative one." Smith's conclusions led to the emphasis on brain size as the key distinctive feature between human and ape.

[2] Louis had a similar relationship with John Napier, and the inequality of the situation did not go unnoticed by their British colleagues. "I agree with you about Napier in being young and still with a reputation to make," Le Gros Clark wrote to Kenneth Oakley in December 1962. "I fancy he is trying to keep on good terms with Leakey & others with a view to gathering some of their material for study & description. The same applies, I imagine, to Tobias. . . ."

when he told Tobias to forget the promised studies of Jonny's Child and the Chellean skull. "This is disappointing," Louis wrote, "but it is the way you have forced our hands."

Tobias managed to patch up the relationship and regain the fossils with a four-page reply, but thereafter he took great pains not to antagonize Louis. Thus, when he began his study of the "pre-*Zinj* child" (as he and Louis referred to Jonny's Child) in April 1962, he reported his initial impressions in careful, almost obsequious language. "[T]he child parietals certainly do seem to have covered a brain somewhat larger than that of *Zinjanthropus,*" he reported in May. He even thought the child "possessed a brain bigger than that of any of the existing australopithecines." However, he doubted that his final estimates of the child's adult brain size would be greater than 650 cc, well short of the *Homo* range. Then, taking the plunge, he admitted that his "present feeling about the child is that it is an australopithecine." Louis, of course, would be furious when he read this statement, so Tobias quickly defused it. "These are rambling thoughts—I am thinking aloud, that is all for the moment." "Thinking aloud" was to be Tobias's refrain for the next year and a half.

During that period, the two men maintained a spirited correspondence, with Louis badgering Tobias about his opinions and Tobias playing coy. Every month, Louis would write to ask if Tobias was ready to name the new hominid, and Tobias would cautiously respond that he was not yet certain. By December 1962, Louis had become desperate.

> I do want to hear what you have to say about the "child" . . . Mary and I are sure (more and more so every time we go over the data) that it is NOT *Australopithecus.* I think only those with "Psychosclerosis," as Le Gros once put it and who cannot bear the idea of two contemporary branches of Hominid, could ever put it into that sub-family. However, you *must* say what you feel and, if necessary, I shall have a "disagreeing" chapter giving the other point of view.

A month passed before Tobias replied, and then he wrote gently, but firmly, "So far, I must say, I have seen nothing which excludes the specimen from belonging to the Australopithecinae . . . ," adding good-naturedly, "I hope my interim feeling . . . is not the result of premature 'psychosclerosis'!!!" He would not budge, nor would Louis, who tried once more to bully Tobias. "I do *NOT* think the pre-*Zinjanthropus* child . . . fit[s] in with the Australopithecines, I think it is *Homo,* or near *Homo* . . . I think that you, like many others, suffer from a prejudice against having two contemporary stocks. . . . Can you tell me *why* . . . ?" At the end of his letter, though, Louis grudgingly admitted that what he really needed were more specimens, particularly a "face and a frontal to be sure. . . ." Then, in an effort to regain the upper hand, Louis once again reneged on his offer to Tobias to analyze the Chellean skull: "We will see after you have completed this study."

By now, however, Tobias was learning to ignore Louis's threats. He agreed that they needed additional specimens, but for the time being all he had were the parietals and the partial lower jaw. When he completed his initial study of the child's brain capacity in mid-June 1963, he immediately dispatched a cable to Louis, and did nothing to soften the blow: CRANIAL CAPACITY PREZINJ BETWEEN 600 AND 700 . . . DIFFICULT TO RECONCILE WITH HOMO . . . TEETH ALSO POINT TO AUSTRALOPITHECINE AFFINITY.

If Louis was the least bit disappointed by this news, he never let it be known. Instead, he took a different tack, belatedly questioning Tobias's ability to determine the child's cranial capacity with only the parietals. He felt, too, that Tobias had not taken sufficient note of their shape, which "does not seem to me to fit at all into the Australopithecines. I admit they are not *Homo* in the most modern sense, but if we are going to limit ourselves to two genera [*Homo* and *Australopithecus*], then I would call them *Homo;* otherwise I would place them in a new genus, possibly Epi-Homo."[3] Besides, Louis added, he had just heard from John Napier, who could find, Louis said, "no significant difference" between the hominid foot that had been found with the child and "present-day *Homo sapiens.*" Similarly, the hand from the same site, though certainly primitive, also had a human look to it. Louis felt that these fossils, as well as the hominid leg and collar bones, corroborated his belief that the child was *Homo.*

In London, John Napier had reached much the same conclusion. His initial study of the hand, published in November 1962 in *Nature,* described it as "short [and] powerful with strong, curved digits, surmounted by broad, flat nails." It also had a "strong and opposable" thumb, and would have had "the physical capacity" to make the "small pebble-tools . . . found on the living floor." Although the hand bones could not be "closely matched with any known hominoid species living to-day," they did resemble those of a juvenile gorilla and adult human.

Napier had parceled out the leg and foot bones of Jonny's Child to his colleagues at the Royal Free Medical School, Peter Davis and Michael Day. Day was then living in a little garret room above the medical school, and there he assembled the foot bones with a "lot of plasticine." "That was the most astounding thing to see," he recalled, "because it was quite clear to

---

[3] In July 1962 the Wenner-Gren Foundation had held a conference for anthropologists to discuss the hominid taxonomy. There was now a plethora of names for the fossils, since almost every one that had been discovered had been assigned to a new species or genus. At the meeting, the scientists agreed to a new, streamlined taxonomy. All the fossils that had been called *Pithecanthropus*—for example, Java and Peking men—were now labeled *Homo erectus,* as Clark Howell had suggested in an influential 1960 paper. The various forms of australopithecines, such as *Paranthropus* and *Plesianthropus,* were divided into two categories: *Australopithecus robustus,* for the large-skulled species, and *Australopithecus africanus,* for the more delicate form. Louis attended this meeting and agreed to the changes, but somehow managed to persuade his colleagues to leave both *Zinjanthropus* and his Chellean Man in their own separate categories. Two years later, however, *Zinj* would be classified as *Australopithecus (Zinjanthropus) boisei,* and the Chellean skull labeled *Homo erectus.*

anybody—you didn't have to be an anatomist—that this was obviously a bipedal foot. It was a foot of someone who could stand and walk like you and me. It was thrilling, and Louis Leakey was delighted." Most important, the foot did not have the grasping big toe of an arboreal creature. Napier, Davis, and Day also found that the leg bones from the *Zinj* site functionally fit the foot, enabling them to study in greater detail how this creature had actually walked. Although the "walking pattern was not identical with modern man, . . . the distinctions are not going to be great," Napier wrote Louis midway through this analysis.

By the end of July 1963, only Tobias remained unconvinced that the fossils represented an early form of *Homo.* He told Louis that he had "by no means made up" his mind about pre-*Zinj,* but that the cranial capacity did not solve the problem. He was now making precise measurements of the jaw's teeth as well as those of the australopithecines. In answer to Louis's accusation of "psychosclerosis," he wrote: ". . . I can assure you I am not in the least bothered by any philosophical preconception about the number of higher primate genera which can have lived side by side. I am trying rigorously to be guided by the evidence before my eyes; if the fossils suggest two genera, I shall not hesitate to say so; if it suggests two species . . . or any other conclusion, I shall equally not hesitate to say so!" It did seem to Tobias that "all these small differences [between the child and the other australopithecines] may add up to a total picture of a new species," but he was not ready to commit himself. He said simply, "The study continues." It continued through the remainder of the summer and well into autumn, and Louis waited impatiently for Tobias to change his mind.

Louis had meanwhile moved into a new office in the Centre for Prehistory and Palaeontology on the Coryndon Museum's grounds. Creating the Centre —a fully autonomous institution overseeing all the archeological and paleontological collections formerly housed at the museum—had been one of Louis's chief ambitions since resigning his position as the museum's director in June 1961. After a year of lobbying and fund-raising, he had gained the approval of the Kenya government and the museum's trustees, and secured a Ford Foundation grant of £2,000 per annum for the next three years to cover the Centre's operating expenses. Louis had also persuaded the National Geographic Society, which was already paying his salary, plus providing the funds for the Olduvai excavations, to contribute £350 annually to the Centre to pay laboratory expenses and an assistant.

In July 1962 the Centre opened in three squat, mustard-colored buildings formerly occupied by the government's Desert Locust Control program. All the many crates of stone tools and fossils that had lined the museum's hallways and basement were transferred here. Louis took the title of director and hired Shirley Coryndon, who had worked with Donald MacInnes on

Rusinga Island in 1950, as his paleontological assistant.[4] Mary moved into another office, and with the addition of a small staff—a secretary and an accountant, two African workers, a laboratory technician, and a deputy director—the Centre sprang to life. In no time at all, it was as madly hectic as Louis's old museum office.

The Leakeys also used the Centre as a place to coordinate their expeditions. Just before a trip, piles of mattresses and bags of cabbages and potatoes would suddenly appear on the Centre's lawn. The supplies were often destined for Olduvai, but other times, they were packed off to Lake Victoria where Louis was excavating a new Miocene site called Fort Ternan, which a farmer named Fred Wicker had discovered in 1959 on his land.[5]

Mary was not enthusiastic about the undertaking. It drew Louis's attention away from Olduvai, and siphoned off money and men as well. She also disapproved of the way Louis chose to manage the dig, almost like an absentee landlord. Heselon Mukiri, Louis's longtime Kikuyu assistant, would run the day-to-day operation, overseeing a crew of five Kamba workers, while Louis checked in periodically. "I regretted a new site being opened up without adequate supervision and thought it best not to be associated," Mary said about her decision not to participate. By the time George Gaylord Simpson, the eminent American paleontologist, visited the Leakeys in June 1961, as he traveled through Africa studying fossil mammal collections, the tension between Louis and Mary was palpable. Nor did it lessen when Simpson sided with Louis. "Mary did not think much of the prospect [for Fort Ternan]," Simpson recalled in his autobiography, "but I encouraged Louis to investigate a bit further. . . ."

Louis and Simpson decided to check on Heselon Mukiri's progress at the end of June, about a month after the digging had commenced. They had already planned a safari to Olduvai and Louis's old Rift Valley sites, and added Fort Ternan to their itinerary. The tour would give Simpson a good background in the geology of East Africa before he settled down to study its fossils. At least that is what Louis thought. But Simpson had another purpose —or "ulterior motive," as he wrote in his autobiography—in wanting to see Louis's field operations. He had spent the last several months in England, where he learned that certain unnamed "conservative" scientists mistrusted much of Louis's work.[6] They were particularly suspicious of the string of discoveries the Leakeys had made at Olduvai in 1960, and Simpson, "a very good field man," had been tapped to "determine whether . . . [Louis's] reports tallied well with the field evidence." None of this was known to

---

[4] Shirley Coryndon's first husband, Roger, was the son of Sir Robert Coryndon, after whom the museum was named.

[5] Louis had actually noticed these deposits in 1932, while exploring the nearby Koru fossil beds with Arthur Hopwood. He had searched the Fort Ternan site for fossil primates then, but was unsuccessful.

[6] Simpson's biographer, Leo Laporte, speculates that some of this concern may have even come from Louis's old ally, Le Gros Clark, since Simpson and Le Gros were close friends.

Louis, who cheerfully set off for Olduvai in mid-June with Simpson, secretly playing the role of Professor Boswell, at his side.

At the gorge, Louis led Simpson up and down the sediments, showing the locality of each discovery. (Louis had taken care to mark and photograph these finds after his run-in with Boswell at Kanam/Kanjera.) Any doubts that Simpson might have had about the quality of the Leakeys' work were quickly dispelled. He was enchanted by the gorge ("There cannot possibly be a more historic spot anywhere on earth," he observed in his journal), and awed by the amount of work the Leakeys had done. "[Their] field operations . . . met the highest professional standards," he wrote, and Louis's "statements of fact were indeed factual."[7]

After spending several days at Olduvai, they drove back to Nairobi, packed up fresh supplies, and headed out to Lake Victoria and Fort Ternan, arriving at the dig on the morning of June 20. Heselon and his crew had removed several tons of earth in a series of stepped trenches, and Simpson wandered off to take a photo of the excavation, while Louis inspected the latest finds. Heselon carefully opened one box of fossils after another, and then, with a bemused smile, handed Louis a small tin. "Eagerly I lifted the lid and examined the specimen inside," Louis wrote later. What he saw made his "heart beat faster," and moments later he was "running full out" toward Simpson, shouting, "George! George! I've got it! It's here! I've got it! George! I've got it." In his hand, Louis displayed two fragments of a primate's upper jaw and a lower molar. "[The jaw] had a sort of protohuman look to it," Simpson recalled. The two men were particularly struck by the fossil's "canine fossa," a small hollow set just below the eye socket in the cheek bone. Such a hollow is found in human skulls, but not in those of apes, and Louis considered it a major distinguishing feature.[8] Although the jaw was not as old as the *Proconsul* skull Mary had found on Rusinga, it did predate Olduvai by more than ten million years, and would, Louis was certain, help fill "an enormous gap in the panorama of man's development."[9]

Louis carefully repacked the jaw in the tin, and then he and Simpson drove to the nearest telephone to call Mary. He did this, Simpson said, with a "bit of 'I told you so' because of her skepticism about the value of work at that site. In fact she was still skeptical until Louis assured her that I had seen the new primate specimen and agreed that it was highly important." That night, en route to Rusinga Island, Louis and Simpson celebrated the discovery with a bottle of Champagne aboard the *Miocene Lady.* They downed some "Leakey Safari Specials," too—equal parts of condensed milk and

[7] Simpson reported his findings to his British colleagues, and did what he could on Louis's behalf to help him become a member of the Royal Society, but to no avail. "It still seems to me unfair, indeed little short of scandalous, that a few carping critics prevented his election to fellowship in the Royal Society," Simpson wrote in his autobiography.

[8] The hollow exists in humans because of the reduced root of the canine tooth.

[9] In March 1962 Garniss Curtis of the University of California at Berkeley dated the deposits of Fort Ternan at fourteen million years. The Rusinga beds date between sixteen and twenty million years ago.

cognac—which "sound horrible," Simpson acknowledged, "but [are] very effective pick-me-up[s] after a hard day in the field."

Back in Nairobi, Louis dispatched his usual bundle of letters enthusiastically announcing the find and begging everyone to keep the news confidential. "Last week, . . . we found the first indications . . . of a higher primate, certainly hominoid, but more than that I cannot say at present," he wrote to Melville Grosvenor, the National Geographic Society's president. "If we get more of this individual, it would be an even greater discovery than those made at Olduvai so far and will fill the biggest gap in the whole history of evolution from pre-human to human. At the moment, what we have is only a small fragment, but very exciting indeed and I must ask you to keep completely quiet about it."

Louis hoped, as always, that further digging would reveal more of this creature's skull, but nothing more than a single canine turned up. Heselon's team did find twelve hundred other fossils, including bones of a new species of giraffe that was as small as a calf, as well as new mastodon and rhinoceros species. He closed the dig in July, and Louis decided to write a brief description of the new jaw. By mid-December 1961, he had completed a draft of his report and sent this to Simpson (who had replaced Le Gros Clark as Louis's Miocene confidant) for his comments. Although Louis told Simpson that he was "very hesitant to create a new genus of higher primates," he had done just that, naming the fossil in honor of Fred Wicker, on whose property it had been found: *Kenyapithecus wickeri,* or Wicker's Kenyan Ape.

Simpson did not quibble with Louis's paper (though others would), and in March 1962, Louis and Mary announced the discovery together at the National Geographic Society. They had been invited to receive the Society's highest honor, the gold Hubbard Medal, for "revolutionizing knowledge of prehistory by unearthing fossils of earliest man and giant animals in East Africa."[10]

Society members and the press had now come to expect Louis to unveil some new treasure, and he did not disappoint them. He had brought a cast and slides of *Kenyapithecus,* "a descendant of the *Proconsul* stock," with him, and told his audience that this "new creature . . . is definitely heading towards man." It was not itself a man, because man would be found with tools, and Louis did not expect to find artifacts with *Kenyapithecus.* The Leakeys did, however, have stone tools at Olduvai with *Zinjanthropus* and Jonny's Child. "These two creatures lived side-by-side at Oldoway," Louis explained, using the old pronunciation for the gorge, and "represent[ed] two different experiments towards what eventually became man." One of

[10] The Hubbard Medal had first been awarded to Robert Peary, discoverer of the North Pole. Other recipients included Sir Edmund Hilary and Louis's colleague from his 1930 expedition, Sir Vivian Fuchs. Fuchs received it for leading the British Trans-Antarctica Expedition. Besides Mary, only two other women, Anne Morrow Lindbergh and Jane Goodall, have been awarded the medal. In 1993, Louis and Mary's son Richard was also given this honor.

the two became "real man," and Louis promised his audience that he would return one day to tell them all about him.

Fort Ternan continued to be a productive site. Louis and Heselon reopened the dig in the fall of 1962, finding several teeth of another *Kenyapithecus,* as well as part of the creature's lower jaw and a limb bone. Louis was elated and once again dispatched a stack of "confidential" letters.

Although Louis was spending most of his time in Nairobi, engaged in various projects and his study of the Fort Ternan fossils, he occasionally paid visits to Olduvai. He happened to be in camp on the morning of October 22 when a workman named Ndibo Mbuika shyly asked him to come look at something he had found. Ndibo led Louis down the footpath toward the MNK (Mary Nicol's Korongo) excavations, then knelt beside a little pile of stones. Underneath them, glinting in the sun, lay a single tooth, still *in situ.* "It was a beautifully preserved hominid premolar," recalled Mary, whom Louis had summoned at once. They quickly searched the surrounding area, turning up several fragments of a very thin skull. These lay on the surface, partially hidden by tall grasses and shrubs.

The Leakeys immediately called their men over to start a new excavation. All the vegetation was removed, and the first layer of soil carefully scraped off and washed through one-sixteenth-inch mesh sieves. "The mandible, two pieces of maxilla [upper and lower jaws], parts of both temporal bones and several other fragments came to light almost immediately," Mary wrote about this discovery, "whilst many small fragments were recovered when the soil was washed." Louis was particularly delighted to see that the lower jaw contained a nearly complete set of teeth. Like the teeth of Jonny's Child, these were rather narrow. The thinness of the skull fragments recalled the child as well, and two days into the dig, Louis sent Tobias an exultant note: "Confidentially—but DO NOT TELL ANYONE AT PRESENT, ONLY THE FAMILY AND YOU KNOW—we have a new hominid lower in Bed II than [the] 'Chellean' skull. Most exciting but no more now. We are still excavating."

From the small size of the teeth, Louis guessed that the new hominid was a female and the Leakeys christened her "Cinderella," or "Cindy" for short. Over the course of the next several days, more bits and pieces of Cindy turned up. Louis, however, could not stay for the duration of the dig. He had previously made plans to work on his study of the Miocene fossil primates at the British Museum of Natural History during November. He would have to learn about any new finds from Mary.

Nearly four weeks passed after Louis's departure, and nothing more of Cindy surfaced. She was "being a proper little bitch," Mary wrote Louis in frustration. Still, Mary and her Kamba crew persisted in the hunt, and finally struck gold. Elated, Mary cabled Louis, announcing the discovery of more of

Cindy's skull as well as some teeth belonging to her "boyfriend." These consisted of two very large molars and one canine; Mary inferred from their size that they came from a male. The very next week, she cabled Louis again, telling him about yet another exciting find: a skull that Mary first believed to be an australopithecine. Another one of Mary's workers, Maiko Mutumbo, discovered this hominid while out exploring. Sadly, the fossil had been badly trampled by Maasai cattle, reducing much of it, Mary noted, to "1,500 fragments of bone and of teeth, many smaller than a grain of rice." But the fragments of the upper and lower jaws, as well as the skullcap, could be pieced together. Again, its size indicated that the hominid had been a male, and Mary christened it "George." The skull's shape indicated that the creature was not an australopithecine, but another individual like Cindy and Jonny's Child. The Leakeys now had remains from eight of these hominids. They had been found in several different geological horizons of Beds I and II, and so spanned a period of time from 1.8 million to 800,000 years ago.

The discoveries delighted Louis. He was certain that the accumulated evidence would convince Tobias about the presence of a new type of *Homo*. Even before Louis had learned about Mary's latest finds, he had written Tobias, saying, "I think our newest discovery [Cindy] strongly confirms . . . my view that the pre-*Zinj* material is not connected with the australopithecines but with something which is closer to Man."

Tobias did not respond to this letter for several weeks, but when he did, he wrote "URGENT" across the top. After his intensive study, he had finally reached the conclusion that the pre-*Zinj* child was not, after all, an australopithecine.

> Taking into account the dental features, mandibular features, cranial capacity [which he now estimated at between 675 and 680 cc, just shy of the "cerebral Rubicon"] and what information I have on the post-cranial bones, it is clear we have here an exciting new form which is intermediate between the Australopithecine and Pithecanthropine *[Homo erectus]* groups. . . . Once again, you have been proved right! I know you felt it was a hominine [in the *Homo* genus] all along; but I feel it is only as a result of my intensive and detailed studies of the specimen that I have been able to convince myself that this is the correct interpretation!

Years later, Tobias would also credit the Leakeys' additional discoveries as a key factor in his decision, although he did not see these until January 1964. "It was that whole cluster of further finds that made it clear that the original find of '61 wasn't a freakish, abnormal exception," Tobias explained in 1984.

> It was indeed a member of a population. There were a whole lot of them, and they all had bigger brains and narrower teeth and a number of other features . . . which showed a nearer approach to the genus *Homo* than to

> *Australopithecus.* So at that point I became convinced that Louis was right, that we did have a separate species and that it should belong with *Homo.* . . . It is not as Sonia Cole says [in her biography of Louis] that Louis talked me into accepting it. I sat on it for three years. . . . The specimens talked me into accepting themselves, it wasn't Louis.[11]

John Napier agreed with Louis's diagnosis as well, and the three men began working on a report about the new species of *Homo* for *Nature.* Tobias also readied his paper on the cranial capacity of Jonny's Child, and the Leakeys wrote a description of the new Olduvai discoveries. All three reports were published in the April 4, 1964, issue of *Nature,* with the announcement of "A New Species of the Genus *Homo* from Olduvai Gorge" appearing last. For the species name itself, Louis had turned (at Tobias's suggestion) to Raymond Dart. "He was a wonderful person with language," said Tobias,

> a man who had invented astonishing names like "*Australopithecus*" and "osteodontokeratic." So I said to him, "Raymond, we've got this new species. We think it's different from *Australopithecus africanus.* We've got good evidence that it's the first stone tool maker." And he went away and came back with this winner of a name, *Homo habilis*—which is our word "ability" if you drop the "h." It means "able" or "handy," the first handyman, which is what he was.

Assigning the fossils to a new *Homo* species extended the age of humankind by one million years. The 600,000-year-old *Homo erectus* fossils from China, formerly known as Peking Man, had previously stood as the oldest direct ancestor. The new species also confirmed Louis's belief in the contemporaneous existence of two different types of hominids. Nearly two million years ago, on the shores of the ancient Olduvai lake, *Homo habilis* and *Zinjanthropus* had looked out over the African savanna together—though not necessarily in harmony. "[I]t is probable," the authors suggested, "that . . . the *Zinjanthropus* skull represents an intruder (or a victim) on a *Homo habilis* living site."

In their report, Louis and his colleagues also changed the definition of *Homo,* primarily because the brain capacity of Jonny's Child (which they listed as the type specimen) fell short of the 700 cc limit. Their new definition was much more comprehensive, and included an erect posture and a habitual bipedal gait, a precision grip, and a brain capacity of only 600 cc. The precision grip was of importance because it implied the ability to make tools. Indeed, the authors credited *Homo habilis* as the maker of all the

---

[11] Tobias said that his final conversion came when Louis showed him the upper and lower jaws of OH 13 (Olduvai Hominid 13, "Cindy"), which held small, slender, but long teeth. "[M]y eyes widened and shone; I turned to Leakey and said, 'Louis, this is *Homo.*' "

cultural material found with the fossils: artifacts, the rough, circular shelter, and the windbreak.

By assigning Olduvai's oldest stone tools to *Homo habilis,* Louis effectively stripped *Zinjanthropus* of his human standing. Though only three years before, he had lauded *Zinj* as "not only a 'true' man, but the earliest known stone-tool making man," he now readily agreed with what his colleagues had argued all along: *Zinj* was only an australopithecine, and so probably incapable of fashioning tools. From now on, *Zinj* would be known as *Australopithecus boisei.* Louis's new interpretation did not disturb him in the least, although it put him in a position that the majority of scientists would find horrifying. "Many people who published something that was later proved to be wrong would regard it as a tremendous setback to their career, to their credibility and to everything else," said Michael Day. "Louis was never like that. He published what he thought at the time. If it was wrong, the next time you met he would say, 'My dear boy, that's all nonsense. Take no notice of it. It's all been changed.' "

There was one additional reason that Louis and his colleagues had changed the definition of *Homo*—one that did not appear in their *Nature* paper, but that Louis addressed at a press conference on April 4, 1964, at the National Geographic Society, where he announced *Homo habilis.* After describing the significance of the new species, he explained that he had also redefined the genus. The previous definition, he said, had described *Homo* as a creature capable of making tools "to a set and regular pattern." But a year earlier, he had received word that "chimpanzees in the wild were making tools" exactly in this manner. "[I]n consequence," Louis said, ". . . our new definition of man is a very much more complex one. We decided we must exclude the chimpanzees from the United Nations." Louis had received this news about the chimps from his protégée Jane Goodall, whom he had sent five years before to live among the wild chimpanzees of Tanzania.

Chapter 17

CHIMPANZEES AND OTHER LOVES

Though Louis occasionally took on a young man as a protégé, his closest professional relationships were with women. Perhaps, as some of his colleagues suggested, this was because Louis was, in animal-behavior terms, an Alpha Male: domineering, competitive, and intimidating to other men. Consequently, he had very few close male friends (indeed, in his writings hardly any men surface as "friends" aside from the paleontologist Donald MacInnes). Instead, he sought the companionship of women, particularly those who were young and talented and bright. They, in turn, found him so magnetic and inspiring that eventually a dozen female disciples were sending Louis reports on everything from primate behavior to aardvarks to early humans in North America.

"Louis was a superenthusiast," said Mary Smith of the *Geographic.* "He was so full of desire to know everything in the world that he just turned people on. He'd pour this electric energy into them as if they were vessels. Those who were receptive, he kept pushing and shoving and urging. And that's what sent the Goodalls to the jungles, the Fosseys to the mountaintops, and the Birutés to orangutanville."[1]

[1] When a young woman of twenty-seven, Mary Smith was herself inspired by Louis. She had gone to Olduvai Gorge as an observer for the National Geographic Society in 1962. After dinner, she joined Louis outside, leaning back in the camp's canvas chairs and gazing at the stars. "It was an absolutely brilliant night in the middle of the Serengeti," she recalled. "We got to talking about astronomy and things, and he had this eclectic interest in everything, and I can't tell you how far-reaching that conversation was. I remember he shut his eyes and put his hands on his head and said, 'You know, you people at the National Geographic with your facilities and your ability to reach people—you must do something about the evolution of early man. You can't have those Bible people hold back

In some ways, these relationships were like those between any student and teacher (and perhaps made up for the students Louis lost after his falling out with Cambridge University). But they also recalled Pygmalion (more cynical observers might say Svengali), for Louis preferred his protégées to be intellectual blank slates. He could then mold and shape their minds, and fire the young women seated at his feet with his desires and ambitions. But like Pygmalion, Louis found that the danger in sculpting the perfect image was that all too often one fell in love—as he did with Jane Goodall, the young woman he dispatched to study the chimpanzees.

Louis met many of his protégées through chance encounters, and his initial meeting with Jane, in the spring of 1957, was typical. A lithe, pretty, hazel-eyed blonde of twenty-three, Jane had come to Kenya from England "hoping to get involved with animals." She had been smitten with the African continent since reading the tales of Tarzan and Dr. Dolittle as a child; after working as a secretary and a waitress, she saved enough money for round-trip passage to Nairobi. Initially, she stayed with a school friend whose parents owned a farm in the White Highlands, then moved into the city where she again worked as a secretary. But she was looking for more than this from her African experience, and when she confessed her love of animals to her friend's parents, they suggested she contact Dr. Louis Leakey at the Coryndon Museum.

"So I rang up the Museum," Jane recalled, "and I said, 'I'd like to make an appointment to meet Dr. Leakey.' And the voice said, 'I'm Leakey. What do you want?' " Louis listened politely to Jane's shy request for a meeting to talk about animals, then penciled an appointment on his calendar. Though he did not say so to her, and indeed kept his thoughts to himself for several months, he had been looking for some time for someone to study the wild primates.

At the time Jane Goodall contacted him, Louis was preoccupied with the fossilized Miocene primate bones he and Mary had unearthed on Lake Victoria's Rusinga Island and other nearby sites. Their collection of primate fossils was so extensive and diverse—they had collected specimens representing 450 individuals of the apelike *Proconsul africanus* alone—that Louis postulated an immense and genetically rich ape population for East Africa in the Miocene era. Here, he argued, was "the main evolutionary centre for the higher Primates and the birth-place of man himself." On Rusinga and Mfangano islands, Louis and Mary had also collected the amazingly well-preserved fossils of plants, seeds, fruits, and insects, enabling them to reconstruct mentally the Miocene apes' forest home. Louis pictured it as an open forest gallery, "with great rivers running down into the lake, while at the

---

the knowledge we're learning. It must be brought to the public. Promise me, Mary, that you'll look into this.' And I said, 'Yes, sir, I certainly will.' " Twenty-three years later, Mary Smith fulfilled her promise by overseeing the production of an issue of the *National Geographic* devoted to human origins.

same time there were fairly extensive open grasslands lying between the rivers." It was, Louis believed, remarkably similar to the habitat of at least one of the Miocene apes' descendants: the chimpanzee, *Pan troglodytes* schweinfurthii, and in particular one group of chimps that lived in the forest bordering the shores of Lake Tanganyika.

Louis had first learned about these chimpanzees in 1945 from a fellow Cambridge anthropologist, Jack Trevor. While stationed in Kenya with the British Army, Trevor had taken a leave to visit a small reserve in Tanganyika that had been set aside for the apes. He and Louis briefly corresponded about the chimpanzees, and later Louis claimed that he sent a young man "who must forever remain nameless" to study the animals in 1946, but who had "failed utterly." No records remain of this aborted effort. Ten years later, in 1956, Louis made another attempt to learn something about wild apes, this time sending a former secretary, Rosalie Osborn, to track the mountain gorillas at Kisoro, Uganda. (Louis had contemplated taking on this project himself, confessing to the owner of the hotel at Kisoro, who had contacted Louis in an effort to find someone to observe the gorillas and accustom them to people, "I only wish I were young and free and I would jump at it myself. . . .") However, Rosalie had only enough funds to stay four months, and Louis did not succeed in raising additional money for an extended study. A few months before Jane Goodall arrived in Nairobi, Louis had even corresponded with the district officer near the chimpanzee reserve about sending Rosalie to study the chimps, but nothing came of this, as she had decided to return home to England.[2]

Unaware of Louis's interest in primates or potential primate-watchers, Jane stepped into his office a few days after her phone call. Louis left no record of their meeting, but he was sufficiently impressed by Jane to offer her a job on the spot as his assistant secretary. He promised that it would not be merely a deskbound job, and the very next day took her for an early-morning drive in Nairobi National Park, querying her about animals along the way. "He asked me all kinds of questions about animals, and as I'd read about them all my life, I could answer him," Jane recalled. "Louis was very impressed because I knew what ichthyology meant, and I knew what ant bears were. I knew all about them. It wasn't just luck. He could see that I really was serious."

As a further test, Louis introduced Jane to Mary. He had suggested that it would be worthwhile for Jane to work at Olduvai that summer, but advised her, "It depends on whether you get on with my wife." Though Jane had not

---

[2] Louis had offered the chimpanzee study to another young woman, Cathryn Hosea Hilker, at about the same time he was considering Jane. "There's a young woman currently working in my office who wants the job, but she doesn't have the background," Louis told Hilker, who had recently graduated from the University of Cincinnati with a degree in anthropology. Hilker, worried about the five-year commitment Louis insisted on, turned him down. Years later, she met Jane at a lecture in the United States, saying, "Jane, I hate you. No, I don't really mean that. Oh, goodness, I had the chance to do this myself . . . and I didn't. I hate myself, really." Hilker now works with the exotic cats at the Cincinnati Zoo, and is deeply involved in the conservation of cheetahs.

heard any negative gossip about Mary, Louis's ambiguous remark alarmed her, and she arrived at the Leakeys' Langata home petrified. Mary and her sons had been out riding, and Mary decided to play a joke on Jane by offering her a ride on a horse that habitually walked backwards. But the moment she sat in the saddle, the pony began to buck, and Jane dismounted at once. "I thought it must be suffering from saddle sores, and I was right," Jane said, revealing a sensitivity to animals that Mary appreciated. "Of course, then I was okay as far as Mary Leakey was concerned, and I was able to go to Olduvai. Ichthyology for Louis and saddle sores for Mary."

Jane traveled with the Leakeys to Olduvai that summer along with another young woman, Gillian Trace, who had grown up in Kenya. For Jane, Olduvai held the promise of all that she sought in Africa: wide, open spaces and the close proximity of animals. Louis had made this possible, and thirty years later she recalled "the way he opened doors for me." But Gillian alerted Jane to the dangers of Louis's charm. "I hope you don't go falling in love with Louis," she said one morning as they walked to the gorge. "Everyone does." The idea startled Jane. "Louis was thirty years older than me, and he was very fat and his teeth were bad. The idea of him touching me was disgusting."

Jane worked at Olduvai for a month, swinging a heavy pick alongside Louis and Mary, searching for fossils, and sorting artifacts at night. In the evenings, there were walks with Louis to see giraffes and rhinos, and after dinner, long talks under the stars about animals, discussions Jane remembers as "intoxicating." Finally, one evening toward the end of their stay, she and Louis walked to the edge of the gorge, and as the stars came out above Mount Lemagrut, Louis began telling her about "a group of chimpanzees that lived along the shores of a lake, very isolated and far away, and how exciting it would be to learn about their behavior. I didn't have the faintest idea that he meant that I should do this." Overcome by the images he evoked, Jane begged Louis to stop telling her about the chimpanzees, saying, "*Please,* don't talk about this. It's what I want to do." And Louis replied, "Why do you think I am talking about it?" If she was willing to devote herself to such a study, Louis promised that he would somehow find the money for it.

Jane was overwhelmed by Louis's unexpected offer and by his faith in her, and she was determined not to let him down. "That's what worried me and kept me awake at nights. I knew he'd sent other people to study the apes and they had failed. If I didn't succeed, he would have been left with nothing, having made yet another mistake."

When they returned to Nairobi in the fall of 1957, Louis began making funding inquiries on Jane's behalf. "I have found a highly suitable [young woman] from the point of view of her personality and interests, and, by the time I send her, from the point of view of her training also," he told Sherwood Washburn, a physical anthropologist at the University of Chicago who had participated in a primate-collecting expedition in Southeast Asia in 1937 and studied baboons in Africa in the 1950s. Louis thought Washburn's support might help him win an initial grant from the Wenner-Gren Foundation or the New York Zoological Society. At this point, he was not envisioning the extensive project that Jane's research eventually became, but rather a four-month study that would "shed some light on the behaviour of *Proconsul.*"[3] Louis sent a similar letter to London's Royal Zoological Society, but, though these organizations congratulated him on his proposed study, none was willing to grant the needed monies.

There were two strikes against Louis's proposal: first, Jane was a woman; second, she lacked any academic credentials. Neither of these concerned Louis, who had decided after his 1946 experience with the nameless young man that women were better suited to long-term animal studies. He believed women were generally more patient than men, perhaps because of his own impetuous nature and Mary's steady one. Louis thought, too, that a woman would be less threatening to the male-dominated ape society. "He thought there was something about male hormones that brought out aggression in animals," explained a colleague. "Whereas a woman has some aura that says, 'I'm one of you. I won't hurt you.' "[4] Though many men have since proved to be successful primate watchers, Louis's bias toward women opened up an avenue for them in field biology that otherwise might not have existed. His preference continues to be influential today; in the United States, the science of primatology has almost equal numbers of male and female researchers.

Louis was also unconcerned about Jane's lack of higher education. "He wanted someone with a mind uncluttered and unbiased by theory," Jane wrote in *In the Shadow of Man,* her popular book about her chimpanzee research, "who would make the study for no other reason than a real desire for knowledge." Actually at the time, in the late 1950s, there was very little theory to cloud anyone's mind. Ethology—the study of animal behavior—was such a new field that even trained zoologists knew little about the best way to habituate animals in the wild (that is, accustom them to the presence of humans). "Sure I was getting my Ph.D.," commented Irven DeVore, then a graduate student of Washburn's who arrived in Kenya in 1959 to study the wild baboons, "but my training wasn't much better [than Jane's]. There weren't any guidelines then for doing animal behavior. You could have put on one small library shelf all the monographs on mammals that had

[3] Louis believed that the study of living animals was essential to understanding those of the past. Before he would allow Alan Gentry, then a young graduate student at Cambridge University, to study the fossilized antelope bone collection in the Coryndon Museum, Louis insisted that he spend three weeks in Nairobi National Park "getting to know the living ones"—an approach that initially surprised Gentry. "It wouldn't have occurred to me to go into the subject like that, but it did give me a good idea of what life was like for a gazelle on the East African plains."

[4] Irven DeVore, now the Ruth Moore professor of biological anthropology at Harvard University, studied the baboons in Nairobi National Park in 1959 and concurred with Louis's opinion, noting that the baboons were "extremely tolerant of my wife and child, but not of me, even though I tried to be gentle and move slowly and so on. . . . A male trying to study primates just starts with an enormous amount of strikes against him in terms of habituation."

been done in the field since, say, 1900." DeVore remembers reading a study on a herd of England's red deer as a way of preparing himself for his research.

Louis's own knowledge of animal behavior, however, was extensive, based on his childhood experiences as a hunter and his years of living in the bush. While waiting for funds to begin the study, he shared much of this with Jane, teaching her stalking and tracking techniques, and most importantly honing her observation skills. In the daytime, Jane continued to work at the museum as his secretary, but Louis introduced her to everyone as his chimpanzee researcher. "I met Jane there in Louis's office," recalled Betty Howell, the wife of the paleoanthropologist Clark Howell. "She had a little bushbaby in her arms, and Louis said, 'I'm going to send this girl out to study the chimpanzees. She's just a natural, and she'll be wonderful at it.' But I thought she looked too young and too thin and too blond. I was very impressed, of course, but she just looked too vulnerable to me."

Nearly two years passed before Louis succeeded in finding a sponsor, and when he did, it turned out to be an unlikely source: the Wilkie Brothers Foundation of Des Plaines, Illinois. Dedicated to the "achievement of unattained horizons in human welfare," the foundation had been started by Leighton Wilkie, who with his brothers ran a tool-manufacturing firm, DoAll, in Des Plaines. Wilkie was so interested in the history of tools that he had started his own traveling museum, called "Civilization Through Tools." He had also attended the 1955 Pan-African Prehistory Conference in Livingstone, Northern Rhodesia (now Zambia), in order to meet the "brave pioneers"—the Leakeys and Raymond Dart—who had discovered some of humankind's earliest tools. They had provided him with specimens for his museum, and consequently he was eager to help when Louis finally turned to him in early 1959. The flaws that had troubled other potential backers—Jane, the untrained female—did not bother Wilkie.

So certain was Louis that he would eventually find the money that he had already sent Jane to England in mid-1958 to study primate behavior at the London Zoo under Professor Osman Hill, and to take informal primate anatomy courses from John Napier, the anatomist at the Royal Free Hospital. Napier had described the limb bones of *Proconsul africanus* and would later join Louis in naming *Homo habilis.*

For Jane, the sojourn in England put some welcome distance between her and Louis. After a year of working as his secretary, and putting up with his annoying habit of listening to her private telephone calls, Jane had finally realized that Louis was in love with her. He poured his heart out to her one night shortly before she departed, a confession that "horrified" Jane. "Louis was a wonderful man in many ways," she recalled about this unwanted declaration. "When he talked to you about his ideas, about something he was deeply interested in, it was intoxicating—particularly if it was *you* he wanted to do a study for him. And I worshipped him. He was a grand old man to me, a wonderful man. But I could never imagine being romantically involved with him. I told him so, and he was very hurt, I think."

Still, Louis did not give up. Desperately in love with his "chimpanzee girl," he pursued her with love letters, giving her the Kikuyu name "Mwendwa" for "My Most Beloved." Jane responded by telling Louis that she could never return his feelings, and defensively began a relationship with a young man in London. Louis's love remained hopelessly one-sided, and by the time he sent news of the Wilkie Brothers Foundation's $3,000 grant in February 1959, his ardor had subsided. Nothing came of Jane's London liaison, and when she finally returned to East Africa in the spring of 1960 ready to begin her chimpanzee research, neither she nor Louis mentioned his earlier passion. Jane never held Louis's declaration of love against him.

> What I was most afraid of was what my rejection of him might mean for my study of the chimpanzees. But Louis was such a super person. Even though I rejected his love, he didn't take back his offer and did everything he could to help. He genuinely fell in love with people. I'm sure he fell in love with all his women researchers, and because Louis was a man of passions, when he fell in love, of course, it would be passionate. It couldn't be any other way with him.

Jane was not the first young woman Louis had fallen for, though she may have been one of the first to reject his advances. His vitality and charisma appealed to many women, and in time, he gained a reputation among his colleagues as a womanizer. Some colleagues were envious, some disapproving ("Louis was a philanderer, you know," commented one associate, lifting his eyebrows, "always after other men's wives"), but all were baffled by Louis's attractive powers. "I remember the first time I met Louis Leakey [in 1959]," said Irven DeVore, who was then a young man of twenty-five.

> He was dressed in one of those awful boiler suits [coveralls], and he had a great shock of unruly white hair, a heavily creased face and about three teeth—he was actually kind of snaggle-toothed. And when my wife, Nancy, and I got back to our hotel, I said to her, "Objectively, he must be one of the ugliest men I've ever met. What do you think?" And she said, "Are you kidding? That's the sexiest man I've ever laid my eyes on." And of course, I could see immediately that all the things I was responding to in Louis—his enthusiasm and infectious energy—were charismatic to women as well.

By the time Louis approached Jane, he had had several affairs. His spying activities for the Criminal Investigation Department during the 1940s and early 1950s, which took him far from home, gave him ample opportunity, as did his trips abroad.[5] The most serious of Louis's affairs began on a sojourn to study the fossil pig collection at the British Museum of Natural History in 1954, where he met a petite, blue-eyed young woman named Rosalie Osborn, who very nearly became, according to Richard Leakey, "the third Mrs. Leakey."

Like Jane, Rosalie first worked for Louis as a secretary, typing his report on the fossil Pleistocene pigs he and Mary had collected at Olduvai and other sites. Rosalie was rather plain and serious, but she was also "full of energy, interested in the world and very, very bright," in the words of a friend. "She admired anyone of intellectual standing, and so got along well with Louis." When her relationship with Louis became something more than secretarial is uncertain; Rosalie herself has declined to comment about it. But Colin Leakey, Louis and Frida's son, noticed something about his father and Rosalie when he met them in Cambridge in 1954.

Colin had only recently met Louis for the first time since infancy, in 1952. "That was Mother's decision," said Priscilla, Louis and Frida's daughter. "She always said she wanted us to meet him when we were grown up, but she felt it was easier for our lives to be quite separate until then."[6] They heard news about their father occasionally from Donald MacInnes, Priscilla's godfather, and other friends. Priscilla learned more by "looking up Father's name in *Who's Who* in my school library," where she discovered she had three half-brothers. But she did not see Louis again until 1949, when she was eighteen, a meeting she describes as "very easy really after the first few minutes." Colin, who was then sixteen, chose not to meet his father on this occasion, but waited another three years. "Priscilla arranged that first meeting," he recalled. "And it was done in this fairly absurd manner. We were both to go to [London's] Holland Park at a particular time, but enter at separate gates, and walk to a particular bench where we were to meet. Presumably, I would greet this man whom I'd not seen in nearly twenty years and say, 'Hello, are you my father?' And he would respond, 'Yes. You must be my son.'" If their initial meeting was awkward, father and son nevertheless quickly discovered that they were at least physically very much alike. "You could easily see the genes at work," said Colin. "Our hand

---

[5] Louis's Kenyan dalliances gave rise to many rumors. Karl Butzer, a paleoarcheologist at the University of Texas, remembers traveling through the Rift Valley in 1967 on his way to the Omo River with some of Louis's Kenyan crew, who cheerfully discussed his indiscretions. "Supposedly, there was a trail of Louis's progeny all the way up the Valley," said Butzer. There is no evidence, however, to substantiate such tales.

[6] Frida never remarried, but continued to live in the Cambridge house she bought when married to Louis. She never asked Louis for money to help raise their children, knowing that although he was generous, he did not have an income. Frida relied on her own small capital, and did some writing and broadcasting to make ends meet. "It was no good being sorry, moaning about it [the divorce]," she said, "one just had to get on. There was no other option. I had two small children, a rabbit warren of a house, and so I lived by my wits for a time. Then the war came, and I became billeting officer for the London-evacuated children, had goats, grew barley and that was that. I knew very, very little of what was going on in Louis's life." Frida died in September 1993 at the age of ninety-one.

gestures and speech-patterns were almost identical." (Even today, people who knew both men say that Colin strongly resembles Louis.)

When they met again, two years later, Colin was a student at King's College, Cambridge, where he was studying botany. "I remember receiving this summons to go to meet my father at the Blue Boar Inn. So I went off, looking forward to having a long talk with him. And when I got there, Louis was sitting on a bench and beside him was this creature—an unexpected young woman. Rosalie. I instinctively sensed their relationship, and was angry with him for making me a party to his duplicity." Colin felt that Rosalie was also embarrassed, but that Louis was not, an insensitivity that dismayed his son. Several years passed before Colin chose to see his father again.

Louis's affair with Rosalie intensified after he arranged for her to move to Kenya in the summer of 1955. Ostensibly, she came to work as his secretary, but it was not long before Mary discovered the true nature of their relationship. "Of course I was aware of what he was doing," she said. "Anybody would be." She had seen this sort of behavior from Louis before, and often had been "furious with him" because he got involved with "such useless sorts of women." But Rosalie's arrival was by far the most serious threat to Mary. Rosalie came during the last years of Kenya's Emergency, a period that had been extremely hard for Mary. "I think Mary was finding Louis's wanderings a bit difficult," said Richard Leakey, "both in terms of Mau Mau and his preoccupations with what she took to be girlfriends. It was very stressful for her. And she made it worse for herself because she took to drowning her sorrows, and Louis had no tolerance for people who drank too much. The more she drank, the more he sought solace from Rosalie."

Louis threatened to leave Mary, and the once-spirited Leakey household began to crumble under the strain. "We had rows about that [relationship]," said Mary. "It was quite serious. But I wasn't going to have the whole family broken up; the children were all quite young." At the time, in late 1955, Jonathan was fifteen, Richard nearly eleven, and Philip six. Of the three sons, Richard apparently suffered the most from his parents' quarrels. "I used to hear them fighting, and I would beg them not to fight," he recalled. "I think Jonathan simply disassociated himself from it. He'd lock himself in his room, and when they'd shout and yell, he didn't want to hear anything about it. But if I heard them, I'd break in and say, 'For heaven's sakes, stop this. Please don't shout at each other.' " Richard begged his father not to leave as well, imploring, "Daddy, please don't leave me. Don't go, Daddy."

At some point during these sad times, Richard was thrown from his pony, suffering a severe concussion. The injury was so traumatic that for a time, the doctor was unsure if he would survive. Richard's fear that his parents might separate apparently compounded his injury, and he lay in bed, a wan, thin child, with his head wrapped in bandages, nervously watching the adults. It was then that Louis broke off his relationship with Rosalie. "I saw his letter to her [after Louis died]," Richard said. "In it, [Louis] says something like, 'I just can't do this. Jonathan can cope, and Philip's too young to be affected, but it's going to ruin Richard if I leave. So we have to sever the knot.' So all my screaming and yelling somehow did some good."

But Louis did not completely abandon Rosalie. Like other women, she had been drawn to him as much by his enthusiasms as his interest in her, and they now used his projects as a way to stay in touch. She continued to work as his secretary until February 1956, then traveled to Rusinga Island to look for fossils and to collect fish from Lake Victoria for Yale University. The distance may have helped mend Rosalie's broken heart, for she gamely took up Louis's offer some months later to help habituate the gorillas on Mount Muhavura in Uganda. "Louis asked me if I'd like to do it," Rosalie recalled, "and I thought it sounded interesting, fun. So I went." Rosalie was one of the first observers to follow the gorillas as they moved through the forest.

When her funds ran out in January 1957, after a four-month stay, Rosalie left the mountain, the gorillas, and Louis, and returned to England. Louis continued to assist her, helping her land a job at the British Museum of Natural History, then arranging for her to study zoology at Newnham College, Cambridge University. Rosalie eventually returned to Kenya as a biology teacher. She never married.

The gorilla project had been proposed to Louis by Walter Baumgartel, who owned a lodge on Mount Muhavura, and who hoped to discover a way of enticing the gorillas into clearings where tourists might view them. This experiment failed, but Rosalie did learn to track the gorillas through the mountain mists, and made some records of their nests, vocalizations, and eating habits. When she left, another young woman named Jill Donisthorpe pursued the gorillas for ten more months on Baumgartel's behalf.[7] Both women's reports encouraged Louis about the possibility of successful long-term primate studies. "Two girls with no training in . . . biology and no previous field experience of this type have secured significant information," he told the New York Zoological Society in December 1959 at a special meeting the society had convened on primate studies. "The fact that they have been able to achieve this is an indication perhaps that to study these animals is not as difficult as we have been led to believe."

Indeed, two zoologists sponsored by the New York Zoological Society and the National Science Foundation, George Schaller and John Emlen, were even then in Uganda engaged in the first comprehensive study of the mountain gorilla—a project that Louis enthusiastically endorsed, even though the observers were male. Researchers had also embarked on studies of the chimpanzee: Junichero Itani had set up camp on Lake Tanganyika, south of the area where Louis was then planning to send Jane Goodall, and

[7] When Rosalie's funds ran out—she was relying on her own savings—Walter Baumgartel ran a newspaper ad that brought Donisthorpe. Donisthorpe published an article about her work in the *South African Journal of Science,* as well as a charming popular account, *Gorilla Mountain.*

a Dutch observer, Adriaan Kortlandt, had started a project in the eastern Congo (Zaire). There was some urgency to these efforts, for the gorillas and chimpanzees were already under pressure from neighboring human populations, and many researchers doubted that the animals would survive in the wild.

Louis shared this apprehension. While searching for funds for Jane's study, he had also been in touch with the game warden of the Gombe Stream Chimpanzee Reserve in Tanganyika, where a population of about 150 chimpanzees lived. From the warden, Louis obtained maps, descriptions of the park's flora and fauna, a rainfall chart—and warnings that the local farmers and fishermen opposed the reserve. Living in a region where land was already in short supply, they coveted the park's untouched twenty square miles. This was particularly worrisome as the colony of Tanganyika was soon to become the independent nation of Tanzania, and Louis thought the new government might open the park to settlement. He hoped that Jane would have at least eight months in the field before politics interfered, and so he had all her camp equipment purchased and waiting for her when she returned to Nairobi from England in May 1960.

"Louis bought my tent and binoculars, and told me to wear dull, drab clothes [so as not to alarm the chimps, a rule that persists to this day]," Jane said. "I had a couple of tin mugs, tin plates and tin trunks, and the binoculars, which were secondhand and terrible. I don't think my eyes have ever recovered from them. Louis always did everything secondhand and cheap, and sometimes that was bad, because whatever it was would fall to pieces."

Louis and Jane's relationship was by now decidedly paternalistic, with Louis playing the role of kindly mentor. Whatever passions he had once felt for her Louis now put aside, concentrating instead on getting Jane to Gombe. Actually, it was a matter of getting Jane *and* her mother, Vanne Morris Goodall, to the park. A vivacious, warmhearted divorcée, Vanne would serve as Jane's chaperone in the forest. The wardens had refused to allow an unescorted woman in the reserve, and only reluctantly agreed to the study when Vanne volunteered to accompany Jane. "I am anxious that no untoward mishap should occur," David Anstey, the senior warden, wrote to Louis just before Jane was ready to depart for Gombe. "The area is isolated and I am not happy that the two ladies should be there." Anstey had such misgivings that a few days later he revoked his permission entirely. The local fishermen were protesting the reserve again, and Anstey simply could not permit the ladies to step into such a volatile situation. Jane would have to wait.

She waited three weeks, though Louis saw to it that she made good use of her time. He knew of a colony of vervet monkeys living on Lake Victoria's Lolui Island, and sent Jane and Vanne there, along with the skipper of his *Miocene Lady,* Hassan Salimu, for the field experience. In the meantime, the Kigoma district commissioner and Warden Anstey sorted out the troubles with the fishermen—apparently, their real concern was Jane's impending arrival. They could not conceive of anyone traveling all the way from England to study chimpanzees, and so had decided she must be a government spy. The tribal headmen agreed to let Jane come only after Anstey promised that they would meet her and could observe her at work.

When Louis finally summoned Jane back to Nairobi, her heart soared. "I could think of nothing save the excitement of the eight-hundred-mile journey to Kigoma [the closest major Tanzanian town to Gombe]—and the chimpanzees," she wrote in *In the Shadow of Man.* Two weeks later, in mid-July 1960, Jane, Vanne, and their cook Dominic had settled into their camp under the tall oil-nut palms of the Gombe Stream Chimpanzee Reserve. The following day Jane saw chimpanzees in the wild for the first time.

"You will be pleased to hear that the research on the Chimpanzees . . . is at last under way," Louis wrote to Leighton Wilkie, a week after Jane had arrived at Gombe, "and that on the very first day . . . Miss Goodall [saw] two chimpanzees out in the open not far from the camp, and the prospects are very good. . . . She has been observing as many as ten chimpanzees at once."

For ten days, the chimpanzees obliged Jane by coming each morning to feed on the reddish fruits of a nearby *msulula* tree. But then the animals vanished and for the next two months, Jane seldom saw a chimp. When she did, they fled. "I was really worried that I would fail, that the money would run out, and Louis wouldn't be able to get any more because I had failed," Jane recalled about this frustrating period. "I remember writing him and saying, 'I just can't do it. I can't habituate the chimps. I haven't got the time.' And he would write back and say, 'I know you can do it. I have faith in you.' The more he said that, the worse I felt."

Yet over the course of the next three months, the chimps gradually grew accustomed to her presence, and Jane learned that they would tolerate her if she did not move closer than sixty feet. She came to recognize many of the chimpanzees as well, acknowledging their individuality by giving them names, such as Flo, Goliath, and Leakey.[8] She made notes on their calls, collected their dung, sampled the fruits and nuts they ate, watched them groom each other and build sleeping nests high in the fronds of palm trees. Then one day, only four months after her arrival at Gombe, she witnessed what no one had seen before: chimpanzees feasting on a wild piglet they had killed. A few days later, she made another unexpected discovery. Her favorite chimpanzee, one she called David Greybeard, sat beside a termite nest "carefully push[ing] a long grass stem down into a hole in the mound. . . . I was too far away to make out what he was eating, but it was

[8] Though the habit of naming study animals is common today, at the time it was considered bad science, as was the concept of individuality. Names suggested anthropormorphism; traditional ethologists preferred labeling their subjects with numbers. Jane was also criticized by a "major [scientific] periodical" for using the pronouns "he," "she," and "who" when referring to the chimps in her first technical paper, written in 1963. The editors had crossed out these words, replacing them with "it" and "which." "I am glad to say," Jane noted twenty-five years later, "that the final version conferred on the chimpanzees the dignity of their separate sexes."

obvious that he was actually using a grass stem as a tool." After David left the mound, Jane pushed one of his discarded stems into the hole, and pulled out a cluster of termites. Later, she would see the chimps modify twigs and stems, stripping away the leaves to make a proper termite-retrieving tool, "the first recorded example of a wild animal not merely using an object as a tool, but actually modifying an object and thus showing the crude beginnings of toolmaking."[9]

Louis was naturally delighted by these discoveries, particularly the chimpanzees' tool use. "He said, 'Yes, of course, early man must have been using twigs and straws and grasses long before he used stone tools,'" Jane recalled. Her observations came at just the right time: the Wilkie Brothers Foundation money was running out, and Louis was desperately searching for new funds. This time he approached the Wenner-Gren Foundation, and they willingly provided a small grant. "Miss Goodall . . . is getting increasingly good results," he proudly wrote to Harold Coolidge, a zoologist at the National Academy of Sciences in Washington, D.C., who helped him secure the grant. "[She] has now had a Chimpanzee family sitting within ten feet of her, behaving in a normal fashion as though she was not there. She has also been 'investigated' by a male Chimpanzee who eventually rushed up to her and knocked her hair and the top of her head with his hand to try and make her move and show what she was. . . . I can say, without hesitation, that her results are far exceeding my wildest hopes and that it is vital to keep on with the work."

By late November 1960, the reserve wardens were confident that Jane could fend for herself, and so did not object when Vanne returned to England. Though Louis's assistant, Hassan, came to watch over her camp, and Dominic continued to provide her with meals, Jane was virtually alone in the forest for much of the next year. The loneliness was wearing, and she found herself greeting the trees in the forest, and saying "Good morning" to the stream where she fetched her water. Her most immediate contact with the outside world was Louis, who wrote to her constantly, advising her on everything from analyzing chimpanzee dung to casting their footprints. He organized her finances, filled out grant applications, bought her a new pair of binoculars, and arranged for her to attend Cambridge University under the tutelage of the animal behaviorist Robert Hinde. But most of all, he was friend and confidant, commiserating with her by mail when heavy rains made life miserable and chimp-watching impossible, rejoicing when she wrote of observing new, unexpected behaviors.

In March 1961, Louis persuaded the National Geographic Society to award Jane a $1,400 grant. As a condition, the society's magazine editors insisted on sending a photographer to her camp, but Jane flatly refused and Louis

---

[9] Other observers had witnessed chimpanzees in Liberia cracking open palm nuts with rocks; and in the Cameroon, a hunter reported watching chimpanzees poke sticks down an underground bees' nest, then lick off the honey. But Jane's observations were the first well-documented reports of this type of behavior.

backed her up. The chimpanzees were used to seeing Jane alone, and an unknown person could destroy their confidence. He suggested that they send Jane's sister, Judy, as the photographer, since she looked somewhat like Jane. When the editors balked at this—Judy had no photographic experience—Louis got the London weekly newspaper *Reveille* to pay her passage instead.

"That was when we started calling him our 'Fairy Foster Father,'" Jane said. "Because of the doors he opened. We were supposed to be his 'Foster Daughters,' but that was a bit too much for me. So I was 'F.C.'—or 'Foster Child.'"

From late 1961 until Louis's death in 1972, Jane addressed nearly all of her letters to him "Dear FFF," and signed them "Love, F.C."—an affectionate acronymic game that Louis joined.

Whether or not Jane was aware of it, her nickname for Louis was coming almost literally true, for by now Louis was intimate with her mother. When he traveled to London, he stayed in Vanne's flat on Earl's Court Road, and in the evening escorted her to concerts and the ballet. "Vanne is very easygoing and good-natured, a fantastic person, really," commented one friend. "Louis was probably very relaxed with her."

Louis's close relationship with Vanne and his reputation for womanizing gave rise to some wild rumors, the strangest of which was the speculation that Louis might actually be Jane Goodall's father. In her biography of Louis, *Leakey's Luck,* Sonia Cole claims that Louis and Vanne knew each other in their teens in England. Jane Goodall dismisses this information, and Vanne refuses even to comment. However, both Colin and Richard Leakey believe the story of Louis's paternity could be true, saying that only a father-daughter relationship could explain Louis's strong ties to the Goodall family. Mary, on the other hand, regards the story as preposterous. At any rate, in mid-1933, or approximately nine months before Jane's birth on March 4, 1934, Louis was in England, married to Frida but passionately involved with his then-student Mary Nicol; it is doubtful he had time for another relationship. Although the rumor made the rounds of scientific meetings and conferences over the years, it did not reach Jane Goodall until the mid-1980s. She called the tale "absolutely incredible. . . . There's no way it could possibly be true. . . . How funny." Jane then described her close physical resemblance to her own father, Mortimer Herbert Morris Goodall, who worked for the Jaguar Motor Company and loved to race motor cars.

Louis's affection for the Goodalls distressed his family. Mary responded by dismissing Vanne as "a fool," but Richard, who was fifteen when Louis sent Jane to Gombe, grew increasingly resentful of the younger woman. Although he and Jane had been "the best of friends," Jane said, when they first met in 1957, and had made plans to catch and tame a wild zebra, six years later he was openly hostile toward her. By then, she was an established scientist as well as an emerging celebrity, with full funding from the National

Geographic Society, and was working toward a Cambridge University Ph.D.[10] In contrast, Richard was a high school dropout, running a safari business and searching for a way to "make my mark in the world." When Louis sent him to Gombe along with Joanne Hess, an observer from the National Geographic Society, to report on Jane's work, Richard was, Jane recalled, "very arrogant." He sneered at her research, taunted her about nearly stepping on a poisonous snake, and was in general "mean and nasty." Richard, in turn, thought Jane's treatment of Hess "appallingly rude." Back in Nairobi, he continued his attack on her, reporting on her research in the worst possible way to his father. His efforts at sabotaging their relationship very nearly succeeded, and for several months Louis and Jane exchanged painful letters, with Louis questioning her research and Jane imploring him to continue believing in her. "If only you came you would see for yourself that things are not as Joanne [Hess] and Richard painted them," Jane wrote. "[N]ever, never think that I will let anything happen to [the chimpanzees] through what I am doing. . . . [A]nd your faith—don't, please don't, lose it now. It hurts you know."

In particular, Louis objected to Jane's use of bananas as a means of luring the chimpanzees to camp, a practice that Richard had witnessed and that Louis, apparently, knew nothing about. Worried that the feeding would unnaturally affect the chimps' behavior and jeopardize not only the validity of her study but her own safety, he advised her to stop at once. But Jane and the National Geographic cinematographer Hugo van Lawick continued to set bananas out for the chimps, arguing that it was the only way to film the animals' behavior. "We couldn't have done it without that," Jane said. "But Louis was right. It would have been far, far better not to feed them. He had a bad feeling about it, and it was a bit sad because many of the findings that I think would have excited him, he tended to push aside. When I tried to tell him about the long-term bonds between mothers and their families, he'd say, 'Well, that's all mixed up with the banana feeding.' And it was such a pity, because it wasn't."

Louis never did go to Gombe, and he remained critical of Jane's work until a few months before he died. Only then did he grudgingly admit that the feeding may not have tainted her study as much as he feared.

Richard, though he believed Jane's feeding to be wrong, subsequently realized that his behavior then was immature. "But I had very mixed emotions about my father's involvement with the Goodalls, and I was a very jealous young man." He was also very ambitious, and had set his sights on becoming far more than a safari guide.

---

[10] When Louis first sent Jane to Gombe, she had little scientific training aside from what he had arranged for her in England. She was also young and beautiful, and many professionals regarded her as "a blonde bimbo," said Mary Smith. Not until the mid-1960s, when she completed her doctoral thesis, was she taken seriously.

Chapter 18

RICHARD MAKES HIS MOVE

By the late 1950s, the great passion Louis and Mary had once felt for each other had waned. Their professional interests, the stones and bones that had brought them together, now dominated their lives, lending their relationship a businesslike air that their colleagues found remarkable. "There were few outward signs of affection and concern," said Betty Howell, the wife of the paleoanthropologist Clark Howell. "It was all prehistory and archeology." Alan Gentry, a paleontologist at the British Museum of Natural History, noted that while Louis always called Mary "Mary-mine," she, in return, "did not give the impression that she was very close to him or even liked him very much." Yet they stayed together, primarily because of their children. The years immediately after Louis broke off his affair with Rosalie Osborn were so harmonious that Richard Leakey remembered them as "some of the happiest times."

Each of the Leakey boys had a pony and belonged to the Langata Pony Club, which Louis and Mary had helped organize for the white, middle-class neighborhood children.[1] On weekends the club members would sometimes meet for jumping, steeplechase, and dressage competitions. But more often, they would simply ride together among the wildlife that roamed the nearby Athi Plains, daring one another to chase a giraffe, rope a zebra, or gallop close enough to a rhino to slap its rump. On other occasions, Louis bundled

[1] Louis and Mary also founded the Dalmatian Club of East Africa in 1949, and were members of the East Africa Kennel Club. Louis became president of that club in 1959. Several of their Dalmatians won prizes, and Victoria was judged Best in Show in 1957.

some dozen children into his Land Rover for a field trip to the handaxe site of Olorgesailie or to spend a day at the museum. But whatever the activity, the Pony Club gathered first at the Leakeys' home. "They were a very open-house family to us kids," recalled Christine McRae, a club member, "and we all used to spend our entire holidays at their home. We'd take a packed lunch, then ride to the Leakeys and then through the forest and across the river to the plains. And of course, there were not many houses out there then. It was all open and we could ride for miles, right to the Ngong Hills."

Richard Leakey was especially fond of horses, in spite of the skull fracture he had suffered when thrown by his pony at age eleven. He had never excelled at the school sports of rugby and cricket, and so was delighted to find "something I was good at." Out on the plains among the animals, free to ride at will, Richard soon found where his heart lay and decided then that he would have a career that gave him "an outdoor life in Africa."

Having spent many of their holidays on their parents' expeditions, Richard, Jonathan, and Philip were competent at bush skills and mature beyond their years. Louis and Mary never coddled their sons, nor worried about entertaining them, believing that children were better off when left to fend for themselves. "They all three became very independent this way," Mary noted. "Quite apart from convenience, I don't approve of playing games with children. They had nannies, but their nannies were really there to prevent them from falling in the fire." By the age of fourteen, each son knew how to drive and repair a Land Rover, to track and shoot game, find his way in the bush, and manage the logistics of running a camp. "The Leakeys' sons were always self-sufficient, extraordinary, pioneer-type boys," said Betty Howell. "You could give them a piece of wood and say, 'Make a bow and arrow for me,' and they'd come up with something that would work. They were very adventurous, and really like our pioneers, only 150 years later."

On their parents' expeditions, they were expected to help excavate and catalogue the finds, as well as assist in the daily camp tasks. "If one wanted to be at Olduvai, one pitched in," said Richard. "Money was always short, and so it was almost a job to pay for one's keep." They were not assigned trivial chores, but rather—to the surprise of many visiting scientists—entrusted with such key responsibilities as servicing the Land Rovers or building the camp's *bandas*. Recalled the archeologist Maxine Kleindienst, "When I was at Olduvai [in the summer of 1960], fifteen-year-old Richard was in charge of the water supply," a task that required towing water trailers thirty miles to a spring near the top of Ngorongoro Crater three times a week. "He was the one who made sure all the trucks were running and the supply line was running—and this was taken for granted."

But there was also time on expeditions for exploring, watching wildlife, trapping snakes and hunting birds, and playing "early men and early women," experimental games to see if they could live as our ancestors had. Sometimes at dusk, the boys, their parents, and assorted visiting scientists would go "springhare-ing"—making a slow drive across the Serengeti until a springhare (a nocturnal rodent that hops like a kangaroo) was spotted in the glare of the headlights. Someone would then leap from the car and make a mad dash to grab the animal's long tail: perhaps quick-footed early humans had also seized small animals bare-handed. (Once Louis demonstrated for a reporter how early humans similarly could have used nothing more than their hands to hunt small antelopes. Camouflaging himself with leafy branches, he slowly advanced toward a gazelle, taking two hours to reduce the original 250 yards to six feet, "at which point," the reporter noted, "the prehistorian brought down the gazelle with a perfectly timed flying tackle.") Other times—in an effort to demonstrate that early humans may have obtained their meat by stealing as well as hunting—Louis and Richard chased carnivores from their kills. "We chased off packs of hyenas, and once we disturbed some lions on a kill, without a vehicle, just carrying our firearms to see if they'd leave," Richard recalled, "which they did."[2]

Their outdoor life left the boys bronzed and freckle-faced, with sun-bleached locks of hair spilling over their eyes. Despite their age differences, they were physically very much alike: tall and skinny, with angular faces, and those "long Leakey legs," as Phillip Tobias noted. Self-assured and independent, they were nevertheless gauche—well versed in African bush ways, but lacking any polish—and not one was a serious student. "It was a surprise, considering their parents' reputation, to see how unacademic Louis and Mary's children were," recalled Gentry. But with lions and rhinos roaming just beyond their front door, it would have been more surprising to find the Leakey boys at home dutifully poring over their schoolbooks. As it was, they regarded school as only a burden, "a sadness of any youth," as Richard once said.

In primary school, Richard had been an excellent student, but he slipped when he entered the Duke of York Secondary School at age eleven. The school was open only to European boys, and run on a strict colonial regimen. Richard never fit in. Sensitive and highly strung, he worried about Kenya's racial inequities, and was promptly branded a "lover of niggers." At the time, in 1956, Kenya Colony was still in the throes of the Mau Mau troubles, and Richard's sympathies for the Africans made him an instant social outcast.

On his very first day at the Duke of York, another boy pointed him out as a "nigger lover," prompting a gang of older students to grab him. "Before I knew what had happened," Richard wrote in his autobiography, "I had been placed inside a wire cage some three cubic feet in size. . . . The hinged lid

---

[2] In an interview with a reporter from the *New York Times Magazine* in 1971, Louis claimed that he "experimented with Richard. We went out naked, picking up some giraffe limb bones and jaws to act as rudimentary weapons, but not such as we could offend or kill anybody with—just protect ourselves a little. And we drove off the hyenas and vultures long enough when they came into the kill. They were furious and after ten minutes, I signaled to my son, 'Get out. It's not safe any longer. They're going to kill us now.' But we got a little zebra. . . ." Richard says that while he and his father did chase off animals from kill sites, they never did so unclothed.

was closed and padlocked. I was crouched like a monkey in this tiny cage, with no way of escape." Delighted with their prize, several hundred boys took turns poking him with sticks, spitting and even urinating on him until they left for the morning assembly. Richard remained huddled in his cage, "very miserable and frightened," but was eventually spotted by one of the teachers, who had to use a hacksaw to cut through the lock. "He had no doubt I was to blame," Richard wrote, "and so, wet through, filthy and stinking, I began my first day of senior school."

The school's many rules were a further aggravation. Richard found himself caned for letting his knee socks slip to his ankles, and for missing morning chapel (a punishment that led him to decide he would never be a Christian). He enjoyed his biology class, but hated Latin and math, and skipped them whenever he could. Far more interesting was his own small-scale trapping business, supplying wild animals to the photographer Des Bartlett for nature films. The money that Bartlett paid Richard for each mongoose, porcupine, wild cat, or bush baby he trapped gave Richard a "marvellous degree of independence" from his parents.

If the trapping business sometimes made him late for school, it also taught him a great deal about handling wild animals and confirmed his desire for an outdoor career. But he was determined that it would be a career that had nothing to do with fossils or archeology. Though Richard loved the adventure of expeditionary life, he was also concerned about trying to "develop my own stature, my own reputation, [and] position in the shadow of my father. . . . So I thought the best way to get out of his shadow was to get out of his business. And so at the time when I could have gone to university . . . I made the conscious decision not to do so, because this way I could get well-removed from this grand ship that was sailing so well in early oceans."

Richard did not graduate from the Duke of York, but simply quit school in December 1960. He had just turned sixteen, and "it was now," he felt, that his "real life would begin."

One Sunday morning in January 1961, Louis told Richard to stay seated after breakfast. Louis was then fifty-eight, forty-two years older than his son, and looked it, with his once-angular features now fleshy and worn, his figure rumpled and slouched. Beside him, Richard was as lean and handsome as a young Gregory Peck. Dark-eyed and sharp-featured, he was also intensely ambitious and "very prickly about being" his father's son.

Louis and Richard had never been close, perhaps because, as Mary suggested, "they were too alike in temperament." Louis may have been somewhat jealous of Richard as well, for outsiders thought that Richard was Mary's favorite. But for whatever reason, Louis had never been "as fond of Richard as [he was] of the other two when they were boys," Mary wrote in her autobiography, "and had treated him with visibly less indulgence. . . ." Nor did Louis pamper Richard now, but told him he must make a decision. Louis would continue supporting him if he pursued his studies; but if Richard did not go back to school, he would have to find a way to support himself. Richard was actually pleased at the choice—"I shall always be grateful to my parents for their willingness to allow me to make my own decisions about the important things in my life," he noted—and told Louis that he was not returning to school. He had other plans and asked to borrow £500 in order to purchase an old Land Rover. It was the last money he borrowed from his parents.

For a while, Richard continued to trap wild animals for Bartlett's films as well as for overseas zoos, and then started an animal skeleton enterprise. Knowing that universities and museums needed such skeletons for their collections, Richard used his Land Rover to transport the corpses of wild animals to the museum, where he cleaned and labeled them. Many animals died in a drought that devastated Kenya and Tanganyika in late 1960 and early 1961, and Richard's small business was soon booming. "Richard and Kamoya [Kimeu, then one of Mary's Kamba assistants] had a great big pot, like one for cooking up missionaries," said Garniss Curtis, a geologist from the University of California at Berkeley who determined the dates of Louis and Mary's fossils. "And they'd put the skeletons in there, boil them up, and then put the bones on anthills to strip them clean. They all came out nice and white. Then they put them in sacks, and when I was there, they had a room about forty by twenty filled with bones." Impressed with Richard's drive and enterprise, as well as his ability to identify each bone ("He had each one labeled, and was unbelievably skillful at this"), Curtis offered Richard the chance to attend Berkeley. "I told him I could get him in with a little remedial help if he wanted. But Richard was just not one for the books." Instead, he wanted adventure, and like other young men who had grown up in Kenya, chose to live by his wits, earning an income from the colony's land and animals.

By 1961 Kenya Colony was beginning to acquire an air of modernity. The economic boom that began after World War II had transformed Nairobi from a homely tin-shack town into a bustling city of palm-lined streets, traffic jams, and swank office buildings. There were soap and paper factories, cement plants, a new international airport with jet-sized runways, and, overseeing all, a "sedate, white-collar community," as one magazine reported. Yet the country as a whole had not lost its frontier aura, and many boys of European descent succumbed to Kenya's spell, choosing careers as safari guides, hunters, game wardens, and bush pilots. Even Richard's elder brother, Jonathan, who had finished secondary school (and who Louis hoped would pursue a career in prehistory), decided instead to live in the bush, collecting the venom of poisonous snakes (which is used to make antivenin) for a living. In time, Jonathan became one of the world's major suppliers of snake venom for the antivenin market, and an expert on African snakes.

Kenya's political climate, too, favored these Kenya Cowboys, as the young, self-sufficient white Kenyans were often called. In January, 1960, Britain's Colonial Office decreed that the colony would become independent as soon as possible. Black Africans would run the country then, likely limiting the role—and career opportunities—of the whites. Many of Kenya's white settlers considered the Colonial Office's decision a betrayal: the Emergency over the Mau Mau troubles had barely ended, and the British government was still encouraging new settlers to emigrate to Kenya. But the nationalist feelings that lay at the heart of Mau Mau had not died away. Although eighty thousand Kikuyu had been imprisoned and "rehabilitated," new leaders from other tribes had continued to call for independence and a government based on one man/one vote. Around the world former colonies were emerging as young, new countries. In Africa, Britain had granted Ghana independence in 1957; the French were preparing to leave Algeria and their West African colonies, and the Belgians were packing up in the Congo. Britain had also scheduled Tanganyika for independence in 1961, Uganda the following year, and Kenya soon thereafter.

Stunned by the announcement of Kenya's new political future, many white settlers decided to leave. By 1962, Kenya's white population had dropped by ten percent and farm values had plummeted—land that had been worth $60,000 in 1959 was barely fetching $15,000 three years later, presuming a buyer could be found.

Although Louis had no intention of leaving, he was also uncertain about Kenya's political future. In the 1950s he had argued for a multiracial government, one based on a fixed number of representatives of Europeans, Africans, and Asians, and had secretly hoped to fill one of these seats himself. "Louis had a real interest in politics," Mary recalled, "and I think he probably would have gone into it. But I hated it; I mean I hate anything to do with politics. They involve people, inevitably, and I don't like people, either. I would have had to go away, live in the bush somewhere." Mary refused to support Louis's ambition of running for office, and the matter was dropped. But he never lost his interest in the subject, or his belief that his connections to the Kikuyu tribe gave him a special understanding of Africans. Louis was not, however, always as perceptive as he thought, and on the subject of independence he was wrong.

In 1960, when the Colonial Office began discussing the possibility of majority rule in Kenya, Louis made a radio broadcast on the Voice of Kenya. In it, he called again for a multiracial government in Kenya. He said that while black Africans should have a role in the government, they were not ready for full democracy. Like other white Kenyans, he believed that was, as he put it, the "road to Inefficiency, Insincerity and Insecurity. . . . It is the road of those who pretend they are ready to take over the functions of Government, but who in their hearts know quite well that there are still too few trained, efficient and responsible Africans who could take over the thousands of jobs now held competently by Europeans." Nor was Louis immune to other fears of the white community. When a crime wave broke out in early 1961, he predicted that another Mau Mau era was at hand. "Kenya today is on the edge of an abyss," he wrote in an article entitled "Time for a Firm Hand in Kenya," published in London's *Sunday Telegraph*. "The situation is too much like what it was in 1951 and 1952 [when Mau Mau erupted]," he continued, "to be looked at with complacency," and then called for strong government action against illegal political organizations.

What most concerned Louis was the impending release of Jomo Kenyatta, who had been convicted of being the leader of Mau Mau in 1953. After his conviction, Kenyatta had served a six-year prison term in the desert near the Sudanese border (he was given one year off for good behavior), and had spent another two years under house arrest in the outpost of Lodwar, several hundred miles from Kikuyuland. Yet even from that distance, he exerted a powerful influence on Kenya's politics. Always a hero to his people, Kenyatta's long exile had made him as legendary a figure as Nelson Mandela is today, and at political rallies cries of *"Uhuru na Kenyatta!"*—"Freedom with Kenyatta!"—filled the air. There would be no freedom for Kenya without Kenyatta, and in August 1961 the British government finally released him. Three months later, Kenyatta was elected president of Kenya's leading political party, KANU, the Kenya African National Union. Though independence would not come for another two years, there was little doubt in anyone's mind who would be the country's first prime minister.

Understandably, Louis feared some type of retribution from Kenyatta. Besides the role he had played as translator at Kenyatta's trial, Louis's activites as a spy and an informer among the Kikuyu during the Emergency had soured relations between him and the tribe. "In fact," said Charles Njonjo, whose parents had been educated at the Leakey mission, and who would help draft Kenya's constitution, "at that time, during the independence, Louis was very suspect among the Kikuyu." It was not improbable that Kenyatta would expel Louis from the country. The British government had offered its citizens a choice of Kenyan or British citizenship. Although Louis wished to become Kenyan, he also made certain that his British papers were in order.

The issue was less problematic for his sons, who never considered leaving. "Kenya was my home and I'd always thought of myself as Kenyan," said Richard, voicing a sentiment shared by Philip and Jonathan. "It never entered our heads to leave. It was absolutely inconceivable that I couldn't have a future here because of the color of my skin." With the money he had made trapping animals and marketing skeletons, Richard formed a safari company, first guiding members of the National Geographic Society and then other wealthy clients who thought that because of the Leakey name, they might be meeting Louis. "They were appalled to find an eighteen-year-old in charge," Richard said, "but by then it was too late. I had their deposit and we were off."

To expand his safari business, Richard decided to learn to fly. In October 1963, three months before Kenya's independence day, he obtained his private-pilot's license and made his first flight as captain, taking a group of clients to Olduvai. On the flight, he noticed an area along the western shores of Tanganyika's Lake Natron that resembled the fossil beds of Olduvai. "I knew what I was looking at because I grew up with a family that spent its life groveling in the sediments," Richard recalled. "It looked familiar country to me, and I thought I'd like to return, [and] take a look at it. I was curious."

Mary had actually spotted the Lake Natron fossil beds in 1959, on her way back to Nairobi from Olduvai after discovering *Zinj.* She, too, thought the sediments would be worth investigating, but it was impossible for her to take time away from Olduvai. Now, four years later, with his parents' approval, Richard set out for Natron with a truck and Land Rover, and two of Mary's Kamba workers.

For Richard, the expedition was a welcome departure from guiding tourists. He had grown weary of tending fussy safari clients, but he was not willing to give up the adventure of bashing about in the remote bush to "new and marvelous places." The journey to Natron fulfilled his every expectation, since the lake was accessible only by following a faint track, and the sediments Mary had described to him were not easily located. "Richard was radioing his parents that he could find nothing but lakebeds and fish fossils," said Glynn Isaac, then a young archeologist who was excavating the Olorgesailie handaxe site for Louis. "So Louis and Mary said to me, 'Well, you'll have to go and work with him because clearly he needs someone with a bit more experience.' " Glynn welcomed the opportunity, and met Richard a few days later on the edge of the Serengeti.

Glynn Isaac was then twenty-eight, a small, wiry, mirthful man who had grown up in South Africa, studied at Cambridge University, and was planning a career as a laboratory archeologist when he met Louis. "Louis was looking for someone to finish the work at Olorgesailie," he recalled. "I remember sitting in a row outside the room where he was interviewing people [at Cambridge]. But when I got in, all that he asked me of any consequence was, could I repair a car if it broke down?" Having maintained a creaky 1934 Austin throughout his college career, Glynn could answer this question truthfully, and so was hired largely on the strength of his mechanical aptitude. Like Richard, he was bush-savvy as well as mechanically minded, and shortly after he arrived in Kenya in 1961, the two became good friends. They had already collaborated on various trackless ventures around Olduvai, and now set out together to find the fossil exposures at Lake Natron.

Glynn could see immediately that the region Richard was exploring had been underwater in prehistoric times—"Anything that fell out of a canoe would have been there"—and suggested that they look for the shoreline of the ancient lake. Here, they were more likely to find the fossils of extinct mammals, stone tools, and possibly the bones of early humans. They took another flight over the Natron basin, and from the air pinpointed a region just north of the modern-day lake that looked promising. "We practically destroyed two Land Rovers getting there," said Glynn, "but eventually we made our way to the Peninj River—a beautiful place." The river ran through a deep gorge and cascaded over a high escarpment before spilling into Lake Natron. Near the river, numerous steep-sided gullies snaked through the barren red earth, and here, "with a sense of triumph," they found a good scattering of mammal fossils. "Richard was really up for fossil hunting now," said Glynn. "But it was because of the adventure. He enjoyed the difficulty of getting to the place, the fun of searching for things and finding them."

Back in Nairobi, Richard and Glynn showed their fossil samples to Louis, who agreed that the material looked worthwhile. Their samples of pig and antelope fossils resembled those from the Bed II level at Olduvai, Louis said, and so would be dated at approximately 1.5 million years. Then he generously advanced them funds from his National Geographic account for a monthlong expedition. They left shortly after New Year's, 1964, with Richard as the logistics leader and Glynn in charge of the science. Accompanying them were Philip Leakey, not yet fifteen; Glynn's wife, Barbara; a National Geographic photographer, Hugo van Lawick (who had worked with and would later marry Jane Goodall); and a staff of six Kenyan workers from Mary's Olduvai crew.

As the expedition leader, Richard decided that rather than trying to drive back to Peninj, it would be easier to cross Lake Natron with a raft, then walk the two miles to the site. Once across the lake, Richard hired thirty porters from a nearby Wasonjo village to help his team carry all their goods. But troubled by his "strong social conscience," he decided to show his "solidarity" with the porters by hefting a forty-pound load and struggling along with it at the rear of the column. It was not a wise decision. "Naturally, with this load, I could not catch up with the head of the column," he wrote in his autobiography, "and before long there was chaos. The line broke, the porters pillaged the loads . . . and then took off. We lost blankets, sheets, pots, pans and a lot of food." In the future, Richard made sure that he was "in command [when] in the lead!"

Glynn Isaac directed the scientific exploration of the site. Although Peninj was not as prolific as Olduvai, there were still well-preserved animal specimens, particularly the bones of antelopes and pigs, as well as a scattering of handaxes and other stone tools. But none of this really excited Richard. "I had enjoyed the challenge of getting the expedition to the camp and everything set up and I looked forward to the organizational aspect of leaving again, but the fossil hunting and archaeology left me rather bored after such an adventure," he wrote. "Thus I took the first legitimate opportunity to return to Nairobi." In so doing, Richard missed the most exciting moment of his expedition.

On January 11, only a few days after Richard left, Kamoya Kimeu was walking down a narrow ravine when he happened to glance up and noticed "a very white object very far up the cliff." "It was difficult to reach it," he recalled, "and when I did there was nowhere to stand." But he could see that it was a fossilized bone, and when he brushed away the earth, a tooth set in a jawbone gleamed back at him. Certain that it was hominid, Kamoya shouted for Glynn, who quickly climbed up the cliff and began shouting as well. "None of us were blasé about these things then," Glynn recalled, "so there was high excitement."

It appeared that an entire hominid mandible (the lower jawbone) was buried in the cliff face. Glynn and Philip Leakey set off at once by raft and Land Rover for the town of Magadi, where there was a telephone, and called Richard. He, in turn, called his parents at Olduvai, and by the next morning, "the clans were assembling" at Peninj, as Glynn put it. Everyone marched out to the gully, and took turns climbing up the steps Kamoya had cut into the cliff to take a look at the fossil. "It was literally sticking out of the cliff," said Glynn, "half the jaw embedded and half sticking out." He then excavated the fossil, producing a beautifully preserved lower jawbone from a hominid resembling *Zinjanthropus,* although half a million years younger in age.

Louis was especially delighted by the fossil, and decided that it supported his theory that australopithecines such as *Zinj* had nothing to do with the ultimate evolution of humans. "The new lower jaw from Natron . . . show[s] that the *Zinjanthropus* type continued to evolve unchanged," he told a press conference in Washington, D.C., on April 4, 1964, "having already overspecialized away from man. While on the other side of the story there was evolving . . . a new species of *Homo.*" This new species was *Homo habilis*—the man with ability—that Louis and Phillip Tobias, the South African anatomist, had argued over for three years. Only a month before the Peninj discovery, Tobias had finally accepted Louis's contention that the smaller hominids from Olduvai represented a new human species. Tobias had then made plans to visit Kenya to see the Olduvai skulls Mary had unearthed at the end of 1963 (these, together with Jonny's Child, would be named *Homo habilis*), and so was in Nairobi when Glynn Isaac arrived with the Peninj jaw packed in a biscuit tin.

During that same week, Louis and Phillip began to draft their paper for *Nature* describing the new species, *Homo habilis.* Louis and Mary were also writing an article announcing the many hominid discoveries her team had made at Olduvai in 1963, and Louis decided to mention the Peninj fossil as well.[3] He believed it strengthened his argument that two hominids—one, the slender handyman; the other, the heavy-jawed *Zinj*—had lived side by side in East Africa two million years ago. Louis then entrusted the jawbone to Phillip Tobias, who would describe its anatomy in detail.

Richard felt cut out of the excitement of describing the fossil as his parents, Tobias, and Isaac debated its interpretation. Louis had not forgotten him, but the scientific work could only be done by specialists—and so, in

---

[3] These fossils—skulls, teeth, and jawbones of *Homo habilis,* the cranial vault of a *Homo erectus* skull, and the *Australopithecus boisei* (or Peninj) mandible—were all announced in Louis and Mary's April 4, 1964, paper, "Recent Discoveries of Fossil Hominids in Tanganyika: At Olduvai and Near Lake Natron."

Louis and Mary's *Nature* paper, while Isaac was praised as the scientific leader of the expedition, Richard was described only as the initial explorer. His parents also thanked him—albeit in a footnote—for handling the expedition's logistics. When Louis announced the discovery in Washington, D.C., at the National Geographic Society, he again graciously referred to Richard as the "leader" of the expedition. But by now, it had become painfully clear to Richard who the real leaders were: the scientists. "Richard saw that he ran the logistics and I ran the science," recalled Glynn Isaac, "and at the end of the day, it was the science that mattered. That was a turning point for him." "On that expedition, I was really just the 'tent boy,' and that didn't please me very much," Richard said about this revelation. "I started to have ego problems again. . . . So . . . I thought, 'No, no, no, no, no. We have to do better than this. Let me go to university.' "

Richard's decision was validated by a second expedition he and Glynn led to Lake Natron in the summer of 1964. Louis had helped them secure a $14,000 National Geographic grant, enabling them to spend three months in the field. Although the hominid fossils eluded them this time, they did turn up some excellent mammal specimens and, to his surprise, Richard discovered that he actually enjoyed digging out fossils. "In many ways it was so much more satisfying than conducting tourists around East Africa's National Parks," he noted in his autobiography. Then, too, there was a pretty blond, blue-eyed archeologist along—Margaret Cropper. She had worked for Mary at Olduvai, and had discovered one of the hominid skulls there in 1963. She had also been Jonathan Leakey's girlfriend at one point. During the second Natron expedition, Margaret and Richard became quite close, and when she left for Scotland in September 1964 to complete her studies at the University of Edinburgh, Richard wondered what he "was going to do next." Suddenly going to England and studying for a degree seemed like an excellent idea to Richard. Five months later, he was on his way. "[F]or once I had made a decision about my future of which my father really approved and, moreover, I would be able to see Margaret."

Chapter 19

A GIRL FOR THE GORILLAS

Independence came to Kenya Colony precisely at midnight December 12, 1963. The British flag was hauled down for the last time, and the green, red, and black banner of the newly independent nation hoisted aloft. Beneath it, smiling broadly, stood Jomo Kenyatta, Kenya's first prime minister. Throughout 1963, as the colony approached independence, Louis had worried that his role in Kenyatta's trial "could easily [bring] serious hostilities against my family." By April, he had become so anxious that he even turned down an invitation from England's Royal Society to exhibit his fossils. But the reprisals white Kenyans so feared never took place. "I have no intention of retaliating or looking backwards," Kenyatta announced in August, three months after his election as president. "We are going to forget the past and look forward to the future."

Kenyatta's magnanimity, which helped ensure the young country's stability, extended on a personal level to Louis as well. When Malcolm MacDonald, Kenya Colony's last governor and a close adviser of Kenyatta's, suggested arranging a meeting between Kenyatta and Louis, both readily agreed. As a pretext, MacDonald told the prime minister that Louis wished to show him some of his Kenyan fossils. "I was at that meeting," recalled Charles Njonjo, who served as Kenya's first attorney general. "I think Louis was nervous, wondering what sort of reception he would get. He was brought in by [Peter Mbiyu] Koinange, the son of the chief, so that gave him some confidence. At first, it wasn't a very happy meeting; it wasn't warm. The old man [Kenyatta] was still holding back. But Kenyatta was wonderful; he was wise [and] big. And so they separated as friends." The two men, who had not seen each other since the trial, spoke together in Kikuyu, and finally Kenyatta gave Louis a hearty embrace.[1]

Louis and Mary thus started 1964 free of worries about their future in Kenya, and, because of their many Olduvai discoveries, assured of international acclaim. Yet the years ahead would not be especially happy for them as a couple. Their stones-and-bones partnership was breaking down, pulled apart by new demands and interests—and the fame *Zinj* had brought them. "It was inconceivable . . . that Louis and I should have regarded the finding of *Zinj* and *Homo habilis* as anything other than a triumph, a series of discoveries to be followed up regardless of the amount of effort required until every possible piece of information had been extracted," Mary later wrote in her autobiography. "That is what we set ourselves to do, and the cost included elements that we had not considered." Louis's time was increasingly spent traveling overseas giving fund-raising speeches, while Mary did most of the research. At first, this arrangement had seemed sensible; they had not foreseen that it would lead them down increasingly divergent paths. Or that the greatest cost of finding *Zinj* would be their marriage.

Had Olduvai remained their sole interest, Louis and Mary might have been spared the anguish of watching their partnership dissolve. But Louis was as intellectually restless at sixty-one as he had been at thirty, when his old nemesis Professor Percy Boswell had branded him a "honey bee, forever flitting from one [thing] to another." His curiosity remained insatiable. "Louis was totally boundless," observed Lita Osmundsen, the former director of the Wenner-Gren Foundation. "It was like the cosmos was his meat."

But his quest to "know all about everything in the world," as Mary Smith of the National Geographic Society, put it, also carried a price: Louis spread himself too thin. By 1964 he was directing the Centre for Prehistory and Palaeontology, which was affiliated with the museum; writing scientific articles and overseeing excavations in Kenya; supervising the Tigoni Primate Research Centre, a facility for studying captive monkeys that he had established near Nairobi in 1958; monitoring Jane Goodall's chimpanzee research; serving as a trustee of the East Africa Wildlife Society; and investigating topics as diverse as the medical properties of zebra fat (certain tribes in Kenya used it as a treatment for tuberculosis) and the prehistory of Israel.

Louis supported these activities almost solely with funds he had raised—or planned to raise—from sources outside of Kenya. Grants from the Ford and National Science foundations paid for his prehistory center (although

[1] In the same spirit of reconciliation, Kenyatta appointed as Kenya's acting chief justice the English judge who had turned down his last appeal, thus sending Kenyatta to prison. When MacDonald asked Kenyatta if it was the same man, Kenyatta replied, "Yes. . . . He's a good lawyer and he'll be a good acting Chief Justice."

the Ford Foundation's annual $7,000 grant would expire in June 1965); and a seven-year, $100,000 grant from the U.S. National Institutes of Health covered most of Tigoni's expenses. Louis himself continued to receive an annual salary of $7,840 from the National Geographic Society, as well as grants from the Society ($33,600 annually) for the excavations at Olduvai and Fort Ternan. But because Louis spent his money as quickly as it arrived—typically on some new scientific enterprise—the Leakeys were perpetually in desperate straits.

"I certainly do hope you can do something about cash," Mary wrote in an urgent letter to Louis in the spring of 1963 (though she noted that she could have written a similar letter anytime during the 1960s), when he was a visiting professor at the University of California at Riverside. "Things are pretty desperate." There was no money to pay her student assistants at Olduvai, or to buy more gasoline, or to pay the shop bills. The telephone bill and Philip's school fees were due as well. Mary was only getting by because Richard had given her a loan.

Louis sent her the needed money from his visiting professor's stipend. It was little more than a Band-Aid on a running wound: not long after receiving Mary's plea, Louis decided to launch yet another project, this time a dig for early humans at Calico Hills in California's Mojave Desert. A young archeologist named Ruth DeEtte ("Dee") Simpson had shown Louis the Calico site, about seventy-five miles from UC Riverside. Simpson had actually met Louis in 1959, when she traveled to London to show him the artifacts she had found at Calico. Louis was then doing research at the British Museum of Natural History, and his secretary allotted Simpson a five-minute meeting. It lasted, however, for four hours. "All the other appointments were canceled," Simpson recalled. "Then he came over to the hotel and had dinner with my mother and me, and stayed until two in the morning talking about how he would like to become involved if this were really an early site in America."

Simpson's story and artifacts intrigued Louis. In Pleistocene times, some one hundred thousand years ago, Calico had been a lush land, its snow-fed mountain streams spilling into Lake Manix, its meadows teeming with game and fowl. Now it was harsh and barren. Only the primitive-looking stone scrapers that Simpson collected along Calico's lower slopes hinted at its past. After examining these artifacts and tramping across the desert hills, Louis came away convinced that "if there was a place where we could find evidence of Upper Pleistocene . . . man [in America], this was it!"[2] Archeologists then generally believed that people first arrived in the Americas only 12,000 years ago. They based the date on the 1927 discovery of a stone arrowhead embedded in the rib cage of an extinct species of bison that was

[2] Louis was cagey about giving a precise date for the Calico site. In public, he talked in terms of 15,000 to 50,000 years. But in private with Dee Simpson he would say, "You know, Dee, man has been here 100,000 years, so you might as well think along those lines."

found in Folsom, New Mexico. But Louis had argued for an earlier date since his days as a lecturer at Cambridge University and was determined to confirm one.[3]

Many archeologists would come to regard Calico as Louis's downfall, his ultimate folly. But when Louis decided to explore the Calico Hills, he was still riding high on the fossil bounty from Olduvai, and at the peak of his career. With the Olduvai fossils, he had proved what he had argued since he was a young anthropology student at Cambridge: that humankind had evolved in Africa. The textbooks had been rewritten, as he had vowed they would be, and the professor who had told him to search in Asia had been shown to be wrong. By 1963, even as he was savoring this vindication, Louis was searching for another challenge. Calico gave him, at age sixty-one, a fresh new fight.

Though Louis did manage to get a National Geographic grant to investigate Calico, he was also scrambling to keep funds coming into Tigoni, his primate research center. As with so many of his scientific projects, Louis had launched Tigoni with a woman—in this case, the Cambridge-trained and newly widowed primatologist Cynthia Booth. A loud, brassy blonde who liked, said Tigoni worker Frances Burton, "bawdiness and 'macho,' " Booth moved from Ghana to Kenya in 1958 at Louis's suggestion. Booth had purchased a farm north of Nairobi, and there she and Louis set up a laboratory for raising monkeys, using modest grants from the Chicago Field Museum and Nairobi's Coryndon Museum—as well as, to Mary's annoyance, money from his own pocket. Louis envisioned Tigoni as primarily a conservation facility, where monkeys would be bred for the medical trade, thereby alleviating the need to trap wild monkeys. But he also hoped it would be used for taxonomic and behavioral studies, a place where students could be trained before studying primates in the wild. While the facilities impressed visiting primatologists, very little was actually accomplished at Tigoni, and in spite of a generous NIH grant from the United States, the center was always desperate for money. "[Tigoni] was a disaster," Mary said, "and Louis used to agonize over it, and make telephone calls to [his supporters in] California whenever the money ran out. And nothing was being achieved. I thought he was wasting his time."

By the time Louis got involved with Calico, his pattern had become all too familiar to Mary: a new scientific scheme, another woman (although, as in the case of Dee Simpson, these were not always romantic attachments), and

[3] In 1929 Louis told the archeological students he was tutoring at Cambridge University that "circumstantial evidence" alone indicated that the New World must have been settled at least 15,000 years ago. After word spread among Louis's professors about his radical lectures, he received a surprise visit from the eminent American anthropologist (and staunch supporter of the 12,000-year date), Dr. Ales Hrdlička. "He burst into my rooms [at St. John's]," Louis told an audience forty years later. "[He] didn't even wait to shake hands—and said, 'Leakey, what's this I hear? Are you preaching heresy?' " Though taken aback, Louis said no, and explained his position, arguing that the distribution of Native Americans and the diversity of their languages required more than "the few thousands of years that you . . . allow. And I shall go on saying that without any evidence; it must be so!"

a financial mess. "It couldn't have been easy being married to Louis Leakey," said Lita Osmundsen. "Here you have a man of passion, flair, and insight, and then you have Mary, a woman who is a thoroughgoing scientist. She was the spine in the operation. For someone as dedicated, principled, and duty-bound as she is, having to get up and cook breakfast every morning for Pegasus would have been pretty devastating."

For Mary, Calico was the last straw. She was suspicious of the site from the beginning, particularly when she learned that Simpson was excavating an alluvial fan—deposits laid down by floods and mudflows. Though Louis had selected the site himself, hammering metal spikes in the earth and announcing to an amazed Simpson (who was overwhelmed by Louis's intuition), "Dee, dig here," Mary was not impressed. "I tried to tell him I thought he was mistaken," Mary said, "but then you couldn't mention it to him. At least, I couldn't." Where once she had felt unquestioning admiration for Louis, Mary now began to feel scorn, and sometimes she openly chastised Louis for what she regarded as his follies. Phillip Tobias, who came regularly to Nairobi to work on the Olduvai fossils, always dined with the Leakeys on his last night in town, and Louis would always bring up some new theory he had saved for the last moment.

> He might say, "Oh, Phillip, I want to hear your views on my new classification of the primates." And then he would say, "Well, we have to have a family for this, and a family for that." But before I had a chance to react to him, Mary would give him a withering look and say, "Really, Louis! Do you really believe anyone is going to believe that? I don't think you'll sell that idea to a soul." And Louis would shut up and never say another word about his new classification.

The Africans who worked for the Leakeys at Olduvai and the Centre for Prehistory and Palaeontology noticed the same lack of respect. "We knew Leakey was a big man," recalled Kamoya Kimeu, "somebody who was known everywhere, who could see anybody, any other big man, like Kenyatta. But when you would go home to his house, you would find he was nothing there. He himself was nothing in his own home. But outside, everybody was listening to him."

Louis reacted to Mary's criticism by spending more time overseas. In America the exposure he had received from the *National Geographic*—between 1960 and 1965, the magazine published four articles by or about him, and in 1966 aired a television special, "Dr. Leakey and the Dawn of Man"—had made him a celebrity. Twice a year he traversed the country, speaking to adoring, sold-out crowds from Washington, D.C., to Chicago, Salt Lake City, and Los Angeles. Louis relished this attention, something he had never received in England. He was invited to the University of Illinois and Cornell University as a visiting professor, given honorary doctorates and anthropology's highest awards, and interviewed on radio and television talk shows. "He loved to be Louis Leakey," said his son Richard, "and he loved to be recognized, and to stimulate people by talking about what he'd done and who he was."

But Louis was also racing the clock. He was now nearly sixty-three, overweight, and suffering from painful arthritis in his hips. "I think Louis had a tremendous sense of his own mortality," said Mary Smith. "And that's what made him want to be everything to everybody and do everything. He knew he had a finite amount of time and it was running out."

Mary did not disapprove of all of Louis's new projects—and, in fact, was involved herself with the excavations at Ubeidiya, a site in Israel's Jordan Valley, where stone tools and fossils reminiscent of those from Olduvai had been found by Israeli archeologists in 1959.[4] Mary visited the site on Louis's behalf in 1963 to evaluate the artifacts, and came away deeply impressed. The stone tools and fossils were unquestionably genuine—and this time, the scientists were men. "[Louis] has been as much impressed as I was," Mary wrote to Phillip Tobias, after Louis made his own visit to Ubeidiya in 1965, "and is now fully charged with enthusiasm to get funds for Steklis [the archeologist in charge of the excavation]. I hope he succeeds."

Mary also supported Louis's plans for building new laboratories at his Centre for Prehistory and Palaeontology (which he did with the aid of Wenner-Gren and National Geographic grants), and for starting a department of osteology, which would house the skeletons of modern animals for comparison with the fossilized forms. A collection of this type did not then exist in Africa, and the Wenner-Gren Foundation readily gave Louis $20,000 as seed money. She was less enthusiastic about Louis's primate studies ("I don't think they can really tell you anything about how the early hominids behaved," she observed), but could only look on in dismay when in the spring of 1966 he announced a long-term study of the mountain gorilla.

By now, Louis's first "ape lady," Jane Goodall, had become an international success, with stories in the *National Geographic,* and a film and popular book under way. In 1960, when she had first gone out to Tanganyika to begin watching the chimpanzees, Jane and Louis had thought she would also undertake the gorilla study, as well as studies of the pygmy chimpanzees (bonobos) and orangutans. "I don't want to spend ten years here [at Gombe] myself," she had told Louis in 1963, "because meanwhile the orang and the pigmy chimp [will] become extinct . . . and anyway, I shall be too old to rush about in Borneo's humidity or the Congo swamps, to do new field studies." But three years later her plans had changed. She had completed her doctor-

[4] Dated at 1.2 million years, these artifacts represent some of the earliest evidence of human culture outside of Africa, and may represent one of our ancestors' migration routes out of Africa to Asia and Europe.

ate at Cambridge University, and had married the wildlife cinematographer Baron Hugo van Lawick, whose work often took him away from Gombe and the chimps.[5] Jane and Hugo lived half the year at Gombe, where Jane had established the Gombe Stream Research Center with funding from the National Geographic Society (and plenty of advice from Louis), and the other half filming various animals on the Serengeti. Happy with this arrangement, Jane also realized that she was not ready to leave the chimpanzees and embark on another primate study. There was still too much to learn at Gombe, particularly about the close ties between chimpanzee mothers and infants—the bonds, Jane had discovered, that held the chimps' society together. Wherever else she might go, Gombe Stream was now her home.

With Jane's success behind him, Louis was very receptive to the tall, striking, dark-eyed young woman who nervously approached him in March 1966 in Louisville, Kentucky, after one of his lectures. Her name was Dian Fossey, and she had actually met Louis three years before at Olduvai Gorge, where she had stopped briefly en route to the Congo. An animal lover since childhood, Dian had borrowed $5,000 to finance a seven-week African safari, and had set out hoping most of all to meet the Leakeys and to see the mountain gorillas. She had accomplished both goals, although with some difficulty: touring Olduvai, she had badly sprained an ankle, and vomited all over a "treasured fossil." Dian's subsequent trek to the gorillas was a painful ordeal, but hobbling along with the aid of a stick, she succeeded in climbing to their mountain home. She spent three days searching for the animals, and on her last day, was rewarded with a breathtaking glimpse of six gorillas huddled together in the dense greenery of the forest. "Surely, God, these are my kin," one of her Congolese guides whispered, voicing a sentiment Dian shared. There and then, Dian decided that somehow she would return to learn more about the gorillas.

At Olduvai, before her accident, Dian had talked to Louis about her hopes of seeing the gorillas, and he had responded enthusiastically. Now, three years later in Kentucky, as she shyly stepped forward to shake his hand, Louis looked at her intently, his eyes narrowing in a squint. Then he said, to her amazement, "Miss Fossey, isn't it? Please wait until I've finished with all these people." Once free of his well-wishers, Louis plied her with questions. Had her ankle healed? Had she seen the gorillas? What was she doing now, and what were her plans for the future? Louis listened carefully to her

[5] A romantic, Louis enjoyed arranging matches for his protégés—including that between Goodall and van Lawick. Van Lawick was a neophyte cinematographer when the National Geographic Society sent him to Gombe, at Louis's suggestion, to film Jane's work. When discussing the assignment with Louis, van Lawick suddenly realized that Louis also had other things in mind. "I don't think he meant for me to realize what was happening," van Lawick said later. "But he made some sort of remark, and I turned around and just caught a smile on his face. And then I began to wonder, 'Hey, wait a minute.'" Neither Jane nor van Lawick objected to Louis's marital engineering ("I didn't have to go along with his fun, did I?" said Jane), and at their wedding thanked him for making it possible. They have since divorced, although they remain friends.

replies, then asked her to come see him the next morning at his hotel. He thought she might be the right person to undertake the study of the mountain gorilla.

Louis interviewed Dian at the Stouffer's Louisville Inn, scratching his impressions of her on a small hotel notepad. He printed her name and address across the top, then her age—thirty-three, looped with a circle—noted that she had been to college, that she loved animals, and didn't mind being alone. She had no fear, she told him, and then said, "That is where I belong." Louis apparently agreed, for at the end of the hour-long interview (during which, Dian noted in her journal, "Dr. Leakey did most of the talking"), he offered her the job. He promised to raise the money to get her started, as well as future fame and fortune, and advised her that since she would be living in an extremely remote region, it would be wise to have her appendix removed.[6] That night, Dian telephoned her mother in California. "Mother! Mother! Dr. Leakey was just here. And he wants me to come study the gorillas. He wants me to do it!" Though her mother was horrified at this idea, she also knew that it was impossible to dissuade her daughter. But unlike Jane Goodall's mother, Kitty Price (who had divorced Dian's father and remarried) never gave the study her blessing.

Louis returned to Kenya at the end of April 1966, and was soon making inquiries about visas, study permits, and funding on Dian's behalf. Though Dian, like Jane, was untrained in animal behavior studies—she had been working as a physical therapist when Louis met her—Jane's success gave Louis a measure of credibility. "I have at last found a girl whom [*sic*] I think is another Jane Goodall," he told Robert Hinde, the zoologist at Cambridge University who had overseen Jane's doctoral studies. "I am hoping to send [her] to continue the study of the Mountain Gorilla which George Schaller started so successfully, but never carried through to its proper end." In other letters, he labeled Dian a "winner," an "indubitably suitable candidate" who was "determined to do something worthwhile with the gorillas." He wanted to get her "launched . . . as soon as possible," and only nine months after interviewing her in Kentucky, he did. Grants came from the Wilkie Brothers Foundation, the National Geographic Society, and the New York Zoological Society, and Cambridge University agreed to consider Dian for its doctoral program, as it had done with Jane.

On December 22, 1966, Dian—*sans* her appendix—arrived in Nairobi. Louis quickly took her in hand, introducing her to his staff and helping her buy last-minute items and a battered, canvas-topped Land Rover for her trip to the Congo. He also invited her to his home for lunch, where Mary looked up long enough to say, "So, you're the girl who is going to out-Schaller

[6] In the 1930s Louis had nearly lost Mary and his friend Donald MacInnes to appendicitis attacks, so his advice was not as whimsical as it might seem. He had also recommended an appendectomy to Jane Goodall, but she laughed at the idea.

Schaller, are you." "It was a statement, not a question," Dian would later note, and an "intimidating thought" to carry with her.

Two weeks later, Dian and the wildlife cinematographer Alan Root headed south to the Congo and the Parc des Virungas, where she planned to set up her camp. Root and his wife, Joan, had befriended Dian on her previous trip to the gorillas (indeed, it was because of the Roots' guide that she first saw the animals), and though he considered her and Louis's plan dubious, he offered his help. En route, they stopped at the Traveler's Rest to see Walter Baumgartel, the hotelier who had first interested Louis in the gorillas. He was pleased that Dian intended to pursue his favorite subject, but worried about her timing. Kivu Province, the region in which she would be working, was in a state of unrest, and he warned her that no one could predict what the military would do. Dian shrugged off his concerns; politics did not interest her. Besides, the park authorities had welcomed her study.

Three days later, on January 6, 1967, Dian, Root, and several dozen porters began the long trek into the Virungas. They reached her campsite, Kabara meadow, in the late afternoon. Here, at eight thousand feet, in a grassy glade surrounded by mossy, gnarled *Hagenia* trees, Dian and Root built what they thought was her permanent camp. Sometimes while they worked, they heard the far-off hoots of gorillas echoing in the forest, but during the two days Root was able to spend at Kabara, they never saw the animals. In what little time he had, Root also gave Dian a short course in tracking, a skill she lacked entirely. Once after finding relatively fresh gorilla tracks, Dian promptly headed down the trail, confident that Root was behind her. But she soon realized that she was all alone, and puzzled, she retraced her steps. "[I] found him patiently sitting at the very point where we had first encountered the trail. With the utmost British tolerance and politeness, Alan said, 'Dian, if you are ever going to contact gorillas, you must follow their tracks to where they are going rather than backtrack trails to where they've been.' "

Root's departure left Dian alone in the forest, except for the company of her camp staff, cook, and gorilla tracker, Sanweke.[7] None of the men spoke English, Dian's newly acquired Swahili was rough, and she was suddenly overwhelmed with loneliness. Watching Root disappear among the Hagenia trees, Dian found herself clinging to her "tent pole simply to avoid running after him." But Dian was, as Louis had quickly perceived, a determined young woman. She forced herself to forget her loneliness, and set about her business of tracking and habituating the mountain gorilla.

Within a week, Dian and Sanweke made contact with a family of nine gorillas, and by the end of the month, she could write confidently to Louis, "To date I have twenty-three hours and seventeen minutes from nine contacts with two 'families' . . . of gorillas." She had discovered that they were

[7] Sanweke had tracked gorillas for Carl Akeley, George Schaller, and the Roots.

"more curious than apprehensive" about her, and particularly enjoyed watching her climb trees, "peel bark and eat leaves." She did not tell Louis that she had been charged several times—something he had warned her would happen—or that she had not always succeeded in holding her ground, as he had insisted she must. After taking several "nosedive[s]" into the surrounding foliage when screaming, yellow-fanged gorillas rushed full speed at her, Dian overcame her fear, though "only by clinging to the surrounding vegetation for dear life." In time, she learned to pacify the animals by mimicking their sounds of contentment, nibbling at wild celery stalks, and staying crouched in a submissive posture.

Louis was elated by Dian's early success. "This is first class and I congratulate you," he wrote after receiving her first letter. He reminded her, too, that she must have patience during the inevitable "blank periods" and cautioned her about a poacher she had mentioned seeing near Kabara. "[D]o not get into difficulty with this man; he could make things very difficult for you and it worries me a little," he wrote presciently.

Louis had sent his reply to Dian from California, where he was lecturing at fifteen different junior colleges in the space of two weeks. He had already been on the road for one month, having left Kenya in mid-January 1967. During that time, he had visited Ubeidiya in Israel, accepted an honorary fellowship from his old college of St. John's at Cambridge, and presented talks everywhere from London and Paris to New York and Boulder, Colorado. Between January 23 and March 3, he gave twenty-three lectures, a very full schedule for a man nearly sixty-four. But because he typically spoke for five hundred dollars or less, Louis booked every engagement he could find.

"We'd watch this man come over, hat in hand, to get money to carry out his work," recalled Helen O'Brien, whose husband, Allen, a California entrepreneur, helped Louis launch his own foundation in the United States in the late 1960s. "And he was only just number one in the whole wide world in his subject! Why a great scientist like that would have to come over and beg was beyond us."

Louis's frenetic schedule took its toll. On March 3 he gave his last talk in California at the University of California, Riverside, and also squeezed in a visit to Calico. He then flew to Chicago, where that city's Academy of Sciences was expecting him as their featured speaker the next evening. But shortly after his plane landed at O'Hare International Airport, Louis collapsed. Although the newspapers initially reported that he had suffered a heart attack, Louis insisted that he "had simply collapsed out of fatigue." Then, too, his exhausting itinerary may have triggered an epileptic fit, although he had not suffered from these in many years.

Mary had accompanied Louis on part of his travels and was waiting for him at his next destination in Arizona when he passed out in Chicago. She and Louis had intended to explore Arizona's San Pedro Valley with the geologist Vance Haynes as part of Louis's interest in the peopling of the Americas. "I'd told Louis that we had a New World equivalent of Olduvai Gorge at San Pedro," Haynes recalled. "It has the same [fossilized] fauna and terraces like Olduvai, but no one had ever found any stone tools there. Louis said that was just because we weren't looking in the right places, so he was going to come here and find the artifacts."[8] Haynes's description of San Pedro had led Louis to believe he could reenact his 1931 discovery of artifacts in Olduvai Gorge. But neither Louis nor Mary visited San Pedro. Instead, she left immediately for Chicago, where she found a tired and grumpy Louis. Worse than the collapse, in his opinion, was the way the ambulance attendants had manhandled him onto the stretcher: now he had a wrenched back.

Injured back, epileptic fit or heart attack—none of these afflictions really worried Louis. As soon as he returned to Nairobi, and in spite of Mary's pleas, he was back to his old schedule. He had to catch up on his correspondence with Dian, who was reporting wonderful news of her work with the gorillas, and then there was a whole new expedition to arrange, this time to Ethiopia. Bed rest was out of the question.

[8] Louis never did find time to visit San Pedro Valley. To date, no one has ever found artifacts with the Pleistocene animal fossils in the valley.

Chapter 20

TO THE OMO

Until the Leakeys' 1959 discovery of *Zinjanthropus,* few paleoanthropologists had considered searching for early human fossils in East Africa; by 1967, few considered looking elsewhere. East Africa's sediments—fossil-rich, abundant, datable, and largely unexplored—were now prime hominid hunting terrain. Beginning in 1960, when the National Geographic Society funded Louis and Mary's Olduvai excavations, and continuing to the present day, hardly a year has passed that has not seen an expedition head off into the East African bush in search of hominid treasure. "This was an unprecedented, absolutely unique time in the whole story of paleoanthropology," said French paleontologist Yves Coppens, who joined the East African bone rush in 1967. "There was work, money, and people—five hundred people working several months every year for more than twenty years along 2,000 kilometers of the Rift Valley—because the Leakeys found *Zinj.*"

Almost all of these expeditions involved the Leakeys. Louis had discovered many of the prime fossil localities in Kenya and Tanzania, giving him—according to scientific custom—first exploration rights; he directed the Centre for Prehistory and Palaeontology, and recommended visiting scientists for Kenya government research permits; and he controlled the Centre's fossil collections, deciding which scientists could examine the material. Few of his colleagues wielded such power, and many came to regard East Africa, and Kenya in particular, as a Leakey fiefdom. "The Leakey family believes that all the archeological and paleontological materials in Kenya and Tanzania are theirs," said the geologist Basil Cooke, who first encountered this attitude in 1948 when he came to Kenya with an expedition led by the brash American entrepreneur Wendell Phillips. "It is their preserve," Cooke said, "and anyone else who wants to come in has to get their approval, their cooperation, and their permission." In fact, most visiting scientists depended on Louis's assistance simply to get their expeditions under way. Clark Howell remembers relying heavily on Louis in 1957 to arrange his excavation at Isimila, Tanganyika. "Louis helped with everything—the purchase of vehicles, what we should take, petrol, permits, insurance, licenses, all of it. He knew about these things and I didn't."

But for all his generosity, Louis could also be petty and small-minded, turning secretive about sites he had worked even with scientists he employed. When Merrick Posnansky, a British archeologist, came to Kenya as warden of the country's prehistoric sites in 1956, he almost immediately ran afoul of Louis. One of his first tasks entailed creating a museum-on-the-spot for the handaxe site of Olorgesailie, which Louis and Mary had discovered in 1942. Though the Leakeys were busy exploring Olduvai and Rusinga Island, Louis frowned at the idea of Posnansky initiating a dig of his own at Olorgesailie. Louis was further annoyed when the young archeologist started an excavation near Hyrax Hill (which Mary had worked during the war years), even though Posnansky had Mary's blessing. "Basically, Louis didn't like anyone who competed with him on the East African stage," Posnansky said. "He was then still desperately seeking an important hominid . . . and was reluctant to let go" of any sites where such fossils might be found.

Stories circulated, too, about Louis deliberately sending expeditions on wild-goose chases. "He cheated Bryan Patterson [a Harvard University paleontologist] in this manner," said Ernst Mayr, the eminent Harvard zoologist who helped arrange Patterson's 1963 expedition to Kenya. "Patterson went to three sites [that Louis recommended], and found nothing. He was very surprised and inquired around Kenya and was told—and this is nothing but hearsay—that Louis had been to all these sites and had found there was nothing, and that was why he sent Patterson to them, to paralyze his expedition."

From letters and records at the National Museums of Kenya, it is clear that Louis had acted in good faith with Patterson. Louis had only made short visits to the three areas and had hoped to return one day for further exploration; instead, he offered them to Patterson, who had the funds and equipment necessary for exploring such remote regions. But the rumors that Louis kept the best sites for himself persisted, and led to his being widely regarded "as the dictator of East Africa," according to Mayr. "It was the way he acted, telling people where they could go [to search for fossils] in Kenya."

Louis's reputation as the godfather of East African fossil sites was further enhanced in 1965 when Ethiopia's Emperor Haile Selassie I asked him to organize an expedition to Ethiopia. The two men—Louis, white-maned and charming, the Emperor, diminutive and imperious—met during a State House luncheon in Nairobi hosted by Jomo Kenyatta in the fall of 1965. "In the course of our luncheon, Haile Selassie said, 'Dr. Leakey, why has my country got no fossils like you find in Tanzania and in Kenya?' " Louis recalled at a National Geographic Society lecture four years after this meeting. "Of course, that was my opening, and I said, 'Well, your Royal Highness, if you would allow us to go and search in your country, I know where we might find something.' [He said,] 'Why haven't you been there?' 'Well, it's always been difficult. Your government has not given us the facilities.' 'All right,' he said, 'I'll arrange it.' "

Louis specifically suggested that an expedition be sent to the southern reaches of Ethiopia, along the Omo River, where the French paleontologist Camille Arambourg had worked in 1934.[1] From Arambourg's collections—as well as fossils that Louis's assistant Heselon Mukiri had picked up while running guns to Ethiopia during World War II—Louis knew that the region held fossil deposits dating to late Pleistocene times. Clark Howell had also spent three months searching the Omo deposits in 1959, but had run into trouble with Ethiopian border guards who demanded that he leave his fossils in their care (the fossils were never seen again). In deference to Arambourg's and Howell's earlier work, Louis asked to include both men on his new expedition. Three months later, on February 10, 1966, Louis was summoned to the Ethiopian Embassy in Nairobi. There the ambassador read aloud a letter from the emperor, requesting Louis to lead an international expedition, comprising Kenyans, Ethiopians, French, and Americans, to the Omo River. The expedition would depart in June of 1967, giving Louis somewhat more than a year to make arrangements.

Louis wasted no time getting started. He sent an "OMO OK" cable to Howell (who was eager to return to Ethiopia) and a letter explaining the plans in detail to Arambourg. He then asked the National Geographic Society for a grant, telling Leonard Carmichael, the chairman of the research committee, "I am of the strongest possible opinion that the Omo Valley will yield fossil remains of Early Man and his cultures of the highest possible importance." Plans were made, too, for an American geologist, Francis Brown (who was then a graduate student at the University of California at Berkeley and whom Louis referred to as "Young Frank Brown"), to spend the fall making an initial survey of the Omo deposits. Louis and Mary planned a short, preliminary visit to the area at the same time, together with their son Richard and his wife of three months, Margaret. As much as Louis loved fieldwork, he could no longer walk the long distances fossil hunting required—the arthritis in his hips was simply too painful. Instead, Louis decided, Richard would lead the Kenyan contingent of the Omo Expedition, assuring Louis of some control should there be any important finds. What Louis did not reckon on was the drive and hunger for independence of his intensely ambitious son.

---

[1] The Omo deposits had been discovered in 1902 by a French expedition led by Comte de Boaz. Boaz died on the expedition, but a colleague brought some of the fossil specimens to Paris, where they later caught Arambourg's eye.

• • •

In the spring of 1966, when Louis received the emperor's approval to proceed with the Omo Expedition, Richard was twenty-two. He had spent six months of the previous year in England, studying biology, chemistry, and physics, essentially completing his high school education—and pursuing Margaret Cropper, the young woman he had fallen for in Kenya. She was finishing a degree in archeology at the University of Edinburgh, and was flattered by Richard's attention and the idea that the two of them might form a team like Louis and Mary. "Not at those exalted levels—or I didn't see it at that level anyway—," Margaret said, "but the same idea, yes." Although five years older than Richard, Margaret was taken with his sense of "always [being] very much in control, in command of a situation. I think it was probably like men in uniform having some kind of special appeal," she said about her initial attraction to Richard. "He fit the image of a big white hunter, not that he was ever a hunter, but he had that glamour. I was young and impressionable, and I wasn't worldly wise enough—I was infatuated."

In England, although Richard's bush glamour paled somewhat—"He was definitely the 'country boy come to town,' though he never would have admitted it," Margaret said—the two saw each other as often as possible. At the same time, Richard was working frantically at his studies, determined to complete his last two years of high school in six months. "I had lots of catching up to do," Richard said about this period of his life. "I was fairly impulsive, fairly insensitive. I think I could perhaps be best described as an angry young man. I didn't believe that anybody but myself knew what was going on in the world. I had lots to do in a hurry."

In quick succession, he finished his studies, passed his university entrance exams (although his scores were not high enough to qualify for Louis's alma mater, Cambridge University), and proposed to Margaret. But instead of applying to other universities, as he originally had planned, he returned to Kenya in January 1966. Soon he was married to Margaret, and busy directing an excavation at Kenya's Lake Baringo (Margaret provided the "academic guidance," and Louis the funds), acting as his father's deputy at the Centre for Prehistory and Palaeontology, and running photographic safaris. He had also purchased nine acres in the forested Nairobi suburb of Karen, where he built a home. And then there was the Omo Expedition to help organize. The idea of returning to England and becoming a student once again when he was already living at full tilt seemed dull and grey. "I thought, 'Ah, that's enough books for me for a lifetime. And no, I'm not going to apply.' So I didn't go back," Richard said.

Because of his position at his father's Centre, Richard knew he could search for fossils even without a degree. Nor was he the least bit bashful about trading on his family name or using his father's connections—he had already done both to launch his safari business. "Richard didn't see anything wrong in using what he had, which happened to be an 'in' in the right place," said Margaret, who was appalled by what she considered Richard's arrogance. "His attitude was that anybody would be a fool not to use [the family name]. He'd say, 'Well, ordinary people have to bother about degrees; I don't need to. I'm here and I can jump on my father's name and do it anyway.' " His father's name might give him a start, but without a degree he would be vulnerable to the charge that he was nothing more than an amateur fossil hunter, someone lacking the theoretical background to interpret what he found.

But no one could make Richard change his mind. Not Margaret, who criticized him for thinking "he was as good as his father"; not Clark Howell, who offered to help him get into the University of California at Berkeley; and certainly not his parents, who knew it would be futile to disagree with their headstrong son.

At the same time, Louis and Mary's youngest son, Philip, also announced his intention to quit school. His parents had hoped that by sending him to a secondary school in England they might change his mind. But Philip had other ideas. "Because I was Kenyan, the British wouldn't let me into their country without a return ticket," he said. "So at the end of the first term, I just got on the plane and came home. And that really upset everybody. It was like coming into winter in Nairobi; the chill had set in in our relationship. And I was given an ultimatum by my parents: either I went back to school, or I got out of their house." Philip was fifteen and one-half at the time, but did not hesitate in his decision. He chose the latter option, took a tent, and moved south of Nairobi to a plot of land on the Mbagathe River, where he staked a mining claim. He built a home there, where he and his family live today.

Of Louis and Mary's three sons, Richard had always been "the ornery one," Mary said, often "getting into rages" when he could not get his way. If his temper tantrums did not work, he devised other means to do what he wanted. Once, on a family expedition, Louis insisted that Richard eat his serving of kedgeree—a mixture of rice, sardines, and hard-boiled eggs, which Richard loathed. He grudgingly complied, licking up every last bit, but then pushed his fingers as far down his throat as he possibly could, "with predictable results," Mary noted. Richard was never forced to eat his kedgeree again. Richard evidently had the same appetite for school, and his parents wisely decided to avoid the subject.

The lack of a university education did not stop Richard from having grand plans—or assuming a larger role at his father's Centre for Prehistory and Palaeontology than Louis intended. Whenever Louis traveled overseas, he turned his office over to Richard, along with written instructions about how to handle the Centre's daily affairs. Often, these involved nothing more than dealing with the staff and accounts, but sometimes other, larger matters were pending. Despite his father's wishes, Richard, who had "always been his own boy and was now his own man," according to Clark Howell, handled these as he saw fit. Inevitably, there were arguments between father and son.

"The first misunderstanding I had with Louis was over the return of Tanzanian [fossil] material," said Richard. "I thought we didn't need it, and it was blocking our storage space." Tanzania had recently opened a new National Museum in Dar es Salaam, and the museum director was eager to have some fossils from Olduvai—the country's most famous prehistoric site—in its collections. Louis and Mary had already returned Tanzania's most famous fossil, *Zinjanthropus,* presenting it to President Julius Nyerere in a public ceremony at Dar es Salaam on January 26, 1965. Phillip Tobias had completed his study of the skull and the Tanzanians had constructed a special air-conditioned vault for its safekeeping, so Louis had no quarrel with its return. But the Olduvai animal skulls and skeletal material were another matter. Louis still hoped to write a monograph about the ancient Olduvai fauna, and was reluctant to part with any of his vast collection, even surface finds.

Louis had intended to handle the negotiations with the Tanzanians in a meeting scheduled for March 31, 1967. But after he collapsed from fatigue at the Chicago airport early that month, he delegated Richard to go in his place, expecting his son to "go and listen, but not speak," said Richard. To Louis's dismay, Richard not only spoke but agreed that some of the Olduvai material should be returned. "It was material that had been found prior to 1959, prior to my mother's rather meticulous work, and material that had been described. It was also in the way, we had nowhere to store it, it was in boxes, and the sooner it got out of Nairobi the better, as far as I was concerned," said Richard. Richard also believed that the material rightfully belonged in Tanzania, and that by demonstrating "good faith" on this issue, his parents would have an easier time retaining their Tanzanian research permits.[2]

None of these arguments appeased Louis, who was furious with his son. The two had a "tremendous row," Richard said, and Louis attempted to break the agreement, even appealing to Melvin Payne at the National Geographic Society. "Richard has obviously fallen for [the Tanzanians'] arguments," Louis wrote, "but he has not had a great deal of background of administrative problems. I fear that I see the whole plan as the first stage of getting us out of Olduvai altogether." In the end, there was nothing Louis could do to alter what his son had done. "Father found it extremely difficult to accept that the decision had been taken without him," said Richard, "but thirty cases of fossils went down [to Dar es Salaam]. And that really marked the beginning of the difficult phase [between us]."

[2] Prior to Tanzania's independence in 1961, Louis and Mary received their research permits from the Tanganyika Geological Survey. These were automatically renewed, and the Leakeys could come and go as they pleased. They also had no difficulty bringing their Kenyan employees with them. But after independence, the Tanzanian government decided to exercise more control over Olduvai. The Leakeys were required to file annually for permits, and were questioned extensively about their Kenyan work force. They resolved this latter issue by training Tanzanians to work as guides at a small museum built on the edge of the gorge. They reached an agreement about permits as well, enabling Mary to work unimpeded at Olduvai until her retirement in 1984.

Father and son quarreled, too, when Richard asked his wife, Margaret, to make an inventory of all the archeological and paleontological collections at the Centre, something that had never been done. Indeed, Louis's record-keeping was so poor that in some instances there was no indication of where the material had come from. Once again Louis resisted Richard's plan. He was angry at the "implied criticism of his earlier work," said Richard, and Margaret's inventory was dropped.

These and similar quarrels led Louis to ask Richard to find a new job. Richard did not have far to look. Immediately adjacent to the Centre stood the National Museum (later Museum*s*) of Kenya, as the Coryndon Natural History Museum had been renamed following Kenya's independence. If the Centre wasn't being properly run, the museum was even in worse shape, at least in Richard's eyes. It was still staffed by British expatriates, many of whom acted as if the colonial era had not yet ended—an attitude that irritated the Kenyan government.[3] "There was a strong feeling among senior government officials that something ought to be done about getting the museum into Kenyan hands," said Richard, who had become friends with several civil servants through his work at the Centre. "So I decided I wanted to have some involvement at the museum, and they supported me."

Early in 1967, Richard formed the Kenya Museum Associates, an organization that included several prominent Kenyan leaders and that offered to help raise money for the poorly funded National Museum. "It was a *tiny* museum in those years," said Richard. "It had something like a £15,000 budget. And we were offering an additional £5,000. Once they started biting at the bait, it was a question of saying, 'Well, you know you really can't do that unless we have some involvement.' " The museum's curator, Robert Carcasson (who had replaced Louis in 1961 when the Centre for Prehistory and Palaeontology was established), responded by giving Richard an observer's seat on the museum's board of directors. Carcasson did not realize it, but in so doing, he had let in the Kenyan equivalent of the Trojan Horse. By June 1967, as he was readying the Kenyan contingent of the Omo Expedition, Richard already had Carcasson in his sights.

Few places on earth are as remote as Ethiopia's Omo Valley. Tucked into the southwestern corner of the country, and bordered by the Sudan to the west and Kenya to the south, the valley is a semidesert, drought-ridden land, peopled by the warring, nomadic Mursi and Sirma tribes. "[Their] reputation as thieves and murderers is . . . very bad," the game warden for the Omo Valley, G. H. H. Brown, counseled Richard in a letter a few days before the

[3] One of the most striking incidents involved a young Kenyan technician Richard had hired at the Centre who stopped at the Museum one day to use the lavatory. But the curator refused to allow him to do so, insisting that the facilities were only available for the senior white staff. Richard was furious, but turned the incident to his advantage, using it, he said, as "ammunition in [his] fight with the Museum management."

Omo Expedition departed. "I think this is exaggeration; but . . . they are all (a) very suspicious of strangers and (b) equipped with firearms. . . . This bit of Africa," he added as a final warning, "has never been governed or administered by anyone at all and primitive reactions prevail." The Omo Expedition could not rely on local food or gasoline supplies but would have to be entirely self-contained. And if something went wrong, they would either have to alert Louis in Nairobi, five hundred miles to the south, by radiotelephone, or hope that someone in the little Ethiopian garrison town of Kalam could help.

None of this worried Richard, who welcomed the chance to test himself in the African bush. "If ever I had imagined myself romantically struggling to survive in tropical Africa, here was my chance to do it for real," he wrote in *One Life,* his 1983 autobiography.

Richard had spent the better part of a year preparing for the expedition. Although he was leading only the Kenyan portion, he had volunteered his help as general quartermaster to the French and Americans. Howell and Arambourg were relieved to have his on-site expertise ("We had to buy everything from the truck to the spoon," noted Yves Coppens), and soon lists of supplies and personnel, as well as letters laying out the agreements among the three parties, were making their way among Nairobi, Paris, and Berkeley. Only once did the elderly Arambourg, who was eighty-two, question Richard's role by cabling Louis to ask if all was in order. "[A]ll preparations for the Omo Expedition appear to be going satisfactorily," Louis wrote in reply, "but the details are in my Son's hands, not mine." The scientific side of the expedition, however, would remain Louis's, Howell's, and Arambourg's concern.

"My father looked at my initial role in this [the Omo Expedition] like I was continuing my photographic safaris," said Richard, twenty years later. "I was the safari leader, and I would take people, scientists, to various areas." Father and son agreed that Richard would call Louis via radiotelephone regularly, and Louis scheduled several field visits during the three-month expedition. Then, too, Louis reminded him, Howell would be close at hand to offer guidance. As wise and as seemingly minor as these restrictions were, they did not sit well with Richard, who would later complain, "I always had to have a chaperone."

By the end of May 1967, all the expedition supplies were assembled in Nairobi. Packed into three one-ton trucks were everything from mess and camping tents, three kerosene-powered refrigerators, camp beds, and blankets, to shovels, picks, and drums of plaster of Paris, to sacks of flour, salt, rice and maize meal—there were even a wooden dinghy and rubber raft for traversing the river.

Although the expedition was always referred to in the singular, in reality it consisted of three separate groups: the Kenyans (of whom, Richard noted, "I was the undisputed leader"), the Americans (led by Clark Howell), and the French and Ethiopians (headed by Camille Arambourg). Each group had their own source of funding: the National Geographic Society for the Kenyans, the U.S. National Science Foundation and Wenner-Gren Foundation for the Americans, and the Centre National de la Recherche Scientifique for the French and Ethiopians. Once in the field, each party would have its own separate camp and separate fossil deposits to explore. At the outset, because the Omo Valley is so vast, covering 150 square miles, and the fossils so plentiful ("The number of fossils is simply staggering," the geologist Frank Brown had reported to Louis at the end of his 1966 survey), this seemed the best decision. But it also lent the expedition a competitive edge, which did not take long to surface, although as Yves Coppens observed, "Nobody was killed, and we are still talking to each other, which is a sort of success."

Shortly before dawn on June 3, 1967, this large—some forty people—and polyglot party piled into eight Land Rovers and three trucks. Louis and Arambourg, who would fly up later to the Omo, had come to the Centre's grounds to see off the expedition. Limping slightly from the pain in his hips, Louis walked about greeting the expedition members and giving last-minute words of encouragement. It was a bittersweet moment for him. "Louis would have loved to have gone," said Howell, "but he was having trouble getting around. You could see that." Still, he enjoyed being the catalyst behind this enterprise, and was pleased to see one of his sons continuing his life's work—although as several of Louis's friends later noted, he did not want Richard to surpass him.

The Omo Valley lay more than five hundred miles northwest of Nairobi, over rough tracks, and Richard, at the head of the expedition's convoy in a blue Bronco, led the way. The slow five-day journey gave him ample time to size up the competition. "Richard had a certain amount of disdain for the French and American groups. He didn't figure they really knew how to run a safari in East Africa," said Paul Abell, an American geochemist who had volunteered to work for Louis in 1966. Abell had become friends with Richard during that summer when they worked together on an excavation at Lake Baringo. Subsequently, Richard invited him to join the Omo Expedition.

Richard was particularly annoyed by what he called the "effete airs" of the French. At the end of each day's long drive, as soon as the vehicles stopped, "the French whipped out a table, chairs and white tablecloth, wineglasses and a bottle of wine," said Abell. "And they sat around and had an evening drink, watched the sunset, while the African labor unloaded the trucks and got the tents up. This rankled Richard, that the French were placing themselves a cut above the Africans. In his group, you worked right alongside the Africans; there was no division of labor whatsoever."

The Americans, with their habit of opening a can of beans, fishing out a spoon, then leaning against their trucks for a hasty meal, merely amused

Richard. Later, when Howell's team pitched their final camp in the midst of waist-high savanna grass some distance from the Omo River—a decision that required them to use a water trailer—and never moved, Richard decided they were lazy.

Some of Richard's rapid judgments probably stemmed from his own worries about how his team would fare against the French and Americans. Howell, Arambourg, and Coppens were all accredited scientists who had brought along other scientists and graduate students to work with them. In contrast, Richard's team was a motley, unspecialized crew. Besides the geochemist Paul Abell, Richard's group included Margaret (the only person with academic training in archeology and anthropology); Allen O'Brien, a wealthy California entrepreneur and adventurer who had befriended Louis; Alex Duff-MacKay, a biologist from Kenya's National Museum who came to collect samples of the local animal species; Bob Campbell, a photographer on assignment for the National Geographic Society; Kamoya Kimeu, the sharp-eyed Kamba fossil hunter; and eleven other Kamba excavators and camp staff. Then, too, Richard had asked for the most remote and difficult fossil area to explore—fifty square miles of sediments that lay across the Omo River, where there was neither a bridge nor a ferry. He planned to cross the river with his vehicles and supplies on a motorized rubber ferry that Allen O'Brien had designed and tested. Still, it would take Richard's team several additional days simply to get into the field, while Howell's and Arambourg's team were already searching for fossils.

The French had been allocated the southernmost fossil deposits, where Arambourg had previously worked, and just beyond the Ethiopian town of Kalam, on a bluff overlooking the Omo River, they set up their camp. Richard and Clark Howell continued north another fifteen miles, when Howell elected to stop. The road now was little more than a track through dense stands of thorn bush, but Richard forced a way through, traveling only forty miles in two days, and finally reached the banks of the Omo River early in the afternoon of June 10. "The atmosphere of the Omo River that day was almost tangible," Richard wrote in his autobiography. "The river was sluggish, the air hot and humid, and thick gallery forest grew all along it; numerous birds were to be seen and in the great fig trees there were countless black-and-white colobus monkeys." There were also many crocodiles in the river, while others—ranging in size from twelve to twenty feet—sunned themselves on the Omo's banks. "They were enormous brutes," said Paul Abell.[4]

To cross the river, Richard's team first built a boat ramp, siting it close to a village of Mursi people—the unpredictable tribe that the game warden had warned Richard about. But instead of being hostile, the Mursi first were

---

[4] After visiting the Omo Expedition in late July, Louis wrote to Melvin Payne that he had "counted 590 crocodiles, none of them less than 7 or 8 feet long, and some of them nearly 20 feet" along the river between Howell's and Richard's camps. Looking down at the river bank from Richard's camp, he counted another thirty crocodiles.

frightened ("They abandoned their village when they saw us," said Bob Campbell), then curious and finally friendly. They had seen only a handful of white people before, and still retained their own distinctive culture. The young women distended their lower lips with lip plugs, three-inch-wide clay disks that distended the flesh into a flat plate, while the men sported fanciful hairdos of packed clay and feathers.[5] They were armed, as the warden had said, but with ancient Italian bolt-action rifles and long fishing spears. Sometimes, Richard's crew saw them fishing, poised immobile at the water's edge, their spears held high, ready to thrust at the slightest ripple. The Mursi warned their visitors about the crocodiles, which were large and bold enough to take cows, sheep, goats—and humans—that wandered carelessly by the river.

It was not long before Richard discovered just how brazen the Omo's crocodiles were. In traveling up and down the river, he had at first stayed close to the river's shoreline, just in case a crocodile became too curious. But the reptiles seemed to ignore the boat, perhaps because of the engine's noise. Just after dawn one morning, as Richard, Margaret, Kamoya, and Paul Abell set out down the river, Richard turned the boat into midstream. "The colobus monkeys had begun their pre-dawn chorus and it was a beautiful time to be about," Richard wrote in his autobiography. "The light was changing rapidly and the pastel colours of the vegetation were screened by wisps of mist rising from the water. . . . [W]e were mesmerized by the extraordinary beauty of our surroundings." Suddenly, less than seventy-five feet away, a crocodile surfaced, swimming against the current, but nevertheless coming straight and fast at their boat. " 'Is that a hippo or a croc?' we asked," recalled Kamoya. " 'Croc,' said Richard." He turned the boat toward the reptile, but the crocodile was far swifter. It shot at them like a torpedo, the curl of a wave cresting on its snout. In the next instant, the crocodile reared up out of the water, its massive jaws wide open, and bit the side of the boat, splintering the wood. Richard made a violent turn in the other direction, nearly lost his passengers, shipped a lot of water, but got free. For a moment, it seemed the crocodile had disappeared. Then it shot out of the water again, and with a great tearing crash snapped its jaws onto the stern, less than six inches from where Richard sat. "We were in a state of panic," said Abell, "but Richard didn't panic and drove the boat straight up onto the bank"—with the crocodile still firmly attached. As soon as they touched the shore, everyone leaped out of the boat, while the crocodile swam back and forth, angry at losing its meal. "Even then," said Kamoya, "for three minutes we couldn't talk."

They were now six miles from camp and on the wrong side of the river. Richard tried once more to continue their trip, but another crocodile charged, forcing them to the shore again. This time they decided to walk

---

[5] Richard's team found lip plugs in the region that dated to 5,000 years ago, indicating a long record for this tradition.

back to camp—though even this journey had its dangers: a baboon troop threatened to attack, and then Richard spotted the fresh tracks of a pride of lions headed in the same direction. Richard slowed his pace, the lions turned off, and late in the day his weary team finally reached camp.

Richard immediately radiotelephoned Louis, asking for a larger, aluminum boat and a forty-horsepower engine. Louis jotted his requests on the back of an envelope, then cabled the National Geographic Society. Without a safe means of traversing the river, Richard's crew was stranded and fossil exploration came to a halt. "The country in our area is very broken up and covered in very thick bush and undergrowth," Richard wrote his parents in a lengthy letter detailing the crocodile attack. "This makes overland travel almost impossible and distances are such that walking is virtually out of the question. I have spent the last four days at it and have made no headway at all—the vehicles are certainly unable to deal with the situation." If there was any satisfaction in it, Richard knew that the Americans ("and to some extent, the French") were having an equally frustrating and exhausting time reaching the fossil deposits.

Howell eventually solved his problem by cutting tracks through the bush to the sediments, while the National Geographic Society cabled funds for a new boat and engine for Richard. "With this . . . we were again able to use the river. . . . [And] we could taunt and tease [the crocodiles] in safety to our hearts' content," Richard wrote in his autobiography. They also moved their camp closer to the fossil-bearing sediments, and only a few days after receiving the boat, on June 29, the Kenyan team made its first hominid discovery, winning the unannounced race among the three parties.

That morning, Richard and his crew had left camp just after dawn. They walked about a mile upriver, then turned inland to the fossil beds, which lay a short distance from the river. Beyond the Omo's lush shoreline, the land grew increasingly arid: the tussocks of long grass and thornbush thickets gave way to pebbles and wispy tufts of grass, then even this bit of vegetation disappeared, and there was nothing but the deep, barren gullies of the Omo Basin. Striped chocolate, yellow, and pink, these undulated across the earth like giant candy-colored ribbons. And everywhere, as the geologist Frank Brown had promised, there were fossils.

By 7:30 A.M., Richard's team had fanned out across this maze. They carefully picked their way up and down the hills, pausing to study the fossilized limbs of an antelope or a partial skull of an extinct elephant. "We were spread out across the gullies," recalled Paul Abell, "when we suddenly heard a shout from Kamoya that he'd found something. And that pulled us back together." Kamoya had spotted several thick fragments of a hominid skull lying on the surface of a yellow-colored sandy soil; other pieces were embedded in the earth, indicating that the skull was *in situ*. Based on the surrounding animal fossils—which indicated the mid-Pleistocene, about 150,000 years ago—Richard knew that the skull would not be as old as those from the lower deposits at Olduvai. Still, it was a hominid, and the first to be found by the Omo Expedition. "We were a bunch of pretty rank amateurs then," said Abell, "so of course we celebrated."

They immediately began an excavation and recovered not only additional pieces of the skull, but parts of its upper and lower jaws, as well as arms, hand, foot, and leg bones, some vertebrae and pelvic fragments—perhaps, Richard hoped, the remains of a *Homo erectus.* It was a dramatic discovery, and Richard radioed his father with the news late that afternoon, reporting that his team had found "Joseph"—a code name they used for discussing hominid finds over the public radiotelephone channels. Louis dispatched a cable at once to Melvin Payne at the National Geographic Society, using the same secret message as his son: "RICHARD REPORTS FINDING JOSEPH." Then he dictated a letter to Payne, explaining just who Joseph was —and noting that the expense of the boat had been more than justified by this discovery.

One week later, after the excavation was completed, Margaret flew to Nairobi with the fossils and Louis immediately began drafting notes on the back of an envelope about his son's wonderful find. A quick glance at the skull and mandible told him that it was not *Homo erectus,* but rather an early form of *Homo sapiens,* our own species. This did not disappoint Louis, who thought that it might be used to resurrect his Kanjera skull fragments —the fossils that Professor Boswell had placed in a suspense account thirty-two years earlier. "Skull is *MAGNIFICENT,*" Louis wrote in his notes, "& will solve for all time the early appearance of *Homo sapiens.*" He planned to compare it with the Kanjera fossils as well as other early *Homo sapiens* fossils from Europe. Louis elaborated on this idea in a letter he wrote that same day to Melvin Payne: "I found the Kanjera skull fragments in 1933 & claimed them as primitive *Homo sapiens.* . . . How loudly at this most of my colleagues scoffed." But with Richard's fossil in hand, Louis noted, "I have very little doubt indeed that . . . it will soon *confirm,* for good & all, the much earlier acceptance of *Homo sapiens* than heretofore accepted . . . I'm delighted."

Louis's assumption was correct—the Omo skull eventually helped overturn the current thinking about the emergence of early *Homo sapiens.*[6] At the time, anthropologists believed that modern humans had first trod the earth 60,000 years ago, after the demise of the Neanderthals. Kamoya's discovery, which was subsequently dated to approximately 130,000 years ago, more than doubled the time span for our species and provided some of the first evidence that *Homo sapiens* and *Homo neanderthalensis* lived at the same time. But the fossil did not, as Louis had devoutly hoped, redeem his Kanjera skull fragments. He could not pinpoint his fossils' discovery site, and they remained where Boswell placed them, in a suspense account.

---

[6] When the skull fragments were pieced together, they revealed a skull that is very modern in shape. The limb bones, too, which are those of a tall, muscular male, are "not significantly different in body form from East Africans of today," Christopher Stringer wrote in a 1993 study of the specimen.

While Louis studied the Omo skeleton, Richard's team continued to search for fossils, though with little success. Richard knew by now that the sediments they were exploring were unlikely to produce any fossils of great antiquity; nothing much older than 150,000 years would be found. In contrast, the American and French teams were searching deposits ranging from one to four million years in age. Only two weeks after Kamoya's discovery, Yves Coppens turned up the coveted prize: the fossilized lower jawbone of a primitive australopithecine, dating to at least two million years ago.[7] The Kenyan team, too, scored again in early August, when Paul Abell picked up fragments of a second early *Homo sapiens* skull ("JOSEPH'S RELATIVE," as Louis called it) that fit together to form a beautiful black-tinted cranium. Yet it wasn't enough for Richard.[8]

"Richard was very much interested in following in his father's footsteps—although he wouldn't say so—by finding the earliest *Homo,*" said Abell. "When he discovered that few of our sediments were much older than mid-Pleistocene, he was very disappointed. He was definitely miffed that his area was not as productive—that is, as old—as the others."

Richard was also irritated by what he considered Howell's patronizing air. Richard, Howell, and Coppens (Arambourg, suffering from the desert heat, returned to Paris at the end of July) periodically visited one another's camps, usually after a hominid or other important fossil had turned up. Each man naturally wanted to see what was being found, and the visits also gave them a chance to offer congratulations, although these were probably less sincere than they sounded.[9] At the Americans' camp, Howell often used these fossil displays as a chance to raise the issue of Richard's unfinished education. "I was a very strong encourager of him going to college," Howell recalled. But witnesses to these talks remembered Richard "chafing at Clark's well-meant suggestions," as Karl Butzer, a geologist on the expedition, put it.

When he returned to his own camp, Richard often entertained his crew with "acid comments about the incompetent French and Americans," said Abell. He was especially appalled at the fossil-collecting techniques dis-

---

[7] The jawbone has since been dated at 2.6 million years, and is thought by some scientists to be an early form of *Australopithecus boisei.*

[8] Paul Abell's discovery, which came from deposits originally assigned to the Americans, infuriated Clark Howell. Only the day before, Howell had suggested that Richard's team explore this region, which was closer to the Kenyan camp. "Clark didn't think there was anything in them [the deposits] anyway," said Abell. But only two hours after setting out, Abell picked up the first fragments. He quietly rolled them into a sock, then handed the sock to Richard at lunch, saying, "This might be something you're interested in." Richard unwrapped the sock and was momentarily stunned. In contrast, Howell "was absolutely livid," said Abell. "He was so put out with himself that something he had just given away to the Kenyans would produce that."

[9] Louis joined in these one-upmanship games as well. After visiting the French team to see their australopithecine fossil, he wrote to Melvin Payne at the National Geographic Society that he had seen "the somewhat weather-beaten and damaged hominid mandible . . . which is their most exciting find. . . . It has no teeth at all and is a bit ragged." In contrast, he could hardly praise Richard's *Homo sapiens* fossil enough, insisting that "Richard's party is the envy of all the others because of his hominid and also because of his having the only early Pleistocene site so far located." In fact, Richard was far more envious of the other teams than they were of him.

played by some of Howell's students. "I remember in particular seeing a crocodile and baboon skull [in the field] before they were collected," Richard recalled. "Then I saw them in camp, and was horrified at what had happened." The specimens were badly broken, convincing Richard that at least he did not need a university education in field techniques. Yet, without a degree, it did not seem possible for Richard to free himself from "these professors," as he began to refer disdainfully to the academics who wished to guide him.

Louis's occasional visits to the Omo Expedition exacerbated Richard's feelings of inadequacy—particularly since his father and the other scientists quickly took control of the fossils that Richard's team was finding. "I was terribly frustrated," Richard said about this period of his life. "I was constantly being reminded that I was a sort of 'tent-boy.'" At another time he remarked, "I already knew how to organize an expedition and how to find fossils. I wanted to have my own show."

In mid-August Richard got his wish. He had gone to Nairobi for a week to check on his safari business. On the return flight, a thunderstorm blew up over the western shore of Lake Rudolf (Turkana), and the pilot changed his course, making a detour to the east. "This took us up to and along the eastern shore which I had never seen before," Richard wrote in his autobiography. The pilot flew low over the land, and to his surprise, Richard realized he was looking at vast areas of sediments, not volcanic rock as maps of the region indicated. It was the kind of country that produced fossils, it was in Kenya, and Richard was "wildly excited" by his observations.

Back at his camp on the Omo, Richard called Clark Howell and made arrangements to borrow a three-seater Hughes helicopter the American party had chartered. At seven o'clock in the morning a few days later, Richard and the pilot took off, headed for somewhere east and south of the Omo River, along the northeastern shore of Lake Rudolf. It was a long flight, and in the helicopter, Richard was tense with excitement. Once again, he spotted extensive sedimentary outcrops along the lakeshore where the maps showed only lava. Noticing an area that looked particularly promising, Richard asked the pilot to touch down. "I jumped out of the helicopter and immediately found some fossils as well as stone artefacts," he wrote in his autobiography. "We explored further, landing at various other outcrops and everywhere we stopped there were fossils." Hopscotching across the landscape, Richard soon had a small collection of stone tools and fossils, primarily the teeth of pigs and elephants, which would give a rough indication of the region's age. He knew already, from a quick glance, that they were considerably older than anything he had yet found on the Omo.

Here in Kenya, in his country, was what he needed: a site of his own, away from his father and the annoying chaperones. Richard quickly made up his mind. He would not go back to the Omo the next year, but would return to Lake Rudolf at the head of his own expedition.

1

Louis as a young man with his brother, sisters, and parents at Kabete. Seated in the foreground are his parents, Mary ("May") and Canon Harry Leakey; left to right in the back row, Gladys, Louis, Julia, and Douglas.

Canon Harry Leakey and his assistant, Muthago, translating the Bible into Kikuyu at Kabete. 2

3

A fully initiated member of the Kikuyu tribe, Louis shows off a warrior's regalia for the camera, about 1920.

Louis's Second East African Archaeological Expedition, 1928–29, at Lake Baringo. Front row: Louis and Penelope Jenkin. Back row: John Solomon, Frida Leakey (Louis's first wife), Tom Powys Cobb.

4

5

The children of Louis's first marriage, his six-month-old son, Colin, and his three-year-old daughter, Priscilla, 1934.

6

Members of Louis's Fourth East African Archaeological Expedition at their Kanam camp, 1934–35. In the second row are (left to right): Sam White, Peter Bell, Allan Turner, Louis, Percy Boswell, Peter Kent. In the third row, third from left, is Heselon Mukiri, Louis's chief fossil hunter.

7

Louis, at right, walks with Professor W. E. Le Gros Clark at Oxford University, which had just awarded Louis an honorary D.Sc. Le Gros's support was vital to Louis and Mary.

8

Mary Leakey, at about age three, sitting with her parents, Cecilia and Erskine Nicol

9

MARY PRESENTS THE SKULL OF *PROCONSUL AFRICANUS* (NOW *P. HESELONI*) TO THE PRESS AT HEATHROW AIRPORT, 1948.

ONE OF MARY'S FAVORITE EXPEDITIONS TOOK THE FAMILY TO STUDY THE ROCK PAINTINGS NEAR KONDOA, TANZANIA. HERE SHE BALANCES ON A LADDER TO TRACE THE ANCIENT ART, 1951.

10

11

LOUIS AND MARY EXPLORING FOR FOSSILS AT OLDUVAI GORGE IN THE 1950S. LOUIS ALWAYS CARRIED A RIFLE TO PROTECT THEM FROM THE LIONS AND RHINOS THAT THEN INHABITED THE GORGE.

12

13

Olduvai finally yields its treasure: Mary points to the teeth and palate of *Zinj, Australopithecus boisei,* the first major hominid fossil the Leakeys found at the gorge after a nearly thirty-year search.

Leakey expeditions were always family affairs, including Dalmatians and fox terrier. Above, Louis, Mary, and eleven-year-old Philip excavate a nearly two-million-year-old occupation site in Olduvai Gorge.

14

Louis at home working on one of his manuscripts, while a pet hyrax perches on his shoulder.

15

MARY AT HER MUSEUM WORKBENCH IN 1967, ANALYZING THE STONE TOOLS FROM OLDUVAI.

LOUIS, AT SIXTY-SEVEN, SHARES A MOMENT WITH HIS PROTÉGÉE JANE GOODALL, THE FAMED CHIMPANZEE RESEARCHER, AT A LEAKEY FOUNDATION MEETING, 1970.

16

17

LOUIS'S GORILLA WATCHER, DIAN FOSSEY, HOLDING THE ORPHAN, COCO, IN 1969.

18

Louis's "Orangutan Girl," Biruté Galdikas, with two orphan orangs, Sugito and Sobiarso, 1972.

The Calico Early Man Site in California's Mojave Desert, where Louis helped launch a search for the earliest Americans.

19

20

Margaret Cropper, Richard, and Philip find their way through the tall grass at Lake Natron, Tanzania, with the aid of a Msonjo guide, 1964. Two years later, Margaret became Richard's first wife.

21

At Louis's Centre for Prehistory and Palaeontology in Nairobi, Richard shows the lower jaw of an *Australopithecus boisei* from Lake Natron to his parents and Melvin Payne, then vice-president of the National Geographic Society.

22

Louis shows some of his Olduvai excavation team how to make a stone tool and skin an animal in ten minutes.

23

Fossils always brought the family together: Louis discusses a specimen with Richard at Lake Turkana.

THE FIND THAT LEFT RICHARD "BREATHLESS"—THE SKULL OF A NEARLY TWO-MILLION-YEAR-OLD *AUSTRALOPITHECUS BOISEI* HE SPOTTED WHILE WALKING DOWN A DRY RIVERBED NEAR LAKE TURKANA, 1972.

24

ARCHEOLOGISTS, UNDER THE DIRECTION OF GLYNN ISAAC, BEGIN THE FIRST SURVEY OF THE KBS TUFF SITE, 1970.

25

26

RICHARD AND MARY DISPLAY THE *AUSTRALOPITHECUS BOISEI* SKULLS THEY FOUND—HIS (KNM-ER 406) FROM EAST LAKE TURKANA; HERS (*ZINJ*) FROM OLDUVAI.

27

KAMOYA KIMEU, HEAD OF THE HOMINID GANG, AT EAST LAKE TURKANA IN 1970.

DINNERS AT KOOBI FORA ALWAYS ENDED WITH LIVELY DISCUSSIONS; HERE, RICHARD'S TEAM GATHERS FOR THE EVENING MEAL AND DEBATE IN 1972.

28

29

Richard holds up two of his prized hominid finds from East Lake Turkana. On the left, *Australopithecus boisei*; on the right, skull 1470, an early *Homo*.

30

Camels at Koobi Fora: Richard, Meave, Kamoya Kimeu, and Peter Nzube head into Lake Turkana's badlands to search for hominid fossils.

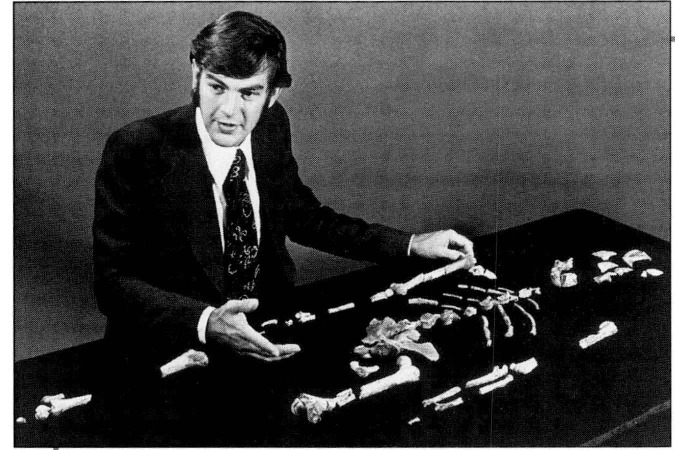

Donald Johanson displays the partial skeleton Lucy, *Australopithecus afarensis*, soon after its discovery, 1974.

31

32

Paleoanthropologist Tim White and paleontologist John Harris with Kamoya Kimeu's Hominid Gang at Lake Turkana in 1975. Left to right: Bernard Ngeneo, Peter Nzube, James Kimani, Musau Mbitti, Harrison Mutua, Kamoya Kimeu, Wambua Mangoa, and Mavudu Maluita.

Paleoanthropologists of all persuasions gather for a friendly photo at the University of California, Davis, in 1976. Left to right: Glynn Isaac, David Pilbeam, Garniss Curtis, Owen Lovejoy, Donald Johanson, David Look, Richard Leakey, and Alan Walker.

33

Mary studying the find of her life: the hominid footprint trail at Laetoli, 1978.

34

35

Anatomist Alan Walker stands beside the skeleton of the Turkana Boy, a fossilized 1.5-million-year-old *Homo erectus* from west Lake Turkana.

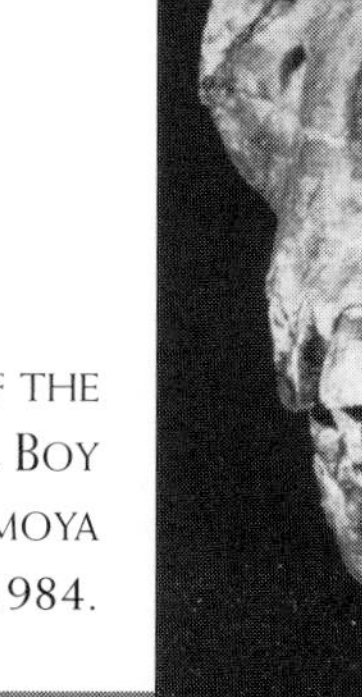

The skull of the Turkana Boy found by Kamoya Kimeu in 1984.

36

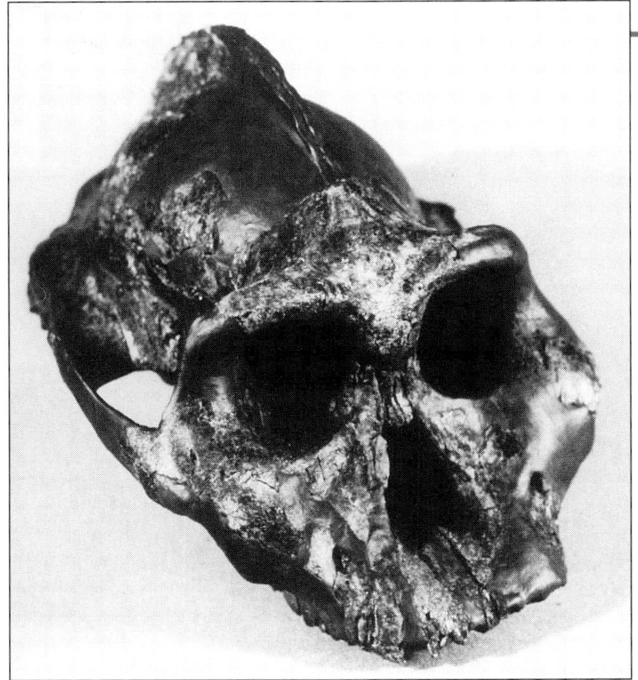

The Black Skull, discovered along the shores of western Lake Turkana by Alan Walker in 1985. Its unusual facial features surprised scientists.

37

Richard, Meave, and their daughters, Louise and Samira, enjoy an afternoon at their Lake Turkana home.

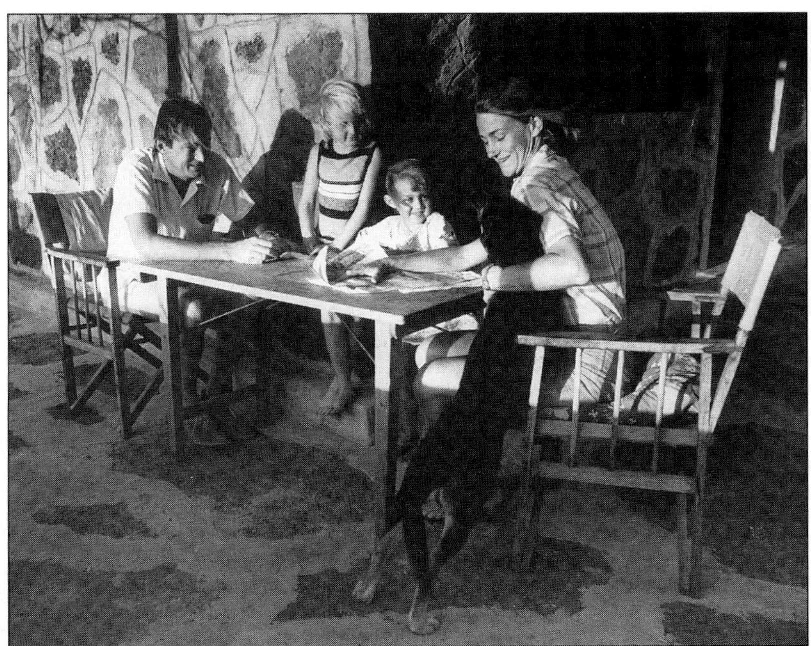

38

39

40

Richard and Meave explore for fossils on the west side of Lake Turkana, 1987.

The legacy continues: Louise and Meave with the Turkana Boy's skull at the Nariokotome River Camp, 1984.

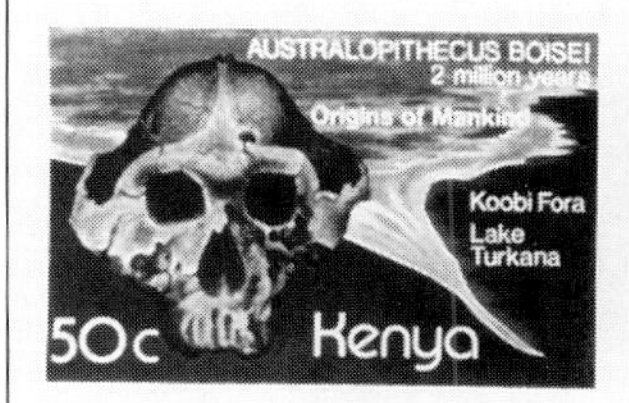

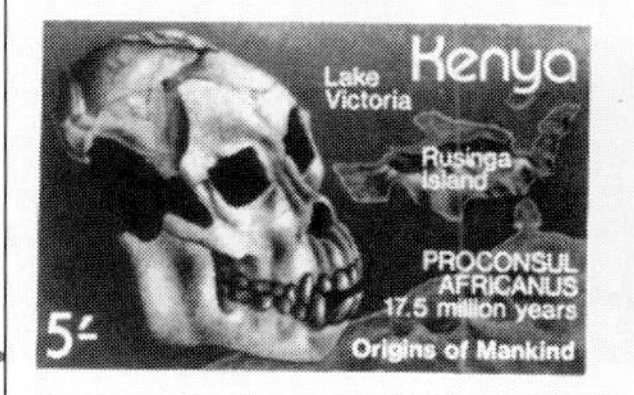

Kenya celebrates four of the Leakey family's discoveries with a set of postage stamps: *Australopithecus boisei*, *Homo habilis*, *Homo erectus*, and *Proconsul africanus*.

41

Mary, eighty years old, at Olduvai Gorge in 1993, where scientists from around the world gathered for an international conference in her honor.

42

President Daniel arap Moi of Kenya ignites elephant ivory confiscated from poachers while Richard looks on, 1989.

43

44

Richard, as director of the Kenya Wildlife Services, flashing one of his famous mischievous smiles, 1990.

Chapter 21

BREAKING AWAY

On August 7, 1967, while Richard searched for fossils along the Omo River and Mary worked on her book about her Olduvai Gorge excavations, Louis turned sixty-four. He was an old sixty-four, almost toothless, overweight and hunched, his thinning white hair swept back from his weathered face. Often wracked with pain from his arthritic hips, he used a cane to hobble around. "I doubt whether, after about 1966, Louis was ever wholly free from pain," Mary wrote in her autobiography, "but so often he contrived to conceal it." Louis simply could not be bothered by physical concerns. If his once-energetic step was less certain, that was no cause to stop or slow down, especially not now, when, if he chose, he could have the world's spotlight focused on him.

"Before Louis, paleoanthropology was only for the scientific world, but Louis and the National Geographic Society opened it up to the public," said Yves Coppens, who learned the value of publicity from Louis, and gave numerous interviews after finding his first hominid skull in 1960. "And Louis was very criticized for doing this—as I was—because it made him a star. He is one of the reasons why the field is so public today and why it is so competitive. Because," Coppens added, laughing at the absurdity of the thought, "we are stars."

Louis had always had a talent for wooing the press. He never traveled overseas without a cast of some important new find tucked away in a pocket, waiting for a moment to spring it on unwary reporters, to dazzle them with a bit of weathered jaw as if it were the Hope diamond. Louis performed the same stunt at his annual meetings with the National Geographic Society's research committee. "We always joked that at granting time Louis would pull something out of the deep freeze," said Gilbert Grosvenor, the society's current president. Louis similarly enjoyed surprising researchers at his Centre for Prehistory and Palaeontology, popping an undescribed fossil jaw or tooth out of his pocket for them like a magician pulling a rabbit from his hat. "Louis always liked having one or two fossils tucked away, ones that no one knew about," said Peter Andrews, who came to Kenya in 1969 to finish his Cambridge University doctoral thesis under Louis. "He loved having surprises. He'd pull out a fossil, then say, 'Look at this!!!' and laugh in that breathless way." Such antics earned him the nickname of the "Abominable Showman" among his more critical colleagues, but endeared him to the press. Newspaper and magazine accounts about Louis in the late 1960s invariably referred to him as a "famed," "revered," "eminent," or "tireless" scientist.

Even when Louis's infirmities prevented him from doing fieldwork, he continued to make discoveries—and headlines. In 1965 he decided to go hominid hunting in the Centre's collections, specifically among primate fossils that dated to twenty million years ago. He and other scientists had collected the specimens in the 1930s and 1940s from various sites in the Lake Victoria basin. Then, in 1951, Louis and Wilfrid Le Gros Clark worked out a classification for these Miocene apes. But Louis had never been satisfied with this analysis. He was particularly unhappy with one fragmentary jawbone that Le Gros insisted was simply a primitive ape. After a "great argument," Louis had reluctantly agreed to this interpretation, believing even then that it actually represented an ancestral member of the family Hominidae—the human family. Now, fifteen years after losing the first round, Louis decided to review the debate. If he was right, then the human line would extend to the early Miocene, about twenty million years ago, as his mentor Sir Arthur Keith had predicted.[1]

Louis had already in 1961 pushed the human line back to the mid-Miocene (about fourteen million years ago) with the fossilized upper jaw of *Kenyapithecus wickeri* from Fort Ternan. Although the jaw was narrow like an ape's, its canines were small like a human's. For Louis and many anthropologists at the time, this was a key trait, marking when the human line began to separate from the ape, as Charles Darwin had first suggested it would in *The Descent of Man*. Without large canines, so Darwin had argued, the apelike creatures would have been forced to begin using tools—one of the first steps on the road to becoming human. Louis did not regard *Kenyapithecus wickeri* as the progenitor of the human line, however. To him it was a "transitional stage" between the "root stock . . . and man and near-men." As such, it was possible to see in *Kenyapithecus* not only the direction hominids

---

[1] Today, based on genetic comparisons of humans, chimpanzees, and gorillas, most paleoanthropologists agree that the human and ape lines separated between seven and five million years ago—at the tail end of the Miocene and not at its beginning as Louis and Sir Arthur Keith had suggested.

were heading, but also something of their earlier origins. Theoretically, then, Louis might be able to find some similarities between this fossil and the jawbone that dated to twenty million years ago—and prove, as he had argued with Le Gros, that it also belonged on the human line. With jawbone in hand, Louis sat down among his Miocene fossils to do just that.

Finding an ancestor of *Kenyapithecus wickeri* would also help Louis bolster its credibility as a distinct family and species—a status that he was in danger of losing. In 1963 two other paleoanthropologists, Elwyn Simons and David Pilbeam, had revised the Miocene primate classification. Deciding that there were striking similarities between the Miocene apes from East Africa and those from India, they had lumped many of the genera together. Thus, according to their interpretation, neither *Kenyapithecus* nor *Proconsul* was valid—a view that infuriated Louis, who carried on a long battle in scientific journals with the two men. At the same time, he was also writing letters and articles defending his new *Homo* species, *Homo habilis,* which other scientists (including Le Gros Clark) had dismissed as nothing more than an australopithecine. Sometimes Louis called his lab technician Ron Clarke into his office and read him the latest volley in these hot debates. "He'd say, 'Ron, Ron! Look at this!' " Clarke recalled. "And he'd laugh breathily. . . . Then he'd read it to me and say, 'The man's a fool! The man's a fool!' And he'd scribble such things on the side of the paper, 'NONSENSE!' and 'RUBBISH!' and 'ROT!' "[2]

In public, Louis could get even more carried away. At an anthropological conference in Chicago in April 1965, he jumped up and shouted at David Pilbeam, who was giving a presentation, to sit down and shut up. "I was still a graduate student," said Pilbeam, who is now director of the Peabody Museum at Harvard, "and there were a whole bunch of worthies at this meeting. By chance, I was actually presenting a paper [discussing the new Miocene ape classification]. I was halfway through it, when Louis started to yell at me, saying that this had all been published, it was all in the journals, and why did we have to listen to it again. He said I was completely out of order and should shut up." Terrified and appalled, Pilbeam looked around the room for someone to come to his defense. But no one—not even Pilbeam's thesis adviser or the conference chairman—spoke up. "One eminent, now-dead person, who was sitting next to me, shrugged and basically turned his back when I asked him what I should do." Louis continued to glare at Pilbeam, but the younger man gathered his nerve and said, 'I'm not going to sit down. You should shut up and let me finish.' And he did." Afterwards, although Louis came up to Pilbeam, shook his hand, and said,

[2] Yves Coppens has Louis's personal, annotated copy of Camille Arambourg's book *Mission scientifique de l'Omo, 1932–1933,* about his 1932 Omo Expedition. "It is annotated in a competitive way with Arambourg," said Coppens. "Leakey was not happy that somebody was working in the same field, as no one is, so he was not agreeing with everything. In the margin there are some annotations done with pencil. Sometimes it is 'Yes,' but other times it is, 'NO! NO WAY! IMPOSSIBLE!' And the pencil is passing through the pages."

"No hard feelings," Pilbeam left the room a shaken man, certain that his career was at an end. Instead, his colleagues congratulated him for having the courage to stand up to Louis, and his career (although not his relationship with Louis) soared.

Shortly after this encounter, Louis began seriously to work on his study of the Rusinga Island fossils, searching for evidence to counter Simons and Pilbeam, and for an ancestor of *Kenyapithecus*. From the broken bits of jawbones and teeth—fossils that Louis and Le Gros Clark had originally classified as species of *Proconsul* and *Sivapithecus,* both early apelike creatures—Louis picked out seven specimens, including the fossil over which he and Le Gros had originally argued. These had been erroneously labeled, he decided after a detailed, lengthy study, and actually constituted a single new species. He named his new find *Kenyapithecus africanus,* and described it as not only an ancestor of *Kenyapithecus wickeri* but "a very early ancestor of man himself." In one swift move, Louis had not only resurrected *Kenyapithecus,* but once again claimed the position of having the earliest hominid fossil. Humankind, in Louis's eyes, now extended to twenty million years ago.

"That's part of the Leakey syndrome," said Elwyn Simons, who quickly dismantled Louis's new species, although it has since been resurrected.[3] "You know, 'The fossils that I find are the important ones; they're all on the direct line to mankind. But the fossils you find are extinct side branches.' "

Louis introduced his new hominid on January 14, 1967, at a press conference in Nairobi. Only the day before, Bryan Patterson, a Harvard paleontologist, had announced his discovery of a 2.5-million-year-old hominid elbow bone in Kenya's northern deserts.[4] Patterson's discovery had made the front page of the London *Times* and the *New York Times,* with both papers noting that it predated the Leakeys' *Homo habilis.* "[U]ntil now [that] was the oldest known manlike fossil," observed the writer for the New York paper. Patterson's moment of glory, however, was short-lived. Whether or not Louis had planned it, his announcement of *Kenyapithecus africanus*—a twenty-million-year-old ancestor—stole the limelight. "Dr. Leakey's claims are potentially much more important for the study of man's ancestry than the discovery reported from Harvard," noted the London *Times* in its article about Louis's discovery. "They concern the stage when manlike and apelike stocks parted company," and not merely the "stage when members of the manlike stock became unquestionably human."

Louis was now, as one reporter described him, "the white-haired man who touched the oldest man-related creature on earth." And no one in his field, it seemed, could beat him at this game.

---

[3] Indeed, fossils discovered by Richard and Meave Leakey in 1985 have verified one of Louis's basic contentions: that there was great variety at both the generic and specific level among the Miocene apes.

[4] This hominid specimen has since been redated to approximately four million years.

• • •

Louis strengthened his case for *Kenyapithecus africanus* with the discovery of a nearly complete lower jawbone on Rusinga Island in August 1967. A few months before, he had arranged for a young American geologist, John Van Couvering and his paleontologist wife, Judy, to prepare a detailed geological map of the island and its fossil sites. When Louis arrived for a visit in August, Judy casually handed him a box, saying she thought she had found a primate. Louis quickly unwrapped the fossils and agreed at once: Judy had uncovered a few loose teeth and a well-preserved mandible of a primitive ape.[5] Back in Nairobi, Louis cleaned the jawbone and teeth, inserting those that had fallen out of their sockets. After a day's work, he had "a most beautiful example of *Kenyapithecus africanus,*" he told Melvin Payne of the National Geographic Society. It was much more complete than any of the other specimens Louis had so labeled, and Louis began drafting a report for *Nature.*

Hoping to find additional specimens of Miocene apes—particularly *Kenyapithecus wickeri,* the fourteen-million-year-old hominid—Louis had also reopened the excavation at Fort Ternan. During the 1966 season, Louis's onsite supervisor, Heselon Mukiri, had uncovered an area with numerous antelope bones, all of them broken. One skull with a depressed fracture particularly impressed Louis. To his eyes, it appeared that the skull and other bones "had been clearly smashed by heavy blows with a blunt instrument," as he would tell an audience five years later. Louis was not completely surprised then, when in the fall of 1967, Heselon found a "peculiar lump of lava [with] battered edges" among the bones. No other lava was present at the site, and Louis concluded that *Kenyapithecus wickeri* had carried the stone there, then used it "to break open animal skulls in order to get at the brain and bones to get at the marrow," as he wrote later in a paper for *Nature.*

Here, Louis thought, was the first evidence for early tool use, but he said little about it to his colleagues, deciding to surprise them at the Sixth Pan-African Congress on Prehistory and Palaeontology. This time the international gathering of archeologists, anthropologists, and geologists met in Dakar, Senegal, in December 1967. Louis was elected to chair the section on the earliest hominids, while Yves Coppens was chosen as secretary. "He asked me very secretly to take over the chairmanship [at one point]," Coppens said, "so that he could make his announcement of the fourteen-million-

[5] Louis misidentified another primate fossil that John Van Couvering found. "It was a piece of jaw with some teeth in it, and I thought it looked like a hominoid, like *Proconsul,*" said Van Couvering. "So when Louis showed up one time, I trotted this out, and very proud of myself said, 'I think I've found a *Proconsul.*' Louis looked at it and said, 'Oh, no, dear boy, not a *Proconsul.* It's a pig. They'll fool you every time.' " In fact, many people, including Louis, have mistaken pigs for hominids and vice versa; and this time, Louis was wrong. The fossil jaw was subsequently identified by Peter Andrews as a new species of Miocene ape. It is now classified as *Rangwapithecus vancouveringi.*

year-old tool. He enjoyed so much being in a position to provoke everyone about such a tool; and I liked the idea, too. So I took the chair, and he made his announcement very strongly as he used to do. But he was a bit disappointed because everyone was so shocked that nothing happened." No one raised a hand to ask a question, no one stood up to criticize his claims. Instead, there was complete silence in the hall. Louis and Coppens looked at each other awhile, then Coppens said, "Too strong. It was too strong. Next time be a bit softer."

Or it may have been, as Elwyn Simons noted, that his colleagues thought Louis's claim "outlandish." Even Mary considered Louis's suggestion "most unconvincing. I can't believe he really thought it was a fourteen-million-year-old stone tool," she said. In more critical circles, Louis's lump of lava added to a growing suspicion that his scientific judgment was slipping. His search for early humans in California had already tarnished his reputation, as had his habit of spouting odd, quirky ideas. "He came up with a lot of innovative theories," said Simons, "but sometimes they were too simplistic. Like after he sighted a white giraffe in Africa, he said that that was how the different races of humans got started—that they would have avoided mating with the other color. So he opened himself up to quite a bit of snickering with these kinds of things."

Louis's theory about his stone tool, in fact, was not particularly radical. David Pilbeam, Simons's coauthor on several papers about the Miocene apes from India, had already suggested that one, *Ramapithecus,* was actually a hominid. He thought that it was probably nearly bipedal, and that it had used its hands "extensively and perhaps tools as well."[6] Pilbeam based his hypothesis on rather slim anatomical evidence: a single jawbone and its teeth. Yet because the canine teeth were small—a trait paleoanthropologists seized on as proof of everything from tool use to upright posture—Pilbeam argued that *Ramapithecus* was well advanced. Louis's contention, then, that the small-canined *Kenyapithecus wickeri* had used tools was not unreasonable. But Louis was the only one who would suggest that such tools could actually be found, and then attempt to produce the evidence.[7]

Louis had a cast made of his fourteen-million-year-old stone tool, and tucked it in his pocket when he set off on his annual pilgrimage to the National Geographic Society in early January 1968. He had an abundance of discoveries to show the members of the Research and Exploration Committee this time: the tool, the new mandible of *Kenyapithecus africanus,* and the early *Homo sapiens* skull that Richard's team had found on the Omo Expedition. The bounty of finds, Louis was sure, would convince the com-

---

[6] In 1982, after discovering a nearly complete skull and the face of *Sivapithecus,* a close relative of *Ramapithecus,* Pilbeam reclassified this once-early human ancestor as a proto-orangutan.

[7] Yves Coppens says that he is "still teaching the reality of Louis's tool. I think it is important to recognize some stones which could have been used by animals in the line to man, even if it is just a hypothesis. This stone exists, it is about the size of a hand with natural sharp edges which are clearly crushed, and it is difficult to believe that such a stone could be there naturally at such a site."

mittee to continue funding his Miocene work at Fort Ternan and the Kenyan portion of the Omo Expedition. Louis also planned to ask for a new grant for Mary to begin excavating Beds III and IV at Olduvai Gorge. But he had no intention of requesting funds for Richard's dream of exploring the fossil exposures at Lake Rudolf, in spite of the promising fossils Richard had shown him. Louis expected his son to remain with the Omo Expedition, "perhaps because of his personal commitment to Emperor Haile Selassie," Richard wrote in his autobiography. "[B]ut I also think he felt strongly that I would benefit from working under Clark Howell and Camille Arambourg," the two scientific leaders of the Omo. This was precisely why Richard, now twenty-three, wanted a project of his own: "I was determined to put an end to my working 'under' people." If his father would not make the request to the National Geographic Society for him, Richard decided he would do it on his own—although he said nothing about his plans to Louis.

As field leader of the Kenyan portion of the Omo Expedition, Richard was invited to the National Geographic meeting along with Louis. He waited quietly while his father described the plans for the coming season at Fort Ternan, Olduvai, and the Omo, where, Louis explained, the disparate teams planned to work more closely together. Richard then respectfully read his own report about the Omo Expedition, but at the end of it—much to his father's surprise—announced that he had no intention of going back to the Omo. He had found a better place to search for hominid fossils, and he thought the $25,000 that was earmarked for the Omo Expedition should be spent at Lake Rudolf. For a moment, the committee members were silent. Then Louis spoke up. "He said something like, 'Young people today are so impatient,' " Richard recalled. " 'I worked at Olduvai for thirty years before making important discoveries. You can't even do it for a year or two before getting fed up and going on somewhere else.' He thought I should stay at the Omo where he knew there were lots of fossils." Richard countered that there was also an abundance of fossils at Rudolf and that they were untouched. It was a polite if strained exchange, and then the committee members asked them to leave the room. The decision would be made—as the committee's decisions always are made—behind closed doors.

Outside, in the hallway, father and son waited together. "Father was disappointed that I wanted to break away from the Omo," said Richard. "Not so much that I wanted to go somewhere else, but more because he wanted to have his hand in the Omo." Louis sighed again about the impatience of youth, but never reproached his son about his unexpected bid for his own scientific expedition.

After the committee's meeting ended, the chairman, Leonard Carmichael, invited Louis and Richard to join them for lunch in the society's private dining room. When all were seated, Carmichael congratulated Louis on his grants for Olduvai and Fort Ternan. He then turned to Richard and said his request had been approved as well. "You can have the money," Melville Grosvenor, the society president, added, giving Richard a stern look, "but if you find nothing you are never to come begging at our door again." The warning did not trouble Richard; he was delighted. Louis's international enterprise—the exploration of the Omo Basin and its hominid fossils—would now be the sole concern of the French and Americans. If Louis was surprised or dismayed by the decision, he did not show it. Instead, he warmly congratulated his son. "Father supported me," said Richard. "We were in public and were part of the same family, and you can't show disunity in public."

At last Richard had what he wanted—a project of his own, independent of his parents. But his triumph had come at a price. He had abandoned a project his father had started, then siphoned off its funds for his own use, leaving Louis rather abruptly cut out of the whole enterprise. And Richard had done it as much to spite his father as to further his own ambitions, as he now admits. "For me it was a gamble and I had no idea of how things would work out," Richard would write about this incident in his autobiography. "I now know that I made the right decision but for quite the wrong reasons."

Fresh from his success with the National Geographic Society, Richard returned to Nairobi in late January 1968 determined to secure a position at Kenya's National Museum. He needed the monthly paycheck, but also had visions of what the museum could become—"an interdisciplinary natural and earth sciences complex, something like the Smithsonian Institution"—and he was certain that he was the only one capable of bringing this about. Few of the museum's senior staff or trustees agreed, however. In their eyes, Richard was merely a twenty-three-year-old callow youth, lacking both experience and academic degrees. It was inappropriate that such a young person be given any responsibilities or authority, argued Robert Carcasson, then the museum's director. Carcasson was even more alarmed when he realized that what Richard was really aiming for was his own job, the directorship of the National Museum.

At the time, in 1968, the museum was a small institution, primarily concerned with the natural sciences, and employing only twenty-seven people.[8] British expatriates, like Carcasson, held the seven senior positions, while the highest-ranking Kenyan was a ticket clerk. The situation was much the same on the museum's board of trustees: of the sixteen seats available, only two were occupied by black Kenyans; the remainder were held by whites, predominantly former colonials. Nor were there any plans to involve more Kenyans or to train any for museum careers. In Richard's eyes, the Kenyanization of the museum—and not his own age or academic qualifications—was the key issue.

---

[8] Louis's independent Centre for Prehistory and Palaeontology oversaw the archeological and fossil specimens.

With a political astuteness seemingly beyond his years, Richard had invited Joel Ojal, the permanent secretary in the ministry that oversaw the museum, to serve as the chairman of the Kenya Museum Associates, which Richard had founded. Like Richard, Ojal "felt that something ought to be done about getting the museum into Kenyan hands," said Richard. But it was not only Richard's ideas that Ojal liked; he was impressed by the younger man's initiative as well—Richard's Museum Associates was now raising ten percent of the museum's budget, and had obtained a Ford Foundation grant to send two Kenyans to the Smithsonian Institution for training as laboratory technicians. Soon, Ojal was listening to Richard—and not to Carcasson or Sir Ferdinand Cavendish-Bentinck, the museum's chairman of the board—about how the museum should be run.

In late 1967, Joel Ojal had a private meeting with Sir Ferdinand and gave him an ultimatum. "Either the Museum took on Richard Leakey and made efforts to Kenyanize senior posts or else all government funding would cease. Not surprisingly," Richard wrote in his autobiography, "my chances of a job were greatly improved." The board concurred and decided to offer Richard a part-time position as executive officer. Richard, however, was uninterested in a part-time position; neither did he like the ring of "executive officer." He wanted to be a full-time administrator, and he wanted the title of director. "Sir Ferdinand was trying to neutralize the damage I could wreak immediately as a twenty-three-year-old who'd never worked anywhere in his life," Richard said. "And I understood exactly what he was doing, and was having no part of it. I wanted to wreak damage; I wanted to change everything. That's why I wanted the job."

Between February and May 1968, Richard's museum job remained in limbo. But he was hardly idle. He was once again running his father's Centre (Louis had traveled to London in April for a hip replacement operation, and would not return until August) and preparing for his East Rudolf Expedition. This was scheduled to depart from Nairobi at the end of May; Richard hoped to resolve the museum issue before then. By now, the museum's board of trustees included several Kenyan members who had been appointed by Joel Ojal, following Richard's recommendations, and they, too, wanted to Kenyanize the museum. "I was then a very young man, as was Richard," recalled Perez Olindo, one of the new appointees, who was soon to become the first Kenyan director of Kenya's national parks. "And many of these older colonial authorities had the name 'Sir' at the beginning of their names, and they looked at us as notorious young boys who wanted to change the face of Kenya overnight. Naturally they had something to protect, but we persisted."

In mid-May the board met again to consider Richard's appointment, and this time—even though Carcasson, the museum's director, threatened to resign—the decision went Richard's way. Beginning in October, he would be the administrative director with full responsibility to the board for all the administration of the National Museum. He was not, however, to concern himself with the scientific side of the museum; this would remain Carcasson's responsibility until his departure, when a scientific director would be hired. (Carcasson had carried out his threat, and handed in his resignation at the meeting.) "This was no threat to me," Richard wrote in his autobiography, "because I knew that whoever was in charge of the finances was in fact in control of everything and once the director left I would be completely in charge."

Although it had required political maneuvering worthy of Machiavelli, Richard had achieved one of his key goals: he had been "supported by the Kenya government as a Kenyan to replace a foreigner!" His next goal was to find a hominid fossil, as he had promised the National Geographic Society. On May 25, 1968, Richard set out for Lake Rudolf to do just that.

Chapter 22

RICHARD STRIKES OIL

On maps of East Africa, Lake Rudolf (Turkana) appears as a long, narrow splash of blue in the surrounding beige desert. But from shore the lake is a lustrous green: the Jade Sea, as it is often called. Fed by the muddy waters of the Omo River, Rudolf is thirty-five miles across at its widest point and 155 miles long. Its northern shores lie just inside Ethiopia, while its southern waters lap the black and red rocks of Kenya's desert volcanoes. These small volcanic hills form a barrier to the lake, and without an outlet, Rudolf lies sparkling and shrinking under the tropical sun.

Ten thousand years ago, the lake had an outlet: the Nile. Rudolf was then two hundred feet higher than it is today. Its waters extended much farther north into Ethiopia, and washed against a range of mountains in the east that now lie twenty miles away. Over the past four million years, the lake has risen and fallen several times as the climate changed, and because of these dramatic fluctuations, its shoreline sediments contain one of the world's richest fossil records.

When Richard set out for Rudolf at the end of May 1968, he did not know how extensive the fossil deposits would prove to be. His brief helicopter visit to the region the year before had shown that there were fossils to be found, but their range and age were unknown. Since that visit, he had made a second reconnaissance flight in a small plane, and from the air chose his expedition's first destination: Allia Bay, about midway up the lake. Aside from a twenty-year-old army track, there were no roads to Allia Bay, but Richard was ready for the challenge. He knew he was about to embark on a "great anthropological adventure. It was virgin ground," he said, "and that's where I wanted to be. Somewhere down there . . . was the key to our existence, and I would be the one to find it."

Unlike the huge caravan that Richard had led to the Omo River in 1967, his East Rudolf Expedition was small and spartan. With his $25,000 National Geographic grant, he outfitted one truck, two Land Rovers, and a Ford Bronco. Kamoya Kimeu and a staff of ten Kamba workers would accompany him, as well as a small scientific team of graduate students. No one in the league of Clark Howell or Yves Coppens would be along, and though Richard had done this by design, it was a great concern to Louis. "My father asked, 'Who are you taking with you?' " Richard recalled about his idea of leading a team of unknowns. "The inference was, 'Who was responsible scientifically?' And I said, 'Why, nobody.' Because it wouldn't have been my expedition if I'd had heavies on board; and anyway, no scientist worth his salt was going to go scampering around North Turkana [Rudolf] with Richard Leakey as his boss."

Richard did rely on his father and his father's colleagues, however, to find his teammates. Through the British paleontologist Robert Savage, Louis had met John Harris, a wiry, dark-haired graduate student at the University of Bristol and Savage's research assistant. Louis told Harris about Richard's upcoming expedition, and suggested that Harris come along to assess the mammalian fossils at Rudolf—an invitation that Harris gladly accepted. To help analyze any hominid fossils his team might discover, Richard turned to the anatomist Michael Day (who had described the *Homo habilis* foot bones for Louis and Mary). Day asked if Richard would include his student, the lanky and good-natured Bernard Wood, on the expedition, and Richard agreed. Archeological discoveries would be the concern of Richard's wife, Margaret. Richard also invited Paul Abell, the geochemist who had worked with him at the Omo, and the photographer Bob Campbell.

Wood was especially surprised and pleased by Richard's invitation to join the expedition. "Richard had never even met me," said Wood, "and yet he sent me a letter saying, 'Yes, do come along.' " Richard's letter actually said a good deal more. He was delighted to have Wood along, provided he could pay his own air fare, and then there was the little matter of security. "There is one point that I must make clear," Richard wrote, "and that is that the area we will be operating in is officially termed 'hostile.' There is a long history of the locals raiding, murdering and mutilating and for this reason, we will have a thirty-four-man fully armed escort. I doubt very much whether anything will happen but in the event of trouble, I cannot be held responsible for loss of life or limb." Although Wood had never been on any trip more taxing than a Boy Scout outing, Richard's warning did not deter him and he eagerly signed on. Due to Wood's school schedule, however, he would not join the expedition until July.

Richard sent much the same letter to Harris, who was equally unconcerned about the potential dangers. Then, too, Harris had done fieldwork in the Sahara, and so was familiar with the hazards of exotic exploration. Yet

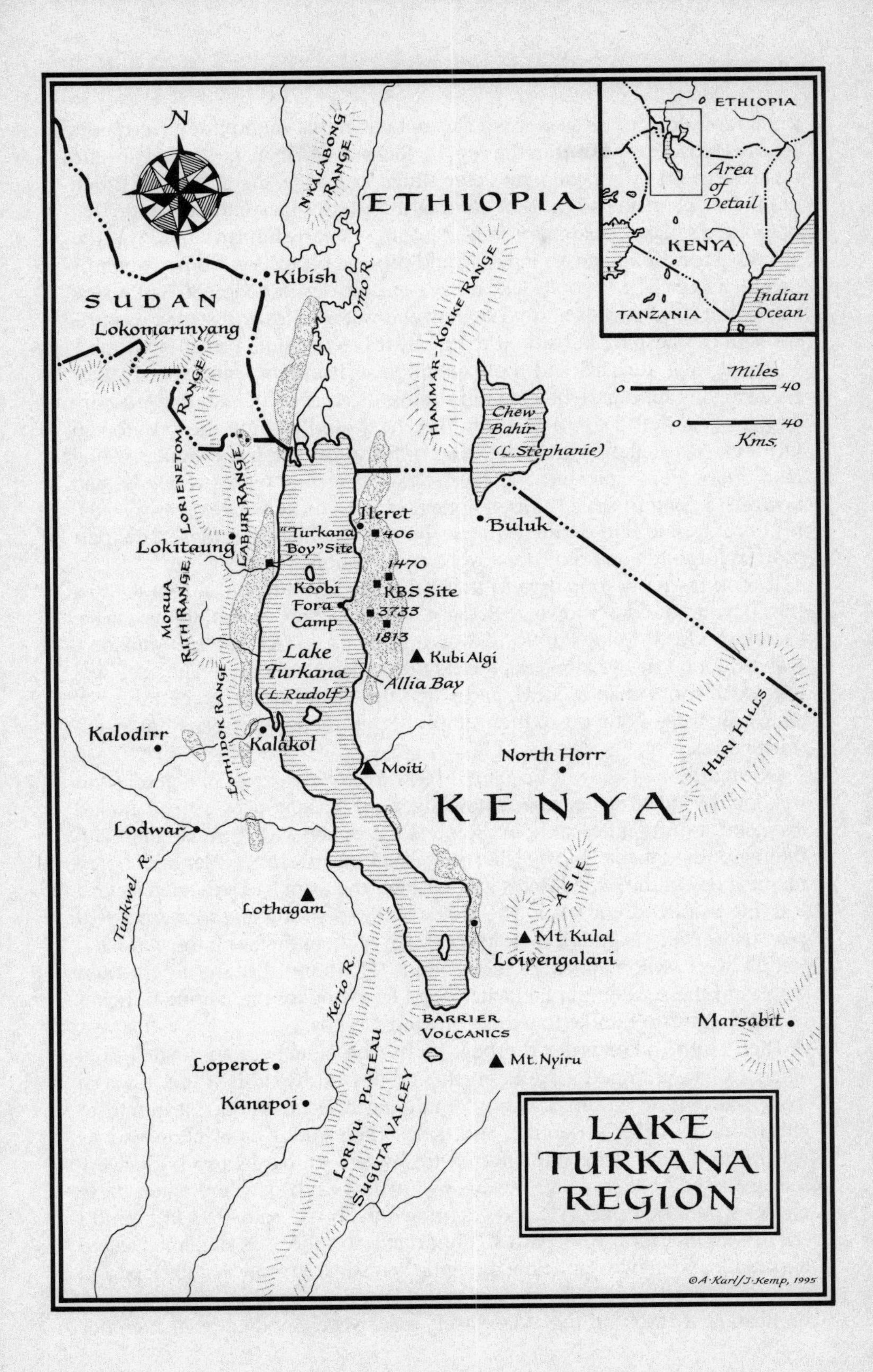

N
ETHIOPIA
SUDAN
KENYA
TANZANIA
Area of Detail
Indian Ocean
Miles
0
40
Kms.
0
40
NKALABONG RANGE
Omo R.
HAMMAR-KOKKE RANGE
Kibish
Lokomarinyang
LORIENETOM RANGE
LABUR RANGE
Chew Bahir (L. Stephanie)
Ileret
406
"Turkana Boy" Site
Lokitaung
Buluk
1470
KBS Site
Koobi Fora Camp
3733
1813
MORUA RITH RANGE
Lake Turkana (L. Rudolf)
Kubi Algi
Allia Bay
LOTHIDOK RANGE
Kalodirr
Kalakol
Moiti
North Horr
HURI HILLS
Lodwar
Turkwel R.
Lothagam
ASIE
Mt. Kulal
Loiyengalani
Kerio R.
BARRIER VOLCANICS
Marsabit
Loperot
Mt. Nyiru
Kanapoi
LORIYU PLATEAU
SUGUTA VALLEY
LAKE TURKANA REGION
©A·Karl/J·Kemp, 1995

the Rudolf desert was so remote and unknown that the Kenya government strictly forbade any travel in the region. Richard had managed to get around this restriction by appealing to Peter Mbiyu Koinange, the childhood friend of his father's, who had become Kenya's minister of state and who approved of Richard's plan to search for important sites of early humans in Kenya. The armed escort Koinange arranged would provide protection from the nearby proud, fierce Gabra people. Like many nomadic desert people in East Africa, the Gabra herd camels and goats between water holes and sparse pasture, sometimes engaging in battle with enemy tribes or raiding small outposts.

By mid-May Richard had the expedition well in hand, and shortly after receiving his appointment as administrative director of the National Museum of Kenya, he left Nairobi for Allia Bay; he would assume the position in October. As usual, he was at the head of his caravan, a lean, angular young man dressed in khaki shirt and shorts, a meerschaum pipe—which he had begun smoking to make himself appear older—tucked in his shirt pocket, his beret beside him on the car seat, the political maneuverings of the past months forgotten and only the adventure of Rudolf ahead.

It took his party four days to travel the 350 miles from Nairobi to Allia Bay. Beyond the dusty town of Baragoi, where Richard picked up his police escort, the team followed World War II army tracks; when these vanished, Richard found his way through the desert by dead reckoning. The pace was not much more than a crawl, and when the team finally reached the bay after a final ten-hour push, they simply set up their camp beds under the star-filled desert sky.

During the night, a wind began to blow from the lake. At first gentle and rustling, by dawn it became a howling gale, blasting across the bay and campsite at thirty miles an hour. Some of the police who were familiar with the region said that it "blow[s] like this virtually all the time," Margaret noted the next day in the expedition's journal, after the team had struggled against it trying to pitch their tents. The winds were so strong and persistent that years later every expedition member spoke of them first, as if the wind and Rudolf were synonymous in their minds. Doubting that the tents could withstand the sudden blasts, Richard had everyone join in cutting bunches of marsh grass to bundle together for wind screens.

There were other tasks: the boat had to be launched, an airstrip built (Richard had arranged for fuel, fresh supplies, and visitors to be flown in from Nairobi; he would also be commuting to Nairobi once a month to check on his father's Centre), and, since Lake Rudolf is alkaline and its waters produce a diarrhetic effect, a freshwater source had to be located. Richard issued orders—even to his own wife—as to how and when these chores would be done. "He was very much the boss," said Margaret, "and I fell in with the troops." Richard's "authoritative posture," as Paul Abell called it, paid off and in two days' time the camp was squared away. Kamoya located a water source in a deep, algae-covered lava pool close to camp; though unpleasant to look at, the water itself was sweet. Another team member discovered an old army airstrip that was easily cleared, and Paul Abell landed a thirty-pound Nile perch on the first boat run up the lake. The next day, June 4—only a week after leaving Nairobi—the real work of the expedition began.

"Richard needed badly to get at least one hominid fragment that first year," said Paul Abell, "and he had to make the decisions as to how the time and effort were to be allocated. So he was pushing us pretty hard." Richard first sent his core team of fossil hunters—Kamoya Kimeu, Peter Nzube, and Mwongela Muoka—along with Paul Abell and John Harris to explore the sandstone outcroppings near the volcanic cone of Sibilot, southeast of camp. These crumbling, beige-colored hills were studded with fossils, many of them broken, but others intact. Indeed, there was such an abundance of animal fossils—remains of ancient pigs, antelopes, elephants, and hippos—that Richard had to continually remind his crew that they were there primarily to make a general survey and to find hominids.

"I had been extremely lucky at [Lake] Natron [where Kamoya found a lower jaw of *Australopithecus boisei*], and I was determined to prove that luck was still with me," Richard wrote in his autobiography. It was. On only the second day of their explorations, Kamoya picked up a few hominid teeth resembling those of the South African australopithecine *A. africanus.* Two days later, Paul Abell spotted a few hominid skull fragments in a dry streambed. "So far [we have] the occipital [the back of the skull], but will sieve tomorrow," Margaret wrote in the expedition's journal, then added with tongue-in-cheek amazement, "Talk about an embarrassment of riches! Just as well—the reputation would be ruined otherwise." The sieving produced a few pieces of the parietal bones, which form the top of the skull. "These were all very, very scrappy bits," said Abell, "but they gave us the first evidence that hominids had lived in the area."

Although the fossils were not *in situ* and therefore could not be dated, Richard was delighted by his team's early success. "There are literally miles and miles and miles of sediments, and I am more than confident of a truly worthwhile season," he wrote to his father, who was in Washington, D.C., for physical therapy following a hip replacement operation, and to complete a book on the wildlife of East Africa for the National Geographic. "[W]e have already found two hominids, . . . [and] I am certain that further discoveries will be made in the very near future."

Richard's prediction was on target. Only two days after writing this letter, on June 15, his team "struck oil," as he told his mother in one of his typical coded radiotelephone messages. This time, Peter Nzube turned up a massive fossilized lower jawbone of *Australopithecus boisei,* the same species as *Zinj.* It was a badly weathered specimen, but had a few teeth in place, and Richard, recalled Paul Abell, "was absolutely tickled silly. That fossil was worth its weight in gold to him. It justified the expedition and guaranteed that we'd be coming back." That night "we ate a better dinner than usual," said Abell, "and the pressure we'd been under let up a bit."

But not for long. Richard was intent on surveying as much of the Rudolf exposures as possible, an area that he estimated extended for nearly one thousand square miles. He intended that his team would cover it all on foot, despite temperatures of 105 and 110 degrees. "Richard drove us unmercifully," said Abell. "We were bushed all the time."

Richard himself rose at five o'clock every morning, then made the rounds of the tents, waking the others. He had set up his camp much as he would one of his commercial safaris, and a little before 6:00 A.M. everyone had gathered in the mess tent to swallow a cup of tea and a slice of bread. In the course of their explorations, the team had discovered that the sediments north of Allia Bay were far more fruitful than those closer to camp. To reach them, the party motored up the lake in the small boat, with the "wind blowing like blue blazes," said Abell. By six-thirty or seven o'clock they were out on the deposits, walking as much as twenty miles a day among the abundance of fossils. Some of the armed guards always accompanied the fossil hunters, keeping a lookout for hostile nomads. But few were ever seen. In fact, the expedition members soon realized that they were very much alone in this part of the desert. "The place had been pretty much depopulated by tribal warfare," said Abell. "But there were animals everywhere. It was like living in the middle of a game park. You never knew if you were going to walk around an outcrop and find yourself face-to-face with a lion or cheetah or giraffe."

On days that the team explored close to camp, they returned to its shade and shelter for lunch around noon; otherwise they knelt against shrubs or under twiggy thorn trees to share a loaf of bread and jar of marmalade.[1] At three o'clock, after the hottest part of the day had passed, they set out again, walking up and down one rocky slope after another, crossing yet another dry streambed, pausing to study the fossilized skull of an antelope or the teeth of an elephant. "The afternoons were the worst," said Abell. "The winds had died down by then and the air was stifling hot." Yet they never carried water with them on their forays, but waited until they returned to the boat or camp for a long, cool drink. "We preferred being thirsty to having the weight of a canteen," said Abell. "And in that country you could go a full day without fluids, as long as you knew you were going to be back by night. You just planned not to get lost."

Dehydrated and footsore, the crew straggled back to camp by six o'clock, revived themselves with a swim in the lake—while keeping a wary eye on the crocodiles usually lined up on shore—changed into fresh clothes, then shared a drink together before dinner. Dinner was served promptly at seven-thirty. The regularity and semiformal structure lent Richard's camp an

[1] Richard's cook, Edward Kandindi, baked loaves of fresh bread in the ashes of a campfire every day, following a long-standing tradition started by Louis, who was famous for his bread. Even when Louis traveled overseas, he carried a jar of yeast with him, so that he could turn out fresh loaves in his hotel kitchenette. Today, Richard's camps are generally regarded as serving the best bread in all of Kenya.

"English air," said John Harris, which carried over to the dinner table.[2] Taking his seat at the head of the table, Richard presided over the meal, sometimes slicing filets of fresh Nile perch for each team member, more often spooning up portions of curry, or carving a roast of wild oryx or topi (an antelope). "Richard was the one who did the hunting," said Abell. "He would go off about sundown and would never shoot an animal in robust health. It always had to be the one that was old or a laggard, so we got some pretty tough meat. He wasn't really a 'live-off-the-land' type, and found it very traumatic to shoot anything. There were days when he wouldn't come back with anything at all, when he couldn't find any animal he could morally shoot."

After the initial flush of hominid discoveries, nothing more turned up in the coastal deposits, and at the end of June Richard moved his camp twenty miles inland to an oasis called Nderati. Here, in the midst of barren volcanic hills, lay a series of man-made wells surrounded by a forest of desert Doum palms. Richard's team pitched their tents in the palms' cool shade, then built a new airstrip, and because water was plentiful, set up canvas showers. But the water supply was also seen as a potential danger, possibly attracting nomadic tribesmen and desert bandits. The few local people that the team had seen—tall, dark-skinned men with feathers adorning their hair, and rifles and spears slung over their shoulders—had kept their distance. The expedition's police escort, however, was not taking any chances; it stationed sentries in the hills above the oasis and patrolled the camp at night.[3]

It was at the Nderati camp that Bernard Wood finally caught up with the expedition in late July—and met Richard for the first time. Both were tall, slender young men, athletic and lithe, but Richard was bronzed from the sun while Wood, just out from London, was "white as the snow," as he recalled. Richard's team, Wood soon discovered, was also intensely competitive, vying with one another for Richard's approval. John Harris, the paleontologist, had already developed a close working relationship with Richard, one that he was not about to let Wood—the new student from England—invade; and "an edge," as Wood put it, quickly developed between them.

A week before Wood's arrival, Richard's team had scored again, with Mwongela Muoka producing another australopithecine jawbone. Like the previous find, this one was badly weathered, but Richard remained confi-

---

[2] There were actually two dinner tables: one for management, meaning Richard, Kamoya, and the scientists; the other for the camp staff, who typically ate a dinner of *posho* (maize meal) and meat. This hierarchical arrangement continues today, and to outsiders, it can feel uncomfortable. Richard says that his usual egalitarianism does not work at the dinner table, since the Kamba workers speak little English and so would feel ill at ease among the scientists. "Dinner is a time for relaxing, and the men would not feel relaxed at my table."

[3] Only once did the escort see any "action." Coming back to camp one afternoon, someone thought he saw a group of men huddled under a tree in the distance. Richard and the police assumed they were bandits, and set off to capture them. But after charging the tree, Richard realized that the bandits were nothing more than lava boulders. "We felt very silly but thankful that we had not gone back for a machine gun and further troops!" Richard wrote in his autobiography.

dent that better specimens would turn up. The area north of Nderati, they had discovered, was "crammed with fossils," said Paul Abell. One site alone was so rich in fossilized hippo remains that the team christened it Hippo Valley, while other areas contained quantities of monkey and elephant bones. Realizing that it was impossible to collect all the best specimens, Richard settled instead for a few examples of each species to impress the members of the National Geographic, and to show off as "trophies" to Clark Howell, who had returned to the Omo for another season.

By August, Richard's team had pushed its explorations as far north as the Ileret River, only ten miles shy of the Ethiopian border. If they thought they had struck oil before, here they hit a gusher. "At Ileret, we were finding hominid fragments virtually every day," said Abell. Many of these fossils were only small fragments, but Kamoya spotted half of an australopithecine jawbone, bringing the season's identifiable total of hominid remains to three, and far surpassing Richard's expectations. The specimens were also of a much greater age than those he had found at the Omo. There the team had searched deposits no older than 150,000 years, whereas at Rudolf, Richard estimated the beds ranged in age from one to four million years.

Richard closed his camp at the end of August, and after a three-day drive his dust-covered expedition arrived back in Nairobi. Louis had recently returned from his four-month sojourn in America and England, and he and Mary delightedly examined their son's finds. Louis thought that Richard's hominid fossils dated to at least two million years ago, making them about the same age as the oldest hominids from Olduvai. There was little question now that Richard had been right in deciding to abandon the Omo project and explore Rudolf instead. Louis congratulated him on this decision—though his voice carried a hint of envy and chagrin. "[Father] told me that he had always wanted to explore the area but lack of funds and equipment in the early years had made it impossible," Richard wrote in his autobiography. "I was never quite sure whether this was really true because he certainly never mentioned it before the expedition; in any event I decided not to believe it because I wanted to feel that it was entirely my site."[4]

In public, however, Louis gave Richard full credit for his discoveries. Worried that his son might be "over modest about his Hominid finds," he even wrote a letter to Melvin Payne at the National Geographic Society

[4] During World War II, Louis had received a sketch map of the northeastern Rudolf area from J. K. R. Thorp, the district commissioner at Marsabit, which was then the nearest administrative post. In reply, Louis asked if he could send "a native collector who has worked at Omo" to look for fossils, but was not given permission to do so; he did not pursue the matter after the war. In 1951 Louis did make an attempt to explore the southwestern region of Lake Rudolf, but found the roads nearly impassable. He also damaged his car and truck, and concluded that only very well-equipped and well-funded expeditions could venture into this region. When American paleontologists asked about where they might search for fossils in Kenya, Louis invariably sent them to southwestern Rudolf.

stressing their importance.[5] "[Richard] has unquestionably got *'Zinjanthropus'* [*A. boisei*] material from three different sites," Louis wrote, "and he also has what I, personally, think is an occipital of *Homo habilis* from the fourth site, which means there are at least four sites which will be well worth detailed study next year. He has also obtained an excellent sample of fossil fauna indicating that a part of the section there is older than Olduvai."

A week after Richard's return to Nairobi, Clark Howell drove in from the Omo. He, too, had had a successful season, finding two australopithecine lower jawbones and an abundance of animal fossils, and spread these out on Louis and Mary's dining room table. In the space of a three-month season, the two teams had collected five major hominid specimens. For Louis, it was surely a bittersweet moment. Both Howell's and his son's expeditions were in the field because of discoveries Louis and Mary had made: they had shown the importance of East Africa in human evolution. Yet it had taken them years of searching to uncover what Richard and Howell had found in a single season.

Perhaps the sight of these fossils touched an old hunger in Louis, for not long afterwards he proposed a new expedition, one that he himself would lead, to an area south of Lake Turkana called the Suguta Valley. A young geology student from the University of London who was helping Louis at his Centre had examined the valley's geology, Louis told Mel Payne in the same letter he wrote extolling Richard's finds. Its deposits dated from five to ten million years ago, meaning that any hominid fossils uncovered would fill "in that gap between Fort Ternan and our early Pleistocene [from fourteen million to two million years], which is so vitally important for understanding the evolution from *Kenyapithecus* to the Australopithecines and *Homo,*" Louis wrote. "[Suguta] will give us one of the most important of the many links still missing in human evolution." In other words, Louis hoped to find here the transitional species between what he considered to be the first hominid, *Kenyapithecus,* and humankind's recent ancestors, the australopithecines and early *Homo.* Payne should be prepared for a grant request to explore Suguta the next year, Louis wrote.

Louis's suggestion, however, that he might lead this expedition was merely wishful thinking. Although his April hip-replacement operation had been a success, he was still in considerable pain and managed to get about only with the aid of two metal crutches. Instead of pursuing a full physical therapy course while he was in America, he had spent only ten days exercising his hip, cutting the regimen short in order to fly to California, where his supporters were organizing the L. S. B. Leakey Foundation to help him raise

---

[5] Louis need not have been so concerned, as Richard was far from being overly humble. "The East Rudolf Expedition was a tremendous success," Richard wrote unabashedly to Leonard Carmichael, the chairman of the National Geographic's Research and Exploration Committee. "Amongst other things, we collected three good specimens of *Australopithecus,* all of which are a good deal older than *'Zinjanthropus.'* There is no doubt that this new locality is the richest and probably the most extensive Pleistocene site in East and Central Africa."

funds for his many projects. Nor did he stay on the strict diet his physicians had assigned, when they told him he must lose weight to reduce the strain on his hip. He had also slipped and fallen on his hip several times after the operation, further injuring it. No matter how much he yearned to return to active fieldwork, Louis would never do so.

Instead, he remained alone at home in the Nairobi suburb of Langata when Mary left for Olduvai Gorge in late September. She had finished her book on her excavations in Beds I and II, and was ready to begin digging in the gorge's uppermost deposits, Beds III and IV. Olduvai was very much her project now, while Louis was more of an interested bystander. As she had done while excavating the lower beds, Mary planned to live at the gorge. "One did not go to Olduvai just to live: one went to work and live," she wrote in her autobiography. She herded her five Dalmatians into her Land Rover, then headed down the grassy drive of their Langata home, calling out a good-bye to Louis, as he hobbled out on his crutches. He stood at their entryway, an old, white-haired man, braced on one crutch, waving the other high overhead as Mary pulled away.

Chapter 23

MINING HOMINIDS AT OLDUVAI

Since closing her Olduvai excavations in 1963, Mary had made only occasional visits to the gorge. She had spent most of the following five years either in her cramped laboratory at the Centre for Prehistory and Palaeontology or in her even smaller study at home, analyzing her vast collection of artifacts and fossils from Olduvai's lower beds. Though it was tiny, Mary favored her home office, with its simple wooden desk facing the one large window. She sat here for hours, one of her Dalmatians asleep at her feet, while she bent over her finds, looking up now and then at the distant Ngong Hills to rest her eyes, then turning back to what would become her classic study, *Olduvai Gorge,* Vol. 3: *Excavations in Beds I and II, 1960–1963.*[1] In this volume, Mary presented her analysis of the physical characteristics of each of the 37,127 Olduvai artifacts, described the twenty fossilized hominids and nearly 20,000 animal remains she had excavated at the gorge, and discussed what these discoveries suggest about early human behavior.

Mary's Olduvai artifacts were the oldest known tools, with those from the

[1] Louis wrote the first volume of the series, *Olduvai Gorge: A Preliminary Report on the Geology and Fauna, 1951–1961,* and Phillip Tobias wrote Volume 2, *Olduvai Gorge: The Cranium and Maxillary Dentition of Australopithecus (Zinjanthropus) boisei.* Louis and Mary had originally intended to collaborate on the third volume as they had on past works, with Mary analyzing the stones and Louis the bones. But Louis's " 'spare time' " efforts "did not achieve results," Mary noted in a letter to Tobias, and she and Richard's wife, Margaret, completed this task as well. Since she had done the bulk of the research and writing, Mary Leakey is credited as the sole author of Volume 3.

lowest sites in Bed I dated at nearly two million years ago.[2] These were simple implements, often nothing more than a rounded cobble with a sharp edge. Yet they had obviously been fashioned with some purpose in mind.[3] Indeed, as Mary sketched and catalogued her finds, measuring each tool's dimensions and noting the material from which it was made, she realized that the hominids must have been making tools for several thousand years before they arrived at Olduvai.

At four of the numerous tool sites in the gorge, Mary had uncovered remains of *Homo habilis,* "the human with ability." From Mary's study, it appeared that the name was even more apt than Louis first realized, for *Homo habilis* had not manufactured just one type of tool, but had created a toolkit with twenty distinct implements. Such a diversity two million years ago suggested that there must have been earlier "more primitive and less organised stages of tool-making," Mary noted in her book. In *Homo habilis*'s toolkit, Mary identified choppers, cleavers, hammerstones, anvils, chisels, picks, scrapers, and awls. She classified other tools by their shape, producing a list of polyhedrons, discoids, spheroids, bifacial points, ovates, and *outils écaillés* (scaled tools). In addition, the hominids had left behind "utilized material," such as cobblestones that had been used for some purpose, and quantities of small, razor-sharp flakes called debitage.[4]

Besides suggesting that *Homo habilis* had created particular tools for particular purposes, Mary's study showed that these early humans had either traveled or traded with other people to obtain the best raw material. Most of the tools in Bed I were made from a green lava found in streams flowing into the ancient Olduvai Lake, or a white quartzite quarried from an outcrop near the gorge. As the hominids' skills increased, they turned to chert (a type of silica) nodules to make smaller, finer tools, digging these rocks out of deposits along the lakeshore. For heavier tools, such as choppers and cleavers, they continued using the more durable lavas and quartzite. But they had also discovered a fine-grained green lava (called phonolite) twelve miles away near the small volcano of Engelosen. This rock took an

---

[2] In 1985 Richard Leakey's West Turkana Expedition discovered an earlier artifact site that dates to 2.4 million years; in 1976 Helene Roche found Oldowan-type tools at Hadar, Ethiopia, dated to 2.6 million years.

[3] The Oldowan choppers are not as simple to make as they might appear—or as the term "simple implement" might suggest. Experimental archeologists who have duplicated the Oldowan choppers believe the toolmaker probably worked in a squatting position. He rested his selected cobble against one thigh and struck it with a hammerstone. Each blow had to be made in an ordered sequence, at the correct angle and with the right amount of force to produce the type of edge he had in mind. Stronger, regular edges were sometimes added later by chipping off a series of small flakes in a process archeologists call "retouching."

[4] For many years, archeologists regarded debitage flakes as merely the chips left behind after making a stone tool. But in 1960, after trying—and failing—to butcher a cow's leg with the large choppers, Louis realized that the hominids must have relied on the smaller, knife-edged flakes for this job. "[C]hoppers were useless for disjointing the leg and cutting through the ligaments that held the bones together . . . ," Mary wrote. "Small quartzite flakes, however, were found to be most serviceable for this purpose and by using them it was not difficult to detach the head of the femur from its socket in the pelvis."

especially sharp edge and in time became the hominids' preferred material. They used other types of stones as well, including quartz, granite gneiss, and a pink feldspar, selecting specific varieties for specific tools—evidence that the hominids understood the different properties of their raw materials.[5]

In some instances, the hominids made tools from animal bones, chipping and shaping the massive limb bones of elephants, horses, and giraffes, as well as the canines and incisors of hippos. Like the stone tools, these were the oldest bone tools in the world—and entirely Mary's discovery.[6] At sites showing the most hominid activity, she had excavated more than twenty thousand fossilized animal bones. Out of these, she selected 110 as having been used or modified by hominids, basing her choice on the bones' "odd or funny" look. While nearly all twenty thousand bones had marks or fractures on them—usually caused by hominids smashing them, by animals gnawing them, or by earth movements—these specimens appeared to have been deliberately chipped and flaked.[7] Unlike Louis, who probably would have held a press conference announcing another earthshaking discovery, Mary was extremely cautious about interpreting her bone specimens. In her book, she noted only that the bones had been intentionally flaked, and suggested that they might have been used as scrapers and awls, though she did not classify them as such.

Mary exercised similar restraint in analyzing all her Olduvai material. Although Louis might "make decisions . . . about things without really any valid reasons that one could put one's finger on," as Mary once observed, her own conclusions were based on detailed measurements and statistics. She had weighed and measured every stone, counted the number of flake scars each one bore, noted the angle and depth of the flakes, counted the types of tools and animal bones at each site, then calculated the percentages of each type of tool uncovered and the ratio of artifacts to faunal remains. The type and number of tools was particularly important, allowing her to assess the hominids' increasing sophistication. For example, the toolkit from lower Bed I contained an average of only six different implements, while in middle Bed II, 100,000 years later, the kit had an average of ten. Mary also found fewer animal bones in the Bed II sites, but a greater number of tools. In the lower bed, the bones of antelopes and small game predominate, while Bed II sites are filled with big game. The number of bones may have decreased because one elephant can feed as many people as several

---

[5] The geologist Richard Hay helped Mary identify the types of stones and locate the sources of nearly all the rocks the hominids used.

[6] In 1955 Raymond Dart had suggested that the South African australopithecines had used bones as tools, but few archeologists accepted his evidence. The bone tools that Mary identified did not resemble Dart's.

[7] Twenty years later, the anthropologist Pat Shipman studied Mary's Olduvai bone tool collection with a scanning electron microscope. When a bone has been used for some purpose—digging in the ground or scraping a hide—it is altered microscopically. Nearly all of the bones that Mary had identified as tools showed this microscopic alteration.

antelopes. Apparently by Bed II times, the hominids had become better scavengers or hunters, capable of at least stealing the remains of (if not killing) such large game as giraffes, hippopotamuses, and rhinoceroses, in addition to elephants.[8]

In Bed II times, the hominids also began making protobifaces (the first tool to have both a point and a sharp edge) in place of the simple choppers. Mary thought that they had learned how to make perfect spheroids as well, and because these round balls occur in such great numbers in Bed II, she speculated that they represented "an important change in tool requirements." In fact, as archeologists know today, the round balls were simply well-used hammerstones.

Mary assigned the simple stone tools from Bed I to the Oldowan Culture, a term Louis had coined in 1936. At the time, Louis believed that the early humans' toolmaking skills had developed in a straight line, leading from the simple Oldowan choppers to the fine handaxes of Bed IV. Mary, however, offered a more intriguing interpretation: an intruder, perhaps *Homo erectus,* had brought the Acheulean handaxe culture to Olduvai. According to her analysis, the Oldowan Culture evolved directly into what she called the Developed Oldowan (the expanded toolkit of Bed II times). But in her eyes none of these implements seemed to foreshadow the handaxe (although the protobifaces appeared to "represent attempts to achieve a rudimentary handaxe by whatever means was possible"). Instead, the elegant, pear-shaped handaxes appeared suddenly in the gorge deposits—and in the same geological horizon as the Developed Oldowan, although this and the Acheulean were never mixed. Later, after she had excavated Beds III and IV, Mary found the Developed Oldowan and Acheulean handaxes at separate but contemporary sites from Middle Bed II to the top of Bed IV, a span of nearly a million years.[9]

Mary and her colleagues puzzled over why the two industries were never found mixed together. The geologist Richard Hay suggested that the hominids may have used the implements in different environments. For example, it seemed that they used Acheulean handaxes only along streams, and the Developed Oldowan tools by the lakeshore; or perhaps, these people had

[8] The most striking evidence for the hominids' meat-collecting skills were the two butchery sites in Bed II, where Mary uncovered the remains of an elephant and a *Deinotherium* among a profusion of tools. It is impossible to determine, however, whether the hominids actually killed these animals or simply stumbled onto their carcasses.

[9] The same two industries—Developed Oldowan and Acheulean—have been found at many early human sites including Olorgesailie, Kenya; Ubeidiya, Israel; and Melka Konture, Ethiopia. In every case, the tools seem to be linked to a particular part of the landscape. Thus the Acheulean handaxes are found close to stream channels, while the Developed Oldowan (which does not include handaxes) tools are found far away from streams and in finer-grained sediments. "You can see this division all through the archeological record," says Rick Potts, an archeologist at the Smithsonian Institution. Today Potts and other archeologists think that only one hominid manufactured both of these industries. "But we still don't understand why that division exists, why we find the Acheulean or Developed Oldowan tools in separate localities."

separate sites for activities that required either handaxes or Developed Oldowan tools. But Mary favored the idea of an advanced people coming to the gorge, bringing the handaxe culture with them, then living side by side with the earlier inhabitants. "[T]he existence of two contemporary cultural streams does not seem in any way impossible, particularly when it is known that *Homo erectus, Homo habilis* and a robust australopithecine [*Australopithecus boisei*] all existed at Olduvai during Bed II times," she wrote. With further fieldwork, she thought the puzzle could be solved.[10]

There were other fascinating questions as well. Was *Homo erectus* the inventor of the handaxe culture? From Mary's excavations, it seemed that these tools first appeared at the gorge at the same time as this ancestor (about 1.5 million years ago). But because the two partial *Homo erectus* skulls the Leakeys had found at Olduvai were not associated with stone tools, Mary regarded this question as unsolved. She hoped that new excavations would finally answer it. Nor was it clear what the handaxes were used for; the tools were large, typically covering the palm of a man's hand, yet they were sharp-edged on all sides, and would cut into the hand if used for chopping, or digging. They may have been hafted onto spear handles, but the shaping again seemed wrong. Perhaps Mary's new digs in Beds III and IV would resolve this matter as well.

Besides starting new excavations, Mary planned to have part of her team scour the gorge for more hominid fossils. Such specimens were still rare, the relationships among them still muddled (few paleoanthropologists were ready to accept *Homo habilis* as a legitimate species, for example), and only new discoveries could provide the answers.

On a more practical level, hominid fossils had also become an expedition leader's meal ticket. Like her son Richard and their colleague Clark Howell, Mary needed to find hominids simply to maintain her funding. "The N.G.S. are now feeling the squeeze," Mary wrote to Phillip Tobias in February 1968, "and have stated that unless we come up with some spectacular finds this year, they are unlikely to give us a further grant in 1969. Since one cannot make finds, sensational or otherwise, without digging, this means that I must embark on Bed IV. . . ."

By early October 1968, Mary and her Dalmatians were settled once again at Olduvai. As before, she employed a large Kamba workforce, several of whom had just returned from Richard's first Rudolf expedition. Perhaps the

---

[10] Peter Jones, an experimental archeologist, carried out various tests with the stone tools at Olduvai during the late 1970s. He was the first to make tools from all the various stones the hominids used at the gorge. Jones discovered that many of the stones, because of the way they break, will only yield crude results—which may explain why the protobifaces of Bed IV times still look like Developed Oldowan tools. He believes that rather than revealing two separate cultures, the Developed Oldowan and Acheulean handaxes simply reflect the nature of the materials from which they were fashioned.

pressure from the National Geographic Society (which Mary likened to having "a pistol . . . held at [her] head") had some value, for only three weeks after renewing her excavation, Peter Nzube picked up an almost complete skull—including the face and nasal bones—of *Homo habilis* in lower Bed I. Though the specimen was badly crushed and encased in a limestone matrix, Mary could see that most of it was intact and that it would very likely settle the question about whether *Homo habilis* was a human ancestor or an australopithecine. Over the next few weeks, the fossil hunters spotted other hominid remains: a single molar of *Homo habilis* from Bed I; then two fragmented *Homo erectus* mandibles from Bed IV; and finally, on December 31, 1968, an arm bone (possibly belonging to *A. boisei*) from Bed II.

"You really do seem to have hit the 'jackpot' at Olduvai," Desmond Clark wrote Mary, congratulating her on her discoveries. "[It's] quite amazing the way the hominids turn up." The National Geographic thought so, too, and when Louis applied for his 1969 grants—including another $13,440 for Mary's work at Olduvai—no one on the Society's research committee hesitated.

While Mary was "mining hominids at Olduvai," as one colleague phrased it, Louis was engaged as a visiting professor at Cornell University in Ithaca, New York. He had flown to London in mid-October for another round of physical therapy on his hip; two weeks later he was on his way to Cornell for a two-month stay. It was a prestigious posting, with a monthly stipend of $3,500, plus travel and living expenses—as well as, for Louis, the pleasure of being among academics who regarded him as a great man. In their appointment, Cornell's board of trustees hailed him as "an explorer of the distant past . . . who has indeed discovered the ancestors of Adam, ordered and named them, giving us and our posterity new progenitors."

If Louis sounded grand and godlike on paper, in the flesh he proved to be informal and charming, as the undergraduates who attended his seminars soon discovered. Sometimes he perched on a wooden crate in front of the class with lumps of flint at his feet, then whacked out a stone tool and used it to crack a cow bone for its marrow. Picking up a bone splinter, he spooned out a bit of marrow and ate it with gusto—all the while holding forth on the importance of tools in human evolution. Other times he described the hunting techniques he had learned as a boy from a Kikuyu hunter, skills that had probably served *Homo habilis* as well. If you knew the properties of certain trees, Louis explained, you could braid a strong cord from their bark, then use this to make a snare for small game—and all this could be done with one's bare hands. Knowledgeable hunters could even catch springhares and small antelopes bare-handed, as Louis had done, relying on stealth and an intimate understanding of animal behavior to get close enough to seize the prey.[11] The earliest people probably survived in a similar fashion, Louis suggested, until one day someone accidentally knocked a flake off a stone —perhaps when using a rock to break a bone for its marrow—and discovered that the flake's sharp edge made a wonderful tool.

Louis's lively seminars and public lectures brought a steady stream of visitors to his campus apartment, where he served tea and scones he had baked himself. The university had offered to house him in a club on campus, where his meals would have been prepared for him, but Louis declined. "I must be independent," he said, explaining that he sometimes rose early or worked late, and wanted to be able to eat when he chose.[12]

Louis spent some of his early morning and late evening hours preparing for his lectures, but most of his spare time was devoted to his ongoing projects: sorting out animal and staffing problems at his Tigoni Primate Research Centre, where researchers were studying the fertility of wild primates; maintaining the search for early humans at Calico, California; finding a photographer to visit Dian Fossey and her gorillas; raising enough funds to keep all his enterprises afloat; and also tending to such mundane matters as whether or not one of his Land Rovers had been repaired in Tanzania. In his weekly letters to Richard, who was managing both his father's Centre for Prehistory and the National Museum, Louis worried about all of these topics, and Richard replied in detail, continually assuring his father that "everything is under control." Richard also sent regular reports about Mary's Olduvai excavations, and the progress in removing the new *Homo habilis* skull from its limestone matrix.

Louis had already left Kenya when Peter Nzube found this fossil, and every bit of news he heard about it left him wanting more. "Your final paragraph intrigues me," he wrote Richard in late November, after Richard sent a letter saying the skull resembled those of the South African hominid, *Australopithecus africanus.* "I am wondering whether you are *really* dealing with something like *A. africanus* rather than a specimen of *Homo habilis.* The teeth, if they turn up, should be decisive on this point." Mary kept Louis abreast of her discoveries as well, telling him when the skull's teeth were

---

[11] Louis once explained the procedure for catching springhares bare-handed to a reporter from the *Saturday Evening Post:* "When you see a hare, it runs straight away and you run straight after it. It has its ears back, but not all the way back. The ears go all the way back when it's about to dodge, a sharp right or a sharp left. If you're right-handed, you always dash to the right anticipating a dodge to the right. That means the odds are fifty-fifty, and you should catch half the hares you chase. If you've guessed correctly, the hare runs by instinct directly at you and you can scoop it up like fielding a fast grounder."

[12] During Louis's tenure as a visiting professor at the University of California at Riverside, in 1963, he had also had a housekeeping apartment. "One caller who volunteered to drive him to an engagement found him ironing a shirt," reported the local newspaper. "Another found he had deposited a bread loaf tin above the gas heater . . . to keep it at an even temperature while the yeast worked. 'It'll be ready for the oven when I get back,' " [Louis] explained. The newspaper's article about Louis carried the headline, "Working Scientist Leakey No 'Head in Clouds' Scholar," and included photographs of Louis ironing and cooking.

uncovered (they were *Homo habilis*) and when the partial *Homo erectus* jawbones and arm bone were found.[13] But aside from their shared intellectual interests and their worries about funding, Louis and Mary had little left in common.

Their relationship had further deteriorated during the last year. The pain in Louis's hip often made him irritable. "We still talked things over, discussed ideas," Mary said, "but a lot of the time there was friction." In particular, they quarreled about Louis's involvement at Calico. Mary had visited the dig in 1967, and although impressed by "the meticulous excavations," she was unconvinced that Louis's team was actually finding implements made by early humans. "I do not think a natural origin [for the artifacts] can be ruled out," Mary told a colleague after her visit.

Calico worried Mary deeply. The basic idea—that Calico would prove that humans had entered the Americas more than fifty thousand years ago—was highly controversial, and Mary feared that Louis's reputation as a scholar would be ruined if he persisted in this search. "Mary said to me over and over, 'I must stop him. Somehow, I must find a way to stop him,'" recalled Marie Wormington, emeritus archeologist at the Denver Museum of Natural History and a highly regarded authority on the peopling of the Americas, who also was unconvinced by the Calico tools.

So desperate had Mary become about Calico that, unbeknownst to Louis, she tried to put an end to the dig. "Mary was trying to protect Louis's image and reputation," said Mary Smith of the National Geographic.

> She thought he was being self-destructive, and she did say to me that this Calico thing must not be published. Which is not why we didn't. There just wasn't anything there; there wasn't enough evidence.

But neither Mary's behind-the-scenes efforts or caustic remarks to Louis about Calico persuaded him to stop. He was determined that he would overturn the prevailing scientific theories about the archeology of North America, just as he had upset previous ideas about human evolution. In 1968, when the National Geographic Society decided not to provide further funding for Calico, Louis turned to the Wenner-Gren and Wilkie foundations, as well as the newly launched, California-based L. S. B. Leakey Foundation for Research Related to Man's Origin. Its very first grant went to Calico.

The prime mover behind the Leakey Foundation (as it is called today) was Allen O'Brien, an entrepreneur and adventurer from Newport Beach, California. He and his wife, Helen, had first met the Leakeys on a safari in Kenya in 1965—in fact, Louis, who was managing Richard's safari company while Richard was at school in England, had been their guide, and had taken them on a personal tour of Olduvai. On that trip the Leakeys and the O'Bri-

---

[13] "The teeth are text-book *H. habilis*," Mary wrote to Desmond Clark, "so that his status should be either established or otherwise." Mary's letters to Louis from this period are not extant.

ens formed a lasting friendship. Whenever Louis was in Southern California, he stayed with the O'Briens, and Allen drove him to all his appointments and lectures. Louis was then receiving five hundred dollars or less for his speaking engagements—a sum of money that appalled the O'Briens. "He was being exploited," Helen said, and Allen thought that was wrong. That's what gave him [Allen] the idea to start a foundation." Several other wealthy and prominent Californians joined in the effort, and on March 26, 1968, the Leakey Foundation was born. But like many projects Louis was involved in at this period of his life, the foundation was on rather shaky ground financially—indeed, it needed Louis as much as Louis needed it. Many of his lecture tours were now centered on raising funds for the foundation; Louis, in turn, expected the foundation to spend its monies only on his projects. Inevitably, there were conflicts.[14]

In time, the Leakey Foundation and its demands on Louis's energy and time came to be another sore point between Louis and Mary. The Calico dispute ended in a stalemate: they simply dropped the subject. But early in 1968, they found themselves unexpectedly at odds again over what should have been a happy occasion: at Phillip Tobias's instigation, the University of Witwatersrand in Johannesburg announced it was awarding them both honorary doctorates.

Louis promptly replied that he had to decline the honor; he was much too prominent a Kenyan citizen to receive awards from a South African university. (Kenya did not then recognize the South African government because of its apartheid policies.) But Mary, who then, as now, was a British citizen, decided to accept the degree—her first such award.[15] "It is a very great honour," she told Tobias, "and although I have never sought academic distinction, it gives me a nice warm feeling particularly coming from an African University instead of from overseas, where values tend to be distorted." She was dismayed at Louis's decision to decline the award, believing that it was "not in any way a political matter, particularly in view of the University's known views on racialism."[16] Louis, however, would not change his mind. He was convinced, as he told Tobias, that if he did accept the degree "the effect on popular and public opinion would be very grave indeed here, and even more so in Tanzania, and it could very easily prevent any possibility of further work by us at Olduvai."

In the end, Mary traveled alone to Johannesburg. On March 30, 1968,

---

[14] In 1969 the L. S. B. Leakey Foundation was saved from almost certain financial ruin by Robert Beck, who had sold his electronics company to the Xerox Corporation and become a very wealthy man. He gave the foundation what it needed most: a million-dollar endowment. The Leakey Foundation is today a thriving organization, providing anthropologists and primatologists around the world with grants for research into human origins and primate studies.

[15] Louis had already received honorary degrees from Oxford University, the University of Guelph, the University of Utah, the University of California, and the University of East Africa, Dar Es Salaam, Tanzania.

[16] The University of Witwatersrand was even then an integrated institution, and Phillip Tobias, who headed the anatomy department, an antiapartheid activist.

dressed for the first time in academia's scarlet cap and gown, she stepped forward to accept her doctorate of science from the University of Witwatersrand's chancellor. It was given in recognition of "her originality of mind and important innovation . . . in archaeological fieldwork," said Phillip Tobias in a brief introductory speech. "Probably nobody has spent more time engaged in actual excavation in Africa than Mary Leakey," he continued, adding that it was "not easy for a wife of a famous husband to gain acknowledgement for her personal achievement, but in the case of the Leakeys the merits of the wife, Mrs. Mary Leakey, stand out so prominently that they cannot be missed."

In Nairobi, Mary's husband spent the day at home, listening attentively to the radio, certain that his wife was about to cause an international incident. But nothing happened, and when Mary returned to Kenya in April, she was *Dr.* Mary Leakey.

Perhaps that title and her nearly completed book gave her more confidence, as did Louis's fallibility over Calico, for not long after she returned from South Africa, Mary's colleagues noticed a distinct change in her behavior.[17] "When I first met Mary in 1961," said the archeologist Glynn Isaac, "you could talk about technical matters with her, but when you turned towards the interpretation, what these stones and bones mean, she always used to say, 'Well, we'll have to ask Louis about that.' " Isaac left Kenya for three years to complete his doctorate, and when he returned in 1968, he discovered a new Mary. "Clearly now, and it seemed quite suddenly, she was going to stand on her own feet; she was not only doing the research, she was assigning meaning to the result." Nor did Mary think of herself now as simply a partner at Olduvai, the one who ran the dig while Louis raised the funds. Olduvai had become her show—she found the hominids, directed the excavations, and, if necessary, she would even get the grants. "To put it in crude terms," said Phillip Tobias, "Louis's role at Olduvai was scarcely necessary any longer."

"There was never any decision to part company," Mary said about the increasing separation between herself and Louis. "It just came about as being the easiest way to live. And there was Calico, which was a disaster. I don't think Louis's brain was what it had been by any means. I think senile is too strong a word, but he seemed to lose his critical faculties. It was very distressing."

When Louis came home from Cornell in early January 1969, Mary drove up from Olduvai to meet him at the airport. They spent a few days together, looking at the new Olduvai hominids, and discussing her excavations and Louis's upcoming travel plans. He had time only for a month in Nairobi to straighten out affairs at his Centre for Prehistory and at Tigoni; then he

---

[17] Clark Howell observed that Louis and Mary always wanted to be "treated as individuals and not as a family. Even in the 1950s, it was not 'Louis and Mary Leakey.' It was Louis Leakey and Mary Leakey. They each had his/her autonomy. It wasn't like so many marriages, where the wife is merely a mirror for her husband. Mary was her own person in all respects."

would fly back to the States for a round of fund-raising talks. His supporters in California needed him to help raise money for the L. S. B. Leakey Foundation, and he wanted once again to visit Calico. As Mary listened to Louis, she realized that she could not save him from what she considered his "destructive behavior." She would have to let him go, while she returned to what was now her own home, Olduvai Gorge. "Had Louis needed me to be here [in Nairobi], I would have given that my priority," she said. "But it was quite clear that he didn't. So the clear-cut path to me was to go on with the work. And I did."

Chapter 24

DEAREST DIAN

By 1969 Mary believed that Louis's scientific acumen and reputation were slipping, but his fans in America thought otherwise. Louis's semiannual fund-raising tours and the many *National Geographic* articles and television programs by and about him had made him a legendary figure, especially on college campuses. Students flocked to hear his impassioned lectures. Ostensibly Louis's talks concerned Olduvai and human evolution, but they were also a means for him to inspire the young. He described humankind's origins, the discoveries of tools and fire, and then argued that in spite of the uncertainties of the present—the threat of nuclear war and environmental pollution—there was hope for humankind's future. It lay with the students seated before him.

"You, the young generation, have got to see to it that our wonderful heritage, and it is a wonderful heritage—art, music, science, religion—is not lost to us," Louis said, concluding one such talk at San Francisco State University. "It mustn't be lost to us. I want my grandchildren and great-grandchildren and great-great-grandchildren after that for another twenty million years to go on developing until they finally reach a point of Omega." With an evangelist's zeal, he challenged them to explore the unknown as he had, to dedicate their lives to science and space exploration, to peace, to Africa. And like an evangelist, though the lecture halls were packed, Louis seemed to speak to each person individually. In response, the students loved him.[1]

[1] Louis also had the rare gift of being able to communicate with children. "Louis was tremendously good with children," said his son Richard. "He had an affinity with young people, formative people of ten, twelve. He would fascinate them, stimulate them. It is an image that I remember and I'm—envious? Proud? One of those. But he would always talk to them and not at them."

"I remember him coming to lecture [at Foothill College, Los Altos], this big man with a shock of white hair, hobbling onto stage with his two canes, struggling really, and the whole audience leapt to its feet cheering and applauding him," said Bob Drewes, then an undergraduate student who went on to specialize in African reptiles partially because of Louis's influence. "He was awesome."

Even those who knew him intimately were astonished by Louis's charisma on stage. Betty Howell, the wife of Clark Howell, attended one of Louis's talks at the University of California, Berkeley, in 1969. Though she had known Louis since 1957, she had never heard him lecture, and was amazed to find herself in tears at the end.

> I was *so* moved by him. It was as though I'd never known the man. He'd taught me to bake bread and taught our son how to do string tricks. And here [at his lecture] was all this beauty and strength and positive direction coming out of him and it just didn't seem like it could possibly be the same person. Then a woman of about fifty-five came up to me and said, "I saw you with Dr. Leakey. Would you give him this for me?" And she put in my hand a chain with a medallion on it; she was *so* moved by his talk she wanted to give something to him. And I thought, "My God, he's like a guru." I had no idea he would have such an effect on people. It was like a religion.

Wherever he went in America, Louis drew crowds of similar admirers, but nowhere was he idolized as he was in Southern California. By 1969 he was traveling twice a year to Los Angeles, to give lectures and attend social events organized by the L. S. B. Leakey Foundation, and to visit the excavation for early humans at Calico. Both enterprises attracted people, primarily women, who were so enamored of Louis that many scientists disparaged them as members of the "Leakey Cult." "We all tended to put Dr. Leakey on a pedestal," said Rosemary Ritter, who was then a college student working on the Calico dig. "We'd heard so much about him and his finds in Africa that took him and his wife thirty years to make. I admired him, too, because he was willing to make mistakes, to put his name on the line and stand up for what he thought was right. And then when I met him for the first time—he had such an aura about him that I wanted to rush up and put my arms around him. He was just the type of person who could bring the good out in everyone."

Though other scientists might have been chary of such unquestioning adulation, Louis responded like a Hollywood ingenue. He had always loved the limelight, and now he not only had adoring crowds of followers, but through the wealthy backers of the Leakey Foundation, invitations to Los Angeles high society. "He was the center of attention," said Sonia Cole, a close friend of Louis's who later wrote his biography. "[A]nd he had to appear at dinners, and go to parties, and he loved the whole thing. A lot of his friends and family criticized this and said, 'Where was Louis the scientist? What was he doing being a sort of film star?' But I always thought if he enjoyed it, why shouldn't he have a little fun towards the end."

There were black-tie dinners in Louis's honor, cocktail parties and luncheons. Louis moved among the movie stars, financiers, land developers, and tycoons as easily as he did among the elders of an African village, dispensing warmth and charm to all. "His charisma was just out of this world, and even chit-chatting with people, he was absolutely spellbinding," said Helen O'Brien. Louis unquestionably enjoyed the social whirl, but his primary motive in attending these events was to raise money. The Leakey Foundation was still struggling to attain a secure endowment, and Louis's charm often helped persuade people to donate. Ironically, although the foundation was started to help lighten Louis's fund-raising burden (Louis estimated that he needed at least $300,000 a year), it seemed to have the opposite effect. Now he was not only asking for money to keep projects like Calico afloat; he was also trying to find funds for the foundation. None of these financial worries, however, stopped Louis from launching new enterprises, or taking on new disciples.

By now Jane Goodall and Dian Fossey, Louis's star protégées, were famous in their own right. It had been ten years since Jane had begun her study of the wild chimpanzees, while Dian had lived among the mountain gorillas for three. Louis often spoke of their research in his lectures, stressing how important these studies were in understanding human evolution. "Louis was really quite dramatic about this," recalled Biruté Galdikas, who was an attentive listener at a lecture Louis gave at the University of California at Los Angeles in March 1969. "He said he had just received a telegram that morning from Dian Fossey, saying that the gorillas were now tying and untying her shoelaces. And he patted his pocket to show that the telegram was still there. It suddenly struck me that this was the man I should talk to. I wanted to study orangutans, and the fact that Louis Leakey had encouraged others, that he had helped Jane and Dian, made me think he would help me."

At the end of Louis's lecture, Biruté joined the circle of people waiting to speak to him. "There was a crowd around him and I had to wait my turn. I forget exactly what I said, but I told him I wished to study orangutans." Louis smiled and nodded at this, but then turned away to answer another question.

> There were all kinds of people coming up to him. So then I started telling him about the steps I'd already taken to get the study under way: writing to Malaysia, contacting zoological societies. And that's when he really lit up. He said he was going to return to Africa the next day, but I was to write him and keep in touch. He stressed that several times. When I walked out of there, I knew I was going to study orangutans, and that Louis was going to help me. I even said to a friend as I walked out, "Well, Louis Leakey is going to do it."

Louis was as serious about his intentions as Biruté thought. With the gorilla and chimpanzee studies well under way, he was ready to back a study of the third great ape, the orangutan, thus completing observations in the wild of humankind's closest relatives.[2] Only a few hours after meeting Biruté, Louis telephoned her anthropology professor at UCLA and asked him to arrange an interview with the pretty, almond-eyed young woman for the next morning.

During his visits to Los Angeles, Louis often stayed at the spacious Westwood home of Joan and Arnold Travis, two wealthy members of the Leakey Foundation, and it was here amid the rustling palms and sweet roses that Louis interviewed Biruté. "He said he was considering helping me," Biruté recalled, "and he wanted me to undergo some tests. These were brain teasers like you find at the back of magazines, like joining some lines in the shortest possible way. I flunked the first one, but passed the second, and Louis was really happy about that."

The two then had a long talk, and although Louis did not make a firm commitment to her, Biruté left their meeting convinced that he would. They exchanged addresses, and that afternoon Louis flew to London en route to Nairobi. By the time he reached home, he had already drafted letters to the governments of Malaysia and Indonesia inquiring about the status of the orangutan. "I think, for all practical purposes, that once Louis had conceived of something, it was already done," Biruté said sixteen years after her first meeting with him. "His mind just raced a million miles ahead of everyone else's. So when he decided that I was his candidate, in his eyes, the orangutan study was essentially done." For Louis, organizing the actual research was just a pro forma affair.

Louis returned to Nairobi at the end of March 1969. He had spent the last two months crisscrossing the United States, giving lectures in Washington, D.C., New York, Philadelphia, Cleveland, Salt Lake City, and Los Angeles, even though, as he himself admitted, he "was lame and moving on sticks and not as fit as [he] used to be." Louis would also soon turn sixty-seven, yet he had lost none of his energy or enthusiasm.

On April 1 he was back at his desk at the Centre for Prehistory and Palaeontology, where he turned his attention to his work there and at the Tigoni Primate Research Centre. He had staffed both with students who came primarily from America and England. Some had responded to adver-

---

[2] From Louis's papers in the National Museums of Kenya, it is unclear whether he had approached anyone about an orangutan study before he met Biruté. He had discussed such a study with Jane Goodall early in her career, but by 1969, she was entirely devoted to the chimpanzees. Nevertheless, when he began seeking funds for Biruté's project, Louis always told the granting agencies that he had made it "known that [he] was anxious to find a girl who was willing and competent to initiate an Orang outan study," and had "interviewed a variety of people" before selecting Biruté—words that made his choice appear less impulsive than, in fact, it may have been.

tisements that Louis placed in *The Times* of London, but more often they wrote to him on their own after hearing him, or reading about his work with Jane Goodall and Dian Fossey. "I had always wanted to work with Louis Leakey, and I wrote him a letter to that effect," said Frances Burton, an attractive brunette of twenty-six whom Louis employed in 1966 to care for the monkeys at Tigoni. "To my surprise he answered, and set up a meeting at my apartment." Burton was then a New York City high school teacher, but she had majored in zoology and dreamed of "working in the path of Jane Goodall," as she wrote Louis, adding, "Above all, the dream I have involves working under you."

As he had with Biruté, Louis gave Burton an intelligence test—this one involved finding the most economical way to link water pipes to several houses, a puzzle that Louis said tested one's "clear thinking"—and after she had successfully completed it, offered her the position. The meeting left Burton, as she would later tell Louis, in "a state of dreams." "I was overwhelmed by [Louis's] lack of elitism," she said about their meeting. "[H]e had, after all, come to my apartment, this famous person. He seemed forthright, certainly taken with himself, and above all enjoying everything. He did the talking. I nodded, smiled, indicated my interest, and occasionally got a question in." Two months later, in early 1966, after Louis sent her a Kenyan residency permit and money for her airfare, Burton arrived at Tigoni.

In much the same informal manner, Louis hired Juliet Ament, a zoologist who had been working in Tanzania on a tsetse-fly-control program, to set up the osteology department at his Prehistory Centre in early 1965. "I'd smashed myself and my landrover *[sic]*, and ended up in Nairobi, jobless and convalescent," she recalled. "On a friend's advice, I went to see if Louis had a job which I could do, [and] he immediately said yes. I was surprised to be accepted so readily." Louis asked Ament, a striking young woman with jet-black hair, to put together a collection of complete skeletons of every vertebrate species in East Africa—not simply one of each, but fifteen, in order to show sex and age differences. Such a collection would be of great use to researchers studying the animals' fossilized ancestors, Louis believed. The Wenner-Gren Foundation had agreed, and granted him $20,000 in seed money. But once he had hired Ament, Louis left the osteology department largely in her hands. When she needed funds or equipment, he provided them, and while he gave little technical advice, he always "encouraged [her] to believe that [she] was doing something worthwhile." It did not take Ament long, however, to realize that one of the reasons Louis had hired her was because she was a young, attractive woman.

"Louis enjoyed the company of presentable young women," said Ament, "so he kept his eyes open for them. Most of us were flattered by his attention and easily led into his schemes, to be left to get on with the job on our own when something else demanded his attention. . . . This didn't suit everybody."

Ament witnessed a steady stream of young people—men and women—

drifting in and out of Louis's Centre. "[They] had been caught up in Louis's enthusiasm and gone along with him expecting him to play the role of fairy godfather," she explained. But after building up their expectations, Louis offered little direction, leaving them to make their own opportunities in Africa. In Ament's eyes, these "were considerable," but she acknowledged that unless a person had "strength of purpose . . . and a good deal of initiative, which [Louis] admired, they drifted off again before long."

For those willing to make their own way, Louis did offer substantial material support. A letter from Louis was often all that was required to help a student land a place in Oxford or Cambridge University's doctoral program, or to obtain a grant from a funding agency. Both Jane Goodall and Dian Fossey studied at Cambridge for their doctorates because of Louis—as did Peter Andrews, a young forestry student who became fascinated by fossils in the mid-1960s. "I came across some fossil sites in western Kenya while I was working in the forestry department," Andrews said, "and I took some of the fossils to Louis." Louis encouraged him to switch fields, and Andrews soon found himself studying anthropology at Cambridge. In 1969, with Ph.D. in hand, Andrews returned to Kenya as Louis's assistant on the Miocene ape fossils. "He gave me the whole collection to describe, which hadn't been worked on in about twenty years," said Andrews.

Louis and Andrews planned to work on the study together, but this was not to be. While Louis did provide Andrews with airfare, salary, and a large Wenner-Gren grant, he had no time to give for scientific guidance. "It wasn't that he wasn't interested," said Andrews, "it was just that he had a zillion other things going on." There was, however, another element at play in Andrews's inability to see Louis: Andrews was a young man, and Louis liked women. "Louis would be working in his office and would always be too busy to see you," he recalled. "That was the advantage of being a young, attractive female—you could actually get into his office. But when you weren't, you had to seize your chances. So I used to have discussions with him as he walked down the corridor, going from one appointment to the next."

While Andrews was not discouraged by Louis's scattered attention, other young people—particularly those working at Tigoni—often were. Louis had launched the primate center with Cynthia Booth in 1958. They shared a vision of breeding monkeys for medical research in order to stop the trade in wild monkeys. Booth was then newly widowed, a blond, brilliant (with double firsts from Cambridge University in zoology and chemistry) young woman, and she and Louis apparently had an affair. This did not last, and by 1966, Booth, now remarried, had grown bitter and frustrated. She complained to students at Tigoni that Louis did not provide her with sufficient funds to operate the research center; at the same time, despite great quantities of data, she never published any of her studies, making Louis's fund-raising efforts extremely difficult. She and Louis were now awkward friends, and whenever Booth saw Louis, she would "always dig at him a bit," recalled one of the Tigoni workers. Perhaps because of this and his own busy schedule, Louis seldom visited Tigoni (although he constantly worried about it), adding to Booth's frustration. As a consequence, Tigoni was something of an orphan. For the first eight years, the U.S. National Institutes of Health supported Tigoni with generous grants. But when nothing was published, the NIH withdrew its funding, leaving Louis with another project to finance through his lectures.

There were, however, a few people who did receive from Louis the guidance and moral support that others longed for. At the top of Louis's list were his star protégées, Jane Goodall and Dian Fossey. Although he never visited them in spite of repeated invitations, he did write hundreds of letters, offering advice on everything from the importance of boiling their drinking water and analyzing ape dung to how to prepare for major lectures. Whatever they asked for—binders, typing paper, binoculars, work gloves and boots—Louis bought, often making the purchases himself. "We'd ask Louis from time to time to buy things for us in Nairobi," recalled Hugo van Lawick, then Jane Goodall's husband. "And we were absolutely horrified to discover that he was doing all this himself; we'd always assumed he sent someone from his Centre. He wasn't in the best of health and he'd be almost on the trot doing this shopping." When Jane and Hugo decided that they would no longer ask Louis for anything, "he got terribly upset," said van Lawick. "He thought we were trying to pull away from him."

Perhaps because she was "a lone woman" as she wrote Louis at one point, Dian never suggested that he might be doing too much for her. She had also suffered greatly in her first year observing the mountain gorillas, and needed his assistance. Unlike the sunny, lakeside home of the chimpanzees where Jane worked, Dian's study area at Kabara in the Congo (now Zaire) was high, dense rainforest: a cold, misty, dripping environment with thickets of stinging nettles and gorilla trails that seemed only to lead uphill. But as depressing and arduous as the weather and terrain might be, Dian had pressed ahead with her work—until one morning in early July 1967, when a party of armed Congolese park soldiers appeared at her camp and escorted her off the mountain, saying she had to leave for her own protection. A few days before, white mercenaries serving the rebel leader Moise Tshombé had attacked the eastern Congo. President Joseph Désiré Mobutu had then broadcast a radio message warning that white foreigners were trying to take over the country—an announcement that many Congolese interpreted as an order to kill all whites.

Dian knew nothing of the radio broadcast; she was simply upset that she had been forced off her mountain only six months into her study, and wanted nothing more than to return. In Nairobi, however, Louis had a far better understanding of events in the Congo. He had read about Mobutu's message and then learned that the Congolese army had shot down two civilian planes from East Africa. The situation was not good. "Frankly, I am exceedingly worried," he wrote Dian on the very day that she was being escorted from her camp, "and have been in touch with the American Ambassador here several times today." Unaware that she had been removed from Kabara, he had already investigated sending a helicopter to rescue her.

Dian's situation quickly worsened—although the exact course of events remains unclear. She herself left differing accounts—a simplified version in *Gorillas in the Mist* and more straightforward reports in her letters to Louis. Later, when she was free, Dian told Louis more lurid, unpleasant details. Detained for sixteen days, apparently she had at first been more or less free to come and go as she chose; she had even left the Congo for Uganda at one point. But she insisted on returning, determined that she would go back to Kabara and resume her gorilla study. Instead, Dian found herself "earmarked" for a Congolese general and locked inside a cage, where she was kept on public display for two days. During this time, she was spat and urinated on, and may have been raped.[3]

Louis tried to help, at one point sending a plane to Uganda to pick her up, but Dian had returned to the Congo before it arrived. He cabled her money as well when she asked for this, and continued to press the American Embassy and State Department for help. In the end, Dian escaped on her own, making a mad dash in her car across the Uganda border to the Traveler's Rest inn of Walter Baumgartel, the hotelier who had first contacted Louis about the gorillas. Though Dian feared that Louis would consider her a failure because she had abandoned her study, he was impressed with her determination, cool head, and great courage. "She has had a pretty thin time, but she is completely undaunted," Louis wrote proudly to Melvin Payne at the National Geographic Society, adding that all Dian wanted to do was "proceed forthwith up the mountain from the Rwanda side, and make contact [with the gorillas] again." Louis, however, had insisted that she come to Nairobi for a rest and discussions. He sent a chartered plane to pick her up, and Dian climbed on board.

It may have been during the five days he devoted to Dian's recovery that Louis first felt the stirrings of love for her. They were, as he would later say in a love letter to her, "kindred spirits in so many ways." Louis liked Dian's pluck and courage, and readily engaged in a "bellowing" argument on her behalf with American Embassy officials who did not want her to return to the gorillas. Louis was attentive and kind, too, taking her out to dinner and on game drives in Nairobi National Park, and praising her "magnificent work." He knew that she had worked hard, and never doubted that she would succeed in habituating the gorillas, even if it meant starting over. To

[3] Biruté Galdikas said that both Dian and Louis had told her that Dian had been raped. "She told me she had been held in a cage and raped. Also, Louis Leakey told me twice that Dian had been raped —once during a conversation when he said, 'She's been through so much.'" Another friend, however, said that while Dian experienced the emotional equivalent of being raped, she had escaped the act itself by screaming at her tormentors, "'You don't have the balls to rape me!'"

this end, he persuaded the Wilkie Brothers Foundation and the New York Zoological Society to renew their grants, and finally sent Dian off to Rwanda in mid-August, equipped with new tents, bedding, and camping gear—and bolstered in spirit by Louis's admiring faith in her.[4]

Though another tragedy—this time the brutal murder of three of her Belgian friends—temporarily halted Dian's efforts to return to the gorillas, by mid-September she had settled into Rwanda's Parc National des Volcans, just across the border from her old study haunts. Again, she was camped high in the mountains, this time on a grassy saddle between the peaks of Karisimbi and Visoke. From their names, she invented a new one, Karisoke, for her research center. Almost immediately after setting up camp, Dian heard the distant sound of a gorilla beating its chest, and the next day found gorilla tracks and trails everywhere. "You will be very happy to know that I've found a utopia—not only for the gorilla but for me as well," she wrote in elation to Louis. "Not only is this area teeming with gorilla, it is beautiful beyond description." The only concern she raised was financial—supplies in Rwanda cost twice as much as those in the Congo. She did not, however, mention the cattle herders or poachers she had encountered on her first day in the park. Nor did she say that she had already taken it upon herself to destroy every poacher's snare she found.[5]

Louis was naturally "delighted" that she had discovered "a good place with plenty of gorilla." He advised her not to bother about the expense of her supplies: "[A]s long as the work is being done, you must not worry about that at all." He would attend to the problem of funding just as he had always done for all of his projects and protégés.

Two years would pass before Louis saw Dian again. During that time, they continued to exchange letters nearly every week, with hers relating both her frustrations and successes with the gorillas. The almost constant rain depressed her at times, but by January 1969 she had once again established close contact with the gorillas. "MEMBERS OF GROUP 8 NOW COMING WHEN CALLED SHARING TREE WITH ME EXAMINING EQUIPMENT AND FOLLOWING AFTER CONTACT," she elatedly cabled Louis. For his part, Louis played the role of confidant and adviser as he had done with Jane, though his schedule and a hip operation prevented him from being as attentive as Dian desired. The money from the National Geographic seldom arrived on time, and her repeated requests for an assistant to help with a gorilla survey seemed to go unheeded. She thought it best if she came to Nairobi to discuss her problems with Louis. The altitude and her poor diet were taking their toll as

[4] Dian's bravery did, however, worry Fairfield Osborne, the president of the New York Zoological Society, who wrote to Louis, "What new breed of human being is this? These young women go out alone to far places, obviously relishing the risks involved. . . . Do you think they are trying to prove they are better than men?" Louis did not respond to this query.

[5] Though intended for small antelopes, these wire snares sometimes caught gorillas, causing cuts, infections, and often gangrene.

well, and by the summer of 1969, she was convinced that she had contracted tuberculosis.

"I shall be most glad to help in any way I can," Louis responded. "As far as I can see, I shall be able to put you up at the house, instead of having to stay at the Devon [Hotel], . . . and then we can talk in the evening, and by day doubtless you will have things to do and so shall I. Mary will be at Olduvai as far as I know." He set aside the week of October 8 for Dian's visit.

It was only after Dian arrived in Nairobi that Louis learned of her lung condition, and he immediately put her in touch with a specialist.[6] Dian had a series of tests done at a hospital, and then instead of tending to their separate affairs as Louis had said they would in his letter, he took her on a game-watching safari. Perhaps he wanted nothing more than to give her a complete rest, to take her mind off her many worries—the poachers; the inefficient park staff; the baby gorillas that had been taken from her care by a Belgian zoo; the American student whom Louis had hired to help with the gorilla census, who then proved an absolute failure—but at some point on their safari, Louis and Dian became lovers.

Overweight and physically disabled, Louis was for all practical purposes separated from Mary. He even admitted to close friends that he had failed in his marriage, and was consequently looking for someone to give him a wife's care and understanding. But although Louis was often surrounded by women who found his energy and charisma as magnetic as ever, and although he liked to give the impression that there was more to some of these relationships than met the eye, at this point Louis had little physical intimacy in his life.[7] Dian's warmth and affection were thus unexpected and overpowering, and Louis responded. They separated on October 17, when Louis had to leave for the States via London. He sent her three letters that day.

> Dian my dearest, dearest love . . . Dian my most beloved. I just HATE going off and leaving you like this. I care so much and in such a short time we have come to *belong* so much to each other and to share so

---

[6] Dian did not have tuberculosis as she feared, but rather emphysema.

[7] Only seven months before his affair with Dian, Louis had fallen in love with Judy Strong Castel, a young, vivacious secretary at the National Geographic Society. He brought her to Kenya, ostensibly to work as his personal assistant, but in reality he hoped she would be his mistress. All he wanted, he told her in a love letter, was for her to "care for me . . . [not] because I'm 'Dr. Leakey' but because I'm a simple man who wants life to be happy & content [with] a shareing *[sic]* & real understanding." As he had Jane Goodall, Louis gave Castel the Kikuyu name Mwendwa, for "My most beloved." But while Castel was happy to be Louis's confidante and later said he had changed her life, she had no intention of becoming his mistress. "Sometimes he would hold me in his arms, and he'd plead with me to move in with him, but I just couldn't love him physically. He was missing too many teeth." Louis apparently gave up his pursuit of Castel after she had accompanied him to his beach house, then left him to babysit for her young daughter while she partied with friends her age. Her behavior had "emphasized the generation gap," Louis wrote in an angry 3:00 A.M. note to her. "I suppose it was meant to." Castel advised other young women who asked her how they might get a job with Louis, "The man is always looking for a girl. If you want to work for him, sleep with him. He desperately needs affection."

much. I'll be writing (and phoning) soon. I love and love and love and love every bit of you and all you stand for and are. Take care of yourself.
Louis

My dearest love, I am so worried of having to leave you at this time of sad terrible worry for you [wondering if she had tuberculosis] and I'll be thinking of you all the time. . . . I love you and love you so and there are no words that can describe the peace and calm—happiness that are springing from our love of each other. I know you love and love most deeply too. Get well quickly and take care of yourself insofar as that is possible. My love I love you so and give myself to you. Louis

Dearest love in such a short short time our love was born and grew and matured into something deep and secure—however much you try to pretend its [*sic*] only temporary. Which it is *Not,* Not, Not. Last night you made me so utterly happy as you suddenly accepted that you too loved me. And last night with you peacefully asleep in my arms . . . was bliss. . . . What a heavenly week we had with so many lovely memories to share while we are apart and for all the future. Take care of yourself my dearest one. I love you so. Louis

Louis continued to pour out his heart to Dian, but waited in vain for a reply. "I have such lovely memories shared with you—sunsets, lion, cheetah, clouds, that last night at Langata [Louis and Mary's home] and on and on. I *love* you so. . . . I must stop now. I've got to get ready to go to the airport to fly to Los Angeles where I'll hear from you." But she did not write or call him, while Louis wrote almost every day. He had a ruby for her he told her at one point, and needed to know her ring size so that he could have a special ring made. "I want you to have it soon as a deep token of my caring."

But even this elicited no response, and when Louis returned to Nairobi in mid-November, he had yet to hear from Dian. ". . . I've had no news of you *at all* since I spoke with you on the phone just before I left," he wrote rather bitterly. "I've written four or five letters to you and only hope one or two of them reached you. There is an East African Airways flight this evening. I wonder if it will bring me a letter from you. I shall keep this until I see tomorrow's mail. . . . [But there was no letter, and Louis added another plea to this letter.] Darling love, your last words to me on the phone were *I love you too,* so I'm hoping to hear from you soon, soon, soon."

Not until December 7, almost two months after they had parted in Nairobi, did Louis hear from Dian, and then she sent a strictly businesslike note along with her field report. Perhaps it was Dian's way of trying to return their relationship to its previous status, but Louis refused to acknowledge the apparent one-sidedness of their affair. "Dearest Dian," he replied at once, "Thank you so much for the 'official' letter of Dec. 7th with the copy of your report. . . . I was most glad to have all the news, even if I would also have welcomed a personal letter to me, my loved one, however brief. . . . Darling I love you so much. I know you love me too. . . . Louis."

Finally, just before Christmas, Louis received a warm, cheerful letter from Dian, yet its tone was decidedly newsy and could hardly be mistaken as a love letter. Louis, however, responded as if it were. "Just as I was leaving this morning I got your lovely letter of Dec 11th. . . . I love you, just quite simply, deeply, whether you are miles and miles away or very very close. . . ." He knew Dian would soon be leaving for Cambridge University to begin her doctoral studies, and he hoped to see her there. "If plans go well I'll be up there [at Cambridge] in February (for 1 night I hope) and again in March. We *might even fly* back [to East Africa] together. . . . Dian my dearest, I'll want to see you in Cambridge, let me know your address there as soon as possible. My heart and mind and thoughts are with you and will be."

Whatever Dian thought about Louis's ardent passion for her, she kept to herself, but the coolness of her own response spoke volumes—even if Louis was unwilling to listen. Once, after yet another love letter arrived from him, she wrote in her diary, "Don't know what to do about L. God—what a mess." Perhaps she found this desperate neediness in a man she had once idolized embarrassing, or perhaps she had nothing to offer in return, having fallen in love herself with Bob Campbell, the photographer the National Geographic Society had sent to Karisoke to film her work with the gorillas. But regardless of how she responded, Louis's love for her did not cool.

In early January 1970, Dian moved to Cambridge, and to Louis's delight, she once again sent a warm letter, telling him how much she hated the cold and crowds of England, and how she longed for her mountain home. As always, her letter sent Louis's spirits and hopes soaring and he wrote back ecstatically, "At last I have news from you—I'm so grateful to you for writing when experiencing such misery and purgatory. I feel for you so much. Yet I know you must go through with some part of it [the doctoral program]." He then mentioned his upcoming trip to England again, and said, "I'll phone you as soon as ever I can after arriving in London and make a plan to meet you. . . . It looks as though I'll arrive either on Feb. 5th or 6th. . . ." They would have dinner and a "quiet talk" together one evening.

Louis wrote this letter to Dian on January 30; the next day he drove to Olduvai where he "simply had to get . . . things done," as he later told his staff at Tigoni. There he apparently felt some chest pains and nausea, which he interpreted as a sign of indigestion. Instead of seeing a physician, Louis then drove the three hundred miles back to Nairobi in his usual nonstop fashion, and boarded his plane for London on February 5. During the flight, the pains returned, and as soon as his plane touched down, Louis, feeling that he was about to collapse, called for an ambulance. He refused to go to the hospital, however, and insisted on being taken to Vanne Goodall's flat. She summoned a doctor, who promptly put him in the hospital. Louis had suffered a mild heart attack at Olduvai, the doctor determined, and shortly after checking into the hospital, Louis had another, very severe coronary. Six months would pass before he was on his feet again.

Dian wrote to Louis immediately, offering her care and concern. But she did not visit him until March 3, almost a month after his attack, although she was only ninety minutes away by train in Cambridge.

Chapter 25

FATHER AND SON

Louis remained on the critical list at London's Princess Beatrice Hospital for five days following his heart attack. His physician prescribed three weeks of bed rest and said Louis could be expected to make a full recovery if he avoided all "agitation, anxiety and anger"—a nearly impossible prescription for Louis.

As soon as he could, Louis had a phone installed at his bedside. When not on the phone, he was either dictating letters or seeing visitors—all in an effort to continue directing everything from his Centre for Prehistory and Palaeontology in Nairobi to the L. S. B. Leakey Foundation in Pasadena, California. Any suggestion—no matter from whom or how carefully worded—that he temporarily put aside some of these tasks only triggered his temper. "[A]t present, I am not to write or move," he told Tynka Robertson, his secretary in Nairobi, whom he suspected of holding back his mail. But, he added testily, his doctors' orders did not mean that she was to keep problems from him. "[I]t is *much more worrying* for me to be kept in the dark about things, than to know what is happening, & then to sort out what must be dealt with." She must send *all* of his mail, in other words.

Louis sent similar appeals to Joan Travis at the Leakey Foundation and to Leonard Carmichael at the National Geographic Society, insisting that he be kept fully informed about all of their decisions. He fretted that he had let down the foundation by missing its speaking engagements, and worried about how he would find the money needed for the excavations at Calico, to keep his Tigoni Primate Research Centre running, and to send Biruté Galdikas to study the orangutans. Though Travis tried to soothe Louis, assuring him that he was *"not* alone," but "surrounded by people who care *deeply* and who are all working together to lighten your load," Louis continued to despair. Only a few months earlier, he had failed to persuade the National Institutes of Health to renew its grant to Tigoni, and had only been able to obtain half of his usual funds from the National Science Foundation for his Centre for Prehistory. Now he would also be short the monies he had anticipated raising from his lecture tour in the States.

In mid-February, two weeks after his collapse, Louis received another shock. Leonard Carmichael, in an otherwise cheerily sympathetic letter, announced that the National Geographic Society was cutting its support of Mary's Olduvai Gorge excavations from $48,440 to $40,000. And this time, the money would be paid directly to Mary rather than to Louis, presumably because of Louis's illness and "the many demands on [his] time and energy. . . . You will be interested to know," Carmichael continued, hoping to soften these blows, "that, with the grants just authorized, the Society has now made more than one million dollars available for your projects and those of your associates [over the preceding ten years]."[1]

The reduction of the Olduvai grant crushed Louis. He and the directors of the Geographic—Leonard Carmichael, Melvin Payne, and Melville Grosvenor—had always had a close friendship, and Louis did not hesitate to draw on this now. "Forgive me, dear Leonard," Louis wrote at the end of a nine-page, single-spaced letter he dictated to Vanne Goodall from his hospital bed, "for writing to you at such length and in this vein, but I felt that for the record certain things should be absolutely clear." In particular, Louis felt compelled to recount his and Mary's major achievements since 1960, when the Geographic first began to support them. Louis did not complain about the money being paid directly to Mary, except to note that it would be "exceedingly difficult for her to operate [the account] from Olduvai," and he hoped that the $40,000 did not include the $9,000 salary that the Geographic usually paid him.

Apparently, Louis's sickbed plea had some effect. Carmichael responded that the Geographic would indeed continue to pay his salary, and hoped that "this action . . . will serve to relieve somewhat your anxiety about finances." Yet the $9,000 was only a temporary remedy, for Louis's projects were a financial disaster. He had accounts upon accounts and was forever borrowing from one to cover a deficit in another. Louis's affairs were hopelessly muddled. Richard, who had visited Louis in the hospital a day after his heart attack and who was now trying to untangle Louis's financial mess in Nairobi, wrote, "You explained to me at some length the circumstances which led to your unfortunate illness recently, and quite frankly circumstances at present are such that I might well suffer from the same illness."

---

[1] The Geographic was also giving $2,000 to Richard Hay for a geological survey of Olduvai Gorge; $12,740 to Dian Fossey for her gorilla study; $25,000 to Richard Leakey for his fossil-hunting expedition to Lake Rudolf; and $25,000 to Jane Goodall for her chimpanzee research.

Some of his comments might "seem harsh," Richard continued, but he hoped that Louis would "not take them this way." However, father and son had been quarreling over Louis's Centre for Prehistory for the past year and a half, and it was inevitable that Richard's intervention would rekindle their argument. Louis had in fact told Richard as much during his son's visit. "He looked shaken, frightened—hurt," recalled Richard, "and he said, Did I realize how much stress and strain I was causing him? He gave the impression that he was there [in the hospital] because of my actions." Richard had gently disagreed with his father, suggesting that his heart attack had more likely been triggered by his "desperate scurrying around for money." But Louis seemed "totally irrational and paranoid," Richard said, "so it was no good defending myself." Louis wanted Richard to look after his prehistory center in his absence, as Richard often did, but he did not completely trust his son—nor anyone else, for that matter. "He was very dominant about everything, and enormously hopeless about relying on other people. He didn't trust them to be loyal to him," Richard said later.

Two months before his father's collapse, Richard had turned twenty-five. Lean, angular, and intense, he had focused all of his ambitions on becoming administrative director of Kenya's National Museum, a position he finally attained in October 1968. Theoretically, he was in charge only of the museum's administration, while the director, Robert Carcasson, oversaw its scientific affairs. But, as Richard wrote in his autobiography, "I knew that whoever was in charge of the finances was in fact in control of everything and once the Director left I would be completely in charge."[2] From the outset, Richard acted as if he already were.

To make his presence known, Richard first had the museum's dusty-red exterior washed (though he had no authority to do so, Carcasson angrily informed him), and then contrived to slow traffic on the museum grounds by building speed bumps. These were too large, and Richard's first victim happened to be Carcasson, whose small car was stranded in mid-bump. "I was not at all popular," Richard wrote in his autobiography, "and had to agree to keep in closer contact with the Director."

But Richard also had substantive successes. Only three months after taking office, he persuaded Perez Olindo, the first Kenyan director of Kenya's national parks, to transfer to the museum control of the country's archeological monuments and paleontological sites.[3] This not only increased the museum's prestige but added to its coffers as well, since the government provided subsidies to run the sites. At the same time, the permanent secretary in the ministry of natural resources, Joel Ojal, a powerful figure in the young Kenyan government and a staunch ally of Richard's, asked him to

---

[2] Carcasson had resigned in protest when Richard was appointed, and only remained through 1970 to complete his contract.

[3] These included the coastal sites of Fort Jesus, a fifteenth-century Portuguese fortress at Mombasa, and Gedi, the ruins of an eighteenth-century town settled by Arab traders, as well as several archeological sites Louis and Mary had discovered: Olorgesailie, Kariandusi, and Hyrax Hill.

prepare a five-year plan for the museum. Richard proposed to replace the existing small, dowdy collection of buildings with a large complex modeled on the Smithsonian Institution. The new National Museum would be internationally recognized, be staffed with Kenyans, play a key educational role in the country, have regional branches throughout Kenya, and oversee all archeological and paleontological affairs. It was a grand scheme, and it put Richard once again at odds with his father.

Louis then controlled Kenya's archeology and paleontology through his Centre for Prehistory and Palaeontology. Though situated on the National Museum grounds, the Centre was run by Louis as if it were an independent entity, using funds he raised overseas. In his Centre, Louis housed the great fossil and archeological collections from Olduvai, as well as the Miocene ape fossils from western Kenya, and the new hominid fossils Richard had discovered at Lake Natron, the Omo River, and Lake Rudolph. Louis decreed who could study this material and recommended who should receive research and excavation permits for Kenya as well, giving him great power not only within his country but within the international scientific community. He could deny entry to researchers for the flimsiest of reasons, and had done so.

"I remember pleading with Louis on one occasion to renew the excavation permit (which he had refused) of a very able but very brash young American archaeologist who had got on the wrong side of Louis and was thus being penalised," recalled Jean Brown, who worked at the museum. "The young man in question was much too new and much too brash to show the father of Kenya archaeology the required deference."

John Karmali, who in 1969 was one of the newly appointed trustees of the museum board, recalled that "the Centre was like a little private empire of his [Louis's] own, and Louis hated giving up that power. He was constantly fighting a tremendous rear-guard action with Richard and us, the new trustees."[4]

Richard had first suggested in February 1969 that Louis might consider merging the Centre and its department of osteology with the museum. But father and son were immediately at odds. For Richard, the issue was one of nationalism, of building the little museum into a grand institution; but for Louis, the museum's reputation, and that of the Centre, rested solely on what he, L. S. B. Leakey, had achieved. The Centre existed because Louis existed. "Father always used to say, 'These things have nothing to do with the museum; the museum is quite separate,'" Richard recalled. "And I kept saying, 'But they *should* have to do with the museum. The strength of its future lies in bringing everything under one place.'" Richard argued as well that such a merger would give the "Centre a more national flavour," and

---

[4] To help him implement changes at the museum, Richard had asked Joel Ojal to appoint several Kenyan trustees to the board, which had previously been dominated by elderly colonial gentlemen. With a majority of sympathetic trustees on his side, Richard was able to push through his new plans.

that as a result the Kenyan government would be more apt to support it, freeing Louis of some of his financial burden.[5] But Louis disagreed.

Richard was further irritated, and embarrassed, when Louis continued to employ an expatriate British woman as his secretary at the Centre rather than a Kenyan. His irritation turned to anger when he discovered that F.D. Castings, a London-based firm that Louis had hired to produce fiberglass casts of the East African hominid fossils, was actually owned and operated by Judy Goodall, Jane's sister. In fact, Louis had helped Judy start the firm, whose acronym stood for "Foster Daughter," the affectionate term she used when signing her letters to Louis.

"I had wanted to get a casting program going at the museum to help us make money," Richard said, "and my father was bitterly opposed to the idea on the grounds that we weren't capable of doing it. Then, when I discovered that the Goodalls were involved, I became extremely upset because it seemed that Judy's casting program was in direct competition with my own endeavors at the museum. And Louis seemed to be siding with her against Kenya and me." Jealous and angry, Richard had his own lab technicians trained to make casts. When their casts rivaled those of Judy's, Richard proudly showed them to his father. "He was furious for a while and probably deeply hurt, but that made no difference to me then," Richard wrote in his autobiography. "Once more, I was doing the right thing but largely for the wrong reasons; in other words, I was acting at least partly out of spite even though the programme I initiated did benefit the Museum and Kenya."

Father and son quarreled again when the government, at Richard's request, revised the Antiquities Law, tightening the rules for engaging in research and excavation in Kenya. Essentially, the new law required scientists to fill out formal applications, rather than writing to Louis. These were sent to the Office of the President, then forwarded to Richard's desk for approval. "It was simply a channel," Richard said about this misunderstanding. "But instead of going to Louis as the sort of superstar of paleoanthropology, the applications were coming to me as director of the museum. I think he took considerable exception to that, and saw it as a deliberate subversion of his position. In fact, it was undertaken with a view to try to establish an organized procedure that superseded personalities, and that provided some regularity and stability."

While Richard professed that nationalism, organization, and efficiency were his priorities, to onlookers it appeared that he merely wished to "stomp on his father," as one visiting researcher to the museum observed. "Most of us who were there then [at the museum in 1969 and 1970] didn't

[5] When Richard was appointed administrative director, the Kenya government provided approximately $45,000 for the museum's budget. In his five-year plan, Richard proposed increasing this to two million dollars, "a preposterously high budget when put against budgets in previous years," he noted in his autobiography, "but I was not particularly concerned with such comparisons." He succeeded in obtaining about half this sum from the government, and the other half from donations and such marketing enterprises as selling fiberglass casts of fossils and operating museum shops.

know that the museum and Centre for Prehistory were separate units, and it looked to all of us like Richard was trying to take over Louis's job," said M. E. Morbeck, then a graduate student in physical anthropology. "And Richard was often rather rude about his father, saying things like, 'He doesn't know anything about such-and-such,' usually some administrative matter. There was no question that it was a power play, and there was a lot of tension because of it, and all I wanted was not to get caught in the cross-fire."

The tension was exacerbated by the location of Louis's and Richard's offices. Both were housed in Louis's Centre for Prehistory, a stuccoed, mustard-colored building, with Louis's office on the ground floor and Richard's on the second. Louis, hobbling on crutches, would come puffing into his office early every morning, usually arriving minutes after his son, who made a point of being at work by 6:00 A.M. Sometimes father and son conversed genially enough, but other times, at the height of their quarrels, they refused to talk to each other. "Louis would yell at his secretary, who would yell to Richard's secretary, who would then give the message to Richard," said Morbeck. "Then Richard's response would be shouted back to Louis through the secretaries. And occasionally we would hear Richard and Louis yell back and forth, just from window to window." Once, in a choleric moment, Louis had all the locks changed on his offices to keep Richard out. "The first time you hear something like that, you think you've ended up in South Kookyland," commented one researcher. Father and son were equally quick-tempered at the museum board meetings, often making very direct, tactless remarks to each other. "It was very tough times," said Thomas Odhiambo, then the chairman of the board. "And that was part of the business of being chairman, to make sure that the father and the son were not sitting close to each other."

Although the museum's trustees held Louis in high regard, they nevertheless sided with Richard in his vision for the museum. "In that period, the main thing was to bring the museum into the research field," said Odhiambo. "To change it from a family affair to a national research institution. And we went a long ways toward that." In December 1969 the board agreed, and Louis reluctantly accepted their decision, to merge the Centre for Prehistory and Palaeontology with the National Museum. The merger would happen gradually, over the course of several years, and Louis would retain his position as honorary director. Still, when the merger was complete, Louis would in essence be working for his son.

By the middle of March 1970, Louis was well on his way to recovery from his heart attack. He had been released from the hospital and was recuperating at Vanne Goodall's flat in York Mansions, a five-story, red-brick building near the British Museum of Natural History. With its sunny sitting room, worn, overstuffed furniture, and grand piano, the flat had an inviting air. Vanne

Goodall's warmth and good humor, too, were a tonic for Louis. "I think that 1 York Mansions is the next best thing to being in Africa," Dian Fossey, who was studying at Cambridge University, wrote in a chatty note to Louis after visiting him at Vanne's during his convalescence. "[Y]ou forget to look out the windows in a home like that one, and I'm so pleased that you have such a home to recuperate in."

The anatomist Michael Day stopped in to visit as well, and reported to Mary that "[Louis] really begins to look his old self again, improved in colour but still taking life sensibly and easily." Louis was incorrigible, though, and did every bit as much work as his cardiologist would allow. He spent long hours every day—albeit propped up in an easy chair in the sitting room or lying in bed—dictating letters to Vanne, reviewing the draft of *Adam or Ape,* a collection of articles about evolution and paleoanthropology he had helped edit, and beginning the second volume of his autobiography, *By the Evidence.* He hoped to pay his medical expenses with this latter book.

But Louis's initial progress was suddenly cut short by another contretemps with Richard. Dian Fossey had just returned to Africa, and from there she wrote Louis that Richard was "interfering in gorilla matters." "[These] have *nothing whatsoever* to do with you or the centre," Louis angrily scribbled at the end of an otherwise calm letter to his son. Richard was "going beyond [his] powers," Louis added, and if he did not stop, Louis would fly back at once "against Doctor's orders." His fate would then be in Richard's hands. Richard, as if on cue, shot back an equally provocative letter, and father and son were at each other again.[6]

In fact, Louis was in no condition to fly anywhere. Shortly after hearing from Dian, Louis also received alarming letters from Juliet Ament, who ran his osteology department, and Rochelle Porter, one of his student protégées, whom he had planned to send to Rwanda to work with Dian. Richard had informed Porter that there were no funds in Louis's gorilla account to send her anywhere, so she was returning to the States, which greatly upset Louis. "I cannot adequately express to you the distress and worry I felt on your behalf over what has transpired," Louis wrote in anguish to Porter. "[I]t was entirely unjustifiable of my son Richard to act as he did. . . . Richard is authorised to make certain day to day decisions on my behalf [at the Centre], but not on major matters affecting staff . . . and I cannot imagine what has got into him to have done so." But like a cold-souled accountant, Richard was eliminating his father's projects and people wherever funds were short. Thus he had told Ament that "money is so scarce . . . it is necessary to close Osteology . . . Richard said [and] I quote," Ament continued in her letter to

---

[6] In a subsequent letter to Louis, Richard explained the problems more fully: "On the gorilla expenses that I asked Dian to settle: you say that had these bills been sent to you, you would have paid them. First, there is no money here for this and secondly these bills go back to October/ November and you have sat on them. Several statements have your writing on them with 'What is this?' or 'To be paid from Gorilla.' and yet nothing was done."

Louis, " '. . . in his (your) absence the affairs here are my responsibility.' " Other students and employees telephoned to pass along additional "bad Richard" tales, and these so upset Louis that he flew into a rage and collapsed in bed, whereupon Vanne summoned the doctor.

A week later, after regaining his strength, Louis dictated a four-page confidential letter to Richard, which Vanne dutifully transcribed. "Since writing my letter of March 25th," he told his son, "I have had two very severe shocks. In fact they arrived simultaneously and hit me so hard that I had another minor heart attack, and we had to send for the specialist again. I do wish you would realise that for a person suffering from my particular heart condition, mental shocks and worries of the kind you are causing me can do every bit as much harm as running up a hill, walking too far or trying to lift something heavy."[7]

Louis expressed his outrage over Richard's interference in his affairs in no uncertain terms. He was especially livid that his son had closed the osteology department ("You must surely realise that this is my very special 'baby' "), and that Richard had examined the gorilla account. "[A]rrangements concerning gorilla studies are entirely my affair, and not yours," Louis insisted. "It is infinitely more likely to cause a really major further setback in my health, if you do not consult me over such things, than if you do consult me—and my specialist confirms this. In fact he is going to write to you stressing this."[8]

Richard received his father's letter and the promised one from Dr. J. L. Grant on the same day. In response, Richard cabled Louis at once, trying to assure him of his good intentions, and promising to send bank statements and documents illustrating the financial problems at the Centre. But to Dr. Grant, Richard was openly hostile. "I have received your letter . . . and I am quite frankly amazed," he began. "Nobody could be more concerned about my father's health than I am and I believe that one of the greatest contributing factors to his recent illness was the fact that nobody had been prepared to curtail the ever increasing amount of activities that he is involved in." Richard was certain that unless these were reduced his father would "suffer a major heart attack or similar demise within a month of his return to Nairobi. The magnitude of the financial problems are [*sic*] beyond description. . . . I am not doing anything more than endeavouring to help," Richard continued, "and it is my belief that everything should be done to prevent a tragedy that will surely come."

For a while after this exchange of letters, father and son drew back from the threat-counterthreat of heart attacks and tried to discuss matters more regularly on the telephone as well as by mail. But misunderstandings continued to crop up, and inevitably Louis felt cut out of the picture. He was always

---

[7] Actually, Louis's physician determined that he had not had another heart attack, though because of the "degree of physical disturbance . . . undoubtedly it had been a possibility."

[8] Louis wrote similar letters to his protégés, accusing Richard of giving him another heart attack. These letters would later lead to rumors that Richard was responsible for his father's death.

"kept in the dark," he told Rochelle Porter, and no longer knew how to get his son's attention.

Louis appealed to Mary as well, but she wisely refrained from taking sides in these father-son disputes, and tended instead to her excavations at Olduvai. "I do not think Richard ever wished to topple Louis from his position," she commented twenty years later, "but he felt Louis had lost his grip on things and [Richard] wanted to do a more efficient job himself."

In early July 1970, Dr. Grant finally gave Louis permission to fly home, but only after admonishing him about the seriousness of his illness. Louis could no longer expect to work as he had in the past, but must rest every day and lose weight as well. As for the short-term memory loss Louis was experiencing (and which greatly irritated him), this symptom might pass with time. Still, Louis's recovery had been remarkable, and Grant believed that it should continue if Louis followed his advice. "One of his [Dr. Grant's] other patients, who has exactly the same trouble as mine, has just gone ski-ing [*sic*] . . . and has resumed playing tennis," Louis reported optimistically to Leonard Carmichael at the National Geographic Society. "[Dr. Grant] sees no reason why I should not be fit to go up and down Olduvai Gorge and do any of my other normal activities, *provided I obey him*. . . ."

But Louis no longer had his old stamina, and lacked the heart to keep fighting with his son. "Louis was very fond of Richard in spite of the jealousy between them," commented Jean Brown. "He wanted a son to follow in his footsteps but not to eclipse him." Richard's star, however, was on the rise. With his administrative victories at the museum and tremendous successes in the field at Lake Rudolf, he was well on his way to consolidating his position as the new godfather of East African paleoanthropology.

Chapter 26

JACKPOT AT KOOBI FORA

"I was always struck by Richard's extraordinary drive," said Bill Richards, a Los Angeles businessman and trustee of the Leakey Foundation. "His determination and energy were right off the scale, way past the redline on the achievement meter, and I remember thinking many times, 'What is it that makes Richard run?' "

Many other friends—even family members—wondered about Richard's drive as well. Though Louis had always been in a hurry, compelled to launch project after project, he did not have the almost demonic urge that seemed to push Richard. Some colleagues thought an Oedipal need to topple his father fueled Richard's ambition; others, that the young man simply sought the glory of international fame. What no one suspected, and what Richard never divulged, was that in March of 1969 he had been diagnosed with a terminal kidney disease. Richard was then twenty-four, and the best prognosis his physicians could offer was that he had between six months and ten years to live.

The disease itself had struck Richard suddenly in late September 1968, three days before he was to begin his job as administrative director at the National Museum. One morning he woke with a painful sore throat and fever; the next day, his face was flushed and puffy, a certain sign of fluid retention, and he sought the advice of a doctor. Richard's throat infection, the doctor advised, had spread to his kidneys and unless the young man rested for six weeks, the disease would ultimately destroy them. Stunned, but disbelieving, Richard chose to ignore the doctor's advice. Besides, his new job—which he had fought so hard for—began in two days' time; if he did not show up for six weeks, the museum board would surely withdraw its offer. "The job was essential to me and crucial to my ambitions and I could not possibly consider the prospect of losing it," Richard wrote in his autobiography. "Consequently, I spent only a morning in bed and thereafter went on with my normal activities. Apart from a dull pain in my back, the obvious signs of kidney problems soon disappeared and I was at my desk as planned on 1 October."

But the doctor's warning continued to nag at Richard, and en route to the United States in March 1969, he visited a kidney specialist in London. The physician confirmed the earlier diagnosis: Richard could expect his kidneys to deteriorate until they failed completely. Richard would then either have to have a transplant or be put on dialysis for the remainder of his life. "This time I believed the doctor and I was a little frightened," Richard wrote. He was also depressed, but he quickly resolved that although the disease was incurable he would put it "entirely out of [his] mind and . . . live a full and normal life as long as possible." He would tell no one—other than his wife, Margaret—about his condition; nor would he follow the doctor's special diet or have periodic medical checkups. He was not an invalid, he decided, and he had no intention of living like one.

With the doctor's diagnosis secreted away, Richard flew on to Washington, D.C., and a meeting at the National Geographic Society. He had been invited to discuss his upcoming expedition to Lake Rudolf and to present his second lecture with his father at Constitution Hall.[1] Three thousand people filled the hall that night, and as he approached the podium, Richard's hands broke into a cold sweat. "Boy, I was nervous," he recalled twenty years later. (Leonard Carmichael's stage introduction—"He is one of the youngest-looking [museum] directors I've ever seen"—did little to bolster his ego.) But Richard's voice never quavered as he narrated a film from his 1967 Omo River adventures. In fact, once his stage fright passed, Richard found the audience's attentiveness and applause to be thrilling.

Richard's polish impressed the Society's Research and Exploration Committee as well. Carmichael and Melville Grosvenor, the Society's president, had been stern with him the preceding year when Richard requested funds to explore Rudolf instead of the Omo. That gamble had paid off—Richard's team had produced three partial australopithecine jawbones—and the committee awarded him another $25,000 grant for 1969. As he had vowed to himself, Richard never mentioned his kidney condition to anyone at the Geographic; besides, in its early stages, the disease did not appear to affect him. Tanned and vigorous, he neither looked nor felt ill, and when he set out for Rudolf at the end of May, no one suspected that behind his urgent ambition lay a deadline.

• • •

[1] Richard had made his first appearance at the hall in 1964, following his Lake Natron expedition.

On his exploratory expedition to Lake Rudolf in 1968,[2] Richard's team had discovered particularly promising fossil deposits north of a sandy spit of land called Koobi Fora. Richard and Kamoya had unintentionally walked to the end of the peninsula that year, and Richard had then decided that it would make an excellent base camp: with water on three sides, the camp would be cooled by the lake's steady breezes and, if bandits attacked, it could be easily defended. Freshwater would be a problem, though, because of the alkalinity of Lake Rudolf; the team would have to haul its drinking and cooking water from the Nderati Oasis, fifteen miles away over rough tracks.

Because of the limited water supply, Richard organized a small team for 1969. Once again, he invited his geochemist friend Paul Abell from the University of Rhode Island, and enlisted Kamoya Kimeu, Mwongela Muoka, and Peter Nzube as his chief hominid hunters. Richard had taken his father's advice about finding a geologist to begin mapping the deposits, and asked a young Harvard graduate student, Anna K. Behrensmeyer (or Kay, as she preferred to be called), to join his team.[3] Don Siegel, a geology student at Pennsylvania State University, would work as her assistant. A *National Geographic* photographer, Gordon Gahan, would also join the team, as would a young zoologist from Louis's Tigoni Primate Research Centre, Meave Epps. "I had hopes of encouraging her to become a palaeontologist," Richard wrote in his autobiography, as an explanation of Meave's presence. This was only a half-truth, for Richard and Meave were also in love.

At times, Richard's life has borne an uncanny resemblance to his father's, and his affair with Meave in some ways mirrored Louis's with Mary. Like Louis, Richard was married when he met Meave; and his wife, Margaret, had just given birth (to a daughter, Anna) as Louis's first wife, Frida, had done thirty-six years earlier. Anna's arrival provided Richard with a ready excuse for his colleagues who wondered why Margaret—the archeological member of his team—was not joining him in the field. Those who knew Richard well were not taken in by this explanation, however, for it was commonly known that his marriage was in shambles. From the beginning, Richard and Margaret had bickered in public, and the marriage struck many as a mismatch.

"Even at the end of their honeymoon [in 1966], they were really being rather nasty to each other," Paul Abell said about Richard and Margaret. "Whenever Margaret found Richard exaggerating or saying something she didn't agree with, she'd come down on him like a ton of bricks. Although she didn't use the term, it was the equivalent of saying 'Horseshit!' " Richard usually responded with an equally unpleasant remark, but he loathed domestic quarrels and would then let the matter drop.

Margaret's pregnancy with Anna in 1968 did not rescue the deteriorat-

---

[2] Lake Rudolf's colonial-era name was officially changed to Lake Turkana in 1974.

[3] Richard had heard from his father time and again how the lack of a geologist's map for the deposits at Kanam/Kanjera in 1932 had sabotaged his early career.

ing marriage. Perhaps because of her conventional upbringing in class-conscious England, Margaret did not approve of Richard's intention to pursue paleontology without a university degree—and told him so. She disliked the way he used his father's name to advance his career, and disparaged his appointment as administrative director at the museum. "Of course Richard thought he could do anything his father could do," Margaret said. "In fact, that was one of his problems: he thought he was as good as his father. He thought he was as good as anybody no matter whether they were kings or what—he was their equal. He couldn't quite see that his father had a kind of stature that he [Richard] simply didn't possess, and he couldn't really understand why he couldn't fill the same position."

The repeated arguments left Richard worried and frightened—they were all too reminiscent of his parents' bitter quarrels that had devastated him as a child. When Anna was born in March 1969, Richard decided that his marriage had to come to an end. "We were often snapping at each other," Richard said, "and I had this *horror* of Anna growing up in a house where there were going to be fights—and it was clearly going to be like that. And I simply wasn't able to comprehend being drawn into this nightmare of parental fighting again. It was terribly distressing. But I couldn't have a partner who didn't support me, didn't believe in me."

Richard had also met Meave, and was taken with her spirited mix of adventurousness and gentleness. "She was open, generous, reflexively kind, and made friends easily," said one friend, who also remarked on Meave's serious, tough-minded side. Spotting a roadkill in Kenya once, Meave stopped the car, picked up the dead animal, and "casually bunged it in the boot [trunk]" for a later dissecting session.

Slender and fair, with sparkling hazel eyes, Meave had first come to Kenya in 1965, after responding to an advertisement Louis ran in *The Times* of London. "I'd just finished a degree in zoology [at the University of Wales] and wanted to get a job in marine zoology," Meave said. "But at the time it was hard for women to get such work. No one would hire me, and I was getting really fed up. Then somebody told me about an advert in the back of *The Times* for a job in Kenya. There was a phone number, and I remember phoning from this phone box with these tiny little coins and not having nearly enough. Louis was on the other end of the phone, and he told me to come have an interview with him in London."

Meave met Louis at Vanne Goodall's flat, where he often conducted these interviews. She told him about her education, that she loved the out-of-doors, knew how to repair cars (at her university she'd maintained a 1936 Morris 8), and had raised lambs and Chinese goslings for extra money. She also intended to study for her doctorate. In response, Louis offered her a position at his Tigoni Primate Research Centre caring for the monkeys. He suggested that she might find a doctoral thesis subject there as well, perhaps analyzing the skeletons in the monkey collection. "Louis was so enthusiastic about everything," Meave recalled, "and he made the work sound exciting and interesting. And even though Tigoni didn't turn out to be what I had expected, that didn't bother me."

Following Louis's suggestion, Meave completed the research for her doctorate at Tigoni (she chose to study the limb bones of colobus monkeys), and after spending two years in Kenya, returned to the University of Wales to write her thesis. Ironically, she never met Richard during her initial stay in East Africa. "Everyone had told me, 'Oh, he's so dreadful; he's so awful,' " Meave said. "I hadn't any desire to meet him. That's all I heard those first two years: 'He's a dreadful fellow.' He was brash and eager to succeed, and he rubbed people the wrong way."

In the autumn of 1968, Meave returned to Tigoni at Louis's request. She had reluctantly agreed to serve as Tigoni's acting director for six months, at the end of which time Louis promised to send her on a field study of warthogs, sable antelopes, or lowland gorillas. But her visions of a Jane Goodall–like career soon vanished, for two weeks after her return, Richard summoned her to his office. Louis had just left for one of his fund-raising tours in the United States and, as usual, Tigoni was out of money.

"Louis's accountant had told me that there wasn't any money for Tigoni, and that these monkey people had to be very careful," said Richard. "So I called this girl, Meave Epps, and she came to my office and I said, 'You've really got to stop spending so much money'—which is exactly what I'd been told to tell her. And she got very annoyed and said she hadn't been spending any money; there wasn't any to spend. So that is how we met; it really didn't amount to very much."

Though irritated by Richard's "long lecture," as Meave described their first meeting, she also instinctively liked him. "He wasn't dreadful at all," she recalled with some surprise. "I really liked him, but because he was married, I didn't pursue it."

A few weeks later, Richard called Meave again, this time to ask if she could help him dissect some monkeys. Richard was then writing a description of the fossil skeleton of an extinct giant colobus monkey, *Paracolobus chemeroni*—from Lake Baringo—in part to demonstrate that he could publish scientific papers without a degree.[4] "I had very consciously decided not to write papers about the fossil hominids because of all the arguments among paleoanthropologists and because Louis was involved, and I had no desire to get into an area where we would be competing," Richard said about his decision to specialize in monkeys.[5] "There are only about six

[4] Richard had already published two papers on fossilized monkey skulls in the *Proceedings of the East African Natural History Society,* a publication of Kenya's National Museum. Louis had asked the British Museum of Natural History to publish one of Richard's papers in its "Fossil Mammals of East Africa" series, but was told that the series only included descriptions of fossils in the BMNH's collections.

[5] As part of this strategy, Richard gave the *Homo sapiens* skulls his team found on the 1967 Omo Expedition to Michael Day to describe, and the australopithecine jawbones from his 1968 Rudolf Expedition to Alan Walker and Jo Mungai.

people in the world who care about extinct monkeys, so it is very uncontentious, and you can write about them without a Ph.D." Needing a better grasp of primate anatomy in order to write about the *Paracolobus* skeleton, Richard turned to Meave.

"Initially, it was all perfectly academic," Richard said. "I wanted to learn how to dissect monkeys, and I thought because she'd done her thesis on colobus monkeys that she'd like to see this fossilized one. So we began to see a lot of each other because of this monkey, and I began to really enjoy being with someone who thought I was quite clever and on the right track."

Margaret, however, was not deceived. "They say that if a marriage is working, another woman arriving on the scene doesn't make any difference," she said. "But anyway, she [Meave] arrived. And Richard suddenly developed a sort of uncontrollable interest in cutting up monkeys. She was teaching him how to do this—sort of 'Come and see my etchings' stuff."

It was during this period that Anna was born, and Richard began pressing Margaret for a divorce. "Margaret said I would have to initiate it, which in Kenya meant committing adultery," Richard said. "So I said, 'Fine. That will be easily done.' "

Not long afterwards, Richard invited Meave to join him on his second Rudolf Expedition.

At the end of May 1969, Richard and his African staff drove to Koobi Fora to establish a base camp. Since no road led to the peninsula, they first had to forge a track from Nderati. This proved to be easier than Richard had expected, and only three days after leaving Nairobi, he and his crew were pitching their tents on a sandy rise overlooking Lake Rudolf. They cut reeds for windbreaks, built a large grass hut for a laboratory, and marked out an airstrip on the mudflats of a dry lake six miles from the camp. Richard anticipated having visitors fly in, as well as making trips himself occasionally to Nairobi to check on museum affairs.

By the time Richard had readied his camp, the geologist Kay Behrensmeyer was on her way to Kenya, and Richard drove back to Nairobi to pick her up at the airport. The next day they headed north together to Koobi Fora.

Behrensmeyer had only met Richard once before—in 1968, when he asked her if she would join his team—but she had heard plenty of rumors, mostly negative, about him. Much of the gossip came from her doctoral supervisor, Bryan Patterson, an eminent paleontologist who then held the prestigious Louis Agassiz chair at Harvard's Museum of Comparative Zoology. Patterson had made his name collecting fossils in the western United States as well as in South America. He had also led five expeditions to Kenya, beginning in 1963. Louis had helped coordinate these expeditions, providing not only field assistants but the fossil localities themselves. At Louis's suggestion, Patterson had explored several Miocene fossil localities south and west of Lake Rudolf. None of these proved as productive as Patterson had hoped, and in 1965 his team discovered a new Pleistocene fossil site, called Kanapoi. Although the Kanapoi deposits were not as old as those at the Miocene sites, they were considerably richer in fossils. And here Patterson's crew turned up their best find, the fossilized elbow bone of a three-million-year-old hominid, possibly an australopithecine.[6]

At the time, Patterson's hominid bone was a million years older than anything Louis and Mary had unearthed at Olduvai. Though Patterson's wife and others would later say that Louis was jealous of the find because of its age, there is no evidence that this was so. Louis was suspicious, however, of Patterson's claim that Kanapoi was devoid of artifacts. He thought it more likely that Patterson's team of geologists and paleontologists—and no archeologists—had simply missed the stone tools. In early 1966, Louis asked his son Jonathan, who was passing through the Kanapoi region on a snake-collecting foray, to see if he could spot any artifacts. With no trouble at all, Jonathan found choppers and flaked tools reminiscent of those from Olduvai at the site. Subsequently, Richard also stopped at Kanapoi and confirmed Jonathan's discovery.

The missed artifacts were one problem, but in Richard's eyes Patterson's team had committed an even graver sin. "There were vehicle tracks all over the fossil beds," he recalled. "Their main road went right over the top of a complete [fossilized] crocodile skeleton, which irritated me a lot." He persuaded the Ministry of Natural Resources that there had to be tighter controls over scientific expeditions in Kenya, and in 1968 the research permit process was changed.

Patterson was one of the first to run up against Kenya's new rules. In the spring of 1968, he wrote Louis, as he had always done in the past, saying that he planned to return to Kenya that summer. He wanted to continue his research at Kanapoi and Lothagam, another Pleistocene site, where one of his team members had discovered a weathered australopithecine mandible in 1967. But a letter to Louis was no longer sufficient, Louis explained in his reply: Patterson would have to fill out a series of forms and file these with the Ministry of Natural Resources. When Patterson wrote to Louis for clarification, he received a letter instead from Richard, who archly informed him that the new regulations were designed not to harass foreigners but to "ensure that the work is in fact scientifically conducted." "There have been occasions," Richard noted in another explanatory letter to Patterson, "when sites have been ruined; all the diagnostic surface material has been collected and no excavations carried out or proper records kept. Perhaps amongst the worst offenders are Geologists. . . ." Since Patterson's team was composed primarily of geologists and paleontologists, his reaction to Richard's letter can be imagined—and reports eventually reached Richard's ears.

"From what I heard," Richard said, "Patterson's response was, 'What the

---

[6] Later this specimen was dated to four million years, making it one of the oldest hominid fossils.

hell is this young whipper-snapper doing? This boy of Leakey's writing me such pompous letters! I'll be damned if I'm going to work there.' " In his written reply, Patterson said simply that he would be unable to come to Kenya for the 1968 season. But he actually had no intention of working in the country again, and advised his colleagues and students that the Leakeys were not to be trusted. "Patterson felt that he had been sort of disciplined by Richard," said Kay Behrensmeyer, "and he felt rather bitter about this treatment because he'd done a lot of work there. He felt the Leakeys were trying to exclude him because he had come up with these new, very rich localities which also included hominid fossils."

Unbeknownst to Richard, stories were soon circulating among American scientists about his treatment of Patterson and his political maneuverings. There was gossip, too, that Richard was having difficulties attracting qualified scientists to join his team; people were supposedly turning him down because they did not want to risk their professional reputations.[7] Even students were advised against going.

"When Richard invited me to join his team, not one of my [Harvard] advisors thought I should go," said Behrensmeyer.

> My advisors, and even people at academic meetings, told me that Richard was not to be trusted; that his expedition was that of an amateur; there weren't any established scientists attached to it; it sounded like a treasure hunt; and that the Leakeys were trying to create a dynasty, an empire over all the fossil collecting in East Africa. But it was apparent that most of these people had never even met Richard. Finally, I just decided that there were so many people with negative impressions—but with very little to base them on—that I would go see for myself.

Behrensmeyer was, however, puzzled about why Richard had invited her, a second-year graduate student, whom he had met only once. She had been part of Patterson's last team to Lothagam in 1968—when he sent only students to collect specimens found the previous summer. At the end of that expedition, Richard asked her to stop by the museum. "He said he was looking for someone to do the geology at Turkana [Rudolf], and asked me if I'd consider it," she recalled. "He was very direct and gave me the feeling that he really wanted me to join his crew. He said, 'We're going to create a young team and not be burdened with all these old guys who think they know what they're doing.' But I still had trouble imagining that he didn't want someone more experienced, so I didn't commit myself."

Though Richard had in fact initially invited Richard Hay, who was mapping Olduvai Gorge, he preferred working with a team of younger people—

---

[7] These rumors started after Richard Hay, the geologist working with Mary at Olduvai Gorge, declined Richard Leakey's invitation to map the deposits at Rudolf. "Koobi Fora was simply too big," Hay said about his decision to turn down this project. "I had all I could deal with at Olduvai."

scientists without any ties to his father—just as he had told Behrensmeyer. "I felt that at least in the first few years, we didn't need any great scientists along," he said. "And since I had to have scientific chaperones, I was bull-headed enough to take chaperones of my own age—teenage babysitters for teenagers."[8]

Behrensmeyer attracted his attention because she had already proved that she was capable of excellent field work in East Africa, and because she was a woman. "It's a funny thing, but I've always gotten along better with women than with men," Richard said.

Although Behrensmeyer's advisers were not enthusiastic about Richard's offer, her peers were; one suggested that it was "the opportunity of a lifetime." On her thirteen-hour journey north with Richard to Rudolf, Behrensmeyer reached the same conclusion. "We drove all night and talked the whole way," she said. "The car stopped at one point, and Richard got out and hit it with a hammer to make it go. I just had this building impression of someone who was incredibly competent and very much in command, and very knowledgeable about all kinds of things in Kenya. I decided that even if the other things were true, he had a genuine commitment to the science—I sensed that right away—and a love of the challenge Turkana [Rudolf] presented, and a love of working outdoors." Behrensmeyer liked it, too, that Richard treated her as a "professional equal instead of as a measly graduate student."

At Koobi Fora, Richard introduced Behrensmeyer to the other team members and showed her around the camp. Richard ran his camp as he had always done, with Europeans and Africans sharing an equal burden of the work. But it was still a hierarchy, and, for many Americans, one with jarring colonial overtones. "I was used to field camps in this country where one does everything for oneself," Behrensmeyer said. "And here was this camp all laid out on the lakeshore with people to wait the tables, bring the food, clean your socks. I was given my tent and someone to look after me—a Kenyan, who would be my 'field accompaniment.' He wasn't exactly an assistant, but someone who would just see that I didn't get in any trouble."

The next day, Richard took Behrensmeyer on a tour of the Rudolf deposits, again impressing on her that this was a solid and ideal scientific situation. He imagined that she would find a thesis topic to pursue, and offered to provide whatever support she needed. Behrensmeyer needed little encouragement. One look at the region had convinced her that Rudolf, with its untouched geology and fossils, was a "fantastic area for geological and paleontological work"—as, indeed, Richard's expedition would prove.

---

[8] Richard notes that selecting a younger team produced another unforeseen benefit. "Many of these young people, like Kay Behrensmeyer and John Harris, were molded by being given an opportunity to make themselves at Koobi Fora," he said. "They are now internationally recognized experts in their fields. And that wouldn't have happened if the older scientists had run the expedition. How many graduate students did their doctorates at Olduvai Gorge? Zero."

• • •

Richard may not have been as Machiavellian as Patterson and others imagined, and he certainly had no intention of keeping Patterson out of Kenya, but he was actively interfering behind the scenes with Clark Howell's Omo Expedition, which was scheduled to return to the field for the summer of 1969.

On his 1968 expedition, Richard had proved that hominid fossils could be found in the Rudolf deposits, but he had not located stone tools or archeological sites—the best clues to our early ancestors' behavior. Nor had he succeeded in relocating the rich fossil and artifact site he had discovered on his 1967 helicopter surveillance of the Rudolf region; in his excitement he had failed to note its location on a map. Perhaps the site lay close to the Ethiopian border—perhaps, Richard worried, it lay inside Ethiopia itself. If so, then the deposits were southeast of where Clark Howell would be working, across the Omo River. What if Howell crossed the river?

Rather than writing directly to Howell about his concerns, Richard contacted the director of Ethiopia's National Museum, Kebbede Mikael, sending him two letters on the same day. In one, Richard asked for permission to cross the Ethiopian border in order to explore the area north of Rudolf; and in the other, which he asked to be kept "in strict confidence," he suggested that Howell's student field assistants were incompetent. They were "in no way qualified to work in an area so rich in fossils both of animals and early man, and I thought it my responsiblity to notify you of this," Richard wrote. His concerns were not completely unfounded. Richard had observed Howell's team collecting fossils on a previous expedition and was appalled when the students damaged two fossilized animal skulls. But Richard also wanted to keep Howell out of the southeastern Omo region. "I wanted it myself," he said later. "And I was then an absolutely miserable fellow. I'm sure I was capable of being most unpleasant and uncharitable; I wouldn't have earned my reputation if there weren't some truth to these stories."

In response to Richard's letters, Kebbede Mikael granted him the permission he sought, then forwarded a list of Howell's crew members to Richard ("Professor Leakey," as Mikael addressed Richard) for approval. "It is obvious that my decision [to allow Howell's team into Ethiopia] will depend on the qualifications and experience of the scientists proposed and you alone could give me the correct and reliable assessment," Mikael wrote.[9] Richard happily obliged, telling Mikael that since Howell's team did not include a "competent Palaeontologist" it would be best if Ethiopia limited the group's activities. "I see no reason why the Americans should not spend their time exploring further and completing the geological and geographical re-

---

[9] The Ethiopians readily accepted Richard as an expert because of their relationship with and admiration for Louis.

searches that were initiated in 1967 and partially completed in 1968," Richard wrote. This would keep Howell's expedition confined to the fossil exposures west of the Omo and well away from the area Richard was eyeing.

Howell, of course, had no idea that his team—and even his letters to Kebbede Mikael—were being scrutinized by Richard. Nor did he suspect that when his expedition arrived in Ethiopia, Mikael sent Richard a telegram asking if Richard thought the Americans should be allowed in the field. "RECOMMEND AMERICAN TEAM WORK OMO AS PER DETAILS MY PREVIOUS LETTER," Richard cabled back. Mikael complied, leaving the way clear for Richard to explore north of Rudolf as he chose.

Howell probably would not have been surprised to learn of Richard's meddling, for paleoanthropology was changing. Collegial bonhomie had given way to backstabbing, as the stakes—fame and lucrative grants—grew higher. Indeed, Howell had already been a victim of such cutthroat ambition—and at the hands of Camille Arambourg, the elderly leader of the French contingent of the Omo Expedition. Though all three teams had agreed in 1967 to share the announcement of important hominid fossil discoveries, Arambourg had held a private press conference as soon as he had returned to Paris. The French team had found the oldest hominid specimen that year, the lower jaw of an australopithecine, and Arambourg claimed it as his own. For Clark Howell, Arambourg's defection marked the end of an era. "And Arambourg was supposed to be the gentleman of French anthropology," Howell said sadly to a colleague. "There goes the *last gentleman.*" Richard's meddlesome behavior was only more of the same.

Ironically, Richard's machinations turned out to be unnecessary. Only six days after arriving at Koobi Fora, Behrensmeyer discovered exactly what Richard had hoped to locate: an early Pleistocene tool site with artifacts *in situ.*[10]

Behrensmeyer had found the artifacts while collecting samples of tuff (volcanic ash) that would be used for dating the Turkana deposits. Based on the fossils collected the previous season, Richard had estimated the age of the exposures at close to two million years, the same age as the lowest bed at Olduvai. An analysis of the tuff—which contains minerals that crystallized at the time they were blown out of the volcano—would give a more precise date.

Richard was in Nairobi for a museum board meeting the day Behrensmeyer made her discovery. She left the site undisturbed until he returned the next day, after a nonstop, all-night drive with Meave. In a story Richard wrote for the *National Geographic,* he described Behrensmeyer rushing up

[10] Richard never did take advantage of his Ethiopian research permit. Nor did he ever relocate the promising site he had found in 1967, which he still believes lies in Ethiopia. "To this day, these sites across the border . . . have not been studied and will, in my opinion, offer some future archaeologist considerable joy," he wrote in his autobiography.

to him with the cryptic message: "You were right—they do exist, and we have them *in situ*.... What's more, there's good tuff directly on the site!" They then leaped into Richard's Land Rover and drove off to the site, where Richard examined the tools in "elation." In his autobiography, Richard was more honest about his immediate reaction. "I confess to having been somewhat sceptical and probably very jealous," he wrote. "[H]ow could a geologist recently introduced to the area find something that I had looked for unsuccessfully during a period of about four months! I consoled myself with the uncharitable thought that Kay was almost certainly mistaken; she had probably found recent stone artefacts of which there were many in the area and in which we had no interest." But, indeed, Behrensmeyer had found early stone tools: choppers and sharp-edged flakes were scattered throughout the grey volcanic ash.

Not only were the artifacts obviously of great antiquity, they were embedded in the tuff itself, an indication that the tools must have been manufactured prior to the volcanic eruption that covered them with ash. With such a clear association between tools and tuff—a paleolithic prehistorian's dream—it seemed that the site would be easily dated. Richard's initial jealousy now gave way to pure pleasure, which was heightened by his own discovery a short while later of another ancient tool site mixed with hippo bones—apparently a group of hominids had stopped here long ago to butcher a hippo's carcass.[11]

The team decided to leave this second site intact for archeologists to excavate later, and then dug a small trench at Behrensmeyer's. In it they uncovered more tools and flakes, and several animal bones—primarily antelope remains—that hominids may have cracked open for their marrow. Behrensmeyer also collected two samples of volcanic tuff, which Richard dispatched to the geochronology laboratory of Jack Miller and Frank Fitch in Cambridge, England. Their lab, called F.M. Consultants, had already dated rocks from Olduvai and other East African sites, and was highly regarded by British geologists. Richard hoped that F.M. could run a date on his samples "as a matter of some urgency." But the initial samples were contaminated with minerals from older volcanic eruptions, and it was not until the end of the season that Richard had the dates he wanted.[12]

---

[11] Richard had actually walked across this site the year before but had not noticed the tools. While studying the tuff at Behrensmeyer's site, he suddenly realized that he had once seen an almost identical outcrop—perhaps there were tools associated with it as well. "It is quite easy not to see things that are later obvious," Richard wrote about his missed tools, "and I am convinced that I search with a predetermined image in my subconscious. If I am searching for fossil bones it is unlikely that I will find stone tools and vice versa." The site he remembered was close to that of Behrensmeyer's, and after a half-hour walk he came to "exactly the place I had remembered and there to my delight were stone flakes too, just like those at Kay's site!"

[12] Ideally, the ash from each prehistoric volcanic eruption would be separated from the next by a layer of sedimentary material. This would create the picture-perfect layer cake seen in geology primers, with the oldest ashfalls on the bottom and the youngest on top. In general, this is what had happened at Olduvai. But at Rudolf, it was a different matter. Rivers and streams brought the volcanic ash down from the surrounding highlands and washed it across the floodplains, forming thick

In the meantime, Richard's team set out to search for fossils and more archeological sites. Now that they had artifacts similar to those from Olduvai, Richard was eager to discover remains of the hominid who had made them. Paleoanthropologists were still undecided about these early toolmakers—was it a species of *Australopithecus, Homo habilis,* or some as yet undiscovered species of *Homo*?

Thinking that his team could cover distances in the desert better on camelback, Richard borrowed four riding camels from the Kenya police, and rented six pack camels from a village of Gabbra people; two Gabbra also came along as camel handlers. Camels, of course, presented a temptation to the desert *shifta* (bandits), and even before the animals arrived at Koobi Fora, Richard's party was ambushed at a waterhole. The previous year, a large police escort had accompanied the expedition, but Richard had managed to cut this down to two men. The two exchanged shots with the bandits, killing one; none of Richard's party and none of the camels were harmed.[13] Later, a pride of lions also attacked the camels; once again they escaped. A few days later, the camels were safely herded into a small thorn-fence corral near the tents.

For the next few days, Richard and his camel-riding team—Meave, Kamoya, and Nzube—tried to learn how to handle the animals, a far more difficult task than they had imagined. "The easiest part of camel riding is thinking about it beforehand," Richard would later note. The camels consistently attracted lions, which hung around the outskirts of camp, and one night crept within three feet of the sleeping camp guards. After that scare, there were nightly lion alerts, with team members rushing out of their tents to toss firecrackers at the beasts.

By early July, Richard was ready to test the camels in the field, and shortly after dawn on July 6, he and his camel safari headed north toward Ethiopia. They planned to spend a few days zigzagging along the border.

Besides their adaptability to desert travel, there was another reason Richard chose camels for this safari. He liked the romantic image of himself as

---

deposits of tuff. But because of the rivers' churning action, the older volcanics and younger ashfalls were mixed. Thus the first sample that Richard sent to Fitch and Miller produced a date of 200 million years—which was obviously incorrect. Fitch responded with a lengthy letter describing what types of tuff to collect to avoid this problem, and Behrensmeyer eventually succeeded in finding what the lab needed. At the end of the 1969 season, Jack Miller also collected tuff from the site for further analysis.

[13] On his last camel safari, toward the end of their field season, Richard had another encounter with one of these armed bands. He had led two of his team's camels to a water hole five miles from camp, when the bush ahead seemed to "explode with people. Dark, wiry, and clad in loincloths, they carried knives, spears and rifles," he recalled in an article for the *National Geographic.* He counted eighteen men, who ran off to a nearby low ridge, then stopped and watched him. They raised their rifles, and in turn, Richard lifted his own above his head, at the same time "ruefully finger[ing] the two rounds of ammunition" in his pocket. He then determinedly set about his business of filling the water containers, while several of the men crept closer to watch. They followed him halfway back to camp, too, then vanished into the desert. Richard never saw them again.

"a camel-riding explorer of the great African desert"—a sort of latter-day Lawrence of Arabia. He played the role to the hilt. Mounted on camelback, with a fringed red-and-black burnoose shielding his bronzed face, his eyes squinting against the desert sun, Richard set off jauntily into the wasteland. Meave, Kamoya, and Nzube wore similarly colorful turbans, and all rode perched on wooden saddles topped with fluffy white sheepskins, with the reins and riding crops gathered in their hands. Behind them walked the two Gabbra camel herders with the pack animals.

"We traveled on these reconnaissance trips with Spartan simplicity, in order to cover as much territory as possible," Richard wrote in the *National Geographic* about his camel safaris. "We rose at dawn, and a cup of tea sufficed for breakfast. At midday we stopped to roast a few potatoes and chew a couple of strips of sun-dried antelope meat."

Stopping in the early evening, they gathered around a small campfire, where Richard cooked vegetable or meat stews and curries. When night fell, the sky seemed blanketed from horizon to horizon with stars, and the only sounds came from the rustling desert wind and the cries of jackals and hyenas—though on certain nights, the thunderous roar of lions chilled the air. Lions crept around the camp as well, but never attacked, perhaps because the camels' loud, distressed snorting woke everyone. Richard would then stoke the fire, and the camp would settle back to sleep. To protect the camels, the team slept in a wide circle around the animals, forming a human barrier. Richard thought this arrangement would also help them to catch any bolting camel, because, as the team discovered, even hobbled camels could run away quickly. The camels were troublesome in other ways: sometimes they intentionally walked under low thorn trees in an effort to brush off their riders, or they would all come to a halt at once and not start up again—despite beatings and shouting—until they decided to go. Worst of all, they walked at a slow, bone-breaking gait of about three miles an hour, leaving their riders stiff and sore.

Despite the camels' shortcomings, Richard organized a second camel expedition in early August, again heading toward the Ethiopian border. After three days, they had traveled only fifteen miles north of Koobi Fora but were already weary of riding their camels. Noticing a small outcrop of fossil deposits, Richard suggested they stop for the day. The exposures actually proved to be rich with fossils and Richard decided they would spend the morning exploring as well.

Early the next day, on August 5, the team set out in pairs, with Richard and Meave heading in one direction, and Kamoya and Nzube in the other; they would meet back at camp at one o'clock. But Richard tired of the search by ten-thirty; it was a hot day, he was still sore from camel riding, and he and Meave decided to return to camp. Looking for the shortest route back, Richard headed down a dry streambed—a place usually devoid of fossils. Indeed, he was not thinking about fossils at all, but about the camp's welcoming shade and a cool drink, when suddenly his heart leaped. "Meave!"

he shouted with such alarm that she would later say she thought he had spotted a snake or a lion. "It's an austr-austr-australopithecine!" he managed to stutter. Staring back at them from the sandy riverbed was the face of an *Australopithecus.* The skull lay twenty feet away, its bony sagittal crest and thick brow ridges clearly visible above the sand. Meave and Richard crept slowly forward, half fearful that the fossil was only an illusion. But the skull did not vanish, and moments later they were "peering at it from just a few inches, marvelling at its completeness and the incredible fact that we had found it. It is the only time that a discovery has left me truly breathless," Richard wrote in his autobiography.

Wishing to share their discovery with Kamoya and Nzube, Richard and Meave built three huge stone cairns beside the skull, then returned to camp. As soon as Kamoya and Nzube arrived, the four set off again, almost running, to find the fossil. Richard pointed it out from a distance, and in celebration, Kamoya and Nzube hoisted him on their shoulders and danced. Later, in the cool of the afternoon, Richard carefully collected the skull—an unusually easy operation since there were no broken fragments. Indeed, the fossil was remarkably complete, with only the teeth and lower jaw missing, and bore a striking resemblance to the *Australopithecus boisei* skull (as *Zinjanthropus* was now called) that Mary had found at Olduvai. Richard wrapped the skull in newspapers and sheepskin, placed it in a small box, and roped it tightly on his saddle.

At sunrise the next day, they were on their way to Koobi Fora. Paul Abell spotted them first, four distant riders moving fast across the desert. He studied them through his binoculars for a moment, then grabbed his gun. "That looks like nothing I've ever seen on earth!" he exclaimed. Richard's camel safari was not expected in camp for another week, and strangers on camelback in Kenya's northern desert could only mean trouble. The team at Koobi Fora nervously watched the approaching riders, until Mary Leakey, who had flown up to see Behrensmeyer's tool site, said softly, "It's Richard."

Soon the riders were in camp. Richard dismounted, untied his box, carefully removed the layers of wrappings and placed the new skull in his mother's hands. "It's beautiful," Mary said. "Absolutely magnificent!" Almost ten years earlier to the day, Mary had discovered the skull of *Zinjanthropus* at Olduvai, and now she sat studying one that was nearly identical although discovered some seven hundred miles away.

Clark Howell was also at Koobi Fora, visiting the camp with Mary, when Richard produced his big, intact australopithecine skull. Howell's Omo team had yet to produce anything comparable, and although he praised Richard's find, Howell looked crushed. "You could just see the disappointment written all over Clark's face," said Paul Abell. "Richard was having phenomenal luck and Clark was obviously chagrined that he had nothing as good."[14]

---

[14] Howell's team did turn up more than two hundred hominid fossils between 1967 and 1974, dating between 3.5 and 1.2 million years old. The specimens included several australopithecine mandibles,

A few days later, Richard and his team—some on camels, some in a Land Rover—returned to the skull site. He hoped that the skull's teeth might be found by sieving the soil, and he also wanted to continue exploring these deposits. Though the sieving produced nothing more, Mwongela Muoka did discover another hominid skull. Found protruding from a low bluff, this one was broken into fragments. It was also missing the face and jaws, but enough of the cranium was present to raise Richard's spirits. With its thinner, smoother shape, it did not resemble any of the australopithecine skulls that Richard had seen. "I felt, with mounting excitement," Richard wrote in the *National Geographic,* "that . . . further study might show this to be no near-man, but perhaps even a species of the genus *Homo.*" If so, his team may have even located that first elusive toolmaker.[15]

By late August, it was time for the first Koobi Fora field season to end. In three months, Richard's team had made extraordinary finds—besides the hominids, the team had discovered a bounty of beautifully preserved animal fossils, including the skulls and skeletal remains of a saber-toothed cat, several primates, elephants, crocodiles, antelopes, and hippos. Richard had also received the second round of results for the date on Behrensmeyer's tool site from Fitch and Miller. This time, the sample of volcanic ash had yielded a date of 2.4 million years, a much more reasonable result. Because the artifacts were embedded in the ash, they were the same age, making them the oldest tools in the world. Delighted by the age and the wealth of fossil and archeological material, Richard decided to plan for a long-term project, one that would be interdisciplinary and involve many scientists.

Back in Nairobi, Richard dashed off a letter to Allen O'Brien, the adventurer-entrepreneur who had joined him in 1967 on the Omo Expedition and had helped found the Leakey Foundation. "I have just returned from a most successful season at Rudolf," Richard wrote, "and although I am not yet able to give specific details, I can assure you that we have 'hit the jackpot.'" The jackpot would prove to be so lucrative that before long scientists would be clamoring to join Richard's team.

---

partial skulls, teeth, and postcranial bones. "But," said Howell, "we never found anything as beautiful as the fossils from Koobi Fora." The Omo fossils were always more fragmentary than those from Koobi Fora because the geochemistry at the two sites differed; while the fossilization process at Koobi Fora made the skeletal material hard and rocklike, at Omo the fossils were extremely fragile and broke into pieces at the slightest touch.

[15] Today, this skull is classified as a female *Australopithecus boisei.*

Chapter 27

MISADVENTURE AT CALICO

Richard's discoveries at Koobi Fora in 1969 delighted his parents. Louis dispatched bottles of wine to the Koobi Fora camp, then sent a letter to Melvin Payne at the National Geographic Society. The australopithecine skull was such a prize that Richard "practically fainted when he realised what he had found!" Louis told Payne. Then, conveniently overlooking his initial opposition to Richard's plans to explore Rudolf, Louis added, "Compared with the continued fragmentary finds of the French and American expeditions, I think we can say that this discovery more than justifies having transferred the Kenya party to work at East Rudolf and leaving the others to work in Omo."

Although pleased by this first discovery, Louis was more jubilant about the second hominid Richard's team uncovered. Only the cranium of this skull remained. Even so, its thin bone and smooth shape suggested a human ancestor to Louis—although not *Homo habilis,* the small, two-million-year-old hominid from Olduvai Gorge that Louis, Phillip Tobias, and John Napier had named in 1964. This new, as yet unnamed, specimen was dated tentatively to 2.6 million years, making it somewhat more than half a million years older than the oldest Olduvai fossils.[1] It was also shaped differently from the *Homo habilis* specimens. Nevertheless, Louis thought the new skull would rekindle the debate about that species, which few of his colleagues had accepted.

Habilis had been controversial from the outset. Many paleoanthropolo-

[1] The date was later revised to 1.8 million years.

gists had rejected it, including Le Gros Clark, the Oxford anatomist who had helped rescue Louis's career in the late 1940s and had worked closely with him on the Miocene primate fossils. "I have allowed myself to be quite outspoken," Le Gros Clark told a colleague in 1964, "because I think it is so important to nip this nonsense of 'Homo habilis' in the bud."[2] Le Gros and several other paleoanthropologists waged a heated battle against Louis, Tobias, and Napier in *Nature, Discovery,* and even the London *Times.* These critics argued that there was insufficient evidence to support the new species, and that the fossils more closely resembled australopithecines than any type of *Homo.* But most of all, they objected to the idea that *Homo habilis* and the australopithecines had lived side by side on the African savanna two million years ago—an idea that contradicted the prevailing theory of human origins.

At the time, in the mid-1960s, the majority of paleoanthropologists subscribed to the single-species theory of human evolution. This theory held that because of their special adaptations—the two-legged stance, the invention of culture, the enlarged brain—only one true human ancestor could have existed at any time. The australopithecines, which dated to about two million years ago, were widely regarded as the transitional species between humans and apes. They were upright creatures, similar to humans in their build, but their brains were small and they apparently did not make stone tools. It was thus perfectly acceptable that three types of australopithecines had once roamed across Africa: the large-toothed vegetarians represented by *Australopithecus boisei* in eastern Africa and *Australopithecus robustus* in South Africa (these were considered "dead-end" species, meaning they left no descendants), and the smaller, possibly meat-eating *Australopithecus africanus,* whose fossils had been discovered only in South Africa. Proponents of the single-species theory believed that the *Homo* line had emerged directly from this last creature, with *Homo erectus* the first human. In this interpretation, *Homo habilis* represented either an early stage of *Homo erectus* or an australopithecine—and was therefore not a separate species. "I think they are two successive stages of one single lineage, culminating in ourselves," John Robinson, a paleoanthropologist at the University of Wisconsin and a leading critic of *Homo habilis,* told a reporter in 1966, voicing a commonly shared opinion.

Louis's hypothesis upset this orderly, step-by-step progression of human evolution. Not only were the australopithecines and early humans contemporaneous, he argued, but there were also a variety of early *Homo*s, all competing with one another in a Darwinian battle for survival.[3] Human

---

[2] Instead of italicizing the name, as is the correct practice with scientific names, Le Gros Clark always placed quotation marks around *Homo habilis*—a sort of punctuated sneer. Le Gros's quotation marks greatly irritated Louis, who fumed against them in his published retorts to Le Gros's attacks and in private.

[3] Unlike many of his colleagues, Louis did not believe that humans descended from the australopithecines. If *Homo habilis* and the australopithecines were contemporaneous, then, as Louis pointed

evolution was not a special, one-time-only event, as his colleagues believed, but, as with other animals, a lavish experiment. And like the lineages of those other "ordinary mammals," such as pigs, horses, and elephants, the human lineage resembled a bush more than it did a ladder.

"The time has come to get away from the 1, 2, 3, 4 idea of man's development," Louis told a symposium on human origins at the University of Chicago in 1965, where he further upset his colleagues by announcing that he had identified a third type of early *Homo* at Olduvai and implying that more would surely follow. In addition to *Homo habilis,* Louis and Mary had discovered fossils of *Homo erectus* in the gorge, although in different geological levels. The *Homo habilis* specimens dated to nearly two million years, while the *Homo erectus* fossils were a little over one million years old. "This is only the beginning," Louis intoned to his squirming audience, and then "with all the ardor of a sea captain speaking from the quarter deck," as one reporter noted, proceeded to describe this new ancestor.

Louis based his third *Homo* on a skull Mary's team had discovered at Olduvai in 1963, which was known as O.H. (for Oldvuai Hominid) 16, or "George." George had eroded out of the same geological levels as *Homo habilis,* making proto-*Homo erectus,* as Louis referred to this new species, a companion of *Homo habilis* as well as *Australopithecus boisei.* In Louis's eyes, only *Homo habilis* led to modern humans, while *Homo erectus* and the australopithecines died out.[4]

"Man developed just like the animals did," Louis declared, "with various species living side by side until the weaker died out or were annihilated, leaving the stronger until eventually modern man emerged." Two million years ago, four species of giraffe had existed side by side at Olduvai, as well as six species of elephants and four kinds of pigs. "Why should there only be one line of man?" Louis asked. "It's a prejudice [which] comes from a religious attitude . . . that man was a special creation." Louis then called on his fellow scientists to "review all their previous ideas about human origins and to substitute for those theories new ones which would be more in keeping with the facts that [are] now known."

Richard's 1969 discoveries at Turkana—where *A. boisei* apparently had shared the ancient lakeshore with yet another type of *Homo*—seemed to confirm Louis's multiple-species theory. If the second fossil was "man, not a

---

out in a letter to Le Gros, "we will have to look for a common ancestor of the hominines and australopithecines way back in the Pliocene [about seven million years ago] and abandon the idea that the hominines went through an australopithecine stage."

[4] Louis never placed *Homo erectus* in the direct human line because of its large brow ridges, a feature he and his mentor, Sir Arthur Keith, considered characteristic of primitive primates. Few of his colleagues supported this idea, and today *H. erectus* is accepted on the family tree as the immediate ancestor of *H. sapiens.* However, the tree's shape, as Louis predicted, has become considerably more bushy. "George" is now classified by some scientists as *Homo habilis.*

'near man,' " as Louis told a Leakey Foundation audience in the fall of 1969, then it, too, would have to be given a specific name.[5]

Louis, of course, would have loved nothing more than to name this new skull, then wave it under the noses of John Robinson and Sir Wilfrid Le Gros Clark. But it was not to be. Richard had his own ideas about the pursuit of paleoanthropology, a field he saw as beset with petty quarrels for which there were no answers. Unlike his parents, he did not christen his fossils with pet names; instead, he always referred to them by their field numbers, a tactic he hoped would lessen the emotional attachment people often felt for the ancestral bones. Nor did he intend to use his fossils to bolster particular taxonomic or evolutionary theories. "I thought that could be done later, after a larger sample had been collected," Richard explained. "In that early stage, I simply wanted anatomical descriptions, not taxonomical arguments." Thus, rather than following his parents' tradition and offering the specimens to Louis and Phillip Tobias to describe, Richard turned to two anatomists at Nairobi University: Joseph Mungai, the Kenyan head of the anatomy department, and Alan Walker, a British paleontologist and anatomist who had worked on Miocene primate fossils from Kenya and Uganda.

Louis was stunned, outraged, and disappointed. "Why did you ask them?" he demanded of his son. Richard had, after all, invited his mother to describe the first artifacts from Rudolf. Furthermore, Mungai, although an experienced anatomist, lacked any paleontological training, while Alan Walker had never studied Pleistocene hominids and was a young, outspoken critic of Louis.[6] "My father didn't say so directly, but the implication of his question was that he had wanted to be involved," Richard said, "and that I had made very much the wrong decision, bringing outsiders in."

Richard, however, had consciously sought outsiders. In particular, he wanted to involve black Kenyan scientists as part of his efforts to Kenyanize the museum and its research projects. Mungai was not only the first black Kenyan scientist but the first black African scientist to be involved with hominid research. But Richard's reasons only exasperated his father, who was certain that his son was "working to eliminate him from any participation" with the new fossils and the museum, Richard wrote in his autobiography. Following his 1970 heart attack and long convalescence, Louis's suspicions escalated into near paranoia as Richard assumed a larger role at the museum and at Louis's Centre for Prehistory and Palaeontology. Nor did Louis's fears ease when he returned to Nairobi in July 1970. His family had rushed to Nairobi Airport to greet him, as they always did when Louis returned from a visit overseas, staging a joyful reunion, which puzzled those who knew of the family's quarrels.

---

[5] Louis thought the discovery of this new species of *Homo* would also help reinstate his Kanam jawbone—but it did not.

[6] Initially, Richard invited only Jo Mungai to describe the fossils. Mungai then suggested that Walker also be involved.

"They always made a terrific fuss whenever Louis came back," said Peter Andrews, a paleontologist who worked with Louis at the Centre.

> Mary came up from Olduvai, Richard came in from the field, Johnny was there, even Philip turned up. And they put on this fantastic family show of solidarity at the airport where they were all matey, hugging and kissing. This was at a time when most of them were not speaking to each other—Mary and Louis were essentially separated; Richard and Louis weren't speaking; Richard and Philip couldn't stand each other. But at the airport, there was a very strong sense of family, an almost Italian attitude, sort of family before all else. As soon as [the reunion] was over, they wouldn't speak to each other again.[7]

So it was that Louis, still convalescent, soon found himself alone in Nairobi, trying to untangle his affairs at the Centre and at Tigoni. "At the moment I am alone here," he told Joan Travis, one of the directors of the Leakey Foundation, three weeks after his return. "Richard is away [at Turkana] till mid-September and Mary is away [at Olduvai] and I have the full responsibility of making decisions and plans." He was trying to rest, as his doctor had ordered, but the workload was daunting and he simply had no time to attend to the Foundation's requests. "I am already doing rather more than my Doctor wants me to do," Louis explained, after describing his busy schedule in great detail. "I feel that I must not overwork further, otherwise I shall set back my present progress." Yet, he was already making plans to travel to the United States in late September for a fund-raising tour, and to host a conference on the Calico Hills Early Man site in California.

While Louis's role at Olduvai, Rudolf, and the museum had dwindled, he was still the overall director of the dig at Calico, seventy-five miles east of Los Angeles. But, as with the naming of *Homo habilis* and Louis's multiple-species theory of human origins, Calico was controversial. Few of his colleagues accepted his contention that the site proved that people had arrived in the New World more than 50,000 years ago. The upcoming Calico Conference, Louis was certain, would change that.

Louis's team of archeologists, many of them volunteers, had been laboring under the desert sun at Calico since November 1964. During their six-year excavation, they had dug twenty pits (one seventy-seven feet deep), recovering several hundred stone tools and flakes—plus thousands of untouched rocks and bits of stone, which were dumped to one side. The work was extremely tedious—digging inches at a time through a cementlike deposit of rock, boulders, clay, and gravel to pick out the promising pieces of chalcedony—but the team kept at it, largely because of Louis. He was the

---

[7] Richard notes, however, that at these family reunions only his parents hugged and kissed.

one who had told Dee Simpson, the on-site director, where to dig; he was the one who had agreed that the chipped pieces of stone were primitive artifacts; and he was the one who prophesied the team's ultimate victory.[8] "I am counting on all of you," he had told his crew in August 1966, "to help achieve a real victory over prejudice in favor of facts about U.S.A. prehistory."

To keep up his crew's morale, Louis tried to visit Calico twice a year, stopping off between lectures and staying for two or three days. These were memorable times for the Calico volunteers. "Louis knew everyone by name," said Simpson, "and he was so good to everybody, making sure that people were rotated in their jobs so no one got bored. He always seemed to feel that we were his friends and he was one of the team."

At night by the campfire, Louis would tell them, after swearing everyone to secrecy, about the latest Olduvai or Rudolf discoveries; by day, he praised their own efforts, stressing the importance of their excavation. "Together you will help to make history," he told them at one point, "and you will some day be able to look back with pride and say, 'I was on that project—I helped.' " With the desert wind blowing through his white hair, Louis would hobble around the site, exclaiming over prized finds, sympathizing with those digging in the rockiest areas, and repeating his favorite mantra, "Patience, patience, patience." "Dr. Leakey had a way of making even the littlest, most unimportant person feel important," said Rosemary Ritter, a member of the Calico crew. "And that's why people were so willing to work for him."

In 1968, at a depth of twenty-three feet, Ritter made one of the most significant Calico discoveries—a semicircle of stones that Louis interpreted as a hearth. "There were three main rocks, one stacked on top of another, and they all pointed toward the center," said Ritter. "It just looked strange, and when I showed it to Dee, she said we'd have to wait for Dr. Leakey's opinion." Louis did not see the "feature," as Simpson first referred to the stone circle, until February 1969, about four months after its discovery. He sat quietly studying the formation for about half an hour, while Ritter and Simpson stood to one side, "holding our breath," Ritter recalled. "Louis used to tell me that he never did anything until he had a chance to go away and think it over as a Kikuyu, and I suppose he spent that time thinking about the hearth as a Kikuyu," said Simpson. Finally, Louis announced, "I like the looks of it. But we mustn't call it a hearth yet, because we don't know for sure. So we'll call it Twiggy [after the famous 1960s model]—you know, flat in front and round behind."

For Louis, who described the discovery to Richard as "a very exciting

[8] Dee Simpson and John Kettl, a member of the Calico crew, showed Louis the first artifacts from the dig in March 1965, four months after the excavation began. They set out fifty possible artifacts for him to examine, but then noticed the state of Louis's glasses. "You could hardly see his eyes through them," Simpson recalled. "What are we going to do?" Kettl whispered to Simpson. "I don't think the man's washed his glasses since he had egg five days ago." Simpson summoned her courage and offered to clean Louis's glasses. Handing them back to him, she asked, "Isn't that better?" "Well, I suppose so," Louis replied, "but I've got used to looking around things." Louis then studied the fifty pieces of stone and selected seven as artifacts.

thing that quite definitely looks like an artificial structure," the rock circle (which he decided was a hearth after it was completely excavated) offered additional proof that people had lived and hunted in the Americas long before 12,000 years B.P. (before the present). If so, then Louis and his crew would have achieved the "major breakthrough in American prehistory" that he had promised his team. All he had to do was convince his fellow archeologists that the artifacts were man-made and not the product of natural forces as many skeptics claimed.

Louis had decided in 1968 that a conference at Calico would be the best way to convince the archeological community about the site's validity—and perhaps even persuade the National Geographic Society to reinstate its funding.[9] He and Simpson had seemed to gain a modicum of respect that year when *Science* published a report on their preliminary findings. Louis was further encouraged when the archeologists Clark Howell and François Bordes agreed that the flakes—though not necessarily the tools—were the work of humans. "I was never 100 per cent convinced about the artifacts," Howell said later, "though when I first saw the material, I was more strongly inclined to accept the possibility than some of my colleagues. I remember thinking, 'My god, this [tool] has criteria a, b, and c—the very criteria we use to separate the artificial from the natural.' So I was willing to entertain the possibility."

To Louis's delight, Bordes went one step further and in September 1969 at a UNESCO-sponsored conference on human origins announced to the gathered delegates that "Leakey is right. . . . Calico is the earliest known site of man in America. . . ." But Bordes was only one voice, and a French one at that. For Calico to be accepted, Louis had to persuade his North American colleagues—many of whom resented the whole enterprise and were irritated that Louis had not familiarized himself with North American archeology in general.

"If Louis had had full command of all the early sites and data here, we might have been more open to what he was saying about Calico," said Emil Haury, an archeologist at the University of Arizona who had directed several excavations at sites in the Southwest dating between ten thousand and twelve thousand years ago. "But he was playing a hunch, and he was a little bit cavalier." "A lot of American scientists disliked Louis because of his somewhat arrogant attitude and because of his ability to get funding," added Dee Simpson. "They saw him thumbing his nose at them: 'You didn't pick up on the Early Man sites, but we will.'" In his hurry to revolutionize North American prehistory, Louis did not stop to worry about the effect his pronouncements might have—or whom he might be offending. Even when such eminent American archeologists as Robert Heizer at the University of

---

[9] The National Geographic Society funded the Calico excavation between November 1964 and the spring of 1968, contributing a total of $100,000. The society withdrew its support after two of its staff archeologists visited the site. They were unconvinced that the artifacts there were man-made.

California, Berkeley, refused to meet with him or to view the Calico artifacts, Louis did not change his style but instead plunged ahead, certain that his conference would force everyone (Heizer included) to change their minds.

Invitations to the conference were sent to nearly one hundred internationally recognized archeologists and geologists (though not to Mary Leakey, who had made it clear that even if she was invited, she would not come), while an additional one hundred places were reserved for members of the Calico team. The meeting itself was billed as the International Calico Conference and was scheduled for October 22 to 25, 1970, at the San Bernardino County Museum in Bloomington, California.

Louis knew that he came to the meeting as an underdog, but that was the very role he relished. All his life Louis had waged campaigns to overturn conventional thinking about human origins, and this conference was simply another skirmish. Nor did he have any doubts at all, as he told Simpson, that he was going to win.

The day before the conference, Louis went around "like a mother hen," as one friend recalled, hunched over his cane, checking the display cases and advising his team on proper scientific etiquette. "He wanted everything done so that the scientists could reach their own objective conclusions," said Simpson. "So we couldn't put any labels on the tools identifying them as scrapers or choppers, and we weren't supposed to offer any explanations about the material unless we were asked."

Like the showman that he was, Louis had also planned a dramatic event for the conference—one, he felt, that was sure to convince any fence-sitters. Rainer Berger, a geochemist at the University of California, Los Angeles, and a friend of Louis's, had asked a Czechoslovakian laboratory to test one of the hearthstones for evidence of heat. The results were positive, and Louis arranged for the Czech scientists to telephone him with the news at an appropriate moment during the conference. Louis could then stun his audience with this apparently sudden development. "Louis thought it would be the high point of the conference," said Joan Travis, a Leakey Foundation director who helped plan the meeting. "He really thought that everyone would be convinced when they saw these beautiful stone tools and heard about the hearth. He pinned a lot of hope on what they would decide."

Louis was so certain of his colleagues' verdict that he concluded his opening-night speech—given in the dining room of the Bloomington Holiday Inn—with a triumphant vision of what the Calico Conference meant to American archeology. This meeting, he suggested, was comparable to the 1859 visit of a team of British scientists to the archeological sites of Jacques Boucher de Crèvecoueur de Perthes along the Somme River. The French archeologist had claimed that he had uncovered artifacts together with the fossilized bones of extinct mammals—the first definitive evidence that humans had a long ancestry. "He had been ridiculed and laughed at," Louis told his audience. "But the scientists came back and said they were satisifed, and from that day on, the prehistory of Europe went forwards. I believe that from this weekend on, a new chapter will be written in the prehistory of America."

While Louis's impassioned speech stirred his Calico team, it left the delegates uneasy. He had asked for their objectivity in judging the site, but many left the dinner feeling pressured, as if Louis were seeking converts. The Calico crew—nervous and protective of Louis—seemed to the visiting researchers less a team of scientists than cultists. "It was like a holy cause," said Vance Haynes, a geologist at the University of Arizona who attended the conference. "And Louis was their Messiah." Even Mary shared this view. "Why he's nothing but a guru to those people!" she had snorted to Glynn Isaac after visiting Calico in 1967.

The following morning, Louis led the scientists on a tour of the site. While the archeologists were impressed by the meticulousness of the excavations, the geologists were less certain about what they were seeing. The site lay in an alluvial fan, a mudslide of rocks and boulders, which looked "pretty damned old," said Richard Hay, the geologist who had mapped Olduvai Gorge for Mary. He was surprised then when Thomas Clements, a geologist from the University of Southern California, told the gathering that the site dated to only 70,000 years B.P. The date seemed doubtful as well to Kenneth Oakley, a prehistorian from the British Museum of Natural History, who had helped solve the Piltdown mystery. But when he asked Clements to substantiate the date, Clements faltered. "Kenneth asked him how did we know the site didn't date to Pliocene times, rather than the late Pleistocene, as he claimed," said Hay. "And Clements couldn't answer that." If the site was Pliocene, then it would date to two million years ago or more—an age that not even Louis could support.

There was further confusion when the archeologists Glynn Isaac and Desmond Clark stopped to photograph a circle of stones they assumed was the hearth. Simpson suddenly appeared behind them, saying, "It's not that one; it's this one over here." "It seemed to us that there could be any number of hearths," said Clark. "It just depended on how many cobbles you took away [when excavating] to make one."

There were questions, too, about the artifacts themselves. Were they natural or man-made? What kinds of natural processes might produce the chipped and shaped pieces of chalcedony that Simpson and her team were uncovering? "One of the problems," said Clark, "was that while there were tools and flakes, there were no cores [the scarred chunks of rock left behind after making a stone tool]. And Glynn had an answer for that—all of those rocks they had dumped to one side were the cores." To Clark and Isaac, and many of the other archeologists, the alluvial fan itself explained the tools and flakes—they had been produced when boulders in the mudslide struck nodules of chert. "They were pressure-induced flakes, so they did look very much like those of human manufacture," said Clark.

Few scientists were convinced by Louis's reasoning about the site. They suspected instead that poor science alone accounted for Calico. "He wasn't willing to look at alternative explanations, to test other hypotheses," said Vance Haynes. "And in this type of situation, you have to give equal weight to the possibility that the artifacts—and all of the things they were uncovering—resulted from natural processes." Indeed, for Haynes and many of the other geologists "all of the [physical] mechanics are there for producing" the artifacts naturally.[10]

Though Louis hobbled around urging everyone to be open-minded and exclaiming over his team's evident hard work and determination, by the end of the first day even he was subdued. "Clements' inability to tie down that date sort of cast a shadow over the whole thing," said Walter Schuiling, a historian at San Bernardino Valley College and member of the Calico team. "You could see the disappointment in Louis's face."

Still, Louis had his trump card: the news about the fire in the hearth—though even this announcement did not have quite the impact he had planned. While Rainer Berger explained his analysis of the hearthstone, Louis sat expectantly at the speakers' table, waiting for the Czechoslovakian scientists' phone call. It never came. Instead, Berger rescued the moment with a simple concluding statement, "In the final analysis, it would appear that in that circular arrangement of stones there must have been a fire." Immediately, the auditorium exploded with applause and cheering, albeit primarily from the Calico team. Louis added to the din by banging his cane noisily against the table. "His grin lit up the whole hall," Simpson recalled.[11]

But neither drama nor Louis's conviction could save Calico. There were simply too many questions about the site's age and the artifacts, and very few of the assembled scientists were willing to grant Calico a place of glory in American prehistory. At the same time, few wished to offend or hurt Louis by voicing their opinion. "We all kept a very low profile," said Desmond Clark, "because we didn't want to upset Louis. He had been ill, so one was careful in one's remarks."

Consequently, at the conference's final session when Louis pressured his colleagues for their opinions, a curious and unpleasant silence settled over the gathering. Many of the scientists sat with their arms folded, their eyes focused on the floor. A very few, such as Glynn Isaac, ventured gentle criticisms, but by and large the audience was silent. "People didn't agree with Louis," said Karl Butzer, a geologist at the University of Texas, "but they didn't come right out and say so. They used every euphemism in the book to try to say they didn't agree without actually saying they didn't agree. And Louis's face was getting redder and redder. It was an exceptionally sorry

---

[10] Haynes described the various geological processes that could produce each of the Calico artifacts in a 1973 paper, "The Calico Site: Artifacts or Geofacts?," which was published in *Science*.

[11] Berger's analysis of the hearthstone was insufficient to convince the delegates that "Twiggy" was a hearth or that there had been a fire. Other stones from the dig and surface rocks should have been submitted to similar tests, argued Vance Haynes, as should stones from a recently made hearth. These would then provide a control against which to evaluate the rocks in the Calico hearth. These tests have still not been done.

sight to see." "The great bulk of us sat on our hands," Clark Howell added. "We were downcast and embarrassed, because there seemed to be only one voice crying in the wilderness—Louis's. And we were friends."

It was neither the revolution Louis had hoped for, nor even a spirited debate, and he left the Calico Conference bitterly disappointed. By the time he arrived home in late November, Mary already knew the outcome of the conference via letters from Desmond Clark and Glynn Isaac. She was thus surprised to hear Louis say to her only moments after greeting him at the airport, "It's all right; they all accepted the evidence." Mary said nothing in reply, and Louis never again mentioned Calico to her. "That's the way Louis was," said Richard Hay. "When he lost an argument with you, he'd never bring the subject up again."[12]

Louis, however, never thought he had lost this argument. To those he was closest to late in his life, such as Vanne Goodall, he always expressed his absolute faith in Calico. "I shall be proved right, of course," he told Vanne after the conference, "but I shall no longer be here." Louis was "silent for a moment," Vanne recalled, "and then came the familiar chuckle . . . the irrepressible twinkle. 'You see,' [Louis] said with a grin, 'they'd have to change *all* the textbooks if they admit I'm right!' "[13]

Two months before the Calico conference, on August 7, 1970, Louis had turned sixty-seven. Though he had shed twenty-eight pounds in an effort to comply with his doctor's orders, he was still overweight and, in spite of his hip replacement, continued to suffer pains in his right leg. His cane had become a permanent feature, and in the mornings when he arrived at his office at the Centre for Prehistory and Palaeontology he was invariably in an ill humor. "Louis was usually red-faced, and he'd bark at a few people, and everyone would scurry around deferentially," said Richard Michael Gramley, then a Peace Corps volunteer working at the museum.

Louis was now at an age when most people retire, but he refused to let up. He continued to launch new projects even as his old ones were coming unraveled.

---

[12] At Olduvai Gorge in 1964, Richard Hay had experienced firsthand this side of Louis's often stubborn nature. Louis had always interpreted every red-colored geological layer in the gorge as Bed III. Hay, however, discovered that some of the red-colored layers were actually part of Beds II and IV, and he dug a trench to illustrate this to Louis. But the two men argued over Hay's analysis before he was able to show Louis the trench. Indeed, Louis never saw the trench. "Mary knew how sensitive Louis was to this sort of thing, and she quickly had her men fill it in," said Hay. "She knew the subject would never come up again, that that was Louis's way of acknowledging defeat—so if it's finished, let's bury it."

[13] Dee Simpson and her team of volunteers continue to dig at Calico, still certain that they are recovering genuine artifacts. In retrospect, Simpson believes that they held the conference too soon, that the material they have uncovered in recent years is more persuasive. In 1991 a team of Russian scientists visited the site and expressed surprise that Calico is considered suspect among North American archeologists. Apparently the Calico artifacts are reminiscent of stone tools these scientists have discovered in Siberia. In America, however, Calico is generally dismissed as Louis's biggest folly. "We knew Calico would eventually die away," said Clark Howell, "which it did. Maybe it's not dead for Miss Simpson and her cohorts, but in the world of science, it is dead. It's become like Sasquatch and the Yeti."

"Everything seems to be going wrong," Louis wrote Dian Fossey on January 12, 1971, detailing his fears that the National Geographic Society was no longer going to support him, and that his Tigoni Primate Research Centre and the Leakey Foundation were on the verge of collapse. "The people who promised money for Biruté [Galdikas to study the orangutans] have fallen down, the money for Tigoni is not available except on conditions which I am not prepared to accept, and by and large everything is in a great mess."

That afternoon he attended a meeting about Tigoni which lasted until 6:00 P.M. "By the end of it," he recalled later in a taped letter he sent to Vanne Goodall, "I was pretty well fagged out and almost dead to the world." He went home to rest, and as soon as he lay down realized he was about to have another heart attack. His servants quickly summoned his physician, who diagnosed an embolism. It would be wise, the doctor thought, if Louis left Nairobi's high altitude for a while and spent a few days relaxing at the coast. The year before Louis had bought a small cottage at Malindi, a sleepy palm-shaded town overlooking the blue waters of the Indian Ocean.[14] This was his retreat from the world where no one could find him (to guarantee this, he had painted the false name "L. S. Tookey" on his gate, though as one friend pointed out, "only an idiot" would be taken in by Louis's ruse), and a few days after his near attack, Louis flew to his refuge.

Louis did not travel alone, but—as was often the case in his later years—was accompanied by several young women and his driver, Clemente.[15] For three days, Louis and his retinue relaxed on the Malindi shore, he lying under the palms, the women swimming and snorkeling. At night, they feasted on fresh fish seasoned with coconut milk, studied the stars, talked, and read, and were as Louis recalled "generally lazy." The coast was exactly the tonic Louis needed, and by Monday, January 18, he had begun to feel so much better that he decided to return to Nairobi the next day. As a parting gift to his young friends, Louis offered to show them the ruins of Gedi, an Arab town dating to the sixteenth century that had been set aside as a national monument. Louis's bad leg prevented him from touring the site himself, so he sent the bikini-clad young women and Clemente off together, while he waited in the shade of a mango tree.

"Then I got up," Louis told Vanne, "and went to look at one of the ruins . . . known as the Great Palace. No sooner had I got there, and I was just

---

[14] Three years before, in 1968, the Emperor of Ethiopia had awarded Louis the Haile Selassie I Award for African Research—a prize that included a gold medal and diploma plus 40,000 Ethiopian dollars. Louis had used some of this money to buy his Malindi cottage. That same year, Louis had also received the Royal African Society's Wellcome Medal for "distinguished service to Africa" (an honor previously bestowed on Dr. Albert Schweitzer); and the president of Senegal had appointed Louis a Commander of the National Order of the Republic for his services to Africa.

[15] Louis was aware of the gossip that circulated about his young women friends, but hopelessly naïve about how to curtail it. Once he told Andrew Hill that he was taking a certain young woman with him to Malindi. "But it's all right," Louis explained, "because I'm taking two other girls along as well!"

having a look at a carving I wanted to examine more closely . . . , than a swarm of bees attacked me *en masse.* It was perfectly *horrible,* [a] terrible, terrifying experience. I was alone. The bees simply came in their thousands upon thousands all around me and mobbed me." Louis hobbled away as fast as he could, brushing bees from his eyes, mouth, and throat, shouting for Clemente and the girls, and in his terror walked straight into a wall. He hit his head, fell, and badly hurt his injured leg again. Somehow he managed to struggle to his feet and tried again to escape the bees. By now, his friends had returned, but in their scanty dress they could not help. Louis, who seemed to be the only one with a clear head in spite of his suffering, told them to grab the blankets from the car and use those to cover themselves while they helped him. Finally, they managed to get him into the car, along with several hundred bees, and after killing the bees, raced back to the cottage. "Drive quickly, quickly, quickly," Louis gasped to Clemente. "I'm collapsing."

The women spent the better part of the evening pulling the bees' stingers out of Louis. He would later say he counted eight hundred stings before he lost track. The bees' poison seemed to paralyze him as well, though his physician later diagnosed the paralysis as a stroke, perhaps one he suffered as a result of his fall. "My brain was working absolutely clearly, but otherwise I was absolutely paralyzed," Louis said, "[and] towards the middle of the night, I was almost certain I was going to die." He lay awake, convinced that he was "in extremis," but unable to move, and then toward dawn began to "turn the corner." By sunrise, Louis knew he would recover, and immediately began making plans to return to Nairobi. At the same time, his son Jonathan arranged for a Flying Doctors' plane to pick him up—though as soon as the plane took off, Louis realized that its unpressurized cabin was endangering his weakened heart. "I was much too weak to protest even or to say anything about it," he told Vanne. "I just grinned and bore it and hoped for the best." By nightfall, Louis was finally in the hospital and under the care of his physician.

For several days, Louis continued to be paralyzed, unable to move either his legs or arms. The stroke had affected his memory as well, and sometimes "after about five minutes of thinking about anything at all seriously, I would suddenly just forget and completely go blank and be unable to remember words and names at all," he said. The bee stings had also injured his eyes, leaving him for a time with only partial vision. Yet by the end of a week, Louis began to recuperate. He made a tape recording for Vanne and asked her to send it on to his daughter, Priscilla. His right hand was still paralyzed, and though he was doing physical therapy and learning to eat and write with his left hand, he expected that it would be awhile before he was able to write letters. The tape would let "the family," as he referred to the Goodalls, know what he had endured, and that he was now on the mend. He thought he might return to the coast to rest and strengthen his heart. "I can't be worried [there] by too many . . . people coming in all the time, [saying]

'Please this, please that, please the other thing,' until I just get absolutely bothered," Louis said petulantly. "I'm just going to have to do less."

This promise lasted less than a minute. In almost the next breath, Louis was planning a visit to Vanne's, followed by another lecture tour in the United States, "because I do need the money so desperately badly. . . ." Louis might be crippled, partially paralyzed, and suffering from a poor heart, but none of these was reason to slow down. Instead, it seemed to him that he had been given another chance. "Quite frankly, I still don't know how I survived at all," Louis confessed to Vanne, his voice slurred and breathless. "I must be tough. I must be very, very tough."

Chapter 28

AN UNSTOPPABLE MAN

Louis remained in the hospital for a month recovering from his bee stings and stroke. He was finally released in mid-February 1971, though he had to return daily for physical therapy. As she always had, Mary had traveled up from Olduvai as soon as she heard about Louis's plight, but she headed back just as quickly when she was assured of his recovery. Louis thus once again found himself alone at their Langata home. White-haired and heavy, with his paralyzed right arm dangling awkwardly at his side and his left arm braced against his cane, Louis hobbled slowly but determinedly through the house—a half-crumpled figure who refused to give way.

"Louis used to joke [that] . . . he had more lives than the cat . . . because he had already had more than nine life-threatening experiences," said Joan Travis, one of the founding members of the Leakey Foundation. "And he survived them all. So he, with bravado, would say, 'You know, I'm going to go on forever'; and then on the other hand he would recognize his frailties and he would say, 'I'm a prisoner in my own body; it's a terrible thing. And I used to play tennis and I used to do this and I will again some day.' He would hold it out [as a possibility] for himself as well as for you. But it was also apparent that each succeeding incident would take its toll."

More than ever now, Louis was forced to delegate his field studies, to find protégées who would embrace his dreams as their own. Besides Jane Goodall, Dian Fossey, and Biruté Galdikas—the "trimates," as they were sometimes called—Louis had a young woman studying the De Brazza monkeys in western Kenya, another beginning to track wild colobus monkeys, and a third looking into (yet again) the classification of the wildebeest. Galdikas was about to leave for Indonesia to watch the orangutans, as soon as Louis found the money. Although financing these ventures was always troublesome, Louis was omnivorously curious; he wanted to launch studies of the pygmy chimpanzee as well as such little-known African mammals as the aardvark, zorilla (a creature resembling the skunk), and warthog—animals he had watched as a child, but whose behaviors remained little-known.

There was certainly no shortage of projects—and an equally long line of eager students to take them on. Most of Louis's recruits, as ever, were bright, attractive young women. "Everyone knew that Louis liked to take on young women who were untrained in academic matters and send them into the field," said Alan Walker, who was then teaching anatomy at the University of Nairobi. "You'd meet some young woman and you'd ask, 'What do you do?' And she'd say, 'Oh, Dr. Leakey's got me doing this.' And it would be something for which she was totally untrained, that Louis was training her to do —sometimes quite important things. I don't think he ever did this with a man; he might point a young man in a direction and help him get his doctorate. Whereas with women, Louis liked to thrust them into the unknown."

As Louis aged, the young women came to serve an additional purpose: that of flattering an old man who craved the admiration and respect he seldom received anymore from his colleagues, or from Mary. Louis had always been attractive to women, and even in his later years when he was overweight and nearly toothless, he exerted a magnetism that younger men envied. Yet as his female coterie swelled, Louis looked less the proud peacock than the old roué. "There was always another young girl with him," recalled Karl Butzer, a geologist at the University of Texas, "and he looked rather silly, limping along on his cane, acting like a guru with this retinue of young girls following him around. Everybody of course saw this, and most were uncomfortable with it."

No one knew what to make of Louis's growing band of young female protégées, most of whom were in their late teens or early twenties. Were his relationships with them sexual, or was it simply that Louis, who loved mysteries, wanted his colleagues to believe this? Either way, people sniggered about Louis and his "harem," and many serious young women who thought that working with Louis would help their careers instead found themselves regarded merely as sycophants, or worse.

This was particularly true in Nairobi, where the relatively small white community gossiped as much in the 1970s as it had in Karen Blixen's day. "If you became involved with Louis, you were tagged," recalled the anthropologist Elizabeth Meyerhoff, whom Louis invited to Kenya in 1970 when she was a sophomore at the University of California at Los Angeles. "You became a 'Leakey or Tigoni Girl'—a term that, in other people's eyes, connoted that you were flirting with Louis or he was flirting with you, and possibly having an affair. And there's no question that Louis was a womanizer or that he'd had affairs with a lot of women. But he was sixty-eight and could hardly walk, and I was just nineteen—the whole thing was completely off-scale."

Sometimes Louis even seemed to stage titillating scenes just to set tongues wagging. Once Louis invited Meyerhoff to dinner at Langata along with the anatomist Michael Day and his wife, both very proper people. When she arrived, Louis presented Meyerhoff, a stunning, blue-eyed strawberry blonde, with a bouquet of roses, then thrust her into the role of hostess for the evening, giving every impression that she was his mistress. "It put me in a very funny position," Meyerhoff said, "and the Days were angry and confused. But Louis loved it. He was old and crippled, and to have this young girl pouring the wine at his wife's house was probably terrific. The sad part was that I didn't have the guts to ignore them [the Days] and just enjoy the scene."

Wherever Louis went, he acquired fresh disciples. He had met Meyerhoff at Joan Travis's home in Los Angeles, and persuaded her to come to Africa, although what she would do in Africa was never defined. "I came over here knowing only that I was going to do something 'exceedingly important' with Dr. Leakey," said Meyerhoff. She was thus surprised and a little baffled to meet another girl on the plane to Kenya who was also "coming over to do something exceedingly important" with Louis. In Nairobi, there were even more young women, all vying for a share of Louis's attention.[1]

As if he were Paris handing Aphrodite the prized apple, Louis led each woman to believe that she—and she alone—had been specially chosen by him for a key project. Sometimes, the gift he proffered was nothing more than the vague promise he had made to Meyerhoff. For these young women, Louis acted as professor, instructing them in African prehistory, anthropology, paleontology, and animal behavior. They formed a pool on which he could draw when the right subject (and funding) came along—and while they waited, they also served as needed confidantes, a private circle of admirers whom Louis could summon at will. Several of the women lived and worked at Louis's Tigoni Primate Research Centre (which locals dubbed "The House of the Rising Sun"), but readily dropped their own research to spend time with Louis. He invited them on safaris, to lessons and dinners at his home, and sometimes for private overnight tête-à-têtes. More than one

[1] The anatomists Alan Walker and Michael Rose, who were both then teaching at the University of Nairobi, tell a story about a young woman hitchhiker they met. They picked her up on their way to explore some fossil beds in western Kenya and gave her a lift to a town in that area. En route Walker and Rose explained their research. "We said things like, 'We work with the Leakeys,'" recalled Walker. "And she said, 'Who are the Leakeys?' and 'What's a fossil?'" She wanted a lift back to Nairobi, and the two men arranged to meet her in a few days' time, but she did not show up. Some weeks later, Louis invited Walker to dinner at Langata. "And who should open the door, but this woman!" said Walker. "She quickly said, 'Shhh. Not a word.' Then Louis introduced us, saying this was the woman who was going to do his orangutan study [this incident occurred before Louis met Biruté Galdikas]. Apparently, she'd got a lift back to Nairobi with Louis and by the time they got to town, Louis had decided she was going to be his orang woman. So she was staying at Langata, being the hostess, and being taught by Louis all the things you need to know to be a primate behaviorist."

visitor to the Leakey home found Louis dressed in his baggy coveralls and perched on the edge of his chair, holding forth on some topic to a wide-eyed young woman (who sometimes wore the same style of ill-fitting coveralls) seated at his feet—the very position that Mary had occupied in Frida's house nearly forty years before. And just like Mary, they had fallen under Louis's spell.

"Louis was exceedingly charming," said Meyerhoff, "and he had this amazing childlike enthusiasm for life that was catching. Everything was new and important to him. He was incredibly intelligent and thoughtful, and at the same time he could laugh like a kid and get thrilled over a rose." Most seductively of all, Louis made his young women friends "feel important," treating them as if they were "intelligent women" rather than naïve, inexperienced girls. When he chose, Louis captivated young men the same way.

"Louis had this knack when he was talking to you that made you feel that you were the only person who mattered to him at all," said Derek Roe, an archeologist at Oxford University, who met Louis when Roe was a graduate student.

> Never mind all the scientific conferences he'd just been to, all the important people he'd talked to; he wanted to tell you what he was doing. He would take a fossil tooth out of his pocket and say, "Look at this; what do you think of this?" And one was deeply flattered. It didn't matter that you weren't the least bit qualified to give an opinion; he really wanted to know what you thought. If he did this to people who were impressionable, the effect must have been explosive. They would become clay in his hands.

Having seduced the young women intellectually, did Louis then seduce them physically? He was, after all and in spite of the crowds of admirers, a lonely old man, sometimes desperately in need of affection—and "understanding," as he told one young woman. "I don't want to be Dr. Leakey the archaeologist," he wrote her. "I want to be Louis—someone whose ideas about God & beauty & how to live are needed & worthwhile." And at times, he apparently tried to spark a romance, perhaps hoping that once again he would find someone like Dian Fossey, someone to hold in his arms and love.

Not long after returning to his home from the hospital, Louis telephoned Gabrielle Dolphin, a twenty-year-old brunette who was newly arrived from the United States, and invited her to Langata for some promised lessons in African prehistory. She was temporarily living and working at Tigoni, a long drive from Langata, so Louis suggested she stay overnight. He sent his driver to pick her up.

At the time, Dolphin barely knew Louis, having met him only five months earlier, in October 1970, after writing him a letter asking if he needed a research assistant.[2] Louis had responded by arranging an interview with her at his hotel in Washington, D.C., an encounter that nearly left Dolphin speechless, so in awe was she of meeting Dr. Leakey. "I remember the exact time he [Louis] walked into the hotel lobby—10:25," Dolphin later wrote in her journal. "I was shamefully enough beside myself—there stood the man about who *[sic]* I had read and idolized as a tiny kid—Dr. Leakey—Olduvai Gorge—*Zinjanthropus*—it was all I could do not to burst into tears. I walked up to him glowing fiery red and I could only utter, 'finally we meet,' with this huge, rather idiotic grin on my face." After another interview and a brief exchange of letters, Louis wired Dolphin money for an airline ticket and she was on her way to Nairobi. She would be working with primates, she thought, though as with Meyerhoff, Louis's plans for her were ill-defined. But shortly before Dolphin arrived in Kenya, Louis was felled by the bees and several weeks passed before he was well enough to see her. Thus the invitation to Louis's home seemed to Dolphin a renewal of his faith and interest in her.

Initially, as Dolphin and Louis discussed her interests over a long lunch, this seemed to be the case. She talked about her dream of a career in anthropology, while Louis dropped hints about a "secret dig" he was planning to an area that was so remote the team would have to be airlifted in. He wanted to include her and so would train her in paleontology. "[Dr. Leakey] then extended his hand in partnership to me as the director in charge of a new dig at Fort Ternin *[sic],*" a stunned Dolphin wrote in her journal. She was "one out of 29 [students]" he had considered bringing to Kenya, Louis told her, and because she was very bright, he thought she should attend Cambridge University for her doctorate. He would be her tutor. That night Louis gave her his book, *Adam's Ancestors,* to read, and promised to take her on an early-morning game drive in Nairobi National Park for "lessons in observation."

Over the next two months, Louis invited Dolphin to his home almost every week. "It's becoming a second home," she noted in her journal at one point. Louis gave her a telephoto lens for her camera, boxes of fossils to study, lessons in paleontology, taught her to make stone tools, and took her on safari. Every animal they saw elicited a story or some tidbit of knowledge from Louis, which Dolphin dutifully recorded. Sometimes he talked about the hunting techniques of early humans, other times he focused on the behavior of the animals themselves, demonstrating an empathy for their instincts that amazed Dolphin (as well as many other friends whom Louis entertained this way). "Louis has this unbelievable power to become an animal," an awestruck Dolphin noted one day, after witnessing Louis's ability

[2] Dolphin had also written letters to such eminent anthropologists as Sherwood Washburn, Irven DeVore, and Phyllis Jay Dohlinow. Louis was the only one who responded with the offer of a possible job.

to predict the behavior of a lion they had encountered.[3] At his home, Louis was often just as engaging. Before the morning game drives, they fixed tea and toast together in the kitchen—simple moments that invariably led to laughter as a sleepy Louis poured his tea in the sugar bowl or suddenly discovered the rubber handle of his cane melting against the stove. Then, as they dined on Louis's veranda, his pet hyrax, Flossy, would appear, begging for a sip of Dolphin's tea or a bowl of rose petals. And always there was "Louis's childlike expansiveness. He was just thrilled about everything he was doing, and he was purely thrilled when talking with me about my possibilities."

But there were other moments that left Dolphin worried and ill at ease. "Louis was crippled and sometimes he'd ask me for help to get out of the bath tub," she recalled. "And I'm a kind person, I wouldn't deny an older person help if they needed it. But it felt like a setup"—especially since his cook was nearby and as capable as Dolphin at providing the needed assistance. Other times, after his bath, Louis would appear at her room, dressed in his robe. "He'd want me to come into the courtyard so he could tell me about the constellations, or there'd be something he'd want to show me, or to have a cup of tea. I could hear him coming; his cane would go clack, squeak, clack, squeak, and my heart would just race. I didn't know what would happen, what he would do, or what he was after." Louis further frightened Dolphin by sometimes stopping in her bedroom to have a chat, then placing his hand affectionately on her knee. "I wanted to believe it was fatherly," she said, "because I was always in a state of denial about what he was doing. I mean, it couldn't possibly be true, because I was there to do something important and he was my doorway. But I also knew that if I didn't watch it, it could turn into something overt. So I was always on guard." Dolphin made certain that she never got ready for bed until she knew Louis had retired, "that he was down and it would be too difficult for him to get up again."

Rumors circulated about the passes Louis sometimes made, but these did not stop young women from idolizing him or wanting to be in his company. Many accepted his behavior as part of his African heritage, that in a way he was acting like the powerful Kikuyu elders who took the most desirable girls for their brides. "Louis was very attuned to the Kikuyu system of patriarchy, so he liked having women around him because it added to his status," said Toni Kay Jackman, whom Louis brought to Kenya in 1972 to begin a

---

[3] Louis's ability seemingly to transform himself into various animals also amazed Bill Graves, then a *National Geographic* editor, who joined Louis on a game drive in 1964. "Gradually Louis fell into a habit I had noticed before," Graves reported in the *Geographic*. "He began to speak for the animals we were watching, verbalizing what seemed to be in each one's mind. He was so absorbed in interpreting their thoughts that he forgot my presence completely. Soon I felt I was actually listening to the animals themselves." With Louis as interpreter, Graves watched a lioness stalk a herd of zebra, only to have her hunt disrupted by a passing vehicle. As the zebras bounded away, "Louis spoke once more for the lioness—a short, sharp monosyllable of disgust."

study of the pygmy chimpanzee. "I never had any problems with him, but I also felt that if he were younger and in better health he could have been a handful. But he was always respectful of me."

By the end of April 1971, Louis had packed his bags for a fund-raising tour in the United States. Not only did he need the money, but he was eager to attend a University of California symposium in San Francisco honoring Pierre Teilhard de Chardin. Louis had been invited as a guest speaker, even though he was working on a book that fingered Teilhard de Chardin as the perpetrator of the Piltdown hoax. (Louis had divulged this to the organizers of the conference, who told him not to worry; they welcomed controversial opinions.) The trip would last six weeks and because Louis was still crippled from the bee stings, he took along his new personal assistant, Cara Phelips, the twenty-five-year-old daughter of a neighbor. She would see to it that Louis did not overexert himself, a task that she finally understood soon after they arrived at their first stop, New York City. "We went to the top of two skyscrapers just for the fun of it," a weary Phelips wrote her parents. "[It was] Louis's idea, not mine." Three days later, they were in San Francisco.

Although the organizers of the Teilhard de Chardin symposium had invited several academic luminaries—Conor Cruise O'Brien, George Gaylord Simpson, Theodosius Dobzhansky—none was as well-known or popular as Louis. San Francisco mayor Joseph Alioto sent his chauffeured car to drive Louis to his hotel; after lunch there were private meetings with Alioto and the University of California chancellor, as well as a special press conference, where Louis discussed everything from fossils to the state of the world in general. That night the university hosted a dinner for the symposium's key speakers, and the following morning, May 1, the conference itself began.

Louis was not the keynote speaker—indeed, his lecture, entitled "The Gathering of Man: From Savannah to City," was not scheduled until midafternoon. But as soon as he stood up, "the audience rose and clapped and clapped," Phelips reported to her parents. "After he had spoken they rose again and clapped and clapped. Nobody else had half as much applause. . . ." Louis never mentioned his suspicion about Teilhard's role in the Piltdown affair, but instead discussed his own interpretation of hominid fossils and his theories about major human discoveries (fire being the most important, he thought, as it allowed people leisure time to invent the arts and sciences).

The meeting drew to a close late in the afternoon and, as at the end of any public gathering, the hall quickly filled with the sounds of people bustling about readying to leave. Louis, seated at the speakers' table on a raised dais, also pushed his chair back—only he pushed too hard, his partially paralyzed right leg stuck under the table, and with a loud crash, he fell backwards onto the stage. For a moment no one seemed to notice what had happened, then a cry of dismay filled the hall. Louis lay on his back in the chair, his eyes closed in pain. "When Louis fell, he hit the back of his head," recalled Joan Travis, "and I rushed onto the stage. . . . He was stunned and surprised." Someone brought a pillow, and Louis rested his head on it, lying where he fell for a full fifteen minutes. He seemed to recover somewhat, though he had sustained a slight concussion and had bad bruises on his arm and thigh. Still, he was certain that nothing was seriously wrong; he would be back at the symposium the next day.

But Louis was not as well as he thought and by late afternoon the following day was back in bed again. "I was foolish enough not to rest . . . but to go to the conference again," Louis later confessed to a friend, "so that I had the most ghastly headache for a day. . . . However, I am alright *[sic]* again, but still have some pain in my back and shoulder."

Louis continued to insist that nothing was seriously wrong. He promised not to overwork, but also intended to complete his lectures—a three-week schedule of talks that had him flying up and down California, giving presentations nearly every day. He did stop long enough in Los Angeles for skull X-rays, but these turned up nothing. The headaches, however, would not go away and at times were nearly incapacitating. "Louis has good days and bad days depending on how much he's doing," Phelips told her parents. "It's impossible to keep him quiet. I have now learnt to refuse point blank to do something when he's tired, like phoning or letters." Louis fumed at Phelips for these enforced quiet periods, but he was really railing against his own body, which once again seemed to be letting him down. Yet he never missed a single lecture, and with his usual enthusiasm gave each audience his all.

Louis's last talk was scheduled for the evening of May 25 at the University of California at Riverside, one of his favorite stops. That morning, however, he received some disturbing news about the finances of the Leakey Foundation and the headache returned full force. "Some of the trustees kept trying to impose [spending] guidelines and controls [on Louis]," said Joan Travis, "and Louis wasn't the kind of person to take this kindly." Nor would he listen to her suggestion that he cancel his last talk. Instead, he stretched out in the passenger seat of Travis's Jaguar and had her drive him the fifty miles from Los Angeles to Riverside. He had brought an overnight bag along since he planned to visit his friends at the Calico excavation the next day. At Riverside, Louis once again delivered an inspiring lecture, though even members of the audience could see that he was unwell. He had chosen to show a film, *Land of the Sonjo,* about Richard's 1964 expedition to Lake Natron when Kamoya found an important hominid fossil, so that he did not have to talk continuously. Nevertheless, as the evening wore on, Louis grew increasingly ashen-faced and began perspiring heavily. From the wings, Travis watched him worriedly, certain that he was about to collapse. Somehow, Louis managed to make it through to the end, and only after the last round of applause let Travis take him by the arm.

"Louis was terribly sick the last time he talked at Riverside," said Dee Simpson, the director of the Calico dig, who had also watched his performance with concern. "He was clear as a bell, but he was so sick that when he finished, he had to be guided off the backstage and into the car. [Then] he looked out the car window and said, 'I'll see you tomorrow.' That's the one time he didn't make it."

Travis scrapped Louis's Calico plans and instead drove him back to her home. By the time they reached Los Angeles, Louis had lost all coordination, and was sadly confused, asking Phelips if she shouldn't be arranging his slides for his next lecture. "Louis couldn't walk or even get out of bed," Travis recalled, "and that's when Dr. Charles Carton [an eminent neurosurgeon and neighbor] said, 'It's getting worse and worse; I'll have to operate. Who is his next-of-kin?' " Mary was out of touch at Olduvai, but someone succeeded in reaching Richard by phone. He immediately gave his permission for the operation. On May 29 Louis was wheeled into surgery, where Dr. Carton removed two blood clots from Louis's brain—one caused by his fall in San Francisco, the other just beneath it, apparently the result of his fall at Gedi when the bees had stung him. With the pressure from the clots removed, Louis regained nearly the full use of his right leg, arm, and hand again.

"Louis was so elated when we saw him in the recovery room," said Travis. "He raised his right arm up and very rapidly moved all his fingers and said, 'See, it wasn't a stroke.' And that was so important to him because people were saying, 'Oh, Louis had a stroke and he's now just a silly old man.' And everything improved—his handwriting and walking—almost from the moment he entered the recovery room."

For the next several weeks, Louis stayed at the Travis home, waiting for the three holes the surgeon had cut in his head to heal. He no longer suffered from headaches and with much of his paralysis gone, he began seriously to watch his weight. It seemed there was a chance he could recover some of his old vitality. "Louis's ailments had had him sidelined," said Travis, "and he often felt that Richard and Mary were excluding him from the world that he loved. And he'd get frustrated and angry when he'd get word from home that Richard was doing something without consulting him. It was bitter for him to feel his own powers declining and at the same to have somebody—particularly his own son—nipping at his heels."

After the operation, however, as Louis's health improved, he seemed somewhat less worried about Richard's activities. In fact, Louis suddenly became, in Travis's words, "mellow," a change in temperament she attributed to Louis's physical improvement. "As Louis became less physically impaired, he also grew more hopeful about taking on fieldwork again, and that hope alone improved his outlook," she said. It was only physical disabilities and not mental ones, Travis insisted, that had led to Louis's "anger and frustration" and caused him to sometimes behave erratically. But his family and other friends thought otherwise. To them, the blood clots were proof of something they had suspected for some time—that Louis had been suffering from some type of brain damage.

"Louis was gaga toward the end of his life," said Jane Goodall, "though for about the last year [after the operation] he was lucid again. But for a while he was very pompous and acted in a way that was nothing like the old Louis. We made a film together for the Leakey Foundation, and I was supposed to explain my research, but Louis wouldn't let me speak. He kept interfering, showing off for the camera, and holding his head in the most arrogant way. It was really too much."

Richard also felt that the blood clots explained some of his father's recent odd and irascible behaviors, as did Mary. Once, not long before his brain operation, Louis had spoiled an evening with old friends at Olduvai. "Louis had gone off to bed," recalled Ron Clarke, who was then working for the Leakeys as a lab technician, "and Dick Hay, Desmond Clark, Mary and I were sitting around drinking and talking. All of a sudden Louis appeared and he was very angry. He shouted at Mary, 'Mary! You said you were coming to bed, and I see you're just sitting here drinking brandy! If you're going to drink brandy, then I am, too!' " Mary's face turned white at Louis's words. "She was clearly frightened," said Clarke, "but then Desmond Clark said very calmly, 'Ah, Louis. Do come and join us.' And then Louis looked very confused and agitated and said something about how he was just looking for his spectacles, but he sat down with us for a while. Mary was aghast."

"That behavior was part of Louis's paranoia, his conspiracy theory," said Richard. "He was worried about people talking about him. And he thought the family were ganging up against him, so if Mary sat up late, talking to friends, they had to be discussing him."

In his later years, Louis saw conspiracies everywhere, as if friend and foe alike were against him, working to bring him down. Richard and Mary were conspiring to take over his Centre, his colleagues were conspiring against his new species, *Homo habilis,* and another group had conspired against him at Calico. Did Louis's conspiracy theories stem from a partly damaged brain, as Richard believed, or as with many leaders, was it the loneliness of his position that haunted him? Perhaps his one-man battles on behalf of unpopular scientific theories simply had dragged on too long.[4] Whatever the source of Louis's dark worries, they apparently lessened after his operation.

By the end of June, Louis had recovered sufficiently to begin his journey home. Dee Simpson joined the Travises in seeing Louis off at the Los Angeles airport. Louis told her he was sorry he had been unable to visit Calico, but promised he would on his next trip. Then, just before boarding the plane, he turned to Simpson and said with a big chuckle, "Everybody always said I had holes in my head, and now I do." "The man," said Simpson, "was unstoppable."

• • •

---

[4] Joan Travis first encountered Louis's embattled attitude shortly after helping to found the Leakey Foundation in 1968, when Louis presented her with a copy of *Adam's Ancestors.* Inside, he had inscribed, "With gratitude for helping me in my fight." "Everything was a fight for him," said Travis, "a battle he had to win."

Louis did not return to Nairobi until mid-July. He had planned a brief stop in Washington, D.C., to press the National Geographic Society about providing a grant of $2,000 to help launch Biruté Galdikas's orangutan study. But his stay was extended when he once again fell ill, this time with a high fever and bladder infection, forcing him back to the hospital and onto a heavy dose of antibiotics. Still he made his appeal to the Geographic, and even though its research committee was reluctant to commit to another long-term primate study, he succeeded in wangling the funds from them. With this grant and matching sums from the Wilkie Brothers Foundation and the Leakey Foundation, Louis finally felt he had sufficient funds to send Galdikas and her husband, Rod Brindamour, to Indonesia. First, however, Louis wanted the couple to come to East Africa for some last-minute training sessions, including trips to Olduvai Gorge and Jane Goodall's camp at Gombe Stream. While they were in Nairobi, they would stay with him at Langata.

Because of his illness, Louis arrived home only a few weeks before his guests, but dropped all of his other activities to spend time with them. As he had with Gabrielle Dolphin, Louis squired them on game park safaris, showed them the sites of his early digs in the Rift Valley, and shared his knowledge about field life in general—but this time, there was never a hint of impropriety, perhaps because Galdikas was accompanied by her husband. Instead, Louis kept the relationship purely paternal. "We were like his children," said Galdikas. "Rod and I could ask him for anything, and he would try to get it for us, even if it meant using his own personal money. He'd buy books he thought we should read, and he paid for our trip to Olduvai."

Louis was unable to accompany Galdikas and Brindamour to the gorge, but he assured them that Mary would look after them. Assuming that part of Louis's generosity stemmed from his "African side, where once you're considered family, they will do anything for you," Galdikas anticipated the same reaction from Mary. Instead, Mary was cool and distant, distinctly uninterested in Galdikas and her orangutan study. It was only after dinner, over several glasses of Scotch, that Mary opened up, talking to Galdikas in a "woman-to-woman" way—and in the process revealed a contempt for Louis and the women who surrounded him. Galdikas was shocked. "She was jealous of Jane [Goodall]," said Galdikas, "and she said many nasty, unrepeatable things about her, and told me that Dian [Fossey] was a 'mad woman, crazy.' She totally misconstrued Louis's relationships and was jealous of the most inappropriate people—even Elizabeth Meyerhoff and Vanne Goodall [Jane's mother]. It was unbelievable."

If Mary had hoped for a sympathetic female ear to tell her sorrows to, she could not have chosen a more inappropriate confidante. Galdikas refused to believe Mary's allegations, insisting even today that Louis's relationship with Vanne was simply that of "two old, old friends who were very comfortable with each other," while his relationships with Jane, Dian, and Elizabeth were "totally paternal"—an interpretation colored by her own experience with Louis. Perhaps their friendships with Louis were "closer" than hers, she said; nevertheless, they were all "Louis's daughters, though Jane was the favorite. That was what Louis really wanted," Galdikas said. "A daughter."

Whether Louis's protégées were surrogate daughters or something else, Mary was clearly weary of her husband's wandering eye. For the last three years, she and Louis had lived almost totally separate lives, yet neither time nor distance had taken away the sting of Louis's philandering—though sometimes Mary could be good-humored about it. Once she told Andrew Hill, then a graduate student in paleontology at Cambridge University who came to work at Olduvai, "Oh, it's harmless, Andrew. Louis just collects them like medals."[5] Still, the women were a key reason behind Louis's and Mary's separation, though they never officially divorced and continued to see each other. "Our separation was papered over to some extent," Mary said, "because I always stayed at Langata when I came to Nairobi." But even these visits had their painful moments. "Louis was not always careful," she said, "so he was always covering up and that irritated me. And there were always these other females coming in"—who all too often, Mary claimed in her autobiography, were of a "deplorably low standard."

Mary was not above letting the young women know how she felt about them. When Penelope Caldwell, a daughter of one of the members of the Leakey Foundation, arrived in Kenya to work at Tigoni, Mary was not the least bit welcoming. "She looked me up and down, up and down, up and down, and finally said, 'Well, everyone to his own taste,' " Caldwell recalled. And when Biruté Galdikas and Mary met again at Langata, Mary was openly hostile. Galdikas and her husband were then still living at Langata as Louis's guests, and were present when Louis gave Mary a big box of chocolates tied with a red ribbon. He said they were from an "anonymous admirer," although neither the gift nor Louis's playful words elicited a response from Mary. After opening the box, Mary passed it to Louis and to two of her visiting students, took a chocolate herself, then pointedly set the box down—ignoring Galdikas and Brindamour. "There was a very uncomfortable

[5] Hill witnessed Louis's technique with women shortly after arriving in Kenya in 1971. Louis had asked Hill to take Mary's new assistant (whom Louis had not met) to Olduvai; Louis was going, too, but he had no room for this student since he had packed his Land Rover full of vegetables for the Olduvai crew. "Louis said I should come round his house at seven-thirty in the morning, and pick up this woman; she would be on the doorstep. So I went round and she turned out to be extremely beautiful. And there was Louis with his arm around her, calling her 'My dear.' Then he said to me, 'Hill. The vegetables are there.' So I had to load up my Land Rover with the vegetables, and Louis drove off with the girl." Later, Hill had his revenge. Louis had invited the young woman to stay at Langata on her return, and Mary suggested that Hill stay there as well—Mary's way of playing a joke on her husband. "It was very rude," said Hill. "I just turned up, saying that Mary told me to come. And Louis was furious and would not speak to me. But he was, of course, a gentleman and could not completely ignore me, so he would talk to me through the young woman [employing the same tactic his sons used against each other in their quarrels]. He'd say to her, 'Does he want a drink? Does he want more meat?' After a couple of days like that we got on very well."

silence," said Galdikas, "and finally Louis said, 'Mary, perhaps Biruté and Rod would like a chocolate as well?' Mary looked real huffy then, and snorted, and pushed the chocolates to me."

After that incident, Galdikas wondered, "Why does Louis put up with her? What is the attraction? Even when she treated him like dirt, Louis never fought back. He continued to be nice to her, to speak well of her, and he tried so hard to please her. And I think it was because he had made the same kind of commitment to her as he had to Jane and Dian. He was crazy about Mary in the same way he was crazy about them. He had the same passion for all of them."

But there is little evidence for this. How passionate can any marriage remain after thirty-five years, particularly one so buffeted by extramarital affairs and financial pressures? Louis, for his part, no longer looked to his wife for comfort or understanding, nor did he seem to comprehend or care how betrayed Mary felt each time he returned to Langata with a new girl or lavished them with gifts. (Louis particularly doted on Dian Fossey, giving her a ruby ring, a portable heater—after she complained about the cold in her mountain cabin—a special bush jacket she admired, and embellished telegrams at Christmas. To other protégées, he gave roses, plane tickets, cameras and lenses, tents, camping gear, books, and took them on safari—all the while that Mary was living an utterly spartan life at Olduvai.) Each new project he launched drained away more of his energies and money—and Galdikas and her orangutans simply repeated Louis's pattern once again.

Mary's visits to Langata seldom lasted more than a few days. She was busy now at Olduvai excavating the handaxe sites of Bed IV, the upper level of the gorge, and had only come to Nairobi during Galdikas's stay because Louis had announced he was going abroad again. Barely six weeks had passed since he had returned from his last harrowing trip, and Mary was concerned, particularly since Louis had filled his schedule as usual with back-to-back lectures. He had to go, he insisted. Tigoni was on the verge of financial collapse, and he had to persuade the National Science Foundation in the United States to renew its grant to his Centre for Prehistory. But there was more to it than finances, Mary thought. "Louis . . . had to prove that his powers were not waning," she wrote in her autobiography. "Hence the refusal to rest, the quest for new projects, the acceptance of every invitation to lecture, to make a film, to visit a university, to receive some wealthy visitor . . . [and], at a more human level, hence his women. . . ."

At the end of September 1971, Louis was off to California—and lecture halls packed with admiring fans—once more.

Chapter 29

ROAR OF THE OLD LION

Shortly before Louis left for his autumn lecture tour in the United States, Richard returned to Nairobi. After three months in the Rudolf desert, he was fit, tan, and loaded down with hominid fossils—partial skulls, jawbones, leg and arm bones of both early humans and australopithecines. In the first two weeks alone of this expedition, his fourth to Rudolf, Richard and his team had scooped up eleven hominids, prompting him to "whoop like a hyena," one observer noted, as the bones practically spilled from the earth. Richard's field season the previous year, in 1970, had been equally productive, yielding what Mary called "a whole galaxy of hominid[s]." These included, she told Phillip Tobias, "a mandible that nobody can seriously suggest is anything but *Homo,* a huge *Zinj* [*Australopithecus boisei*] mandible, bigger than any yet found, a small, nearly complete skull that may be a female *Zinj*[1] *and* post-cranial material [bones other than the skull]."

Altogether, during their four summers at Lake Turkana, Richard's team had amassed forty-nine hominid specimens; in comparison, it had taken Louis and Mary thirty-six years to find thirty-six hominid fossils at Olduvai. No one had ever before discovered early hominid fossils in such quantity and diversity—or at such a rate—and the scientific world was dazzled.

[1] The "female *Zinj*" was virtually intact, and Richard upon seeing it immediately guessed that it must be a female *Australopithecus boisei*—the first recognized female specimen. "It was clearly a *boisei,* but without a big [sagittal] crest," he explained about his hunch. Richard's team had actually discovered the first female *A. boisei* in 1969, but that fossil was incomplete; Louis and Richard had initially interpreted it as a new species of *Homo.* Following the 1970 discovery, however, this 1969 partial skull was also classified as a female *A. boisei.*

"Your discoveries . . . are surely the most exciting thing to happen in recent years in the field of human paleontology!" Clark Howell wrote enthusiastically to Richard after viewing his finds, while the usually staid British journal *Nature* praised Richard's "remarkable flare [*sic*] for knowing exactly where to look for fossils."

Nature also readily published Richard's articles announcing his discoveries—though there was so much to report each year that the articles began to sound like Hollywood sequels: "New Hominid Remains and Early Artefacts from Northern Kenya"; "Further Evidence of Lower Pleistocene Hominids from . . . Kenya"; and "Further Evidence of Lower Pleistocene Hominids from . . . Kenya, 1972." There were large photo stories, too, featuring Richard and his fossil trophies in the *National Geographic* and *Life* magazines, as well as a steady stream of invitations to lecture at American universities and colleges. (Richard, who still lacked a university degree, often began these talks by quipping that the only time he had attended college was to lecture.) In a scientific field where success is measured as much by the luck of the hunt as by rigorous research, Richard had, at age twenty-six, taken the lead.

Richard's sudden, explosive success posed a dilemma for Louis. He had raised his sons to be competitive and ambitious, and had always hoped that one of them would follow in his footsteps. But Richard was no longer merely following; he was surging ahead. While proud of his son's achievements—and delighted that the hominid fossils at Rudolf supported his interpretation of those from Olduvai—Louis also felt threatened. "Louis sometimes spoke of how few great men had great sons," recalled Alvin Gittins, an artist the Leakey Foundation engaged to paint a portrait of Louis in 1970. "He understood that for his sons to fight their way from underneath his shadow required a certain toughness, and so he seems to have raised them as animals on the African uplands are raised—to be aggressive, to be tough, to keep him on the defensive, and ultimately to bring down the old lion himself. And he encouraged them!"

"Look, here is a buck," Louis said to Gittins one day, while the two were touring the Nairobi game park. "He's got his does around him, but he's on the decline. There's a young buck over there, and he's attracting the does away; and over there are the peripheral, disinherited old bucks. They have nothing now, except their own company." With every fossil hominid Richard produced, Louis felt himself pushed further toward that periphery—but it was a position he refused to accept. He might be physically on the decline, but he continued to view Richard, in spite of his successes, as far too young and inexperienced to lead the herd.

Besides, Richard's success appeared to have come so easily. Fossil hominids seemed to be everywhere at Rudolf; Richard need only bend down and pick them up, while Louis and Mary had struggled to find the specimens that had brought them fame. The disparity made Louis slightly defensive, and in a report to the National Geographic's research and exploration committee about Richard's 1970 expedition, Louis claimed that he never would have gone to Olduvai in 1931 if he had had the vehicles to get to Rudolf. "However, in those days," Louis wrote, "quite apart from financial difficulties, we had not got such modern equipment as 4-wheel drive Land Rovers, helicopters and light aeroplanes, nor was there any reasonably sized motor vessel on Lake Rudolf, and it was not practical for me although I did endeavour to go to the area to explore it, unsuccessfully. . . ."

Louis was even more ambivalent about Richard's success at persuading the Kenya government to support the museum (something Louis had never been able to do), and his son's ability to draw large and enthusiastic audiences overseas. Piqued by his father's lukewarm approval, Richard responded with a mixture of hurt pride and disdain. Louis, he would later say, had become "a sick old man at the end of his career" who "found my successes difficult. I was not old enough or mature enough to respond to that adequately." Consequently, father and son seldom saw each other, except at the museum, and whenever Richard reported the discovery of a new hominid. Then he made a special trip to Nairobi to share the fossil with Louis—and for those moments, at least, they were father and son.[2] With the dirt-encrusted bones spread across a long table at the Centre for Prehistory, the two would sit shoulder to shoulder, examining the specimens, comparing jaws and teeth, the angle of a joint, the shape of a brow. "Fossils gave Father enormous pleasure," Richard said later about these meetings. "He'd given his life to finding these things, and so was always very pleased to see our newest finds—and they were such beautiful specimens."

Outside of these few meetings, Richard and Louis seldom discussed the fossils or their implications for the study of human origins. "It was a very sad thing," Richard recalled,

> but Louis and I didn't exchange a lot of opinions at that time [between 1970 and 1971] because we were quarreling about the excavation permits [whether Louis at his Centre for Prehistory or Richard as director of the museum had the ultimate authority for issuing these]. So we didn't get on and didn't have a lot of time together talking. Consequently, I don't have a complete idea of what he thought about human evolution, or about a lot of things—I really wasn't his intellectual confidant. Had I been, I might have been a lot smarter.

In an effort to have some influence on his son, Louis befriended the anatomist Alan Walker, whom Richard had invited to describe his fossil

[2] Richard sometimes tested his father's skill at identifying fossil bones. Once he and some of his teammates returned from Koobi Fora with a skull that "looked like a crushed loaf of mouldy bread!" recalled the archeologist Gayle Gittins. "Richard said in a low voice, 'Let's test the old man's genius!' But Louis stood straight up, rather enjoying the situation, went into the next room, asked for a chair and sat down. Here was this 'thing'—it hadn't even been cleaned. Louis sat silently while Richard and the others stood in the background. He scrutinized it from every angle for about four minutes. Then he quietly said, '*Equid* [horse].' Richard's and his friends' faces just fell, even though they were terribly impressed at Louis's quick and correct identification. . . . And Louis laughed—just like a child."

hominids. Initially, Louis had disliked Walker intensely.[3] But later Louis discovered that Walker had attended St. John's College, Cambridge, Louis's alma mater, and had studied with his colleague, the anatomist John Napier. "So he could hardly imagine that I was poorly trained," said Walker. Richard had also become a close friend of Walker's, and Louis turned to him as an intermediary. "I used to stay with Alan when I first came out to Kenya," said Andrew Hill, a paleoanthropologist who joined several of Richard's expeditions. "And Louis had adopted Alan. Since Louis wasn't going to talk to Richard, he had to have somebody to shout his ideas at and show new things to. And it was always Alan who was phoned up in the early morning to come round and see Louis. I think Louis also knew that it was a way of talking to Richard—that Alan would tell Richard what was going on."

The ongoing quarrels between father and son were now legendary, and for every fossil discovery that drew them together, some new administrative issue cast them further apart. Soon enough, the fossils themselves became an arguing point. Not only had Richard's team discovered an unprecedented number of hominid specimens; it had also turned up nearly one thousand well-preserved fossils of all types of animals: extinct species of elephants, pigs, monkeys, antelopes, giraffes, horses, hippos, and crocodiles were all represented. Fossils in the barren Rudolf desert were so common that one could "hardly walk ten paces without seeing" one, Richard wrote in his autobiography. There were archeological riches as well from the two sites that Richard and Kay Behrensmeyer had discovered in 1969. Originally dated to 2.4 million years ago, the artifacts were then the oldest-known tools in the world—some 600,000 years older than those from Olduvai. (The date was subsequently revised to 1.8 million years ago, making them roughly the same age as the oldest Olduvai implements.) It was a fantastic amount of material—hominid fossils, faunal specimens, and artifacts—and promised, as Richard told Gilbert Grosvenor of the National Geographic Society, to resolve "the perennial queries about human evolution."

Answering these questions, however, would require more than a roomful of stunning specimens. The fossils would have to be studied, measured, carefully compared, and subjected to the detailed type of analysis that Phillip Tobias had carried out on *Zinj* and the Olduvai *Homo habilis* fossils. Often, this type of research was done far from Kenya—in England or South Africa —primarily because Louis's Centre for Prehistory lacked the proper facili-

[3] Only a few years before, when Walker was teaching at Uganda's Makerere University, Louis had advised the country's then-president, Milton Obote, to stop the young man from searching for fossils. "I had shown Louis some fossils from a new Miocene site I'd found in Uganda," said Walker, "and I thought he would be interested. But he said, 'No, no. I'm not interested. We've got lots of Miocene sites in Kenya; we don't need any more.' " Louis was apparently more concerned that he had no control over the young scientist's explorations, and wrote immediately to President Obote —whose staff passed the letter on to Uganda's director of antiquities, a friend of Walker's, who then apprised him of its contents. "The letter basically said, 'Dear President Obote. This is just to inform you that there's a young unqualified, untrained Englishman ruining fossil sites in your country, and it ought to be stopped,' " said Walker. "And it was signed, 'Louis.' But nothing happened, and I just carried on."

ties. "The hominid room, where the fossils were kept, was in one of the old museum buildings," recalled M. E. Morbeck, then a graduate student in physical anthropology who visited the Centre in 1970 to study Louis's Miocene primate fossils. "And nothing was kept in a safe. All of the hominid fossils from Olduvai and other sites—fossils like the Peninj jaw [the *A. boisei* lower jawbone Richard's team discovered at Lake Natron in 1964]—were kept in cabinets in Dinky Toy boxes or cigar boxes, and these were just piled one on top of the other. There was one bare lightbulb hanging in the middle of the room—and that was the only light, aside from what came in the windows, for studying the fossils. It was just incredible."

For Richard, the poor storage facilities were acutely embarrassing. His Rudolf fossils and artifacts also ended up in crates and shoeboxes stashed in a wooden-floored hut, with no protection from either fire or theft. "It was a disastrous picture," Richard said, "and it struck me as just outrageous that such important things should be kept like this." To make matters worse, Richard had also launched a campaign to have all the Kenyan fossils that were then being studied abroad returned; but there was nowhere to store them and he was soon under pressure from his father to stop. Although Louis shared Richard's concern about the storage problem and, Richard said, "often complained bitterly about the appalling conditions," he did nothing to improve the situation, feeling that there were always more pressing matters. In late 1970 Richard decided to tackle the problem himself.

"I told Louis that I thought it would be very easy to raise money to build an air-conditioned vault like they had in Dar es Salaam for *Zinj,* around which we could then build research rooms," Richard recalled. "I planned to raise the money myself; I thought it was something useful I could do." But Louis strongly opposed Richard's idea. "Louis told me that it was a waste of money," Richard said, "that we didn't need such fancy plans, that the younger generation had no idea of how money should be spent, that we wanted carpeted rooms and fancy chandeliers, when money really needed to be spent on other things. He couldn't conceive of raising a million dollars to build a building, when he was fighting to raise $5,000 to run Tigoni."

In spite of his father's opposition, Richard presented his plan to the museum's board of trustees and with their approval launched a fund-raising campaign in the spring of 1971. To prospective donors, Richard explained that he had drafted plans for an Institute of African Prehistory—a scientific complex that would be named "in honour of Louis Leakey who has without question done more than anyone else in this field," as Richard explained to the executive director of the Leakey Foundation. Richard had, however, said very little to Louis about his plans—and certainly nothing about tapping his father's connections at the foundation—only announcing them to his father on the evening before Louis left for his 1971 accident-ridden lecture tour in the United States. Louis was not impressed.

"Richard suddenly turned up at the house [last night]," Louis wrote indignantly to Joan Travis about this visit, "having not been in my house for nearly five months, and told me that there was a plan . . . to do more than merely add buildings to my Centre, but instead, to set up something to be named after me." Louis was instantly suspicious, imagining that rather than honoring him, Richard's plan was really a ruse to oust him from his Centre. He was certain that his son intended "to take over the whole of the collections and functions of my Centre, [which would then] be run by Richard . . . with me retiring and getting out. . . . Obviously, in due course, Richard and Mary will have to take over when I leave off, but I do not wish to be pushed out into the cold just yet."[4] As always, Louis was angry that Richard had not consulted him. "I must be kept much more in the overall picture," he complained to another Leakey Foundation member, "rather than being told about things after each step has been taken, which frankly, I find very hard to swallow from my son."

Louis wrote these letters prior to his fall in San Francisco and subsequent brain operation, so it may be that they reflect the paranoia that several observers noted about Louis. But Richard himself was not well—though Louis never knew this. Three years had passed since Richard had been diagnosed with terminal kidney disease, and while he continued to appear healthy, never missing a day's work, his kidneys had begun to deteriorate. "I used to have a lot of pain from the kidneys," Richard said. "I'd get very bad backaches, swollen ankles and a swollen face, and terrible headaches." Two or three times a week, he would wake with a blinding migraine, brought on by high blood pressure, another symptom of the disease. He treated the migraines with Panadol, swallowing as many as six a day to dull the pain. But he did not treat his blood pressure, and its constant pounding left him "very tense, and very easily irritated. I tended to be very fractious," Richard recalled. "I was very uptight and could get upset by the smallest things. Overall I was probably less tolerant and patient of others than society normally expects."

Richard knew his clock was ticking. Three years earlier his physicians had predicted his kidneys would fail in ten years' time. Did he now have only seven years left? If so, how many of his dreams and ambitions could he realize? "He wanted to do certain things before he died," explained Meave, whom Richard had married in October 1970, a year after divorcing Margaret. "He'd say, 'You know they say my kidneys are going to pack up. But I'm going to live a long life.' But I think he really felt his time was short, and so if people objected to what he wanted to do or how he wanted things done, he felt he was wasting time and he'd get very uptight."

The museum was high on Richard's list of objectives. He wanted to transform it from a small, provincial museum into an institute of international

[4] Others shared Louis's view of Richard's actions. "I'm sure that Richard did propose the institute as a way to take over Louis's interests," said the paleoanthropologist Andrew Hill. "He did it in a way that would have made Louis look silly if he complained, because it was all done in Louis's honor. And of course that made Louis furious."

standing. But to do that, Richard did, in fact, need control of the fossil collection—just as Louis had suspected. Particularly important were the fossil hominids, which were in his father's care. At the end of 1969, Louis had agreed in principle to merge his Centre with the museum one day, and now at every museum board meeting, Richard pressed him to set a date. Further, Richard began urging his father to relinquish Tigoni, the primate center. "The idea was to develop the museum so that it would become rather like a consortium of several different research centers," explained Thomas Odhiambo, then the chairman of the museum trustees. "The Louis Leakey Memorial Institute of African Prehistory was going to be one; the Primate Centre another; a Material Culture Centre the third. It was all part of a plan to have Kenya become preeminent in these fields; to make it a major center for research in the whole of Africa." But it was one thing, Odhiambo added, to argue that the Centre for Prehistory should fall under the museum's jurisdiction (on paper, it already did), and quite another to ask for Tigoni.

"Tigoni was Louis's own child; even the title was private," Odhiambo explained. "So it was an imposition to ask him to kindly hand that over to the museum even though it [the primate center] was badly managed. And Richard was not kind about this. He was not adverse to saying to his father, 'What will happen when you die?' "

"I didn't do it in any sense to shock," Richard said about his blunt question. "It was just a straightforward statement, and as far as I can recall, father was always very direct and to the point with people. And it seemed to me that we had a big crisis coming on our hands, if this matter wasn't cleared up, and he wouldn't see it."

After recovering from brain surgery, Louis did grudgingly agree to hand over the Centre for Prehistory to the museum in June 1972. He would stay on as honorary director, while Richard assumed the role of administrative director. As for the Louis Leakey Institute, Richard let the matter drop. He would wait until the Centre was fully under his control, then raise the funds he needed to build a modern research center. Tigoni, too, remained Louis's. To make certain that no one had any doubts on this issue, Louis always distributed a memo to his staff before leaving on his lecture tours. "If any major crisis should occur at Tigoni please consult me. *TIGONI IS NOT THE CONCERN OF RICHARD LEAKEY OR THE MUSEUM TRUSTEES OF KENYA* AND HE SHOULD NOT BE ASKED TO DO ANYTHING FOR TIGONI." It was something of a last roar from an old and failing lion.

"The last time I saw Louis," recalled Mary Smith of the National Geographic Society, "I looked at him and thought, 'You're going to die.' He just looked so bad—terribly tired and very, very fragile for a man who was so big and robust-looking. I think he sensed that he didn't have any more time left. And he was still frantically trying to look into the pages of every book that had ever been written. That's really what he was trying to do: to find out everything there was and then tell everybody."

Louis had seen Smith in 1971 in Washington, D.C., at the beginning of his autumn lecture tour. He may have looked exhausted to her, but, as Louis told Dian Fossey, "I am very very much better, in fact better than I have been for nearly two years." He felt so good that he was now "going full steam ahead again"; he even thought he would visit Dian and her mountain gorillas "in the not too distant future." It was purely fantasy, of course. Louis's right leg and hand remained partly paralyzed (he never was able to write properly again), and when in Nairobi, which at five thousand feet is only half the elevation of Dian's mountain camp, Louis's doctor was constantly packing him off to the coast to relieve the stress on his heart.

Yet Louis's successful brain operation had left him feeling so invigorated that he could scarcely keep still. Between the fall of 1971 and the spring of 1972, he was constantly on the road, lecturing, raising funds, attending conferences, finishing books and starting new ones, listening to and assisting the dreams of students, and trying to launch a special one of his own: an expedition to the Suguta valley, a region south of Lake Rudolf that he had been eyeing for some years.[5] "With the enthusiasm and stubbornness that have marked his career," wrote a reporter for the Washington, D.C.-based *Science News,* "Louis S. B. Leakey says he can prove that true ancestral man existed seven million years ago." He was busy "making plans for an expedition next summer" that would "bring back the earliest evidence for the origins of man's ancestors."

In fact, Louis was simply hoping to stir up enough interest (particularly at the National Geographic Society) to raise the money he needed to equip such a venture. But few at the Geographic shared Louis's optimism. Besides, the United States was in a recession. Even at the Geographic, funds were tight; and some on the grants committee thought the Society was already giving the Leakeys (father, mother, and son) more than enough money.

The recession hampered all of Louis's fund-raising efforts. People packed the halls to hear him lecture, but did not readily open their wallets. Louis's supporters at the Leakey Foundation tried to help with rounds of parties and lectures at nearly every California college campus. They also persuaded a more well-heeled crowd to ante up forty dollars each to listen to a debate on human aggression between Louis and Robert Ardrey, the playwright-

[5] Louis had arranged for his son Philip and the Spanish paleontologist Emiliano Aguirre to make a three-week survey of the Suguta valley in 1969. They actually never reached the valley itself—which is far from any roads and in virtually waterless country—but explored some deposits to the south. They found a variety of fossils, including some from a previously unknown species of three-toed horse, and a pig's tooth, which Louis considered a promising sign. He was further encouraged by a geologist's discovery of a ten-million-year-old primate's tooth from a site called Ngorora. Although Ngorora lay about fifty miles to the south of Suguta, Louis often cited this tooth as evidence of what could be found.

author of *African Genesis,* a best-selling book, at the California Institute of Technology in Pasadena. "Well, we were a great hit," Ardrey recalled some years later. "Yet... in a burst of modesty so uncharacteristic of the man, [Louis] touched my hand and said, his nose wrinkled up: 'I don't think we were worth it. Do you?' "

The money dribbled in, seven hundred dollars for one talk, a thousand for another. Somehow the Leakey Foundation managed to scrape together enough funds to bail out Louis's Centre for Prehistory and Tigoni, and to help Louis get his fourth primate project, a study of the pygmy chimpanzee, under way.[6] But overall the foundation was something of a disappointment to Louis. The great man its members so admired and so wanted to help still "went around without teeth or a nickel in his pocket," said Joan Travis.

Searching for money, Louis traveled back and forth to Nairobi, London, and the United States. Only after spending February as a visiting professor at Cornell University in Ithaca, New York, did he slow down—and then only because he dropped a log, destined for the fireplace, on his left big toe, fracturing it in four places. "It was agonizing," he wrote to Mary from Joan Travis's home in California, where he was recovering. He had done nothing for his foot at Cornell except to wear a loose shoe, and now he was flat on his back with a swollen leg and foot, and taking drugs to ward off phlebitis. Typically, though, he was also going out that evening to a fund-raising dinner, and then there were "some lectures" to give "over the next eight to nine days." Mary could expect him home in April.

Mary had tried many times to persuade Louis to cut back on his work and travels, but to no avail. Once she and Jean Brown, an archeologist at the museum, visited Louis together in the hospital at Nairobi after one of his minor heart attacks and urged him "to slow down a bit. [B]ut he said, 'I'm not taking any notice of either of you bossy women,' " Brown recalled. Louis took equally little notice of his physician's orders.[7]

Louis was back in Nairobi and still toying with the idea of visiting Dian Fossey ("My only problem is whether my heart and lungs would bear the altitude of your camp," he wrote her in May) when he received news of an

---

[6] Louis selected Toni Kay Jackman, a graduate student at Cornell University, for this project in early 1972. Through the Leakey Foundation, he raised nine hundred dollars for her to come to Kenya and work at Tigoni while he trained her in observation techniques. Louis also sent her to work with Jane Goodall at Gombe Stream for three months in the summer of 1972. Jackman would then, Louis planned, travel to an area in Zaire to begin her project. But Louis died shortly before Jackman left Gombe and she was unable to follow through on his plans. The animal she was going to study, *Pan paniscus,* the pygmy chimpanzee, is actually not any smaller than the common chimpanzee, *Pan troglodytes.* In 1973, the Japanese primatologist Takayoshi Kano launched a study of *Pan paniscus* that continues to this day.

[7] "My wife and I were in Nairobi after one of Louis's heart attacks," recalled the geologist Basil Cooke, "and he was supposed to lie down each afternoon. We'd invited him out to dinner for his birthday, and Louis said he would meet us at the restaurant after he'd taken his rest. When we got there, he was already waiting for us and looking very bright." The Cookes asked if Louis had had a good rest. "And Louis said, 'Oh, I didn't rest. I wrote up the whole wildebeest chapter for my book for the National Geographic Society.' "

unexpected windfall: the Guggenheim Foundation had awarded him $100,000 for his Suguta expedition.[8] At last, Louis would be in the field again. "He was so elated that he had that money, that there was enough to engage a pilot so that he could fly to the site," said Joan Travis. "He just loved the whole idea of exploring for fossils again. It was going to be his last hurrah."

It was too late to organize the expedition for the summer of 1972, so Louis postponed it to 1973. Besides, he had already planned to spend August in London with Vanne Goodall, polishing the second volume of his autobiography and beginning a book about the Piltdown forgery—which, he told his literary agent, "is liable to be very controversial, sell very well and probably have a big potential as a serial to run in one of the major newspapers in England. . . ."[9] Then he had his annual autumn lectures to prepare, along with a series of special seminars on human evolution for the California Institute of Technology.

"May I remind you that we will need to know your outside dates for the Fall as soon as possible," Joan Travis had written Louis in the spring of 1972, "so that we can start making arrangements." She thought he should plan on spending most of September and all of October in the United States. "You know," she wrote, "ten days for Dee Simpson [at Calico], three to four days for Seattle, three to four days in San Francisco, a few days in Chicago, a few more in New York, a few in Minneapolis, Los Angeles and so on. Films, cassettes, TV, McMaster, whew! Seriously, do try to allow for these, plus lectures."[10]

It is not surprising that Louis began closing many of his letters with the refrain, "Forgive the brevity of this letter . . . but I am very pushed for time."

---

[8] The grant was a direct result of Louis's meeting with Robert Ardrey. After their debate, Louis had told Ardrey about the Suguta valley and his belief that it contained a wealth of hominid fossils. Ardrey then shared Louis's plans with an old friend, Dr. Henry Allen Moe, a former director of the Guggenheim Foundation, who persuaded the foundation to award Louis the grant.

[9] Louis believed that the Jesuit priest Pierre Teilhard de Chardin had planted the fake fossils in the Sussex gravel beds as a sort of "practical joke, which went wrong." Louis discussed this idea with several of his colleagues, none of whom was convinced, although more recently the Harvard paleontologist Stephen Jay Gould revived this theory for many of the same reasons that Louis held. The mystery of who staged the hoax has yet to be solved, though Charles Dawson, who reported the discovery, has long been regarded as the primary culprit. Louis, however, thought that Dawson had acted in concert with Teilhard de Chardin. "Perhaps you know," Louis wrote intriguingly to Glyn Daniel, an archeologist at Cambridge University, "that when Teilhard was asked to comment [about the hoax], on my instigation (because I sent the reporters to him), he would only say something like . . . 'I do not wish to comment, but I can tell you *I know* that Dawson was not *responsible.*' Please note that he did not say 'Dawson did not do it,' nor did he say that Dawson was not implicated, but 'Dawson was not *responsible.*' " In *Leakey's Luck,* Louis's biographer Sonia Cole claims that Louis was "polishing" his book about the Piltdown hoax shortly before he died. But Mary Leakey insists that Louis "never got further than writing the notes and rough draft," which are now stored in the archives of the National Museums of Kenya. Mary regarded Louis's ideas about Teilhard de Chardin "as pure fantasy and I was reluctant that they should be made public—it would have been good publicity for whoever published them—again, I wanted to avoid cashing in on this."

[10] The Leakey Foundation was preparing a series of sixteen-millimeter films showing Louis holding forth on various subjects. The foundation intended to distribute these to universities for a fee. McMaster University in Ontario, Canada, had invited Louis to present its prestigious Redman Lecture in October.

• • •

With little fanfare, Louis's Centre for Prehistory and Palaeontology was transferred to the National Museums of Kenya on June 30, 1972. Richard was away at Rudolf on his fifth expedition at the time, and Mary was at Olduvai. Louis ignored the change, coming into his office early in the morning as he always had. But those who worked for him felt that it must have been a blow. "Louis was an Alpha male type," said Peter Andrews, a paleontologist on the centre's staff, "and he liked to be right, to be the center of attention and to be running things—which is why it was so awful when Richard chucked him out. Suddenly Richard was the head of the museum, and his father was just the head of this one rather small section. It was extremely unpleasant." Others regarded the transition as inevitable. "I remember saying to Ron Clarke [a technician at the centre] in 1966 that even though Richard was then to some extent still the wayward son, one day we would be calling him 'Sir,'" recalled Alan Gentry, another paleontologist. "And certainly before we left Nairobi, my prophecy did come true."

Richard could at times seem affected, pompous even, but nevertheless always had "the charisma of a natural born leader," said the geologist Kay Behrensmeyer. Already in the four years since he had been appointed museum director, he had gained the ear of the young Kenyan government and persuaded it to do what neither his father nor Robert Carcasson had ever been able to: increase the government's contribution to the museum's annual budget from £23,000 to more than £200,000. By 1971 Richard had further enhanced the museum's prestige by establishing two regional branches—one in Kitale, in western Kenya, and one at Mombasa, on the coast—so that the National Museum of Kenya was now properly called the National Museums.[11] He had tripled the size of his staff as well, and arranged training for many of his employees in museums overseas.

Richard's concern for order and organization would now be brought to bear on Louis's Centre—which was as crammed with books, papers, fossils, and artifacts as Louis's head was with ideas. Even Louis's office reflected this creative chaos.[12] Dusty bits of bone and broken handaxes lined the windowsills, files were stacked on top of cabinets, and papers were piled in baskets and on his desk. In contrast, Richard's office was spartan and tidy, without pictures or decorations of any kind.

Richard brought the same order and appetite for work to his expeditions. "He ran a tight operation," recalled Kay Behrensmeyer. "We were always up and out at dawn, and very regular in our whole schedule, which Richard put

---

[11] Today there are seven regional branches scattered across the country.

[12] Dian Fossey retained a "very special keepsake" from Louis's office. "It is one of his huge, white desk blotters that I have mounted and framed," she wrote in a letter two years before her death. "The blotter is filled with doodles—squares inside of squares and all types and shapes of geometrical forms and whirls inside of whirls." For her, Louis's jottings represented "the balance and sanity of Dr. Leakey's ever inquisitive mind."

a lot of stock in." By 1972 such organization was imperative, since the size of his expeditions had exploded. At times more than seventy people—scientists, graduate students, Kenyan fossil hunters, and camp staff—were in the field. "We'd gone from being a small, exploratory outfit, to a busy, multifaceted expedition," said the archeologist Glynn Isaac, whom Richard had invited to be coleader in 1970. "We'd all be in the field at the same time—paleontologists, geologists, archeologists. So there was a lot of sharing and interacting and a feeling that we were an organized team."

Richard still based his team at Koobi Fora, the sandy spit of land that stretched into Lake Rudolf, but his tented camp was a thing of the past. Only a hundred feet from the water's edge, he and his staff had constructed an airy, thatched communal dining room, wooden huts for a kitchen, sleeping quarters and laboratories, and a home with wide verandas for himself and Meave. Richard equipped the kitchen with a propane-fueled refrigerator, and since he commuted to Nairobi every other week to attend to museum affairs, the camp was well-stocked with fresh fruit and vegetables. There was almost always fresh fish from the lake as well, or Richard might bring up a crate of frozen chickens, or a slab of beef or leg of lamb.[13] Some nights, he surprised his team with bowls of strawberries or succulent peaches. "It was like going to Club Med," said Betty Goerke, an archeologist who was astonished to find such pleasures in the midst of the broiling desert. "And it was such a contrast to Glynn Isaac's field camp, where we were stuck in tents far from the lake, and all we had was corned beef, beans, rice and maize meal. You looked forward to visiting Koobi Fora."

After dinner in the glow of kerosene lanterns, Richard, who always sat at the head of the table, discussed the day's work with his crew and planned the next. As the expedition leader, Richard was the one who set the agenda. He largely ignored the objections of others. "I remember a walk on the Koobi Fora beach once with Richard and [paleontologist] Vince Maglio," said Behrensmeyer, "and Vince was a little uneasy about the way Richard was doing certain things. He felt the collecting and documentation of the fossils wasn't tight enough, and he implied that because of his university background he knew more than Richard did. And you could see that Richard did not appreciate the advice. He was irritated that someone from a university would try to tell him what to do in his own fossil beds."

In spite of his lack of formal scientific training, Richard was a quick study. "He was like a sponge, the way he listened to others, their ideas and theories, and made them his own," said the physical anthropologist M. E. Morbeck. "Richard had a phenomenal memory and grasp of detail," added Paul Abell, the geochemist who had been part of Richard's team since 1967.

[13] Richard had abandoned flying in 1967 after crash-landing at Olduvai—"a stupid incident . . . in which I very nearly killed several good friends," he wrote in his autobiography. But as the Turkana expedition and Richard's role at the museum expanded, it quickly became clear that flying was the best way to commute between Koobi Fora and Nairobi. Richard regained his license and nerve in 1970 and resumed piloting his own plane.

"He'd bring some specialist to camp and after a few days he'd wrung every drop of information out of them. He could talk just as expertly as the 'expert' did."

Surprisingly, this talent rarely alienated his team of scientists, perhaps because above all, Richard had a gift, like his father, for making people believe in themselves and their abilities. "He would build people up, make them feel that he was their friend and comrade-in-arms, and created such an amazing team spirit, that people would say, 'I'll do anything for him,' " said Behrensmeyer.

In time, this devotion to Richard would lead to charges that he demanded blind loyalty from his team members. But in the early years of the expedition, there was only zestful enthusiasm for the young leader and his scientific enterprise. "We were finding everything that Clark Howell and his team had gone to the Omo to find," said Glynn Isaac. "We had spectacular hominid specimens, wonderful fauna, and stratified archeological sites that could be dug and studied and argued about. So it was by far the best place to be."

While Howell and Yves Coppens and their crews toiled once more in the Omo deposits, finding hominid teeth and fragmentary mandibles and skull fragments, Richard's team was rolling in ancestral bones. Two years before, he had asked Kamoya Kimeu to head a small, mobile team of men to search primarily for hominid fossils. Dividing the vast Turkana region into a series of sections, the Hominid Gang, as Kamoya's team came to be called, spent their days walking up and down the deposits, eyes alert for the signs of an important find—the glint of tooth enamel, the puffy arch of a hominid's brow, the smooth curve of a bit of cranial bone. While Richard, Meave, and other team members occasionally found a hominid bone, the Hominid Gang was responsible for most of the specimens.[14]

By 1972, burdened with museum work and the logistics of the expedition, Richard seldom searched for fossils anymore. But he enjoyed collecting them and asked Kamoya to notify him of each discovery. Because of the fragile nature of many of the specimens, Richard "made it a rule," he wrote in his autobiography "that all fossil hominid finds . . . [were to be] left undisturbed" until he had visited the site. He would then inspect the fossil, photograph and record it, determine the best strategy for excavating it, and remove it himself—a task that often required hours of work as he lay stretched full-length or squatted on his knees under the desert sun.

To keep abreast of the Hominid Gang's work, Richard and Kamoya stayed in touch via radiotelephone—and hardly a week went by when the receiver at Koobi Fora did not crackle with the news of another discovery. Barely two weeks after the 1972 field season began, Richard flew north to an area

[14] In 1970 Richard and Meave independently found the first hominid femurs (thigh bones) in Kenya. Meave found hers first; her discovery triggered a subconscious memory in Richard. Retracing a route he had taken some days earlier, he came back a few hours later with a fossilized femur of his own. "I was amazed and wondered how he did it," Meave wrote in her diary, "but he simply remarked that he had to maintain his position!"

called Ileret, where he inspected the first find—fragments of a skull that he thought might possibly be those of *Homo erectus.*[15] The next week, Kamoya reported "some hominids," as Meave noted in the Koobi Fora journal; two days later, there was news of a femural shaft (leg bone), ankle bone, and more skull fragments; and three days later, a report of an australopithecine mandible. It began to seem that all one had to do was walk across the deposits at Turkana and hominid fossils appeared. At the season's end, Richard's team had amassed another thirty-five specimens.

Ironically, the most important hominid fossil found that year—an almost complete *Homo* skull from deposits then dated at more than 2.6 million years—was at first scarcely given any attention at all. The precise date of its discovery is not even recorded. But a few days prior to July 27, Bernard Ngeneo stopped to study a small pile of fragmented fossils in a steep gully dotted with scrubby thornbushes. Other people, including Kamoya and the paleontologist John Harris, had noticed this same pile before, but seeing primarily bits of antelope bone they had passed it by. Ngeneo, however, knelt down for a closer look. "I was not sure it was a hominid and so I spent a long time checking it," he recalled. "Then I found a piece that had these suture marks, and I thought then it might be a hominid because I'd seen those marks on the skulls of human beings." He called Kamoya over to look and Kamoya agreed. "So we called Richard [in Nairobi] to come and check," Kamoya said.

Richard left his office in suit and tie, flew to Koobi Fora, and at dawn the following day, with the anatomist Bernard Wood in tow, piloted his Cessna to a region designated "area 131." Kamoya and his crew were waiting for him at a short airstrip they had scraped in the desert. As soon as Richard landed, everyone set off at a fast pace to see the new find. "What a place to find a hominid!" Richard exclaimed as he bent over the scrappy pile of light-colored fossil bone. He and Wood picked up a few of the larger pieces, and studied them intently. "None of the fragments," Richard noted in his autobiography, "was more than an inch long, but some were readily recognizable as being part of a hominid cranium." There were other good signs: some of the pieces obviously came from the back of the skull, while others were from the top and sides; there were even bits of facial bone. Perhaps there were enough fragments to reconstruct a complete skull. Since the skull was so badly broken, Richard knew that most of it would be uncovered by sieving, which would take weeks. This time he would entrust the work to Kamoya. But Kamoya's crew also had other work to complete, so the sieving operation was delayed for several weeks.

"1470 [as the skull came to be known, after its field number] seemed so insignificant when it was first found," recalled Meave. "Nobody took any note of it. We didn't even sieve it for ages. And then, when we did, we had it all."

---

[15] Today these fragments are classified as *Homo,* species undetermined.

By August 20, Kamoya's crew had collected a small bag of fragments, which Richard brought to Koobi Fora. After Meave washed them and laid them in the sun to dry, Richard and Bernard Wood began trying to fit the pieces together. Sitting at a worktable on the veranda, they quickly joined together several of the larger pieces from the top of the skull. "In no time at all . . . we realized that the fossil skull had been large," Richard wrote in his autobiography, "certainly larger than the small-brained *Australopithecus* such as we had found in 1969 and 1970." But they could not proceed without additional pieces. These came out of the ground slowly, some bits no larger than a thumbnail. Richard soon grew frustrated with the slow reconstruction and turned it over to Meave, who had been so fond of jigsaw puzzles as a child that after fitting one together with the picture side up, she would turn it over and do it again with only the cardboard side showing. The skull, however, was an even greater challenge. "We have a jigsaw puzzle with no edge pieces," she said wearily after one particularly frustrating day.

Meave had given birth to her and Richard's first child, Louise, in March, and had brought the four-month-old baby to camp. Now with Louise resting in a basin of water—to keep her cool in the desert air—Meave puzzled over the skull. There were some days when she would find only one or two matching pieces. Still, in a little more than three weeks, she had fitted enough of it together for its general shape—and importance—to be seen.

"It was larger than any of the early fossil hominids that I had seen," Richard wrote, "but the question was, how large was the brain?" He and Wood ventured a guess by sealing the gaps in the skull's vault with Plasticine and tape, then filling it with sand. They measured the sand in a rain gauge, coming out with a volume of eight hundred cubic centimeters—which placed the skull at the low end of the cranium capacity of *Homo erectus.*[16] But this skull came from deposits that had been dated to more than 2.6 million years, making it at least 1.5 million years older than any known *H. erectus* fossil.[17] "Historic moment!" Wood wrote elatedly in the Koobi Fora camp journal at the conclusion of their experiment. "Tremendously exciting—quite memorable." In their hands, they knew they held the largest hominid skull of its age that had ever been found.

Meave soon had much of the face joined together as well, and Richard decided—to Meave's annoyance—to take the skull to Nairobi to show to Louis. "The skull was very fragile and I wasn't done with it, and I didn't think he needed to be flying off with it like that," Meave said. But Louis was soon leaving for London and a lecture tour in the United States, and Richard wanted him to see the skull before he departed. "I knew how excited Louis

---

[16] Later, when more accurate methods were used, the volume of 1470's brain was determined to be 775 cubic centimeters. By comparison, the skulls of *Homo erectus* range from 750 to 1,100 cc; the craniums of average modern humans contain about 1,400 cc.

[17] Later, these fossil deposits were correctly dated to about 1.88 million years ago, while the 1470 skull is now dated at about 1.9 million. *Homo erectus* is dated to about 1.8 million years in East Africa.

would be to see it, especially knowing of his difficulties with *Homo habilis,*" Richard said. "I just felt that it was desperately important for him to see the skull; that it would give him enormous pleasure at a time when he was not getting much pleasure out of life."

On September 26 Richard flew south to Nairobi with the skull packed carefully in a wooden box. Mary, who had come to Koobi Fora for a few days, traveled home with him. Later that morning, they walked into Louis's office and Richard opened the box, then placed 1470 in his father's hands. Louis's face lit up at once. Here at last was absolute proof of what he had long argued: that a large-brained *Homo* had walked the African savanna more than two million years ago. "The feeling of delight and enthusiasm [on Louis's face] as he handled this skull was almost tangible," his secretary later recalled, "and his hands, and the expression on his face can never be forgotten." Looking up at his son, Louis said, "They won't believe you!" Then he paused and added, "I give you 100 per cent, Richard, I give you 100 per cent."

Louis, Mary, and Richard sat in Louis's office discussing the skull until well past noon—one of the few times in several years when they were all together. "[It] was a long and extraordinary discussion," Richard wrote in his autobiography, and they decided to continue it over dinner at his parents' home. "Louis was excited, triumphant, sublimely happy," Mary wrote about this dinner. "It was almost like old times." They laughed and talked, gossiped and joked, and Louis exclaimed again and again over the wonderful skull. For once, the tension between father and son disappeared. "In many ways I felt closer to [Father] that day than I had since my early childhood," Richard wrote. "I had a distinct feeling . . . that we had finally made a real peace." But he also felt worried about his father's upcoming trip, and wished the evening would last longer, "that we had more time to talk."

Driving his father to the airport late that night for an early-morning flight, Richard's premonition grew stronger. "I hadn't taken him to the airport in years," Richard said, "and I can remember vividly having a feeling that this was his last trip; that it was the last time I would see him." He urged his father to take care of his health as Louis walked through the departure gate, and Louis nodded vigorously. Then he was gone.

Richard left the next day for Koobi Fora and Mary drove south to Olduvai. Five days later, on October 2, Richard landed again at Nairobi, this time bringing Meave, Louise, and Glynn Isaac home. "We'd finished the field season," Richard said. "And we'd unloaded the airplane and were starting toward our car, when somebody—a man—came across the tarmac and called to me." Richard paused and looked at him. Then the man said, "Have you heard about your father?"

Chapter 30

AN END AND A BEGINNING

Louis—"the man with the million year mind," as one newspaper characterized him, who had "thrown a flood of light on man's ancestry"—was dead.

For Richard, standing on the tarmac at Wilson Airport, the news was a shock. Although he had half expected that his father would not return from this last journey, the actual words left him stunned. Only moments before, he, Meave, and Glynn Isaac were rejoicing at the close of another successful fossil-hunting season. Now they walked somberly to Richard's car and headed for Langata and Mary. Louis had died the day before in London, on October 1, and Richard, it seemed, was the last of the family members to know.

Mary had been at Olduvai when the news first came. Louis's daughter, Priscilla, had placed a call from London to the Langata house, and by chance Philip happened to be there. After making a radio call to Jonathan at Lake Baringo, Philip hired an airplane and flew to Olduvai. Light aircraft stopped so infrequently there that Mary knew there must be some emergency as soon as she saw the plane signal its intention to land. She was not surprised, then, to see Philip step down from the plane. "Another heart attack?" she asked. "Or is he dead?" "Dead," said Philip. Mary quickly gathered a few belongings and flew back to Nairobi with her son.

From Philip, Mary learned what little was known about Louis's death. Louis had traveled safely to London, where he stayed, as usual, in Vanne Goodall's flat. They had spent three busy days together working over a final version of the second volume of his autobiography.[1] But Louis lacked his usual energy and Vanne persuaded him to see Dr. J. L. Grant, the physician who had attended him after his heart attack in 1970. This time, Grant gave Louis a series of electrocardiogram tests, but they revealed nothing abnormal. Louis had, however, been treated for alarmingly high blood pressure in Nairobi, and as a precaution Grant suggested that he reschedule his flight to New York from October 2 to October 5. Louis never made the plane. Early in the morning on the day after his EKG tests, and while getting dressed in Vanne's apartment, he collapsed.

Vanne urgently summoned Grant, who had Louis rushed to the intensive care unit of London's St. Stephen's Hospital. But there was little to be done. Louis was dying—and apparently knew it. Vanne sat by his side, then left at nine o'clock in the morning, and a nursing sister took her place. Half an hour later, Louis died. He was sixty-nine.

"I spoke to the nursing sister . . . who was with Louis when he died," Priscilla later wrote to Mary, "and she said it was quite peaceful and without pain at the end. He looked very peaceful and serene. . . . The sister said his last spoken words were regret that he would not be here at the time of Richard's announcement [about the 1470 skull] next week." Priscilla had seen her dead father at the hospital and offered "a prayer and farewell—I hope not presumptuously—on behalf of us all."

Louis had spent so much of his later years in the company of other women that Mary was not surprised to learn that he had suffered his fatal heart attack in another woman's home. But it added a bitter edge to her grief, and perhaps for that reason she decided to handle the funeral expediently—Louis's body would be cremated and the ashes brought back to Nairobi. When Richard arrived at Langata, these plans were already under way—and Richard was aghast. "I was very upset at the way the whole thing was being handled, and being a pillar of tact, I let them all [Mary, Philip, and Jonathan] know what I thought," Richard said. He was particularly alarmed by what he regarded as their "insensitive" plans for the cremation, which is a taboo in Kikuyu culture. "Louis was a Kikuyu and they greatly admired and loved him, and it seemed to me absolutely essential that his body be brought back and buried in Kikuyuland," Richard said. "And I was very nasty, very mean about this and said things that should never have been said."

Determined to have his way, Richard canceled the cremation and arranged instead for Louis's body to be flown home in a coffin the next day. Then there was the "ghastly waiting" at the airport, as Mary wrote in her autobiography, "while crate after crate of ordinary cargo was unloaded before the coffin could be reached."

After Richard's outburst, the family decided to hold a private graveside funeral for Louis at the little church in Limuru, where Louis's parents are buried. Set on a hillside high above Nairobi in a grove of eucalyptus trees,

[1] This was published posthumously as *By the Evidence: Memoirs, 1932–1951*.

the church, with its chiseled stone walls and soaring spire, is decidedly English. Yet it rises from Kikuyu soil, making it a fitting resting place for a man who had straddled these two cultures. White and black Kenyans, including a representative of President Kenyatta, were also in attendance—though Mary insisted that the number of guests be kept to a minimum. For the final scene of Louis's life, at least, she wanted privacy. But Louis, well-loved and very public, had his devotées, and "somehow far too many [of these] other people" showed up at the funeral as well, Mary wrote in her autobiography. Tense and irritated by the whole affair, Mary was further distressed when Jonathan, who was driving her to the church, made a wrong turn en route, delaying their arrival by nearly half an hour. By then, "nobody was in a mood to say any words to anybody," Richard said, "and the burial was dealt with very quickly and expeditiously."

Two days later, a larger, open memorial service was held in Nairobi's All Saints' Cathedral. People thronged this event, packing the church and spilling into its aisles—as if Louis himself had come to speak rather than to be remembered. Kenyan and foreign dignitaries, the museum staff and young women from Tigoni, Nairobi shopkeepers and several of Louis's Kikuyu contemporaries, as well as his friends and family, came together "to pay [their] respects to the memory of one of the great sons of Kenya," eulogized Charles Njonjo, Kenya's attorney general. Njonjo reminded the assemblage that Louis had come into the world at Kabete, and from there had gone on to win "for himself a world-wide lustre of affection and respect.[2] It was always his great gift to merge himself, quite unaffectedly, into any segment . . . of mankind. And he would wish to be remembered, not in terms of any race or nationality or tribe, but as a member of the human species."

There were other formal addresses, including one by Louis's brother-in-law, Leonard Beecher, later the Anglican archbishop of East Africa. Then, this service, too, drew to a close and the Leakey family, Mary said with relief, was once again "left more or less in peace to sort ourselves out."[3]

Louis's sudden death made headlines around the world, with obituaries appearing in newspapers, magazines, and scientific journals; many letters of condolence were sent to the family as well. Such an outpouring of sympathy and affection might have prompted some families to immediately erect a monument to a man of Louis's achievements. But after his funeral, the Leakeys seldom even visited his grave. Mary left for Olduvai a few days after the service at All Saints' Cathedral; Richard was busy at the museum and planning a trip abroad; Jonathan returned to his home and snake venom business at Lake Baringo; and Philip was again involved in his safari business.

[2] Charles Njonjo's parents had been educated at the Leakey mission school at Kabete.
[3] Another memorial service was held for Louis on October 20 at St. John's College Chapel in Cambridge. His first wife, Frida, together with their children, Priscilla and Colin, and many of Louis's British friends and colleagues attended.

Although they grieved for Louis—"I still find it hard to realise that his tremendous vitality and personality have gone," Mary wrote a month after the event to Joan Travis—they were not sentimental about his passing. Perhaps for this reason, Louis's grave lay unmarked for weeks, then months, and eventually a year—although the family did have plans for a tombstone.

"Mary didn't want a marble headstone or a cross," Richard said, "so we decided to put a large piece of quartzite [the stone from which early humans made many of Olduvai's finest handaxes, and which Louis often used to demonstrate their craft] from Olduvai on the grave. But it was quite a long time before anything was done."

Jonathan finally drove to the gorge, selected a chunk of the white rock, and brought it back to Nairobi. But when he went to Limuru to place the slab on his father's grave, he was amazed to discover another headstone already in place. Made of marble, it bore an inscription:

Louis Seymour Bazett Leakey
1903–1972
"Wakaruigi"
"Son of Sparrow Hawk"

You live on
In the minds you inspired
In the projects you pioneered
In the lives you improved and created
In the hearts that loved you
You cannot die.

A coded message was carved across the very bottom: ILYEA.

Who had interfered in the family's affairs? No one at the church seemed to know. Baffled by the mystery, Jonathan called Richard, who was as perplexed as his brother, until Jonathan mentioned the acronym, ILYEA. "I told Jonathan that I couldn't be certain, but I thought that Rosalie Osborn must have placed the headstone there," Richard recalled. "She used to sign her letters to Louis like that—ILYEA—I'll Love You Ever Always." Richard had discovered a packet of Osborn's letters to Louis among his father's papers, and many bore this acronymic signature.

Much later, Rosalie admitted to putting the marker on the grave. She had never stopped loving Louis, and after the end of their affair in the early 1950s, she had continued to live in Kenya, working as a schoolteacher, and hoping that somehow she might reenter his life. Louis, too, apparently retained a special affection for her. Once he confided to Elizabeth Meyerhoff that Rosalie had been "the great love" of his life. But to the Leakey family, her presence evoked such painful memories that Richard had made a point of asking her not to attend Louis's funeral. She had gone alone to the gravesite after the ceremony. Then, unlike Louis's family, she had continued to make regular visits to the Limuru cemetery to place flowers on his grave. After a year of such visits, and seeing that "nothing had been done," Rosalie ordered the headstone. Even in death, it seemed, Louis would have his other women.

As scandalous as the headstone was, with its message of undying love, it seemed far more scandalous to remove it. "I told Jonathan that we could hardly move it," Richard said. "We'd sat around doing nothing and somebody else had marked the grave; we should be grateful. But, the nerve!"[4]

Rosalie's tombstone still stands at the head of Louis's grave. And the ILYEA inscription continues to puzzle visitors who make the pilgrimage to Limuru. More than one has subsequently said to Mary, "I love your headstone. Could you tell us what ILYEA means?" Her answer is brisk and brusque: "I don't know. I didn't put the headstone there." "Mary hates that cemetery," Richard said, "and she has insisted with all of us, 'I'm absolutely *not* to be buried there.' " She has instructed that she be cremated and her ashes scattered at Olduvai.

Louis had never prepared for death. He apparently had discussed the idea of a will, but he never wrote one or named an executor for his estate—leading to rumors that he had left nothing but debts.[5] "Actually, Father's affairs weren't nearly as bad as they might have been," said Richard, who nevertheless spent weeks untangling Louis's finances. "The lack of a will turned out to be quite immaterial because there weren't any assets, aside from his house at the coast, that a will could have dealt with." The family sold Louis's Malindi cottage to pay Kenya's death taxes, while Richard and Mary covered the "few hundred pounds" of unpaid domestic bills. However, Louis's other bank accounts—those holding money for his many private projects, from Tigoni and Fort Ternan, to Dian Fossey's gorilla study and Toni Kay Jackman's promised pygmy chimpanzee research—proved difficult. A probate court would have to decide the fate of these funds, which could take as long as eight months, Richard told Jackman; in the meantime, the accounts were frozen. Although this left Jackman without airfare home, "Richard didn't seem to care," she said. "He told me we were all on our own."

Adding to the confusion and ill-will were Louis's assurances to Jackman

---

[4] Jonathan suggested placing the family's slab at the opposite end of the grave. But Richard thought it best to bring the rock to the museum, where, he said, "it sits to this day. Maybe I'll use it for my grave."

[5] Although Louis had a peculiar way of handling his finances, he always paid his debts. "He could get credit in any shop in Kenya just because he was Louis," said Richard. "He was trusted as a man of integrity, who kept his word, his promise. . . . And I think that's the thing I most wanted to achieve for myself: to be perceived as having that sort of integrity."

and his other protégées that he had made provisions for them in his will.[6] Jackman refused to believe Richard when he told her that there was no will; she suspected that Richard had appropriated the funds for her study for his own purposes. Louis's other protégés had much the same experience. "The National Geographic Society's financial assistance left for my gorilla research project in Dr. Leakey's account was absorbed by Richard," Dian Fossey wrote pointedly in a letter two years before her death. Elizabeth Meyerhoff, the California student Louis had brought to Kenya in 1970, lost not only her funding but even her camping gear. With Louis's assistance, she had just launched a study of the Pokot people in Kenya's northern desert when news arrived of Louis's death—together with instructions that she was to return everything Louis had given her. "Mary and Richard were incredibly pissed off with what Louis had done with his money," she said. "Everything became quite cold and callous, and I realized what my position was vis-à-vis the Leakeys. Because of my intimacy with Louis, I was now an outcast."

It was not so much the money that Louis had spent on his protégés that upset the Leakey family, but rather the effort and cost to his health that the money represented. In their view, Louis had killed himself trying to raise money for dubious projects, such as Tigoni and Calico. "There was never enough money to do these things properly," said Richard, "but there was always just enough so that they didn't die. So these projects lingered on, and Louis felt *terribly* responsible and kept chasing around, looking for money." Louis's hectic lecture tours for the Leakey Foundation were a particular sore point with Mary, who regarded the foundation's members as little more than cultists. Although Mary subsequently became friendly with many of these people, she initially blamed them for Louis's death. "Mary thought that we had worked Louis to death," said Helen O'Brien, whose husband, Allen, had started the Leakey Foundation. "We supposedly killed him off by wearing him out. Hah! Excuse me, but Louis wore himself out. He never did anything he didn't want to do. And Louis loved nothing more than the roar of the crowd."

It seemed everyone close to Louis blamed someone else for his death, perhaps because each privately believed that he or she had contributed, either directly or indirectly, to the stresses that had claimed him, or at least had stood by while he worked himself to death. Most of the fingers were pointed at Richard—by Leakey Foundation members who had heard Louis's bitter complaints against his son, by the young women at Tigoni, by Dian Fossey. All believed that Richard's maneuverings at the museum were responsible for Louis's death. Few of these accusations reached Richard's ears, but Dian met with Richard one day at Nairobi's Thorn Tree Café to discuss

[6] There was also a rumor that Louis had left his inheritance to "some California woman," Richard said. "We don't know who she was. But Allen O'Brien often told the story that this woman in California was out to get the inheritance; that Louis had said he was leaving everything to her. But nothing came of it."

her gorilla project and there publicly declared what everyone else was whispering. "It was a very unpleasant meeting," Richard said. "She was very upset and told me that I had caused my father's death—that it was all because of my ambition and behavior." Richard did not wait to hear more. Pushing away from the table, he dismissed her charges with an angry "Rubbish!" then stalked away.

Dian could not have been completely guiltless. She had carefully kept her distance from Louis since his declaration of love, seeing him for the last time in the spring of 1971 during his convalescence from the bee stings. Only ten months before his death, in early January 1972, he had written despondently to her about his financial worries. He had no funds either for Tigoni or his own research, and worst of all no one to turn to. "Life is being most *awful* for me. . . . I'm in the depths of despair and so terribly alone with no one who seems to understand how important *work* is to me." But Dian was different, Louis wrote; she did know. "Just now I wish you were here near to me to calm the awful uncertainty and give me peace." If Dian answered Louis's plea, her letter apparently has not survived. Nor did she at first seem affected by his death, barely making note of it in her journal. "We go out to find Group 8," she wrote on October 6. ". . . Fog moves in after 2 and we leave. Cable here on return saying Leakey has died. I'm sorry for his sake. . . ."

It was not surprising that Fossey and others cast Richard as the responsible party in Louis's decline and death. To Louis, Richard had been rebellious, at times even disrespectful, and to his father's protégés, cold and abrasive. But Richard had hoped that their relationship would change. "For me, his death couldn't have come at a worse time," Richard said. "He died too soon, just when we were beginning to put our misunderstandings behind us and move onto a better footing—and as friends we could have done so much. People didn't understand, but his death was a great loss to me."

To ease his anguish, Richard revived his plans for an international institute of prehistory. The building would be of modern design and spacious enough to house the growing archeological and paleontological collections. Such an institute would easily attract foreign researchers, adding to Kenya's —and Richard's—prestige. And it would do what he had tried to do while Louis was alive but Louis had never permitted: honor his father. "[I]t seemed an appropriate way to commemorate and continue the work of a pioneer prehistorian," Richard wrote in his autobiography. He would call the new building The International Louis Leakey Memorial Institute for African Prehistory (TILLMIAP).

Richard also planned to do what he could to save Louis's Tigoni Primate Research Centre. He asked all but one of the few remaining foreign researchers to leave, and stripped the Kenyan staff to a minimum. "Many of the men have been given notice," Louis's secretary Pat Barrett wrote to a former Tigoni worker, "and now, of course, the inevitable is happening.

Despite the extensive loans they got from Louis . . . they are all trying in one way or another to get money, and don't appear to realise the fountainhead has gone." In his place stood his son, with the accountant's reports in hand.

While Richard was preoccupied with sorting out his father's affairs, Meave and the anatomist Alan Walker continued to work on the reconstruction of the hominid skull 1470. Meave had glued much of the skull together in the field, revealing a hominid with a large braincase and smooth brow. By late October, she and Walker succeeded in joining the upper jaw to the face and Richard decided it was time to reveal his team's most dramatic discovery to the world—although he was still uncertain about what he should call it.

"What a curious fellow!" Richard wrote to Bernard Wood, who was now in London and who had helped with the early stages of the skull's reconstruction. "There are no doubts in my own mind (. . . Alan Walker agrees too) that the skull is not *H. erectus* and is new and should have a new name. However, to launch a species on one skull is unwise. . . ." Near the skull Richard's team had also uncovered two femurs and two lower leg bones that were remarkably humanlike. Richard speculated that these fossils belonged to the same species as the skull and suggested that by lumping together all the specimens he might persuade his colleagues that he had indeed found a new species. "I know that it would be a strange approach," he told Wood, "but it might serve to confound the expected critics."

Wood, however, advised caution. "One wants to be hypercareful," he replied, "because people will be just waiting for a repeat of the 'launch' of *Homo habilis* which wasn't entirely auspicious. . . . I am sorry to be chicken . . . but [the skull] could be known by its [field] number, and as *Homo sp.,* without any loss of importance as a fossil. . . ." Wood thought that there was "something very and unnecessarily emotive about naming a new species; people think that because you give it a name you have all the answers—I can't truthfully say we have all the answers." Further, the fossil's importance might be obscured if there were "prolonged wrangling in the press about nomenclature."

Richard accepted Wood's advice. Although he "wanted the limelight," as he admitted ten years later, he also wanted to avoid Louis's mistakes, particularly the contentious battles Louis had waged over naming his fossils. Richard would thus introduce his new skull to the world simply as 1470, the earliest example of the *Homo* lineage; he would not assign it to a particular species.

As much as he wanted publicity, Richard yearned even more for recognition from the scientific establishment. To be regarded seriously, he would have to announce 1470 in a scientific forum before presenting it to the general public. A symposium that London's Zoological Society was convening in early November seemed to offer the perfect opportunity, particularly since Richard had already been invited as a speaker. He contacted the Zoological Society's directors about his plans, suggesting that he describe the fossil to his colleagues in the morning, then hold a press conference in the afternoon. The Society might even benefit from the attention, Richard thought. Instead, its directors were alarmed—apparently as much by the idea of Richard on stage displaying a major new find as by the threat of a mob of reporters.

With some misgivings, the directors agreed to Richard's plan, although they insisted that he speak to the press elsewhere. Their meeting, they explained, had been organized in memory of the eminent anatomist Sir Grafton Elliot Smith, and they did not want the media intruding.

Richard accepted the society's terms and on November 9, 1972, arrived at the symposium with slides and casts of his fossils. He was scheduled as the second speaker, following a lecture on Elliot Smith by the society's secretary, Sir Solly Zuckerman. An imposing, white-haired figure, Zuckerman had a formidable reputation as a zoologist and chief scientific adviser to the British government. At one time, he and Louis had been friends, but when Louis pointed out an error in one of Zuckerman's articles, Zuckerman turned against him. Rumor had it that Zuckerman had successfully blackballed Louis's name every time he had been nominated for membership in the Royal Society, Britain's elite scientific fraternity.[7] Now Zuckerman—who had led the opposition against Richard's appearance—did what he could to discredit Louis's son.

Ostensibly, Zuckerman's lecture was structured as a defense of Elliot Smith's role in the Piltdown hoax.[8] Elliot Smith and other leading anatomists had identified the skull in 1912 as the oldest human fossil—precisely what Richard would be claiming for 1470. Zuckerman asserted that Smith could not possibly have been one of the hoaxers; he had been hoodwinked by an "amateur scientist," something that could happen again unless professional scientists were careful.

The allusion was not lost on Richard. After Zuckerman concluded his talk, Richard stepped to the podium. "I am an amateur," he declared, fixing his eyes on Zuckerman. "And I propose simply to present to you the facts of our new discovery and leave it to you and others to judge what it is." Zuckerman's speech had gone far beyond its allotted time, leaving Richard with only thirty minutes to present his fossil. There wasn't sufficient time to read his written text, and instead, Richard discussed his discoveries ad lib, illustrating his points with slides. Toward the end, he produced a cast of

---

[7] "In spite of the fact that I have been nominated a Fellow of the Royal Society on four or five occasions," Louis wrote to Joan Travis in 1971, "... I am given to understand that in every case my nomination has been blocked by Zuckerman who seems to have a personal spite against me. I believe this stems from the fact that I drew his attention privately (and not in public, as I could have done) to a very important error in one of his science papers. Up to that time he and I used to correspond as 'Dear Louis,' and 'Dear Solly,' and I had been his friend for more than 30 years. He now opposes everything I have done ..."

[8] Ronald Millar had published a book, *The Piltdown Men,* a few weeks before the conference, fingering Sir Grafton Elliot Smith as the mastermind behind the fraud.

1470 along with casts of the leg bones. These, he noted, "have astounded anatomists . . . because they are practically indistinguishable from the same bones of modern man."

The fossils, Richard continued, were all nearly one million years older than any previous hominid find. Furthermore, the skull's large brain and smooth brow placed it firmly in the *Homo* line. Together, the specimens seemed to prove what Louis had long argued: that early humans and australopithecines were contemporaneous. And because they had lived side by side for such a lengthy period of time, it was unlikely that *Australopithecus* was a human ancestor. "It would seem," Richard concluded, "that *Australopithecus,* as known, can be excluded from our own line of ancestry."

Immediately, the audience was abuzz. Richard's fossils had, as he had hoped, confounded his colleagues and, at the end of his speech, he took his seat to a loud, enthusiastic round of applause.

Zuckerman, however, was not to be outdone. Seated on the speaker's dais, he was the first to jump to his feet. Then, inclining his head toward Richard, he intoned, "Mr. Chairman, may I first congratulate Mr. Leakey, an amateur and not a specialist, for the very modest and moderate way he has given his presentation." A murmur of protest rippled through the audience at the word "amateur," but Zuckerman was undeterred. "May I also express my personal gratitude," he continued, "and certainly the gratitude of many others who have worked with him and his father, for the work they have done, not as anatomists, as Mr. Leakey pointed out, not as geochemists or anything else, but just as people interested in collecting fossils on which specialists can work."

A stony silence settled over the audience at Zuckerman's gibes, but Zuckerman ignored this hostility and, much to everyone's surprise, began to praise the fossil itself. He had never accepted the idea that australopithecines were human ancestors; 1470 would finally put an end to what he considered a tedious argument. "You may not have intended to," Zuckerman told Richard, "but you have demolished all that [that is, the theory of an australopithecine ancestor] with your skull." Richard smiled in spite of the earlier insults and replied, "I am quite pleased I have." Other scientists then rose to congratulate him on his discovery, and the applause "went on and on until Richard blushed," said the anatomist Phillip Tobias.

For a moment, it seemed that the Leakeys and Zuckerman might be reconciled. But when the meeting adjourned, the doors swung open to a crush of reporters and cameramen. "I was as surprised as Zuckerman seemed to be furious," Richard wrote in his autobiography. Apparently Richard's press announcement about 1470 had been released too early in the United States, and the media was now clamoring for interviews. Zuckerman, of course, would have none of it and shuffled Richard off to his office, where he was held temporarily "against [his] will." The standoff ended only when Richard arranged for the press to meet him at Kenya's High Commission later that afternoon.

By the next morning—and in spite of Zuckerman's best efforts—Richard was a celebrity.[9] Newspapers around the world trumpeted his discovery, many of them featuring Richard and his fossil as a page-one story. "Skull Pushes Back Man's Origins One Million Years," the *New York Times* declared, while London's *Evening Standard* erroneously celebrated Richard as a "Briton" who had found the "Missing Link." Popular magazines, too, carried lengthy articles about Richard and the new skull, emphasizing its great age. Although he had wanted the publicity, Richard would later say he was surprised at the amount of attention 1470 garnered. "I can only imagine that the world just happened to be quiet that day because it is unthinkable that under normal circumstances a fossil skull would make the front page in so many papers all over the world." Yet only thirteen years before, the news of *Zinjanthropus* had catapulted his parents to fame, if not fortune. It was now Richard's turn to face the spotlight.

---

[9] In spite of his antipathy toward the press, Zuckerman did comment on Richard's skull, telling reporters the discovery "might lead to a controversy as great as the notorious 'Piltdown man' hoax. . . . The skull may be old," he added, "but it remains to be seen what it is."

Chapter 31

THE BEST BONES

After announcing the 1470 skull, Richard did not linger in London to bask in the limelight. He was determined to raise a million dollars for the National Museums' new institute of prehistory (a goal his father would have deemed preposterous and unreachable), and immediately returned to Nairobi to lay plans for a major fund-raising tour in England and the United States. Two months later, in January 1973, with the news of 1470 still fresh in everyone's minds, Richard was on his way, searching for benefactors as Louis had done so many times before. And like his father, Richard drew crowds of admirers.

At St. John's College, Cambridge, Louis's alma mater, Richard spoke to a "fantastically overcrowded" lecture room, the archeologist Glyn Daniel reported to Mary. Equally enthusiastic audiences greeted Richard in London, New York, and Washington, D.C., where the National Geographic Society awarded him its Franklin L. Burr Prize "in recognition of outstanding contributions to science."[1] From Washington, he traveled to Chicago, Wichita, San Francisco, and Los Angeles, often lecturing college audiences that had once applauded his father. In Berkeley he stopped long enough to meet with Glynn Isaac and other members of his Rudolf expedition, and to discuss research objectives for their 1973 field season. Then he was back on the road, relating his tales of fossils and adventures in a wry, relaxed, self-deprecating style that delighted his audiences.

Like Louis, Richard seemed to have "an innate flair for communicating

[1] Richard shared the $2,500 award with the six members of his hominid team.

with people, whether there were two or two thousand in the audience," said Gilbert Grosvenor, the president of the National Geographic Society. He also had a gift for getting people to open their wallets. "Father and Richard both had that aura, a way of making people believe in them, of convincing people to give them the money they needed," said Philip Leakey. And with a charm also reminiscent of his father's, Richard fielded questions with grace and confidence. But there was an essential difference. Louis did not don a stage personality as Richard did. "Richard is much more circumscribed and calculating than Louis," said Lita Osmundsen, the former director of the Wenner-Gren Foundation. "He has the charm, the charisma of his father, but it seems learned rather than instinctive. It is like a cloak he wears." Paul Abell, who had worked with Richard at Rudolf, remarked, "The Richard I knew in the field was a different person. You could kid him, give him a hard time. He didn't have the smoothness, the pushiness he had on the lecture circuit. But we understood that he had to slip into another persona when meeting with someone who might give money."

Although he enjoyed the standing ovations and unabashed admiration, Richard had no desire to be idolized as his father had been. Nor did he cultivate protégés as Louis had. He was on the lecture circuit for one purpose only—to raise money for his institute—and all of his energies went to this goal.

Perhaps because of his intense focus, Richard succeeded in raising nearly $200,000 on this six-week tour, whereas Louis's best catch on a tour of similar length was barely $25,000.[2] He also secured another $45,000 grant from the National Geographic Society for his fifth Rudolf expedition. It was scheduled to depart at the beginning of June, only two months after Richard returned to Nairobi from his lectures. But the shortness of time did not worry him. Since 1970, Koobi Fora had become his second home, much as Olduvai was Mary's, with a year-round staff looking after his personal effects. It was thus with a sense of ease and anticipation that on June 2, 1973, Richard once again flew Meave and fourteen-month-old Louise to Koobi Fora and the fossil-laden deposits of Lake Rudolf.

Over the next few weeks, the rest of Richard's team—archeologists, geologists, paleontologists, and Kamoya's crew of hominid hunters—trickled into camp. Primed by their previous successes in the field, plus the ongoing media attention (Richard's story about the discovery of the skull 1470 had just appeared in the *National Geographic*), the scientists did not take long getting back to work. "Everyone arrived with great and pleasant expecta-

---

[2] Half of Richard's $200,000 came from the Guggenheim Foundation, which agreed to grant Richard's institute the $100,000 it had previously earmarked for Louis's hominid-hunting expedition to the Suguta valley, an area south of Lake Turkana.

tions," said Paul Abell. Nor did it take long for the hominid fossils to start rolling in once more.

Richard first dispatched the fossil hunters, along with paleontologist John Harris, to the Karari Ridge, where 1470 and nine other hominid specimens had been found the previous year. Some thirty miles northeast of Koobi Fora, the uplifted sediments of the ridge lay inland from the cooling breezes of the lake. The heat and glare from the sun were especially intense here, making fossil hunting a grueling task in the afternoon. Despite the suffocating temperatures, Kamoya "really worked his crew," said Paul Abell—and in no time at all, they were making discoveries.

Maundo Muluila found the season's first hominid fossil only a few days into the hunt: four small fragments of a cranium. Not enough remained of this specimen to determine the species, but every hominid fossil, even those which were badly weathered, boosted the team's spirits—they were all proof that Rudolf's treasure box was not yet empty. Indeed, only a few days later, Wambua Mangao found a piece of lower jawbone eroding out of deposits as old as those that had yielded 1470; and about a mile away, John Harris discovered another hominid mandible, this one with many of the teeth set like sepia-tinted pearls in the bone. Richard would later assign both of these specimens to the *Homo* lineage.

Other tantalizing bits and pieces of bone turned up over the next weeks, and then at the beginning of July, Paul Abell spotted the top of a skull emerging from the ground. "I scratched around enough to be sure that it was hominid," Abell said, then marked the site with a cairn of stones. Much of the rest of the skull was embedded in the ground, and Richard excavated these pieces, revealing one of the most puzzling hominid fossils ever discovered. Like a robust australopithecine, this creature had a sagittal crest running the length of its cranium, yet its teeth were small like those of *Homo habilis*—a combination that Richard noted was "in contrast to all the [australopithecine] specimens previously recovered from East Rudolf." Perhaps it was a new species—and in Louis's hands, it probably would have received a new name. But Richard continued to be cautious about his discoveries, writing later in *Nature* that "KNM ER-1805 [as the skull was numbered] is undoubtedly important, but its interpretation is enigmatic."[3]

While Richard was busy excavating this fossil, Meave took a few steps down the hillside and found yet another hominid specimen. Only the lower jaw and teeth remained of this creature, but unlike 1805, the teeth were large, and Richard unhesitatingly labeled it *Australopithecus.*

Over the next few weeks, hominid fossils continued to pour in: more jaws and teeth, fragmented skull caps, leg and arm bones, and the most complete skeleton of *Homo erectus* that had yet been discovered.[4] Fragments of the

[3] KNM-ER 1805 continues to present a mystery. No other hominid fossils of this type have been found, and the skull is simply labeled *Incertae sedis* (of uncertain taxonomy).

[4] Kamoya Kimeu found this specimen as well as a second, far more complete *Homo erectus* skeleton in 1984. (See chapter 40.)

skull and jawbones remained, as well as portions of both the leg and arm bones, which were oddly thickened. "The interesting thing is that this fellow was diseased," Richard would later say, after the anatomist Alan Walker had studied these fossils in detail. "The bones look almost rheumatoid. It's a pathological condition you see in people who have ingested too much vitamin A, usually from eating too much raw liver." Whatever the specific source of the liver, *Homo erectus* and the other hominids apparently had a wide range of animals to feast on. In their search, Kamoya's crew had turned up fossils of everything from rodents and aardvarks to hyenas and camels. So rich had life been along the ancient lakeshore that animals evolved rapidly, producing an almost profligate display of size and form: four species of giraffe, nine different pigs, and twenty-four types of antelopes had all called Rudolf home.

With such an abundance of animal life, even on days when hominid bones were scarce, the fossil hunters could count on uncovering prizes. One typical morning's search brought in a well-preserved monkey's jaw and femur, an elephant's molar and two tusks, plus the skull of a baby elephant. Still, it was the hominid fossils that everyone wanted—although finding them required as much luck as skill, and even among the scientists superstitions grew concerning how one ought to look for the ancestral bones. "You couldn't go out wanting to find a hominid," explained Kay Behrensmeyer. "If you had that in mind, then you were bound not to find one." To keep their luck—and windfall of fossil hominids—intact, Richard forbade reference books on hominid bones or human anatomy in camp; not even a chimpanzee skeleton was permitted. "He wanted his fossil hunters to depend on their own abilities," said Behrensmeyer. "And there was also the idea that if you were trying that hard, reading about and studying the bones, then they wouldn't come out of the ground for you; they wouldn't let you find them."[5]

Richard's own time for fossil hunting continued to dwindle. More than ever, he was an administrator, preoccupied with museum affairs and camp logistics. Simply keeping the seventy-odd scientists and staff supplied with food, fuel, and spare vehicle parts required weekly trips between Nairobi and Koobi Fora. Sometimes the expedition's seven-ton truck managed this commute without difficulty. But when it was broken down and provisions were running low, Richard ferried up the goods in his Cessna. For the expedition, Richard and his plane were the one certain link to the outside world.

---

[5] Richard did, however, encourage his hominid team to study comparative anatomy intensively at the museum, a decision he reached after the first few Rudolf seasons. "It had always struck me that one only finds what one is looking for," Richard said. "And the hominid team were finding only teeth and jaws, and there had to be a lot more than that. So I arranged for them to spend three weeks at the museum studying human skeletons and comparative animals." The intensive instruction paid off; in subsequent seasons, the team found hominid arm and leg bones at about the same rate as skulls and teeth.

On his trips to Rudolf, Richard first landed at Koobi Fora to unload supplies, then took off again immediately for Kamoya's bush camp. "He was always anxious to see what had turned up in the way of fossils," said Abell. En route, Richard often flew low over the encampments of other team members who were working away from the main camp, lobbing down small bundles of letters and canned goods as he passed overhead. "Christ, you would have thought he was in the RAF, the way he'd fly along over the exposures, then suddenly peel off and dive down," said Ralph Holloway, an anthropologist who accompanied Richard on one of these flights. "I think he enjoyed my sheer white coloration during those descents."

For Kamoya's crew, Richard also always brought up a side of beef or leg of lamb, a welcomed break from their regular diet of posho (maize meal) and tinned beef—though sometimes the men did scavenge a hunk of antelope or zebra meat from a lion kill, after chasing away the lions with the Land Rover.[6] The lions were always disgruntled about losing their dinner, though only one ever gave chase in return. It raced after the vehicle—an open-air Land Rover crowded with eight terrified men and one fly-infested carcass—and almost succeeded in jumping aboard. After that, Kamoya made certain the lions had eaten their fill and had moved away before taking what remained.

Wherever he was camped, Kamoya and his team would clear and mark a short airstrip for Richard, enabling him to land nearby. Then, moving across the desert with such long strides that Kamoya's men had to run to keep up with him (they dubbed him "the Ostrich"), Richard would inspect the most significant fossil sites and excavate any hominid fossils the team had discovered. But he seldom paused to do any exploring himself, and in 1973 found only one hominid specimen, four weathered fragments of a lower jawbone. "It takes four or five days at Turkana just to get your eye in, to actually start seeing the fossils," Richard said. "So it was much more efficient to have Kamoya's crew do the prospecting, then come up and collect what they had found."

Indeed, by the end of the field season in September, Kamoya's team had amassed twenty more hominid fossils, including another nearly complete skull. Kamoya himself turned up the first pieces of this specimen, eroding from sandy deposits only slightly younger than those that had yielded the skull 1470. At the time, Richard attributed a date of 2.6 million years to 1470 and two million years to the new specimen, which was numbered KNM-ER 1813. But like Paul Abell's earlier find, 1805, this new fossil puzzled Richard and his anatomists. With its small brain, small teeth, and delicate jaw, it was reminiscent of *Australopithecus africanus,* the slightly built, gracile creatures from South Africa. Only a year before, Richard had argued that this

[6] In the early days of the expedition, Richard had sometimes hunted the antelope for his crew. But by 1973 this was no longer permitted, as he had successfully lobbied the Kenyan government to declare East Rudolf a national park. In 1974 it was renamed East Turkana National Park.

species had probably not inhabited ancient East Africa, although other scientists disagreed. (In their eyes, the fossils Louis had labeled *Homo habilis* were actually small australopithecines.) Now Richard wondered if Louis had indeed misclassified at least one of the Olduvai hominids, an almost complete skull that had been found in 1968. "The size and morphology of the teeth of [1813 and OH 24, the field number of the Olduvai fossil] are alike," Richard wrote in *Nature,* "and the cranial capacities may also be comparable."[7] If so, then three, not just two, hominid species—one a large-brained *Homo,* represented by 1470; the second a strong-jawed, small-brained creature, *Australopithecus boisei;* and the third a gracile type, like *Australopithecus africanus*—had shared the shores of Lake Rudolf between three and one million years ago.

Casting aside his usual caution, Richard went one step further. There was another unusual hominid specimen from Rudolf, a lower jawbone that shared some similarities with hominid fossils from the Ethiopian Omo region. Found in 1972, this fossil had been assigned the number KNM-ER 1482 but listed as taxonomically uncertain. Perhaps, Richard suggested in *Nature,* 1482 and the Omo specimens represented a fourth hominid form, one that was also contemporaneous with the other three. This creature, which he declined to name, may have been "a remnant of an earlier population that disappeared during the early Pleistocene," Richard wrote in *Nature.*[8] All four lineages, Richard thought, probably evolved from an original ancestor that "may be traced back well beyond the Plio-Pleistocene boundary." In other words, the common ancestor of *Homo* and the australopithecines had yet to be found.

Richard's four-hominid hypothesis would have gladdened his father's heart. It was precisely the kind of bushy lineage that Louis had proposed on several occasions. Indeed, on a visit to Koobi Fora only two months before he died, Louis had predicted that his son would find evidence of three hominid species at Rudolf. Further, Richard's hominid lineage effectively excluded the australopithecines from anything but a tangential role in human evolution, precisely as Louis had argued. Since many scientists regarded *Australopithecus africanus* as the immediate ancestor of the *Homo* line, Richard's article was certain to reignite old fires, as an anonymous correspondent in *Nature*'s "News and Views" section pointed out.

"If one accepts that gracile australopithecines and *Homo* coexisted in the early Pleistocene," this unidentified writer noted, "then one must also accept that [*Australopithecus africanus*] does not, indeed cannot, represent the

---

[7] OH24, nicknamed "Twiggy," was a very crushed and flattened cranium. In his 1991 two-volume study of *Homo habilis,* Phillip Tobias asserts that Twiggy is a female member of this species. He regards KNM-ER 1813 as another female *H. habilis,* and KNM-ER 1470 as a male *habilis.* Richard, however, continues to believe that KNM-ER 1813 and Twiggy represent an altogether different type of hominid.

[8] The French members of the Omo team had found the first specimen of this hominid in 1967. Camille Arambourg and Yves Coppens created a new genus for it, *Paraustralopithecus aethiopicus,* meaning a creature closely related to *Australopithecus.*

ancestral hominid group. The gracile australopithecines are widely accepted as the basal hominid stock and yet Leakey's current classification of the early hominids implies that the known members of this group coexisted with the genus *Homo* and were in fact too late to provide the ancestral population."

The similarity between Louis's and Richard's ideas would in time lead to charges of a "Leakey line" about human evolution. "It's all part of the Leakey Syndrome," said Elwyn Simons, who had carried on a lengthy debate with Louis about the taxonomy of the Miocene apes. "You know, the earliest *Homo* has yet to be discovered, and all the fossils which they [the Leakeys] find are important and on the direct line of mankind, but the fossils everyone else finds are extinct side branches." Nevertheless, Richard maintains that it was the fossils themselves and not his father's opinions, or an inflated ego, that led him to his conclusions. "Obviously, there is some continuity in our thinking, but it doesn't amount to anything like Scripture, or to a fixation on large brains versus small brains," Richard said. "I think both Louis and I were looking for more or less the same thing, and that is, when did our species, *Homo,* begin?"

With the mountain of hominid fossils stashed at the museum, Richard worried very little about criticisms of his theories—which *Newsweek* reported had "caused near apoplexy among many experts." "I had the best bones going," he said, "and so commanded recognition. Ultimately, it didn't really matter what the fossils said. The fact that they were saying something was what was important to me in terms of my position at the museum and my ability to raise money."

Thus, when Clark Howell and others from the Omo team began raising questions about the geology and dating at Koobi Fora, Richard just shrugged. He was "Paleontology's Daring Young Man," as one magazine characterized him and, for the time being at least, at the head of the pack.

Richard's star had risen so far and so fast that by age twenty-nine he was everything that students of paleoanthropology longed to be. Magazine photographs of a turbaned Richard on camelback exploring the African wilds had so glamorized the science that university enrollments in anthropology soared.

"There was a whole group of us from the late sixties who were inspired by Richard," said M. E. Morbeck, then a graduate student in physical anthropology. "He was young, handsome and dashing. He had the limelight with all the glamor and world attention. It was an image many students wanted for themselves. They wanted their [hominid] fossil. They forgot how much work it takes; that you have to know anatomy and spend hours looking at monkey, ape and human bones before you ever look for fossils. All they wanted was to hunt for fossils, as if it were searching for treasure."

In the profession itself, Richard's bonanza of Rudolf bones had overshadowed the South African hominids, as well as those from the Omo and even Olduvai. "The older guys, like Phillip Tobias and Michael Day, were working on the Olduvai material, which by 1973 wasn't important any more," said Morbeck. "All the younger guys wanted to work with the Turkana [Rudolf] fossils, which were *really* important because there were so many of them."

Yet Richard had his detractors as well, and while they may have envied the quantity of his fossils, they were not impressed with the quality of his science. As Clark Howell suspected, something seemed to be wrong with the dating at Koobi Fora.

The first hint of this problem came in 1971 at a Wenner-Gren anthropological meeting in Austria. There Basil Cooke, a specialist on fossil pigs, which he had studied at South African sites, the Omo, and Olduvai, presented his preliminary data on the pigs from Rudolf. "I was working on the fossil pigs, not because I'm interested in pigs as such," Cooke later explained, "but because of their geological value. They are a very good dating tool."

At all of the African sites, paleontologists had discovered several different lineages of extinct pigs. As the pigs had evolved, their molar teeth had changed, increasing in length and height. These differences were so distinctive that Cooke found he could use the pigs as a kind of paleontological clock, providing another way to date the fossil deposits. Thus, if fossils of the pig *Nyanzachoerus pattersoni* turned up at a site, Cooke knew that it must date to about six million years ago, whereas fossils of *Mesochoerus olduvaiensis* indicated deposits of two million years.

The Rudolf sites, identified by Richard's team as 2.6 million years old, contained pigs that were barely two million years old, according to Cooke. In other words, Richard's prized finds, the 1470 skull and artifacts reputed to be the world's oldest, were possibly not any older than the oldest hominids and tools from Olduvai.

Richard, however, had relied on one of England's most prestigious geochronology laboratories, F.M. Consultants Ltd. in Cambridge, to date the volcanic ash from his fossil sites—and he was not about to abandon their results because of Cooke's pigs. The F.M. geochronologists, Jack Miller at Cambridge and Frank Fitch at Birkbeck College in London, had also used the latest dating technique (the argon-40/argon-39 method) to secure the age. But the technique was still in its infancy, and as other scientists would later point out, the age spectrum that the mass spectrometer machine produced was not always easy to understand. Nevertheless, Fitch and Miller confidently supplied Richard with a date of 2.61 ± 0.26 million years as the age of the KBS (for Kay Behrensmeyer Site) tuff, as the sample came to be known. The date was, as Fitch said, a "golden oldie."

Richard had relied on Fitch and Miller's date in 1972 to emphasize the importance of the 1470 skull, citing their dating results in his report in *Nature.* But he made no mention of Cooke's pig study, even though he was well aware of Cooke's analysis, having invited him to describe the Rudolf pig fossils. "I felt that Basil had some prejudice against me," Richard said later, "because I was a cheeky young man and self-confident and operating way above my station. I thought he was trying to bring me down, and therefore I didn't trust him." (Cooke, who in the 1950s had helped demolish one of Louis's pet ideas about dating and climates in East Africa—the pluvial hypothesis—admittedly did enjoy being the thorn in Richard's side. "People ask me if I've worked with the Leakeys," he quipped, "and I say, 'Yes, but more often against them.' ")

Between 1971 and 1973, Cooke continued to add to his pig data. He and Vince Maglio, a paleontologist on Richard's team, also discovered that the fossil elephants from Rudolf and the Omo did not correlate; nor did a variety of other animal fossils, including those of the horse and several species of antelope. Trying to explain the discrepancy, Richard's team came up with the suggestion that the environments in the two regions had been so different two to three million years ago that faster-evolving races of animals had developed at Rudolf. "We would sit at Koobi Fora with our binoculars and look across to the Omo and say, 'Well, what the hell would have made the ecology there so different from here?' " Richard said. They speculated that some zoogeographical barrier had kept the animals separated, and took heart from the fact that even today certain species of birds and antelopes at Koobi Fora are not found at the Omo.

In spite of these examples, Richard's paleontologists knew that such an idea (which came to be called the ecological hypothesis) seemed to make little sense, particularly since modern species of pigs are the same on both sides of the lake. Nevertheless, the team seized it as the one weapon they had for fending off Cooke's data.

In early September 1973, Richard's team joined with forty other scientists —including Cooke and Clark Howell—in Nairobi for a two-week conference on the Lake Rudolf Basin. Howell was also now convinced that the date for the KBS tuff was wrong, and made his point by reading two lists of fossil animal species, one from the Omo and one from Koobi Fora. The lists were identical, except that the species from Koobi Fora were dated some 600,000 years older than those from the Omo, although the fossils came from equivalent deposits. It took Howell fifteen minutes to read his lists, and then he looked up at his audience. "They are not congruent," he said, and left the stage.

Richard's team did not blanch. "We already knew that the Omo team didn't want the correlations to be the way we did," explained Behrensmeyer. "But Clark's paper certainly made it public."

Cooke, too, raised the issue, presenting his pig data in even greater detail. The discrepancy between particular species "is considerable," he noted, " and cannot be ignored." Richard and Isaac tried to counter his results with their ecological hypothesis, but the Omo scientists scoffed at this suggestion.[9]

Still, Richard and his team refused to back down. "There'd always been a rivalry between us, but now it was becoming more like a football match," said Isaac, recalling the mounting tension as the scientists sat glaring at one another. The stakes were high. "We had spectacular hominid specimens and archeological sites," Isaac explained,

> whereas at the Omo, although it is a magnificent stratigraphical sequence, the preservation is poor. So they had little in the way of hominids or archeological sites, and were probably a little envious. At the same time, Richard had, in 1470, "Man Much Older Than You Think," while I was the owner of the oldest archeological site in the world—and we were reluctant to see all that go down the drain. So all this undoubtedly did give the dating issue a charge. It wasn't purely a group of disinterested scientists trying to get it right.

Richard, too, thought that Howell's attack—like that of Cooke's—had been driven primarily by a desire to bring Richard down. Further, he had no reason to doubt Fitch and Miller, who continued to insist that their dating techniques were impeccable.[10] Besides, his team had just recovered "20 more hominids . . . including two rather magnificent skulls," as he wrote to Phillip Tobias. What did the Omo team have to show?

"I didn't feel that we needed any help from Clark [Howell] or anyone else to sort this matter out," Richard said later. "We were doing very well. Why would you ask for help when you think you're winning?"

By the end of the meeting, the dating issue remained unresolved. But it was a weak spot in Richard's success, one that he knew others would continue to try to exploit.

Meanwhile, on the sidelines, watching the two teams sparring, sat a fresh contender, Donald Carl Johanson. One of Howell's students, he had spent three seasons with the Omo expedition and had just completed his doctorate in paleoanthropology at the University of Chicago. Dark-haired and dark-eyed, with a decidely nonacademic taste for designer suits and fine wines,

---

[9] Alan Gentry, another paleontologist, also reported on the antelope fossils from Rudolf and the Omo, demonstrating again that the fauna indicated a problem with the date. Richard was initially impressed by Gentry's report, since he was an independent scholar, not attached to either team. But Richard also thought Gentry's evidence was too slim, and so did not agree that it invalidated Fitch and Miller's date.

[10] Other scientists were also impressed with Fitch and Miller's dating technique. In a postconference report in the *South African Journal of Science,* the geologist Karl Butzer noted that "Interpretation of the geology proved to be most controversial," but then added his support for Fitch and Miller's work. "These East Rudolf argon-40/argon-39 determinations are age spectra datings," he wrote, "the best of their kind."

he was a year older than Richard and just as ambitious. "Don really wanted the big-time fossils," said M. E. Morbeck, who met him when they were both graduate students in physical anthropology. "He used to tell me when we were students that it was too bad that Clark Howell had never found a fossil, and that he was going to find the fossil that Clark never got." Johanson had a site already picked out for his discovery, the Afar desert of western Ethiopia. After the Nairobi conference, he and a small party of French and American scientists were planning to explore it. But at the conference itself, he kept a low profile. "I had no fossils," he would write later in his 1981 book, *Lucy,* "only a stakeout and a hope. Faced with the glittering finds of Richard Leakey, I felt rather insignificant."

Richard and Johanson did exchange a few words at the meeting about Johanson's plans for the Afar. "Do you really expect to find hominids there?" Richard asked. "Older than yours," Johanson replied, adding, "I'll bet you a bottle of wine on that." "Done," said Richard.

Richard planned to send his own team the next season into deposits at Rudolf that dated to four million years. Here they might uncover fossilized animal specimens to settle the dating controversy. And, if Richard's hominid lineage was right, they might find evidence of an even older species of *Homo.* But like Professor Percy Boswell, who had brought a halt to Louis's meteoric rise thirty-eight years before, Basil Cooke would not go away.

Chapter 32

THE GLADIATORS' CLASH

Rumors ran rife after the September 1973 Lake Rudolf Basin Conference. Instead of resolving the debate over the date of the KBS tuff, Richard's efforts at the conference seemed only to have inflamed the issue. Feeling that his team was being "unfairly knocked about," Richard sided even more strongly with his geochronologists, Fitch and Miller, acquiescing to their demand that no other lab be allowed to date the Rudolf deposits—a virtually unprecedented edict that stunned and angered the scientists who ran up against it.

"Richard took me on a field trip to Koobi Fora in the fall of 1973," recalled Richard Hay, the geologist who had mapped Olduvai Gorge for Louis and Mary, "and on the flight up he said to me, 'Do not take a sample big enough to date.' He said he wanted to keep the home team happy, that Fitch and Miller were very territorial." Hay had no intention of collecting any such samples, but Richard's admonition gave him pause. "As a geologist, I'd just never been told not to take something bigger than your thumbnail. And that was part of the problem. If you have only one team working on a problem like that, and they're in error, things can get way out of kilter. When the crash finally comes, there's bound to be a great deal of noise."

Garniss Curtis, a geochronologist at the University of California at Berkeley, who had dated samples from Olduvai in 1960, asked Kay Behrensmeyer for a sample of the KBS tuff and was rebuffed. A flinty, rigorous scientist, Curtis was as intrigued as he was angry. He smelled a cover-up and so did others. Gossip to that effect was soon circulating—particularly in the ivy-covered halls of Berkeley, then the leading center of paleoanthropology in the United States. Not only was Curtis at Berkeley; so were Richard Hay, Frank Brown, Clark Howell, and many members of the Omo team. Somewhat alone in their midst stood Glynn Isaac, staving off as best he could the criticisms of Richard.

But Isaac had grown increasingly concerned about the dating issue after meeting with two geophysicists at Stanford University, Brent Dalrymple and Alan Cox. They were outsiders, unattached to either the Berkeley or Nairobi camps, and, Isaac thought, could discuss the problem objectively. "We had a long and interesting discussion of the whole problem of dating the Rudolf tuffs," Isaac wrote to Richard a few days later, "and to cut a long story short, both of them are highly critical of much of the F & M [Fitch and Miller] procedure and of the presentation of results. . . . Now all this does not at all amount to the suggestion that we have an erroneous chronology," Isaac hastened to add, knowing that any intimation of this sort would put Richard on guard, "only to the fact that geophysicists with no axe to grind find their [argon-40/argon-39] data unsatisfactory." Dalrymple and Cox suggested that because the Cambridge lab's measurements were so erratic and difficult to interpret, Isaac and Richard should have another lab date the deposits—but this time using the more "conventional" potassium/argon method.[1] Garniss Curtis was a specialist in this type of dating, and, shortly after his Stanford meeting, early in 1974, Isaac made his way across the Berkeley campus to Curtis's lab.

Richard did not object to the idea of involving another lab in the dating dilemma, but he was suspicious of Curtis, who had worked closely with Clark Howell and the Omo team, making him seem a part of the rival camp. Curtis was also an old friend of Louis's and Mary's—and perhaps partly for that reason, Richard did not trust him. Indeed, Richard was certain that even his own mother was eager to see him fall.

Since Louis's death and their quarrel over his burial, Richard's and Mary's relationship had been strained. "I said some things then that hurt very badly, and that estranged me from Mary for a very long time," Richard said. Their relations took an even icier turn when Mary violated one of her cardinal rules: "Every Leakey has to have his or her own empire," she was fond of saying, "and heaven help any other Leakey who sets foot in it uninvited." Yet two years before, in the summer of 1971, she had done exactly that, turning

---

[1] At the Lake Rudolf basin conference, Fitch and Miller had presented forty-one age determinations for the KBS tuff. These ranged from 223 million to as little as .091 million years. Only one of their measurements actually yielded 2.61 million years, although six others came within a quarter million years of that date. Yet there were also eight measurements giving a date of 1.9 million years—almost exactly the date that Basil Cooke insisted on. In their paper, Fitch and Miller had attempted to explain the confusing mix of dates by arguing that the tuff had been both contaminated and "overprinted"—that is, some geological event had altered the tuff's radiometric clock. In a 1974 review of their paper, Brent Dalrymple scoffed at this explanation. "These two mechanisms could be used to explain anything," he wrote to Glynn Isaac and Clark Howell, "as their effects on the potassium-argon technique are exactly the opposite." In other words, the tuff could have been either contaminated or overprinted, but not both.

to Curtis for help when she feared that Richard's scientific enterprise at Rudolf was foundering because of "incompetent geologists."

Inevitably, word of his mother's meddling, however well-intentioned, trickled back to Richard, and he reacted angrily. It was just another example, he felt, of Mary's cool contempt for him and his work. "There's no such word as 'matronizing'—it's 'patronizing'—but in this case 'matronizing' would be better," Richard said. "She was dismissive of many of the things I did; she never approved of my efforts to build the Louis Leakey Memorial Institute, and she was very dismissive of the Koobi Fora project. She thought that I really didn't know what I was doing."

As Richard's problems at Rudolf grew, so did his troubles with Mary. She'd begun making tart, dismissive remarks about his geologists to her friends—who were only too happy to pass these on to Richard, who in turn complained to her, "It's embarrassing for me when people raise their eyebrows and say, 'You should have heard what your mother said about you last night.' " If she wanted to behave that way, he added, then "we should see as little of each other as possible."

In spite of her son's threats, Mary continued to align herself with what Richard regarded as the "enemy camp," and Richard responded as he had when quarreling with his brothers: as much as possible, he cut his mother out of his life. "In public, we always tried to maintain a tolerable show of unity," Richard said. "But other than that, I just kept out of her way; I was very angry."

Whom could he trust? By choice, most of the scientists on his team were as young and inexperienced as he. "At the Omo, they had far more established workers, people like Basil Cooke and Clark Howell," said John Harris, the paleontologist on Richard's crew. "And I think that was basically the crux of the problem—we were not as sure of ourselves, not as experienced. Fitch and Miller kept reassuring us that there was nothing wrong with the date whatsoever. And being gullible, we believed them."

Fitch and Miller also continued to be "adamant" that theirs was "the only appropriate method" for dating the Rudolf deposits, as they told Richard and his colleagues at a February 1974 meeting of the Rudolf team at the University of Rhode Island.

"If Fitch and Miller had just allowed the possibility of an error," Richard said later, "if they had said, 'Well, maybe we're wrong and maybe we should check this with another lab,' the whole thing would have been resolved immediately. But they didn't and, as a matter of principle, I thought it was not correct to abandon one's senior colleagues who were convinced that they were right."

Isolated and embattled, Richard turned increasingly to Glynn Isaac, his coleader, for advice. The two men had been friends since their expedition to Lake Natron in 1964. There and at Rudolf, Isaac's good humor and ready quips had often provided a foil to Richard's more direct, hot-tempered ways.

Isaac also had a better understanding of their fellow academicians—even Clark Howell and Basil Cooke, he repeatedly assured Richard, were less interested in defaming the Rudolf team than in seeking the truth about the date. It was an issue he wanted resolved as well, since his interpretation of the archeological sites hinged on having them accurately dated. And, like other team members, Isaac had wearied of Fitch and Miller's resistance to involving another lab.

Mulling over the impasse, Isaac devised what Richard would later call "a typical Glynn compromise": have the tuffs from both Rudolf and the Omo redated. Fitch and Miller could check the Omo dates, while Garniss Curtis verified those at Rudolf. Richard agreed. Curtis could now have a sample of the KBS tuff.

Richard was still convinced that fossil evidence supporting both the 2.6-million-year date and the ecological hypothesis (the idea that animals evolved at a faster rate at Rudolf than at Omo) would be found at Koobi Fora. To that end, when the 1974 field season got under way in June, Richard dispatched his Hominid Gang into some of the oldest deposits at Lake Turkana (as Kenya's government renamed Rudolf in 1974), fossil beds dated between three and four million years old that had previously been explored only cursorily.

As in previous seasons, the camp staff and Kamoya's crew were settled in the field by late May, while most of the scientists made their way to Koobi Fora by mid-June. Richard again flew many of them up to Turkana, but he cut his visits short, because Meave was expecting their second child. On June 12, 1974, she gave birth to another daughter. They left the baby girl unnamed for a few days, then chose the Kenyan Moslem name of Samira for Louise's sister. Meave planned to bring both children to Turkana later in the season, which had gotten off to a good if slow start.

After only a few weeks' searching, Kamoya's men had a handful of hominid cranial fragments and teeth—boosting everyone's spirits. Their first ventures into the older beds also produced some interesting pig specimens, which John Harris suspected might be new species. If so, perhaps they would put an end to Basil Cooke's theories. But just as the team was hitting its stride, tragedy struck. Only a few days after Samira's birth, Richard received an urgent nighttime radio call from Koobi Fora: Jeff Hammock, a graduate student from Rhode Island, was missing; they needed to launch an immediate aerial search of the desert. At dawn the next day, Richard took off in his Cessna for Turkana.

"It has given me nightmares for years," said Paul Abell, Hammock's colleague who had parted from him the morning before, pointing him toward tuff deposits about a mile from Glynn Isaac's camp. "I told him to keep the [Karari] ridge on his left, and he had a set of aerial maps showing the camp and the sites, so I thought he'd be okay." Hammock planned to sample three deposits, and was expected back at camp at noon along with the rest of the crew. He had been in the field for ten days and seemed to be fully acclimatized, so that when he did not arrive on schedule, the other scientists went ahead with their lunch, thinking that he had found something interesting to explore. But by two o'clock, as the desert's temperature soared, they knew that something had gone amiss. Hammock should have been back, if not for lunch, at least for a long drink of water, for like all the heartiest members of Richard's team, he had gone into the desert without any.

Abell led the other scientists and crew to where he had last seen Hammock. From there, the group fanned out on foot, calling and whistling, but there was no response. By nightfall, Hammock was still missing, and Kamoya drove south to Koobi Fora to call Richard. The rest of the group parked a Land Rover with the headlights switched on and tended fires along the Karari Ridge through the night, hoping that these might serve as a beacon. But by the time Richard arrived the next morning, Hammock remained lost. Kamoya had already expanded the search, engaging a missionary aircraft as well as members of the National Park Ranger Force. Nearly seventy people were now looking for the young man.

Neither the intensive search on foot nor the aerial survey produced anything. Hammock seemed to have simply vanished into the desert. Perhaps, Richard feared, he had been killed and eaten by a lion. As the days wore on, it grew increasingly unlikely that Hammock would be found alive, and on the morning of the fifth day, Richard decided to end the search. Sick at heart, he flew back to Nairobi to confer with the American ambassador, while the search party disbanded.

And then, miraculously, Hammock was found. Abdi Mohammed, the East Turkana National Park warden, had started back to his post at Alia Bay on a seldom-used track when he came across a row of boulders rolled across the road. "And there was Jeff sitting under a tree," said Abell. "He was conscious, but he was just skin and bones, he'd lost so much weight." Mohammed drove him back to camp, where the others administered first aid and tried to discover what had happened to him; but Hammock was delirious from dehydration.[2] That afternoon, Richard flew the young man to Nairobi, where he was placed in the city hospital's intensive care unit. The scientists at Turkana celebrated, but Hammock soon died in the hospital. Five days without water in the torrid desert had damaged his internal organs beyond recovery.

For Richard, Hammock's death was a bitter blow. He had always prided

[2] Hammock's journal was found with him and in it he kept a log of his travail. From these notes, Abell believes that Hammock was probably never more than six miles from the Karari Ridge camp, though the group could never fully determine what his movements had been. Once he apparently crossed the main road to Illeret, "but he didn't recognize where he was," said Abell. He also heard and saw the planes searching for him, but if he made any attempt to attract their attention, he failed. "It wasn't until the third day that he really admitted that he was lost," Abell said, still distraught over Hammock's death twenty years later. "He wrote that sometimes he'd doze at night, but then he'd wake up with hyenas nibbling at his shoes. It was a terrible ordeal."

himself on running a safe, efficient camp, and enjoyed playing the role of the infallible leader. With an almost Victorian sense of duty, Richard had accepted responsibility for everything that happened at Turkana, and he now shouldered the burden of Hammock's death. "Richard was never critical of me or my role in this," said Abell, "although we, all of us, did talk extensively about what had happened and what we should have done." Prior to Hammock's disappearance, Richard had often advised his team not to work alone in the desert; he now made it an absolute rule.

Hammock's death left the expedition disheartened and depressed. Although the team members resumed their work, a general feeling of malaise hung in the air. "That whole season was fraught with low morale and problems," said Kay Behrensmeyer, who arrived after the tragic event. "It started on the wrong foot and those of us who came later didn't fully appreciate what the others had been through, how exhausted they were from the search." She and the other newly arrived members were "gung-ho to get back to work," and their high spirits irritated their colleagues, leading to petty quarrels. Then the supply truck broke down—not once, but time and again, and the usually bountiful pantry at Koobi Fora shrank to something less than C-rations. "For a while there was only canned tripe, Marmite [a yeast extract spread], and rice to eat," said Behrensmeyer.

The hominid fossils dwindled, too. There seemed to be little more than scrappy bone and teeth fragments in the older beds, leaving Kamoya and his crew despondent. The only hominid of any significance that season was found by the archeologist John W. K. (Jack) Harris. While excavating a new archeological site, he uncovered the lower jawbone of an australopithecine next to a scattering of stone tools and broken bones—the only hominid fossil ever discovered at Turkana in an archeological dig. The fossil's presence raised many questions, and for the first time since Hammock's death, the dinner table at Koobi Fora was once again animated with theory and debate. Like the *Zinj* skull at Olduvai, this mandible was surrounded by tools that Isaac had previously attributed to early *Homo*. Perhaps the australopithecines had fashioned tools after all—or, as Louis had been fond of imagining, perhaps this fellow had served only as an entrée at his cousin's dining table. "There is no way to tell," Isaac and Harris would later write in a monograph about their excavations at Turkana.

Although hominid fossils were in short supply, the fossil hunters did find new species of extinct mammals. In particular, they discovered several new species of pigs, raising Richard's hopes that his team now had the necessary faunal evidence to support Fitch and Miller's date. Indeed, after the paleontologist John Harris made an initial study of the specimens, Richard grew even more convinced that the material would settle "the prevailing debate about dating and fauna differences with the Omo . . . once and for all," as he wrote to a friend in the autumn of 1974 at the close of the field season.

Eager to present the new data as soon as possible, Richard asked John Harris to prepare a paper for a conference on East Africa's Rift Valley that was scheduled to meet at the Geological Society in London in February 1975. Clark Howell and "his henchmen," as Richard derided the members of the Omo team in a letter to Isaac, would also be present, and Richard anticipated a lively battle. "There is every indication that Berkeley is sending a 'team' to do us in at London on the issue of dating," he wrote to Jack Miller, two months before the meeting. "I am more than prepared to settle issues regarding fauna and you will have a chuckle. I am sure that you will deal with the geo-physical issues and I merely urge that we keep it cool and 100% effective!" Miller readily concurred. "I am in one hundred per-cent agreement with you that we should be completely cool about the job," he responded. "I have always thought that the thing was getting far too hysterical."

Richard's team gathered at Cambridge three days before the London meeting to finalize their strategy. By then a new worrisome rumor had surfaced: Garniss Curtis had finished his test of the KBS tuff, producing a date of 1.8 million years; he would present his results at the conference. If Fitch and Miller were shaken by this news, they did not show it. "They continued to assure us that their work was first class, that there weren't any problems with it," said John Harris, and that Curtis's test was "irrelevant." For Richard, the rumor only stiffened his resolve. He had come for a "gladiators' clash," as Isaac phrased it, and the new pigs, Richard was certain, would deal a final, "powerful blow."

On February 19, 1975, the geologist Walter W. Bishop struck the gavel in the Geological Society's massive greystone Burlington House, opening the three-day East Africa Rift Valley conference. Although only a fraction of the papers being presented concerned the Turkana dating issue, it was nevertheless the meeting's most talked-about topic. "It was a very dramatic time in the science, like a very important, indeed the most important, football match," said the geologist Richard Hay. "Things were up in the air. What *is* the age of it? It was all much more exciting than something you get right the first time around." The tension was further heightened, Hay added, by the fact that "everyone knew that Garniss Curtis had gotten KBS tuff samples, and that it was the first time anyone aside from Fitch and Miller had anything from Koobi Fora to date. So there was great interest in Curtis's work." There was also gossip. Although untrue, rumors were soon circulating that Curtis had obtained his sample illicitly, that a graduate student from Berkeley had gathered the tuff in secret and smuggled it out. (The rumors persist even to the present day.)

Curtis was scheduled to give his paper on the afternoon of the second day, immediately after Fitch and Miller's presentation. Curtis's talk was directed specifically at Fitch and Miller, rather than the general gathering, and delved into the intricacies of geochronological testing. His own test of the KBS tuff had produced two dates—one of 1.6 million years, the other 1.8 million years. "I thought that pair of dates would shake up Fitch and Miller," Curtis later wrote to Richard, "and that they would finally understand that they had a much bigger problem than they thought." Instead, "they failed to see its [his paper's] application to them and gave it condescending praise."

At the end of Curtis's speech, Miller took the floor. "I'd heard that Miller was a butcher's son," said Richard Hay, "and he was like a butcher himself, the way he went after Curtis. He cut him up, right there." In his archest English, as Curtis recalled, Miller sniffed that he knew that Curtis was doing *"extremely* good work at Berkeley—considering the limited amount of equipment he has. But he really needs to bring his act up to date with 40/39 dating."

Richard himself was livid, but because of Fitch and Miller, not Curtis. "They'd assured Richard in Cambridge that their dates were irreproachable," said Paul Abell, "but in their presentation, they hedged and put in a number of weasel words that left Richard high and dry, and he was furious." He was also angry with Basil Cooke, who had further refined his fossil pig data, incorporating most of the Turkana specimens. But he had not shared his results (which once again pointed to a date of no more than two million years) with Richard prior to the meeting. "I felt that Basil, as a member of 'my' team, should not have used the Koobi Fora data in the way he did [to question Fitch and Miller] without giving me a full report before the meeting," Richard later wrote in his autobiography. But Cooke had never considered himself a member of anyone's team. "I was an outside consultant," he said, "called in to work on the pigs."

Richard's hopes that his team could quash their critics once and for all now dwindled. He himself attacked Cooke, charging that since Cooke was the only expert on African pigs, there was no way to verify his data. Richard then pulled out his trump—the new pigs that Cooke had not yet seen and that, Richard claimed, supported his team's ecological thesis. Kay Behrensmeyer and John Harris also gave papers defending their hypothesis, but few in the audience were convinced. "There was a rather heated debate about all this," recalled Cooke, "and Richard was getting very hot under the collar. So I got up and said, 'Look, I think we ought to call a halt to this.' " His blue eyes twinkling mischeviously, Cooke then pointed to his tie, which had a picture of a pig on it and the letters MCP underneath. "My wife gave me this tie," he told the gathering, "and she is convinced that 'MCP' means 'male chauvinist pig.' But what it really means is *'Mesochoerus* [one of the disputed pigs] Correlates Perfectly.' " Cooke's joke brought laughter—even Richard forced a smile—and temporarily eased the tension. But that evening tempers flared again.

Richard had invited many of his colleagues to a dinner party at the flat where he was staying in Hyde Park Square. Members of both teams were present, including Clark Howell, Frank Brown, Glynn Isaac, and Bernard Wood, and the conversation inevitably turned to the matter of the KBS tuff. It did not take long to turn into a shouting match.

In the middle of the brawl Richard, with a Louis-like flourish, produced a fiberglass cast of a hominid pelvis Kamoya's team had found just weeks before at Turkana. It was a beautiful specimen, and, if nothing else, served as a pointed reminder of who could find hominid fossils and who could not. Handing it to Howell, Richard noted that even though the pelvis looked very modern, it came from deposits that had been dated at three million years—proof once again of an ancient *Homo*. But Howell was skeptical. "Well, it's beautifully preserved," he told Richard, "and if you'd clean it up better you'd see that it looks just like the one Mary found at Olduvai in Bed IV." That pelvis dated to only one million years and had been attributed to *Homo erectus*—hardly the ancient species Richard had in mind—and he quickly disagreed. "It proves we have *Homo* at 3 million years," he countered. "Yes, it's similar to OH 28 [Mary's specimen], but I assure you it's different. I've looked at it." "Well, look again," Howell growled in reply. "I just don't believe your 3-million-year-old *Homo*. And I don't believe your 2.6 million KBS tuff."[3]

"It didn't take but a minute more," Howell recalled, "and we were all shouting at one another. There were loud arguments, people getting white around their mouths, glaring at each other. It was a tremendous fracas."

Both sides refused to back down. The three-day meeting had resolved nothing, though it had, Howell noted, "really got the issue out in the open." Curtis's paper had driven in another wedge, and it now seemed only a matter of time before the KBS tuff controversy was finally resolved. Mary, for one, already considered it over. "I must confess that I am now a lot happier with [the date for] both tools and hominids," she wrote Curtis, after he ran a second test, again yielding both the 1.6 and 1.8 million year dates.[4] "It now remains to convince Richard to admit [his] error—a thing Louis always did, no matter how it hurt. Let us see if Richard is as good a man as his father!"[5]

Richard was damned if he was going to concede defeat. "My intention is not to give them quarter or satisfaction on any issue," he wrote Isaac. "Louis was persecuted in the 1930's and we are perhaps seeing the same situation but with the refinement of modern methods." Isaac, who was wavering, should "stand firm," Richard urged, "and repel such attacks. . . . I say stand firm, remain objective and let's give leadership in times of stress! I reject the present proposals and see no reason to abandon our original position where the KBS is 2.61 ± 0.26 million years."

But Isaac was impressed with Curtis's results, which, unlike Fitch and

---

[3] Today this hominid pelvis is dated at two million years, making it slightly older than the 1470 skull. Some paleoanthropologists attribute it to *H. habilis,* others to *H. erectus.*

[4] Other labs later analyzed the KBS tuff, finally agreeing on a date of 1.88 million years.

[5] Reviewers of Fitch and Miller's 1973 paper defending their 2.6-million-year date criticized the team's science. But despite the comments of geochronologists Ian McDougall and Brent Dalrymple, Fitch and Miller's paper was published in 1976 without major revisions in *Earliest Man and Environments in the Lake Rudolf Basin.*

Miller's, were consistent. By the end of June, after a flurry of letters passed between Richard and Isaac, Richard could only write in dismay to another friend, "Glynn has totally abandoned the ship." Four more years would pass before Richard would admit that Fitch and Miller were wrong, and Curtis right.

Chapter 33

ON THE TRAIL OF *HOMO ERECTUS*

While Richard was busy amassing fossils at Koobi Fora and arguing over dates, Mary quietly carried on with her work in Tanzania. Even after Louis's death in 1972, Mary had no desire to return to Nairobi from Olduvai, where she had lived since the mid-1960s. "Olduvai had become my home, and it was a much nicer place in which to live," she wrote in her autobiography. Indeed, she had headed back to the gorge as soon as possible after Louis's funeral, and in the following three years had seldom left. Only anthropological conferences overseas or family gatherings in Kenya (she now had six grandchildren from her three sons) could pull her away from Olduvai's grandeur and isolation.

Here, at the edge of the Serengeti Plain, Mary and her Kamba staff labored to complete the excavations in Beds III and IV, and the Masek Beds, the gorge's upper geological levels, which she had started in 1968. Her previous excavations, in Beds I and II, had created a portrait of the earliest stone toolmakers, *Homo habilis,* and their shadowy cousins, *Australopithecus boisei.* Now she intended to do the same for *Homo erectus,* who was then thought to have appeared on the East African savannas about 1.4 million years ago. Together, her excavations would document human prehistory from 1.89 million to 200,000 years ago—the longest and most continuous record of stone age cultures that any archeologist had undertaken.[1]

[1] The Ndutu and Naisiusiu Beds lie above the Masek level. Although no occupation sites were found here, Mary's team did find middle and late Stone Age tools, as well as beads made from ostrich eggshell, all dating to 17,000 years ago. Interestingly, the first human remains discovered at Olduvai —Hans Reck's 1913 Olduvai Man—have been dated to the same period. Still later in the gorge's

When she started on the upper beds, Mary was fifty-five and utterly transformed from the passionate, rebellious young woman who had run away with a married man to Africa. Only her fondness for cigars and Scotch, or an unexpected ribald remark, hinted at that past. Her years of disciplined research had given her a steely, rigorous air. Her dress, too—often a simple white blouse and a blue skirt—was as stern as a headmistress's and matched the look in her eye. It was not that Mary had grown old and crotchety or lacked a sense of humor—indeed, her brown hair showed few strands of grey, her step was light and quick, and she was known for her dry wit. But she had come to measure other people by her own demanding standards, and very few made the grade.

Those who did not—a list that included acquaintances, colleagues, employees, longtime friends, even Louis at the end of his life—could find themselves suddenly scorned, dismissed, and cut out of her life. "She can be ruthless," said Micky Day, a close friend of Mary's and the wife of the anatomist Michael Day, "and you always think, 'Any minute now the chop is going to come.'" "One easily fell in and out of favor," added the geologist Richard Hay. The slightest infraction—asking silly questions, talking too much, being lazy, complaining about the heat or camp food, or simply rubbing Mary the wrong way—could lead to ostracism. By the early 1970s the annals of Olduvai were littered with the names of those who had fallen out with Mary. She called them "Stinkers."

Whether Mary was genuinely misanthropic, or simply a demanding and particular taskmaster, is a matter of debate among those who knew her best. Her son Philip believed she disliked people generally. "She was allergic to them, and hated being around them," he said. Yet with her closest friends, Mary could be jolly and charming, her usually steely blue eyes alight with mirth. "I've always loved her, from the word go," said Helen O'Brien, whose husband launched the Leakey Foundation. "Feisty or not, she was for me; she was fun." To O'Brien and others, Mary's crusty exterior was simply a protective shield, masking her innate insecurities. "She's basically a shy person," said Richard Hay. "She has a fear of other people at some level; she doesn't trust them, doesn't feel comfortable with them, and so she's prickly."

The privations of working and living at Olduvai had only accentuated Mary's bristly manner. "Whatever her duties were—raising the children, working in hard terrain, directing the digs—Mary never shirked," said Lita Osmundsen, the former director of the Wenner-Gren Foundation. "But all that took something out of her. It cost her personally, as well as in time and energy—and therefore she had no patience with people who were softies." In particular, she disliked weak women and always scorned any suggestion

---

prehistory, Neolithic peoples occupied the area. Louis found a ground stone axe of their making at the very top of one of the gorge's gullies. Dated to approximately 2,600 years ago, the axe marked the last phase of Stone Age occupation.

that her life or career had been obstructed by men. When an interviewer for the BBC asked her if she had ever encountered any difficulty in her field because she was a woman, Mary sniffed, "Not as though I'd have noticed it. I've been very busy." "Yet it is a fact that if you are in this kind of field, you come across subtle opposition [from men]," said Elizabeth Vrba, a paleontologist at Yale University and a friend of Mary's. "Mary didn't have any time to spend on it, but that kind of thing changes you. You may start off being gentle and not wanting to irritate anybody, but you soon realize that if you're going to get any recognition, you have to be better than a man all the time. So it is not surprising if she has a few firm sides to her character."

By the mid-1970s, Mary had become an almost legendary scientific recluse. The gorge was her sanctuary, and she jealously guarded her solitude, dictating who could visit and how long they could stay. "[A]bove all I wanted to be left in peace to get on with my work . . . , choosing my visitors carefully when I wanted any at all," she wrote in her autobiography.

Richard Hay and a few other scientists were invited every summer to pursue some program of research in Olduvai's beds. "She always needed a geologist, and I always produced," said Hay, explaining how he had managed to stay in Mary's good graces. "And I kept my own counsel and avoided those topics that I knew would ruffle her fur." Others who were less astute lasted but a day. "Mary belongs to a world where you don't ignore distinctions of age, generation and seniority," observed Glynn Isaac, "and so she expects people to defer to her in her own sphere. Students, especially American students, who haven't understood these social niceties have gotten on rather badly. There have been notable situations where she has wanted people out of her camp." One scientist was dismissed for seeming to criticize one of her dogs, another for hitting Mary's mischievous monkey Simon with a stone. "Mary did not want to see that *mzungu* [white person] again," said Mwongela Muoka, one of her excavators. "She said, 'Take him away.' He even did not have lunch."

The only visitors Mary could not readily exclude were the tourists who began flocking to Olduvai in the late 1960s. All the publicity—the *National Geographic* articles, television specials, and Louis's lecture tours—had made the gorge renowned, and since Olduvai lay midway between the safari lodges of the Serengeti and those at Ngorongoro Crater, it became part of the standard tour itinerary. Louis had welcomed the steadily increasing flow of tourists, and when at the camp, he happily trotted around the gorge with them. But Mary considered them a blight.

When she could, Mary hid from the tourists (whose numbers swelled to 20,400 in 1973), or if Richard Hay was in camp, sent him out to talk to them. She also built a small museum overlooking the gorge, and stocked it with postcards, pamphlets, and guides, hoping these would satisfy the tourists' curiosity. Still, the intruders seemed to pop up everywhere with their "clicking shutters and peering eyes," as Mary wrote in her autobiography. One even snapped a picture of her through the window of her hut washing her hair.[2]

The tourists, like the scientists, came primarily in the summer months, which left Mary relieved and alone with her staff at Olduvai the rest of the year. Occasionally her nearest neighbor, George Dove, who owned a safari camp twenty-five miles from the gorge, stopped by to visit, and her sons sometimes traveled down from Nairobi. Otherwise, she was on her own, managing the camp and excavations, and running her morning health clinic for the local Maasai people as she had always done. "It was very much a one-woman show," said Richard Hay, "with the occasional geologist [meaning himself] thrown in."

Mary's relations with her eight Kamba workers were strictly as employer to employee; she never mingled with them socially, even those she had known for more than a decade. And so in the evenings she still sat alone at the head of her dining table, waiting for her cook to serve her, while her men gathered around their own campfire, laughing and talking over their *posho* and beans. For companionship she turned instead to her four Dalmatians and the shy, wild creatures that lived at the gorge.

"The dogs were a real life line," said Judith Shackleton, an artist who spent several months with Mary in 1970, designing the displays for the Olduvai museum. "I don't think she could have lived that solitary time without them. They were trained not to disturb mice as they came from their burrows in the straw walls to carry off dog biscuits. Wagtails and other birds could also come safely [to camp] for water and crumbs. If at twilight the dogs were missing, there was general panic until they were found; Mary [was] as agitated as she'd have been over her children."

Mary built a birdbath just beyond the camp's main building, attracting flocks of colorful weavers, red-cheeked cordon-bleus, and firefinches, while the gorge's scrub robins grew so tame that they hopped into the dining room, then perched on a chair to beg for cheese. A pair of white-naped ravens frequented the camp as well, coming for the softened bits of dog biscuits Mary set out for them. When she was alone, the ravens would fly right to her for their treat. Sometimes she brought orphaned wildebeest calves to camp after chasing them at high speed across the plains, and housed them in a grass hut next to her own. "They are the sweetest things to me, and I wanted to save them," she said about her sometimes reckless pursuits.[3] So great was her love of animals that one researcher fumed to

---

[2] Mary's disdain for the tourists lessened over the years. For example, during a 1984 visit to the gorge, a group of American tourists spotted her. But instead of turning away, Mary graciously introduced herself, saying, "Well, if they've come all that way to see Olduvai, then I must say hello. It's only fair."

[3] The most famous of Mary's wildebeest calves was Oliver, who chased across the plains with her Dalmatians and terrorized visitors at camp, charging into their tents if he heard someone bathing. Girls were his preference, "apparently on the ground that they could be counted on to give a satisfying scream," Mary wrote in her autobiography. "He clearly knew which tents contained girls, and he also knew from the sounds of the water when was the right moment to attack." More than

another that Mary "treated animals better than she does people." "But then," replied the other scientist, "the animals probably treat her better than people do."

Mary's dogs in particular were everything that life had taught her people were not: loyal and true, unwavering in their love and affection. In turn, Mary was devoted to them. "Once when [Mary's son] Philip was young, he threw a bucket of cold water on the dogs," said Micky Day. "Mary, in turn, picked up a bucket and doused Philip so he would [she said], 'Know what it felt like.' That's the way she is: Dogs first and people—even children—second."

The dogs could do no wrong. "No, no, Smudge," Mary might chide, if one wandered into the middle of a dig. "But people who touched a nail or string at an excavation were immediately shouted at," said Peter Jones, who worked as Mary's assistant in the late 1970s. The dogs, too, received a daily ration of fresh meat, while Mary's workers were allotted meat only once a week. "The Kamba servants just laughed about it," said Jones. "They'd say, 'Look at how she worries over her dogs. But if they were starving, they would eat her.' "

Mary's Kamba workers were by now extremely skilled excavators. Each also held a second camp job—as cook, the dogs' caretaker, mechanic, or driver. Several had been in Mary's employ since 1961 and were used to her imperiousness. "She could be very harsh," said Mwongela Muoka, "and she used to shout very loudly if you did something wrong. But in a way that was good; it made you think clearly about your work." Otherwise it was quiet on the dig, since Mary forbade both singing and idle talk. Only the sounds of the men's ice picks striking the earth, the soft whoosh of dirt being sieved, or the rattle of a wheelbarrow broke the silence.

And so over a six-year period—eleven months of each year, six days a week, from 7:00 A.M. until 5:00 P.M.—Mary and her crew inched their way through the soil and slowly brought to view the life and times of *Homo erectus.*

In later years, Mary would refer to *Homo erectus* as that "dim-witted fellow" —annoyed by his apparent one-million-year obsession with fashioning the same style of handaxe over and over again. But in 1968, when she first began uncovering traces of this ancestor's handiwork, she was intrigued and mystified. Like her research in Beds I and II, these excavations indicated that the early human story was neither simple nor straightforward.

Mary's previous digs, from 1960 to 1963, in Olduvai's lower beds had revealed that about two million years ago *Homo habilis,* the first human ancestor, had made regular use of particular parts of the gorge, possibly as

---

one young woman was sent bolting through the tent flaps with a hastily grabbed towel, and Oliver in pursuit.

campsites, or merely as preferred places for cutting up carcasses, shaping stone tools, or resting. These ancestors lived beside freshwater streams that fed a large alkaline lake. Their tools (which Louis had named the Oldowan industry) were simple choppers, small scrapers, and flakes. Later, they had expanded their toolkit to include heavier scrapers, awls, spheroids, and protohandaxes; Mary named this stage in our ancestors' culture the Developed Oldowan. She found these more sophisticated artifacts throughout Bed II, from approximately 1.7 to 1.2 million years ago, and speculated that *Homo habilis* had also manufactured them.[4]

When Louis and Mary first found the Developed Oldowan tools, they surmised that these artifacts had led in time to the more advanced Acheulean industry, whose hallmark was the tear-shaped handaxe. But Mary's excavations seemed to reveal a more complicated relationship. Instead of finding a tidy progression, with the Developed Oldowan giving way to the Acheulean, she discovered that the two industries had been contemporaneous. Indeed, Mary uncovered Acheulean handaxes at sites dating to 1.5 million years ago, indicating an overlap of at least 200,000 years. In one area of the gorge in Bed II, the two industries were found within a few hundred yards of each other and in the same geological horizon—but they were never found intermingled. Seeking to explain this division, Mary had speculated in her book *Olduvai Gorge,* Vol. 3, *Excavations in Beds I and II, 1960–1963* that the two dissimilar sets of tools represented two different cultures, or perhaps even the work of two different species of hominids. She hoped her new excavations in Beds III and IV would reveal whether this pattern persisted, and how it might be explained.

Mary began her study of the gorge's upper beds with a systematic search of their geological deposits. She already knew of several promising sites that had been identified by members of Louis's 1931 expedition. Two of these in Bed IV had been partially excavated (in 1931 and 1962), yielding hundreds of handaxes and cleavers, and Mary hoped that they might be occupation sites. But she was soon disappointed. The prolific spread of tools did not lie on undisturbed living floors like those she had found in Beds I and II, but rather had been moved into a concentrated mass by rushing rivers.

Mary next turned to a site called FLK (for Frida Leakey's Korongo), in cliffs overlooking the area in Bed I where she had found *Zinj* and her son Jonathan had found the first specimens of *Homo habilis.* In late 1968 Edward Kandini, one of her workers, spotted a fragment of a *Homo erectus* mandible here, lying in a gritty sandstone deposit high in Bed IV. Although only a few

---

[4] Mary found fossils of both *Australopithecus boisei* and *Homo habilis* at Oldowan and Developed Oldowan sites, but attributed the artifacts to the latter because the ability to manufacture tools to a set and regular pattern was considered a distinctive human trait. "The potential capability of *Australopithecus boisei* for making tools is purely conjectural and largely a matter of individual opinion," Mary wrote in her 1979 book, *Olduvai Gorge: My Search for Early Man.* No one had discovered any hand bones of *A. boisei* or the South African *A. robustus,* she noted, which would help determine their manual dexterity.

tools lay beside the fossil, Mary decided to sink an exploratory trench, and this time was rewarded with the discovery of a "fine Acheulian site about 400,000 years old," as she wrote in her autobiography. Almost all of the tools here had been fashioned from a white quartzite and were extremely well-made. In particular, there were beautiful white handaxes trimmed to fine, tapering points, with their cutting edges in "exceptionally sharp condition," Mary noted. She was especially taken with five of the largest specimens whose style and dimensions were so similar that she thought they must all have been the work of one man. Mary had also uncovered tools of white quartzite during her excavations in Bed I, and she and Richard Hay had traced the source to an outcrop on the Serengeti Plain, only a mile away from the gorge. Apparently the stone had proved so useful and desirable that 1.3 million years—and many thousands of generations—later, the hominids were still trekking to the same quarry to gather this material for their tools.

Early in 1970, Mary and her crew moved to another site called WK (for Wayland's Korongo) in the middle of Bed IV, where once again dozens of handaxes and cleavers littered the ground. This, too, proved to be undisturbed and it was not long before her crew hit paydirt. Amidst a scatter of tools and broken animal bones, they unearthed a hominid femur (thigh bone) and part of a pelvis. Since the hominid's skull was missing, Mary could not readily identify the species. But given the size and shape of the bones, and the age of the deposits (500,000 years old), she speculated that the fossils belonged to *Homo erectus*—an opinion that the anatomist Phillip Tobias shared. If so, Mary had discovered the first pelvis bone of this ancestor, and Tobias, who was then in Nairobi working on his study of the Leakeys' *Homo habilis* fossils, could barely contain his excitement.[5] "[I]t may well be the first evidence of the pelvis structure of the first hominids to walk perfectly upright!" he enthused in a letter to Mary. "As such, it is an eviable discovery. Bully for you, my dear. I'm thrilled!!!"[6]

Mary's discovery also marked the first time that the remains of *Homo erectus* had been found *in situ* with artifacts, apparently confirming this ancestor as the maker of the Acheulean handaxes—something that Louis had suspected for many years. But Mary made only the briefest allusion to the possible relationship between the fossils and tools in her report about her find in *Nature*. "It is not certain whether this association implies that *H. erectus* was responsible for the industry, but it seems likely . . . ," she wrote with characteristic caution.[7]

Mary hoped that the hominid bones and tools at WK marked some type of ancient campsite—much as similar collections of stones and bones had

---

[5] Richard Leakey's team found a more complete *Homo erectus* pelvis on the east side of Lake Turkana in 1975, and a second specimen on the west side of Turkana in 1984.

[6] Some paleoanthropologists say that the australopithecines and *Homo habilis* also walked "perfectly upright."

[7] Louis and Mary had found other bones of *Homo erectus* in association with the Acheulean tools, but this was the first time the bones and stones had been found together *in situ*.

defined the living floors in Beds I and II—and she directed her crew to uncover an additional 230-square-foot area. In the process, they unearthed 2,600 specimens, while Mary sat nearby numbering and mapping each item: handaxes, cleavers, scrapers, awls and pitted stones, cobbles and hammerstones, and a variety of animal bones. The last group included fossils of catfish, hippopotami, crocodiles, and the three-toed horse, *Stylohipparion* —the collection perhaps representing some of the remains of *Homo erectus*'s meals.

By now Mary was "hot on the scent after *Homo erectus!*" as she wrote a colleague in the spring of 1971, and she turned her attention to yet another Bed IV site, called HEB (for Heberer's gully). Here, she uncovered five artifact-bearing levels with Acheulean tools, ranging in age from 700,000 to 200,000 years. In one of these horizons, almost all of the handaxes and cleavers (some 75 percent of the tools) had been made from large pieces of fine-grained green phonolite, a lava that Mary and Richard Hay traced to the small volcano Engelosen, about eight miles away on the Serengeti Plain. Like the white quartzite tools, these handaxes were beautifully fashioned, although Mary was puzzled by the lack of accompanying waste flakes. Perhaps, she reasoned, *Homo erectus* had blocked out the tools at the Engelosen quarry, then carried the rocks to their camp or work sites for completion. These hominids also crafted many bone tools, producing a particularly fine handaxe of elephant bone, and turning an elephant pelvis into a mortar.

Unfortunately, aside from classifying the types of tools and animal fossils and noting the association between the tools and the fossils of *Homo erectus,* there was little Mary could say about these Bed IV sites. Although they were not disturbed like the first two she had excavated, they nevertheless lay in old streambeds, and in spite of an intensive search Mary could find no evidence of postholes or traces of any type of shelter. In fact, all the sites in Bed IV lay either beside or directly in old streambeds, perhaps because of a shortage of water. Earthquakes, volcanic eruptions, and climatic shifts had altered the hominids' world, and instead of living beside a large lake, they now dwelled alongside seasonal streams that flowed through grassy plains into small—and sometimes dry—lakes. That their artifacts lay in the streambeds suggested to Mary that *Homo erectus* may have even lived in the beds themselves during the dry season where they could have found water by digging—much as many nomadic tribes in East Africa do today.

Yet the region had remained lush enough during *Homo erectus*'s day to support a population of hippopotami, which together with catfish seemed to be the staple item in the hominids' diet. Mary found fossilized hippo bones in every Bed IV excavation, and even turned up a nearly complete skeleton of the extinct, goggle-eyed *Hippopotamus gorgops* at a site she dubbed Hippo Cliff. Perhaps a group of *Homo erectus* hunters had chased the beast into a swamp and killed it, or had simply been lucky and stumbled onto its carcass. Either way, they had apparently butchered some of the animal, then departed, leaving behind most of the hippo and several of their hefty handaxes and sharp-edged cutting tools.

Two other sites lay close to this ancient Bed IV abbatoir, and Mary's workers sampled them by again digging exploratory trenches. At both sites, they found artifacts at the same geological level (dating to approximately 500,000 years ago) as the *Homo erectus* fossils they had uncovered at WK. But this time, the tools were not of the Acheulean type and thus, Mary thought, not likely to have been made by *Homo erectus.* Instead, she realized that she was once again looking at Developed Oldowan artifacts—tools that she had initially attributed to *Homo habilis.* Finding these distinctive industries once again close together and at sites only 500,000 years old was "quite against expectation," Mary wrote in her autobiography. Unsure about how to interpret her discovery, Mary merely noted in *Nature* that these tools were "at variance with the WK industry." In letters to her archeological colleagues and talks with Richard Hay, however, she was more open and daring, proposing that two lineages of early humans might have lived side by side for as much as one million years.

The key difference between the Acheulean and Developed Oldowan industries lay in the tools' fabrication: to make an Acheulean handaxe (a tool that Mary never found at any of the Developed Oldowan sites) the hominids had to know how to strike a boulder so that they could remove a large flake, one that measured at least eight by four inches and was an inch deep. For a while Mary thought that perhaps the makers of the Developed Oldowan tools had not succeeded in making large flakes because they were using different raw materials for their artifacts, materials that could not be shaped into Acheulean handaxes. "[B]ut the Bed IV sites have put paid to this idea," she wrote to the archeologist Desmond Clark. "They were all using the same materials from the same sources." Hay, who spent many evenings at Olduvai sharing Scotch and cigars with Mary and debating the issue, argued that it seemed more likely that the hominids had used different tools in different settings. "If you plot the Acheulean sites," he explained, "you see that they occur farther away from the lake edge in fluvial [river] settings, whereas the Developed Oldowan tends to be found in lakeside settings. To me, it's more plausible that you have different toolkits in areas where the food supply is different."[8]

Mary hoped to prove her two-hominid theory by discovering *Homo habilis* fossils at one of the Developed Oldowan sites in Bed IV. But in spite of the many tons of earth she and her team removed, no hominid fossils ever came to light. "I would dearly have liked to find hominid remains at one of these Bed IV Developed Oldowan occurrences, to see whether my idea that a different hominid type was their maker held good," Mary wrote wistfully

---

[8] Or, it may be, as the archeologists Kathy D. Schick and Nicholas Toth suggest in their 1993 book, *Making Silent Stones Speak,* that "[a]t sites closer to the lake, the most desirable raw materials for these [Acheulean] bifaces may have been too far away, and the lower-quality bifaces of the Developed Oldowan were the result."

in her autobiography. "But it was not to be, and the reason for the contemporary occurrence of such different toolkits still cannot be stated with certainty."

Mary was equally baffled by a strange group of deep pits her crew unearthed in 1972 at the sole Bed III site worth excavating, JK (for Juma's Korongo).[9] The site had been partially excavated in 1962 by another archeologist, who uncovered Acheulean tools here as well as a hominid femur and tibia. These fossils were badly abraded, making it impossible to determine the species, but because of their presence, Mary decided to explore further. "We were digging a trench, and going down quite rapidly with picks and shovels," she explained during a visit to the gorge in 1984, "when we suddenly encountered a layer of pink siltstone and the pits." Mary had never seen anything like the pits before, and so some were damaged before she realized that they might be significant; she then stopped her crew and directed them to dig a parallel trench. "Of course, we then used only toothbrushes and dental picks and charted the pits with care," she said.

Irregularly shaped and with fingerlike scrapings around their sides, the pits seemed to Mary a certain sign of human activity—although what the hominids were digging for was a mystery. "[I]t's a bit difficult to get inside the mind of *Homo erectus!*" she confided to one friend about her difficulty in interpreting the site. All that could be stated for certain was that the pits were not natural. "You couldn't explain them geologically," said Hay. "Right away, I disowned any part of them. Someone did suggest that an antelope could have done something like that; but so could *Homo erectus.* But I don't know of any natural agency that could have caused those pits and ditches."

More than a dozen pits, some measuring three feet in diameter and a foot deep, had been scooped into the pink siltstone deposit, which had subsequently turned rock-hard. A confusing pattern of narrow channels connected the pits; in some cases, several channels converged on a single pit. There were also stone flakes and bone chips in the pits ("evidence that man was present," Mary noted in *Olduvai Gorge: My Search for Early Man*), as well as numerous small round depressions both in the bottoms of the pits and in the surrounding ground. Mary thought that some of these marks may have been caused by animals trampling the earth, "but others," she wrote in a report to the National Geographic Society, "look very like marks made by the end of a stick."[10] She also found "beautifully preserved . . . tool

[9] With its bright red coloring, Bed III is the most striking of the geological horizons at the gorge, but it is also the least productive archeologically or paleontologically. In his 1951 analysis of Olduvai's geology, Louis argued that the lack of stones and bones in Bed III was a result of a past climatic change; the region had simply grown too dry for the hominids. But Richard Hay showed that it was actually a lack of vegetation and fresh water that was responsible for the scarcity of fossils and artifacts—the Bed III sediments had simply not preserved many traces of the past.

[10] "Some of the small pits could have been made by animals such as hippos or antelopes walking over the surface when it was muddy and picking up clods of wet mud on their feet," Mary wrote in *Olduvai Gorge: My Search for Early Man.* "I have observed similar holes around the shores of Lake Masek [Tanzania] made by animals walking over a muddy area covered by a thin crust of dried soil."

or finger marks . . . round the sides" of the pits, and in one, "a nice little footprint of a child or very small woman." All in all, Mary thought the pits "just about the most exciting item that Olduvai has yet produced," as she told Glynn Isaac.

Yet no one could make sense of what Mary had found. Not Desmond Clark, Glynn Isaac, Phillip Tobias, or Richard Hay, who all came to survey the site, or her son Richard, or even Louis, who was then still alive and loved to speculate about such things. Louis, then living in Nairobi, and Richard flew in for a visit together one day in April 1972. For some moments, they simply stared at Mary's pits, but neither man uttered a word. Finally, Richard looked up and said, "You know, Mother, when there is complete silence in this family it means only one thing: some member of the family is in serious trouble." "It was me he meant," Mary wrote in her autobiography. How would she ever convince her colleagues that the pits had been made by humans, let alone explain their use?[11]

Mary pursued the story of the pits, as well as additional Bed IV sites, through 1973, finally stopping in the spring of 1974. She had at last reached the goal she had set herself fourteen years before: tracing the successive Stone Age industries from the bottom of the gorge to its upper levels. Now she planned to begin her analysis of this latest material, although her excavations had not revealed the definitive answers she had initially hoped for. She had found only a few additional fossils of *Homo erectus,* making it nearly impossible to say anything about his physical characteristics. "Virtually no information is available," she would later write in *Olduvai Gorge,* ". . . other than the fact that [one individual] had particularly massive brow ridges" and that the "walls of [two] skulls were exceedingly thick."

Likewise, she had discovered far less about *Homo erectus*'s living conditions and hunting or scavenging methods than she had about those of *Homo habilis.* Nor had she solved the mystery of the Developed Oldowan and Acheulean industries. Indeed, even the Acheulean perplexed her, since many of the most skillfully crafted handaxes were found in deposits that were older than those containing more primitive-looking, crudely made tools. Instead of advancing in his toolmaking skills, it seemed to Mary that *Homo erectus* had gone backwards. But as she and her assistant Peter Jones studied the tools, they decided that the difference between the types of tools was best explained by the rocks themselves. To make a sharp-edged implement from the fine-grained phonolite required a great deal of work, whereas a basalt handaxe could be turned out in less than a minute. Thus, while a *Homo erectus*'s basalt handaxe might look primitive, it was actually sophisticated—representing an advance in manufacturing efficiency and a greater understanding of the raw materials at hand.

---

[11] In 1984, flying over the desert in northern Kenya, Mary's son Philip noticed similar pits and channels, which were made by a local tribe and are used for salt extraction. Mary believes that they may be a modern version of the JK pits and hopes that one day they will be studied.

Above all in *Homo erectus*'s tools, Mary recognized a basic trait of humankind: the need for tradition. The artifacts from each of the Acheulean sites seemed to have been "made by a group of people who had their own tradition in toolmaking . . . from which they did not deviate to any appreciable extent," Mary wrote. Perhaps the groups represented clans or tribes or even families—whatever the case, 700,000 years ago, people apparently designated their social group by the style of tools they fashioned.

By the fall of 1974, Mary was deep into her analysis of the implements from Beds III and IV, although she was already impatient with the vast number of handaxes *Homo erectus* had left behind. "All of those handaxes over and over again," she fumed later. "It wasn't like the material from Beds I and II. Those tools were interesting and I thoroughly enjoyed writing about them." So she was pleased when her neighbor George Dove showed up at her camp one day with a handful of fossils, including zebra and antelope teeth. He had found them in a load of sand that had recently been delivered to his safari lodge. The sand had come from the bed of the Gadjingero River, which drains through an area called Laetoli, thirty miles south of Olduvai. "It seemed well worth a pleasant day out to go over and see if we could locate the area of sediment from which George's fossil teeth had been washed out," Mary wrote in her autobiography, little dreaming that at Laetoli she would make the most important discovery of her life.

Chapter 34

MOTHER AND SON

Laetoli, the new site that Mary planned to explore, was only an hour's journey by car from Olduvai. She and Louis had actually visited the area twice before—once in 1935, after a Maasai tribesman showed them fossils he had found in the area, and again for three weeks in 1959. They collected fossils on both of these excursions, finding fragments of extinct elephants, antelopes, rhinos, and giraffe. On their first journey, Louis had even turned up a hominid's tooth, but because of its extremely primitive shape, he misidentified it as a monkey's. They had no way of knowing then (since the potassium-argon method of dating had yet to be developed) that the fossils were older than those from Olduvai. But even if Louis and Mary had known this, it is doubtful that they would have soon returned to Laetoli, for only three weeks after their 1959 visit, Mary discovered the skull of *Australopithecus boisei (Zinj)* at Olduvai. Since then, Mary had visited Laetoli only once, on a day's excursion with Richard Hay in 1969.

Yet Laetoli intrigued Mary. She knew that another explorer, the German Dr. Ludwig Kohl-Larsen, had discovered a fossilized hominid maxilla there in 1939, as well as elephant specimens that were considered very primitive.[1] The handful of fossilized zebra and antelope teeth Mary's neighbor George Dove showed her in the spring of 1974 only heightened her interest. Hoping to trace the beds from which the teeth had eroded, Mary and some of her

[1] Kohl-Larsen gave the mandible to the German anatomist H. Weinert to describe; he named it *Meganthropus africanus.* Don Johanson, Tim White, and Yves Coppens reclassified it as *Australopithecus afarensis* in 1978.

staff made several weekend excursions to Laetoli in the fall. They found the original site of Dove's fossil teeth, then followed the deposits farther up the Gadjingero Valley, a dry, scrubby dale bordered with acacia trees, and discovered a series of beds that Mary thought had never been explored before. Here, one morning on their third visit, Mwongela Muoka turned up a hominid's tooth. "This of course intensified our interest," Mary wrote in her autobiography, "and other visits followed."

Mary expanded the search over the Christmas holidays, adding the keen-eyed hominid hunters of Kamoya Kimeu and crew to her own team. Her son Philip, his wife, Valerie, and young daughter, Lara, joined in as well, and by the end of the year Mary had a surprising bounty of hominid finds to report: an adult and a juvenile mandible (lower jaw), a partial maxilla (upper jaw), and several more teeth. Altogether, and in the short space of twelve days, Mary's team had turned up fossils of thirteen individuals—although Mary was not at all certain about what type of hominid they were finding. The primitive-looking jaws and teeth clearly did not belong to *Homo habilis,* the oldest hominid from Olduvai; nor did they particularly resemble any of Richard's finds at Lake Turkana. Mary valued Richard's and his wife Meave's opinions, and on January 1, 1975, sent the fossils to Nairobi with Philip and Valerie.

Richard and Meave, who was an expert on fossilized monkeys and a paleontologist at the museum, agreed that Mary's fossils seemed to represent some type of previously unknown hominid, perhaps an early form of *Homo.* But they also agreed with Mary that additional and better specimens were needed before she named a new species. Further, she did not yet know the date of the Laetoli Beds,[2] having only just sent a box of lava samples to Garniss Curtis at Berkeley for dating. In the meantime, she decided to plan a full-fledged, two-month expedition for the summer of 1975. Based on the site's primitive fossilized elephant teeth, Mary thought that Laetoli must predate Olduvai, and given that hominids had once lived in the area, she hoped now to find their tools.

At Olduvai, Mary had already established that hominids were making a variety of stone tools nearly two million years ago. Even older artifacts, dating to 2.2 million years, had been found in Ethiopia's Omo Basin. Although all of these early implements were simple, they were so diverse (Mary recorded twenty types of tools in the lowest Olduvai beds) that Mary was certain that hominids had been manufacturing stone tools at an even earlier date. But how much earlier? Hoping to find a clue to this question, she and her Kamba crew and a small party of scientists spent much of July

---

[2] The geologist Peter Kent, who had visited Laetoli with Louis and Mary in 1935, named the fossil deposits the "Laetolil Beds"—Laetolil being the anglicized version of Laetoli. Later, the Tanzanian authorities asked Mary to drop the ending "l" in the interests of accuracy. However, the rules of geological nomenclature require scientists to use "Laetolil" whenever naming that particular stratigraphic layer. Thus, while the site is known as "Laetoli" the deposits themselves are always referred to as "Laetolil."

and August carefully searching the Laetoli deposits. Stone artifacts and waste flakes covered the ground, but these were only surface finds, and in spite of the team's best efforts no tools were ever found *in situ.* Nor did Mary add any more hominid fossils to the thirteen that had been found over Christmas.

Even without tools or additional hominid fossils, Mary had scored a coup at Laetoli. Shortly after the close of the expedition in September 1975, Garniss Curtis sent her the results of his dating analysis. Laetoli's hominids ranged in age from 3.35 to 3.75 million years old, making them the oldest hominid fossils in the world. If the fossils belonged to the genus *Homo,* as Mary and Richard believed, then Mary had found the world's oldest human ancestor. Only a year before, Richard had held claim to this title with his fossil skull 1470, then dated to a little more than 2.6 million years. Now Richard would have to relinquish the crown to his mother. The Leakeys had always delighted in each other's discoveries and would put aside their quarrels (at least temporarily) when a new hominid was found. Indeed, it was a fossil—the 1470 skull—that had reconciled Richard and his father only a few days before Louis's death.

Mary may have hoped for a similar rapprochement with her son. Instead, their three-year-old quarrel—a mix of arguments about family issues and Richard's handling of the Turkana dating controversy—was only further inflamed.

Shortly after Curtis released the Laetoli dates, Richard flew to Tanzania with the anatomist Alan Walker and paleontologist John Harris to look at the new site's stratigraphy. Like the deposits at Olduvai and Turkana, Laetoli was complex, and although the geologist Richard Hay had spent two months mapping the fossil beds, Mary was uncertain about some of his conclusions. "We plodded around after Mary as she explained everything," said Walker, "but I suppose Richard was skeptical because not much was said. It was a rather silent field trip." Back at Mary's Olduvai camp, tensions mounted between mother and son until, Richard said, "We got into one of those enormous arguments, and then I went to bed"—but not before telling his mother that she should come back to Nairobi and retire.

"Richard told her, 'Mary, listen, forget it,' " said Peter Jones, who worked as Mary's assistant at Olduvai. "He said, 'Why don't you just calm down and stop trying. Don't get yourself in trouble. You've done all you can do. You're on the shelf.' And that remark really infuriated her. She felt that he wanted her out of the picture."

Mary, of course, had no intention of retiring. Though she was now sixty-two and had undergone a hysterectomy only five months before, she had also just led the "most arduous" expedition of her career, she would later write in her autobiography. Maybe Laetoli's deposits did puzzle her, but she had a "sound wicket" when it came to the date—which was more than Richard could say about Turkana. She was hardly ready to return to Nairobi, and indeed planned to lead a second expedition to Laetoli. All she needed was another National Geographic Society grant—and given the age of her hominids, that seemed a likely prospect.

One month later, on October 30, 1975, Mary convened a press conference at the National Geographic Society in Washington, D.C. It was her first solo performance since 1947, when she presented her first major discovery—the skull of the apelike creature *Proconsul africanus* (now *Proconsul heseloni*)—to reporters in London. That had proved such an ordeal that Mary had gladly let Louis face the press in the intervening years. "I played second fiddle to him because I really didn't like all that fuss and attention," she said. But Louis was gone, and Mary now stood alone at the podium. She had dressed simply for the occasion, wearing a tweed skirt and beige sweater with a strand of beads, her slightly grey hair fluffed back from her face. Then, with little ado, Mary announced the discovery of the thirteen hominid fossils. "They were very fine discoveries," she told the gathering, "but we didn't appreciate their significance until just last month when we learned how old they were. That put them in a whole new light." The 3.35- to 3.75-million-year-old dates themselves, she added, were "watertight"—a reference clearly meant to separate her from the dating troubles of her son.

Mary was less definite about the type of hominid her team had found, saying only that the specimens belonged to the genus *Homo* and so represented the earliest human ancestor. She did not believe them to be australopithecines, those shadowy creatures that Louis had never accepted as part of the human family tree. "I think our new discoveries confirm the view that *australopithecus* was an offshoot of the hominid line, a contemporary of early man who died out," she said. In contrast, the fossils were the remains of people "not unlike ourselves." But beyond this, Mary refused to speculate about her discoveries. "We can't infer anything about the way they [the Laetoli hominids] lived from the present findings," she told reporters who pressed her for details about the hominids' behavior. Until she found stone tools or some other evidence, there was simply no way of knowing anything about these ancestors' hunting skills or where and how they had lived. With luck, Mary said, she might find some such material on her 1976 expedition.

The National Geographic Society awarded her $40,000 for her next expedition, and with the promise of this grant and a gold medal from the American Society of Women Geographers—as well as front-page headlines around the world proclaiming her discovery—Mary returned to Olduvai. Retirement was the furthest thing from her mind.

Gossip soon spread about Richard and Mary's quarrel, but, in fact, they had been fighting since Louis's death, when Richard had challenged his mother's plan to cremate Louis rather than bury him. The KBS dating controversy—in which Mary sided with "the winning team," as Richard phrased it, against him—only exacerbated their dispute. And although Mary says that she never rubbed her son's nose in this error, she expected him at least to admit that he had been wrong. But Richard believed that Garniss Curtis and Mary had conspired to undercut him and devalue his work, and he had no intention of admitting any error. Mary interpreted this bullheaded silence as a symptom of Richard's great ambition. She told Curtis that she feared that "personal ambition has been a more important factor [behind the dating controversy] than scientific knowledge and accuracy."

Whenever Richard caught wind of Mary's opinions, which he inevitably did, he became further convinced of her disloyalty. Richard sensed, too, that Mary did not fully approve of his efforts to build the Louis Leakey Memorial Institute of African Prehistory. After his father's death, Richard had launched a fund-raising campaign for the institute, and by 1975 had raised $200,000 with pledges of $200,000 more. Although this was about a third of the construction budget, Richard was so confident of raising the remainder that he had already begun drawing up plans. The new center would have proper storage facilities and laboratories for studying the fossils that Louis had left packed in shoeboxes and crates, and stashed in disarray in dingy rooms lit with single bare bulbs.[3]

But while Mary's disapproval of his plans was one cause of Richard's angry behavior, another was his failing kidneys, which his mother knew nothing about. By 1975, the seventh year of his renal disease, Richard's damaged kidneys had sapped his energy and made him snappish and petulant. His blood pressure was dangerously high, and his head pounded constantly, giving him a very short temper. "I think Mary and I quarreled at Olduvai during a period when my blood pressure was very bad," said Richard. "I was finding life extremely difficult then because of the disease and I was hypersensitive to any form of criticism. And I think the recognition that one may not actually have that much more to give of one's life—that it may be ending or on dialysis—made one doubly spotty and prickly."

When Richard was diagnosed with kidney disease in 1968, his physicians had given him approximately ten years to live. Although he had vowed not to worry about his shortened lifespan, Richard could not ignore his increasingly violent headaches and sickly appearance. It was not just the headaches that worried him. Always lithe and lanky, Richard had begun to look unhealthily thin. His skin, too, had a sallow cast, and his face looked surprisingly weary and haggard for someone only thirty-one. He and Meave did not discuss it, but they knew that Richard's failing kidneys were beginning to poison him.

None of this escaped Mary, either, although she did not know the cause. Instead, she and many others thought Richard was suffering from stress. After their quarrel at Olduvai, when Richard stormed off to bed, Mary had turned to Alan Walker and urged him to try to slow down her son. "She

---

[3] Mary never accepted an office in the institute, but kept her old one in the now slightly seedy buildings that Louis had constructed in the 1960s. Nor did she mention the institute in her autobiography, *Disclosing the Past*. When asked why she had not, she merely said that the institute did not concern her.

asked me to use my influence with Richard; she had tried and hadn't got through to him," said Walker. "She thought he was going to kill himself if he drove himself at that rate, and she was desperately concerned. I said, 'Well, Mary, you know your son so much better than I do. If you can't get through to him, I don't think I can slow him down either. We can tie stones on his legs, but I'm not sure that's going to help.' "

Richard drove himself despite his illness—or perhaps because of it. "I probably worked at a faster pace on a number of projects because I knew that I might be interrupted," Richard later told a reporter at *The Nairobi Times.* "I just wanted to get at least some of the things done in what could turn out to be a short life." Chief among these projects was transforming the museum from a mere repository of fossils into an international center of learning. He had already more than doubled the museum's budget and staff, adding fossil preparators and laboratory technicians, and revamping the exhibition halls. He had drawn up plans for two more regional museums as well, and had begun transforming Louis's old Tigoni Primate Research Centre into a facility for biomedical studies. But it was through his memorial to his father—the international prehistory institute—that Richard intended to realize his dream fully. Consequently, raising the funds for the institute had become as much a part of his life as searching for fossils at Turkana, and he went after the money with an equal intensity.

"Richard was always a guy in a hurry," said Bill Richards, a Los Angeles businessman who helped him raise money for his institute. "He had extraordinary drive, energy and enthusiasm for this project, and that in itself was captivating." Not everyone was so charmed. In particular, the members of the Leakey Foundation regarded him with such distrust and suspicion that only a year after Louis's death, Richard quit the foundation and set up one of his own. "Richard wanted to take over the [Leakey] foundation," explained Joan Travis, then a member of the board and one of several trustees, Richard claims, who misread his motives. "He wanted to change its emphasis so that it only funded paleoanthropology in Africa and nowhere else. He also didn't want it to fund the primate studies of Jane Goodall, Dian Fossey, and Biruté Galdikas. It was to be only as he saw it. And none of the things that the rest of us had worked so hard for would matter anymore."

In fact, Richard said, he did not want personal control of the Leakey Foundation. "I wanted the foundation that was honoring my father to be totally directed to fossils and archeology, and to leave alone the apes and monkeys," he said, admitting that he had also long disliked and envied Jane Goodall.

> But to a greater extent I felt very strongly that paleoanthropology never had a better chance to raise money than it did with my father's death. I said to a lot of people, "Look, we've never had money. Louis broke his heart trying to get money for many things. But what he really tried to do in his life was to document human origins in Africa. Now with his death,

there's a great chance to do something. So let us focus; let's take a rifle rather than a shotgun approach."

What Richard really wanted was for the foundation to support his institute exclusively. He envisioned raising an endowment of between fifteen and twenty million dollars, and drawing on its income to pay for everything from stipends for visiting scholars to scholarships for graduate students. Only after the institute was thriving should the foundation consider dispensing grants to other fields, such as primatology. Richard outlined his plan at a Leakey Foundation board meeting on March 15, 1973, in Pasadena, California. He had previously lobbied many of the trustees and, armed with assurances of their support, went into the meeting confident of winning.

Instead, Richard was met with a frosty silence. "You could have cut the atmosphere with a knife," he recalled. "I'd expected other people to join with me, but on the contrary there was this long silence with people looking at their blotters. And I thought, 'Whoops! This is not working.' "

Richard had miscalculated by not lobbying the board's two women trustees before the meeting, concentrating instead on the men. "I'm not a feminist," Travis asserted, "but I was very put out when I heard that the guys had had a secret meeting with Richard in Newport Beach the night before. He thought he'd taken care of it all with them." Travis and Tita Caldwell, the other woman trustee, argued that it would be a disservice to Louis's memory not to preserve the foundation's diverse interests, particularly the primate studies. Their objections opened the door, and soon even the men were complaining about Richard's assertions that the existing foundation "was a failure"; nor did they like what trustee Gordon Getty, the oil heir, termed Richard's "strong-arm tactics."

But the final blow came from Richard's mother. Mary met with the board five days after Richard's meeting, and although she expressed support for Richard's institute, she also questioned his plans for the Leakey Foundation, especially his intention to abandon primatology. "I don't think there is any doubt that if Mary had come here and spoken strongly in favor of turning the foundation over to Richard, the trustees would have voted that way," said Ned Munger, then the president of the Leakey Foundation. "And that would have been the end of that." Mary, however, sought the middle ground. Several of the board members then recalled Louis's own warnings not to turn the foundation over to his son, and when the final vote was taken, Richard's proposal was voted down.

More determined now than ever, Richard flew to New Jersey, where he met with David Look, a banker.[4] Look introduced Richard to Charlie Jaffin, a New York attorney, and the three laid plans for Richard's own foundation. Six months later, in the fall of 1973, the Foundation for Research into the

[4] Richard had met David Look through Meave. She had spent a year working as an au pair for the Looks' children.

Origins of Man, or FROM, was in business. Richard's annual six-week lecture tours were now arranged by his FROM associates and the money he raised was funneled into his institute via FROM's coffers.[5] By creating a rival organization, which would be competing for a small pool of dollars, Richard had almost ensured the Leakey Foundation members' lasting enmity.[6]

Meanwhile, Richard was making other enemies. Many paleoanthropologists had grown suspicious of the Leakeys while Louis, as the director of the Centre for Prehistory and Palaeontology, had wielded power over who could or could not search for fossils in his country. Richard modified the research guidelines so that scientists now had to apply to a government science council for permits—a step he believed would defuse charges that he alone exercised control. Yet as one of the few Kenyans capable of evaluating scientists' proposals, Richard was asked to sit on the council. The perception of him as the key power broker in the permit process remained. "He can tell you that it's a government council making these decisions," explained one scientist who preferred not to be identified, "but everyone knows that if Richard thinks you shouldn't get the permit, you won't get it." Already fearful of the power Richard held, many scientists were reluctant to see him expand his base through the proposed Louis Leakey Memorial Institute. His decision to remove himself as the institute's director did little to disabuse his colleagues' belief that he was consolidating his position as the godfather of East African paleoanthropology.

Many of Richard's fellow scientists were envious, too, of his jet-setting among the social elite. Even more than Louis, Richard was at ease among the wealthy and powerful. He counted among his friends members of the British and Dutch royal families, and often entertained ambassadors and diplomats at Koobi Fora—relationships that gave him access to sources of money far beyond the reach of the average scientist. When, for example, he applied to the U.S. National Science Foundation for a grant to help build his institute, he bypassed regular channels. Through the help of a U.S. senator he presented his proposal directly to NSF's director—and received a $100,000 grant. "Richard was very smart, very politically adept," said Nancy Gonzales, an anthropologist who began working at NSF when Richard received his award. "He knew how to use his contacts, and certainly there was a lot of resentment toward people who were like this."

By 1977 Richard had succeeded in raising more than one million dollars for his institute—although at some cost to his reputation. In his haste to beat what he feared might be a literal deadline, Richard came to be regarded as calculating, arrogant, and ruthless—a man whose ambition had no limits.

• • •

[5] Some of the money Richard raised on his lecture tours also went to pay for his Turkana expeditions.
[6] Richard closed FROM in 1984 and merged it with the Leakey Foundation. He is now a life trustee for the foundation, but has never requested funds for any of his own research or expeditions.

There was one part of his life that Richard decided to cut back on in 1975 and, ironically, it was the one closest to his heart: the Koobi Fora Research Project at Lake Turkana. Richard had grown "increasingly despondent," he wrote in his autobiography, about the amount of time he spent organizing the expedition. "I seemed constantly to be worrying about money, the maintenance of vehicles, the deterioration of the camp and the need to repair buildings. I had less and less time to do what I enjoyed—to go into the field and look for fossils." He and Glynn Isaac agreed to scale back the size of their operation from eighty to forty people, and advised visiting scientists that they must be self-sufficient. Isaac even decided not to go into the field; his archeological team had already excavated fifty sites north of Koobi Fora and he needed time in the laboratory to evaluate the results. Overall, Richard and Isaac felt that "a pause for taking stock is essential," as they wrote the expedition members in the fall of 1974.

"That was the beginning of the third phase of work at Koobi Fora," explained Glynn Isaac. "We'd gone from an exploratory phase to becoming a big, multifaceted expedition. But by 1975, we'd explored most of the area, it had become familiar and we were able to do much more exacting science. Things began to wind down."

Koobi Fora itself, once wild and exotic, now seemed to Richard and Isaac a little tame. "My first season there [in 1969], Richard gave me a pistol, saying, 'Carry this always,' " Isaac recalled.

> So I carried this wretched thing in my rucksack all summer wondering what it would do against the locals' rifles. And that year, too, we had a long, difficult drive in a convoy with armed guards just to reach Koobi Fora. Then it became a routine matter to fly up, and the road became a well-beaten track. The county road crew even ran a bulldozer over it. So while Koobi Fora was still beautiful, that exploratory wildness and adventure were gone. We'd shed the excitement of the unknown.

The number of hominid fossils also seemed to be dwindling. In 1974, the Hominid Gang had discovered only sixteen specimens—most of these scrappy tooth fragments—compared with twenty-two hominid fossils in 1973 and thirty-eight in 1972. Perhaps, Richard thought, all the important surface finds had been made, and it would be some years before the winds and rains at Turkana exposed new material.

For the 1975 expedition, Richard had the team explore the badlands between Koobi Fora and Ileret (a distance of about twenty-five miles). Richard also wanted to resolve the dating quandary and the problem with the mismatched animal fossils from the Omo and Turkana—a task that the paleontologist John Harris and a graduate student, Tim White, had begun the previous season. An intense, slightly built young man with a jutting jaw, White was a prize doctoral student of the anthropologist Milford Wolpoff at the University of Michigan. "Tim was active and interested, and wanted field experience," said Wolpoff, "so I asked Richard as a favor if he would take Tim on and Richard said, 'Sure.' "[7] Although somewhat brash and abrasive ("Tim had a log, not just a chip on his shoulder, for some reason," said Wolpoff), White had gotten along well with Harris and Richard, and so found himself in the enviable position among young anthropologists of being invited back as part of the Koobi Fora team.

As always, Richard set the tone and the pace of the expedition, although his visits were often confined to the weekends. He, Meave, and their daughters would fly north in his blue-and-white Cessna Super Skywagon on Saturday mornings. "Once the plane touches down on the grassy landing strip at his camp," wrote a reporter who accompanied Richard on one of these visits, "Richard goes for a quick swim, wraps a colorful African print sarong about himself and pads barefooted to the head of a dining table under a grass roof. There he presides like a firm but benevolent chieftain over the team of loyal and hardworking scientists delighted to work at as rich a fossil ground as Richard Leakey commands."[8] Richard came to Koobi Fora, too, when potential donors to his institute visited Kenya, bringing them along for a swim in the lake, a tour of the Turkana fossil beds, and a dinner he often cooked himself. And he flew to camp whenever Kamoya radiotelephoned with news of the discovery of a new hominid fossil.

Although Richard had worried that the best hominid specimens had already been collected, it did not take long for the Hominid Gang to make their first find. In early July, Maundu Muluila spotted the thick shaft of a hominid thigh bone. A few weeks later, the team discovered two hominid mandibles, and then late in July, Kamoya reported a skull. It was badly weathered and incomplete, but enough remained of the cranium and one eye orbit for Richard to assert that it was "strikingly similar" to the skull 1470—the fossil that had made him famous.

Although many paleoanthropologists placed 1470 with the small australopithecines, Richard had argued that because of its large cranium, 1470 was actually an ancient type of *Homo*. But until he had further specimens there was no way of resolving this controversy. While Richard knew that the new skull would not convince all of his critics, he nevertheless felt that he now had at least one more fossil to back up his claim. "I think 1470 can now be

---

[7] White had actually been trying to join one of Richard's expeditions for several years. He had first written to Richard in the spring of 1972 when he was a senior at the University of California in Riverside, where he had majored in biology and anthropology. He explained that he had attended one of Richard's lectures at UCR and was impressed that Richard had pursued a career outside of academia. White asked about a possible position on Richard's staff, and suggestions about how best to pursue his interest in early hominids. Richard received many such inquiries from students, and as with this one, asked his secretary to respond. She advised White to continue his studies and suggested that he might try meeting Richard at one of his future lectures.

[8] "Richard ran his camp very much like an [English] boarding school," said the anatomist Bernard Wood. "There was a code of conduct that was very much unwritten, but was never changed, regardless of who was in camp." Once Prince Philip came to Koobi Fora, and at dinner Richard offered him the seat to Richard's right, saying, "I'm sorry, but I always sit at the head of the table." But Prince Philip quickly corrected him: "No, where I sit *is* the head of the table."

said to be not a freak but a real entity, a relatively large-brained form of *Homo* that lived in eastern Africa somewhere between two and three million years ago," Richard later told a press conference.

Richard was in camp when the Hominid Gang made their next discovery on August 1. The hominid hunters were then working at an area not far from Koobi Fora, and they pulled in shortly before evening to announce that Bernard Ngeneo (who had found 1470) had spotted some hominid maxillary (upper jaw) fragments and a brow ridge. "Our curiosity was roused when we asked Kamoya if the brow ridge was attached to a skull and he replied, 'Well we cannot say because we can only see a very small piece, most of it is buried,' " Meave wrote in her journal. "Might it be a complete skull?"

The next morning, Meave and Richard drove to the site with Kamoya and his crew. On the sloping ground of a small gully lay two fragments of the hominid's upper jaw. Nearby, poking just above the ground, were the fragile arches of the bony eye orbits—as if someone had been buried feet first, and was now struggling to push free of the earth. So little of the bone was visible that Meave wrote in her diary, "how he [Ngeneo] spotted it I cannot imagine." Richard immediately set to work excavating the fossil, while Meave collected tiny pieces that had broken off and lay on the surface.

By noon, Richard knew that there was more to the fossil than just the brow ridge—although it was difficult to determine how much of the skull was intact since the fossil was extremely fragile and small plant roots weaved in and out of the hairline cracks. Using fine camel-hair brushes, dental picks, and surgical forceps, he carefully uncovered the top of the skull, pausing now and again to brush the fossil with Bedacryl, a lacquer that filled in the cracks and hardened the bone's surface. Richard finally stopped work at 5:00 P.M., having exposed much of the cranium. "It would be terrible if an oryx [a large antelope] trod on it now!" Meave said, looking at the fragile brown bone that rose from the earth like an exotic mushroom. To give the fossil some protection, she and Richard placed a metal basin and thorn bush over it, then reluctantly returned to camp.

Richard worked at the skull over the next three days, then had to fly back to Nairobi to attend to business he could not put off. He returned on August 9 and after lunch finished the excavation, lifting the skull from the ground. The fragile face bones had broken off, but the cranium itself was intact and Richard was elated. "What a find it was!" he wrote in his autobiography. "There was no doubt that this was not *Australopithecus* nor even *Homo habilis* but rather *H. erectus,* a more immediate ancestor of ourselves. Words are inadequate to describe our feelings because for months we had suspected that *H. erectus* had lived in Africa more than a million years ago and here at my fingertips was proof, a perfectly preserved skull found *in situ* in sediments over 1.5 million years old."

The new fossil—catalogued in the National Museum as KNM-ER 3733—was also the oldest example of *Homo erectus* that had ever been found.

Prior to Ngeneo's discovery, *Homo erectus* was primarily known from fossils uncovered in Java and China, which were then thought to date to approximately 500,000 to 700,000 year ago.[9] Fossils of this ancestor—crania, upper and lower jawbones, a partial pelvis, and an arm bone—had also been discovered at Olduvai. These came from a variety of sites in Beds II, III, and IV, and ranged in age from about one million to 500,000 years ago. Richard's team had also unearthed a *Homo erectus* pelvis at Turkana only seven months before finding the skull—although because it was found in deposits that had been erroneously dated to nearly 3 million years, Richard initially claimed the pelvis as an example of a more ancient type of *Homo*. It was subsequently redated to 1.9 million years old.

Yet skulls remain the most compelling of all the hominid fossils—perhaps because in them we recognize something of ourselves—and 3733 included far more than just the cranium. After thoroughly sieving the site, Richard's team recovered nearly all the fragile facial bones and upper teeth. These could not be readily reattached to the skull because the inside of the brain case was completely filled with hard calcified rock. The rock made the skull very heavy, and Richard feared that if it was mishandled when the facial bones were glued on, the skull's weight would crush them. Eventually, in January 1976, Alan Walker solved this problem by boldly breaking the skull into three large fragments with a chisel and hammer. (When Richard learned of Walker's plan, he excused himself from the room—in fact, unable to bear the thought that the skull might be shattered, he left the museum for the entire day.) Once he had split open the skull, Walker was able to remove the rock from the skull's inner surface. He then glued the skull back together, and fitted the delicate facial bones in place, a painstaking task he likened to "doing a jigsaw puzzle in three dimensions, with half the pieces missing and no picture of the whole to use as a guide."

The restored skull revealed all the classic *Homo erectus* features: fat, beetle brows; a flat, low-slung cranium that swept back to a curious bulge; and a face that was flat and tucked under the brain, much like our own, rather than pushed forward like that of a chimpanzee. To Richard, it was "a glorious profile . . . I think it is one of the prettiest skulls I have had the pleasure and privilege of seeing," he told a press conference in the spring of 1976.

For Richard, 3733 was particularly important because it finally laid to rest the single-species theory of human evolution—the idea that only one hominid species had existed at any one time. Richard—and Louis before him—had argued that two types of hominids, *Homo* and *Australopithecus,* had lived side by side at both Olduvai and Turkana. But their critics had always argued that what Richard and Louis interpreted as early types of *Homo* were, in fact, australopithecines. The new *Homo erectus* skull would

[9] Indonesian (Javan) *Homo erectus* fossils were redated in 1994 to 1.6 to 1.8 million years, making them almost as old as *H. erectus* in East Africa.

silence these doubters, Richard knew, because it had been discovered *in situ,* and in deposits that had previously yielded specimens of *Australopithecus boisei.* There was also no question that 3733 was a skull of *Homo erectus*—everyone remarked on how much it resembled its Peking counterparts, although it predated them by more than one million years.[10] With one fossil discovery, Richard had ended one of paleoanthropology's longest debates.

"Richard and Alan were kind to me about that," said Milford Wolpoff, who had been one of the leading proponents of the single-species hypothesis.

> They didn't pull it off as a big surprise in an [academic] meeting, saying to all of our colleagues, "Ha, ha, ha. Milford's got it wrong." Instead, Richard let Alan bring a cast of the skull to me [at the University of Michigan]. And Alan was so excited, he couldn't even wait—he opened up the package in the airport bar—and I understood the implications immediately. So there were never any bad feelings about that. They let me see the skull early in the game, and gave me a gentlemanly way of backing out of my position without making me look like a fool.

The skull 3733 was not the last hominid fossil Kamoya's team found during the 1975 season. Before Richard closed the camp in September, the Hominid Gang added another three hominid specimens to the Turkana fossil trove.

Richard announced his new hominid treasures at a press conference in Washington, D.C., on March 9, 1976. Uncharacteristically, he shared the podium with another paleoanthropologist, Donald Johanson, who had recently returned from Ethiopia with a spectacular hominid haul of his own. Although Johanson had presented his discovery (an astonishing collection of 150 fossilized bones) to the press in January, Richard had invited him to this conference to add further details about his fossils. Richard's and Johanson's joint appearance would also, they hoped, silence rumors that they were "competitors in the search for humanity's history," as a reporter for *Science News* phrased it. "Their appearance together," the reporter continued, "was aimed at dispelling that impression and emphasizing the degree of cooperation that is now a function of research into human evolution."

Richard then unveiled casts of the *Homo erectus* skull, as well as casts of a partial skull the Hominid Gang had found that was, as Richard put it, a "dead-ringer" for 1470, and the *H. erectus* pelvis, explaining the implications of each find. Johanson next displayed casts of his fossils, which included bones of five adults and two children. Most of these were postcranial bones (bones other than the skull), and Johanson speculated that the group of hominids (which he would later call "the first family") had been killed together in a flash flood or similar catastrophe 3.5 million years ago. With

---

[10] The 3733 skull also looks remarkably similar to a new skull from Java, which was found in 1993.

Richard exclaiming, "He [Johanson] has hands, he has feet, he has all sorts of wonderful things," Johanson then exhibited a composite hominid hand he had fashioned from thirty separate hand bones. "Our preliminary observations," said Johanson, "suggest to us that there is nothing in the anatomy or morphology of the bones which would preclude the kind of movements that we are capable of with our own hands today." The hands were designed for tool use, he added, linking them to a later descendant—*Homo erectus.*

Similarly, a broken femur in Johanson's collection appeared to tie Johanson's and Richard's discoveries together—and supported Richard's ideas of an ancient form of *Homo.* Although only the head of the femur remained, it was that of a bipedal walker, evidence that Johanson and Richard claimed as pointing to *Homo.*[11] "No *Australopithecus* had a femur like that," said Leakey, who was "obviously pleased with confirmation of his ideas from another site," wrote the *Science News* reporter.

Together, it seemed, Richard and Johanson were building a case for an ancient form of *Homo,* one that extended back more than three million years. For a year or two, the two men publicly agreed on this interpretation. But there were other forces at play that Richard was unaware of, and it was not long before their camaraderie dissolved into perhaps the most bitter enmity that paleoanthropology has ever known.

[11] Johanson would later reinterpret these fossils as not belonging to *Homo* at all, but representing a new taxon he and Tim White would create, *Australopithecus afarensis,* making this the fourth species of australopithecine in addition to *A. boisei, A. robustus,* and *A. africanus.*

Chapter 35

A NEW CONTENDER

At the time of their 1976 joint press conference about their fossil hominid discoveries, Richard and Donald Johanson had been friends for three years. "I liked Don very much then," Richard said. "He was a good colleague, and part of my age group. We were all young and enthusiastic, and excited about the science—and everybody seemed to get on so well. I remember kidding the older scientists about this, saying, 'We young guys don't have all these squabbles and petty jealousies. We're all buddies.' "

Only a year apart in age—in 1976, Richard was thirty-two, Johanson thirty-three—Richard and Johanson seemed to revel in their public displays of good-humored camaraderie. One reporter, engaged in writing a profile of Richard, once found the two together "sipping 'Tusker' beers on the veranda of Leakey's wooded suburban home. Though they represent expeditions that may be competing for world funds, they were uncommonly brotherly in their concern for the science they share," the writer noted with some surprise.

In the same spirit, Richard and Johanson shared ideas and fossils, exchanged casts of their hominid finds, and joined each other on the FROM (Richard's new foundation) fund-raising circuit. Whenever Johanson passed through Nairobi, Richard always insisted that he be his house guest. And once, at the end of Johanson's 1976 field season, they spent several days at Richard's home on Lamu Island just off the north Kenyan coast, sailing together on the balmy East African seas. Richard had built the home—actually, a rustic series of huts made of palm thatching—in 1977 as a retreat.[1]

---

[1] In 1982 Richard lost this thatched home to a fire; he has since rebuilt with stone.

"Richard nurtured Don," said Kay Behrensmeyer. "He gave Don a lot of opportunities—Richard introduced him to people, took him into the field, spent a lot of time with him, and had him in attendance, like a lieutenant, when Richard came to the States. He told us he wanted to set Don up as the 'American Richard Leakey.' "

But Johanson had other, secret aspirations, and his greatest ambition was to surpass "paleoanthropology's certified supernova," as Johanson described Richard in *Lucy.* The surest way, Johanson knew, was to find better and older hominid fossils than his rival's.

The son of Swedish immigrants, Johanson first heard about the Leakeys when he was a high school student in Chicago. His "imagination [had been] inflamed," he would later write in *Lucy,* by Louis and Mary's account of their discovery of *Zinj* in the *National Geographic.* Johanson had initially intended to study chemistry, but Louis "was proof that a man could make a career out of digging up fossils," Johanson wrote. He decided to follow the same path—a decision that was reinforced when as an undergraduate he attended a seminar Louis gave at the University of Illinois and fell, as so many had before him, under the great man's spell. "Louis not only made the subject of paleoanthropology interesting," Johanson told a San Francisco audience in 1987, "he made it come alive through his infectious imagination." After his encounter with Louis, Johanson wanted only one thing: to join the hunt for hominid fossils in East Africa.

In 1970 Johanson got his wish. He was by then a graduate student at the University of Chicago studying physical anthropology under Clark Howell, and Howell invited him to Ethiopia as a member of the Omo Research Expedition. En route to the field, they stopped in Nairobi, where Howell introduced Johanson to Richard and Mary. Richard has only vague memories of this meeting; to him, Johanson was just another one of Howell's students. But for Johanson, Richard was everything he longed to be: "a glamorous figure, well known to the world's press," as Johanson wrote. Richard had just flown in from Koobi Fora, and set out for Howell's and Johanson's inspection several new australopithecine fossils—"another thrill," wrote Johanson, "for papers on them had not yet been published. I was filled with a sense of being on the cutting edge of a science," he continued, "of talking to people who were shaping it, who had found famous skulls and had themselves become famous." He hungered to join their ranks.

In a field of science where luck counts as much as (if not more than) scientific reasoning, Johanson needed a break—a big break—if he was to find important hominid fossils of his own. He was, after all, still working on his doctorate when he joined Howell on the Omo; the possibility of leading an expedition of his own seemed years away. But then, after spending one more summer in the field with Howell, Johanson met a young, ebullient French geologist, Maurice Taieb, who had a tantalizing tale to tell.

Since 1965, Taieb had been piecing together the geological history of the remote eastern deserts of Ethiopia, a region known as the Afar Triangle. The

Afar appealed to him, he told Johanson, because it marked the northern extension of Africa's Great Rift Valley—one of the world's great continental cracks. "People were just beginning to understand the plate tectonics theory, and so I thought I would study this in the Afar for my dissertation," recalled Taieb, who is now the director of geological research for France's Centre National de la Recherche Scientifique (the French equivalent of the National Science Foundation). He had only limited funds, and so he sometimes explored on foot, packing his supplies on donkeys and staying with villagers along the way; other times, he borrowed a Land Rover and ventured farther into the badlands. On one of these latter expeditions in 1968, Taieb discovered a site both "exceptional and fantastic," as he wrote in his book, *Sur la terre des premiers hommes (In the Land of the First People).* It was named Hadar and its sediments, which stretched for thousands of square miles, were laden with fossils, many of them extraordinarily well-preserved. Over the next few years Taieb and an American geologist, Jon Kalb, explored the area whenever they could, collecting samples of the fossils and mapping the terrain.

But now, Taieb told Johanson, their plans had taken a dramatic turn. Only a few weeks before, in December 1971, Taieb and Kalb had had the good fortune to meet Louis and Mary Leakey at the Pan-African Congress on Prehistory in Addis Ababa, Ethiopia. The two geologists had shown the Leakeys their fossils, and Louis, after remarking that the specimens appeared to be older than those from Olduvai, had immediately agreed to help them launch an exploratory expedition; he and Mary would write letters of recommendation and serve as technical advisers. Louis also urged Taieb to find a paleoanthropologist—someone who could recognize hominid fossils—to join their expedition.[2] Several of the scientists on the French team of the Omo Expedition suggested that Taieb consider Johanson for this role. Would Johanson be interested in exploring the Afar for two months, beginning in April 1972? Taieb asked.

"The allure of that invitation was tremendous," Johanson wrote in *Lucy*. He knew that other regions in the Rift Valley—including the Leakeys' prized sites at Olduvai Gorge and Lake Turkana—had yielded spectacular hominid treasures. What if the Afar turned out similarly? "Richard had taken a chance and struck gold. Shouldn't I?" he asked himself.

Although Johanson had not yet completed his doctoral dissertation, and was advised against going to the Afar by Clark Howell, he nevertheless took the chance. Taieb had also invited Yves Coppens, a noted paleontologist and leader of the French Omo team, and with the money raised with Louis's help the four men set out for the Afar badlands in early April. Taieb and Kalb guided their party to a variety of sites—each one, Johanson later noted,

---

[2] Taieb asked Louis if he thought they would find hominid fossils. "And the first words Louis said were, 'You have to find tools before you find hominids,'" recalled Taieb. "Because that is what he did at Olduvai. But at Hadar [at the site where the skeleton "Lucy" was found], we have not yet found tools, although we have found hominids."

"much better" than the fossil deposits in the Omo Valley—but they saved Taieb's best site for last. When they finally stopped at a cliff overlooking Hadar, they were all momentarily speechless. Hadar appeared to stretch to the far horizon, a barren, rumpled landscape of apparently endless dry gorges, "all of them well stratified, all of them seeming to ooze fossils," Johanson wrote. "It was a place paleoanthropologists see only in their dreams."

"We knew that this was a place we could really sink our teeth into," Kalb said. "So we formalized our partnership. We signed an agreement and called ourselves the International Afar Research Expedition. Taieb was the overall leader." In that agreement, Taieb and Kalb initially included Louis and Mary Leakey as advisers. Johanson, however, balked at mentioning the Leakeys. "He wanted nothing to do with them," said Kalb. "Don said over and over again that he was the anthropologist on the expedition, and he was going to be the only one. He was afraid that if the Leakeys got involved they would control things. But I thought that we'd made a commitment to Louis and Mary, and shouldn't just drop them; it was embarrassing to me. But Don pushed them out of the picture. He was always jockeying to enhance his position."[3]

Despite his desire to limit the Leakeys' involvement, at the end of this first exploratory expedition Johanson and Taieb traveled to Nairobi to report their discoveries to Louis. "Louis was very happy with the fossils, and he took us to lunch at his home, and then to Nairobi National Park to see the lions," said Taieb. "He was very helpful to us as he was with many young people, you know, talking about the specimens, developing ideas, and offering to help us raise funds."

Four months later, Louis died and his involvement in the Afar Expedition was no longer an issue. But Johanson and Taieb continued to use his letters of support in their grant applications—and they each received funding (Johanson from the NSF, Taieb and Coppens from the CNRS, Kalb from the Fund for Overseas Research Grants and Education in Connecticut) for a full-fledged undertaking. They would depart in the autumn of 1973.

It was then that Richard's and Johanson's paths crossed again. Just before heading into the field, in September 1973, Johanson traveled to Nairobi to attend the Lake Rudolf [Turkana] Basin Conference—a gathering of all the scientists elucidating the prehistory of that region. There Johanson bet Richard a bottle of wine that he would find hominid fossils in the Afar that were older than Richard's. It was a friendly wager, and the two men promised to keep in touch.

Johanson had far more than a bottle of wine riding on this expedition. His two-year grant was based largely on the assumption that he would

---

[3] On the expedition agreement, Don Johanson and Yves Coppens were both listed as paleoanthropologists and were to share equally any hominid discoveries. Johanson apparently did not perceive Coppens as a threat.

discover hominid fossils. The team had turned up many beautifully preserved fossils of elephants, hippos, crocodiles, and pigs, but ancestral bones remained elusive, and Johanson began to bully his colleagues. "Johanson was obsessed with finding hominids," recalled Kalb. "He wanted to monopolize the expedition, to make the search for hominids its only purpose. He'd get upset if other work took precedence over his."

By the end of November, two months into the expedition, Johanson had yet to find any evidence of hominids and he was close to despair. Fretting about his future, he one day idly kicked at what he thought was a hippo rib. The bone fell over, onto the ground, and with a start Johanson realized that it was actually the upper part of a primate's shinbone, the proximal tibia. A few yards away, he spotted a primate's lower thighbone (the distal femur) and was astonished to find that the two bones fit together, forming the knee joint. Johanson studied the angle of that joint and suddenly, as he wrote in *Lucy,* "it dawned on me that this was a hominid fossil." If so, then it was the first three-million-year-old knee joint ever discovered—and an indication that even then hominids were bipedal, capable of walking upright.

Johanson could scarcely believe that he had found something so important and so completely new to science. Desperate to verify his find, he and his assistant, Tom Gray, removed a femur from a nearby grave of an Afar tribesman—an act that incensed Taieb and Kalb, since the fierce, nomadic Afar had only reluctantly given the expedition permission to explore their lands. ("We could have been killed for that!" Kalb later exclaimed.) That night, Johanson compared the two bones: except for a major difference in size (the fossil was from a small, three-foot-tall creature), they were identical. "I now had a unique hominid fossil of my own," Johanson wrote in *Lucy,* although without a skull, it was impossible to say what genus or species it represented.

A few days after his discovery, Johanson wrote to Richard, dropping a hint about the fossil and suggesting that the two meet in Nairobi in late December. Richard quickly agreed. "I was there when Johanson came to Nairobi with his knee joint," recalled M. E. Morbeck, "and we took out all the tibia and femur [hominid fossils] from Lake Turkana to compare with Johanson's specimen. We laid everything out on the table, and then Richard came by and said, 'You're not allowed to have all the things out at once. You can only have one fossil out at a time.' And I was surprised because I'd laid out whole skeletons before." Richard, it seemed to Morbeck, was flexing his administrative muscle, showing Johanson just who controlled the fossils in East Africa. The atmosphere was "definitely competitive," she said.

Johanson, Coppens, and Taieb (in a separate power struggle, the three had ousted Kalb) had planned their second expedition for the autumn of 1974. But Johanson was low on funds; he had spent nearly all of his two-year NSF grant on their first foray. By now, Johanson had completed his doctorate and received an assistant professorship at Case Western Reserve University in Cleveland, Ohio. Several friends there offered to assist him with fund-raising, and Johanson turned to Richard and Mary for letters of recommendation—which they quickly supplied. "I see no objections at all to your using my name," Richard wrote Johanson in an accompanying cover letter, "and as you know I have the strongest feelings about the continued success of your project. As I said before if I can be of any assistance I would be only too glad. . . ."

Mary, too, readily dispatched the letter Johanson had requested, noting in it that "Dr. Johanson obtained remarkable results during the short season he was in the field last year, with promise of much more in the future. I trust, most sincerely, that he will obtain sufficient financial support to enable him to conduct a bigger expedition next season and to remain in the field for a longer period."

With the help of Richard's and Mary's letters, Johanson raised the additional funds he needed, and in October 1974 once more returned to the Afar. This time, he was even luckier than before. Only two weeks after pitching their camp—and in the space of three days—the team found four partial hominid jawbones, one a palate with all sixteen teeth still in place. Johanson puzzled over this latter fossil, which was not only remarkably well-preserved but also presented an odd mixture of primitive and modern features. For example, as in modern humans, the front teeth were large in comparison to the back teeth; but, as in apes, the palate itself was shallow, an apelike gap separated the canines, and the teeth rows were parallel rather than curved. "Strange, strange jaws," Johanson wrote in *Lucy.* "Their peculiar blend of *Homo* and australopithecine traits, with a whiff of something more primitive, was utterly baffling." Johanson decided that the jaws represented a new type of extinct hominid, most likely a new species of *Homo,* and he and his colleagues called a press conference in Addis Ababa to announce what they had found.

By now, fifteen years after Louis and Mary had been catapulted to fame by their discovery of *Zinj,* paleoanthropologists were well aware of the power the media wielded in their field. They knew that discoveries of early human remains fascinated the press and general public, and an announcement of a new find, if done dramatically enough, could lead not only to headlines around the world, but to funds for future expeditions. Johanson, Taieb, and Coppens were thus not shy about proclaiming the significance of their fossils. The jawbones, they declared in a statement prepared by Johanson (and in rhetoric worthy of Louis), were "an unparalleled breakthrough in the search for the origins of man's evolution. . . . We have in a matter of merely two days extended our knowledge of the genus *Homo* by nearly 1.5 million years." Furthermore (and again with echoes of Louis), Johanson added, "All previous theories of the origins of the lineage which leads to modern man must now be totally revised. We must throw out many existing theories and consider the possibility that man's origins go back to well over four million years." The jaws were "perhaps the most provocative human fossils ever discovered on the African Continent" and their location suggested the "revolutionary postulate" that the cradle of humankind actually lay outside of Africa.

A week later in a letter to Johanson replying to an invitation from him to visit Hadar, Richard said he was pleased by the invitation, but thought Johanson's press conference claims "a trifle strong." In particular, Richard was annoyed by the suggestion that humans might have originated somewhere other than Africa, since it had long been a fundamental Leakey tenet (indeed, one that Louis spent much of his life proving) that humans evolved there. Johanson's claim that the Afar team had pushed back the time of human evolution by 1.5 million years was also a strike at Richard and his most famous discovery, the skull 1470. "[T]here has been no evidence yet to alter our belief that a date of 3.1 million years [for 1470] is reasonable," Richard noted in his reply (although 1470 would later be redated to a little less than two million years). If Johanson thought that he had won their bet with his three-million-year-old jaws, Richard disagreed. Besides, as Richard pointed out to one reporter, without evidence of a relatively large brain the jawbones could not be assigned to the genus *Homo*.

Richard briefly raised these objections in his letter to Johanson, but then added, "These are minor points! . . . For your records, I am delighted and fully endorse your . . . assertion that the new discoveries are of the greatest importance." He also accepted Johanson's invitation to visit the Afar Expedition camp. He planned to fly Meave, Mary, and the paleontologist John Harris to Ethiopia in late November.

Three weeks later, on November 21, Richard touched down at the Afar, and Johanson and Taieb ushered their guests into camp. Mary had brought along casts of some of the Olduvai hominids, and the group compared these with the Afar jawbones. All agreed that there were distinct similarities between the specimens; the teeth and jaws did seem to fall into the *Homo* line even though they had apelike features as well. Then the conversation turned to the dating problems that Richard faced at Lake Turkana, and almost at once Richard and Johanson were at odds. "It was very difficult for me to follow because they were speaking very fast," said Taieb. "But you could see that they were fighting; it was friendly, but the fight was starting." Added Harris, "Don seemed to think we'd made a mistake about the dating and were doing anything we could to avoid facing it. But it wasn't a desperate cover-up as he implied [about this conversation] in *Lucy*. We were just trying to figure out what the hell was going on."

Richard and Harris had hoped to find some indication that the extinct fauna in Ethiopia correlated in time with the specimens at Turkana as support for their three-million-year date for 1470. Instead, the animals that paleontologists rely on as marker species (for example, they can tell that East African deposits are about two million years old if a certain species of *Equus*, the horse, is present) supported the contention that something was wrong with the dating at Richard's site as Basil Cooke had argued. (At Turkana, *Equus* appeared in deposits that had been dated at three million years

—but at all other East African sites, including Hadar, *Equus* is unknown until two million years ago, when it first arrived from Asia.) Johanson could see the uneasiness on Richard's and Harris's faces as they studied the fossils, and as soon as their party left Johanson turned to Taieb and said, "I'm going to beat the Leakeys." He was confident that Hadar was securely dated at three million years, and equally certain that Turkana was not. And if he was right and Richard wrong, then Johanson could claim title to the oldest human ancestor.

The very next morning after Richard's departure, Johanson made another stunning discovery. Only four miles from camp, he spotted a fragment of an armbone poking from a slope. At first glance, because of its small size, he thought the bone was that of a monkey, but it lacked the monkey's distinguishing bony flange. "My pulse was quickening," Johanson wrote in a subsequent article for the *National Geographic.* "Suddenly I found myself saying, 'It's hominid!' " His eye caught other bone fragments higher up the slope and then "the realization struck us both [Johanson was exploring with his assistant Tom Gray] that we might have found a skeleton. An extraordinary skeleton. . . . There, incredibly, lay a multitude of bone fragments—a nearly complete lower jaw, a thigh bone, arm bones, ribs, vertebrae, and more! The searing heat was forgotten. Tom and I yelled, hugged each other, and danced, mad as any Englishmen in the midday sun."

It was a spectacular find: by far the most complete extinct hominid specimen yet discovered (at the time, the next best example came from excavations at Neanderthal sites, dating to 100,000 years ago). From the shape of the pelvic bone, Johanson determined that the skeleton was probably that of a female. That night as the team celebrated with beer and roasted goat meat, Johanson played the Beatles' song "Lucy in the Sky with Diamonds" over and over again—the musical high (the title is said to allude to LSD) echoing his own. By morning, the skeleton itself was known as "Lucy."

With Lucy's skeleton arrayed before him on a camp table, Johanson knew that he had at last succeeded in transforming himself from "an unknown anthropology graduate [into] a promising field worker," as he later wrote. Even more importantly, he considered the fossils "dazzling enough to match those of . . . Richard Leakey." The two men now stood, at least in Johanson's view, side by side on the same mountaintop.

In Nairobi, Richard learned about Johanson's new discovery from the newspapers and according to Meave was happy for his friend's success. In some ways, Richard was even relieved. He continued to believe that part of his troubles at Turkana stemmed from the jealousy of fellow scientists envious of his many discoveries and the publicity he had received. Now the spotlight would turn to Johanson, and if "Lucy" was splashed across the front pages of the world press, then so much the better, Richard told Glynn Isaac. It would not hurt to have "public attention refocussed for awhile." From

Johanson's early reports, too, it seemed that the new hominid might fall into the genus *Homo.* If so, it would lend support to Richard's contention that the human line extended back more than three million years.

But not all of Richard's friends believed he was truly enthusiastic about Johanson's discovery. "Window dressing," scoffed Paul Abell. "Richard likes to keep his political fences mended, but in private he certainly was not happy about Lucy." Other colleagues also sensed that Johanson's stunning discoveries might change the nature of the two men's friendship. "Some people have been talking about Don 'out-Leakeying Leakey,' " Frank Fitch, the coleader of Richard's dating team, wrote him in December. "I told John [Heminway, a writer] that my money would be on you if it ever came to playing a game of that sort."

Richard, of course, was highly experienced in taking on and beating competitors in the early-human game. But he had seriously underestimated Johanson's ambition. Nor was Richard aware that while Johanson continued to act as if he were Richard's friend, he also secretly continued to demean Richard—and the Leakeys in general—to others. So often did Johanson discuss the Leakeys that several of his colleagues thought he had become obsessed.

"Donald, I think, wanted to be a Leakey," said Taieb, who remembers often listening to Johanson outline his ambitions. "One day, he asked me, 'Maurice, is it possible for me to be as well-known as Richard?' He was worried that he could not, because he does not have a famous mother and father, or the family name." Taieb brushed aside Johanson's fears, dismissing them philosophically. He himself would never be famous, he told Johanson, "because I have an Arab name. Besides, we are just doing some research, which maybe in the next hundred years will amount to nothing. But Donald would not listen; he wants to see his name in lights. . . . I think he decided that since he could not be a Leakey, he had to beat them; he had to be against them," said Taieb, who found Johanson's attitude distasteful. After all, Taieb pointed out to him, "We are all here in East Africa because of the Leakeys' work. They opened the country to scholars, and you must respect them for this, just as you must respect your teachers."

But Taieb had little time for defending the Leakeys against Johanson. His own expedition was at stake, since, after discovering Lucy, Johanson began "to act as if he is the leader," said Taieb. "He wants everything for himself, and it was all because he wanted to pass Richard." For a while, the bickering between the French and American parties threatened the future of the expedition, but before the season ended, Johanson and Taieb patched up their differences. They would return again as a team in the autumn of 1975.

With Lucy in hand, Johanson had little trouble securing funds. He landed a *National Geographic* grant, and the promise of an article in the Society's magazine. And once again, only a few weeks after setting up camp, the team made a spectacular discovery—this time, a virtual hillside of hominid fossils. ". . . I had the unnerving experience of picking up, almost side by side, two fibulas—the smaller of the two shin bones in the leg," Johanson later wrote. "Another Lucy? No, these were both right legs, indicating the presence of more than one individual. Meanwhile, others were shouting over finds of their own, all of them hominid. Fossils seemed to be cascading, almost as from a fountain, down the hillside. A near-frenzy seized us as we scrambled madly to pick them up."

Johanson was ecstatic. If Lucy had placed him on the mountaintop with Richard, this find, Johanson was certain, would push Richard off. Johanson could barely contain his glee, and as he held aloft leg, arm, and hand bones for the camera, he called out, "Hey, Richard, look at this one! This one's a good one! I've got you now, Richard! I've got you now."

Altogether, Johanson and the Afar team retrieved from this single site 197 hominid fossils—a trove that included remains from at least thirteen individuals and comprised jaws, teeth, leg, hand and foot bones, vertebrae, ribs, and skull fragments from an adult and an infant. Johanson speculated that the hominids had been buried together at the site, perhaps when a flash flood swept over them while they were sleeping.[4] If they were a single group of hominids, then they may have had family ties, Johanson suggested. He called this new collection "The First Family."

While reporters were enthusiastic about Johanson's discoveries, they nevertheless seemed always to turn to Richard for an expert's opinion. Johanson's fossils, which he had suggested represented an early type of *Homo,* something slightly more primitive than the 1470 skull, were also viewed as supporting Richard's "revolutionary theory of human evolution," as a 1976 *Newsweek* article stated about Richard's contention that the earliest forms of *Homo* had shared the African plains with other types of hominids. Further, the *Newsweek* reporter considered Richard's 1.5-million-year-old *Homo erectus* skull, 3733, "the most impressive" of all the two men's finds. Nor was Johanson able to claim title to the oldest human ancestor: Mary had beaten him to that with her discovery in 1974 of the 3.75-million-year-old hominid jawbones at Laetoli.

Perhaps none of this would have mattered if Johanson had been content to be one of a triumvirate. But he apparently could not rid himself of his desire to be the world's leading expert in human origins. There was another route to the top, one used regularly in science by younger men intent on making a name for themselves: one disagreed with the expert. Thus, if Johanson's fossils were found to challenge rather than support Richard's theories, then Johanson might at last be able to free himself—and his fossils—from the Leakeys. A different interpretation might even put Johanson at odds with Richard, thereby creating a controversy and bringing Johanson more attention. "Johanson talked about this sometimes with Tom Gray," said Taieb, "and Tom told him he should fight with the Leakeys."

[4] Johanson has since said that the flood scenario is unlikely, noting that the accumulation of hominid bones cannot be explained until further excavations are carried out.

Shortly after the close of his 1975 field season, Johanson traveled to Nairobi to show his new fossils to Richard and Mary, and to spend Christmas at Richard's home. Johanson unveiled his bounty of bones at the museum. The Leakeys were particularly impressed by the mixture of human and australopithecine features. Tim White, a young American paleontologist who had worked with Richard for three seasons at Koobi Fora, also joined in the discussions. Impressed by White's energy and intellect, Richard had recommended him to Mary when she was seeking someone to describe the Laetoli hominids. He was now engaged in that research, and as he handled Johanson's fossils, White remarked, "I think your fossils from Hadar and Mary's fossils from Laetoli may be the same." The suggestion surprised everyone, although there was a slight murmur of agreement as the group took a closer look. Johanson was particularly intrigued. The two sites were separated by one thousand miles and nearly a million years in time. Could the bones possibly have come from the same type of creature? Johanson asked White to stay in touch. Johanson may not have realized it at the time, but White had suggested a way to push past the Leakeys and snare the limelight on his own.

Chapter 36

THE NAME GAME

Donald Johanson's fossil discoveries appeared to support several of the Leakeys' fundamental ideas about human origins. In particular, Johanson's Afar hominids seemed to demonstrate the same general pattern seen at Olduvai and Lake Turkana: an early form of *Homo* living alongside an australopithecine. The pattern differed only in the type of australopithecine—at Olduvai and Turkana, these creatures were robust, with the huge, nutcracking jaws of *Zinj (A. boisei),* while at the Afar, they were represented by the tiny, gracile Lucy. Perhaps Lucy, with her upright posture and softball-sized brain (too small for the *Homo* line), was a northern version of the small australopithecine, *A. africanus,* from South Africa. If so, then three hominids—two species of australopithecines and a species of *Homo*—apparently shared the East African savannas three million years ago.

In 1975, Richard demonstrated that early *Homo* and *A. boisei* had lived side by side.[1] The evidence that another hominid inhabited the plains alongside these two only increased Richard's suspicions about the evolution of the human family tree: that the model was less like a ladder with a step-by-step progression of hominid species and more like a bush. Such a shape would make hominid (including human) origins similar to those of the majority of other animals, and further erode the idea of a special creation. Indeed, Richard suspected that he had already unearthed three contempora-

[1] As long ago as 1960, Louis and Mary had actually disproved the single-species hypothesis with their discovery of *Homo habilis* and *Australopithecus boisei* at Olduvai. But *habilis* as a species was controversial well into the 1970s, and not until Richard's team found a skull of *H. erectus* in deposits of the same age as *A. boisei* did the single-species theory finally fall.

neous hominids (one *Homo* and two australopithecines) at Turkana, and even suggested this in a 1976 article in *Nature*. He did not name this third creature because he felt he had not yet amassed enough evidence.[2]

Mary also thought that two types of hominids (one *Homo,* one australopithecine) were represented in Johanson's collection. But it was the great age of his fossils and those she had discovered at Laetoli that struck her most forcibly. Laetoli had been dated at 3.75 million years and Johanson's site at between 3.5 and 3 million—and both had yielded remains of early humans, proof for Mary that Louis had been correct in his belief about the antiquity of the *Homo* line. "For myself," she told Charles Oxnard, a physical anthropologist at the University of Chicago, "I believe that there can no longer be any reasonable doubt that *Homo* goes back very much further than anyone believed a few years ago." Furthermore, the contemporaneity of early *Homo* with a species of australopithecine supported Louis's (and Richard's) contention that the australopithecines were not ancestors of humankind, but another type of hominid that had diverged from *Homo* at some distant point in the past. The new discoveries from Koobi Fora, the Afar, and Laetoli confirmed "that the australopithecines are really out on a limb," Mary wrote to Garniss Curtis, the UC Berkeley geochronologist who had dated Olduvai and Laetoli.

In 1976 Mary was sixty-three years old and the discoverer of the world's oldest *Homo* fossils. Mary and others (including Phillip Tobias) thought the Laetoli jawbones and teeth she had discovered in 1974 probably represented a new species, in part because her specimens were nearly one million years older than the oldest *Homo (H. habilis)* fossils from Olduvai and differed from them in some key features. But she was reluctant to name this new creature until she had more and better specimens—a decision she would come to regret.

It may have been the experience of watching Louis battle over ancestral names that made Richard and Mary so cautious about naming new species; or perhaps they were simply more careful and precise in their science than he. (Certainly, in Louis's hands, both Mary's Laetoli fossils and Richard's small hominids would already have been named.) Whatever the cause, their caution left an opportunity for others—for once the fossils were described, any scientist could name them, in essence stealing the glory of the fossils for themselves. This was a greater risk for Mary than for Richard, since her Laetoli fossils unquestionably represented something new. Nor did it take long for others to see this, or to take advantage of her caution. Ironically, Mary's very reluctance would in time plunge her and Richard into an even bigger and longer (and still ongoing) naming war than they had ever imagined—or than Louis had ever fought.

---

[2] Richard based the third species on the skull KNM-ER 1813 and fragments of other specimens. These all "have the small endocranial volume (less than 550 cm) and generally gracile characteristics that seem to distinguish them from the robust species," he wrote in *Nature* in 1976.

• • •

Although Mary's 1975 Laetoli expedition had not turned up any additional hominid fossils, her team had recovered the remains of so many new animal species that she felt another field season was in order. The National Geographic Society agreed and awarded her a $70,600 grant for further exploration. So it was that at the end of June 1976, Mary once again put aside her monograph on the artifacts from Beds III and IV in Olduvai, and set up camp with her dogs and crew at Laetoli.

Laetoli lay only thirty miles from Olduvai, yet it differed vastly from the gorge in climate and landscape. Olduvai, with its layers of sediments rising skyward, was like a well-planned geology lesson, whereas Laetoli was more a meander through the past, a site where wind- and water-eroded channels led one gently into sandstone gullies that suddenly deepened and widened, then narrowed again. Here there was neither the geological color nor drama of the gorge, but rather an ancient plateau, tilted against the flanks of the volcanoes Lemagrut, Sadiman, and Oldeani. Earthquakes and rivers had laid bare sections of the plateau, exposing claystones, sandstones, and thick deposits of volcanic ash. It was in these eroded areas, bordered by stands of acacia trees, tall grasses, and lava boulders, that the fossils were found.

To help with the hominid hunt, Mary had invited a mix of scientists, including such established figures as Garniss Curtis and the geologist Richard Hay, as well as younger people, like Peter Jones from England, who was then only nineteen but already a skilled experimental archeologist, capable of duplicating ancient stone tools, and Mary Jackes, a lively Australian who was pursuing her doctorate in anthropology. Two young geologists, Celia Kamau and Marc Monahan, came as well. To manage the camp, Mary enlisted her son Philip and his wife, Valerie; and Hay, too, brought his wife, Lynn, along. Altogether, with scientists, wives, and African crew, Mary's team totaled nearly thirty people—a number that fell far short of Richard's operation at Turkana but nevertheless marked a great departure from Mary's customary solitude. Yet the previous summer she had directed an equally large and diverse group, and enjoyed the camaraderie. But Mary was not prepared for the tumultuous drama of the next three months.

At the outset, the expedition got off to a shaky start when, only three weeks before her team was due to arrive, Mary discovered that a band of Maasai *morani* (young warriors) had burned down her camp huts from the previous summer. The vandalism (which was done, she told Phillip Tobias, "just for the sake of devilry") dismayed Mary, since she had cultivated a friendship with the Maasai at Laetoli and Olduvai, running medical clinics for them at both sites. A new camp had to be constructed at once, and Philip Leakey and Mary's Kamba staff shouldered this task. Philip sited the huts and tents this time closer to the fossil beds, in a protected hollow shaded by spreading acacia trees. "This proved a good site, though it had its drawbacks," Mary wrote in her autobiography, "one of which was the extraordinary number of puff-adders nearby. Sometimes we killed two in a single day, and still more came." The countryside was also so heavily infested with ticks that the group found it "simpler to remove them from our clothing with a brush several times a day than to pick them off singly." More alarming were the temperamental and highly aggressive African buffalos, and the elephants, which "made a considerable nuisance of themselves [by] chasing people," Mary noted in her autobiography. No one was hurt, but several people fell ill with tick typhus, a debilitating, though curable, disease.

Perhaps it was living at such close quarters with the wildlife that created such a wary edge around the camp—but the scientists themselves only exacerbated the unease with their intrigues, plots, and what Mary termed "emotional sagas of every conceivable kind." "There was a lot of internal conflict at the camp," Peter Jones said simply. "People were not getting on, and some people were out of favor with Mary." In particular, several of the wives and other women had landed on her Stinkers' List.

Yet somehow, despite the elephants, snakes, and melodrama, Mary saw to it that her team stuck to the schedule at hand, rising at dawn, working until five—and discoveries were made.

Only two weeks after beginning their exploration, several people picked up hominid teeth lying on the surface exposures. A few days later Edward Kandini found the nearly complete cranium of an archaic *Homo sapiens* skull, dating to approximately 120,000 years ago. Mary Jackes next discovered the skull fragments and milk teeth of a hominid child; later in the season, during further excavations, parts of the child's skeleton, including ribs and finger and arm bones, were also recovered. Then, on July 24, four other scientists—Kay Behrensmeyer, Dorothy Dechant, Andrew Hill, and David Western—arrived in camp to visit Mary and to see the fossil deposits. They were guests, and little dreamed that within the next twenty-four hours they would stumble onto a site that would eventually lead to one of the most remarkable of all hominid discoveries.

"Philip took us on a little tour the next morning," recalled Behrensmeyer, "and after a while David got bored. And apparently when he's in need of something interesting, he likes to tease people by casting elephant dung at them." As manure missiles shot through the air, the group scattered through the brush, laughing and shouting. "We were having a good time pelting each other with dung," said Behrensmeyer, "and then, Andrew and I jumped down into a flat gully where the soil was washed away, and a hard layer of ash was exposed. We were actually looking for more dung supplies, when he said, 'My goodness. Do these look like elephant tracks to you?' " Behrensmeyer said yes, and the two immediately fell to their knees to examine the tracks more closely, then called for the others to come see what they had found. These were not fresh tracks, but fossilized footprints preserved in some of Laetoli's oldest deposits. Scattered around the elephant's trail were tracks of other animals, including antelopes, buffalos, giraffes and birds, and a sprinkling of the marks of ancient raindrops. "It was very exciting," said

Behrensmeyer, who noted that night in her diary that they had found a "fantastic site for footprints."

Mary and everyone else at camp were equally elated—although Richard Hay was annoyed with himself for not noticing the footprints before. "Jesus, we must have walked over those things a hundred times and yet we never saw them," he said. "Certainly, it was a fortuitous find, but sometimes it just takes a new eye."

Mary quickly redirected the focus of the expedition to the fossilized prints, and the team soon discovered that these were not confined to a single gully, but were found wherever a particular layer of volcanic ash, or tuff, was exposed; eventually they identified eighteen sites with tracks worth recording. The footprint tuff itself was about six inches thick. Grey in color and as fine-grained as sand, it looked somewhat like ancient slabs of cement that had been spread haphazardly in the desert. "These were some of the most curious ash layers I'd ever seen in my life," said Hay, who subsequently spent several summers studying how the footprints had been formed and preserved.

As Hay would reveal, the events—geological and climatic—that produced the prints were as fortuitous and unlikely as their discovery. Some 3.5 million years ago, the volcano Sadiman, one of several volcanoes bordering the southern Serengeti Plain, began a series of light-ash eruptions. The ashfalls occurred for several weeks as the dry season gave way to the wet. "You can see that seasonal transition in the types of animal prints in the different layers," explained Hay. "In the lower layers, which were laid down first, there are giraffe, rabbit and antelope prints, and in the upper layers there are elephants and rhinoceros, a sign that the animal migration of the rainy season had begun. There are even impressions from some of the drops from the first showers." The ash contained an unusual mixture of minerals, which, when mixed with the rain, formed a soft, cementlike layer that retained the animals' tracks as they walked or ran across it. When this soft layer dried, it turned to rock, preserving the prints. Altogether, Sadiman showered the plains with ash eighteen times during those few weeks, and each time a different set of animals left their prints behind. Then Sadiman erupted with a final, substantial puff that sealed the prints beneath a thick ash blanket. Over time, this coarse covering layer broke away in places, exposing the footprints for the first time in more than three million years.

The largest of the eighteen footprint sites that the group discovered measured 2,400 square feet, and Mary and Hay later calculated that approximately 18,400 tracks—including prints left by an insect, an ostrich, hares, rhinoceroses, guinea fowl, pigs, antelopes, and a large (possibly sabertoothed) cat—crisscrossed its surface. In all, the team identified the tracks of more than twenty different animals. Mary's crew had already found the fossilized remains of many of these animals in deposits only slightly older than the tracks. Such a serendipitous combination of tracks and bones was almost unique in the fossil record, and would eventually lead to studies of not only the skeletal structure of many extinct creatures, but to their strides and gaits as well.

Hominids, of course, had been present at Laetoli 3.75 million years ago, too—and it did not take Mary long to begin pressing her group to keep an eye out for their tracks. "Mary had the idea that we should look for hominid prints," said Peter Jones, "and it made sense. We had everything else; why shouldn't they be there, too?"

Perhaps they were. But by the end of August, nothing remotely bipedal had been spotted. Then, one day in September, shortly before the camp was to be closed, Jones and Philip Leakey stumbled onto a promising trail. "I'd been taking photos of the footprints and was following an elephant's track, when I noticed these funny, triangular prints," said Jones. "I thought they looked bipedal—there was a left and a right print, then another left and right. But I didn't want to think that they were actually hominid prints." Philip then appeared and said, "Hey, look at that. Those look bipedal." They called Mary to the site, and together all three carefully studied the impressions. The tracks were short and wide, and each bore what appeared to be the characteristic imprint of a humanlike big toe. But Mary wasn't convinced. "She didn't like them," said Jones. "They were ambiguous and there weren't many of them."

But after studying this trail in further detail during the summer of 1977, Mary gradually became convinced that these tracks were "genuine," as she told a colleague in November 1977. She decided to announce them to the world at a press conference in Washington, D.C., in February 1978, along with some other surprising finds: the first fossilized prints of a chalicothere, a peculiar herbivore that sported claws on its hooves; and the possible track of a knuckle-walking ape. Mary also planned to give a series of lectures around the United States, just as Louis used to do. In the last few years she had "conquered [her] phobia" about speaking in public, as she told a close friend, and had begun to enjoy the attention and admiration that now came her way.

Mary's trip would also give her a chance to visit Donald Johanson and Tim White. White and Johanson had met several times in 1977 to study Mary's Laetoli and Johanson's Hadar fossils, and were now certain that the two sets of fossils represented only one hominid species—instead of the two that Richard, Mary, and even Johanson had first recognized. Indeed, only the year before, in March 1976, Johanson had written in *Nature,* "On the basis of the present hominid collection from Hadar, it is tentatively suggested that some specimens show affinities with *A. robustus,* some with *A. africanus,* and others with fossils previously referred to *Homo.*" But by the fall of 1977, White had managed to persuade Johanson that the fossils were all from one species—and a new one at that. They even had a name for it: *Australopithecus afarensis,* or southern-man-ape-from-the-Afar. Because the species was based in part on Mary's hominids—and because the type specimen (that is, the fossil they chose to represent the new species) was also one of Mary's—they hoped that she would join them in their announcement. But first they had to convince her that they were right. So far, their letters had failed to persuade her, but Mary, White, and Johanson all thought they could reach some agreement when they met in March.

Like Richard, Mary was at that time a colleague and friend of both Johanson and White, and she looked forward to seeing them. She, together with Louis, had been an early supporter of Johanson's research in Ethiopia and had, over the years, written him several letters of support, helping him win grants for his project. She enjoyed his company, too, his wit and charm, and on several occasions had welcomed him as her guest at Olduvai and Langata. Johanson was equally fond of Mary. "I kidded her a lot," he wrote in *Lucy;* "we got along very well." Once at Olduvai, as they sat sipping drinks in the African twilight, Mary had offered Johanson her ultimate accolade: "You know, you're like a Leakey; you can find fossils. You know where to look for them and where to find them. That's a Leakey trait." "You're like us in another way, too," she had added, after he thanked her for this high praise. "You're lucky."

Johanson's Afar Expedition had been lucky again in 1976, finding a hominid's lower jawbone and teeth, and a site with stone tools *in situ.* The site was dated at 2.5 million years old, giving Johanson title to the world's oldest artifacts. But by 1977 world events had overtaken Ethiopia, and it became engulfed in revolutionary turmoil. Field research there was no longer possible. Since he could not return to the Afar that summer, Johanson instead began a careful study of the hominid fossils he had already collected.

Tim White was also free in the summer of 1977, having recently completed his doctoral dissertation at the University of Michigan. Talented and ambitious, White had already made a name for himself in paleoanthropological circles. He had worked on Richard's Lake Turkana expeditions for three summers, and was tapped (along with the paleontologist John Harris) by Richard in 1975 to begin a thorough study of the regions' pig fossils. Richard was so impressed with White's research that he recommended the young man to Mary when she was seeking someone to describe her Laetoli hominids, even though White was then still a graduate student. "Instead of giving them to Tobias, Wood or Walker [the anatomists who had described the bulk of the Leakeys' fossils] I told Mary, 'There's a very bright young man who has done excellent work, and I'm sure he should be given these,' " recalled Richard. "It was like what I had done before when I invited Walker and Jo Mungai to describe the first Turkana fossils, and I thought White had much to contribute."

White also quickly gained Mary's trust and regard. "He is a very able, clear-thinking young man," she told Lita Osmundsen, then director of the Wenner-Gren Foundation, "and has written really nice, factual descriptions [of the hominids] for *Nature* and the *AJPA [American Journal of Physical Anthropology].*" Indeed, so highly did Mary think of White that when her team found a hominid child's partial skeleton at Laetoli in 1976, she suggested that they nickname it "Timothy." And just as she had for Johanson, Mary wrote letters of support and recommendation for White, helping him secure grants and his first academic post at the University of California, Berkeley. She had invited him to join her team in the field at Laetoli for the 1978 field season, too, suggesting that he might be the one to find the best hominid fossil there yet.

But it seemed to many that White had a darker, bitter side to his personality that Mary had apparently not yet seen, a way of turning against his benefactors that left people bewildered. Although White never told Mary or Richard, for three summers he had worked on Louis's Calico excavation in California. Dee Simpson, the director of the dig, had encouraged him in his ambition to be an archeologist, but he grew sullen during his last summer there, and Simpson never heard from him again. White went to UC Berkeley for his undergraduate degree, and then moved on to the University of Michigan, where he became Milford Wolpoff's student. Wolpoff, in turn, introduced White to the Leakeys. Yet not long afterwards, Wolpoff, too, felt the brunt of White's wrath. "I'm not a psychologist, so I'll not pretend to understand," said Wolpoff, "but there's something in Tim that makes him turn on the people who have helped him. He apparently can't speak about me today without shaking."

As a student, White had been bright and aggressive, the kind of person that professors expect to go far. If he had a flaw, it was his obsession with scientific accuracy at all costs—a trait that often made him short-tempered and quick to judge others. "Tim knows the 'Right' way,' " said Wolpoff, "and that's with a capital 'R' and maybe fluorescent letters. I used to think that once he got a job and was treated with professional respect, he'd calm down a bit. But I was wrong."

In the field, on Richard's expeditions, White had also been highly regarded.[3] "He was bright, energetic, and very much a part of the team," Richard recalled. At times, though, White's self-righteous stance surfaced even here, leading him to be "unspeakably rude and arrogant to others." Like Wolpoff, Richard assumed that White would eventually outgrow this behavior. Instead, Richard himself became a target.

When White and Harris finally finished their fossil pig study (which Richard had asked them to undertake and which confirmed Basil Cooke's two-million-year date for the Turkana KBS tuff) in the summer of 1976, the two

---

[3] During his second season at Koobi Fora, White started "grumbling," said Kamoya. "He would not listen to orders." Kamoya—and Richard—was particularly annoyed when one of Kamoya's crew found a possible hominid bone. According to Richard's instructions, Kamoya had marked the bone for Richard to excavate when he next came to Koobi Fora. But White insisted on excavating it himself. In doing so, "he broke a small bit off," said Kamoya. The bone turned out to be a hominid femur, and when Richard saw what had happened, "he was shaking with anger," Kamoya recalled. "I never saw him so angry." Richard lectured White about his actions in front of the Hominid Gang, and White was furious. Later he said to Kamoya, "How could he treat me like this, like a child?" Yet Richard forgave White, and invited him back for a third season.

went to Richard's office to present him with a copy of their final draft, which they planned to submit to *Nature.* "Their paper was stamped sort of 'for your files,' " Richard said. "And that annoyed me, because I had never even seen a draft, and I thought, given the importance of this paper, that as a courtesy to the leader of the expedition, they should have first given me an opportunity to comment." Richard grew even more irritated as he read the paper. It dealt not only with the pigs, but referred to several new hominid specimens that had not yet been published—a violation of one of the expedition's basic tenets. "We had long had a rule that unpublished fossil hominids were not to be incorporated in other people's papers," said Richard. "It was done as a courtesy to the fossil group [Alan Walker, Bernard Wood, and Michael Day], so that their research efforts weren't wasted."

Apparently, Richard had not realized that Harris and White planned to apply the results of their pig study to the hominid fossils; in particular, the paper made it clear that the 1470 skull was no more than two million years old. Richard was not ready to concede defeat on this issue, but nevertheless, he insisted that it was the inclusion of the unpublished hominids—and not the dating issue—that upset him.

When Richard explained his objections to White, "he started shouting at me, called me a dictator, said it was a disgrace that I should be in charge—all this rubbish," Richard said. "He said he wanted to have nothing more to do with me, and finally walked out of my office and slammed the door."

Richard was stunned—but again attributed White's outburst to his youth. A week later, he received an anguished letter from White. Four pages long, handwritten in a dense scrawl, and covering the paper from margin to margin, White's letter went through Richard's objections point by point, concluding that he no longer wanted to be part of Richard's team.

The letter further amazed Richard, who responded by first urging White "to take a grip on yourself." He thought White's letter indicated "mental and physical exhaustion" and therefore did not "consider it a serious document. It is my continued wish that you remain a part of my research group," Richard added, "and I reiterate my willingness to bring you out to Kenya next year for further work." But White's letter was actually a statement of how far apart the two men were—and would remain. Richard, as leader of the Koobi Fora Research Project, believed he had acted in the best interests of his team, ensuring that none of the scientists encroached on someone else's territory. In White's world, however, there was no room for social niceties—there was only the pursuit of science, which, in his eyes, Richard had betrayed.[4]

---

[4] Surprisingly, Richard had never before had a run-in with other members of his Koobi Fora team about seeing their papers prior to publication—perhaps because of his strong personality. As Kay Behrensmeyer noted, "The Leakeys are such powerful people and they have so much vision about what they want to do scientifically that they just have the power to get people to conform to their leadership. And people who otherwise might be very independent and powerful in other contexts will bend—did bend under Richard—and came to feel that being part of the team was much more important than expressing their own individuality."

In time, White and Harris did rewrite their paper according to Richard's wishes, and it was eventually published in *Science* (*Nature* rejected it). To Richard's mind, the matter was then closed. "As far as I was concerned, there was no reason *not* to be his friend anymore; in fact, I thought we were friends. Otherwise, I certainly wouldn't have been so warm about recommending him to Mary."

But unbeknown to Richard and Mary, White continued to fume to others about what he regarded as Richard's heavy-handed censorship. The quarrel gave White a means of distancing himself from his latest benefactor. "I think Tim wanted that break," said Wolpoff. "By criticizing Richard, by saying that Richard was standing in the way of 'The Truth,' Tim separated himself from Richard—or as Tim would put it, he defeated Richard."

Whether or not Johanson and White consciously set out to create a species that would challenge Mary, they both knew as soon as they put the tag *Australopithecus* on the Laetoli hominids that Mary would be unhappy. Foreseeing this, Johanson had written a long letter to Mary in late December 1977. In it, he explained why he and White thought that all of the hominids from Laetoli and Hadar (including Lucy) belonged in one genus, and why these specimens should be described as australopithecines rather than early *Homo*.

Although the fossils were markedly different in size, Johanson argued that they were morphologically alike. He and White attributed the size variation to sexual dimorphism; that is, the males were substantially larger than the females. As for assigning the fossils to *Australopithecus* rather than *Homo*, Johanson explained to Mary, "the L/H [Laetoli/Hadar] material does not show the hallmark of the genus *Homo* as indicated by Mayr, [Louis] Leakey, Tobias, Napier, and others—the brain is still small and has not yet commenced enlargement." "What about [creating] a new genus?" Johanson then asked. "This would imply that the L/H were significantly different from later hominids in their adaptations." Thus, according to Johanson, the most logical conclusion, based on the fossils' mixture of australopithecine and early *Homo* features, was to assign the specimens to the genus *Australopithecus*; for the species name, he and White had created the term *afarensis*, based on the Afar region of Ethiopia. Johanson then drew a small diagram showing how *A. afarensis* fit into the hominid phylogeny.

Johanson's sketch depicted a Y-shaped lineage, with *A. afarensis* as the stem, while the *Homo* species formed one arm of the Y and all other australopithecines formed the other. In other words, Johanson and White interpreted *A. afarensis* as the ancestral hominid, not just of later australopithecines, but of virtually every hominid ever discovered: *Homo habilis*, *H. erectus* and *H. sapiens*, and *Australopithecus africanus*, *A. robustus*, and *A. boisei* all owed their origin to *A. afarensis*, according to this scheme. Their conclusions also meant that the earliest species of *Homo* had probably already been found: *Homo habilis,* which appeared about two million years ago. Prior to that, Johanson and White would argue, there was only the australopithecine line.

Mary received Johanson's letter in early January and responded immediately. In a previous letter, after he had first suggested the idea of naming the fossils, Mary had noted that she was "doubtful about a christening ceremony at this stage." Since that time, however, she had had a slight change of mind. "A number of highly respected persons have suggested to me that the L/H material deserves a separate name," she now told Johanson. "I am somewhat in agreement, but the difficulty of arriving at a correct label has seemed to me insurmountable. And I do not think *Australopithecus* is correct. It is a lousy term, based on a juvenile [the Taung skull from South Africa]. . . . Nor is it a direct ancestor of *Homo,* as all of us people agree." Mary also thought that Johanson and White had missed the true import of the fossils. "In my opinion the exciting thing about the L/H material is its ancestral relationship to *Homo*. . . ." She then offered a possible solution to their differing points of view. "If you can find an alternative to *Australopithecus* or make it clear that I disagree on that issue . . . I'll join you and Tim, otherwise I must stand down."

A month later, in February 1978, Mary met White at Berkeley where she reiterated her concerns: essentially, she was unhappy with the appellation *Australopithecus* and did not want to be associated with their naming venture if that was the term they chose. But a paper without Mary's name on it —without the Leakey seal of approval—would probably not receive the same amount of attention, and so Johanson (whom Mary and White had telephoned) suggested another solution. First, they would all name the new species "in a straightforward way." Then he and White would write a separate paper, discussing the implications of the new species and its relationship to other hominids; by not joining in this paper, Johanson explained, Mary would not be giving her approval to the idea that *A. afarensis* was ancestral to *Homo*.

It was a very fine (and dubious) distinction, and although Johanson and White would later insist that Mary accepted this plan, Mary disagrees. "I never thought the Laetoli specimens were *Australopithecus,* and I objected to that term because I thought that anything that was right on the *Homo* line should be called *Homo*. I told them that all along," she said.

But believing that they had Mary's blessing, Johanson and White drafted a paper, describing *A. afarensis* as a new species. Johanson forwarded a copy to Mary, and then in May 1978 set off for Stockholm for a Nobel Symposium on anthropology sponsored by the Royal Swedish Science Academy. Mary would be at this meeting, too—and as a special guest of honor. Sweden's King Gustav planned to present her with the Golden Linnaean Medal for her contributions to the biological sciences. Named for the great Swedish botanist Carolus Linnaeus, the award represented a high tribute to Mary, who would be the first woman to receive it.

Prior to leaving for Stockholm, Mary showed Johanson and White's paper to Richard, and asked his opinion. She was still debating whether or not to add her name. "I think you'd be making a mistake to be coauthor of a paper that puts all these fossils into one species," Richard remembers telling her. "I think it would be wiser to treat the Hadar fossils separately from Laetoli."

Mary was able to gauge other scientists' reactions to *A. afarensis* at the Nobel meeting, where Johanson announced the new species for the first time. Surprisingly, it caused little reaction; there were few questions from other scientists perhaps because, as Richard suggests, the paper was difficult to follow and the implications were not immediately apparent. But Mary was furious—not because of the name but because Johanson had discussed her discoveries at Laetoli in great detail. "Don devoted at least half of his paper to describing Laetoli—the stratigraphy, the dating, the fossils, the fauna—and it put Mary in a very embarrassing position because she was the next speaker and everything that she was going to say had just been said," Richard recalled. "And this was a prestigious meeting with a king in the audience, and here's this old lady who's going to get up and make an ass of herself, repeating everything as if she hadn't heard. She turned around to me and said, 'What am I going to say now? Why did he do that?' And I told her, 'Well, you've got to say something. Extemporize. Say it all again. It doesn't matter. Just get up and speak.' But it was very hard for her."

Johanson, Mary felt, had made her look a fool—and she did not soon forget. "All the things I was going to say about my site that I had been funded for, Don talked about as if they were his own," she said, still irritated nearly ten years later. "It was for me to announce those things; not him. And yes, I was furious; it was so impertinent."

Johanson would later say that he had not meant to encroach; but the Nobel meeting marked the first rift in his relationship with Mary. It did not take long for the break to widen into a gulf.

Chapter 37

FOOTPRINTS FOR THE MANTELPIECE

In January 1978 Mary had announced in a press conference at the National Geographic Society her team's discovery of the Laetoli hominid footprint trail left in volcanic ashes 3.6 million years ago. Scientists were divided about the prints. Some believed that the broad wedges with what appeared to be the definitive mark of a big toe were unquestionably those of a human ancestor, while others were more skeptical.

"Mary showed me a cast of the prints [in London, when she was en route to Washington, D.C.]," recalled the anatomist Michael Day, who had described the fossilized footbones of *Homo habilis* for Louis and Mary, "and I told her they were *not* hominid, which greatly offended her. But I can understand why. She had organized this meeting with the National Geographic, and was going there to deliver her great new find, and I threw cold water on it."

When the prints were discovered the previous summer by Peter Jones and Philip Leakey, Mary had also been unconvinced. But after discussing them with several colleagues, she decided that a hominid had probably made them. She had hoped that Day—as much of a hominid footprint expert as anyone could be, given the paucity of such fossils—would agree, or at least entertain the possibility. His outright dismissal irritated her, but may have also revived her own doubts, since in her announcement, Mary described herself as only "75% certain" that the trail had been left by an early human. The four prints indicated that "the creature", as she referred to this putative ancestor, had walked with a "slow, rolling gait," rather like a bipedal chimpanzee. Such a plodding style seemed to preclude the chance that these hominids were hunters, swiftly chasing after game, Mary noted. Instead, she thought they had lived primarily as scavengers. "It probably ate anything and everything it could find: berries, fruits, scavenged meats," she told the reporters. "In those days, I would have too."

During her visit to the United States, Mary met Louise Robbins, a footprint expert at the University of North Carolina. Unlike Day, Robbins thought there was a good chance the tracks had been made by hominids, but she explained that it was difficult to resolve the issue without studying the prints firsthand. Mary was pleased and immediately invited Robbins to Laetoli to examine the trail, and Robbins accepted.

Mary had also invited Day and two of his students—although because of his earlier criticism, Day found himself on Mary's "Stinkers' List" and relegated to a section of Laetoli far from the fossilized prints. "Louise was Mary's new footprint expert," said Day, while Tim White, "in his cowboy hat and compass and bandoliers," as Day put it, "was Mary's great anatomist that year." Although Mary and White disagreed about his interpretation of her Laetoli hominid fossils, she nevertheless regarded him as brilliant and treated him as a protégé.

The geologist Richard Hay returned to continue his study of the volcanic ash that had preserved the footprints, and the geochemist Paul Abell, Richard's longtime friend, came as well. Mary had asked him to find a technique for removing a section of the animal tracks to display at the Olduvai Museum. To run the camp, Mary turned to Peter Jones, an archeologist who had been living at Olduvai for the last two years.

Although separated in age by more than four decades (in 1978 Jones was twenty-one, Mary sixty-five), Jones and Mary got along surprisingly well, largely because of Jones's adaptable nature. "Mary had a routine," Jones explained. "After work, we'd meet at the living area and we'd have drinks at seven, dinner would start at eight and be finished at eight-twenty. And for the first two years at the end of every dinner, she'd say, 'Well, we don't have puddings down here.' *Two* years! But I learned that the routine was very important and has a value in an isolated situation like that. Things get done, if you say, 'Dinner is at eight on the dot.' "

By the time of the 1978 Laetoli season, Mary had grown so dependent on Jones that she fondly called him her "fourth son"—although Jones noted the affectionate term quickly vanished whenever any of Mary's real sons appeared. Nevertheless, he and Mary "worked very closely together, coordinating projects and people in the field," said Louise Robbins. "Peter could speak fluent Swahili [Mary, like many Europeans in East Africa, used a pidgin, called "Kitchen Swahili"], and knew the thinkings of the native people and interacted with them very well. He knew Mary's feelings too, knew if she was feeling out of sorts, so he could channel things to minimize conflicts in the camp. He was her right-hand man."

Mary's field activities that summer were limited. She had broken her ankle the year before at Olduvai, when stepping off her front porch, and it still pained her. Consequently, she spent most of her mornings in camp, working on her papers and correspondence, and venturing out to see new discoveries in the afternoon. Once, one of her Kamba workers turned up a single hominid molar, but nothing of earthshaking significance was found until July 27, when Paul Abell came sauntering into camp with a smile on his face. He was certain that he had found another footprint that was indisputably hominid.[1]

"I'd finished my project [removing a sample footprint block], and was spending my last day at Laetoli following Dick Hay around the exposures," said Abell. "We were exploring the footprint level, just following it wherever it cropped up." Venturing ahead of Hay, Abell spotted a new section of the hardened ash. "And there right on the edge was what certainly looked like the remains of a [human] heelprint," Abell recalled. "I thought, 'Gosh, that looks promising.' " Searching for more prints, Abell wandered on, then returned, just in time to see Hay—who, with only tuff correlations on his mind, failed even to notice the heel impression—pull out his geological hammer and begin "hammering this footprint to bits. I said, 'Dick, don't do that! I think it's a footprint.' " Hay pulled back and took a second look. "And it *was* a footprint," Hay said later, the memory of the moment still triggering a note of surprise in his voice. Chagrined at his hasty action, Hay carefully fitted the broken pieces back together, and the two were soon telling Mary their news.

The site lay a little less than a mile from camp, and before hiking this distance with her bad ankle, Mary decided to send one of her most skillful workers, Ndibo Mbuika, to dig an exploratory trench—with luck, a complete footprint might turn up beneath the overlying ash. But Ndibo's excavation added a twist. Yes, he told the scientists later the next day, there were more hominid footprints under the ash layer, but one of them was huge. He held up his hands twelve inches apart to indicate the size. This seemed extraordinary and Mary snorted in disbelief. "We all said, 'No, that's ridiculous,' " Jones recalled. "We thought he was exaggerating." Eager now to see for herself whatever it was that Abell and Ndibo had discovered, Mary headed off for the site with several other scientists in tow.

To the group's surprise, Ndibo was right. He had uncovered two distinctly human tracks, one of which was surprisingly long. Only someone six feet tall could have left such a print, yet based on the fossil record, the scientists knew that the hominids of 3.6 million years ago were small creatures, barely four feet tall. Perhaps the print had somehow been distorted before it was preserved, or perhaps the hominid had a curious way of dragging its feet. But as perplexing as the long print was, the mystery took nothing away from the immensity of the discovery itself. Here were unquestionable human

[1] The first set of tracks Mary's team discovered remain enigmatic, although all who have studied them agree that they were left by a creature walking on two legs—perhaps a now extinct species of bear, or an as-yet-unidentified hominid.

▲

tracks, as modern in their appearance as any beachcomber's, and as startling to the scientists as Friday's footprints were to Robinson Crusoe. "We were all amazed, in awe, of how human they were; they were so completely different from those of the previous season," said Robbins. Nothing like this had ever been found before, and Mary was elated. "There was unqualified excitement, everyone was glowing, and Mary, who usually keeps her feelings to herself, even had a big grin," said Hay. "She kept saying, 'Oh, we found it; we found it.' "

To find the two prints—only one of which was huge, the other diminutive—Ndibo had removed a two-foot square of ash and earth about six inches thick. This overburden had sealed and protected the tracks from erosion, so that they looked like prints that had been cast and dried in cement. It seemed likely that additional tracks would be found once more of the overburden was excavated, and Mary put her whole crew to work at the new site, which was labeled the G-1 and G-2 Trails.

With great care, the team dug away the overlying earth, exposing large sections of the hard grey ash underneath, and there, just as they had hoped, were more of the tracks. The prints themselves were filled with dirt, pebbles, and tiny roots, which the crew cleaned away with dental picks and soft brushes. Mary, too, joined in this task, sometimes kneeling next to the prints on a tattered foam cushion, at others pushing it aside to simply stretch out full-length on the earth, her nose and eyes mere inches from her work. The group soon found that they could only excavate the prints in the low light of the early morning and late afternoon; the sun's midday brightness cast too harsh a light for the tracks to be seen clearly. But even with these time limitations, it did not take more than a few days before the team had uncovered two parallel and distinctly human trails.

Apparently, 3.6 million years ago, two early humans—one tall, one short—had been striding across the plains, one of them slightly ahead of the other, when the volcano Sadiman erupted, showering the hominids and earth with ashes. A light rain was falling at the same time, and as the hominids continued on their way each step—left, right, left, right—was preserved in the mixture of muddy ash, leaving behind a kind of Pleistocene version of the Sidewalk of the Stars at Hollywood and Vine. Perhaps, the team speculated, the prints had been left by a male and female, whom they dubbed "Adam and Eve." That might explain the size difference, although the larger prints were also "blurred," as Mary later noted in an article for the *National Geographic,* "as if he had shuffled or dragged his feet." Or the ground may simply have been "loose and dusty, hence the collapsed appearance of his prints."

Yet even with this mystery unresolved, Mary had no doubts about the tracks' importance. After excavating one particularly well-preserved print, she sat up, lit a cigar, exhaled slowly, and announced, "Now this is really something to put on the mantelpiece." Usually, though, Mary worked with little show of emotion, simply pressing ahead at what Louise Robbins described as "a well-thought-out, measured pace. I think she was as amazed as the rest of us at how very old and how very human they were; I certainly felt that every day that we went to work. But she was also a hard-core scientist; she tempered our enthusiasm and told us not to rush our way to the end of the trail. I recall her specifically saying one day, 'These footprints have been here three and a half million years. We don't need to excavate them all in one summer.' "

In letters to colleagues and friends, where it was perhaps easier for her to display her feelings, Mary was forthrightly enthusiastic. "We are having a wildly exciting time here, with footprints that might have been made today," she told Allen O'Brien of the Leakey Foundation. "[They] are extraordinarily well preserved, with toe impressions and all." And to Dick Hoojier, a Dutch paleontologist who had described the rhinoceros and equine fossils from Olduvai and Laetoli, Mary wrote, "We have found magnificent hominid footprints this year, in a dual trail 23 metres [75 feet] long! (confidential) It is a most important find and demonstrates that 3 1/2 m.y. ago the fully bipedal, striding gait had been completely developed." Hoojier instantly recognized the significance of the discovery and replied, "The fully bipedal, steady gait completely developed 3.5 m.y. ago! I shall of course keep this information confidential but look forward to the publication. Few papers shake the world, but this one will be among them I am sure."

Hoojier's assessment was correct: Mary's paper, published in *Nature* in the spring of 1979, did jar the scientific world and finally settled a longstanding dispute about whether or not our earliest ancestors had walked erect. In spite of a variety of fossils—leg, knee and foot bones—that scientists had amassed at sites as far afield as South Africa and Ethiopia, anatomists had been unable to agree about the early hominids' posture and gait. The bones could be interpreted any number of ways, depending on which feature anatomists chose to emphasize, whereas Mary's footprint trail was graphically clear: the early hominids stood tall and walked as easily on two legs as any *Homo sapiens* today. Further, since their brains were still the size of chimpanzees', it was apparently this upright stance, rather than enlarged crania, that first separated these creatures from other primates.

Yet the prints are so modern looking that even Mary could not refrain from attributing some human sensibility to the hominids who had made them. At one point in the second trail, the smaller, possibly female, individual stopped, turned momentarily to the left, then continued on her way. "It looks as if she saw something or heard something over to her left and turned to have a close look," Mary told a press conference in March 1979. "It gives the whole thing a very human aspect." Perhaps this ancestor had stopped to watch a flock of birds, or heard the rumbling of a fresh eruption from Sadiman and turned to study it—or perhaps, as Mary wrote in the *National Geographic,* she had simply "experienced a moment of doubt" and stopped to look over her shoulder at something behind her.

By early September, Mary and her team had uncovered seventy-five feet of the fossilized trail, revealing a total of forty-seven hominid prints. Twenty of these had been made by the larger individual and twenty-seven by the smaller, possibly female, creature. The track, which headed in a straight line to the north, came to an end at a small canyon, where seasonal streams had eroded the ancient ash. But after a quick survey, the team found the tuff and footprints continuing just beyond this fault, and Mary decided to extend the excavation the next season with much the same group of people. She even invited Tim White to return, although their earlier disagreement about how to interpret the Laetoli hominid fossils had finally erupted into a full-scale quarrel.

At the heart of Mary and White's dispute lay two questions: how should the Laetoli hominid fossils be classified, and were they the same species as those that Donald Johanson had discovered in Ethiopia's Afar desert? In spite of Mary's reservations, Johanson and White had classified both sets of hominids in the same species, *A. afarensis.* They had written a paper describing *A. afarensis* for *Kirtlandia,* the scientific journal of the Cleveland Museum of Natural History (where Johanson worked), which was scheduled to be published in the late summer of 1978. On his draft copy, Johanson listed Mary and Yves Coppens (the French coleader of the Afar Expedition) as coauthors. Mary insists, however, that she had never agreed to this arrangement. In a letter to a colleague, she later explained, "Tim showed me a draft of the paper when I was in Berkeley and, although I did not agree, I could not veto its publication; but I never imagined they would include my name, in view of my objection" to *A. afarensis.*

Yet Mary apparently did not make her wishes as clear as she might have. At the Nobel Symposium in Stockholm, Johanson was still under the impression that she had agreed to be a coauthor. However it came about, at this point a misunderstanding arose. Johanson left the Swedish meeting still believing that Mary was a coauthor on the *Kirtlandia* paper—in fact, in a subsequent letter to her, he explained that he had even discussed this with her in Sweden. But Mary maintained that she never gave Johanson permission to add her name to the list of authors.

What was now foremost in Mary's mind was the embarrassment she had suffered at Johanson's hands in Sweden—and all because of a name she had never liked, *Australopithecus afarensis.* "That name rankled her," said Paul Abell. "She was fit to be tied that her fossils were being lumped together with those that Johanson had found in the Afar." Perhaps Johanson's speech had brought home to Mary just who stood to benefit the most from the new name: Johanson and White. The name alone—*afarensis*—assured that attention would be focused on the hominid specimens from Ethiopia, not Laetoli, and their discoverer, Don Johanson. "It was kind of a form of scientific theft," said Richard Hay. "Johanson's choice of *'afarensis'* [for Mary's fossils] really says, 'These are my beasts.' "

Johanson's selection of one of Mary's fossils—a mandible, referred to as LH-4 (for Laetoli Hominid, number four)—as the holotype, or type specimen, only added to this impression. Type specimens are important in taxonomy because they serve as the reference model for the species; if a scientist wishes to include other fossils in the new species, he will compare and contrast these with the "type." But it was highly unusual, if not unprecedented, to name a species for one site, then choose the type specimen from another site, as Johanson had done—particularly when the sites were so far apart in time and space. "It was a horrible thing that Johanson did—taking the type from Tanzania and the name from Ethiopia—and made everyone's hair stand on end," said Ernst Mayr, the noted biologist and taxonomic arbiter. A species' name and the type specimen are intricately intertwined, Mayr explained; but with the name pointing to one site and the type to another, there was bound to be confusion, the last thing a taxonomist wants.

When White arrived at Laetoli in July 1978, he found Mary fuming about *Australopithecus afarensis* and Johanson's speech in Sweden. Apparently White attempted several times to persuade her that he and Johanson were right about their interpretation. But Mary was unswayed, saying only that she had never liked the term *Australopithecus* and thought it a poor choice for the Laetoli hominids. Trained as an archeologist, not a physical anthropologist, Mary had always been uncomfortable discussing hominid fossils in detail. (Indeed, in her correspondence, she only described her ideas about hominid phylogeny once—and then insisted to the recipient that her comments remain confidential.) She enjoyed bantering about the fossils over dinner, but she never entered into the fray at scientific gatherings or in print. "She felt modest about her own anatomical expertise," explained Richard Hay, "and always wanted to get [Phillip] Tobias or some other expert to describe her hominids." White, of course, was one of those experts, and he undoubtedly knew the anatomy of the Laetoli specimens far better than Mary. But he couldn't hide his arrogance, and instead of trying gently to remind Mary about their previous discussions in the States—as Peter Jones might have done—he treated her with barely concealed contempt.

"White was talking strenuously with everybody about Mary, and I'm sure some of this must have gotten back to her," said Abell. "He set himself up as the anthropological expert who knew everything, while Mary was an old fogy who hadn't kept up with the times." To her face, he criticized her Kamba workers, arguing that he needed to teach them the finer points of the human skeleton so that they could find hominid fossils. "Needless to say, whatever hominid fossils were found that season were recognized by my staff and not discovered by White," Mary said later. And after Abell discovered the hominid footprints, White insisted on directing Mary's workers during the excavation. "He started taking over, telling my men they didn't know how to dig," Mary said. "I could never convince him of [their] experience and expertise."

The constant chafing between Mary and White finally erupted into a full-scale quarrel on August 21. The issue again was *Australopithecus afarensis,* with White pressing as before for Mary to accept his and Johanson's interpretation. The two stayed behind in camp to discuss the issue. "It was a very heated argument," Mary recalled, "and it came to a head when he told me, 'Oh, well, it's too late. Your name's on the paper. You can't do anything about it; it's in print.' " White's words were like oil on embers, and Mary abruptly ended the conversation. "That really raised my hackles," she recalled, her eyes blazing again at the thought of this young man telling her what she could and could not do. Mary was also angry because it was only then, she said, that she realized that she had been included as a coauthor on the final version of the *Kirtlandia* paper. And, in fact, Johanson had never sent her the galleys of the article.

Mary was not about to stand by idly while White and Johanson used her name in what she considered to be "their career strategy." Early the next morning, without a word to anyone, she headed straight from her camp to the village of Ngorongoro, an hour's drive away. She drove directly to the post office and there sent Johanson a two-line telegram: "PLEASE OMIT MY NAME FROM PAPER ON THE NEW SPECIES. REGARDS. MARY."

Johanson, however, was not in Cleveland, where Mary had sent the cable, but in Sweden, attending a friend's wedding. When word of her request finally reached him, he was uncertain what to do. The issue of *Kirtlandia* was already printed and ready for distribution, he explained in a letter to her, and to reprint it in its entirety would be very expensive. He suggested including an insert or pasting a label over the list of names, explaining that she had withdrawn; and he apologized at length for the misunderstanding. White also suggested the same course of action to Mary and she agreed, writing to Johanson, "I trust you will see that this is carried out *without fail.*" But she no longer trusted either Johanson or White; as soon as White left Laetoli in mid-September, she wrote a letter to Clark Howell, White's senior colleague at the University of California at Berkeley. In it, she explained what had happened and noted that White and Johanson had promised to correct the mistake; she hoped that Howell would see to it that they did. "I would be so grateful if you could use your influence to ensure that this is published as soon as possible and given due prominence," she wrote. "I am sure that this situation has arisen through misunderstanding and for no other reason, but I am greatly disturbed by it."

Ultimately, a new title page was printed for the *Kirtlandia* paper and the whole issue rebound; it was released in late November 1978.

White stayed on at Laetoli for two weeks following his row with Mary—a time that Mary remembers as "difficult." Yet a few days before his departure, they had another private meeting and managed enough of a reconciliation that Mary invited White to return the next season. Together with Ron Clarke, who had come to Laetoli to make casts of the hominid footprints, Mary and White climbed to the top of O'ndolenya Hill, a promontory overlooking the surrounding plains. "That's a Maasai word meaning 'to go away from' or 'to avoid,' " explained Clarke, "because there had once been some cattle disease nearby. And it's always seemed ironic to me that Tim and Mary should have gone there after their reconciliation—because the next year that's exactly how Mary felt about Tim. She wanted to avoid him."

But on that last day atop O'ndolenya, it seemed that Mary and Tim had weathered their misunderstanding successfully. They were enjoying the view, imagining where the hominids might have walked millions of years ago and anticipating the continuation of the excavation the next season. The naming debate then seemed but a passing cloud over what Mary termed their "state of euphoria" about the footprints. It was only after White left Laetoli for the United States that Mary mentioned *A. afarensis* again—and it was clear that she meant her words to be the last ones on the subject. "Call it what you like," she'd said. "Call it *Hylobates,* call it *Symphalangus,* call it anything, but don't call it *Australopithecus.*" But scientific nomenclature cannot just be wished away, and Mary was fated to hear the name *Australopithecus afarensis* for many years to come.

Chapter 38

BATTLING OVER BONES

Mary closed her Laetoli camp in mid-October 1978. She and her crew took one last look at the hominid footprint trail, then carefully covered it with layers of plastic, fine sand, and small lava boulders. "We are in the sad process of re-burying the footprints," Mary wrote Don Johanson from her camp. "With rain, elephants, buffalo and cattle one feels apprehensive about their welfare until next July." Johanson had just written Mary assuring her that her name had been removed from the *Kirtlandia* article, and their relationship—at least in their correspondence—resumed its previous close and friendly tone. Likewise, Mary and Tim White carried on an amicable exchange of letters through the fall and winter of 1978.

But underneath Mary's surface goodwill, her anger about the name continued to simmer, and by early November she was fuming once again. She had finally received a copy of *Kirtlandia,* with Johanson and White's paper.[1] After reading it, she was at first simply relieved that the article did not bear her name. But she was also frustrated that the new species was now formally published—and that meant that unless the name or species was found to be scientifically invalid, it would forever be the name of Mary's Laetoli hominid

[1] Yves Coppens, a coleader of the Afar International Research Expedition, was also listed as an author, although a few months later he published an article stating that there were actually two species and not just one in the Afar hominid collection. "I asked him, 'Why did you put your name on the [*Kirtlandia* paper], when you must have already known you didn't agree with it?" Richard recalled. Coppens replied, "You know, one of the papers says there could be one species; and the other says there are two. Once could be right, one could be wrong. This way we are probably right with one of them." Because of Coppens's contradictory stance—and his limited role in writing and defending the *Kirtlandia* paper—he is seldom mentioned as a coauthor.

fossils. Determined to unravel the knot that Johanson and White had so neatly tied, Mary immediately wrote to the anatomist Phillip Tobias, telling him in no uncertain terms what she thought of the younger scientists' research.

"It is a slovenly piece of work and I hope that you and many others will join issue," she wrote. Johanson and White, she said, had made several errors. "As an example, the date for the Laetoli specimen is given as '3.6–3.7 m.y.b.p' on p. 3, and as '3.59–3.77 m.y' on p. 5!" Mary regarded the name *Australopithecus afarensis* as scientifically incorrect (". . . there is no justification for the use of *afarensis,*" she wrote another friend, "since Laetoli is older and the original specimen [found by Ludwig Kohl-Larsen] came from there [in] 1939"), and she urged Tobias to take action to squelch the name.

Mary sent much the same letter to the anatomist Michael Day, although she was more direct about what she wished him to do. "Perhaps you would like to write to *Nature* pointing out the wholly unscientific nature of [Johanson and White's] paper?" she asked. Mary needed Day's and Tobias's assistance in her battle since she lacked the anatomical background to challenge *Australopithecus afarensis.* But if her colleagues did choose to write articles attacking the new species, she would "gladly add [her] name," as she told Day.

Mary's appeal did not fall on deaf ears. Neither Tobias nor Day was happy with the new species, and they willingly pledged their support. Tobias thought *A. afarensis* could be squashed by arguing that the Laetoli and Afar fossils actually represented two subspecies of *Australopithecus africanus,* the South African hominid that had first been identified by Raymond Dart in 1925. In fact, Tobias had planned to announce these subspecies—using the names *Australopithecus africanus aethiopicus* for Johanson's fossils and *Australopithecus africanus tanzaniensis* for Mary's—at the Nobel Symposium in Sweden, but Johanson had effectively (and unwittingly) scooped him by presenting his paper first. (According to the International Code of Zoological Nomenclature, the first scientific name attached to a specimen takes precedence over all others.) Tobias nevertheless thought his paleoanthropological colleagues might be sympathetic to his approach because it would untangle the confusing snarl Johanson and White had created by mixing fossils that were separated by a thousand miles and a half million years.

Puzzlingly, Mary did not object to Tobias's strategy, even though he planned to use the name *Australopithecus* to fight her battle. In all of her discussions with Johanson and White she had strongly objected to their suggestion to call her fossils australopithecines; she always referred to them as an early form of *Homo.* Mary did not raise the same issue with Tobias apparently because his tactic would accomplish what she most wanted: to divorce her fossils from Johanson's. But Mary's mixed opinions about human origins—telling one group of scientists she loathed the term *Australopithecus* while not raising the merest objection to the name with another colleague—only added to later suspicions that personalities and personal biases lay at the heart of her quarrel with Johanson and White. "I've often thought that if Johanson and White hadn't named their species *'afarensis'* but *'laetoliensis'* [after Mary's site, Laetoli], she would have accepted it," said the geologist Richard Hay. "But it was that anomaly—using the name from the north, and the type from the south—that was the real sticking point for her."

In the meantime, Johanson and White bolstered their position with a lengthy article in the January 26, 1979, issue of the journal *Science*. Entitled "A Systematic Assessment of Early African Hominids," it was a companion piece to their *Kirtlandia* paper, which had simply named the new species. In their *Science* article, Johanson and White explained why they had settled on the genus *Australopithecus* and hypothesized that their new species significantly altered the human family tree. They argued that the fossils were undoubtedly australopithecines because of their mixture of primitive and advanced features—humanlike, upright bodies crowned with small, apelike heads. The fossils were not, however, of the South African *A. africanus* type, the two men insisted, because the teeth were different, and the specimens predated those from South Africa by more than one million years (an age difference Johanson and White were able to claim by including Mary's 3.75-million-year-old Laetoli fossils; the Hadar hominid remains were then dated at a little less than three million years).

Having hypothesized that the Hadar and Laetoli fossils represented a new species—*Australopithecus afarensis*—Johanson and White then went on to suggest that *A. afarensis* was the pivotal ancestor, the one from which had sprung the human line (beginning with *Homo habilis*) and all the australopithecines, including the South African *A. africanus*. Thus, instead of a many-branched lineage leading back to an as yet unidentified ancestor as Richard had posited three years before, Johanson and White suggested that there were only two branches on the human family tree (*Homo* and *Australopithecus*) and that they both began about four million years ago with *Australopithecus afarensis*.

Until this paper appeared, the new species had caused barely a ripple of interest outside the paleoanthropological community. But *Science* featured Johanson and White's article prominently as its lead piece and with an illustration of one of the Ethiopian fossils for its cover—and soon news of the new species was broadcast in newspapers, magazines, and television reports around the world. *Australopithecus afarensis* was, after all, the first new hominid species to be named in fifteen years, since Louis's *Homo habilis*. Johanson and White's paper also presented a new interpretation of the origins of humankind, one that was certain to be controversial, arguing as it did that the human species, as defined by the term *Homo,* was little more than two million years old. If they were correct, then Mary's 3.75-million-year-old Laetoli fossils could not possibly be regarded as human.

The paper also raised questions yet again about the 2.6-million-year date

Richard still ostensibly claimed for his skull 1470, and which he had assigned to early *Homo.* (In 1976 his dating team had recalculated their results, suggesting a date of 2.4 million years for the KBS tuff.) Other scientists argued that 1470 was more precisely dated at two million years—a date that Johanson and White chose to use in their paper. (Richard did not capitulate until 1980, when two other geochronology labs dated the tuff to 1.88 million years.)

Few readers of their *Science* article could fail to realize that Johanson and White were on an intellectual collision course with Richard and Mary—which may have been precisely what the two men wanted. In *Lucy,* Johanson recalls a journalist friend saying to him, "It seems to me that you and the Leakeys are headed for a big shootout." "I guess we are," Johanson replied.

Only a few days after Johanson and White's *Science* article appeared, Richard arrived in the United States for one of his Foundation for the Research into the Origins of Man (FROM) lecture tours. He had celebrated his thirty-fourth birthday the month before with Meave and their daughters at their grass-and-thatch home on Lamu Island—the one place he told a reporter "where we can totally get away from . . . our work and be quite private." Privacy and peace, Richard had found, were harder to come by since the publication of his and Roger Lewin's two popular books, *Origins* and *People of the Lake.* (Published in 1977, *Origins* alone had sold more than 500,000 copies and had been translated into ten languages by 1979.) He had also been featured in a television documentary about his expeditions, and in November 1977, *Time* magazine splashed Richard's picture on its cover (an issue, Richard liked to note, that outsold one featuring supermodel Cheryl Tiegs). He was the hot young scientist who had "found more and better pre-man and early man fossils than any other anthropologist," *Time* reported, adding that Richard's revolutionary work had already "upset many longheld ideas on evolution and . . . forced science to write a new scenario" for human origins.

All the publicity had led to more requests for Richard to lecture, particularly in the States. At the same time, he had a desk full of projects under way: administering Kenya's National Museums; guiding the newly completed International Louis Leakey Memorial Institute for African Prehistory (TILLMIAP), which oversaw the museum's archeological and paleoanthropological collections and research; fund-raising for FROM; keeping the Turkana Expeditions afloat; and chairing two environmental and wildlife organizations. So much of his life was now centered on boardrooms, meetings, and lectures that he had little time left for fossil hunting; indeed, Richard told a reporter, he had only "about 5 per cent" of his time free for studying human evolution.

Richard did not say that even this small allotment was on the verge of disappearing. The year before, in early 1978, his persistent kidney disease had suddenly worsened, sending his blood pressure soaring and causing such severe headaches that he was often unable to do a full day's work. Richard now looked and felt ill. Not only did he suffer from the constant headaches, but he had little appetite, leaving him fatigued and gaunt. His skin, too, had taken on a sallow hue, he had a peculiarly unpleasant body odor, his legs and ankles were prone to swelling, making walking difficult, and his eyes were sunken, with the brittle, expectant look of the chronically ill. There was little Richard's doctor could do for him, other than increase his medication for high blood pressure. Yet the worse his disease became, the more Richard pushed himself, intent as ever to accomplish as much as possible. As much as he loved fossil hunting, Richard deliberately set aside his fieldwork to pursue larger, more institutional goals—in particular, launching TILLMIAP and providing FROM with a firm financial foundation.

In doing so, Richard transformed himself from young revolutionary to judicious diplomat. "He was no longer going to have his temper raised over these fossils," said Glynn Isaac, recalling the heated debates Richard had eagerly entered into in the early 1970s, "but consciously and adroitly moved to become a sort of elder statesman in the field." Instead of tight-lipped confrontations with his paleontological colleagues, Richard fostered a public image of gentlemanly goodwill. He did this in part because he thought people were more willing to open their wallets for cooperative rather than bickering scientists, and partly because of what he most feared: that without such cooperation, much of what he had built (the expeditions, the National Museums, TILLMIAP and FROM) might collapse if he became gravely ill or died. Thus, when asked by a reporter in 1978 what he considered to be his greatest contribution to paleoanthropology, Richard replied, "the development of the large interdisciplinary international team" at Koobi Fora, rather than the discovery of any particular fossil. With a team, Richard added, there was less opportunity for "prima donnas" to develop, so that even "if a particular figure falls into disfavor or changes his job or falls sick or dies," the work would continue.

On his FROM lecture circuit and at press conferences, Richard preached the same message: the rivalry that had once characterized paleoanthropology was a thing of the past. To prove his point, Richard had taken special pains to demonstrate that he and Johanson—whom the press often identified as rivals—were actually very good friends and colleagues. Whenever Johanson was in Nairobi, he stayed with Richard and Meave, and in the United States, he served as an adviser on the scientific committee of Richard's foundation. Richard and Johanson exchanged regular, newsy letters as well, although only in one, written in the summer of 1977, did Johanson mention his new ideas about the fossils from the Afar and Laetoli. The two men also often saw each other at conferences, and joined together in FROM fund-raising events. And when Johanson approached Richard in the spring of 1978 about searching his old site at Lake Natron, Tanzania, for hominid fossils, Richard graciously relinquished his rights to the region.[2] Johanson needed a new site because war and revolution in Ethiopia had kept him out of the Afar since 1977.

Of course, in all of these exchanges, Richard was still the leader. Whether dispensing exploration rights, extending invitations to his home and FROM, or stepping aside to welcome Johanson on the podium, Richard ran the show. Few thought the friendship could last. "I even said that to both of them," recalled Mary Smith. "I mean, how long could it last? Not very long. Their egos are too big for both of them to stand on the same platform." But when he arrived in the United States in late January 1979, Richard did not, as many in the press seemed to expect, take Johanson and White to task for their article in *Science.* Instead, he was all diplomacy and tact. "I am perfectly prepared to accept that Lucy is a new species of *Australopithecus,*" Richard said in his first reported comments. "But I believe that the common ancestor [of humans and apes] has not yet been found."

In late February 1979 Richard gave a lecture in Pittsburgh, to raise money for FROM, and the next day held a press conference. He had asked Johanson and four other scientists to join him in answering the media's questions, and they sat together at a long table facing the reporters. "I wanted all of the [symposium] speakers there because I wanted to emphasize that FROM was not just 'my' show, but that it represented the field," said Richard.

Most of the queries Richard fielded were of a general nature—What is FROM? for example—but then Boyce Rensberger of the *New York Times* took a more pointed tack. Why, he asked, had Richard not discussed the new species *Australopithecus afarensis* in greater detail in his lecture? Was it because Richard had a different opinion from Johanson about the new species, and if so, could Richard explain his viewpoint? "I thought that was a highly inappropriate question for our FROM function," Richard recalled, "because we were trying to present the field at a popular level, not dwell on the differences of our positions. So I said to Boyce, 'You know, I don't think that's particularly important. Any other questions?' " But Rensberger was not about to be put off, and turning Richard's remark to his own advantage, he quickly replied, "You don't think Lucy is important? Dr. Johanson, you've just heard what Mr. Leakey has said. Perhaps you would like to tell us what the differences are between you?"

Rensberger's ploy surprised Richard. "I was slightly taken aback; I hadn't expected this," he said. "But I was trying to be polite, so I said, 'By all means, Don, go ahead.' " Johanson nodded at Richard, then leaned into the microphone, saying, as Richard recalls, "Yeah, I did hear him [Richard] say that Lucy wasn't important. We all did. And I disagree." He then produced from inside his jacket pocket a statement explaining Lucy's and *A. afarensis*'s significance and read this aloud. "It said something like 'Lucy is the most

---

[2] Richard's team had turned up the first hominid fossil of his career (the lower jawbone of *Australopithecus boisei*) here in 1964.

important fossil this side of Noah's flood,'" Richard said. "I mean, it was a very strong stand, and the whole thing took about ten minutes." Throughout, Richard sat idly at the table, "wondering when this was going to end." When Johanson had finally finished, Rensberger asked Richard again if he would like to give his opinion, but Richard declined, suggesting instead that as it was late, they should all go to dinner.

Before the group disbanded, Rensberger persuaded Richard and Johanson to pose for a photograph, and briefly interviewed Richard as well. "I think Don was right the first time," Richard told Rensberger, referring to Johanson's initial interpretation of the fossils in *Nature,* where he had suggested that some of the specimens represented early *Homo,* the others an australopithecine. "They're sampling two different populations," Richard continued, "*Homo* and *Australopithecus.*" Richard also objected to the hypothesis that *A. afarensis* was the common ancestor of all later australopithecines and early *Homo.* In fact, he told Rensberger, his team had recently discovered eight hominid teeth that would prove that another type of hominid was living at the same time as Johanson's new species. But because the fossils were as yet unpublished, Richard could not discuss them in detail. "The material I've got is very insignificant," Richard admitted, "but there's enough to challenge Don with. It gives me the right to offer my opinion."[3]

Perhaps, under the circumstances, whatever Richard had said would have sounded defensive, but his reference to the eight fossil teeth had a particularly feeble ring; they seemed a very small sword with which to skewer Johanson's bold hypothesis. Nevertheless, Richard's words did constitute a challenge, and Rensberger reported them that way in a page-one story the next morning in the *New York Times.* "Rival Anthropologists Divide on 'Pre-Human' Find," read the headline; while above it ran a picture of Richard facing Johanson, the two apparently ready to slug it out over their fossils. Adding to the prizefighter image were two brief biographical sketches outlining the scientists' vital and career statistics: "Richard Leakey: The Challenger" and "Donald Johanson: Professional with Lots of Luck." Richard had to admit he had been outmaneuvered.

However, while Richard would have preferred that the *Times* feature an article about his foundation (rather than burying the sole reference to it in a sentence), he decided that even this kind of attention might do paleoanthropology some good. "I thought there'd probably be some chat about this, that it might stimulate some interest," he said, "and then the thing would die a natural death."

---

[3] Richard's team had discovered the teeth in 1978 in the southern region of Koobi Fora, from deposits dated at 2.7 million years. The anatomist Alan Walker has called them "dead ringers" for the teeth found at the Makapansgat Cave in South Africa. If so, they would have to be classified as variants of *Australopithecus africanus.* Because they were close in age to the *A. afarensis* specimens from Ethiopia (then dated at 3.2 million years), they would cast doubt on Johanson and White's assertion that *A. afarensis* was the ancestor of *A. africanus.* Nothing more has come of Richard's challenge, however, because additional, conclusive specimens (such as a skull) have not been found.

But Rensberger's article had all the elements that the American public seemed to love in a science story: the young, up-and-coming native son taking on the staid and privileged foreigner (who spoke with an English accent). It was a David-against-Goliath tale, with Richard cast as the giant. The substance of the dispute, and whether Richard actually entered the debate or remained distantly aloof, did not matter. The story that went out from the *New York Times* touted a match between two bitter rivals—and that is how reporters would write about paleoanthropology for much of the next decade.

Only a month after Rensberger's article appeared, Mary added to the controversy. At her press conference at the National Geographic Society in Washington, D.C., where she announced her team's discovery of the 3.6-million-year-old Laetoli hominid footprint trail, reporters asked her opinion about *Australopithecus afarensis.*

Mary stonewalled the gathering first, refusing to discuss Johanson and White's study. "But when pressed," one journalist reported, "she criticized Johanson and White for lumping together fossil finds [from Laetoli and Hadar]. . . . 'They [Johanson and White] are entitled to their opinion,' she said. 'But it is unfortunate they have used the Laetoli specimens as [a basis for naming] *afarensis.*' " Mary then delivered the coup de grace. Overall, she said, she considered Johanson and White's work to be "not very scientific."

"I winced when she said that," recalled Mary Smith, who was present at the press conference. "What she meant was that Johanson was behaving like an amateur [by mixing the fossils]. But you just don't say that to a scientist—that's the kiss of death."

Mary's were fighting words, and in response Johanson came out swinging. Mary wasn't much of a scientist either, he countered to a reporter who read him her quote, and "really shows a poor appreciation of what evolution is all about"—meaning that anatomical similarities, not geography, ruled taxonomy. The name *Australopithecus afarensis* was valid, he added, and it described not only Mary's Laetoli hominid fossils but the creature who left the footprints as well. "The footprints would have to be from *Australopithecus afarensis,*" he said, appropriating Mary's most prized discovery.[4]

Mary herself refused to name the hominids that had left the ancient tracks—just as she had stopped short of assigning a name to the hominid fossils from Laetoli. "They were direct ancestors to man," she told the reporters in

---

[4] Johanson and White did not mention the Laetoli hominid footprints in their article in *Science* because Mary had not yet announced them. But in a subsequent study, published in *Science* in 1980, White wrote that the trails had probably been left by *A. afarensis.* Although it is impossible to prove what creature left the prints, it is generally assumed they were made by the same hominid whose fossils were found at Laetoli. And because these fossils have been labeled *A. afarensis,* the footprint trail is usually attributed to this species as well.

Washington, D.C., and they seemed to display "a lot of human characteristics." Nevertheless, she was "doubtful" that they represented true *Homo*. At the same time, Mary did not want them classified as australopithecines. Whoever had left the tracks and fossil bones behind was "not like [the] other *Australopithecus* . . . which died out," she said, referring to the large, robust form, but without giving her reasons. "We'll leave it [the naming of the species] open until we have more material. It's a very knotty problem."[5]

Mary's blunt and derogatory comments about Johanson and White had upped the ante as far as the popular press was concerned. Although neither Mary nor Richard had yet responded to the two Americans in a scientific journal, the debate was now being reported as "a wonderful, scientific donnybrook, the kind where, if it took place among rowdies in a tavern, barstools would be flying through the air, beer mugs would be crashing against walls and fists would be smashing into noses," wrote Paul Galloway in the *Chicago Sun-Times*. In the midst of the mayhem stood Don Johanson, who, Galloway noted, had "a taste for a good fight" and so was "enjoying it immensely." Johanson had become the new paleoanthropological darling of the U.S. media.

Mary's announcement of the hominid trail did far more than simply heat up the debate with Johanson and White—it stunned the world. *Nature* devoted six pages and its cover to the footprints, the *National Geographic* carried a popular story by Mary,[6] and articles and photographs appeared in the press worldwide. The trail was even mentioned "briefly on the Walter Cronkite news!" Mary told a friend. So ancient and yet so like our own, the barefoot tracks particularly enthralled anthropologists and archeologists. "The footprints seem to have created rather a sensation," Paul Abell wrote to Mary. Abell had found the first hominid heel print in 1978, and now was on sabbatical at Cambridge University. "Everyone is entranced! I was sitting next to the archaeology museum curator at dinner last night [who said] without knowing I had had any part in it, [they were] 'one of the most fascinating archaeological finds ever made.' "

Mary had hoped that the footprints' importance and all the attendant publicity would spur the National Geographic Society to award her a larger grant for her next field season, since operating at a site as isolated as Laetoli was expensive. Instead, and with little explanation aside from financial constraints, the grants committee reduced her funds by $5,000, awarding her

---

[5] Scientists have since compared the fossilized foot bones of *A. afarensis* to the prints—with differing results. Some say the prints are a good match with *afarensis,* while others argue that *A. afarensis*'s toes were too long and curved to have left such a track. They suggest that the prints were probably made by an early form of *Homo*.

[6] At least the article's byline says "Mary D. Leakey." However, in a letter to a friend, Mary wrote, "It is stated to have been written by me, but I assure you I did not write a single word!"

$75,000. But a grant from the Leakey Foundation made up the difference, and Mary's plans to open her Laetoli camp in early July 1979 went ahead without a hitch.

Mary had invited a large and international crew of scientists, including several members from the previous year. But she continued to have mixed emotions about Tim White, whom she had again invited. She was pleased with his descriptions of the Laetoli hominid specimens, but she was angry over the dispute about her name, her fossils, and her footprints. Worried that she and White would quarrel again, Mary wrote him in late April, insisting that when he was in camp, he follow her rules. "As regards yourself Tim," she said, "in order to avoid clashes in the field, I must ask you to appreciate that the direction of this project is in my hands and my decision must be final, even if you do not agree." White's reply was not altogether encouraging. He insisted that he had always followed her orders, but that if he disagreed with her, he would say so. Surely she did not merely want a yes-man in the field? At the end of May, Mary added White's name to the Laetoli camp list—but she did not stop worrying about how she and White were going to get along.

"She was very, very apprehensive about his arrival," said Ron Clarke, indeed, so anxious that "she began thinking up schemes to make life difficult for him once he was there [at Laetoli]." Incapable of simply telling White not to come, Mary instead devised a plan that would inevitably have led to the very confrontation she hoped to avoid. She knew that White intended to help excavate the footprint trail as he had the previous season; but this year, she told Clarke and Jones, she would "send White to an area where she knew there were no footprints. Mary thought this would be a great joke," Clarke continued. "White could excavate there and he wouldn't find a thing. And Mary laughed about it; she thought that would be really funny."

But to Clarke and Jones, Mary's plan seemed only mean-spirited and vindictive. Perhaps she thought her intentional misuse of White's scientific talents would bring him down a peg, or teach him a lesson about who was the real director of the dig. Jones, however, thought her ploy would only have led "to crazy things and incredible tensions in camp." Before the fireworks began, he decided to intervene. He met White at the Arusha, Tanzania, airport and, without a word to Mary, advised him not to come to the field. White spent the summer instead at the National Museum in Nairobi, completing his descriptions of the Laetoli hominid fossils, and corresponding amiably with Mary. Both pretended that he had changed his mind about coming to Laetoli because of the amount of work he had to do on the fossils and because of the size of her team—she already had so many workers, he understood that she did not need his help. But the pretense solved nothing. Mary was undoubtedly relieved not to have White in camp, but her clumsy attempt to wound him only increased the ill will between them. Where once she had had an uneasy truce, she now had an enemy.

The Laetoli season got off to a good start. Mary, Ron Clarke, and two of the African staff were busy at the hominid trail site, where they had uncovered new tracks that were more deeply buried and therefore better preserved than those of the previous year.

Once again, the trail revealed what seemed to be the footprints of two individuals, one tall, one short. And as before, Mary and her crew were at a loss to explain the eleven-to-twelve-inch tracks of the larger creature. "They were "clearly absurdly long," Mary wrote to Phillip Tobias. She enclosed a Polaroid snapshot of the prints, and pointed out another surprise: how close the two trails lay to each other. "You will see that the two individuals were keeping step and walking very close together," Mary noted in her letter. "The only way we have achieved similar trails has been by walking arm-in-arm, as close together as possible. This puts the Pliocene boys and girls into a very romantic light; perhaps the site should be re-named 'Lover's Lane!' " Or, it may have been, as Mary suggested to another friend, that the larger creature had a disability of some sort—was arthritic or crippled—and "was being helped by the smaller person. Two of the big prints have double heel-prints, as though he hesitated and then slid his foot forwards."

For much of the summer, the team tossed around ideas about how the trails had been formed. Had the two hominids walked side by side and arm in arm; or had one come along a little later, and simply walked close to the trail of the first, matching its tracks step for step? "We talked about it and talked about it," recalled Paul Abell, "and we tried all kinds of ways to duplicate the prints. And it simply wasn't possible to walk that close together, unless we put our arms around each other." But Mary and her friends could only joke about such a human image. As her elderly mentor, Gertrude Caton-Thompson, noted, "Boys arm-in-arm with their girl-friends are a common sight in the village, but to see such behaviour in the Pliocene is really shocking!"

Not until early September, after uncovering another twelve feet of the dual trail, was the mystery of the tracks solved—and then by the wildlife cinematographer Alan Root, who had photographed chimpanzees and gorillas in the wild. Root had come to Laetoli to film Mary and the track (which now extended an impressive eighty-nine feet) for a BBC television show on human evolution. Looking at the large prints, Root said the puzzle was easily solved: there had actually been three hominids at Laetoli when the volcano erupted, and that one had deliberately walked in the footsteps of the other. He had often seen chimpanzees play this game of "follow-the-leader" for fun or when they were nervous, with the one behind "holding on to the hips of the one in front," as Mary explained to Melvin Payne. Root's clever insight explained the peculiar length of the largest prints and the curious double-heel marks Mary had noticed in some of the tracks.

By carefully studying the double prints, Mary and her crew deduced that the second hominid had been somewhat shorter than the one in front, while the creature walking so closely beside them was the smallest of the trio. Again, the trail conjured a startlingly human image, that of a family. "If one is prepared to believe that these hominids were already on the road to man, and were not primates with a tendency to ape-like behaviour," Mary later told an audience at London's Royal Society, "then it is plausible to interpret the tracks as being made by a male whose foot had broad, splayed toes, closely followed by a female, with one or the other leading a juvenile." On the other hand, as Mary had learned from chimpanzee watchers, "all chimps tend to hold on to one another when disturbed or frightened." Perhaps the trio had been a family, anxiously making their way together across the African plain when the volcano erupted; or perhaps, they were unrelated hominids who grabbed hold of each other for reassurance. Either way, Mary's team had caught an unexpected glimpse of hominid behavior—something that fossilized bones seldom revealed.[7]

There were other finds that season. Edward Kandini, one of her Kamba workers, spotted a single hominid tooth, bringing the total of Laetoli hominid fossils to thirty; Mary dutifully dispatched the tooth to Tim White in Nairobi to describe. Paul Abell turned up more promising track sites, although the team only had enough time to study and measure the prints at nine of these. Altogether, they recorded 9,525 animal tracks, the most common being those of the extinct hare *Serengetilagus.* Surprisingly, of the large animals, the rhinoceros proved to be the most abundant, leaving behind 254 tracks. Elephants, giraffes, gazelles, large antelopes, and hyenas had also left their marks in the ash—as had guinea fowl, an ostrich, and a solitary beetle. There were trails, too, of the three-toed horse, *Hipparion*—a mare and her foal had even crossed the hominids' track, moving together across the plains at a brisk gait.

All in all, it had been "a wonderful season," as Mary wrote a friend after closing her camp in mid-October. And yet, Mary had been quietly distressed the entire summer. Not long after opening her camp in July, Richard had flown to London to consult a kidney specialist. The news was not good, and Richard had finally felt compelled to tell his mother about his illness. He was now in "end-stage renal failure," as the physician phrased it, and unless he received a kidney transplant or was put on dialysis for the rest of his life, Richard would die.

[7] In her lecture at the Royal Society, Mary described the hominid footprints in greater detail: "It is now evident that the trails were made by three individuals walking together. There is a single trail on the west, made by the smallest individual; the prints to the east . . . are clearly double, made by two individuals walking in tandem, with the second placing his or her feet in the footsteps of the leader. All three individuals were walking in step and most likely holding one another, since any deviation in course, no matter how slight, is closely followed by all three and the stride length is virtually the same, in spite of difference in size of the prints." Perhaps, the smallest one held the largest by the hand, while the middle-sized one grasped the largest by the hips.

Chapter 39

RICHARD REBORN

In March 1979 Richard had returned to Kenya from his U.S. lecture tour—and unexpected confrontation with Don Johanson—weary and gravely ill. He was slowly being poisoned by the accumulated fluids, salt, and protein by-products that his malfunctioning kidneys could no longer excrete, leaving him with the classic symptoms of the uremic: high blood pressure, headaches, swollen ankles and legs, nausea and fatigue. Sapped of his usual vigor, Richard "found the days extraordinarily long," as he wrote in his autobiography. Yet he refused to give in, believing that all he needed was a short vacation or a different type of medicine to restore his energy. He also could not afford the time to be sick: a plot was afoot at the museum and new prehistory institute to depose him as director. Richard's disagreement with Johanson paled in comparison, and he devoted what strength he had to fighting his enemies at home.

Troubles had surfaced at The International Louis Leakey Memorial Institute of African Prehistory (TILLMIAP) not long after Richard had officially opened it in July 1977. Built with funds ($1.5 million) Richard had raised in a four-year blitz of lecture tours and international negotiations, TILLMIAP was designed as an international center for the study of African prehistory and human origins, and was loosely affiliated with the National Museums. The handsome three-story institute was also a tribute to Louis, and at its entryway Richard had erected a bronze statue of his father chipping a stone tool. Louis had objected mightily to his son's plans for the center, but even he would have probably conceded that TILLMIAP's centerpiece—an ample, air-conditioned, walk-in vault for storing and studying the hominid fossils—

was a vast improvement over the rickety shack he had used. The institute also housed a comfortable auditorium (Nairobi's largest), spacious offices, and well-lit laboratories for the staff and visiting scientists. Finally, Kenya had a building worthy of its role as keeper of humankind's earliest heritage —one that Richard imagined drawing students and scholars of human origins to his country, as well as inspiring a fresh crop of Kenyan paleoanthropologists.

Yet despite Richard's role in designing, funding, and creating the prehistory institute—and even though it included the museum's departments of paleontology, paleoanthropology, and archeology—he was not its director. Richard had declined that position because many foreign scientists had believed he was building the institute to expand his scientific empire. Thus, instead of being under Richard's direct control, TILLMIAP functioned as a semiautonomous department of the museum, and had its own director and budget. During the institute's fund-raising phase, this had seemed a sensible solution because it placated Richard's critics overseas. But at home, in Kenya, it served to undermine him.

"The institute really was potentially good," said Andrew Hill, a paleoanthropologist who had transferred from the museum to the new center. "We could look at anything involving the study of humans, from the time of *Homo erectus* back to the Miocene. So it was really more than just a building; it was an idea." But it was apparently not an idea shared by the institute's director, Bethwell A. Ogot—although, at first, he had seemed an excellent choice.

Respected for his studies in African folklore, Ogot had been a professor of African history at the University of Nairobi. Richard and other members of the museum's board had known him for years, and although he was not a prehistorian, he was well-connected in Kenya and, Richard thought, would bring the institute stability and prestige at home, while a staff of local and foreign scientists would add to its collegial, international air. Ogot, however, seemed power-hungry, for within months of becoming TILLMIAP's director he began maneuvering to sever all ties between it and the museum. In this he was aided by John Onyango-Abuje, an archeologist who had once been Richard's star protégé at the museum.

"Abuje was very bright and capable and ambitious," said John Harris, a paleontologist who had also transferred from the museum to TILLMIAP, "and, at one time, Richard had been grooming him as the museum's assistant director, with the intention of turning over the directorship to him one day." Richard had recognized Abuje's talent in the early 1970s, and had helped arrange a scholarship for him to study for his doctorate at the University of California at Berkeley. On Abuje's return to Kenya, Richard appointed him head of the museum's archeology department and, when away overseas, often had Abuje take the museum's helm as acting director. Abuje initiated a research project of his own as well, investigating the prehistoric archeological sites of the Lake Victoria Basin. But the relationship between the two men had soured by the time the institute opened, at least partly because Abuje, like other men Richard had befriended, could see no easy way to rise above him. It was Richard's project at Lake Turkana, not Abuje's at Lake Victoria, that garnered the media's attention; and it was Richard, the museum's director, not Abuje, one of Kenya's first black archeologists, who directed Kenya's prehistoric research.

Richard's enthusiasm for Abuje had also cooled, largely because of what John Harris termed Abuje's "over-excess with alcohol; I think Richard decided Abuje was not to be trusted." Nevertheless, Richard did not abandon his former protégé altogether: when TILLMIAP opened, he supported Abuje's appointment as a senior research fellow and head of the archeology department.

Although much of what transpired at the institute is unclear—tangled in a web of political intrigue—one thing is certain: Abuje and Ogot soon banded together to oust Richard. The two men had much in common: they were the senior black African scientists at TILLMIAP, and both hailed from the Luo tribe. Apparently they intended to draw on tribal loyalties to stage their coup, since they filled most of the institute's junior and mid-level positions with fellow Luo—a stacking of the deck that worried other employees from different tribes as well as the foreign scientists. "It quickly became clear that instead of being an international institute of prehistory and paleontology, it was going to be a Luo-based empire for history and archeology," said Harris.

To further distance the institute from the museum, Ogot began terminating the foreign scientists' contracts, beginning with that of John Harris—then Richard's brother-in-law—in January 1979. It was a calculated slap but Richard refused to be drawn in. He wanted the institute to succeed, and insisted to anyone who complained that Ogot should be given time to find his way. And so, undeterred, Ogot continued to "Africanize" the institute.

Shortly thereafter, in the spring of 1979, not long after Richard had returned from his U.S. lecture tour, Abuje attempted a power play of his own. He complained to Kenya's president, Daniel arap Moi, that Richard was a racist and ran the museum for the benefit of white expatriates only. Such accusations were as emotionally charged in Kenya as they are in the United States, and Richard could not simply shrug off this attack. He urged the museum's board of trustees (which also oversaw the institute) to fire Abuje; if they did not, he would "step aside" as director, he told the chairman of the board, since he would be "unable to continue to serve . . . whilst such indiscipline, disloyalty and sabotage is rampant." Richard also played his trump card: politically well-connected, he asked for a private meeting with Moi, who, after listening attentively, assigned Charles Njonjo, Kenya's attorney general and a longtime friend of the Leakeys, to investigate the troubles at the museum and prehistory institute.

"One does not lightly go for help to the head-of-state," Richard said, "but I was having a lot of difficulty because I was sick, and I told the president,

'I'm not sure I can cope with all these problems.' " He was especially bitter because Abuje had also suggested that the museum be totally separated from the institute and then broken into smaller departments—a dismemberment in Richard's view. Richard had built the museum from a staff of twenty-three to one of nearly three hundred, while increasing the budget from £23,000 to more than £600,000, and adding five new regional museums. "All of this had been achieved with considerable effort," Richard wrote in his autobiography, "and my great ambition had been to complete the Museum expansion before I succumbed to what I knew was inevitable: kidney failure."

Aside from turning to President Moi for assistance, there was little Richard could do; under Kenyan law, he would have to wait for the decision of the museum's board of directors—some of whom supported Abuje and Ogot. Richard was by now also far too ill to do any more behind-the-scenes politicking. Indeed, in late March 1979, shortly after his meeting with Moi, Richard checked into Nairobi Hospital: his blood pressure had soared so high that it was further damaging his kidneys. The battle to save the museum—and Richard's position as its director—would have to be fought without him.

Until he was hospitalized, Richard had never mentioned his kidney ailment to anyone aside from Meave and the anatomist Michael Day, who, ten years before, had helped arrange Richard's first visit to a kidney specialist in England. Other family members and friends could see he was ill, but none of them knew the cause, or how sick he was. "Richard had amazing stoicism and self-control," said Joan Karmali, a friend of the Leakeys and a volunteer at the Museum. "Sometimes I was in his office and he would be talking, and he would stop for a moment, grit his teeth and close his eyes tightly. And I would say, 'Are you all right?' " Richard would immediately answer, "Oh, yes. I'm fine. Fine." "He must have suffered enormously," Karmali continued, "but he *never,* ever let on; he never showed it. I don't think any of us have an inkling what he was suffering, what a really bad state he was in."

Through sheer willpower alone, Richard had staved off the doctors for more than a decade; admitting that he now needed their care came as a "terrible blow," he wrote in his autobiography. He was also distressed that everyone knew he was sick, that he appeared weak and vulnerable at a time when the museum and his staff needed him most. Rumors were soon circulating that Richard had entered the hospital because he was dying—and Abuje walked the museum's and institute's halls boasting that it was all because of his own powers with juju ("witchcraft"). Abuje let it be known that anyone who sided with Richard would suffer the same fate, a claim that terrified some of the less-educated members of Richard's staff—as did the telltale signs of juju (usually charred ashes from some type of sacrifice) that began appearing in the museum's offices and hallways.

Richard's alarmed workers urged him to see their own "witchdoctor" at once to counter the evil spell. "I confess that I was so ill that I did not really care what medicine I took," Richard wrote about the shaman's subsequent visit to his bedside, "although, needless to say, I did not tell my own doctor of these goings on." The shaman cast protective charms in the hospital and at the museum, and left a green-colored brew for Richard to drink—all of which at least improved the morale of Richard's staff. They liked to think that their efforts also helped him recover, since a few days later Richard was released from the hospital with his blood pressure restored to an acceptable level.

But Richard's physician warned him that this was only a temporary reprieve, that his blood pressure could soar dangerously high again without warning. To keep it low, she insisted that Richard be extremely careful about his diet, particularly restricting the amount of salt and protein he ate. Reluctantly, he began an admitted "half-hearted attempt" to do so. That was his only concession to his illness. Like Louis when he had been ill with his heart troubles, Richard had no intention of acting the invalid. As soon as he was out of the hospital, he returned to his usual schedule at the museum: at his desk by 6:00 A.M., one hour's break at lunch, then to work again until 6:00 P.M. or later, if any of the many boards he sat on held a meeting. He seemed determined to show his critics that nothing—neither kidney disease, nor false accusations, nor juju—could keep him down, a posture that continued to confuse people about how ill he really was.

"Richard was by this point very, very thin, but puffy and quite yellow, and he had burst blood vessels in his eyes," said Kathy Eldon, then a volunteer at the museum. "But if you asked him how he was doing, he always said, 'Fine.' " Eldon realized that Richard was an extremely sick man only after she accompanied him in May to Nairobi's main Barclay's Bank to receive a donation for the museum's upcoming fifty-year celebration. "The bank was in an old building and there was a flight of stairs to the lobby," she recalled. "And when we got to the bottom of the stairs, Richard said, 'I think you're going to have to help me.' " Grasping Eldon with one arm and balancing on the banister with the other, Richard slowly made his way up. "But he just didn't have any strength," said Eldon, "and so we stopped repeatedly. We literally inched our way up. And when we finally got to the top, I sort of sat him down on a sofa. But I just had no idea he was that ill, because he never let on."

Yet only a few days later, when the museum's festivities were under way, Eldon spotted Richard out at daybreak, cutting bundles of grass with a hand scythe to spread over the grounds. It had rained the night before and he did not want his guests slipping on the mud. "He just would not, *would not* accept that he should slow down," said Eldon. "Anyone else would have been in bed, getting transfusions." Just as he continued to brush off inquiries about his health, Richard ignored any suggestion that he slacken his pace. "I would try to slow him down, to get him to not do things," said Meave. "But I think he felt that if he admitted to himself or anyone else that he was really sick, then that was it—he couldn't keep going. . . . He knew that if he stopped, that would be the end."

And so it was that barely three weeks after he had been out of the hospital, and although terribly weak and obviously ill, Richard flew off to Paris to begin scouting locations for a six-part BBC television series on human origins, "The Making of Mankind."

Characteristically, Richard had not told the BBC's producers about his illness, believing that if they knew, they never would agree to the series. He also thought that by making the show (and writing a companion book) he might generate enough money to provide a secure endowment for his foundation, FROM. Besides, he was sure the new drugs he was taking—he was now up to twenty-six pills a day—would sustain him long enough to complete the project. But Richard's plan to help scout the French countryside was dashed at the first prehistoric rock shelter the group visited. After scrambling down an incline of a few hundred feet, they realized they had taken the wrong trail to the cave and turned to retrace their steps—except that, just as at the Barclay's Bank, Richard did not have the strength to do so. "[T]o my utter dismay," Richard wrote in his autobiography, "I found I simply could not keep up with the others. Because of my hypertension and the pains in my legs from fluid retention, I was unable to walk more than a few yards at a time without a rest." Once again, he was forced to tell others about his illness, although he couched it in terms of a "kidney complaint," and insisted that any need for dialysis or a transplant was years away. Somewhat reluctantly, the producers agreed to stick by their filming schedule.

At the end of the scouting tour, Richard was back in Nairobi—and back in the hospital. Tests this time indicated that his kidneys were barely functioning and his doctor urged him to see a kidney specialist in England as soon as possible.

The museum and flailing prehistory institute had to be protected in his absence, and Richard's allies in the government quickly acted on his behalf. In a matter of days, the museum's board fired John Onyango-Abuje and arranged for John Karmali, one of the museum trustees, to oversee the museum during Richard's absence. Charles Njonjo then transferred the museum and prehistory institute from the Ministry of Culture to his own jurisdiction to better monitor their affairs and to continue his inquiry. Richard was assured that he would still have a job as director when he returned. Then, on July 14, Richard and Meave left for London.

As they drove to the airport that night, Richard finally gave way to his feelings—although silently, so that no one, not even Meave, would know. "Tears were quietly trickling down my face," he wrote in his autobiography, "and any attempt to say a word would have given away the fact that I was weeping. I do not know exactly what Meave was thinking but I realized that I might well never return to my beloved home or see again my children or friends."

• • •

"I think that kidney business may have been the making of Richard in an odd way," said Mary Smith. "I think Richard thought he was going to die, and that made him face at a young age what Louis had [faced] later in life: that he was going to do all these things and be famous and then he was going to be dead. Richard had to weather the idea that he wasn't going to be able to finish everything he had set out to do; that his life might be short."

In their first meeting, Richard's new physician, Dr. Anthony Wing, drove this idea home point-blank. Richard, at age thirty-five, was in end-stage renal failure, Wing said bluntly, and there were only two ways for him to stay alive: either via hemodialysis (being attached to an artificial kidney machine to cleanse his blood) or by having a kidney transplant. Richard had secretly suspected Wing's diagnosis, but hearing it spoken aloud and in front of Meave left him stunned. He had insisted to everyone before leaving Kenya that he was merely traveling to England for new tests, that he would be home in a matter of days. Wing, however, forbade any more travel. "It was an incredible blow," Richard later wrote. "I wanted to see my children, and I so wanted to see Kenya just once more!"

Wing added Richard's name to the list of patients at St. Thomas's Hospital who should be given dialysis as soon as a machine became available—although Richard eventually hoped to have a transplant. "We never thought of dialysis as a long-term solution," said Meave, "because you need electricity and sterile conditions, so Richard never could have traveled around Kenya. Going to Koobi Fora anymore would have been out of the question. And he would have hated to live that way, tied to a machine. He hasn't got the temperament for it." But a transplant carried other risks: Richard might die during the operation, or afterwards from an infection, or his body might reject the new kidney altogether. Then, too, it placed Richard—who prized his independence and thrived on control—in a very awkward, dependent position. Someone else would have to give up a kidney to give Richard back his life.

Already, all three of his brothers (Jonathan and Philip in Kenya, and his half-brother Colin in England) as well as several friends had offered him a kidney, and had sent blood samples for testing. Richard's brothers' kidneys offered the best chance for a successful transplant, and of the three, Philip's tissue proved to be the best match (which did not surprise those who knew both men; they look remarkably alike, even though Philip is fairer, with light brown hair and blue eyes). So close was Philip's tissue to Richard's that the doctors termed it an excellent match. Yet as wonderful as this news was, it presented Richard with a huge ethical dilemma: how could he accept Philip's kidney when he had not spoken to his younger brother for more than a decade? In fact, Richard had gone out of his way to shun Philip, even pointedly ignoring him in public. "I can remember us saying, 'Oh, we hope it's Jonny's kidney and not Philip's,' " said Meave. "It would have been much, much easier in terms of the relationship."

Four years apart in age, Richard and Philip had been at odds with each other since Philip's late teenage years. Like Richard, Philip had dropped out of high school, but instead of pursuing a career, he had decided, he said, "to experiment." In 1964, a few months shy of his sixteenth birthday, he had moved from his parents' home to the Kitengale, a vast open plain south of Nairobi, and built a hut overlooking the Athi River, where his closest neighbors were nomadic Maasai. He filed a mining claim there, then left to roam the country in search of precious gems and minerals, living with the local peoples wherever he went and picking up their language and culture.

"I rushed around, all over East Africa," said Philip. "And I made money, and I didn't make money, and I made money again. And when I made it, I spent it. And I lived some of the most invigorating years of my life." Where, at the same age, Richard had been serious, anxious to be seen as an adult and given adult responsibilities, Philip was wild and carefree. His dimpled, roguish smile seemed to charm everyone, and he had an uncanny ability for blending with whoever was at hand. "That one is like a chameleon: he can be anything he wants," said Teresia Ng'anga, Richard's secretary at the museum. "When he is with the Kikuyu, he is just like the Kikuyu. When he is with the Maasai, he is a Maasai. Whatever he needs to be, he can be; he can change just like that," she said, snapping her fingers. Philip tried his hand at anything, too: prospector, exotic plant collector, location scout for Hollywood, safari guide, and (for a few weeks for his father) hominid hunter. ("But I decided not to go into that field; it's too small," he said later.)[1]

But Richard disapproved. He thought his brother "unprincipled and irresponsible," and told him so on several occasions—not that Richard's words made any difference. Reveling in his far-flung adventures, Philip merely laughed at his elder brother's tiresome finger-wagging and went on having fun.

In 1969 the rift between Richard and Philip had widened because of a dispute over the Kitengale where Philip lived and which Richard (and Louis, who was still alive) wanted to see added to Nairobi National Park. When Philip refused to move, Richard broke off all contact with his brother, as well as his wife and children. "[F]or more than a decade, none of them had been invited into my house," Richard wrote in his autobiography.

But now that his brother had unhesitatingly offered him a kidney, Richard agonized over his past behavior. It seemed immensely selfish and venal on his part to use Philip "just because," as Richard later wrote, "I wanted to

---

[1] Among other discoveries, Philip (together with Peter Jones) found the first footprint trail at Laetoli, plus postcranial bones there. He also led Louis's 1971 expedition to the Suguta valley.

live." Then, too, Richard really did not wish to apologize for his previous actions. "I had my pride," he noted in his autobiography. Being indebted to Philip—and for the gift of life—was something Richard had never imagined happening; he, after all, had always been the one bailing out Philip. With the tables turned, Richard struggled to find a way to accept what Philip had so generously offered. For days, Richard worried over his decision, talking at length with Meave, their friends, and several kidney specialists, and reading about the experiences of other kidney recipients. Finally, Richard wrote a letter to Philip, laying out all of his concerns, and asking if "in the cold light of day" Philip was still willing to give up a kidney for him. Philip's reply was an unequivocal yes; he had made the offer and he stood by it. Later, Philip would tell a reporter from *The Nairobi Times* that he had only done what was natural. "He is my brother," Philip had said, "and we are very close. If the situation had been reversed and it was I who had needed the transplant, I am sure he would have done the same for me."

When he made his offer, Philip was thirty and just embarking on a new career as a politician—a move that everyone in his family applauded since he seemed a natural at it. He had tried running for a seat in Kenya's Parliament four years before (making him the first white Kenyan to do so since Kenya's independence in 1963), and had lost by only a small margin. He was now trying again, campaigning for a hotly contested seat in one of Nairobi's predominantly black districts. Since it appeared that he had a good shot at winning, Philip hoped that the operation could wait until after the elections—although the election date itself had not been scheduled. "They [the Kenyan government] kept setting an election date and changing it," Meave recalled, "so that we had no idea when Philip would be coming. So it was a long wait, longer than we ever imagined."

Meanwhile, Richard's condition continued to deteriorate. Since he was not a British citizen, which meant he was low on the waiting list for dialysis machines, and since he continued to insist that he was fine, Richard had not yet been given dialysis. "I think one of the reasons the doctors waited so long to put Richard on dialysis," said Meave, "was because he had this great show of appearing well. He'd go see his doctors and they'd ask, 'How are you?' He'd say, 'Oh, fine.' And I knew he could hardly stand up, that he was making this huge effort to appear well all the time. I'd say, 'Why don't you tell the doctor exactly what you feel?' And he'd say, 'No way!' "

But in a friend's London flat where Richard and Meave were staying and where there were no doctors to impress, Richard did little more than lie propped up on a settee, floating in and out of a drugged sleep. He was now passing so little urine that if he did not sleep in an upright position, his lungs filled with fluid and he began choking violently. Once a day, he dragged his thin, poisoned body outside to take a prescribed daily walk to a park bench fifty yards from the apartment—a stroll that always loomed before him as a "major undertaking." "I can remember that he could just make it to the bench," said Meave, "and that was it; then he had to sit. And you're terribly cold when you have uremia. So he would be sitting on the bench in his sheepskin coat on really hot summer days, and shivering." Usually Meave accompanied him on these excursions, but sometimes he went alone. After one of his unaccompanied walks, Richard managed to drag himself back to the apartment building, ride the elevator to their floor, and stagger to the door of the flat. But he did not have the strength to turn the key in the lock. "Meave found him sitting outside the apartment, in tears," said Alan Walker. "He'd got the key in the door, but he couldn't turn it. That's how sick he was."

Finally, at the very end of August, Richard was given his first dialysis treatment, and virtually overnight began to improve. "I spent ten hours on the machine that first time," Richard wrote in his autobiography. "[A]s far as I was concerned it could have been for longer because for the first time in months my uraemia was gone. That terrible taste in my mouth was practically forgotten and there was the beginning of a new warmth in my body." After several such sessions, Richard felt so much better that he wrote exuberantly to his mother (who was then excavating the hominid footprint trail at Laetoli), "It [dialysis] . . . has really put me back on my feet."

By early September, Richard and Meave had moved into an apartment of their own, and brought their daughters, Louise and Samira, from Kenya. Having his family with him and benefiting from the dialysis treatments gave Richard the semblance of a normal life. Three days a week he underwent six hours of dialysis at the hospital, using the time to catch up on his correspondence and to write the story of his life for his children (this was later published as his autobiography, *One Life*). If he did not survive the transplant operation, his daughters would at least have these memoirs. Yet Richard seldom dwelled on this possibility. The Kenyan government had finally scheduled the elections for November 8, and Philip planned to fly to England a few weeks later.

Mary would arrive then, too. All summer long, as she and her team excavated the footprint trail, Mary had quietly worried about Richard and Philip, confiding her fears only in letters to close friends. "I am told [by Richard and Philip] 'not to worry' etc.," she had written Allen O'Brien at one point, "but one cannot overcome such inborne feelings as one has for one's family."

In early November, Mary had returned to Nairobi from Laetoli—just in time to watch Philip defeat ten other contenders (including the incumbent) for a parliamentary seat. He was now Kenya's first white member of Parliament, a success that was heralded in newspapers worldwide. But to his friends' and supporters' amazement, instead of staying in Kenya to bask in the glow, three weeks later Philip flew to England with his wife, Valerie. Like his brother and their mother, he had chosen to say little about the pending operation to anyone, believing, as he later told a reporter for *The Nairobi Times,* that "it was a family affair."

As soon as they arrived in London, Philip telephoned Richard, and Richard invited Philip and Valerie to lunch. "It was the first meal the four of us had had together—ever," said Meave. "So it was rather tense." Shuffling ten years of hard feelings under the carpet left everyone ill at ease. "I think we were all conscious of this," Richard wrote later, "and looking back on it, it is hardly surprising that we were somewhat stiff." Neither brother saw any value in airing past grievances, however, and chose instead to focus on the details of the operation. The transplant was to take place in four days' time, on November 29, after Philip had undergone a series of tests. But first, Philip calmly informed Richard, he needed a bodyguard because since winning the election, there had been numerous threats against his life; further, he had good information that an assassin would try to kill him while he was in the hospital. And, Philip added, since he and Richard looked so much alike, the assassin might mistakenly kill Richard instead. This unexpected news, of course, did not make Richard happy, but since Philip was adamant (the operation would not go forward, he said, without police protection), Richard began calling "a number of good friends" in London, and bodyguards were arranged.

On the afternoon of November 29, Richard and Philip were wheeled into the anesthesia room at St. Thomas's Hospital. Just before the nurses rolled Philip away, Richard asked him once more if he still wanted to go through with the operation. His brother's reply was "short, blunt and unprintable," Richard would later write, but it made all of the nurses laugh. Richard laughed, too, and then the anesthesia took hold.

The operation lasted a little more than three hours, while Meave, Valerie, and Mary sat outside, waiting anxiously together. "It was hard for Valerie," Meave said, "because the operation was actually far worse for Philip than for Richard. They had to cut him [Philip] all the way around one side, to get the kidney and all its connecting tubes out. While for Richard, they just made a little cut and poked the kidney in, then joined it up with the arteries in his leg."

Some days after the operation, and still in great pain, Philip croaked to his visitors, "God, I must have been ill when I said I was going to do this." But by then everyone was telling jokes laced with black humor, the favorite being that Richard could no longer say that he hated his brother's guts, since Philip's kidney was now at work in Richard's body. Richard had known this as soon as he awoke from the anesthesia. Hanging above his bed was a bottle containing a slightly blood-stained liquid that Richard recognized as "my blood mingled with my urine—my own urine." It was the first time he had managed this simple act in three months' time, and the sight filled him with joy. "Although I knew there could be problems ahead," Richard wrote in his autobiography, "at least I was alive, and the transplanted organ had begun to function. I felt that one life was over and I now faced a completely new one—I had just been reborn."

• • •

Richard stayed in the hospital for only thirteen days after receiving Philip's kidney. The grafted organ was working perfectly, although Richard's doctors warned him that his body could still reject it. Yet the days went by and his kidney function grew stronger. "I began to be more and more optimistic that I had one of those rare but possible 'perfect' transplants," he wrote in his autobiography. Good health seemed at last within his reach, and he and Meave began making plans for their return to Kenya.

But less than a month later, on Christmas Eve 1979, and only five days after celebrating his thirty-sixth birthday, Richard's old symptoms returned: the puffiness, headaches, and fatigue. Most alarming of all, he was barely passing any urine. Two days later, he was back in the hospital for tests that proved disheartening: his immune system had turned against the kidney, attacking it as if it were a foreign object that needed to be expelled. In response, Richard's new kidney had stopped working and, Dr. Wing said, might be permanently damaged. "The shock was unbelievable," Richard later wrote, after hearing this news. It seemed impossible that he had endured all of the summer's agony only to find himself once again without a kidney.

To try to save Richard's new organ, Dr. Wing ordered massive doses of additional immunosuppressants—drugs that completely shut down Richard's immune system, forcing it to stop its attack. And gradually, over the next few days, the kidney again began to function; it had not been damaged after all. But Richard could only keep it alive, Dr. Wing explained, by continuing to take the stronger immunosuppressants, drugs that essentially left him without an immune system and so vulnerable to every virus and disease. Just how vulnerable quickly became apparent. Discharged from the hospital on December 29, Richard was back again on New Year's Day, fighting off a viral infection. This time, he was bedridden in the hospital for ten days, and he was released only after he begged to be sent home.

The strain of Richard's ill health began wearing on his family. Just before the transplant operation, all of five-year-old Samira's hair fell out in great handfuls. And when Richard returned from his latest hospital stay, her elder sister, Louise, announced, "You know, Mummy, if Daddy has to go back to the hospital once more, I'm just going home." "It was as if she were actually saying, 'I just can't take it anymore,' " said Meave.

On the evening of January 20, ten days after his last release from the hospital, Richard took another turn for the worse. "He'd been weak, but he suddenly grew weaker," Meave recalled. Complicating matters, Richard had invited to dinner Anthony Marshall, a former U.S. ambassador to Kenya and then the acting director of Richard's foundation, FROM. It was more than a social call: Marshall, Richard knew, intended to question him about charges that he was misappropriating money from his BBC television series on human evolution.

Richard had first conceived the series and companion book, *The Making of Mankind,* in 1978, with the idea that the money from both would provide a secure endowment for FROM—a plan he had announced publicly. But in the United States, gossip had spread that he was actually intending to pocket the proceeds. When the story reached the ears of Anthony Marshall, who was struggling to keep FROM afloat, he became alarmed. "People were attacking him [Richard] when he was down," Marshall recalled, "and I didn't like that." Anxious to get the facts "so I could protect Richard," Marshall flew to London from New York City to question Richard in person.

Marshall, of course, knew about Richard's transplant, but he had no idea that Richard was still ill—since, true to form, Richard had never mentioned any of his recent viral bouts or hasty return trips to the hospital. Only when Richard and Meave greeted Marshall at their flat did Marshall realize that Richard was not at all well. "I'd never have gone over if I'd realized how sick he was," Marshall said later, "but once I found myself there, it would have been just as bad not to have talked to him because Richard would have been concerned. And then we talked too long and too much. . . . And I'm sure it didn't do him any good. But it did give me ammunition to put down his critics."

Richard, however, interpreted the lengthy interrogation as a loss of faith in him, and this proved devastating. "After Tony left, Richard couldn't get up the stairs to bed," said Meave. "Then I was really worried, but he wouldn't let me phone the doctor. He said that I was overreacting, that I was making a fuss." Only with Meave's support was Richard able to climb the stairs and get into bed. She then lay beside him listening as his breathing grew more and more labored, wheezing and slow, and then turned to short, sharp pants. Once or twice Meave suggested they call the doctor, but each time, Richard rebuffed her. "Finally at three in the morning, I said, 'Richard, we've *got* to do something, or you're not going to make it,' " Meave recalled. "He could barely breathe by then." And this time, albeit reluctantly, Richard agreed.

At once, Meave summoned an ambulance. Through London's dark, early-morning streets, Richard was once again whisked to the hospital, where the doctors diagnosed pneumonia and pleurisy, a painful and often deadly inflammation of the lung membrane. To fight the dual infections, the doctors prescribed a variety of antibiotics and placed Richard in an oxygen tent. Dr. Wing then turned to Meave. "You'd better stay the night," he told her, "because if these antibiotics don't work, that's it." She took a seat by Richard's bedside and through the rest of the day and all that night talked to her husband—imploring, urging, and cajoling him to fight, and finally berating him for giving up.

"That was the worst," Meave said. "That was the only time I saw him give up. He really did because he couldn't breathe. He hated the oxygen and I had to keep trying to make him use it. And I could see he had given up and that's when you really have to push a person to keep going. But I wasn't having him die at that point, after *all* we'd been through. And we'd just been told we could go home at the end of February, and that was only three weeks away; in three short weeks, we could be gone, and all was going to be well. And here he was, dying on me! It was bad, it really was."

Richard knew he was dying, too. "Pleurisy is very, very painful," he said later. "Every breath is an agony." Dying seemed easier, and at one point in that long night, he nearly did. "I had an out-of-body experience," he recalled. "I was suddenly floating above my body and looking down on it—and there I was stretched on a gurney with doctors and nurses bustling about. And I thought, 'I wonder why they're all so worried, what the fuss is.' I think I was considering whether I should stay or go." Meave's words brought him back. "She was awful to me," Richard said, "she wouldn't stop; she wouldn't let me rest." Over and over, her words came: "Breathe Richard; don't stop Richard; Richard, listen to me. Breathe!" And so through the night, Richard fought for every breath, while Meave, her fists clenched, her mouth inches from his ear, insisted that he take the next.

At dawn, Richard was still alive. The crisis had passed and the doctors thought he would survive. His new kidney, too, was still functioning, and slowly, once again, Richard began to heal. Later, he would write in his autobiography about the "magnificent effort by doctors and nurses," who had toiled through the night for him. "But I should say," he added, "that had it not been for Meave who gave me the strength and hope to fight, I would not have lived."

That was Richard's final bout; he did not again suffer through a rejection episode or a viral infection. On February 23, 1980, although still gaunt and weak from his long ordeal, Richard flew home to Kenya with Meave, Louise, and Samira.

Chapter 40

"HOW VERY HUMAN"

Eight months had passed since Richard had last seen Kenya, and the sight of its red earth and green hills filled him with joy. He was healthy and he was home, and as he wrote to Kay Behrensmeyer, "[I]t certainly feels good!" After all that Richard had endured, probably no one would have denied him an extra few months of rest. But Richard was eager to show that he not only felt well but was as vigorous as before. Although he told a reporter for *The Nairobi Times* that he would now "probably liv[e] at a slower pace" than he had in the past, Richard was back to work almost as soon as the plane touched down in Nairobi—and at as fast a pace as ever.

Not only did Richard immediately resume the directorship of the National Museums of Kenya, but he began laying plans for a new exploratory expedition to desert regions north and west of Lake Turkana; perhaps in these sediments he would find fossilized hominid remains older than those he had unearthed at Koobi Fora. He launched a foundation to raise funds for kidney patients in Kenya, too, and began untangling the affairs at TILLMIAP following the unsuccessful effort to depose him. The BBC television series on human evolution needed Richard's attention as well, since, as the series' host, he had to be filmed on location at important sites overseas. And so he mapped out a schedule that would take him abroad (to the Middle East, South America, China, and Europe) five times between May and November 1980.

Not long after returning to Nairobi, Richard decided to close the failing TILLMIAP and to merge it with Kenya's National Museums. Within Kenya, Richard's swift action was deemed wise and sensible. "There was no alternative," said John Karmali, who had overseen the museum during Richard's absence. "We had to destroy this idea that the institute was (Bethwell) Ogot's private empire." Outside Kenya, however, the move was viewed with alarm. When Richard closed the institute, "It seemed as if [he] was saying, 'I'm having it all,' " said Andrew Hill, a paleoanthropologist who had served as TILLMIAP's acting director.

Richard's decision also offended the countries, organizations (such as the Leakey Foundation), and individual donors whose contributions had helped build the institute. But Richard defended his action in much the same way that John Karmali did—that it was what was best for Kenya and Kenya's National Museums. Hill, however, thought that Richard's decision also involved Richard's ego, that it was in part a maneuver to demonstrate that despite his long absence he was still in charge.

Whether right or wrong, or good for Kenya, one thing was certain about Richard's decision: it further eroded his reputation overseas, particularly in the United States. Now not only were his critics claiming that he had misappropriated funds from the BBC television series, but that he had defrauded those who had helped build the memorial to his father. Others even suggested that his maneuver was simply a racist power play designed to keep black Kenyans out of paleoanthropology. "Richard's move offended a lot of people," said David Pilbeam, who had been TILLMIAP's acting scientific director. "I'm absolutely convinced there was wide-scale [financial] funny business going on at the Institute [when under Ogot's directorship]. It was an absolute can of worms. . . . But Richard's decision was also shortsighted; it was a major strategic mistake that cost him a lot of goodwill."

TILLMIAP and the gossip aside, Richard was even more saddened and depressed to discover, when he was ill, that many of the scientists he had considered friends were not friends at all. Instead, they seemed to have carefully hedged their bets—and their careers—as to whether Richard lived or died. "All of these rumors that were being spread when I was so sick—people who knew me could have spoken out, said these things weren't true, and stopped this nonsense, if they'd wanted. But many did nothing," said Richard. "They were waiting, I suppose, to see how things turned out; to see if I made it. They didn't want to risk offending Don Johanson, who, if I died, might have been the only one left with any fossils. That's when I really found out who my true friends were."

Curiously, perhaps because of Richard's strengths as a leader, people found it easy to confide in him, and to seek his advice. What they seldom got in return, however, were confidences of his own. Other people might need him, but Richard did not need them. "He had a sort of arrogance, a confidence about him that he didn't need friends," said the anatomist Alan Walker. "But when he was sick, he was weak and vulnerable, and he discovered that he couldn't go it alone; he needed support." It was only then that Richard discovered that at the heart of many of his "friendships" lay nothing more than the quid pro quo of a business transaction.

Richard's unhappiness with his fellow scientists grew by bounds when, in the fall of 1980, he opened an advance copy of Don Johanson's book *Lucy: The Beginnings of Humankind* (coauthored with Maitland Edey). It was largely an account of Johanson's discovery of the small, bipedal hominid from Ethiopia, and Johanson's efforts to persuade his colleagues about his and White's interpretation of the fossils. In *Lucy,* Johanson presented himself as the underdog, struggling (with Tim White's help) to overturn what he portrayed as the antiquated ideas of his chief rivals, the Leakeys—essentially the same plot line that had already emerged in the press about the differences of opinion between Johanson and White and the Leakeys. By itself, this did not surprise Richard. But what upset him (and Mary) was what Richard viewed as Johanson's disregard for or distortion of basic facts about the Leakey family. Richard wrote to Johanson, Edey, and their editors, as did Mary, but none of the mistakes was ever corrected.[1]

If *Lucy* further depressed Richard, it also opened his eyes to the true nature of his relationship with Don Johanson. Before reading this book, Richard had thought their rivalry was largely an intellectual one, a sparring over the significance and interpretation of hominid fossils. The person with the personal animosity toward him and Mary was not Johanson, Richard had previously believed, but Tim White. Many people knew that White lashed out against Richard and Mary to anyone who would listen. (Students in his physical anthropology course at UC Berkeley called it "Anti-Leakey 105.") "Tim has just been psychopathic on the Leakeys," said Mary Smith, who had heard his grievances more than once. "And on the subject of Mary—well, he runs her down so badly that she just couldn't . . . *nobody* could be that bad."

Richard and Mary were well aware of the stories White circulated about them. What they had not realized was that Johanson shared his antipathy. After all, Johanson had written chatty, friendly letters to them both throughout 1980. He was concerned about Richard's illness; asked for the Leakeys' advice about resuming his research in Ethiopia (the country's civil war had kept him out since 1977); thanked Richard for his offer to have Kamoya and other members of the Kenyan Hominid Gang join Johanson's team in Ethiopia; and thanked Richard again for his help in securing a new grant from the National Science Foundation. But all of Johanson's words struck Richard as hollow after reading *Lucy.*

"That's when I realized there was a personal side to our disagreement [about the fossils' interpretation]," said Richard. "Because *Lucy* is a cheap, journalistic slap at Mary and me."

---

[1] Among the errors Richard cited: Louis did not die of a stroke, but a heart attack; Mary did not irresponsibly leave one of her finds, Hominid 16, exposed, so that it was trampled in the night by a herd of cattle—it had been trampled years before it was ever discovered; Louis did not utter the words "It's nothing but a goddamned australopithecine" when he first saw *Zinj.* ("He was a missionary's son," said Richard, "and never swore—and, if he had, certainly would never have used an American swear word.")

Although some reviewers found the book too personal (the review in *Science* was titled "Paleoanthropology without Inhibitions" and complained that Johanson had relied on "hurtful hearsay"), with its good yarns and gossip, *Lucy* climbed the best-seller charts. Increasingly, Richard found himself on the defensive. "I am fine although I confess to feeling rather 'got at' at present with so many nasty things being printed in America about how dreadful I am," Richard confided to a friend in March 1981.

Once again, newspapers and magazines were reveling in the melodrama of paleoanthropology. "Bones and Prima Donnas," "Battle of the Bones," and "Battle of the Top Bananas," ran the headlines. Richard tried to stay above the fray, insisting, as he had in 1979, when the media first suggested that he and Johanson were quarreling, that their intellectual differences were minor. "We're not talking about the future of Kenya or the U.S.," he said with some exasperation to a reporter from *Newsweek*. "We're talking about old bones. They're not that important."

Few reporters bought Richard's line. For one who had so determinedly pushed to acquire his own pile of hominid fossils, it seemed out of character. Yet Richard's words were more sincere than they might have seemed, given his recent life-and-death struggle, which many reporters were apparently unaware of. He was no less ambitious, only more judicious about how he would spend his energy. "The one thing I think has perhaps come solidly home to me as a result of looking back on my life . . . when I was very ill and trying to assess whether it was worthwhile, is that many of the things that one got very excited about, that one committed onself to with a great deal of fuss and time, probably were not so important as one thought," Richard had told a reporter for *The Nairobi Times* not long after he returned home from his transplant operation. But his sentiments were to be tested in a way that Richard could not have anticipated.

Soon after arriving in New York in April 1981 for a meeting with the board of FROM, Richard was asked to appear on "Cronkite's Universe," a television science series hosted by Walter Cronkite, the former CBS News anchorman. When Richard heard that Don Johanson had also been invited, he declined. "Don had been after me for some time to have a public debate with him about *afarensis,*" he recalled. "I didn't want any part of that; after all, what was the point? He had his views and I had mine—but it was silly to debate these with so little [fossil] material to go on." Richard had just published *The Making of Mankind,* however, and his publisher thought the show would help sell the book. Richard was told that Cronkite's aim was to discuss human evolution and "creationism." An ardent anticreationist, Richard agreed to go on. Excusing himself from his board meeting, he caught a cab to the American Museum of Natural History, where the program was to be filmed—and where Johanson and Cronkite were already waiting.

Richard and Johanson shook hands and spoke briefly. "I asked him if he knew what Cronkite was planning," Richard said, "and Don said no. Then I said, 'Listen, I don't want this to turn into a debate about Lucy. We should use this opportunity to show the errors of the creationists.'" "I don't know what's going to happen," Johanson replied. "I suppose it's up to Cronkite." Johanson's vague reply left Richard feeling uneasy. "I kept thinking this is not something I want to be a part of . . . I should just leave."

Johanson was eager for a debate; he had been hungry for a public confrontation with Richard and Mary ever since publishing his 1979 *Science* article that had proposed *A. afarensis* as the pivotal ancestor of humans. "Our announcement of *Australopithecus afarensis* fell into the category of the startling," Johanson had written in *Lucy*. "It discovered and defined a new hominid, and it redesigned the human family tree." Thus, he continued, he found Richard's refusal "to review [the fossils] . . . feature by feature" with him to be "disappointing." (The anatomist Alan Walker, however, regarded Richard's decision as wise. "After all, Richard had not studied their [Johanson's and White's] fossils in detail, whereas Don could cite all sorts of figures and details. He would use taxonomic terms, discuss systematics, and yes, it would be embarrassing for Richard because that's not Richard's business. He's an administrator, an organizer and he leaves the fine interpretation of fossils to others.")

In fact, Richard had not been completely silent about Johanson and White's hypothesis that *A. afarensis* was the progenitor of *Homo*. He and Walker had written a short rebuttal letter to *Science* (published in March 1980) attacking the Americans on phylogenetic grounds. Richard and Walker questioned whether a single hominid had given rise to all later hominids 3.5 million years ago. They also suggested that certain fossils from Koobi Fora and Olduvai (such as 1813 and OH 13) might represent another type of small australopithecine that had coexisted with *Homo habilis* and *A. robustus* about two million years ago. If three humanlike creatures were alive at that time, then it seemed likely to Richard and Walker that Lucy *(A. afarensis)* had not been the sole hominid striding across the African plains less than a million years earlier. Johanson and White's was "much too simple a pattern for my taste," Richard later wrote.

But Richard saw little point in debating the matter. "[We] can argue about this material and call it one thing or another until we're blue in the face. But until more fossils between three and four million years are found, there's going to be plenty of room for disagreement."

Mary, together with several coauthors, had also attacked *A. afarensis* in a letter to *Science,* but simply on the grounds that Johanson and White had violated the rules of scientific nomenclature. Aside from these two letters, Richard and Mary had been silent on *A. afarensis*. Johanson, it seemed, found their silence frustrating—it was the dismissive flick, and it apparently irritated him. He had even written in *Lucy,* "I greatly regret that the Leakeys so far have not addressed themselves directly to what is emerging as the most interesting paleoanthropological issue of the 1980s. Perhaps they will in due course."

So, of course, Johanson must have hoped that Cronkite would steer Richard into a debate about human origins. "Richard had been claiming that the 'rivalry' between us was a myth, largely the invention of the press. I thought that was misleading, and I welcomed the opportunity to meet with Richard on the record," Johanson wrote in his 1989 book, *Lucy's Child.* Johanson had been invited to the show several days before Richard, and arrived with props: a gorilla skull, a cast of an *A. afarensis* skull, and a chart divided into halves. On one side was an illustration depicting his (and White's) Y-shaped version of the human family tree with *A. afarensis* at the stem; the other half was left blank.

After Richard and Johanson exchanged greetings, Cronkite invited them to join him at a small table. Johanson placed the *afarensis* skull on the table and set his chart beside his chair. Richard sat down empty-handed. And the cameras began to roll. "We brought Donald Johanson and Richard Leakey together in the American Museum of Natural History to discuss their different ideas of man's ancestry," Cronkite began. Richard said later, "I realized at once that I'd been set up; it was going to be a debate. And I was furious with myself for getting trapped like this."

After some introductory remarks, Cronkite then invited Johanson, who was only too ready to oblige, to explain his interpretation of human origins. "There has been a controversy that has been going on now for nearly three years between Richard and myself, and it specifically focuses on the family tree," Johanson said. He then explained that he had presented his hypothesis and supporting evidence in 1979, but that Richard had not accepted it. Slumped in his chair, fuming inwardly, Richard stumbled for an answer.

> I've heard it all before. I think it is marvelous what you've done, Don. I just don't agree . . . I am not . . . I'm not willing to discuss specifics of why I think a bone means this or doesn't mean this. I've been around thirty-five years in a family that has seen lots of controversy. I've seen fossils in favor, out of favor, back in favor. Let us be . . . Let us stand back from it. Of course, it is important, Don. I wouldn't minimize it. But I'm not going to say whether you are right or wrong.

Richard paused briefly then, tossed his head back and gave a short laugh. "But I think you're wrong."

Johanson tried another tack. "I've brought along a sort of portrayal of how I look at the family tree," he said, producing the chart with its neatly drawn diagram depicting *A. afarensis* as the common ancestor of humans and later australopithecines. "And I've left a spot for . . ." "For me!" Richard said, completing Johanson's sentence with mock enthusiasm. The camera focused on the blank half of the chart. "No, I don't think we can do this," Richard said. "I haven't got crayons. I haven't got cutouts. I'm not an artist." But Johanson was prepared and produced a red marker. For a moment, Richard sat silent, looking blankly at the chart and marker. "I felt such a bloody fool. I'd walked right into the trap," he said later. But when Johanson began to speak for him, to explain how the Leakey family would draw the tree, Richard interrupted and asked him to hold one edge of the chart.

"I think in probability I would do that," Richard said, marking out Johanson's drawing with a thick red X—a move that Johanson admitted took him "aback." "And what would you draw in its place?" Johanson asked, looking shaken. "A question mark!" Richard replied, and then boldly drew one in the white space. Johanson quickly dropped the chart out of sight behind his chair.

Johanson tried once more to portray Richard as holding to a particular view of human origins, while Richard continued to say that he and Don and other paleoanthropologists needed to find more fossils before anyone's interpretation could be proved. "I would love to prove him right," Richard said at the end. "But I might just prove him wrong."

Richard then hurriedly left the set. "Richard was so mad he could have bitten nails," said Milford Wolpoff, who happened by chance to be doing research at the American Museum that day and had watched the whole show. "There was Don, trying to make a fool out of him, saying 'Look at this man. He's considered a spokesperson for anthropology, but he doesn't know as much as me.' And Richard was trying to get out of it in as gentlemanly a way as possible—which I think he did. But he had every reason to be angry: he was told it was not going to be a debate, and then it was." Richard, however, was most angry with himself: against his better judgment he had let himself be talked into a situation over which he had no control. He vowed that he would not meet again with Don Johanson.

Shortly after the Cronkite show, the National Geographic Society—the Leakeys' long and trusted supporter—refused Richard's request for funds for Koobi Fora and for new explorations north and west of Lake Turkana. It was a stinging rejection, one that had "no good explanations," Richard told Glynn Isaac, aside from the bad publicity generated by Johanson's book.

Perhaps *Lucy* did have an effect, but Richard had also alienated several members of the research committee. "He had a cavalier attitude," explained Mary Smith. "About once a year, he'd come in late with his grant request, sort of toss it in front of the committee, then make sure everybody knew he'd spent the weekend with the Grosvenors [Gilbert Grosvenor, the president of the National Geographic], and more or less say, 'Fill 'er up.' It put people's noses out of joint." This time there had also been a misunderstanding about whether Richard should appear before the research committee in person; when he did not, his grant was denied.

The National Geographic's rejection was a bitter blow. "It was after that when I seriously began to consider leaving paleoanthropology for good," Richard said later. "I was very depressed about the whole business." As he had during other paleoanthropological crises, Richard turned to Glynn Isaac for advice. Although Isaac had also grown somewhat cynical about the science, he counseled Richard to stick with the field, to not abandon it at this low point. "If I'd left then," Richard explained, "I'd have been quitting when I was at the bottom. It's much better to do that when you're at the top."

For his first expedition after his kidney transplant, in 1981, Richard decided to explore a region slightly northeast of Lake Turkana called Buluk. Near the Ethiopian border, Buluk's fossil-rich sediments had been discovered in 1973 by Ron Watkins, a geologist and a member of the Koobi Fora project. Watkins and the paleontologist John Harris had returned to Buluk in 1975 with members of Kamoya Kimeu's Hominid Gang—and "within ten minutes of arriving" had struck pay dirt, said Watkins. He, Harris, and one of Kamoya's men almost simultaneously spotted the fossilized jaw fragments and a limb bone of a monkey. Based on the geology and other fossil animal bones, Watkins and Harris knew this monkey must date to Miocene times, at least fifteen million years ago. They knew, too, that often where such fossils are found, other primates—ones more closely related to humans—are likely to be discovered as well.

Richard had not spent much time exploring Miocene deposits since he was a child. Four years after he was born, Mary had discovered on Lake Victoria's Rusinga Island the tiny and nearly complete skull of the eighteen-million-year-old ape, *Proconsul africanus* (now called *P. heseloni*)—a find that had helped renew Louis's career. Altogether, Louis and Mary had uncovered several genera and species of Miocene apes, one of which, Louis had speculated, had given rise to the later great apes and humans.

"The Miocene was my father's main interest when I started in this business," Richard explained. "All the key Miocene sites—Fort Ternan, Rusinga, Songhor, Mfangano—were his." As a way of distancing himself from his father, Richard had chosen to focus on the more recent past. But the primate fossils from Buluk intrigued him. Perhaps the site held the remains of Miocene apes as well. Richard invited the anatomist Alan Walker, who had published many articles about the Miocene monkeys and apes, to return to Buluk with him in 1981 to see if additional hominoid fossils might be found.

Richard's failure to have his National Geographic grant renewed had dashed those plans. But not for long, Richard hoped. "[P]erhaps [we will go in] January/February [1982] or the summer," he told Walker. "I apologise for this change but it would seem that I am being tested!" But eighteen months would pass before Richard raised sufficient funds on his lecture tours for this project and to explore the sediments on the west side of Lake Turkana. (Turned down by the National Geographic Society for a grant, he never again applied to the Society, although it has helped support the West Turkana expeditions through funds granted to Alan Walker and Meave Leakey.)

Since he had never been to Buluk, Richard flew up for a quick visit in

January 1983, during a family holiday at Koobi Fora. "I wanted to find out where the road and water were, and decide where to put a camp," he explained. Kamoya and his Hominid Gang met Richard and his family at Buluk, and after sorting out the camp's logistics, the group headed into the sediments to explore. As they always did when out in the field, everyone kept their eyes open, just on the chance that someone might spot something —a tactic that paid off when Meave called out that she had found the jawbone of a hominoid (quadrupedal ape). Instantly, everyone turned back and clustered at her side. "We'd been walking single file down this footpath," said Richard, "and Meave was the last in line. That's when she spotted this big [hominoid] jaw and picked it up. I had certainly walked past it and so had everyone else."

The jawbone—half of the mandible with canines in place—was colored a rich brown, and had been lying on the surface among stones of a similar dark hue; a glint from a bit of tooth enamel had caught Meave's eye. Meave's find, together with Watkins and Harris's monkey fossils, suggested that Buluk was particularly rich in these fossilized primates, and that the upcoming summer expedition might be productive.

Since Buluk was a small site with a limited number of fossil exposures, Richard decided to keep the size of this expedition small; besides, he no longer wanted to organize and be responsible for a huge team of scientists in the field. Thus, when he headed back to Buluk in September, Richard was accompanied only by his family, Walker, and Kamoya's hominid hunters. "It was a nice, small group, so everybody got on well," said Richard. Yet Buluk itself was wild and remote enough, with a hint of danger from armed nomadic tribes, to evoke some of the adventure and excitement Richard had encountered in his first years at Turkana. And there were fossils. In Buluk's deposits, the team picked up everything from extinct species of giant hyraxes to previously unknown species of elephantlike animals, large bearlike carnivores, strange antelopes, and huge crocodiles. "It was terribly exciting," Richard said. "Everything we found was new—or at least new to me, since I hadn't really looked at Miocene fauna—and you could envision this entirely new world you'd never been to before."

Best of all, the team found additional fragments of Meave's hominoid jawbone during a short excavation—although they did not find any other fossilized primates. The new pieces provided enough material to give Richard and Walker a better idea of what this ape had looked like. Instead of being similar to Louis and Mary's Victoria Basin hominoids, this creature was oddly reminiscent of *Sivapithecus,* an extinct ape previously known only from Pakistan. Dating to between nine and thirteen million years, the Asian *Sivapithecus* had often been suggested as the distant forebear of the great apes and humans. (Subsequent discoveries in Pakistan, however, have shown that it is actually an ancestor of the orangutan.) The Buluk fossil was even older: seventeen million years. Richard always enjoyed pointing to the importance of Africa in human evolution, and he now used this new fossil to reassert Africa's claim as the center of hominid origins. "[W]e view [the Buluk fossil] as an early species of a genus from which both African apes and hominids could be derived," he and Walker wrote in an article about their find in *Nature* in 1985.

Ironically, however, the African *Sivapithecus* would last only two years before Richard's team found new evidence that disproved it—and threw the entire history of the Miocene apes into turmoil. Still, it was that expedition to Buluk, Richard would say later, that "had relit [his] fire" for paleoanthropology. Now, besides searching for fossils in the three- to four-million-year range that might prove that Johanson's version of hominid origins was too simple, Richard also planned to investigate the Miocene. "Once I started looking into what had been done [research in Kenya's Miocene], I found it to be a mess," he said, the year after his team discovered the Buluk fossil. "Everything is very patchy—partly because Louis and others didn't record things properly. There's no proper faunal record, and there are bits and pieces of 'hominoids' that no one can really say are hominoids. It needs to be sorted out—and that's what I'm going to do."

But it was the western shores of Lake Turkana that truly called to Richard. Before he was ill, he had often gazed "across the jade-green waters of Lake Turkana" from Koobi Fora, "wondering what secrets" the western sediments held, he wrote in *Origins Reconsidered,* a 1992 book about his West Turkana expeditions. In the summer of 1984, he decided it was time to take a closer look.

Richard had already launched a preliminary survey of West Turkana. For a few months during the preceding three summers, he had dispatched Kamoya's crew and the geologist Frank Brown to map and record fossil-rich sediments from the lake's northwestern end to its southern tip. "I'd asked Brown to take on this work because we didn't want to make the same mistake we had at Koobi Fora," Richard explained. There, fossils had been collected before the geology was fully understood—an error that caused many later troubles. Lacking a proper geological map that correctly correlated sediments and volcanic tuffs from one end of the desert to the other —and so established the age of each—the paleontologists simply had no way of knowing what time period the fossils they were finding actually came from. And not realizing how complicated the geology at Turkana was, the first geologists had seriously underestimated the number of distinct tuffs. Frank Brown had almost single-handedly resolved the problems at Koobi Fora, and Richard had sought his expertise before collecting any fossils from the lake's western deposits.

Frank Brown's work was now well along, and Richard told Kamoya to keep a sharp eye out for fossils in the eight- to four-million-year range. "That's where we have a huge gap in the [hominid] fossil record," he explained, "and at some point in that time, I think we'll find the beginnings of bipedalism." Richard knew that there were hominids to be found at West Turkana: in 1983 one of Kamoya's men had picked up a fragment of a 3.3-

million-year-old hominid mandible. Not enough of the fossil remained to be able to say what species it belonged to—but where there was one hominid fossil, there were bound to be more.

As he had in the later years of the Koobi Fora project, Richard relied on Kamoya's team of professional hominid hunters to actually search the deposits for fossils, while he handled the administrative affairs of the museum. He and Kamoya stayed in contact by radiotelephone, with Kamoya calling in every day to report where they were and what progress had been made (reports that also reassured Richard that the men had not been attacked by bandits).

For most of the summer of 1984, these were their only exchanges. But on the morning of August 22, Kamoya had real news: they had found fragments of two hominid skulls at two different sites. "You might want to see them," he joked. Richard laughingly agreed and the next morning, he and Walker (who was at the museum, cleaning and studying the fossilized bones of several *Proconsul* skeletons he had found earlier that year on Rusinga Island) flew north to a dry riverbed about three miles west of Lake Turkana, called Nariokotome.

"We have many bones to show you," Kamoya said as they unpacked the plane. "You will like the hominids." "Skeletons?" Richard teased. Everyone laughed—not realizing how prophetic Richard's joke would prove to be. In fact, Kamoya had only a tiny bit of fossilized hominid bone—no larger than a matchbook—to show Richard. It was "a small piece of hominid frontal [bone from the front of the skull], about 1.5 by 2 inches, in good condition," Walker wrote in his field diary, and it had come from a site that looked as unpromising as the bone itself. Kamoya had pitched his camp under the palms and spreading thorn trees along the dry Nariokotome River. Directly opposite the camp, a slope covered with black rocks and pebbles slumped down to the sandy gulch. It was here amid the dark stone that Kamoya had retrieved the almost equally dark bit of bone. "How he found it," Walker wrote in his diary, "I'll never know."

Although he did not know what part of the skull the bone came from, Kamoya could tell from its thickness and curvature that it was hominid and not some other creature. He had also guessed, correctly, that it belonged to *Homo erectus,* the hominid that preceded *Homo sapiens,* since Brown's dating had established these sediments as being 1.6 million years old. It wasn't as old as Richard had hoped for, but it was part of the human record and Richard decided that the crew should search the slope for additional bits of the skull—although he was far from optimistic that anything would be found. Richard wrote in his diary, "Seldom have I seen anything less hopeful." He was far more eager to see a partly buried hominid skull fragment that the paleontologist John Harris had found in deposits dating to two million years—that, he wrote in *Origins Reconsidered,* could be very important.

As it turned out, all the fossil action was at Kamoya's site that day (only fragments were found at Harris's).[2] And as soon as Richard returned to the Nariokotome, he realized how mistaken he had been. "We've found more bone! Lots of skull!" Kamoya's crew shouted as Richard drove up. Immediately he, Harris, and Walker jumped from the Land Rover and ran "to where Kamoya was sitting, his treasure arrayed before him, like jewels plucked from the dry earth," Richard wrote. In a layer of soft earth, just beneath the surface stones, the men had uncovered more fragments of the frontal bone, as well as the right temporal, and left and right parietals of what Walker was now describing in his diary as a "beautifully preserved (if broken) *Homo erectus.*"

Within a week nearly all of the skull and facial bones had been recovered. Most astonishing of all, Richard and Walker had unearthed small bits of rib, too. They gazed at these uneasily: were they human or merely antelope? It was impossible to say, since by themselves most mammal ribs carry few clues about which creature they belong to. But three days later, with a mounting pile of bones—two thoracic vertebrae, a lumbar vertebra, more ribs, part of the left scapula, the right pelvis, part of a tibia, and the fibula—the two men began to think the unthinkable. Perhaps, Kamoya *had* found a skeleton. If so, it would be the first skeleton of *Homo erectus* ever discovered. "[R]ight beneath our feet might be the discovery of a lifetime," Richard wrote in *Origins Reconsidered,* recalling the "breathtaking moment" when he first realized that Kamoya may have made such a find.

Although fossils of *H. erectus* were known from sites as far afield as Olduvai and Java (and ranging in age from 1.7 million to 400,000 years ago), most of these specimens were skulls or jaws or random limb bones. "[A]s a result," Richard later wrote, "every bone we turned up [at Nariokotome] was the first of its kind to be seen by human eyes." "This is the first thoracic vertebra of *Homo erectus* known to science," began Walker's litany one day. "This is the first lumbar vertebra of *Homo erectus* known to science," he intoned a few moments later. "This is the first clavicle of *Homo erectus* known to science." When a *National Geographic* photographer sat idly nearby, Walker chided, "The first thoracic vertebra you or anyone else has ever seen and only the second pelvis, and you're not taking pictures!"

"You'd have to go to Europe, to Neanderthal graves, to see fossil skeletons as complete," Mary remarked, several days into the excavation. As curious as ever about human origins, even though she was now retired, Mary had badgered Richard to fly her to the site as soon as she got wind of the new find. She arrived shortly after Walker and Meave had finished gluing the skull and facial bones together. It was only the second face of *Homo erectus* ever found—Richard's team had found the first at Koobi Fora—and Mary peered into the skull's sockets as if she could make out what the creature was thinking. "I wonder what he saw," she mused. "It was an Eden, of that

---

[2] Harris's fragment also came from the frontal bone. According to Alan Walker, "It is indistinguishable from that of modern humans."

I'm sure, not at all like the deserts of modern hunter-gatherer bands. Theirs was a rich, bountiful land."

By now the excavation was very much a Leakey family affair, with Richard's wife, daughters, and mother all present—and all taking part. While Meave and Alan pieced the skull together, ten-year-old Samira ran messages between the camp and dig, twelve-year-old Louise helped with the excavation, kneeling on the hard earth beside her father, and Mary supervised, snorting her disapproval at the ragged hole her son had dug. (As an archeologist, her excavation would have featured precise right angles and tidy squares.) But it was all good-natured joking, and Richard glanced up once from the dig to reflect on how happy he was. "There is nothing I like better than digging in the sediments, excavating a hominid bone," he said.

Although Richard had not heard from Don Johanson in some time, he expected to after this discovery was announced. His relationship with Johanson had further deteriorated in 1983, after Johanson suggested to other paleontologists that Richard was actively trying to keep him from returning to Ethiopia. ("Such nonsense!" Richard said, irritated at these charges. "As if I had any way of influencing another government's decisions!")[3] Concerned about the bickering among various American expeditions, the Ethiopian government had declared a moratorium on fossil hunting in 1982, leaving Johanson out in the cold—he had not made a fossil discovery since 1977. But when he made overtures to Richard in the spring of 1983, seeking permission to look at some of the new fossils that had been recently unearthed in Kenya, Richard, still smarting from *Lucy* and the unpleasant rumors, sent a scornful reply.

"I consider your actions and public statements over the past few years to be distinctly unfriendly and lacking in a number of rather important respects," he wrote, after noting that only people he trusted were shown unpublished fossils. "I regret this because there is nothing more unpleasant than to live with the knowledge of such antipathy," Richard continued. "All the same, I do not trust you in the least as a result and so I intend to distance myself." When Johanson protested that he could not understand Richard's hostility, Richard happily fired off another scathing reply. "The reason that my previous letter was written in the tone that it was is simple: I consider you a scoundrel." After that, Richard's secretary answered Johanson's inquiries about seeing hominid fossils, directing him to the government office that handled these requests. Richard's correspondence with Tim White had ended on much the same cold note.

But the skeleton of the Turkana Boy—as Richard and Walker had begun referring to their *Homo erectus* find, having discerned the fossil's sex from the shape of the pelvis—would, they knew, bring more such inquiries from

[3] But the Ethiopians did turn to Richard in 1981 for advice about drafting a new antiquities law. And Richard admits that he tweaked Johanson's nose with this news—telling others just enough information about the Ethiopian delegation's visit to make it seem that Richard's power was as extensive as Johanson feared.

Johanson and White, as well as other scientists. It was certain to reignite the paleoanthropological world, because nothing as dramatic as this skeleton had been found since Lucy. Since it was such a complete specimen, it would answer many questions scientists had long asked about *Homo erectus*. How tall, for example, had these ancestors been? Earlier estimates, based on the study of single leg and arm bones, had suggested they were stocky people of medium height. But the Turkana Boy was clearly "a strapping youth," said Richard. From his teeth, he and Alan knew the boy was about twelve years old when he died (later, more detailed studies of his teeth suggested he was actually between nine and eleven), and his legs gave him the stature of a budding basketball player—he was all of five feet three inches, and would have stood six feet one inch as an adult. Such a build—tall and thin, with long limbs—also characterizes the Nilotic peoples of today, who live in much the same hot, dry, open environment as the Turkana Boy did nearly two million years ago. The similarity is not genetic, Walker noted in a subsequent study, but results entirely from the fact that bodies shaped like the Turkana Boy's (for example, a modern Dinka's) are better adapted for heat loss in these types of environments.

By mapping the precise location of each bone and studying the ancient sediments on which the fossils rested, Walker had also uncovered a few clues to the boy's death. Perhaps, as Walker later suggested, the boy was suffering from a viral infection or sepsis from a newly erupting molar. He found his way to the quiet edge of a lagoon, or the slow-moving oxbow of a meandering river, and there lay down and died. For several days, his body floated face-down in the water; soon the flesh rotted and several of his teeth fell out. A large animal, perhaps a hippo, then trampled the skeleton (the indentations of the animal's prints were still visible in the sediments), snapping the boy's right leg in two pieces, while causing the lighter bones to float closer to shore. The skull and lower jaw also separated, and the cranium floated a little farther away. A million and a half years later, a thorn tree seed sprouted in its cavity—the single tree that had grown on the slope Kamoya had found so interesting.

By the end of September, nearly the entire skeleton, aside from the hand and feet bones, had been unearthed. Perhaps these bones were still buried, but the team would have to level nearly an entire hillside to find them—a project they had neither time nor money for. Reluctantly, Richard decided they would have to resume the search in 1985. "Everyone is a little saddened," Richard wrote in his diary as the team began packing up the camp. "We all had high expectations, but a fully complete skeleton has eluded us."

Yet what they had uncovered was extraordinary. Assembled into a standing position back at the museum, the Turkana Boy took Richard's words and breath away. " 'Remarkable' was all I could say," he later wrote. "Remarkable. . . . How very human he looked . . . !"

Two weeks later, with great pleasure, Richard unveiled to the world what he considers one of the most significant hominid fossil discoveries "of all time." With television shows and newspapers celebrating the Turkana Boy, Richard found himself once again the center of worldwide acclaim. He had made it back to the top.

Chapter 41

GRANDE DAME OF ARCHEOLOGY

In the fall of 1984 the *New York Times* devoted nearly an entire page to praising the Turkana Boy and the Leakey family. "It is the Leakeys who, in a significant way, have given modern paleontology its place among the sciences," wrote John Noble Wilford. Their name, he added, was "synonymous with the study of human origins," noting that it was the Leakeys who had "proved beyond doubt the African origins of man." Richard's new skeleton was only "the most recent in the long line of Leakey discoveries" and because of such finds "even [the family's] severest critics" had to admit that the Leakeys' "towering reputation [was] generally well-deserved." Indeed, even Don Johanson lauded the new fossil as "quite extraordinary." But Johanson had defined himself—and made his fame and fortune—by being at odds with the Leakeys, and could not, Wilford wrote, "resist taking a swipe."

"I'm a little surprised that everyone is so surprised that this was a large individual," Johanson said, referring to Richard's remarks about the boy's unexpected height. "We've had pieces from China and Java that suggest that these [*H. erectus*] individuals approached six feet in size. I don't think this aspect of the discovery is so astonishing." But Johanson himself had described *H. erectus* as being only "medium-sized" in *Lucy*.

For once, though, the press seemed more excited by the skeleton and the boy's striking resemblance to ourselves than by the quarrels between the two anthropologists. That fall on his lecture tour, when audiences leaped to their feet to applaud a photograph of the Turkana Boy's skeleton, Richard no longer doubted the public's "deep-rooted" interest in human origins. To

Richard's great amusement, on one occasion he even drew a larger audience than the rock star Sting. Richard was asked, too, to pose for an ad promoting Rolex watches ("Some men merely make history, Mr. Leakey redefines it," read the copy line), and to dine at the White House with President Ronald Reagan.[1]

Ironically, however, while Richard's popularity with the general public grew, he slowly began to withdraw from the science itself—not from the fieldwork, but from the conferences and meetings where his paleoanthropological colleagues gathered to discuss their latest findings and hypotheses. It was not that Richard was uninterested, but since returning to Kenya he was occupied more with administrative affairs—running the National Museums and the new Institute for Primate Research, as Louis's Tigoni primate center had been rechristened in 1974—than with the detailed analysis scientific research requires.[2] Thus Alan Walker—and not Richard Leakey—would write the anatomical description of the Turkana Boy's skeleton and present details about its bones at scientific meetings. This particular arrangement was not new; Richard and Walker had shared such a partnership for years. But Richard had usually joined in the debate about a fossil's interpretation at paleoanthropological gatherings. Now, however, he was seldom even present, and he made a particular point of shunning any conference where there was the slightest chance of meeting Don Johanson.

Richard's absence from the paleoanthropological stage perplexed and annoyed many of his colleagues. He had particularly angered several American scientists in the spring of 1984, when he snubbed "Ancestors: Four Million Years of Humanity," a huge exhibition and symposium at the American Museum of Natural History in New York. This grand affair was billed as the first display of humankind's most important ancestral fossils. The organizers had requested museums around the world to lend the originals of such specimens as South Africa's Taung Baby, Ethiopia's Lucy, Tanzania's *Zinj,* and Kenya's 1470. But Richard had proved something of a party spoiler. Not only had he chosen not to attend (partly because Johanson had been invited as the keynote speaker), he had also refused to send any of Kenya's original hominid fossils for display, saying that such an exhibit was too risky in a country where creationists were active. (Mary voiced the same opinion. In her lecture at the "Ancestors" symposium she praised the American Museum staff for the "well-organized" event, then blasted them for endangering irreplaceable fossils by transporting them from such distances and exhibiting the specimens in a single room where "a [religious] fundamentalist could come in with a bomb and destroy the whole legacy.")

Many paleoanthropologists perceived Richard's refusal to participate as

---

[1] Richard had also dined at the White House with President Jimmy Carter.

[2] Louis had never allowed the monkeys at Tigoni to be used for invasive medical experiments, which was one reason he had such trouble raising funds to keep it going. Richard took a more practical view, however, and the Institute for Primate Research is dedicated to biomedical and fertility research, using monkeys bred at the institute.

arrogant and selfish. Although absent, he was still a power in their midst—none of them, after all, could carry out any research in Kenya without his permission. (Richard was not the sole voice of approval but sat on a council that reviewed research requests. But a "nay" vote from him was all it took to squelch a request.) And because only a handful of paleoanthropologists knew Richard as a friend, most were also uncertain about how much power he actually wielded. Consequently, when other museum directors (from Tanzania, Ethiopia, China, the Netherlands, and Indonesia) also refused to send their hominid fossils to the American Museum exhibit, many scientists believed that Richard, acting as the instigator, was responsible.

"The extent of your power (as perceived by various persons here) is amazing," the geologist Frank Brown wrote to Richard from the "Ancestors" symposium. "Not only did you not bring the Kenyan fossils, but singlehandedly prevented the Tanzanians & Ethiopians from bringing theirs. No doubt you managed this while on the phone to the royal Dutch family to prevent those specimens from coming & on a separate line to Jakarta to keep the Indonesians away." Overall, Brown added, "there is a good deal of anti-Leakey (particularly REFL [meaning Richard]) sentiment."

Compounding the scientists' grievances, a joint team of Kenyan and Harvard researchers chose to announce their discovery of the oldest known hominid fossil, a jaw fragment of a 4.5-million-year-old unnamed creature on the very day that the "Ancestors" exhibit opened. It was only a coincidence, insisted Andrew Hill, one of the Harvard scientists. But in New York, many scientists thought that Richard had masterminded the news conference as a way of diverting the spotlight from the "Ancestors" exhibit to Kenya, where researchers were busy actually finding fossils.

As powerful as Richard appeared to his fellow scientists, his rule over paleoanthropology actually extended only as far as Kenya's borders. He had occasionally served as a consultant to Ethiopian and Tanzanian officials about such issues as drafting antiquities laws and training museum technicians. But Richard had no say about which scientists might be given research permits to work in those countries. Nor did Mary. Although she (and Louis) had made some of the most important discoveries in paleoanthropology at Olduvai Gorge in Tanzania, she was as dependent as any other researcher on obtaining research permits from the Tanzanian government. But after 1983 Mary no longer needed one. She retired from active fieldwork in the fall of that year and moved back to Nairobi, leaving the gorge and all of its fossil treasures behind for others. The decision about who might now explore Olduvai was up to the Tanzanians.

By the time she left Olduvai, Mary had lived at the gorge for nearly twenty years. Over that time, she had weathered floods; droughts; prowling lions; invasions of vipers, grasshoppers, tourists, and students; ash falls from volcanic Mount Lengai; drunken cooks and mischievous Maasai, who sometimes broke into camp huts or damaged excavations. She had watched, too, as the numbers of leopards and rhinoceroses dwindled to near-extinction on the Serengeti from poaching—both animals, so abundant when she and Louis made their first joint exploration of Olduvai in 1935, no longer lived near the gorge. Mary had endured long stretches of loneliness as well, particularly after Tanzania closed its border with Kenya in 1977 over a political dispute. While Mary received special permission from the Tanzanian authorities to travel between the two countries (as did researchers who came to work with her), her family in Kenya were unable to journey to Olduvai as they formerly had for holidays.

Overall, the border closure had made the hard life at Olduvai even harder: everything from mail delivery to obtaining sufficient food and fuel supplies had become more of an ordeal. Then, alone at camp one afternoon in November 1982, she awoke from a nap to discover that she could not see out of her left eye. "This was very frightening and alarming and I had no idea of the reason," Mary wrote later in her autobiography. She slept uneasily that night, and at dawn, finding herself still one-eyed, placed a radio call to Kenya. By chance, her son Jonathan had just turned on his own receiver at his home at Lake Baringo, and immediately recognized her voice. "Good morning, Mother, how are you?"—Jonathan's voice came crackling over the airwaves. "Jonny!" Mary shouted in relief. "Thank goodness. Listen, I seem to be in real trouble. . . ." That afternoon Mary was on a plane bound for Nairobi to see a physician.

From the doctor, Mary learned that she had suffered a thrombosis—a blood clot had settled behind her left eye, damaging it permanently. "At first . . . I was very depressed and sorry for myself," Mary wrote after hearing this news. But her family came to her rescue. Meave and Richard insisted that she stay at their home, where Richard alternately plied her with "pep talks," fine wines, and haute cuisine meals he himself had cooked. The treatment worked—so well that two months later Mary wrote to Melvin and Ethel Payne at the National Geographic Society, "[I've] almost forgotten what it's like to have two eyes. Two things still daunt me: I cannot draw . . . and I cannot judge the depth of pot holes in the local roads. Nor would I like to drive in Nairobi traffic." But even this fear vanished a few months later. Fed up with her hired driver, Mary set out alone one morning to drive to the downtown market near Nairobi's central mosque. On return, Mary briskly announced, "Well, I did it. There were Muslims everywhere, all over the road, and I didn't hit a one!"

Still, Mary's eye trouble left her feeling vulnerable, particularly about living alone at Olduvai. Her family, too, began pressing her to think about retiring. She had finished the fieldwork at Laetoli, was putting the finishing touches on her autobiography (written with the aid of a ghostwriter), and had nearly completed her analysis of the artifacts from the gorge's upper beds (research that would be described in the fifth Olduvai monograph, *Olduvai Gorge,* Vol. 5, *Excavations in Beds III, IV & the Masek Beds*). It also seemed unlikely that any new major hominid fossils were to be found in the gorge until more of the slopes and gullies eroded. Tanzania's escalating economic and social problems, brought on by years of government mismanagement, only tipped the scales further, as did the decision of her camp manager, Peter Jones, to leave.

In May 1983 Mary accepted the inevitable: she would move out of the little hut at the edge of the gorge that she had for so long called home. "It will be a wrench to leave Olduvai," she wrote to Richard Hay, the geologist and Mary's chief scientific colleague during her years at the gorge. "[B]ut the sense of insecurity in Tanzania makes life impossible, nowadays. Crossing the border into Kenya is blissful, instead of depressing, as it used to be."

That fall, with little fanfare, Mary moved north to the home she and Louis had built at Langata, outside Nairobi. She was only a few miles from the homes of two of her sons, Richard and Philip, and close to numerous friends as well. "We worried at first that she might just shuffle about alone at Langata," said Valerie Leakey, Philip's wife, not long after Mary retired, "but instead she's become quite social. She goes out almost every night to meetings and parties; she's very much in demand." Even more important to Mary, as she told many friends, by living in Nairobi she could "see more of [her] multitude of grandchildren." (She now had ten: nine granddaughters and one grandson, an imbalance that Mary often fumed about.)

Mary was not, however, even in her seventies, the cozy, comforting kind of grandmother, but more of a grand Victorian matriarch. Grey-haired and weathered, she coolly puffed on her short cigars, enjoyed her whiskey, and delivered her tart opinions without a semblance of tact. Her blue eyes (even her blind one) could still snap fire or chill a person to the quick. And although she had gained a gentle plumpness around the middle, she had not gone soft at all, but carried herself "soldier-straight," as Alan Walker put it. Even her high, reedy voice held a note of command; she was, after all, the most famous woman archeologist in the world, as a host of awards and honors attested.

By 1985 Mary held a plethora of honorary degrees from universities, including Oxford, Yale, and Chicago; she had also been elected an honorary foreign member of the American Academy of Arts and Sciences, a member of the British Academy, and had received the Gold Medal from the Society of [American] Women Geographers. Before the decade was out, she would be elected a member of the U.S. National Academy of Sciences, and Prince Philip of England would confer an honorary doctorate on her at Cambridge University. This latter degree, coming as it did from the institution that had spurned Mary and Louis in the 1930s, so touched her that Mary later told a friend she had "fairly skipped down the aisle" to receive it.

Given her lustrous reputation, it was thus all the more galling that, in the spring of 1985, Mary heard from a graduate student in the United States that Don Johanson and Tim White were about to descend on Olduvai Gorge.

• • •

Like Richard's, Mary's relationship with Johanson and White had gone from bad to worse since the publication of *Lucy.* In it, Mary said, Johanson had written "such lies" about her and her work that she had once appealed to the geologist Maurice Taieb, who was Johanson's coleader on the Afar expeditions, to explain why. "I saw her at Olduvai and we were drinking bourbon and she said, 'Maurice, can you believe what Donald has written about me? I have opened my house to him, here and at Langata. I have taken care of him as my son. And yet he writes so many bad things about me!' And she was crying, weeping," recalled Maurice. "I told her, 'Mary, it is not important. Forget what you did for Donald. He wants to do something exciting to sell more books, to make more money, and this is one way to do it.' " Mary then pressed Taieb, "But why is he not telling the truth?" And Taieb replied, "Because Donald is at a young stage and he wants to be the first."

Since the publication of *Lucy* in 1980, Mary had had no contact with Johanson, although she had carried on a brief—if hostile—correspondence with White during the early 1980s concerning his published anatomical descriptions of the Laetoli hominid fossils her team had found. Mary had ended their angry exchanges with a three-line note saying there was no point in continuing the relationship.

News of Johanson's and White's imminent arrival at Olduvai Gorge was thus "terribly upsetting" to Mary, Richard said. "It was like a complete violation." Equally disturbing, Johanson, as leader of the expedition, had not observed the most basic scientific courtesy by notifying Mary about his plans ahead of time; instead he waited until shortly before he arrived at Olduvai to write her a terse letter of announcement. "It's an unwritten rule of science," said the geologist Kay Behrensmeyer, "but you always write the previous researcher at a site to ask what his plans are, if his own research is finished, and to explain your interest in the site, before moving into it yourself. It's the courteous thing to do, but it's particularly important for a place like Olduvai where there is such a lengthy record of research; maintaining that continuity is important."

In fact, as Richard pointed out, if Johanson had written Mary a letter outlining his research plans and asking for her approval, no one could have objected. Even if Mary had disliked the idea, she would have been left with no grounds for complaint. Instead, Johanson's behavior outraged many scientists in the paleoanthropological community. Some even refused to join his expedition, Glynn Isaac reported to Mary.

Mary's longtime Olduvai partner, Richard Hay, also found the news of the expedition distressing. It had left him feeling "sad for some time now," he told Mary. "I know Olduvai belongs to all of science," he added, trying his best to be objective, "but to me Olduvai is *ours,* and I can't imagine going back in these changed circumstances."

In *Lucy's Child,* his book about his work at Olduvai, Johanson tried to justify his action by saying that he had not written to Mary about his plans until he was absolutely certain that he was going to the gorge. "I saw no reason to upset her by parading our hopes," he wrote. Furthermore, he felt that "since Olduvai was a Tanzanian site, it was up to the Tanzanians to contact Mary."

Don Johanson, Tim White, and three colleagues arrived at Olduvai in mid-July 1985 for ten days—"just enough time to sweep quickly through the Gorge and take stock of what it had left to offer," Johanson later wrote. "What we found in those brief few days was more than enough to convince us that the famous old site was far from exhausted." They would return for a lengthier expedition in 1986.

In the meantime, Mary did what she could to soften the blow. During their preliminary expedition, Johanson's crew had stayed at a camp run by the Tanzanian Antiquities Department. But Mary knew that Johanson intended to use her camp upon his return, and she quickly arranged for Tanzanian friends to move out her furnishings. "She just could not abide the idea of one of them sleeping in her bed, as would have happened," Richard said. Instead, her staff packed up her simple iron bed, wooden nightstand, and chair from her hut and the overstuffed couches, chairs, and dining table from her main lodge and trucked them over Ngorongoro Crater to a friend's farm. Not long thereafter, Johanson arrived, striding into the camp compound and asserting his possession with a simple act: he pushed open the door and dropped his bag in, as he later wrote, "what had been Mary Leakey's hut." The "enemy forces," as Mary had begun calling Johanson and crew, had arrived.

Chapter 42

A NEW CHALLENGE

For Richard, Don Johanson's arrival at Olduvai Gorge in the summer of 1986 was doubly irritating. Not only was it an insult to Mary, but it also gave Johanson the platform he craved. Once again, he was seen publicly to be in direct competition with the Leakeys, a role that Johanson loved as much as Richard detested it. Do what he would, Richard could not free himself of Johanson.[1]

Although Richard and Mary had both stopped attending conferences where they might encounter Don Johanson, inevitably their paths crossed. One such occasion was an awards ceremony the National Geographic Society convened in Kamoya Kimeu's honor in November 1985. Kamoya had by then discovered more hominid fossils than any other person (among them the 1984 Turkana Boy skeleton), and the society wanted to honor him with its John Oliver La Gorce Medal. As part of the celebration, the society invited all the key figures in the science—including Johanson and White—to a gala dinner. The invitation to their unfriendly rivals dismayed Richard and Mary —although Mary imagined taking advantage of the situation by "edg[ing] one or other (or both) into the pool in Explorers Hall." At the dinner itself,

[1] In 1985 a reporter for a San Francisco magazine wrote Richard a letter asking several questions about the two men's relationship. In one, she explained that Johanson had complained to her that Richard would not engage in a public debate with him about the fossils. "Instead, [Johanson] says, your public attacks on him are on a personal level," the reporter wrote. "What do you say?" Richard replied, "If Don Johanson can show you any evidence that I have attacked him personally, please publish it. The accusation is groundless and I have been most careful throughout the era of disagreement to avoid any personal references."

however, she and Richard simply stared icily past "the two problems," as she described Johanson and White to a friend.

Meanwhile there was gossip that the Leakeys were working behind the scenes to keep Johanson and White out of Tanzania. The rumors took a decidedly more malicious turn in December 1985, when Richard was told by one of the members of the board of Johanson's Institute of Human Origins that Johanson, White, and Tom Hill, a friend of Johanson's, had once again been gossiping about Richard—but this time at an important retreat where wealthy donors were present. White was reported to have accused Richard of refusing to return important hominid specimens to the Tanzanian government, while he and Hill were said to have resurrected the old gossip about what had happened to the profits from Richard's book, *The Making of Mankind,* and the television series based on it. According to Richard's source, Hill claimed that Richard had diverted money owed him into Richard's personal bank account. Hill had once offered to help market the series in the United States, with the understanding that he would not charge a fee for handling the negotiations. But according to Richard, Hill's final proposal did include a hefty sum for his services, and Richard angrily dismissed him. Since that time, Hill had often alleged that Richard had earned a one-million-dollar profit from the series, but had donated only a pittance to paleoanthropological research, although he had promised to donate all the proceeds.

This gossip was patently untrue, Richard insisted in an angry letter to his attorney, Charles Jaffin in New York City. "I can confirm that I have not personally benefitted from the project," Richard wrote, "this *in spite* of the fact that there is no reason why I should not have [since] the film and book were my idea. . . ."

Richard had put up with these rumors now for six years, and this time he was determined to squelch them for good. He sent Jaffin a copy of the letter he had received from Johanson's board member; Jaffin was aghast. "Tim White's comments are obviously distressing but the comments by Tom Hill are dangerous and I would suggest should not be ignored," he told Richard, adding that Richard had grounds to sue for slander.

"If you feel that there are grounds for a slander action, let us take action," Richard wrote in reply.

An attorney in Jaffin's office then notified Hill and White that a lawsuit was pending against them; they, in turn, appealed to their own lawyers, and the matter was resolved. "My information suggests that Mr. White is being more circumspect at present anyway, perhaps because of your letter," Richard told the attorney in Jaffin's office who had handled the case.

Meanwhile Richard thought that Alan Walker might have found a fossil hominid that would disprove Johanson and White's hypothesis that *A. afarensis* was the pivotal ancestor of humankind.

• • •

Walker had made his spectacular find, a 2.5-million-year-old skull of a robust australopithecine, at the end of August 1985. That summer Richard had again dispatched his hominid hunters and a team of scientists including Walker, John Harris, and Meave, to the west side of Lake Turkana. They pitched their main camp along the dry Nariokotome River where Kamoya had found the Turkana Boy *(Homo erectus)* the previous summer. Hoping to uncover the skeleton's missing bones, the team again tackled the sediments with ice picks and brushes. But what had been an exciting search in 1984, when almost every day the digging produced at least one new bone, had by now turned into drudgery.

The work was dusty, hot, and frustrating. Day after day the excavators knelt on the hard earth, pried it up an inch at a time with their picks, then swept up the soil, washed it, and sieved it—and found nothing. By the end of August, after nearly six weeks of hard labor, the crew had only an arm bone and part of a rib to show for all the tons of earth they had moved. The site was proving "frustratingly barren," Richard later wrote.

Then, on August 30, just as the season was ending, Richard's office phone rang a little after 6:00 A.M. It was Kamoya shouting into his radiotelephone that Richard should fly north at once—Alan had made a big discovery. The call came in on a bad channel, making it difficult for Richard to hear all that Kamoya was telling him. But "I got the salient points," Richard would later write, and he quickly canceled the rest of the week's appointments. "If what I'd just heard through the acoustic fog of Kamoya's call turned out to be correct, I could see that Alan's fossil was going to cause quite a fuss," Richard wrote. At dawn the next morning, he was flying north to Nariokotome.

Two days before, Walker and several of the other team members had taken a break from the excavation and visited an area called Lomekwi, about twenty miles south of the Turkana Boy's site. They had explored Lomekwi a few weeks before, turning up, among other finds, the nearly complete skull of an extinct hippo and a monkey's arm bone. Walker planned to collect both specimens now, and he and his wife, Pat Shipman, an anthropologist, first tackled the hippo skull. Unlike the sediments at the Turkana Boy's site, the earth here was fine-grained and sandy. It was so easily swept away with a brush that by noon the two had exposed almost the entire skull. While Shipman began painting a thin layer of Bedacryl over the fossil to harden it, Walker set off in search of the pile of stones the team had left to mark the monkey's arm bone. He hiked across several ridges, spotting stone cairns here and there—all of them the wrong ones. Then, just as he began walking back to the hippo site, Walker noticed one more small cairn, with an odd-looking, dark-colored fossil lying beside it. One of the team members had marked the bone, but apparently thought it not worth reporting.

Intrigued, Walker stooped to pick up the fossil. His first glance took in an upper jaw that held such huge tooth roots he thought the fossil must be some type of extinct antelope (which apparently was the conclusion of the field worker who had not mentioned the bone). "Then I saw another piece from the front of the skull and thought it was a big monkey," Walker later said. "But then I turned it over and saw a frontal sinus—hominid!" The frontal sinus, a system of cavities within the bone above the nose, is a sure sign that a fossil is a hominid. Other pieces of the skull lay scattered over the ground, suggesting that the rest of the fossil was still buried.

Walker put the fragment back on the ground, then returned to his wife and the hippo skull, only a few minutes' walk away. Shipman sat brushing a last coat of Bedacryl on the fossil, and Walker knelt down casually beside her. "When you've finished that Bedacryl, I'll show you a hominid," Walker said, his voice giving no hint of what he had just found. "Great! What part?" Shipman replied. "Oh, skull," Walker answered as nonchalantly as he could. Shipman dropped her brush at once, and moments later the two were back at the new skull site, picking up pieces of the fossil. These were large and heavy and stained a rich blue-black from manganese in the soil. From the pieces they collected—a part of the frontal bone with massive brow ridges, part of the back of the skull, and most of the upper jaw—they could see that the skull was that of a robust australopithecine, a creature in the same lineage as *Zinj (Australopithecus boisei).* Mary had discovered that fossil at Olduvai almost exactly twenty-six years before. But *Zinj* was a little less than two million years old, while this new skull lay in deposits dating to at least 2.5 million, which pushed back the time period for these ancestral cousins.

It was not the fossil's age, however, but the size of the upper jawbone's tooth roots that got everyone's attention back at the Nariokotome camp that evening. "Look at the size of those tooth roots," Kamoya said over and over, after Walker handed him the fossilized palate. The roots would have secured teeth four to five times the size of a modern human's back molars; Walker thought they were even somewhat larger than *Zinj's* massive grinding equipment. It was that bit of information that Kamoya had tried to convey to Richard over the radio phone the next morning. To most paleoanthropologists, the big teeth of the robust australopithecines (a lineage that includes the East African *Australopithecus boisei* and the South African *Australopithecus robustus*) were adapted for chewing hard nuts and grasses—a specialization that scientists thought had developed in a progression that led from the smaller-toothed *Australopithecus africanus* through *robustus* to *boisei,* and then reached a dead-end about a million years ago. But here was a creature at the beginning of that lineage with the largest teeth of all. Could it be, Richard wondered as he flew north, that Walker had found a new species, a creature that "would make the human bush even bushier?" And if it did add another branch to the bush, would that be enough to weaken Johanson's claim that *A. afarensis* was the stem?

While waiting for Richard, Walker had begun gluing the fossil back together and so had a partly reconstructed skull to show him when he arrived. "Like it?" Walker asked, moments after Richard strode into camp. "Like it?" Richard replied. "It's wonderful, fascinating. Wait till they [other paleoanthropologists] see this." Like Walker and Kamoya, Richard was immediately struck by the prodigious tooth roots and small braincase. Both features were hallmarks of a robust australopithecine, but in this instance the traits were even more exaggerated. Studying the fossil, Richard was certain that the skull would "force many people to change their minds" about the coexistence of multiple hominid species three million years ago. He later expanded on this idea in his diary: "I am even more convinced that the three hominids of 2 million years ago are going to be traced back beyond 3 m.y. Also, it is likely that Johanson-White can be shown to be very wrong in their scheme. We shall see. It's going to be interesting for some."

Over the next week, Richard put most of his team to work on what they were now calling the "Black Skull" site, because of the fossil's distinctive dark patina. By poring over the earth for every fragment and bone sliver, then sifting the soil handful by handful, they eventually recovered nearly the entire skull—with the exception of the lower jaw, a portion of the braincase, and the teeth. At camp, Walker and Meave worked together, as they had on the Turkana Boy's skull, fitting and gluing the broken fossil fragments back together. Within a matter of days, the skull had taken shape. Like other robust australopithecines, it sported a huge sagittal crest running the length of the braincase; its cheekbones flared widely, and it had a wide palate. But it also possessed features that scientists considered more like that of an ape: it had a projecting muzzle, rather than the flat face characteristic of later australopithecines and *Homo;* a braincase no larger than that of an orangutan; and a very apelike jaw. The mix of features struck Richard and Walker as so peculiar that for a while they considered giving it a new species name: *Australopithecus kamoyensis,* in Kamoya's honor.

By the middle of September, the team had recovered as much of the new fossil as time allowed, and Richard brought the season to a close. Now he and Walker had to decide how to classify the Black Skull. They had three choices: name it something new; place it in an already existing taxonomic category; or wait until more fossils of this type were found. Although the first choice was appealing, the two men were no longer so certain that the fossil was unique and warranted a separate species; it would also be difficult to defend. "Naming it something new would have created a controversy," Walker explained, "and we decided not to do that."

Waiting for more fossils of this type to surface was equally unacceptable because Richard and Walker knew that as soon as they announced the fossil, someone would stick a label on it. Ultimately, they decided to settle for the most conservative choice: they would lump it with the rest of the East African robust australopithecines and call it *Australopithecus boisei.* But they would also add a tag identifying it as a "hyper-robust" example of that species. The extra label would, they hoped, silence any critics who might argue that the fossil stretched the definition of *A. boisei.*[2]

While Richard and Walker debated the finer points of taxonomy, Kamoya's crew continued their hominid hunt. Just a few days before closing the camp, Mwongela Muoka came back with a prize: a hominid's lower jawbone from an area only two miles southeast of the Black Skull site. The mandible was thick and heavy, like the Black Skull, and although it came from somewhat younger deposits, it seemed to Walker and Richard another example of the same type of hyper-robust australopithecine.

Wanting to announce the new skull and jawbone as quickly as possible, Richard and Walker began drafting a paper to *Nature* almost as soon as they returned to Nairobi. For Richard, as thrilling as the Turkana Boy had been, this new find was in some ways even more exciting. Here, at last, it seemed he had proof that our ancestral past was far messier than Johanson and White's tidy Y-branch. At Turkana 2.5 to two million years ago, at least three hominids had shared the lakeshore: *Australopithecus boisei,* an as yet unnamed gracile australopithecine (based on certain fossils from Koobi Fora and Olduvai), and *Homo habilis.* The three had evolved in parallel, just as various species of antelopes, pigs, and elephants were doing at the same time. To believe that all three hominids sprouted from Johanson and White's single stem *(A. afarensis)* only three million years ago seemed more and more unlikely to Richard. He and Walker planned to raise this issue in the *Nature* article.

In the meantime, Richard closed the Nariokotome camp. But before he sent the Hominid Gang home, he asked Kamoya and his crew to explore another area farther south that Richard had known about since 1969. That year Richard's team was exploring the fossil deposits on the east side of Lake Turkana, and sometimes he and Meave would drive to Ferguson's Gulf on the west side, then take a motorboat across the lake to the camp at Koobi Fora.

On one such drive, they had spotted a promising-looking area about fifteen miles inland from the gulf. "We said, 'Those look quite fascinating fossil beds; we should go have a look at them sometime,' " Richard recalled. He thought perhaps it was the same area that his father and Donald MacInnes had explored in the 1940s, and had reported as being rich in Miocene fossils. But during the Koobi Fora years, Richard had never had sufficient time to stop—nor, perhaps, sufficient reason, since he was then engaged in searching for our more recent ancestors. But after Meave's discovery of a

---

[2] Richard and Walker also considered classifying the fossil as *Australopithecus aethiopicus.* This species name described the single specimen—a toothless lower jawbone—that had been discovered on the 1967 International Omo Expedition. The fossil was as massive as the Black Skull and came from deposits only 100,000 years older. But since the species was represented by this sole specimen—and this bone, the lower jaw, was not found with the Black Skull—Richard and Walker thought it was risky to call the new fossil *A. aethiopicus.* Other scientists have argued that this would be a more correct classification, but the Black Skull is still listed as *A. boisei,* hyper-robust.

seventeen-million-year-old jawbone of a Miocene ape in 1981 north of Lake Turkana, Richard was curious whether similar creatures would be found in these southwestern deposits.

In early October 1985, Kamoya pitched a new camp about sixty miles south of Nariokotome, along a dry riverbed called Kalodirr. And as Richard and Meave had guessed, the deposits, dating to between sixteen and eighteen million years, were rich with fossils. A few days later, Kamoya was back in Nairobi, handing Richard a partial skull, including the face, of an unknown type of hominoid. "It was a total surprise," Richard later said. "It was an ape, but unlike anything that we'd found or seen—it had a long muzzle and wasn't like any type of hominoid we knew from East Africa. It was completely new to science." This time, certain that the fossil represented not only an unknown species but a new genus as well, Richard did not hesitate to give the fossil a new name. In a subsequent paper in *Nature,* he and Meave called it *Turkanapithecus kalakolensis.*

Deciding that it would be best to return to the site to sieve the sediments for any missing pieces of the skull, Richard and Meave together with the Hominid Gang headed back to Kalodirr for several short expeditions between December 1985 and January 1986. These produced additional pieces of the *Turkanapithecus* skull, plus another partial *Turkanapithecus* skull, as well as a skull and lower jawbone of yet another entirely new hominoid, one the size of a chimpanzee. Again, the skull represented both a new genus and a new species, and Richard and Meave named it *Afropithecus turkanensis.*[3]

As if finding two new genera in one season was not sufficient, Richard's team turned up a third. But the team was able to recover only part of this creature's face and a lower jawbone, not quite enough material, Richard and Meave felt, to give it a new name.[4]

Overall, the 1985–1986 West Turkana Expedition had returned with an astounding haul. In six months' time, Richard's team had expanded the ape and human family trees with four major discoveries—three Miocene hominoids and the Black Skull. "I was very satisfied with that," Richard would later write in *Origins Reconsidered.* The only disappointment came at the Turkana Boy's site. Despite moving tons of earth and digging a hole that resembled an open-pit mine, "we never did find the boy's hands and feet," Richard wrote.

Richard and Walker's description of the Black Skull was published in an August 1986 issue of *Nature*—and instantly put both men on the front pages of newspapers worldwide. With its bronzy sheen, craggy brow, and nearly

[3] Richard and Meave also reclassified the skull she found at Buluk from *Sivapithecus* to *Afropithecus turkanensis.*

[4] By 1994 Richard and Meave had discovered a total of four new genera of Miocene apes at this site.

inch-high sagittal crest, the fossil was by itself magnificent. Beyond that, it was also a seminal find. "[It] is the most exciting find since Lucy," Eric Delson, an anthropologist at the City University of New York's Lehman College, proclaimed to a *Time* magazine reporter. He expected it would cause paleoanthropologists to get out pen and paper and begin redrawing the family tree, just as Richard had hoped.

Richard and Walker had not offered any new trees themselves in their article, but used the Black Skull to question the accepted view of how the australopithecines had evolved, and to cast doubt on Johanson and White's theory that the human lineage arose from *A. afarensis.* But they had done so with reserve. "Whatever the final answer," they wrote, "these new specimens suggest that early hominid phylogeny has not yet been finally established and that it will prove to be more complex than has been stated."

In a press conference, however, Richard was more direct. "This throws cold water on the notion that as recently as three million years ago there was only one species [of early humans] which gave rise to the others," he told reporters gathered in Nairobi. Other scientists were inclined to agree. As Clark Howell of the University of California at Berkeley noted, the Black Skull "implies there were at least two separate lines going at a time that Johanson and White think there was just one."

Johanson and White were then trudging over Olduvai's gullies and unavailable for comment. But later, after having a chance to study a cast of the fossil, they refused to budge an inch. "[T]here was a lot of hype about the Black Skull when it was first announced," White told a reporter in 1987, "and it has been used as an inappropriate vehicle to attack *afarensis.*" He and Johanson argued instead that the fossil bolstered their position that *A. afarensis* was the common ancestor of all later hominids, saying that the skull's mix of primitive and advanced features was just what they would have predicted. They would leave their tree as they had originally drawn it, with *A. afarensis* as the stem, but sprouting three branches instead of two: the *Homo* limb; a branch beginning with *A. africanus* and leading to *A. robustus;* and a separate *A. boisei* limb. They would not admit that Richard Leakey had in any way altered their hypothesis. But Johanson would write in *Lucy's Child,* "nobody really places a great deal of faith in *any* human tree" now.

Richard gave paleoanthropology another good jolt only three months after announcing the Black Skull. This time, he and Meave had published two papers in *Nature* describing the three new genera of Miocene apes from Kalodirr—finds that again left their colleagues stunned and the media dazzled. "Not since his main international rival, Donald Johanson . . . , found the 3.5-million-year-old 'Lucy' . . . have so many new fossils been uncovered," noted a reporter for the *San Francisco Chronicle.* Together the fossils had put paleoanthropology "in a state of nearly unprecedented ferment." As David Pilbeam noted, "Leakey's discoveries more than double the diversity of Miocene apes. And that raises the question of how many more genera and species remain to be found. Clearly, we've only scratched the surface."

The new African Miocene apes provided proof, too, of exactly the kind of diversity Louis had hypothesized long ago. "The old man was right; in fact, he was right more often than most of us gave him credit for in the '60s," said Pilbeam, who had once waged war with Louis over his multispecies classification of these extinct creatures.

For Richard, the fossils were further evidence that the story of human origins was older and far more complex than often thought. "This could push back the point of splitting that led to modern species," he told one reporter. "I can't be more specific than that, but all this ancient diversity will surely shift the arguments a bit." He was even more convinced of this when, seven months later in May 1987, Johanson and White announced their own major find from Olduvai Gorge.

On July 21, 1986, only three days after Johanson had dropped his bag in Mary Leakey's former hut, he and White stumbled onto the partial remains of a 1.8-million-year-old hominid. "The Leakeys took thirty years to find a hominid at Olduvai Gorge," White would later gloat. "We got one in three days." White had spotted the first piece of fossilized bone, the elbow end of the hominid's forearm, while walking with other members of their team across the bottom of the gorge. "Whoa! This is a hominid!" he had shouted to the others. The word riveted the group and all eyes fell to the ground. "There's part of a humerus [arm bone] right next to it!" Johanson cried. They had found the fossils late in the day, and so waited until the next morning to begin a proper survey and excavation—a task that lasted five weeks and ultimately produced 302 pieces of bone.

Many of the fragments were badly weathered unidentifiable bits, but the team managed to piece together three of the arm bones, part of a femur, and a small section of a lower leg bone. Based on the shape of the lower jawbone and teeth, Johanson and White decided the remains were those of *Homo habilis,* the handyman ancestor Louis and Phillip Tobias had named in 1964. If so, it marked the first time that a part of a *Homo habilis* skull had been found with skeletal remains; the fossils would give scientists their first definitive clues to this ancestor's physique.

News of Johanson and White's discovery traveled quickly among their paleoanthropological colleagues. Not only had the Americans discovered an important fossil; in doing so, they had made Mary Leakey look inept—something that others suspected was one of Johanson and White's motives in going to Olduvai anyway. The fossil, after all, was found on the surface, only twenty-five yards from Olduvai's main road. In all likelihood it had been lying there for several years, but Mary's team had not spotted it. "The twists of fate can be very cruel at times," the anatomist Michael Day wrote in a sympathetic note to Mary, after hearing rumors of the new fossil.

Johanson himself had written Mary about their discovery in September. He did not boast or gloat, but merely announced the find, then asked her what field number should be assigned to the new fossil. Mary's answer was terse. "My congratulations on finding a partial skeleton in Bed 1 at Olduvai. The last hominid number allocated by me was O.H. 61 Your skeleton thus becomes O.H. 62." Beyond that, Mary had nothing to say about the fossil. An archeologist, not a physical anthropologist, Mary had made a strict rule never to comment on hominid fossils publicly.

In fact, the magazine and newspaper stories that followed Johanson and White's announcement of their discovery were notable for what was conspicuously missing: comments from either Richard or Mary. Richard had decided, he told reporters, to offer no opinion because he had not seen the actual specimen. But his decision had another, important effect—this time, and for the first time in years, there were no articles discussing the rivalry between the Leakeys and the Americans. Indeed, if the Leakeys were mentioned at all, it was only a historical reference, noting that Louis and Mary had made many famous discoveries at Olduvai Gorge.

In private, though, Richard had a lot to say about Johanson and White's find and their interpretation of the fossil—which, he conceded in *Origins Reconsidered,* had roused his "competitive instincts." For starters, Richard and Walker found the Americans' assertion that the fragmentary remains were a "partial skeleton" a bit of a stretch. Walker dismissed the find as "*really* scrappy," a judgment he based on several equally incomplete specimens that had been found at Koobi Fora. Because these were fragmentary remains, Richard had never gone to the trouble of calling a press conference to announce them.

Beyond that, Richard and Walker disputed—again privately—the Americans' interpretation of the fossil. To their eyes, it was unlikely that this hominid was *Homo habilis.* Based on the length of the creature's femur, Johanson and White had calculated that the hominid would have stood only three and one-half feet tall. Because it was so small, they deduced that they were looking at a female skeleton. Not only was she tiny; she had a surprisingly apelike build, with arms so long in proportion to her legs that her hands would have dangled at her knees. If the fossil was really that of *Homo habilis,* then according to the lineage accepted by most paleoanthropologists, this tiny, apey creature gave rise to strapping *Homo erectus* (the Turkana Boy)—and within a mere 200,000 years. This, at least, was the scenario that Johanson and White were painting. "We sort of expected *Homo habilis* to show a much more even continuum evolution in body type from previous species, leading up to *erectus,*" White said at their press conference. "Now we think the change came all of a sudden."

Richard and Walker, however, thought it much more likely that Johanson and White were looking at an as-yet-unnamed new species: some type of gracile australopithecine. As far back as 1978, in a *Scientific American* article, Richard and Walker had toyed with the idea of naming such a species, using certain Koobi Fora and Olduvai fossils that did not fit the *Homo habilis* mold (even though some of them, such as Olduvai Hominid 13, had been classified as such). "That's part of the problem," Richard would later say. "*Homo habilis* is such a grab-bag mix of fossils; almost anything around two million years that doesn't fit the robust definition has been tossed into it." For example, OH 13, which included the palate, base, and back of the skull, was strikingly similar to the Koobi Fora hominid 1813. But 1813, a nearly complete skull dating to 1.6 million years ago, had never been classified, so confusing were its features: it had the large face and small cheek teeth of early *Homo,* but also the small brain of *Australopithecus.* Such a mix of features baffled paleoanthropologists, some of whom regarded 1813 as a female *Homo erectus,* while others argued that it was a female *Homo habilis.*

Richard and Walker thought it more likely that 1813—as well as OH 13 and Johanson and White's new OH 62—represented something like the South African *Australopithecus africanus.* If they were correct, then, as they had often argued, three hominids had lived side by side at Koobi Fora and Olduvai two million years ago: this gracile australopithecine, *Australopithecus boisei,* and *Homo habilis.*[5]

But Richard did not discuss any of these ideas with reporters who were then writing about Johanson and White's discovery, saving them for *Origins Reconsidered.* Other paleoanthropologists, however, raised the same issues, with most agreeing that *Homo habilis* was a taxonomic designation in need of some work. Johanson and White countered that the differences between the fossils in the *H. habilis* bin could be explained by sexual dimorphism—differences between the sizes of males and females. They had used the same reasoning to explain the size variations among the fossils from Hadar in Ethiopia. In their view, the new hominid represented a pattern they had seen before: large males and tiny females. The pattern just persisted for a longer period of time in early human ancestry than had been previously thought, they argued.[6]

Richard, as usual, said that there were too few fossils for a satisfactory debate. He continued to avoid reporters' questions about his differences with Johanson and White. Nevertheless, an article in the *San Francisco Examiner's* Sunday magazine, *Image,* was entitled "Bone Wars—Two of the Biggest Egos in the Study of Human Origins Are Slugging It Out over a 3.5-Million-Year-Old Woman Named Lucy." Published on August 23, 1987, it appears to have been the last such story about Richard's and Johanson's combative relationship.

---

[5] Richard and Alan Walker contend that until additional specimens are found and the fossils attributed to *Homo habilis* sorted out, there is not sufficient material to name this new species.

[6] In a 1988 study, Bernard Wood, Daniel Lieberman, and David Pilbeam questioned the idea that sexual dimorphism—big males and little females—could explain the size and shape differences among the fossils of these early hominids. Using fossils from Koobi Fora that supposedly represented a male and female, the scientists found that the creatures would have been more sexually dimorphic than gorillas are—an unlikely case, since gorillas have the greatest sexual dimorphism of all the apes. They concluded that the extreme sexual dimorphism that Johanson and White had hypothesized in the early hominids was "highly improbable." However, other scientists using different measurements agree with Johanson and White.

That year, Richard turned forty-three. Seven years had passed since his kidney transplant, and in the interim he had changed physically and emotionally. The steroids he took made his face puffy and swollen, and since he made little effort to curb his love of gourmet cooking, he was overweight. His hair, too, had begun to grey at the temples, and he had begun to hunch slightly as his father had in his later years, though he still walked across the desert sediments with a long, easy stride. ("He's beginning to look just like Louis!" Mary once commented on the changes in Richard's physique.) But perhaps the most noticeable change, said those who had known Richard both before and after his transplant, was his temperament.

"The man you see today and the man of ten years ago are almost two entirely different people," noted anthropologist M. E. Morbeck, who had known Richard since 1970. "When he was younger, he was very high-powered, determined to always be number one; now he's a mellow family-government man, and much more interesting." Others observed the same transformation. "His ego is certainly no longer involved to the same degree it once was in human paleontology," said Harry Merrick, an archeologist at Kenya's National Museums.

Richard himself thought the change had less to do with his operation than with the mere accumulation of years. "I think everyone mellows to a certain extent when they get into their forties," he said, waving aside the suggestion that a new kidney had anything to do with his fresh outlook on life. But he also no longer had to suffer with the nearly incapacitating headaches and hypertension of his kidney disease, which had once left him fractious and weary. Now, too, he finally had managed to shake himself free of what he called "the incredible nastiness" of paleoanthropologists, simply by avoiding the conferences and refusing to comment on Johanson and White's activities. "These fights do real injury to the science, but I'm glad I didn't get out," he told a reporter in 1986. "I am never quite so happy as when out in the field."

Richard was once again enjoying himself. He was executive director of the National Museums of Kenya, which now included seven regional branches as well as the thriving Institute for Primate Research. He sat on the boards of several colleges, both in Kenya and overseas, and other scientific institutions, and was so highly regarded by the Kenyan government that he often acted as a behind-the-scenes adviser to President Daniel arap Moi. At least twice a year, he traveled abroad to lecture, using most of these earnings to pay the expenses of half a dozen Kenyans studying for their doctorates in paleoanthropology and archeology overseas (a simpler method, he had decided, than trying to do this through a foundation). His eldest daughter, Anna, by his first marriage, was just beginning to study ecology at the University of London, while Louise and Samira were in secondary school. He and Meave were building a new home and starting a vineyard as well on fifty acres south of Nairobi in the Ngong Hills. They had time, too, to search for fossils in the summer and winter, and to sail at Lamu Island, where at Christmas they hosted reunions of the entire Leakey clan.

Richard and Mary had settled their previous quarrels during Richard's transplant. "She was forced to show emotions then that she obviously always had, but had never shown before," said Richard, who had often wondered how much his mother actually cared for him. "She came over to England, spent time at the hospital, and worried, and lost weight and went grey-haired. And that meant a great deal." Mary now came nearly every week for Sunday dinner at his home.[7]

Yet despite his full life and hectic schedule, Richard was increasingly restless. "I need a new challenge," he told a friend in 1987, confessing that the daily grind of administering the museum had grown old and stale, while paleoanthropological debates increasingly bored him. He was considering leaving the museum for some new enterprise. Perhaps he would launch a new kind of secondary school for Kenyans, build a Pan-African museum, or become involved at a high level with an international wildlife organization. (Richard had served as the chairman of the Wildlife Clubs of Kenya Association since 1969, and was also chairman of the East African Wildlife Society, an organization that Louis had helped to found.)

But all of these ideas as well as his involvement in a planned 1989 West Turkana Expedition fell to the wayside that spring. On the evening of April 20, it was announced on the radio that President Moi had appointed Richard "the new Director of the Department of Wildlife Conservation and Management with immediate effect." At the time, Kenya's wildlife and national parks were in desperate straits. Poachers armed with automatic weapons were gunning down hundreds of elephants annually, and killing park rangers and tourists as well. And Moi, who admired Richard's leadership at the National Museums and valued his advice on numerous topics, now asked him to do what previous wildlife directors had seemed unable to do: stop the poaching and restore peace and order to Kenya's national parks. Richard immediately resigned his position at the National Museums, although he stayed on as chairman of the board, and turned over the West Turkana Expeditions to Meave, who was head of the museum's paleontology department, to direct.

Moi had given him what he had hoped for: a new challenge. Richard faced it with characteristic Leakey determination. In an explanatory prologue about his career switch in *Origins Reconsidered,* his book about the Turkana Boy, he wrote simply, "I plan to succeed."

Richard was now in charge of Kenya's wildlife, and although he did not know it at the time, he probably would never search for hominid fossils again.

---

[7] Richard was clearly by now her favorite son, something she mentioned to several close friends. "Richard has returned in very good form," she wrote, for example, in 1987 to one friend, after Richard came home from a trip abroad. "It's good to have him back. I'm so spoilt, having all my family round me that I feel sad when one is missing—especially Richard."

Epilogue

For five years, from 1989 to 1994, Richard directed the Kenya Wildlife Service, as the country's national parks system was renamed following his appointment. During his tenure, he built KWS's formerly dispirited ranger corps into an elite fighting force that finally put an end to elephant and rhino poaching. Richard also helped enact an international ban on the ivory trade, and attracted several hundred million dollars in foreign grants and loans to rebuild Kenya's ravaged park system. His firm stand against the corruption and bribery that had once pervaded the wildlife department made him a hero in Kenya, where he was often greeted by people on the street with shouts and cheers.

But Richard made enemies as well—particularly among high-placed Kenyan officials who had been publicly accused of involvement in the poaching and illegal ivory trade. To assure his own safety, Richard arranged a twenty-four-hour armed guard of handpicked rangers to patrol his home, and wherever he went a car bristling with armed bodyguards followed.

They could not protect him everywhere, however. On June 2, 1993, shortly after he took off from Nairobi airport for a trip to a nearby national park, Richard's single-engine Cessna began to lose power, and Richard realized he would have to make a crash landing. Besides himself, four members of his park staff were on board. "It occurred to me that if I did not handle the crash correctly, there would be no survivors," he told a reporter a month after the accident. "So I told the passengers in as lighthearted a way as possible that they were going to have to find their own way from this point on, looked for friendly trees to hit, turned off the ignition and tried to come in level. Unfortunately, I could not see one stout old mango tree." The left wing of Richard's plane clipped the tree, and the plane flipped end-over-end and broke apart. Although the cause of the accident was never discovered, many people in Kenya believe it was an attempted assassination.

Richard and his passengers survived, but he sustained the worst injuries. He had a severe skull fracture, and his legs and ankles were shattered. In spite of numerous surgeries in England to save his legs, Richard lost both of them below the knee. He treated the amputations with his usual humor and determination. Prior to the surgeries, he asked the physicians to preserve his legs so that he could take them back to Kenya for a burial service ("the only part of my funeral I'll be able to attend alive"), and jokingly announced that the title of the second volume of his autobiography would be *Two Feet in the Grave.*

At home in Kenya, the accident did not win him any reprieve from his enemies. Instead, politicians who believed he had accrued too much power and run the KWS in an arrogant manner launched an all-out attack against him in Kenya's parliament and the government-run press. When President Moi authorized a special committee to investigate the alleged wrongdoings at KWS, Richard resigned. "I always said I would stay at KWS as long as I had the support of our head of state," he said, "and when it seemed I no longer had that, I quit." David Western, a conservation biologist with Wildlife Conservation International, was named Richard's successor at KWS.

Richard's fall from grace also had repercussions at the National Museums of Kenya, where he was chairman of the board of trustees. He had resigned as the museum's director when appointed to KWS, but before leaving he had helped pick his replacement, Mohammed Isahakia, then director of the museum's Institute of Primate Research. By 1994, however, Richard was unhappy with the way Isahakia was running the museum. Before he could do anything about Isahakia, however, Richard was relieved of his museum position by the government in the fall of 1994. He is no longer officially affiliated with the National Museums of Kenya.

Fitted with artificial legs, Richard learned to walk again only two months after the second amputation. He now walks without a cane, and has also regained his pilot's license. Committed to continuing his conservation work, he founded a pan-African wildlife consulting agency, Richard Leakey & Associates. As he did before his accident, Richard travels abroad twice a year to lecture and raise funds on behalf of Kenya's wildlife, paleontological research, and the Kenyan students he is supporting.

Although Richard essentially withdrew from paleoanthropology when appointed director of KWS, he has not lost his love for the research or for fossils. He cannot search for fossils as he used to, but he still enjoys excavating them. "I take off my legs and get to work," he explained. In fact, the desert heat, not the rough terrain, causes him the most trouble. Without his legs to help shed heat and regulate his body's temperature, Richard finds the desert's high temperatures extremely uncomfortable. "I just can't be in the desert now for more than a few days at a time," he said.

Meave, who heads the National Museum's paleontology department, now directs the Turkana expeditions, and Richard occasionally joins her in the field. Under her leadership, the Hominid Gang has continued to make discoveries. Their most significant find—fossils of a new hominid species dating in the four-million-year range—will be announced in the fall of 1995. Mary Leakey has completed her analysis of her Olduvai artifacts, and recently published the fifth volume of the *Olduvai Gorge* series, *Excavations in Beds III, IV & the Masek Beds.* She is eighty-two, still smokes cigars and drinks Scotch, and lives in the Langata home she and Louis built.

Meave may not be the last of the Leakey fossil dynasty. Louise, Richard and Meave's eldest daughter, graduated in May of 1995 from the University of Bristol with joint honors in zoology and geology, and is considering pursuing a doctoral degree in paleontology at Cambridge. She has also already raised funds for and directed small expeditions of her own in Kenya. Thus, when Meave left the Turkana expedition in 1993 to be at Richard's bedside following his plane crash, she never hesitated about turning over the field operations to her then twenty-one-year-old daughter. "Louise is running the expedition," Richard proudly told a friend from his hospital bed. "I think she's going to be the one who picks up the family banner."

Appendix

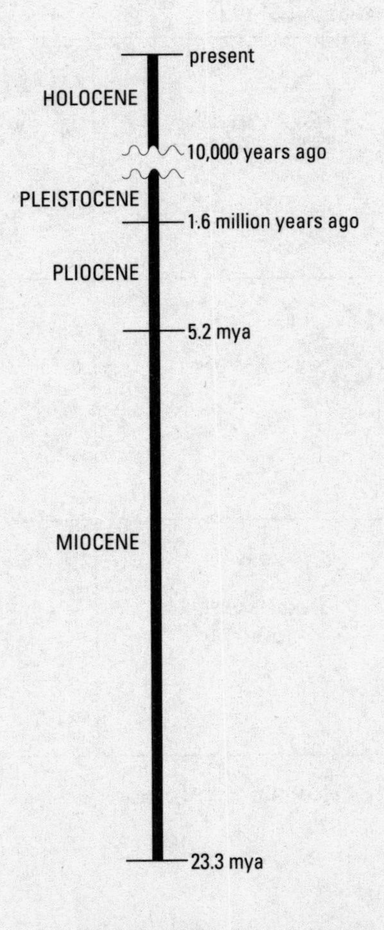

Geologic Epochs

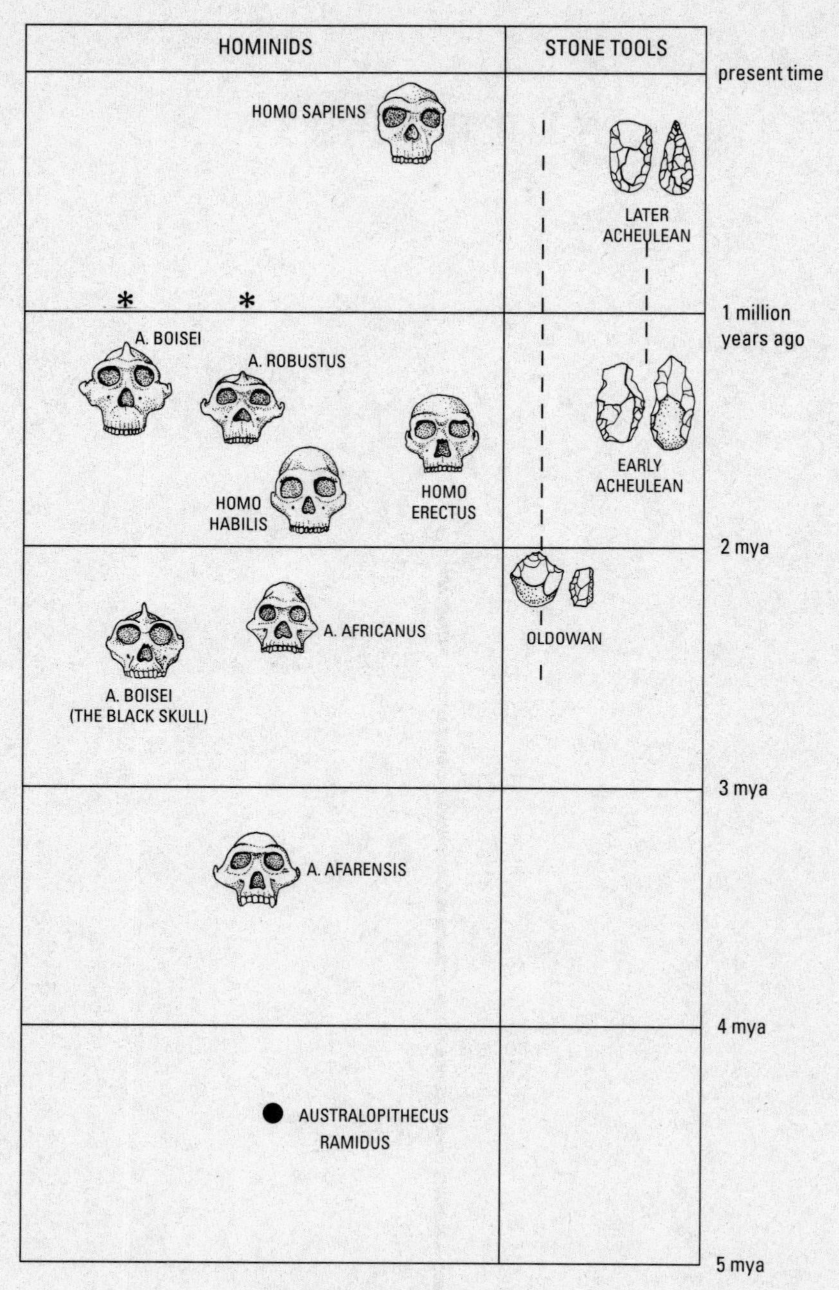

Hominid Evolution

Notes

Key to Abbreviations

BMNH	British Museum of Natural History
KNA	Kenya National Archives
NMK	National Museums of Kenya

Chapter 1. Kabete

14 "wild men," "cannibal feast": L. S. B. Leakey, *White African,* pp. 16–17.
14 "We all traipsed," "Father was . . . carried us along": Julia Barham, interview, London, June 16, 1984.
16 "that miserable scrap-heap": Ewart S. Grogan, as quoted in Norman Wymer, *The Man from the Cape,* p. 144.
16 "tin-pot mushroom," "most lawless" : Francis Hall, *Letters,* Vol. 3, p. 105.
17 "making a hearty meal off his moustache": T. F. C. Bewes, unpublished diary, June 20, 1929 (courtesy of Sarah Howard).
17 "May flower": unpublished reminiscences of Mary Leakey, recalling the nickname Harry Leakey's sister Julia gave her, p. 3 (NMK).
18 "quick, quick, quick": Julia Barham, interview, London, June 16, 1984.
18 "It was nothing . . . at all": *White African,* p. 8.
18 "best-washed baby": Louis quoted in Melvin M. Payne, "Family in Search of Prehistoric Man," *National Geographic,* February 1965, p. 200.
19 "white man's country . . . as we will": Charles Eliot, *The East African Protectorate,* p. 80.
20 "very keen on natural history": Julia Barham, interview, London, June 16, 1984.
20 "We were very animal-minded . . . when they grew up": ibid.
20 "slept with his head . . .": ibid.
20 "nervous and highly strung": ibid.
20 "Gladys was always . . . collected information": ibid.
21 "think and even dream": "Family in Search of Prehistoric Man," op. cit., p. 200.
21 ("We called him. . . . and teased": Julia Barham, interview, London, June 16, 1984.
21 "My mother . . . a plant": ibid.

22 "like a fish . . .": *White African,* p. 27.
22 "happy years . . . incidents": ibid., p. 28.
22 "trying to keep. . . . among the Africans": Julia Barham, interview, London, June 16, 1984.
23 "in language . . . Kikuyu": *White African,* p. 32.
23 "perfectly *awful*": Julia Barham, interview, London, June 16, 1984.
23 "freedom," "possess things": *White African,* p. 33.
23 "great friendship": ibid., p. 60.
23 "patience . . . and observation": Louis Leakey, quoted in "Family in Search of Prehistoric Man," op. cit., p. 206.
23 "hunting passion . . . logic": *White African,* p. 65.
24 "light of the Kikuyu": Chief Koinange quoted in Sonia Cole, *Leakey's Luck,* p. 35.
24 "very unselfish". . . ."dogmatic": Julia Barham, interview, London, June 16, 1984.
24 "kind of hero-worship": *White African,* p. 67.
24 "born with . . . the other": Arthur Loveridge, *Many Happy Days I've Squandered,* p. 1.
24 "chaotic condition": ibid., p. 45.
24 "He lived. . . .'broken bottles' ": Julia Barham, interview, London, June 16, 1984.
25 "throw away . . . piece": L. S. B. Leakey, *Adam's Ancestors,* p. 3.
25 "very, very oldest people": Heselon Mukiri, interview, Nairobi, October 30, 1984 (Kamoya Kimeu, translator).
25 "holes in the ground": Jomo Kenyatta, *Facing Mount Kenya,* p. 32.
25 "I'd thought . . . at me," "certainly implements": *White African,* p. 69.
25 "delighted beyond words": ibid.
25 "I firmly . . . Stone Age there": ibid., p. 70.

Chapter 2. From Cambridge to Olduvai

26 "keen to go": L. S. B. Leakey, *White African,* p. 80.
26 "Louis always knew. . . . to suck eggs' ": Julia Barham, interview, London, June 16, 1984.
27 ("I had . . . over two years"): *White African,* p. 83.
27 "fantastic," ". . . made me very bitter": ibid., p. 89.
27 "Oh, he hated it. . . . spell of his life": Julia Barham, interview, London, June 16, 1984.
27 "little boys of thirteen. . . . its helplessness": *White African,* p. 88.
28 "simply shrugged. . . . achieving what I wanted": ibid., p. 91.
28 "Louis, wherever he went. . . . about him": Julia Barham, interview, London, June 16, 1984.
28 "had examined himself in Kikuyu": *White African,* p. 157.
28 "senior wangler": Sonia Cole, *Leakey's Luck,* p. 52.
28 "Fancy! Tennis in shorts!": Beryl Leakey, interview, Cambridge, June 2, 1984.
28 "indecency": *White African,* p. 180.
28 "My husband. . . . keen enough": Julia Barham, interview, London, June 16, 1984.
28 "overcharged": *Leakey's Luck,* p. 42.
28 "Louis's fellow students . . . with a big ego": E. Barton Worthington, letter to author, May 20, 1986.
29 "very happily and very quickly": *White African,* p. 94.
29 "greatest ambitions": ibid., p. 97.
29 "I little thought . . . my whole career": ibid., p. 98.
29 "My luck . . . in a most unexpected manner": ibid., p. 99.
30 "on a great adventure": ibid., p. 107.
30 "little eccentricities and pretensions": Loris S. Russell, *Dinosaur Hunting in Western Canada,* p. 32.
30 "prided himself on going . . . simplest of food," "unnecessary luxuries": obituary by C. W. Hobley, *Nature,* September 12, 1925, p. 406.
31 "I think them chipped . . . grass fires": Cutler's field diary, June 11, 1924 (BMNH).
31 "yelling like a madman . . . a trooper": *White African,* p. 121.
31 "Leakey very low. . . . aching bones": Cutler's field diary, June 16, July 30, August 31, 1924.
31 "late Mr. C. . . . theoretical study": L. S. B. Leakey, *Adam's Ancestors,* p. 4.
32 "I wonder how. . . . his trumpet hard": Cutler's field diary, December 26, 1924.
32 "Thankful to say . . . Leakey is missing": ibid., January 22, 1925.
32 "murderous bread": ibid., March 20, 1925.
32 "Owing to fever. . . . went to bed": ibid., August 21, 1925.
32 "Maybe if he . . . to carry on his work": C. W. Hobley, *Nature,* op. cit.

32 "blew it to a thousand pieces": *White African,* p. 128.
32 "audience of friends": ibid., p. 153.
33 "much thought," "... native customs," "really be ... 'part time job' ": ibid., p. 161.
33 "At last. ... were coming true": ibid., p. 160.
33 "a real love of teaching": A. Hingston Quiggin, *Haddon, the Headhunter,* p. 119.
33 "Cinderella of the Sciences," "Colleague ... Comrade": ibid., p. 116.
33 "freely, wisely ... 'except higher mathematics' ": ibid., p. 123.
33 "to win souls ...," " 'You can travel ... piece of string' ": ibid., p. 128.
33 "suitable piece of string": Mary and Louis Leakey, *Some String Figures from North East Angola,* p. 9.
34 "very warm friendship," "studying ... of East Africa": *White African,* p. 177.
34 "Half jokingly ... visit Oldoway together": ibid., p. 177.
34 " 'everyone knew he had started in Asia' ": ibid., p. 2.
35 "very reluctant," "very disappointed," "... think of digging": ibid., p. 73.
35 "knives and weapons": ibid., p. 188.
35 "search for the Gumba": L. S. B. Leakey, *The Stone Age Cultures of Kenya Colony,* p. 243.
35 "the oldest culture in East Africa": *Adam's Ancestors,* p. 5.
36 "surprise," "... in East Africa too": ibid., p. 5.
36 "struck by the mineralised condition": L. S. B. Leakey, *The Stone Age Races of Kenya,* p. 58.
36 "important": *White African,* p. 194.
36 "the site must be examined in detail": *The Stone Age Races of Kenya,* op. cit., p. 58.
36 "large and airy": *White African,* p. 195.
37 "ideal places ... to live in": ibid., p. 200.
37 "We'd gone out. ... if they were down": Frida Leakey, interview, Cambridge, June 2, 1984.
37 "lively mind ... love of talk": Penelope M. Jenkin, letter to the author, November 29, 1985.
37 "The supper table ... archeology till dawn": Frida Leakey, interview, Cambridge, May 27, 1984.
38 "Louis was obsessional. ... always an immediacy": ibid.
38 "It was a very daft. ... completely mad" to "appalled. ... more conventional": Frida Leakey, interview, Cambridge, June 2, 1984.
39 "magnificent": *White African,* p. 214.
39 "Camp conditions ... almost luxurious": Elisabeth Kitson, letter to author, September 1, 1985.
39 "[The camp] consists of. ... just bare earth": Elisabeth Kitson to her parents, November 11, 1928.
40 "But then. ... flock of pelicans": Penelope Jenkin, letter to author, November 29, 1985.
40 "extraordinarily clever with his hands": *White African,* p. 210.
40 "lively little ... sense of humour," "good-looking ... brunette": John D. Solomon, letter to author, October 30, 1985.
40 "as I planned ... not come too": *White African,* p. 210.
40 "enthusiastic and vivid ... in time and energy": Penelope Jenkin, letter to author, November 29, 1985.
40 "schoolboyish": Elisabeth Kitson, letter to author, September 1, 1985.
40 "three corrugated ... vast plain," "quite the back ...," "... exciting drive": Elisabeth Kitson to her parents, November 28, 1928.
41 "L. said we. ... brought a gun": ibid.
41 "... in the middle of. ... brought a gun": ibid.
41 "to provide ... so tame and disappointing": ibid.
41 "raid by the Maasai," "...'look after the women' ": *White African,* pp. 221–23.
41 "almost as humdrum ... suburb of London": ibid., p. 221.
42 "rough and plain": Elisabeth Kitson, letter to author, September 1, 1985.
42 "We lived by the gun. ... shooting the animals": Frida Leakey, interview, Cambridge, May 27, 1984.
42 "Louis was the mainspring ... he was wonderful": John D. Solomon, letter to author, October 30, 1985.
42 "... he put the work ... happiness of his workers": Elisabeth Kitson, letter to author, September 1, 1985.
42 "If you were part ... human beings": Frida Leakey, interview, Cambridge, May 27, 1984.
42 "[W]e were gradually ... who made them": *Adam's Ancestors,* p. 7.
42 "By that time. ... monomaniac interest in the thing": Frida Leakey, interview, Cambridge, May 27, 1984.

43 "a rather typical experiment!": Elisabeth Kitson, letter to author, October 30, 1985.
43 "Louis possessed a special. . . . for making a fish harpoon": Frida Leakey, interview, Cambridge, May 27, 1984.
43 "[I]t would not be . . . waves north and south": L. S. B. Leakey, East African Archaeological Expedition, 1928–29, Second Report, November 1, 1928.
43 "Nobody could possibly. . . . maximum possible antiquity": John D. Solomon, letter to author, October 30, 1985.
44 "important changes of climate": *White African,* p. 201.
44 "prove conclusively that . . . by arid periods": *Stone Age Cultures of Kenya Colony,* p. 9.
44 "In retrospect . . . location of sedimentation": John D. Solomon, letter to author, January 16, 1986.
44 "If I remember. . . . were usually correct: ibid.
45 "The discovery was . . . Great Rift Valley": *White African,* p. 234.
45 "A near-fatal step. . . . in the wall": Melvin M. Payne, "Family in Search of Prehistoric Man," *National Geographic,* February 1965, p. 200. "It was Dr. . . . from the cliff": *Visitor's Guide to Kariandusi Prehistoric Site,* p. 2.
46 "I hear you. . . . to make a splash": A. C. Haddon to L. S. B. Leakey, May 1929 (quoted in *Leakey's Luck,* p. 69).
46 Harm is done . . . in all friendship: E. J. Wayland to L. S. B. Leakey, May 1929 (quoted in *Leakey's Luck,* p. 69).
46 "Raymond Dart didn't agree. . . . in Livingstone": J. Desmond Clark, interview, Berkeley, February 5, 1986.
47 "noncommittal comments," ". . . evolutionary story": Raymond Dart, *Adventures with the Missing Link,* p. 54.
47 "Louis did not . . . respect for him," "Nobody could seriously . . . the material obtained": John D. Solomon, letter to author, October 30, 1985.
47 "Almost as soon . . . my main objective": *White African,* pp. 272–73.

Chapter 3. Laying Claim to the Earliest Man

50 "a creature with . . . ape-like face": Arthur Smith Woodward, "Missing Links Among Extinct Animals," *Report of the British Association,* p. 783.
50 "a blend of man and ape": ibid.
50 "Sir Arthur Keith . . . father in science": Frida Leakey, interview, Cambridge, May 27, 1984.
50 "one-man 'court of appeal' ": anonymous, "Lives of the Fellows" (Royal College of Surgeons), 1970, p. 222.
50 "alphabet by which . . . man and ape": Arthur Keith, *An Autobiography,* p. 170.
50 "an ancient form . . . is usually believed": Arthur Keith, *The Antiquity of Man,* p. 732.
50 "ancient population made up . . . species and genera": ibid., p. 711.
51 "Amongst this complex. . . . because of his brain": Arthur Keith, *New Discoveries Relating to the Antiquity of Man,* p. 33.
51 "human remains . . . stone implements," "As to the . . . can be [no] doubt": Arthur Keith, *An Autobiography,* p. 501.
51 "helpful and kind teacher": L. S. B. Leakey, *White African,* p. 207.
51 "enormous quantity of artifacts": Frida Leakey, letter to author, June 9, 1986.
51 "[I]f Mr. Leakey's . . . contemporaries in Europe": *New Discoveries Relating to the Antiquity of Man,* p. 170.
51 "weeks of work": *White African,* p. 273.
52 "unwrapping a skeleton": *The Times,* December 28, 1929, p. 12.
52 "He was a man . . . judgments about things": Keith, *An Autobiography,* p. 501.
52 "Mr. Leakey believes . . . black than white": *New Discoveries Relating to the Antiquity of Man,* p. 173.
52 "resigned himself to his fate": *White African,* p. 287.
53 "I was now certain . . . culture at Oldoway," "diligently," "bones of fossil animals," "I . . . made a small . . . arriving there": ibid., p. 274.
53 "hither & thither . . . ," "joint adventure": Frida Leakey, letter to author, June 9, 1986.
53 "a whole-hearted . . . thing with her": Penelope Jenkin, letter to author, November 29, 1985.
53 "I was awfully. . . . interested in Louis": Frida Leakey, interview, Cambridge, June 24, 1984.
53 "Mr. Leakey's . . . get into difficulties": A. C. Haddon to the Master of St. John's College, September 23, 1929 (courtesy of St. John's College Library).

54 "[W]e didn't write. . . . & sorted flints": Glyn Daniel, letter published in *The Times,* October 12, 1972, p. 18.
54 "innumerable measurements with calipers," "mechanically minded": L. S. B. Leakey, *By the Evidence: Memoirs, 1932–1951,* p. 25.
54 "astonished on seeing . . . from Elmenteita": M. C. Burkitt, *South Africa's Past in Stone and Paint,* p. 168.
54 "first class pioneering . . . Archaeology and Anthropology": A. C. Haddon to the Master of St. John's College, September 23, 1929 (courtesy of St. John's College Library).
54 "Edwardian dress allowance": Frida Leakey, letter to author, June 9, 1986.
55 "made a note . . . the 1931–2 season": L. S. B. Leakey, *The Stone Age Races of Kenya,* p. 9.
55 "It makes me. . . . gone to bed": *By the Evidence,* p. 17.
55 "Oh, my head . . .": *White African,* p. 154.
55 "moments of loss of consciousness": Frida Leakey, letter to author, July 23, 1985.
55 "I was always . . . if I overworked": *White African,* p. 154.
55 "His attacks are. . . . driving a car: quoted in Sonia Cole, *Leakey's Luck,* p. 79.
56 "big expedition to Oldoway": *White African,* p. 276.
56 "completely tired out," ". . . train ticket," "proper doors . . . served its purpose": ibid., p. 279.
56 "from the bottoms. . . . a liberal thinker": Frida Leakey, interview, Cambridge, May 27, 1984.
56 "Louis would always. . . . misread the Bible' ": Julia Barham, interview, London, June 16, 1984.
56 "early Stone Age man," "genial": L. S. B. Leakey, *Adam's Ancestors,* p. 8.
56 "essentially a hunter . . . was most plentiful," "[T]he question of . . . reward us or not": ibid., p. 9.
57 "There was a road . . . the Serengeti plains": *White African,* p. 283.
57 "practicable route to Oldoway": ibid.
58 "native staff," ". . . ten of my excavators": L. S. B. Leakey, East African Archaeological Expedition, 1931–32, Second Field Report, September 1–30, 1931, p. 2 (NMK).
58 "[P]rogress was slow," "very dusty," "appalling": *White African,* p. 285.
58 "a serious drain upon our water supply": ibid., p. 287.
58 "Professor Reck could . . . great scientific discoveries": ibid.
58 "visibly moved," ". . . banks of the Gorge": ibid., pp. 287–88.
58 "a scientist's paradise": ibid., p. 296.
59 "awful visions . . . conclusion," "long-planned expedition": ibid., p. 291.
59 "guide fires," ". . . many of them lions": ibid., p. 294.
59 "prospect of great . . . at Oldoway": ibid., p. 292.
59 "soon after dawn," ". . . in my joy": ibid.
59 "four main objects," "numerous fossil remains": L. S. B. Leakey, "Earliest Man in East Africa," *East African Standard,* October 10, 1931, p. 1.
59 "There it is . . . my four markers": Sir Vivian Fuchs, interview, London, November 22, 1984.
59 "Personally I believe. . . . need re-examination": East African Archaeological Expedition, Second Field Report, p. 4.
60 "almost beyond question . . . of *Homo sapiens*": L. S. B. Leakey, "Earliest Man, Discoveries in Tanganyika," *The Times,* March 9, 1932, p. 11.
60 "batty": Frida Leakey, interview, Cambridge, May 27, 1984.
60 "It was all good fun . . . from a baby": Wilfrida Leakey, "Priscilla in Darkest Africa," *The Boat Train,* p. 120
60 "When Priscilla arrived . . . at the time": Frida Leakey, interview, Cambridge, June 2, 1984.
61 "mystical feeling", "great sense of timelessness": Alan Gentry, interview, London, May 23, 1984.
61 "veritable paradise": *White African,* p. 295.
62 "a large number . . . side of the cliff": L. S. B. Leakey, East African Archaeological Expedition, 1931–32, Fourth Field Report, November 1931, p. 1 (NMK).
62 "lying mixed up with the tools": ibid., p. 2.
62 "a horde of . . . on the spot": ibid., p. 3.
62 "was received . . . frank incredulity": *White African,* p. 296.
62 "green scum": field diary of V. E. Fuchs, October 23, 1931.
62 "fine black dust. . . . dust, dust, dust!": *White African,* p. 296.
62 "pitched battle . . . ," "point blank": ibid., p. 297.
62 "heard first one. . . . skipped to the fire": field diary of V. E. Fuchs, October 23, 1931.
63 "evolutionary stages . . . hand-axe culture": *White African,* p. 295.
63 "very simple," "pebble . . . ," "Oldowan culture": *Adam's Ancestors,* p. 105.

63 "[W]e were well content . . . past two months": *White African,* p. 299.
63 "my young friend," ". . . highest importance": *New Discoveries Relating to the Antiquity of Man,* pp. 155, 172.
63 "In the light. . . . inhabitants of East Africa": ibid., p. 24.
63 "I have in . . . culture at Oldoway": L. S. B. Leakey, East African Archaeological Expedition, 1931–32, Fifth Field Report, January 5, 1932, p. 4.
64 "the relics of his meat feast": L. S. B. Leakey, *The Times,* March 9, 1932, p. 11.
64 "When [the test] proved . . . the older deposits": John D. Solomon, letter to author, March 30, 1985.
64 "rubbish" ". . . OLDOWAY SKELETON," ". . . convince you all!!": Louis Leakey to Sir Arthur Keith, February 26, 1932.
64 "great hopes that . . . discoveries at Oldoway": L. S. B. Leakey, East African Archaeological Expedition, 1931–32, Eighth Field Report, March 1932, p. 1.
65 "fragments of . . . exposure," "no doubt at all": ibid.
65 I've been working . . . "Anti-Oldoway-Man" group!!: Louis Leakey to Arthur Hopwood, March 18, 1932 (BMNH).
65 "Whereas the constant . . . much of it," "We used to . . . during one meal": *White African,* p. 306.
65 "so muddy . . . ," "wheels of the car": *Adam's Ancestors,* p. 205.
66 "all the necessary . . . ," "pushing and pulling . . . ," ". . . head to foot": ibid., pp. 204–205.
66 "almost vertical cliff," "dislodged a . . . matrix": *The Stone Age Races of Kenya,* p. 10.
66 "noticed these teeth . . . fragment": *Adam's Ancestors,* p. 206.
66 "the Kanam mandible . . . of *Homo sapiens*": ibid.
66 "definitely *in situ* . . . to hold his hand!!": Louis Leakey to Arthur Hopwood, May 20, 1932 (BMNH).
66 "[T]he fact remains . . . else but recent": Arthur Hopwood to Louis Leakey, October 3, 1932 (BMNH).
67 "Everyone admits that. . . . 'central African lakes' ": L. S. B. Leakey to Miles Burkitt, March 19, 1932 (courtesy University Museum of Archaelogy and Anthropology, Cambridge University).
67 "I can foresee. . . . make the world believe": Louis Leakey to Arthur Hopwood, August 14, 1932 (BMNH).

Chapter 4. Louis and Mary

68 "without a porthole," "We had to find. . . . my dress allowance": Frida Leakey, interview, Cambridge, May 27, 1984.
68 "green fields . . . and brambles": L. S. B. Leakey, *By the Evidence,* p. 15.
68 "awfully penurious," "Louis loved Cambridge. . . . was most grateful": Frida Leakey, letter to author, April 18, 1986.
69 "magnificent series": L. S. B. Leakey, *Stone Age Races of Kenya,* p. v.
69 "a combined emergency. . . . of the night": *By the Evidence,* p. 16.
69 "pigheaded": Peter Andrews, interview, London, November 23, 1984.
69 "He made. . . . of an 'enthusiast' ": John D. Solomon, letter to author, October 30, 1985.
69 "disprove the assertion": Arthur T. Hopwood to L. S. B. Leakey, August 16, 1932 (BMNH).
69 "So long as . . . triumph for you": ibid.
69 "genuineness . . . in East Africa": L. S. B. Leakey, "The Oldoway Human Skeleton," *Nature,* May 14, 1932, p. 578.
70 "oldest . . . modern man": L. S. B. Leakey, *Adam's Ancestors,* pp. 206–207.
70 "unpalatable": Phillip V. Tobias, "A Re-Examination of the Kanam Mandible," *Proceedings of the 4th Pan-African Congress of Prehistory,* p. 343.
70 "most prominent . . . in Great Britain": *By the Evidence,* p. 26.
70 "decidedly dogmatic": Sir Peter Kent, letter to author, October 19, 1985.
70 "accepted our views . . . horse to run," "indiscreet as ever . . . claims forthwith!": P. G. H. Boswell, unpublished autobiography, p. 217 (University of Liverpool and Imperial College of Science and Technology archives).
70 "lay before . . . a real *Homo*": *Adam's Ancestors,* p. 2.
70 "some difficulty . . . sufficiently cautious": P. G. H. Boswell, unpublished autobiography, p. 218.
71 "a most startling discovery": quoted in "Early Human Remains in East Africa" (report of a conference at Cambridge convened by the Royal Anthropological Institute), *Man,* 1933, no. 65, p. 210.

71 "congratulated Dr. Leakey . . . for the future": ibid., p. 67.
71 "true ancestor": *Adam's Ancestors,* p. 206.
71 "plausibility . . . of the human race": "Man and His Ancestors," *The Times,* October 21, 1933, p. 12.
71 "reached the age. . . . could be": *By the Evidence,* p. 17.
71 "I had . . . retreated from archaeology": Frida Leakey, letter to author, July 17, 1986.
71 "the latest . . . the Stone Age," "the remains . . . modern man": Leakey, preface to *Adam's Ancestors,* p. xi.
72 "a very stuffy . . . want to go": Mary Leakey, quoted in Melvin M. Payne, "Family in Search of Prehistoric Man," *National Geographic,* February 1965, p. 220.
72 "Louis could talk . . . it just right": Beryl Leakey, interview, Cambridge, June 2, 1984.
72 "unperturbed . . . or grandeur": Frida Leakey, interview, Cambridge, June 2, 1984.
72 "[Some people have made. . . . Nothing special": Mary Leakey, interview, Nairobi, July 9, 1984.
72 "in 1933 . . . meet Mary Nicol": *By the Evidence,* p. 24.
72 "the best representations . . . ever seen": ibid., p. 25.
73 "perhaps just . . . strictly necessary": Mary Leakey, *Disclosing the Past,* p. 43.
73 "[F]rom the time. . . . soon and frequently," "several girlfriends": ibid.
73 "solace": ibid., p. 42.
73 "He appreciated women. . . . responded to him": Betty Howell, interview, Berkeley, February 4, 1986.
73 "Women came to him. . . . human that way": Lita Osmundsen, interview, New York City, March 16, 1984.
73 "known to many others": *Disclosing the Past,* p. 42.
73 "I may have . . . the last to know": Frida Leakey, interview, Cambridge, June 2, 1984.
73 "to the core . . . to change [Mary's] mind": *Disclosing the Past,* pp. 43, 44.
73 "surreptitious lunches," ". . . became clear": ibid., pp. 112, 43.
73 "simply dazzled. . . . notice I took": ibid., p. 43.
74 "very intelligent and very attractive": Mary Leakey, interview, Nairobi, July 9, 1984.
74 "as an equal and a colleague," "complete failure": *Disclosing the Past,* p. 42.
74 "felt bad over . . . treating Frida": ibid., p. 45.
74 "This was not. . . . on around me": Frida Leakey, interview, Cambridge, November 29, 1984.
74 "see the . . . African collection": *Disclosing the Past,* p. 44.
74 "not a marriage. . . . understanding between them": Mary Leakey, interview, Nairobi, July 9, 1984.
74 "living separate lives": *Disclosing the Past,* p. 44.
74 "increasingly awkward": Mary Leakey, interview, Nairobi, July 9, 1984.
74 "Louis was sitting. . . . curious": Barbara Waterfield, interview, Cambridge, November 29, 1984.
74 "Tension mounted. . . . position to speak": *Disclosing the Past,* p. 45.
75 "the son Louis had so wanted," "fallen in love . . . her to Africa": Frida Leakey, interview, Cambridge, November 29, 1984.
75 "told us with . . . each of us," "a cad . . . worthless hussy": *Disclosing the Past,* p. 45.
75 "I don't think . . . his children like that": Frida Leakey, interview, Cambridge, November 29, 1984.
75 "along [their] chosen . . . the consequences": *Disclosing the Past,* p. 45.
75 "a tough cookie": Thurstan Shaw, interview, Stapleford, England, November 30, 1984.
76 "to drown": *Disclosing the Past,* p. 19.
76 "the only child . . . of grown-ups": ibid., p. 18.
76 "first began to . . . archaeology": ibid., p. 24.
76 "open nature . . . to make friends": ibid., p. 17.
76 "sheer instinctive joy . . . ," ". . . sorts of good things," "world of their makers": ibid., p. 25.
76 "the Uncooked Dumpling": ibid., p. 26.
76 "He loved to take. . . . great companions": Mary Leakey, interview, Nairobi, July 9, 1984.
76 "She had been. . . . that life completely": ibid.
77 "We so enjoyed. . . . this with me": ibid.
77 "near perfection": *Disclosing the Past,* p. 29.
77 "shattered . . . person in the world": ibid.
77 "I was extremely. . . . very lonely": Mary Leakey, interview, Nairobi, July 9, 1984.
77 "could not find . . . or the nuns": *Disclosing the Past,* p. 32.
77 "was quite loud . . . some of them": ibid.

77 "The abbé kindled. . . . do anything else": quoted in Melvin M. Payne, "Family in Search of Prehistoric Man," *National Geographic,* February 1965, p. 218.
78 "She was . . . top of the tree": Mary Leakey, interview, Nairobi, July 9, 1984.
78 "Meaning the . . . Sex Weary": J. B. S. Haldane, quoted in Ronald W. Clark, *The Life and Work of J. B. S. Haldane,* p. 80.
78 "a good word to say": *Disclosing the Past,* p. 45.
78 "When we met . . . fish & fisheries": E. Barton Worthington, letter to author, May 20, 1986.
79 "rather incisive voice": John D. Solomon, letter to author, December 19, 1985.
79 "[Louis] was . . . at work": P. G. H. Boswell, unpublished autobiography, p. 229.
79 "to spirit . . . for weekends": *Disclosing the Past,* p. 47.
79 "in shock. . . . be put right": Frida Leakey, interview, Cambridge, November 29, 1984.

Chapter 5. Disaster at Kanam

80 "old coal burner," "it was de rigueur . . . made an appearance!": Stanhope White, letter to author, December 18, 1986.
80 "old war-horses": E. Barton Worthington, letter to author, May 20, 1986.
81 "the general massiveness . . . whole mandible": L. S. B. Leakey, *The Stone Age Races of Kenya,* p. 23.
81 "a human stage . . . close to *Homo sapiens*": ibid., p. 9.
81 "suggested . . . new species": "*Homo Sapiens* and His Ancestors: Dr. Leakey on East African Finds," *The Times,* October 21, 1933, p. 12.
81 "geologists and palaeontologists . . . of these fossils": ibid.
81 "congratulated him on his most valuable results": M. L. Tildesley, "The Status of the Kanam Mandible and the Kanjera Skulls," *Man,* December 1933, vol. 33, pp. 200–201.
81 "bursting with information": Sir Peter Kent, letter to author, March 2, 1985.
81 "horror": L. S. B. Leakey, *By the Evidence,* p. 30.
82 "I am not. . . . Near enough": Sonia Cole, *Leakey's Luck,* p. 100.
82 "to show the . . . [fossil's] site": L. S. B. Leakey, "Fossil Human Remains from Kanam and Kanjera, Kenya Colony," *Nature,* October 10, 1936, p. 644.
82 "[W]e tried . . . the position exactly": L. S. B. Leakey, East African Archaeological Expedition, Fourth Season (1934–35), Third Monthly Field Report, December 24th–January 24th, p. 3 (NMK).
82 "impossible to relocate . . . jaw had been found": *By the Evidence,* p. 30.
82 "raised eyebrows": Stanhope White, letter to author, January 16, 1987.
82 "fix the position," "within reasonable limits": *By the Evidence,* p. 30.
82 "early examples," "very primitive . . . *Homo sapiens* type": *The Stone Age Races of Kenya,* p. v.
82 "primitive," "brain capacity . . . human beings today": *By the Evidence,* p. 20.
82 "for the first . . . culture were like": "The Status of the Kanam Mandible and the Kanjera Skulls," op. cit., pp. 200–201.
82 "it appeared not . . . the human family": ibid., p. 200.
83 "to carry him through": Stanhope White, letter to author, January 16, 1987.
83 "completely flabbergasted," "dismayed": ibid.
83 "within ten yards or so . . .": L. S. B. Leakey, Field Diary, January 16, 1935, quoted in Sonia Cole, *Leakey's Luck,* p. 98.
83 "rough sketch map . . . best they could?": Stanhope White, letter to author, December 18, 1986.
83 "I was saving . . . visit to Oldoway": "Fossil Human Remains from Kanam and Kanjera, Kenya Colony," op. cit., p. 643.
83 "Louis would tell . . . our own jobs": Stanhope White, letter to author, January 16, 1987.
83 "as always, Louis. . . . from our beds": Stanhope White, letters to author, December 18, 1986, and January 16, 1987.
84 "with swarms of insects. . . . no insects, no food": P. G. H. Boswell, unpublished autobiography, p. 234.
84 "Notwithstanding the close. . . . decisive verdicts": anonymous, "Early Man in East Africa: Further Investigation," *Nature,* November 10, 1934, p. 730.
84 "Grand Inquisitor," "got on very well": Sir Peter Kent, letter to author, March 2, 1985.
84 "a smallish . . . sort of man": Stanhope White, letter to author, December 18, 1986.
84 "dogmatic," "occasionally irascible": Sir Peter Kent, letter to author, October 19, 1985.
84 "a man of . . . vision," "prone to jump to conclusions": Stanhope White, letter to author, December 18, 1986.

84 "hasty visit": L. S. B. Leakey, East African Archaeological Expedition, Fourth Season (1934–35), Second Monthly Field Report, Nov. 24th–Dec. 23rd, p. 3 (NMK).
85 "one large fragment . . . surface," "promising": ibid., p. 1.
85 "bumped over the plains," "roving herds": P. G. H. Boswell, unpublished autobiography, p. 233.
85 "waterless country," "the toughest meat ever!," "red . . . dust," "the water into soup": ibid., p. 234.
85 "dated geologically . . . archaeologically": *The Stone Age Races of Kenya,* p. 9.
85 "as meaningless . . . without foundation": Phillip V. Tobias, "A Re-Examination of the Kanam Mandible," *Proceedings of the 4th Pan-African Congress of Prehistory,* p. 343.
85 "very pronounced": *The Stone Age Races of Kenya,* p. 19.
86 "simian shelf": *By the Evidence,* p. 21.
86 "had a foreboding . . . well": P. G. H. Boswell, unpublished autobiography, p. 218.
86 "was . . . bony reaction", Phillip V. Tobias, "The Kanam Jaw," *Nature,* March 26, 1960, p. 946.
86 "good fossil remains," "I hoped. . . . recent age," "a half . . . ape": L. S. B. Leakey, East African Archaeological Expedition, Fourth Season (1934–35), Third Monthly Field Report, Dec. 24th–Jan. 24th, p. 2.
86 "The country . . . site to another," P. G. H. Boswell, unpublished autobiography, p. 234.
86 "the most complete . . . *Hippopotamus gorgops*": L. S. B. Leakey, East African Archaeological Expedition, Fourth Season (1934–35), Third Monthly Field Report, Dec. 24th–Jan. 24th, p. 1.
86 "insist[ed] that Leakey . . . or question": P. G. H. Boswell, unpublished autobiography, p. 234.
86 "on a big scale": L. S. B. Leakey, East African Archaeological Expedition, Fourth Season (1934–35), Third Monthly Field Report, Dec. 24th–Jan. 24th, p. 1.
87 "plaintively": Stanhope White, letter to author, December 18, 1986.
87 "unlikely": P. G. H. Boswell, unpublished autobiography, p. 235.
87 "increasingly irritated . . . could not identify it": Sir Peter Kent, letter to author, March 2, 1985.
87 "all attempts to find . . . at the right spot": P. G. H. Boswell, unpublished autobiography, p. 235.
87 "a born scientist": L. S. B. Leakey, *White African,* p. 190.
87 "a native boy": P. G. H. Boswell, "The Search for Man's Ancestry," *Nature,* November 18, 1950, p. 839.
87 "Kanjera [skull] . . . ," "But . . . satisfaction," "spent the rest . . . fragments which fitted)": L. S. B. Leakey, Field Diary, January 16, 1935.
87 "Juma found . . . second bit": L. S. B. Leakey, Field Diary, January 17, 1935.
88 "I realised . . . the same": Stanhope White, letter to author, December 12, 1986.
88 "some distance away," "He was utterly . . . say the least": ibid.
88 "admitted the error," "bad mistake": P. G. H. Boswell, unpublished autobiography, p. 235.
88 "mistakes . . . happened": Stanhope White, letter to author, December 18, 1986.
88 "Professor Boswell . . . though very regrettable": L. S. B. Leakey, Field Diary, January 18, 1935.
88 "I am quite . . . my head on it": Stanhope White, letter to author, December 18, 1986.
89 "If this site . . . Kanam West gullies": Stanhope White, letter to author, January 16, 1987.
89 "[He] did not . . . wild goose chase": Sir Peter Kent, letter to author, March 2, 1985.
89 "within a few feet": *By the Evidence,* p. 35.
89 "uncertain," "whether . . . or not," "Thus . . . useless": P. G. H. Boswell, unpublished autobiography, p. 235.
89 "very down": Stanhope White, letter to author, January 16, 1987.
89 "of the utmost . . . consideration": Arthur Keith, "Early East Africans," *Nature,* February 2, 1935, pp. 163–64.
89 "unfortunate mistake," "unfortunately . . . unresolved": L. S. B. Leakey, East African Archaeological Expedition, Fourth Season (1934–35), Third Monthly Field Report, Dec. 24th–Jan. 24th, p. 4 (NMK).
89 "making the best of a poor show": Sir Peter Kent, letter to author, April 17, 1985.
89 "I had been compelled . . . return to England": P. G. H. Boswell, unpublished autobiography, p. 235.
90 "the Leakey business . . . world" to "the Leakey story . . . horse's mouth' ": ibid., p. 236.
90 "divorce was non-U in Cambridge then": E. Barton Worthington, letter to author, May 20, 1986.
90 "He had crowed . . . business!": Sir Peter Kent, letter to author, March 2, 1985.
90 "some of the older . . . deceptive": Sir Vivian Fuchs, interview, London, November 22, 1984.
90 "Ever since then . . . both of them": *By the Evidence,* p. 27.

91 "rich . . . anthropoid apes": L. S. B. Leakey, East African Archaeological Expedition, Fourth Season (1934–35), Fourth Monthly Field Report, Jan. 24th–Feb. 23rd, p. 1.
91 "parts of . . . apes," "over 30 species," "very old Kikuyu men," ibid., p. 4.
91 "I have been. . . . in this case": A. C. Haddon to L. S. B. Leakey, March (n.d.) 1935.
91 "A letter . . . writing to *Nature*": L. S. B. Leakey, Field Diary, March 16, 1935.
91 "let [Louis] down . . . business": P. G. H. Boswell, unpublished autobiography, p. 235.
91 "scathing": E. Barton Worthington, letter to author, May 20, 1986.
91 "to find . . . on a map": P. G. H. Boswell, "Human Remains from Kanam and Kanjera, Kenya Colony," *Nature,* March 9, 1935, p. 371.
92 [I]n view of. . . . a "suspense account": ibid.
92 "[It] gave rise . . . had been debunked!": P. G. H. Boswell, unpublished autobiography, p. 237.
92 "There is no reason . . . million years old," David Pilbeam, telephone interview, February 23, 1987; "They were probably. . . . [Oldoway Man] skeleton," Tom Plummer, telephone interview, October 17, 1994.
92 "Oldest Fragment . . . Claim": *New York Times,* March 8, 1935, p. 23.
92 "Modern Man . . . Place": *Morning Post,* March 7, 1935.
92 "Got a copy . . . I'd regret afterwards": L. S. B. Leakey, Field Diary, March 28, 1935.
92 "Professor Boswell . . . returned to England": L. S. B. Leakey, East African Archaeological Expedition, Fourth Season (1934–35), Fifth Monthly Field Report, Feb. 24th–March 23rd, p. 1 (NMK).
93 I most certainly. . . . bad faith: L. S. B. Leakey, East African Archaeological Expedition, Fourth Season (1934–35), Ninth and Tenth Monthly Field Reports (combined), p. 1 (NMK).
93 "colossal blunder": Sir Peter Kent, letter to author, March 2, 1985.

Chapter 6. Olduvai's Bounty

94 "be left kicking her heels": Mary Leakey, *Disclosing the Past,* p. 49.
94 "Louis wanted me . . . to see Olduvai": ibid.
94 "rather sleazy": ibid., p. 53
94 "reported sites": L. S. B. Leakey, East African Archaeological Expedition, Fourth Season (1934–35), Sixth Monthly Field Report, March 24th–April 23rd, pp. 2–3 (NMK).
95 "nightmare journey": L. S. B. Leakey, *Kenya: Contrasts and Problems,* p. 55.
95 "in a hurry," "It was. . . . job": Mary Leakey, interview, San Francisco, May 5, 1987.
95 "going out into the blue": L. S. B. Leakey, East African Archaeological Expedition, Fourth Season (1934–35), Sixth Monthly Field Report, March 24th–April 23rd, p. 3 (NMK).
95 "several bunches . . . sardines": Mary Leakey, *Olduvai Gorge: My Search for Early Man,* p. 14.
96 "On some occasions. . . . luggage": L. S. B. Leakey, *By the Evidence,* p. 43.
96 "That night . . . and Sam Howard!!!": Stanhope White, letter to author, January 16, 1987.
96 "The clouds . . . most exciting moment": *By the Evidence,* pp. 43–44.
96 "In a few . . . world": *Disclosing the Past,* p. 55.
96 "two . . . lines": ibid.
96 "I shall never. . . . am nearly home": ibid.
97 "cast its spell," "would . . . again": ibid., p. 63.
97 "From the tourist . . . difficulties": *By the Evidence,* p. 44.
97 "[M]ost of . . . thrill": ibid., p. 45.
97 "a few years . . . Leakey Landrover": Stanhope White, letter to author, February 12, 1987.
97 "the places . . . erosion": *By the Evidence,* p. 46.
97 "incredibly beautiful . . . excitement": *Disclosing the Past,* p. 55.
97 " 'What the hell . . . on the top": Stanhope White, letter to author, January 16, 1987.
97 "pluck[ed] . . . caps' ": Stanhope White, letter to author, December 18, 1986.
98 "If we found . . . ourselves": *Olduvai Gorge: My Search for Early Man,* p. 15.
98 "Every day we . . . brushes,": *By the Evidence,* p. 46.
98 "every step . . . needle points": E. J. Wayland, "The Age of the Oldoway Human Skeleton," unpublished manuscript, September 2, 1932, p. 3 (NMK).
98 "Louis liked . . . identified": *Disclosing the Past,* p. 56.
98 "something over 6 feet": L. S. B. Leakey, East African Archaeological Expedition, Fourth Season (1934–35), Seventh Monthly Field Report, April 24th–May 23rd, p. 2 (NMK).
98 "rather thick": Mary Leakey, interview, Nairobi, July 9, 1984.
98 "very excited": *Disclosing the Past,* p. 56.
98 ". . . I still am . . . plentiful": L. S. B. Leakey, East African Archaeological Expedition, Fourth Season (1934–35), Seventh Monthly Field Report, April 24th–May 23rd, p. 1 (NMK).

99 "all polite": Mary Leakey, letter to author, February 19, 1986.
99 "strongly adverse," "Our joint objections . . . syndrome,'" "pleasant . . . with": Sir Peter Kent, letter to author, March 2, 1985.
99 "she was . . . confident," "fitted in at once," "completely incredulous": Stanhope White, letter to author, December 18, 1986.
99 "They couldn't . . . down for them," "shocked wonderment," "Probably . . . as before": Stanhope White, letter to author, January 16, 1987.
99 "virtual scientific colleague," "somewhat . . . worthy of Louis": *Disclosing the Past,* p. 57.
99 "genuinely . . . complaining": ibid.
99 "An attempt . . . used to": *Olduvai Gorge: My Search for Early Man,* p. 16.
100 "lying . . . canvases": *Disclosing the Past,* p. 57.
100 "eagerly . . . container": *Olduvai Gorge: My Search for Early Man,* p. 17.
100 "We sat . . . sick," "we had . . . rebelled!!": Stanhope White, letter to author, December 18, 1986.
100 "[A]fter which . . . a mile away": L. S. B. Leakey, East African Archaeological Expedition, Fourth Season (1934–35), Seventh Monthly Field Report, April 24th–May 23rd, p. 3 (NMK).
100 "great interest" to "That decided me . . . the new area": *By the Evidence,* p. 50.
100 "in excellent condition": *Disclosing the Past,* p. 58.
100 "crude pebble tool": L. S. B. Leakey, East African Archaeological Expedition, Fourth Season (1934–35), Eighth Monthly Field Report, May 24th–June 23rd, p. 3 (NMK).
100 "surface scatter": *Disclosing the Past,* p. 58.
101 "I was hopeful . . . ten days": *By the Evidence,* p. 57.
101 "greater part . . . Bed I": L. S. B. Leakey, East African Archaeological Expedition, Fourth Season (1934–35), Combined Ninth and Tenth Monthly Field Reports (n.d.), p. 2 (NMK).
101 "finally . . . revolting": *Olduvai Gorge: My Search for Early Man,* p. 16.
101 "muddy, ammonia-tainted water": *By the Evidence,* p. 56.
101 "in toilet paper . . . Bronco": *Olduvai Gorge: My Search for Early Man,* p. 24.
102 "I was . . . yet": Arthur T. Hopwood to L. S. B. Leakey, March 18, 1935 (BMNH).
102 "I am more . . . hidden there": L. S. B. Leakey, East African Archaeological Expedition, Fourth Season (1934–35), Seventh Monthly Field Report, April 24th–May 23rd, p. 4 (NMK).
102 "very scarce": L. S. B. Leakey, East African Archaeological Expedition, Fourth Season (1934–35), Combined Ninth and Tenth Monthly Field Reports, p. 2 (NMK).
102 "we never . . . ourselves": Mary Leakey, *Africa's Vanishing Art: The Rock Paintings of Tanzania,* p. 15.
102 "unsophisticated," "Once . . . lipstick!": ibid.
103 "It was . . . few months": *Disclosing the Past,* p. 62.
103 "extraordinarily naturalistic," "immense . . . high": L. S. B. Leakey, East African Archaeological Expedition, Fourth Season (1934–35), Combined Ninth and Tenth Monthly Field Reports, p. 3 (NMK).
103 "absolutely entranc[ed]," "return . . . art": *Disclosing the Past,* pp. 62–63.
103 "Louis . . . marry me": Mary Leakey, letter to author, November 4, 1986.
103 "We had . . . also there": Stanhope White, letter to author, January 16, 1987.
103 "let the cat out of the bag": Stanhope White, letter to author, December 18, 1986.
103 "I have told. . . . conduct": Peter Kent, letter to L. S. B. Leakey, August 3, 1935, quoted in *Leakey's Luck,* pp. 114–15.
103 "Since you told. . . . divorced?": Harry Leakey to L. S. B. Leakey, August 11, 1935, quoted in *Leakey's Luck,* p. 115.
104 "produced . . . mentioned": *Disclosing the Past,* p. 63.

Chapter 7. Consequences

105 "This meant . . . books elsewhere": L. S. B. Leakey, *By the Evidence,* p. 69.
106 "Mary, . . . properly": Mary Leakey, *Disclosing the Past,* p. 66.
106 "trial marriage": Stanhope White, letter to author, January 16, 1987.
106 "blissfully happy": *Disclosing the Past,* p. 65.
106 "If I can't . . . do that yet": L. S. B. Leakey to Sir Henry Dale, Secretary, Royal Society, November 1, 1935 (Royal Society archives).
106 "to relieve . . . already made": Sir Henry Dale to L. S. B. Leakey, June 28, 1936 (Royal Society archives).
106 "Losing . . . bounce back": Mary Leakey, interview, San Francisco, May 30, 1987.

107 "Whether they were . . . toward herself": Jacquetta Hawkes, *Mortimer Wheeler: Adventurer in Archaeology,* p. 136.
107 "a milestone," "That book . . . in Africa": J. Desmond Clark, "Louis Seymour Bazett Leakey," *Proceedings of the British Academy,* 1973, p. 456.
107 "surest way of determining": L. S. B. Leakey, *Stone Age Africa: An Outline of Prehistory in Africa,* p. 17.
107 "established . . . continent": "Louis Seymour Bazett Leakey," op. cit., p. 456.
107 "simply . . . a 'potboiler' ": *By the Evidence,* p. 73.
107 "unique and enthralling": anonymous, "Science in East Africa: Mr. Leakey's Work," *The Times,* April 30, 1937, p. 21.
108 "I hope . . . in due course": L. S. B. Leakey, *White African,* p. 3.
108 "As the whole . . . *sub judice*": *Stone Age Africa,* p. 165.
108 "In spite of . . . was obtained": L. S. B. Leakey, "Fossil Human Remains from Kanam and Kanjera, Kenya Colony," *Nature,* October 10, 1936, p. 643.
108 "man . . . Upper Pliocene": P. G. H. Boswell, "The Search for Man's Ancestry," *Nature,* November 18, 1950, p. 839; "during the past . . . incorrect terms": P. G. H. Boswell, unpublished autobiography, p. 4.
109 "Dart . . . fighter": George Findlay, *Dr. Robert Broom, F.R.S.,* p. 54; "strode over . . . wavered": ibid., p. 52; "return to the charge": ibid., p. 54; "adult Taung ape": Robert Broom, *Finding the Missing Link,* p. 50.
109 "on or near . . . risen": Robert Broom, "A New Fossil Anthropoid Skull from South Africa," *Nature,* September 19, 1936, pp. 476–77.
109 "to which . . . return": *Disclosing the Past,* p. 66.
109 "I was born . . . tribe": L. S. B. Leakey to the Secretary, Rhodes Trust, August 29, 1936 (Rhodes Trust archives).
110 "elucidate . . . Africa": General Secretary of the Eugenics Society to the Secretary, Rhodes Trust, September 30, 1936 (Rhodes Trust archives).
110 "a critical . . . problem," "biologically unsound," "approach . . . scientist": L. S. B. Leakey to the Secretary, Rhodes Trust, September 8, 1936 (Rhodes Trust archives).
110 "Leakey . . . him": Lord Lothian to H. A. L. Fisher, September 2, 1936 (Rhodes Trust archives).
110 "I think . . . *Nature*": H. A. L. Fisher to Lord Lothian, September 9, 1936 (Rhodes Trust archives).
110 "He [Louis] . . . importance": R. Coupland to E. F. Millar, Rhodes Trust, October 5, 1936 (Rhodes Trust archives).
110 "It was. . . . he took it": Mary Leakey, interview, San Francisco, May 30, 1987.
110 "[We] had already . . . to hand": *Disclosing the Past,* p. 67.
110 "on the ground . . . Mary Nicol," "until . . . improved": *The Times,* June 16, 1936.
111 "we had no . . . from him": Frida Leakey, letter to author, March 21, 1987.
111 "cannot possibly remember": Mary Leakey, interview, San Francisco, May 5, 1987.
111 "reluctant witnesses": *Disclosing the Past,* p. 67.
111 "whatever might await": ibid., p. 68.
111 "good land lying uncultivated": John Ainsworth, quoted in Elspeth Huxley, *White Man's Country,* Vol. 1, *1870–1914,* p. 113.
112 "From that time . . . unknown to us": L. S. B. Leakey, *The Southern Kikuyu,* Vol. 1, p. 33.
112 "an exceedingly progressive man": L. S. B. Leakey, *Kenya: Contrasts and Problems,* p. 12.
112 "dirty windowless hut," "well-built stone": ibid., p. 13.
112 "asked Canon Leakey . . . most wonderful thing": Matthew Njoroge, quoted in Carl G. Rosberg, Jr., and John Nottingham, *The Myth of "Mau Mau": Nationalism in Kenya,* pp. 41–42.
112 "a sober . . . of his tribe": District Commissioner from Kiambu, "Chiefs' Character Book," 1937 (KNA).
113 "so that . . . Kikuyu had lived": *By the Evidence,* p. 77.
113 "solemn promise," "act . . . interest": ibid., p. 78.
113 "the most complete . . . written": L. S. B. Leakey to the Secretary, Rhodes Trust, May 28, 1937 (Rhodes Trust archives).
113 "Louis put. . . . to do that": Mary Leakey, interview, San Francisco, May 30, 1987.
113 "[I]t's going to be. . . . I can see!": L. S. B. Leakey to the Secretary, Rhodes Trust, May 28, 1937 (Rhodes Trust archives).
113 "[W]hat I wanted . . . prehistoric": *Disclosing the Past,* p. 69.
114 "I should not . . . this to me": ibid.
114 "by the grace of God": *By the Evidence,* p. 80.

114 "delighted. . . . longing": *Disclosing the Past,* p. 70.
114 "lying in heaps . . . bodies": *Visitors' Guide to the Hyrax Hill Site,* p. 9.
115 "I felt . . . wander too far": *By the Evidence,* p. 55.
115 "[B]eing still new . . . wished": *Disclosing the Past,* p. 59.
115 "The next minutes. . . . as Mary was": *By the Evidence,* p. 55.
115 "I told you so": *Disclosing the Past,* p. 59.
115 "locate each and every one": *By the Evidence,* p. 86.
115 "I don't feel . . . it must be)": L. S. B. Leakey to the Secretary, Rhodes Trust, May 28, 1937 (Rhodes Trust archives).
115 "They felt . . . not desert you": Lord Lothian, Secretary of the Rhodes Trust, to L. S. B. Leakey, June 29, 1937 (Rhodes Trust archives).
115 "exciting . . . tiring": *By the Evidence,* p. 92.
116 "determined . . . vehicle through": ibid.
116 "So many people. . . . surprise to them": Mary Catherine Fagg, interview, Oxford, November 27, 1984.
116 "exciting cave," "[I]mmediately . . . beads": *By the Evidence,* p. 86.
117 "I had got . . . Kikuyu": L. S. B. Leakey to Carl van Riet Lowe, May 3, 1938 (Government Archives, Pretoria, Republic of South Africa).
117 "There were hundreds. . . . her life": Mary Catherine Fagg, interview, Oxford, November 27, 1984.
118 "Predynastic . . . stuff": L. S. B. Leakey quoting Professor Beck in a letter to Sir Arthur Keith, January 21, 1939 (Royal College of Surgeons archives).
118 "non-Negroid," "mausoleum . . . opal-mining": "Prehistoric Man in East Africa/Opal-Mining in 4,000 B.C.," *The Times,* May 22, 1938, p. 13.
118 "Plans for. . . . strongly": L. S. B. Leakey to Carl van Riet Lowe, July 23, 1938 (Government Archives, Pretoria, Republic of South Africa).
118 "remarkable discoveries. . . . North to South": General Jan Christiaan Smuts to Lord Francis Scott, August 3, 1938 (Government Archives, Pretoria, Republic of South Africa).
119 "I hope . . . the trick": Carl van Riet Lowe to L. S. B. Leakey, August 5, 1938 (Government Archives, Pretoria, Republic of South Africa).
119 "the reasons . . . community": *Kenya: Contrasts and Problems,* p. 108.
119 "A vital question. . . . doubts": ibid., p. x.
119 "on the smell of an oil-rag:" Nellie Grant, quoted in Elspeth Huxley, *Nellie: Letters from Kenya,* p. 127.
119 "[W]e are stranded. . . . exporting it": L. S. B. Leakey to Carl van Riet Lowe, January 2, 1939 (Government Archives, Pretoria, Republic of South Africa).
119 "Certain subversive elements": *By the Evidence,* p. 104.
120 "our rent for awhile": *Disclosing the Past,* p. 76.

Chapter 8. Cloak-and-Dagger

122 "My boys . . . fossils!": L. S. B. Leakey to Archdeacon Owen, Feb. 24, 1942 (NMK).
122 "I spent . . . to bring": Donald MacInnes to L. S. B. Leakey, 1942 (n.d.) (KNA).
122 "Louis loved . . . had been": Mary Leakey, interview, San Francisco, May 5, 1987.
122 "I disliked . . . truth myself": L. S. B. Leakey, *By the Evidence,* pp. 129–30.
123 "one of those . . . frailty": Mary Leakey, *Disclosing the Past,* p. 79.
124 "in fine shape": *By the Evidence,* p. 154.
124 "I quite liked . . . archaeologist": *Disclosing the Past,* pp. 79–80.
124 "[I]t seemed . . . miss": *By the Evidence,* p. 154.
124 "[S]he had spent . . . a rest": ibid.
124 "patting his little back": ibid., p. 155.
125 "freedom . . . war": ibid., p. 145.
126 "When I saw. . . . (round stone balls)": ibid., pp. 159–60.
126 "as if they . . . makers": *Disclosing the Past,* p. 82.
126 "just how . . . satisfactorily": *By the Evidence,* p. 160.
126 "was openly upset": Mary Leakey, interview, Nairobi, September 7, 1987.
126 "The one thing . . . by her loss": Dora MacInnes, letter to author, March 11, 1987.
126 "How does one . . . just carried on": Mary Leakey, interview, Nairobi, September 7, 1987.
126 "I am afraid . . . of my work": L. S. B. Leakey to Archdeacon Owen, April 20, 1943 (NMK).
127 "a real joy to wake up": *By the Evidence,* p. 136.

127 "perfectly magnificent site," "second Oldoway": L. S. B. Leakey to Archdeacon Owen, December 23, 1942, and April (n.d.) 1943 (NMK).
127 "actual . . . floors": L. S. B. Leakey, "A Pre-Historian's Paradise in Africa: Early Stone Age Sites at Olorgesailie," *Illustrated London News,* October 5, 1946, p. 382.
128 *"not one,* but . . . culture": ibid.
128 "had . . . been found anywhere," "way of life . . . culture": L. S. B. Leakey, "Early Man in Kenya, Acheulean Camp Sites," *The Times,* October 4, 1946, p. 3.
129 "It has always. . . . weapon": ibid.
129 "to capture . . . baboon": *By the Evidence,* p. 166.
129 "I personally . . . myself": ibid., p. 167.
129 "a great success. . . . stuff": Mary Leakey to Archdeacon Owen, October 20, 1943 (NMK).
129 "any . . . information," "I fear . . . hand axes": *Disclosing the Past,* p. 83.
130 "to examine . . . fossils": L. S. B. Leakey, *Adam's Ancestors,* p. 3.
130 "there is not . . . impossibility": Sir Arthur Keith, *The Antiquity of Man,* p. 734.
131 "The shape . . . belonged": Georges Cuvier, *Recherches sur les Ossemans Fossiles des Quadrupeds,* 4 vols. (Paris, 1812), quoted in John Noble Wilford, *The Riddle of the Dinosaurs,* pp. 21–22.
131 "ancestral to the Chimpanzee": A. Tindell Hopwood, "Miocene Primates from Kenya," *Journal of the Linnaean Society of London,* 38, 1933, p. 460.
131 "If I interpret. . . . existence": *Adam's Ancestors,* p. 177.
131 "The subdivision . . . allowed": ibid., p. 203.
131 "real ancestors . . . Pleistocene": ibid., p. 202.
132 "their common . . . Miocene": ibid., p. 227.
132 "By the greatest . . . later": L. S. B. Leakey, "African Ancestries, Interpretation of Fossil Finds in Kenya," *The Times,* August 23, 1946, p. 5.
132 "most complete . . . discovered": L. S. B. Leakey to Archdeacon Owen, October 20, 1942 (NMK).
132 "As you will see . . . apes": L. S. B. Leakey to Sir Arthur Smith Woodward, June 8, 1943 (NMK).
132 "I see . . . humanity": Sir Arthur Keith to L. S. B. Leakey, January 28, 1943 (NMK).
132 "common . . . gorilla": ibid.
133 "near . . . stem": L. S. B. Leakey, "A Miocene Anthropoid Mandible from Rusinga, Kenya," *Nature,* September 18, 1943, pp. 319–20.
133 "[I]n *Proconsul* . . . evolved": "African Ancestries, Interpretation of Fossil Finds in Kenya," op. cit., p. 5.
133 "Bausi. . . . *photographs*": L. S. B. Leakey's wartime reports, n.d. (NMK).
133 "I do C.I.D. . . . as well!": L. S. B. Leakey to Kenneth Oakley, December 6, 1944 (BMNH).
134 "state a case. . . . rent free": L. S. B. Leakey to Museum Trustees of Kenya, March (n.d.) 1943 (KNA).
134 "We certainly. . . . rains!": H. William Gardener to Sir Charles Belcher, March 8, 1941 (KNA).
134 "[T]hey would undoubtedly . . . ants": *Disclosing the Past,* p. 80.
134 "unfit for human habitation": L. S. B. Leakey to the Commissioner of Police, Nairobi, July 31, 1945 (KNA).
134 "I am directed. . . . years": Secretary of the Museum Trustees of Kenya to L. S. B. Leakey, March 25, 1943 (KNA).
135 "[A]t first. . . . 'overscented' ": *By the Evidence,* pp. 115–16.
135 "increased by leaps and bounds": ibid., p. 117.
135 "neither . . . Cambridge": Mary Leakey, interview, San Francisco, May 30, 1987.
135 "The trouble . . . world": L. S. B. Leakey to Gordon College, September 30, 1945 (NMK).
136 "It is perhaps . . . opinion": ibid.
136 "While . . . qualifications," "I feel it . . . present": L. S. B. Leakey to Coryndon Natural History Museum Trustees, November 16, 1945 (KNA).

Chapter 9. Race for the Miocene

137 "[My years] . . . *you*": Cecilia Nicol to Mary Leakey, June 14, 1945 (NMK).
138 "babies are boring": Mary Leakey, interview, San Francisco, May 5, 1987.
138 "I don't wonder. . . . adorable": Kathleen D. Frere to Mary Leakey, March 16, 1945 (NMK).
138 "tough guy": Cecilia Leakey quoting Mary in a letter to Mary Leakey, October 18, 1945 (NMK).
138 "strange city": Mary Leakey to Louis Leakey, May 30, 1946 (NMK).
138 "I can't get. . . . near": Mary Leakey to Louis Leakey, May 1, 1946 (NMK).
138 "I can't decide. . . . thing": Mary Leakey to Louis Leakey, May 7, 1946 (NMK).

138 "Really . . . *yourself,*" "it made me . . . safaris": Mary Leakey to Louis Leakey, May 9, 1946 (NMK).
139 "Life . . . everything": Mary Leakey to Louis Leakey, May 24, 1946 (NMK).
139 "I was delighted . . . congress," "from . . . south": Carl van Riet Lowe to Louis Leakey, December 29, 1944 (Government Archives, Pretoria, Republic of South Africa).
139 "[S]cientists were. . . . Prehistory": L. S. B. Leakey, *By the Evidence,* p. 193.
139 "reputation . . . maverick": J. Desmond Clark, interview, Berkeley, June 17, 1987.
139 "entrepreneurial, showbiz manner": Frida Leakey, interview, Cambridge, May 27, 1984.
139 "This is . . . Leakey": E. J. Wayland to Archdeacon Owen, February 10, 1936 (NMK).
140 "Jonathan & I. . . . days": Mary Leakey to Louis Leakey, May 16, 1946 (NMK).
140 "I get. . . . *terribly* badly": Mary Leakey to Louis Leakey, May 28, 1946 (NMK).
141 "very little hope": L. S. B. Leakey to Wilfrid Le Gros Clark, December 31, 1947 (KNA).
141 "[W]e only . . . typewriter": Mary Leakey, *Disclosing the Past,* p. 92.
141 "When I arrived. . . . occasions," "watertight compartments": J. Desmond Clark, "A Personal Memoir," in *A History of African Archaeology,* ed. Peter Robertshaw, p. 193.
141 "We had. . . . ideas": Basil Cooke, interview, New York, April 9, 1984.
142 "the major . . . Africa": "A Personal Memoir," op. cit.
142 "energy and enthusiasm": anonymous, "The Pan-African Congress on Prehistory," *Nature,* February 15, 1947, p. 218.
142 "It was. . . . surprise": Joan Karmali, interview, Nairobi, September 19, 1987.
142 "Hyrax Hill. . . . pioneer research": J. Desmond Clark, interview, Berkeley, May 16, 1984.
142 "I am afraid . . . to man," "man . . . making": " 'Missing Link' Found in Africa," *New York Times,* January 23, 1947, p. 25.
143 "I am now . . . form": Sir Arthur Keith, "Australopithecinae or Dartians," *Nature,* March 15, 1947, p. 377.
143 "tremendous": *Disclosing the Past,* p. 96.
143 "English doubters": G. H. Findlay, *Dr. Robert Broom,* p. 126; "I say. . . . reputation": Robert Broom to L. S. B. Leakey, June 1, 1945 (NMK).
143 "I have become . . . investigation": W. E. Le Gros Clark to L. S. B. Leakey, February 17, 1947 (KNA).
144 "What a lad . . . Uncle Sam": Sonia Cole, *Leakey's Luck,* p. 153.
144 "undertake . . . EXPEDITION": Wendell Phillips to L. S. B. Leakey, January 22, 1947 (KNA).
144 "I am . . . work": L. S. B. Leakey to Wendell Phillips, February 3, 1947 (KNA).
144 "I told . . . pigeon' ": L. S. B. Leakey to Basil Cooke, August 10, 1947 (KNA).
144 "Louis couldn't. . . . American": Mary Leakey, interview, San Francisco, May 30, 1987.
144 "Phillips . . . promoter": Theodore McCown to L. S. B. Leakey, April 17, 1947 (KNA).
144 "principally . . . assistance": ibid.
144 "At one time . . . anthropoids": ibid.
145 "We are . . . America": Wendell Phillips to L. S. B. Leakey, September 2, 1947 (KNA).
145 "If we have . . . areas": L. S. B. Leakey to W. E. Le Gros Clark, April 26, 1947 (KNA).
145 "and, in fact . . . funds": L. S. B. Leakey to His Highness the Aga Khan, April 26, 1947 (KNA).
145 "I have approached . . . Consultant": L. S. B. Leakey to W. E. Le Gros Clark, May 27, 1947 (KNA).
145 "a special Navy ship": Wendell Phillips to L. S. B. Leakey, September 2, 1947 (KNA).
145 "achieve maximum results": L. S. B. Leakey to W. E. Le Gros Clark, May 27, 1947.
146 "the very best of luck": W. E. Le Gros Clark to L. S. B. Leakey, June 14, 1947 (KNA).
146 "a good augury . . . expedition": L. S. B. Leakey, Journal of the British-Kenya Miocene Research Expedition, 1947, July 12, 1947 (NMK).
147 "very pleased": Kenneth Oakley, quoting W. E. Le Gros Clark in a letter to L. S. B. Leakey, October 13, 1947 (KNA).
147 "It does look . . . corner": W. E. Le Gros Clark to L. S. B. Leakey, September 24, 1947 (KNA).
147 "I would . . . play fair": L. S. B. Leakey to W. E. Le Gros Clark, December 29, 1947 (KNA).
147 "[No] reputable . . . expedition": W. E. Le Gros Clark to L. S. B. Leakey, December 24, 1947 (KNA).
148 "a wise . . . scientists": L. S. B. Leakey to W. E. Le Gros Clark, December 31, 1947 (KNA).
148 "virtually unexplored": L. S. B. Leakey to Basil Cooke, August 10, 1947 (KNA).
148 "[T]heir lack . . . members": L. S. B. Leakey to W. E. Le Gros Clark, December 31, 1947 (KNA).
148 "A well equipped . . . scientists": "Fossil Apes in Kenya/Finds Near Lake Victoria," *The Times,* December 30, 1947, p. 3.
148 "under British auspices": Charles Boise to L. S. B. Leakey, January 6, 1948 (KNA).
148 "urgently needed" to "these fossil apes": L. S. B. Leakey to Charles Boise, January 12, 1948.

148 "I am sorry . . . homecoming": L. S. B. Leakey to Kenneth Oakley, September 9, 1948 (KNA).
149 "was very cooperative . . . collectors": Basil Cooke, interview, New York, April 9, 1984.
149 "[T]he party . . . work": L. S. B. Leakey to P. G. H. Boswell, March 3, 1948 (KNA).
149 "proprietary air": Basil Cooke, interview, New York City, April 9, 1984.
149 "threatened . . . fire arms": Charles Camp to L. S. B. Leakey, December 6, 1948 (KNA).
149 "In fact . . . 'my site'!": "A Personal Memoir," op. cit., p. 196.
149 "possibly . . . hoped": L. S. B. Leakey to Charles Boise, September 21, 1948 (KNA).
150 "For some time. . . . to locate it": *By the Evidence,* p. 226.
150 "I have never . . . far away," "Could it be?" "I was . . . running": *Disclosing the Past,* p. 98.
150 "the greater . . . visible": *By the Evidence,* p. 227.
150 "This was. . . . *Proconsul* face": *Disclosing the Past,* pp. 98–99.
150 "We were exhilarated . . . care": ibid., p. 99.
150 "intense frustration": *By the Evidence,* p. 228.
150 "the size of a match head": L. S. B. Leakey to W. E. Le Gros Clark, October 10, 1948 (KNA).
150 "Once I. . . . we got it": *Disclosing the Past,* p. 99.
151 "The ape skull . . . ape men": L. S. B. Leakey, Field Journal from the British Kenya Miocene Expedition, 1948, October 4, 1948 (NMK).
151 "almost infantile . . . often": ibid., October 2, 1948.
151 "We [have] . . . lifetime" to "I feel . . . you": L. S. B. Leakey to W. E. Le Gros Clark, October 10, 1948 (KNA).
151 "good friend . . . Taimur," "I am . . . world": "The Great Iam," *Time,* September 30, 1966, pp. 110–12.

Chapter 10. A Life in the Sediments

152 "Your beast. . . . wonders": Robert Broom to L. S. B. Leakey, October 10, 1948 (KNA).
152 "Well done, indeed!": Henry Field to L. S. B. Leakey, October 27, 1948 (KNA).
152 "My dear Leakey. . . . World of life": Sir Arthur Keith to L. S. B. Leakey, October 23, 1948. (KNA).
152 "marvelous find," "it is . . . here": W. E. Le Gros Clark to L. S. B. Leakey, October 15, 1948 (KNA).
152 "I do very . . . next few years": L. S. B. Leakey to W. E. Le Gros Clark, October 15, 1948 (KNA).
153 "something . . . unaccustomed," "everyone . . . skull," "precious burden": Mary Leakey, *Disclosing the Past,* p. 100.
153 "BOTH . . . AIRPORT": Mary Leakey to L. S. B. Leakey, October 31, 1949 (KNA).
153 "She is . . . find": "Miocene Ape's Skull at Oxford, Mrs. Leakey Hands Over Her Find," *Evening News,* November 1, 1948.
153 "resemblances to the human condition": L. S. B. Leakey, press communiqué on *Proconsul* (n.d.) (KNA).
154 "Professor . . . to tell us": "That Skull," *Daily Herald,* November 19, 1948.
154 "the Miocene . . . interests": *Disclosing the Past,* p. 96.
154 "more human . . . apelike": L. S. B. Leakey to W. E. Le Gros Clark, October 10, 1948 (KNA).
154 "that it was . . . great apes": L. S. B. Leakey to Charles Boise, October 12, 1948 (KNA).
154 "human-like . . . apes": W. E. Le Gros Clark to L. S. B. Leakey, November 2, 1948 (KNA).
154 "I [took] . . . otherwise," "nasty," "It was. . . . famous!": W. E. Le Gros Clark to L. S. B. Leakey, November 20, 1948 (KNA).
155 "*Proconsul* . . . other": L. S. B. Leakey to W. E. Le Gros Clark, December 15, 1948.
155 "approximate[s] rather closely to": W. E. Le Gros Clark and L. S. B. Leakey, "The Miocene Hominoidea of East Africa," *Fossil Mammals of Africa,* No. 1, BMNH, March 1951, p. 114.
155 "It would seem . . . available": Chief Secretary files, n.d., 1948 (KNA).
156 "We always . . . affair": *Disclosing the Past,* p. 96.
156 "real adventures": Richard E. Leakey, *One Life,* p. 21.
156 In those days . . . dawn": ibid., p. 22.
157 "atmosphere . . . expectation": ibid., p. 23.
157 "nothing . . . sediments": Richard Leakey, interview, Rusinga Island, February 12, 1983.
157 "I remember. . . . flies!": ibid.
157 "firmly instructed" to ". . . something to do!": *One Life,* p. 28.
157 "My bone . . . brushing," "was . . . alert": ibid.
158 "furious and deeply upset," "contributed to . . . fossils!": ibid., p. 29.
158 "Father . . . technology," Richard Leakey, interview, Nairobi, July 4, 1984.
158 "This . . . vertebra": Jean Brown, letter to author, May 17, 1988.
159 "I do think . . . Kenya": W. E. Le Gros Clark to L. S. B. Leakey, September 27, 1954 (KNA).

159 "what seemed. . . . seeds": *Disclosing the Past,* p. 103.
160 "wholly . . . alive": L. S. B. Leakey to Kenneth Oakley, January 1952 (quoted in Sonia Cole, *Leakey's Luck,* p. 182).
160 "It was . . . perfect": Jean Brown, letter to author, May 17, 1988.
160 "[O]ne day . . . exciting": L. S. B. Leakey to W. E. Le Gros Clark, January 11, 1955 (KNA).
160 "About 1932. . . . to it": Major G. L. O. Grundy to L. S. B. Leakey, November 29, 1943 (quoted in *Leakey's Luck,* p. 183).
160 "very limited": L. S. B. Leakey to Donald MacInnes, February 21, 1951 (KNA).
160 "CLOSING . . . COSTS": L. S. B. Leakey to Donald MacInnes, February 28, 1951 (KNA).
161 "link the art," "The important . . . well!": Carl van Riet Lowe to L. S. B. Leakey, December 21, 1950 (KNA).
161 "Those . . . Africa": *Disclosing the Past,* p. 105.
161 "It was. . . . fun": Mary Leakey, interview, Nairobi, July 9, 1984.
161 "During . . . satisfying": Mary Leakey, *Africa's Vanishing Art: The Rock Paintings of Tanzania,* p. 8.
161 "to propitiate . . . powerful": *Disclosing the Past,* p. 109.
161 "bottle washer": L. S. B. Leakey, *By the Evidence,* p. 255.
161 "at first . . . mass," "details. . . . freely": *Disclosing the Past,* p. 106.
162 "Poor . . . me!": *Africa's Vanishing Art,* p. 18.
162 "magnificent paintings": *By the Evidence,* p. 254.
162 "I know . . . Kenyatta": *Disclosing the Past,* pp. 102–103.

Chapter 11. Louis and Kenyatta

164 "When someone. . . . presence": Fenner Brockway, *African Journeys,* pp. 87–88.
164 "I am . . . myself": L. S. B. Leakey, *Mau Mau and the Kikuyu,* p. *viii.*
164 "really . . . country' ": L. S. B. Leakey, *Kenya: Contrasts and Problems,* p. x.
164 "to . . . domination": ibid., p. 166.
165 "a sort . . . *grise":* David Throup, interview, Nairobi, November 1, 1984.
165 "grim significance," "[O]f course. . . . taken place": L. S. B. Leakey, *By the Evidence,* p. 238.
165 "just . . . movement": Carl G. Rosberg, Jr., and John Nottingham, *The Myth of "Mau Mau": Nationalism in Kenya,* p. 276.
165 "been a corruption . . . *muma":* ibid., p. 331.
166 "Men and women. . . . maintained": E. J. Baxter, "Mau Mau—The Terror That Has Come to Kenya and What It Means," *East African News Review, Ltd.,* Nairobi (n.d.) (KNA).
166 "constant . . . flames": L. S. B. Leakey, *Defeating Mau Mau,* pp. 100–101.
166 "the supreme . . . connections": Mary Leakey, *Disclosing the Past,* p. 111.
166 "There were police. . . . figure that out": Richard Leakey, interview, Nairobi, July 5, 1984.
166 "with pistol . . . ambush": Jean Brown, letter to author, May 17, 1988.
166 "very overcharged . . . reason": ibid.
167 "difficult," "so used . . . turned it": Catherine Garnett, letter to author, July 26, 1987.
167 "It was . . . all of us": *Disclosing the Past,* p. 111.
167 "Hero of Our Race," "Saviour," "Great Elder": Jeremy Murray-Brown, *Kenyatta,* p. 271.
168 "a bogey-man . . . and a tail": Elspeth Huxley, *Out in the Midday Sun: My Kenya,* p. 196.
168 "up to his. . . . team": David Throup, interview, Nairobi, November 1, 1984.
168 "evil campaign": *Mau Mau and the Kikuyu,* p. 55.
168 "probably . . . unscrupulous few": ibid., p. 85.
169 "I keep getting. . . . Kikuyu": "Likely to Be Long Hearing Says Magistrate," *East African Standard,* December 5, 1952.
169 "It is impossible. . . . it will fit": *East African Standard,* December 12, 1952.
169 "She probably" to ". . . criminal proceedings": Alan Clarke, "Pritt Protest Again Halts Mau Mau Trial," *Daily Herald,* January 8, 1953, p. 2.
170 "his face drawn and ashen-gray": ibid.
170 "mistranslating . . . mind": David Throup, interview, Nairobi, November 1, 1984.
170 "This attack . . . disgust": Sir Evelyn Baring to British Secretary of State, January 13, 1953 (KNA).
170 "splendid work. . . . honorarium": Attorney General to L. S. B. Leakey, May 27, 1953 (quoted in Sonia Cole, *Leakey's Luck,* p. 202).
171 "Louis said" to ". . . being their friend": Julia Barham, interview, London, June 16, 1984.
171 "sociological . . . Mau Mau," "a dangerous obsession" to "rehabilitated": Commissioner for Community Development and Rehabilitation, "Rehabilitation," unpublished, January 6, 1954.

172 "the leading . . . Mau Mau": "Mau Mau at Close Quarters," *The Times,* November 25, 1954, p. 7.
172 "was in fact" to ". . . murdering maniacs": *Defeating Mau Mau,* p. 52.
172 "help to defeat": ibid., p. v.
172 "mental unstability": ibid., p. 127.
172 "a state of . . . co-operation": ibid., p. 128.
173 "acts . . . depravity": ibid., p. 77.
173 "If the bolt . . . disastrous": *Disclosing the Past,* p. 112.
173 "large revolver . . . middle": Jean Brown, letter to author, May 17, 1988.
173 "A lean. . . . Missing Link?": "The White Kikuyu," *The Observer,* July 18, 1954.
173 "This particular . . . research": *Disclosing the Past,* p. 105.

Chapter 12. "Our Man"

175 "absolutely certain," "We. . . . importance": Mary Leakey, interview, Nairobi, July 9, 1984.
175 "All one realized. . . . it was": ibid.
176 "one . . . prehistory": Kenneth Oakley, "The Earliest Tool-makers," *Antiquity,* March 1956, p. 7.
176 "of *Homo sapiens* . . . primitive": L. S. B. Leakey, *Olduvai Gorge: A Report on the Evolution of the Hand-Axe Culture in Beds I–IV,* p. 160.
176 "extraordinarily . . . artifacts": L. S. B. Leakey, *By the Evidence,* p. 249.
177 "We have not . . . cut them up": L. S. B. Leakey to W. N. Edwards, August 11, 1952 (BMNH archives).
177 "living . . . Man": L. S. B. Leakey, "The Giant Animals of Prehistoric Tanganyika," *London Illustrated News,* June 19, 1954, p. 1047.
177 "We were. . . . intervals," "traveled . . . gorge": Jean Brown, letter to author, May 17, 1988.
178 We had . . . day," "wonderful . . . Olduvai": ibid.
178 "august director," "I would work. . . . money' ": ibid.
178 "scurried . . . feet," "frequently . . . neck!": ibid.
178 "turned . . . overnight": Mary Leakey, *Disclosing the Past,* p. 119.
178 "to nag . . . wife": Jean Brown, letter to author, May 17, 1988.
178 "charged. . . . completely": ibid.
178 "some precious find," "most . . . buttons," "often . . . toes": ibid.
178 "a wonderful . . . bush": ibid.
179 "horrified . . . methods" to ". . . going on": ibid.
179 "crushed . . . happening": ibid.
179 "meticulous excavator," "not very good. . . . miss a lot": J. Desmond Clark, interview, Berkeley, May 16, 1984.
179 "became . . . found": Jean Brown, letter to author, May 17, 1988.
179 "Recently . . . Gorge," "so . . . appreciated": L. S. B. Leakey, "A Giant Child Among the Giant Animals of Olduvai?," *Illustrated London News,* June 28, 1958, p. 104.
179 "very, very massive": ibid.
179 "some . . . race," "victim . . . activities": ibid.
180 "dealing . . . remains": L. S. B. Leakey, "Recent Discoveries at Olduvai Gorge, Tanganyika," *Nature,* April 19, 1958, p. 1102.
180 "We decided. . . . else": Mary Leakey, interview, Olduvai Gorge, September 23, 1984.
180 *"13th day of* . . . Bed I": L. S. B. Leakey, Field Journal, Olduvai Gorge, June 24, 1959 (NMK).
180 "I'm sorry. . . . worse": L. S. B. Leakey, "Finding the World's Earliest Man," *National Geographic,* September 1960, p. 431.
181 "was not lying. . . . surely," "They were . . . lot of it": *Disclosing the Past,* p. 121.
181 "I've got him!" to ". . . his teeth!": "Finding the World's Earliest Man," op. cit., p. 431.
181 "I became . . . sticking out": *The Legacy of L. S. B. Leakey,* National Geographic film, January 9, 1978.
181 "to look at. . . . human": "Finding the World's Earliest Man," op. cit., p. 431.
181 "Louis wanted. . . . disappointed": Mary Leakey, interview, Olduvai Gorge, September 23, 1984.
181 "Why . . . australopithecine": Donald C. Johanson and Maitland Edey, *Lucy: The Beginnings of Humankind,* pp. 90–91; "never swore": Mary Leakey, interview, Olduvai Gorge, September 23, 1984.
182 "systematic tool-making," "foresight . . . memory": "The Earliest Tool-makers," op. cit., pp. 4–5.
182 "[T]he most . . . Tool-maker": ibid.
182 "aberrant branches," "that these . . . ate": "Recent Discoveries at Olduvai Gorge," op. cit., p. 1102.

183 *"Mary got it!"* to "... *Australopithecus?"*: L. S. B. Leakey, Field Journal, Olduvai Gorge, July 17, 1959 (NMK).
183 "My fresh ... line": ibid., July 18, 1959 (NMK).
183 "The zygomatic arches" to "... but he is": ibid., July 19, 1959 (NMK).
183 "three-dimensional jigsaw puzzle": "Finding the World's Earliest Man," op. cit., p. 433.

Chapter 13. Fame, Fortune, and *Zinj*

185 "major differences": L. S. B. Leakey, "A New Fossil Skull from Olduvai," *Nature,* August 15, 1959, p. 492.
185 "I am not ... *boisei":* ibid., p. 491.
186 "We ... have got" to "... yet": L. S. B. Leakey to Peter Davis, July 26, 1959 (NMK).
186 "Louis said.... to see it": F. Clark Howell, interview, Berkeley, March 21, 1985.
186 "Most people.... colleagues": ibid.
186 "[I]n fact ... lodged": Sonia Cole, *Leakey's Luck,* p. 231.
187 "glory box": Phillip V. Tobias, interview, Johannesburg, August 3, 1984.
187 "I knew.... spine": Phillip V. Tobias, interview, New York City, April 5, 1984.
187 "personally," "very ... comparisons": "A New Fossil Skull from Olduvai," op. cit., p. 491.
187 "murderous ... cannibalistic": Raymond Dart, *Adventures with the Missing Link,* pp. 106–107.
188 "I'm so ... people": *Louis Leakey and the Dawn of Man,* National Geographic film, November 1966.
188 "wear out" to "so am I": G. H. Findlay, *Dr. Robert Broom,* p. 89.
188 "I am convinced.... which man arose": L. S. B. Leakey to Ralph von Koenigswald, May 12, 1950 (KNA).
189 "Brooms material.... Great": Ralph von Koenigswald to L. S. B. Leakey, January 10, 1951 (KNA).
189 "talked volubly.... australopithecines," "had not.... Johannesburg," "He had ... them": John Robinson, letter to author, February 8, 1988.
189 "did not ... views": ibid.
189 "No, no, no ...": Clark Howell, interview, Berkeley, March 21, 1985.
190 "They were sitting.... I'll show you": J. Desmond Clark, interview, Berkeley, February 5, 1986.
190 "Louis has.... anyone": J. Desmond Clark, interview, Berkeley, February 5, 1986.
190 "He loved ... business": J. Desmond Clark, interview, Berkeley, February 5, 1986.
190 "remarkable discovery": W. E. Le Gros Clark to Kenneth Oakley, August 4, 1959 (BMNH).
190 "I am.... different genus": John Robinson to W. E. Le Gros Clark, August 31, 1959 (BMNH).
190 "I do hope ... massively": W. E. Le Gros Clark to Kenneth Oakley, September 3, 1959 (BMNH).
191 "to harmonize ... zoology": Ernst Mayr, "Taxonomic Categories in Fossil Hominids," *Cold Spring Harbor Symposia on Quantitative Biology,* Vol. XV, *Origin and Evolution of Man,* p. 109.
191 "simulates ... exist": ibid., p. 115.
191 "If individuals ... primates": ibid., p. 109.
191 "The number of.... absolutely necessary": Sherwood Washburn, "Analysis of Primate Evolution," *Cold Spring Harbor Symposia on Quantitative Biology,* Vol. XV, *Origin and Evolution of Man,* p. 76.
192 "I think ... the new view": Ronald Millar, *The Piltdown Men,* p. 220.
192 "To such ... book": Arthur Keith, *An Autobiography,* p. 328.
192 "It was put.... slides": Phillip V. Tobias, interview, Johannesburg, August 3, 1984.
193 "about halfway.... thing here": F. Clark Howell, interview, Berkeley, February 4, 1986.
193 "do quite ... Man": L. S. B. Leakey, "Remains of Man with Oldowan Culture at Olduvai," *Actes du IVe Congrès Panafricain de Préhistoire,* p. 364.
193 "Each one.... dark night": Phillip V. Tobias, interview, Johannesburg, August 3, 1984.
194 "I have no ... run away!" L. S. B. Leakey, speech on human origins at the Israel Academy of Sciences and Humanities, August 2, 1965 (NMK).
194 "an enormous set of 'nutcrackers,'" Phillip V. Tobias quoted in the discussion section of "Remains of Man with Oldowan Culture at Olduvai," op. cit., p. 363.
194 "the connecting ... know him": "Finder Says Fossil Links Ape and Man," *New York Times,* September 4, 1959, p. 1.
194 "oldest ... anywhere": "A 'Stupendous Discovery': The Fossil Skull from Olduvai," *Illustrated London News,* September 12, 1959, p. 217.
195 "television ... Wheeler": "A Skull at Least 600,000 Years Old," *The Guardian,* October 8, 1959.
195 "It was ... life": "Missing Link Proof Goes on Show," *News Chronicle,* October 8, 1959.

195 "the experts . . . discoveries": ibid.
195 "My father. . . . importance of it": Gilbert Grosvenor, interview, Washington, D.C., February 12, 1985.
196 "lived more than 600,000 years ago": L. S. B. Leakey, "Finding the World's Earliest Man," *National Geographic,* September 1960, p. 422.
196 "One thing . . . old, old, old!" Garniss Curtis to L. S. B. Leakey, May 20, 1961 (courtesy of Garniss Curtis).
196 "It was considered. . . . came like locusts": Yves Coppens, interview, Paris, June 6, 1984.

Chapter 14. Mary's Dig

197 "most generously": Mary Leakey, *Olduvai Gorge,* Vol. 3, *Excavations in Beds I and II, 1960–1963* p. xix.
197 "The Kikuyu. . . . Kamba": Mary Leakey, interview, San Francisco, May 5, 1987.
197 "[T]hese boys . . . can": Vivian Fuchs's field diary, 1931.
198 "What . . . bones": Kamoya Kimeu, interview, Nairobi, October 30, 1984.
198 "We didn't. . . . much," "People. . . . worried": ibid.
198 "He spoke. . . . with him": Kamoya Kimeu, interview, Kothiai River Camp, Kenya, August 26, 1987.
198 "You see. . . . believe that": Joseph Mutaba, interview, Nariokotome River Camp, Kenya, September 16, 1987.
199 "very much": Mary Leakey, *Disclosing the Past,* p. 125.
200 "probably like a pygmy": John Hillaby, "Fossil Gives Clue to the 'First Man,' " *New York Times,* October 8, 1959, p. 44.
200 "Does any animal. . . . must be hominid": *Disclosing the Past,* p. 126.
200 "Even early . . . small bodies": Henry McHenry, telephone interview, August 22, 1994.
200 "whether . . . speech": L. S. B. Leakey, "Finding the World's Earliest Man," *National Geographic,* September 1960, p. 435.
200 "You dig. . . . a surprise": Mary Leakey, interview, Olduvai Gorge, September 25, 1984.
200 "That frightened . . . graves": Peter Nzube, interview, Kothiai River Camp, August 30, 1987 (Kamoya Kimeu, translator).
200 "We made. . . . singing!' ": Kamoya Kimeu, interview, Kothiai River Camp, August 30, 1987.
201 "This. . . . dig": ibid.
201 "Mary. . . . teacher": Kamoya Kimeu, interview, San Francisco, November 13, 1985.
201 "like. . . . Mary": Peter Nzube, interview, Kothiai River Camp, August 30, 1987 (Kamoya Kimeu, translator).
201 "Mary was . . . much": Kamoya Kimeu, interview, San Francisco, November 13, 1985.
201 "She knows. . . . quiet": ibid.
201 "This work. . . . hard": Kamoya Kimeu, interview, Nairobi, October 30, 1984.
202 "[W]e have . . . quiet!)": L. S. B. Leakey to Phillip Tobias, March 7, 1960 (NMK).
202 "Trial Trench": *Olduvai Gorge,* Vol. 3, p. 21.
202 "new . . . research": J. Desmond Clark, Foreword to *Olduvai Gorge,* Vol. 3, p. xv.
203 "food debris": *Olduvai Gorge,* Vol. 3, p. 277.
203 "the residue of human faeces": ibid., p. 259.
203 "That's the main . . . magnet was": Richard Potts, telephone interview, November 7, 1994.
203 "Louis put. . . . I need": Irven DeVore, interview, New Brunswick, October 20, 1990.
204 "the most . . . Olduvai]": "Finding the World's Earliest Man," op. cit., p. 428.
204 "like a French person": Betty Howell, interview, Berkeley, February 4, 1986.
204 "Tea. . . . on with it": ibid.
205 "[I]t looks . . . body": L. S. B. Leakey to W. E. Le Gros Clark, August 31, 1960 (NMK).
205 "He sort. . . . fragments," "They were. . . . *Telanthropus*": Maxine Kleindienst, interview, San Francisco, April 21, 1986.
206 "Let me add . . . disturbing": L. S. B. Leakey to W. E. Le Gros Clark, September 30, 1960 (NMK).
206 "fit into . . . and so on": L. S. B. Leakey to W. E. Le Gros Clark, November 15, 1960 (NMK).
206 "very puzzled . . . jaw": W. E. Le Gros Clark to L. S. B. Leakey, November 25, 1960 (NMK).
206 "STAGGERED . . . EVIDENCE": L. S. B. Leakey to W. E. Le Gros Clark, n.d. (NMK).
206 "How badly. . . . jaw": W. E. Le Gros Clark to L. S. B. Leakey, November 29, 1960 (NMK).
207 "I think . . . difficulty," "quite . . . hominid": L. S. B. Leakey to W. E. Le Gros Clark, December 8, 1960 (NMK).

207 "premonition . . ." to "This . . . skull": L. S. B. Leakey, "Exploring 1,750,000 Years into Man's Past," *National Geographic,* October 1961, p. 581.
208 "I fell. . . . Chellean Man": ibid.
208 "What. . . . Chellean *skull":* ibid.
208 "unbelievably massive": Mary Leakey, *Olduvai Gorge: My Search for Early Man,* p. 81.
208 "a direct . . . *Homo":* L. S. B. Leakey to Phillip V. Tobias, December 8, 1960 (NMK).
208 "I have just. . . . foot?": Peter Davis to L. S. B. Leakey, December 21, 1960 (NMK).
209 "You may . . . before": L. S. B. Leakey to W. E. Le Gros Clark, December 8, 1960 (NMK).

Chapter 15. Murder and Mayhem

210 "a bear of a man": Mary Smith, interview, Washington, D.C., February 13, 1985.
211 "Here you see . . ." to "Here . . . tooth brush": L. S. B. Leakey, "Finding the Earliest Man," February 24, 1961, unpublished speech at Constitution Hall (courtesy of the National Geographic Society).
211 "He was . . . *extraordinary* find": Melville Bell Grosvenor to Armand Denis, January 31, 1961 (quoted in Sonia Cole, *Leakey's Luck,* p. 242).
211 "an exciting . . . support": Mary Leakey, *Disclosing the Past,* p. 125.
211 "It was. . . . trees": Mary Smith, telephone interview, March 31, 1989.
212 "Louis was not. . . . his frauds": Ernst Mayr, telephone interview, May 10, 1988.
212 "White thinks. . . . success": Kenneth Oakley to W. E. Le Gros Clark, January 1, 1958 (BMNH). "If I accept. . . . rather wait": L. S. B. Leakey to Kenneth Oakley, January 1958 (quoted in *Leakey's Luck,* p. 222).
212 "to be moderate": W. E. Le Gros Clark to L. S. B. Leakey, April 19, 1960 (NMK).
212 "In particular . . . disparaging way": W. E. Le Gros Clark to L. S. B. Leakey, November 17, 1960 (NMK).
212 "earliest man," "Beyond . . . jersey dress": "Katwinkel's Heirs," *Time,* March 10, 1961, p. 44.
213 "[The press]. . . . murder' ": *Disclosing the Past,* p. 128.
213 "I think. . . . foul": ibid.
213 "British . . . event": " 'Murder Verdict' on Hominid": *The Times,* February 26, 1961, p. 9.
213 "More Secrets . . . Gorge": *Punch,* March 8, 1961.
213 "It would be. . . . a little ridiculous": Geminus, "It Seems to Me," *New Scientist,* March 2, 1961, p. 552.
214 "It would seem. . . . from the break": L. S. B. Leakey, "Exploring 1,750,000 Years into Man's Past," *National Geographic,* October 1961, p. 575.
214 "I think it . . . blunt instrument' ": ibid.
214 "probably was. . . . those days": Tony Irwin, "The Life and Times of Dear Boy," *The Sunday Post,* May 22, 1960.
214 "The fact that . . . by a blow": Mary Leakey, *Olduvai Gorge,* Vol. 3, p. 229.
214 "I concluded . . . overstated": Phillip V. Tobias, interview, Johannesburg, August 3, 1984.
215 "I think they. . . . the continent": ibid.
215 "This isn't. . . . home": not for attribution.
215 "[They] . . . rang": *Disclosing the Past,* p. 117.
215 "But don't . . ." to ". . . keep them in," "I suddenly. . . . with this' ": Betty Howell, interview, Berkeley, February 4, 1986.
216 "I knew that. . . . in any way": Alan Gentry, interview, London, May 23, 1984.
216 "It was like. . . . didn't care either" to "wouldn't cook. . . . wonderful": Betty Howell, interview, Berkeley, February 4, 1986.
217 "More . . . blood!": Jean Brown, letter to author, May 17, 1988.
217 "Louis thought. . . . with needlepoint": Mary Smith, interview, Washington, D.C., February 13, 1985.
217 "She is . . . choice words": Betty Howell, interview, Berkeley, February 4, 1986.
217 "semi-Victorian . . . upbringing" to "Father . . . salt?": Richard Leakey, interview, San Francisco, February 17, 1989.
217 "What could. . . . themselves": Mary Leakey, interview, Nairobi, September 1, 1989.
217 "There was no. . . . irascible": Clark Howell, interview, Berkeley, February 4, 1986.
218 "He would frighten. . . . way to school": Christine McRae, interview, Nairobi, October 26, 1987.
218 "Louis was. . . . get help": Joseph Mutaba, interview, Nariokotome River Camp, Kenya, September 16, 1987.

219 "Most of our. . . . a little exasperating," "Oh, I was. . . . I was": Mary Leakey, interview, Nairobi, July 9, 1984.
219 "really the . . . to date": Mary Leakey to Phillip Tobias, May 12, 1961 (courtesy of Phillip Tobias).
219 "I doubt . . . field": L. S. B. Leakey to Professor L. H. Wells, March 27, 1961 (NMK).
219 "wiser": L. S. B. Leakey to Phillip Tobias, October 21, 1961 (NMK).
219 "inseparable . . . Australopithecines": Le Gros Clark to L. S. B. Leakey, June 15, 1961 (NMK).
219 "I do . . . stage": Le Gros Clark to L. S. B. Leakey, July 5, 1961 (NMK).
220 "If I . . . problem": L. S. B. Leakey, "The Juvenile Mandible from Olduvai," *Nature,* July 22, 1961, pp. 417–18.
220 "Louis said . . . Mary": Glynn Isaac, interview, Cambridge, Mass., April 13, 1984.
220 "I particularly . . . was all": Alan Gentry, interview, London, May 23, 1984.
220 "Coming out . . . servants": ibid.
221 "It was . . . imagined": ibid.
221 "the narcotic hour": Mary Leakey, interview, Olduvai Gorge, September 22, 1984.
221 "Why . . . there": T. Dale Stewart, telephone interview, May 11, 1989.
221 "a heavenly mountain": Mary Leakey, interview, Olduvai Gorge, September 22, 1984.
221 "Every professional . . . nasty": Glynn Isaac, interview, Cambridge, Mass., April 13, 1984.
221 "bad-mouthing . . . me off": Richard Hay, interview, Urbana, Illinois, February 4, 1985.
222 "Their roaring . . . ground": Mary Leakey, interview, Olduvai Gorge, September 22, 1984.
222 "The stupid . . . either": Mary Leakey to Louis Leakey, November 9, 1963 (NMK).
222 "The leopards . . . louder": Maxine Kleindienst, telephone interview, June 12, 1989.
222 "Louis told . . . bowl": Helen O'Brien, interview, Newport Beach, California, April 18, 1985.
222 "Louis took . . . rhino?": Yves Coppens, interview, Paris, June 6, 1984.
223 "I had . . . them": Mary Leakey, interview, Olduvai Gorge, September 23, 1984.
223 "It rained . . . working": Maxine Kleindienst, interview, San Francisco, April 21, 1986.
223 "If you . . . drowning": ibid.
223 "Confidentially . . . 'child' ": L. S. B. Leakey to Phillip Tobias, December 13, 1961 (NMK).
223 "He had . . . right": Mary Leakey, interview, Olduvai Gorge, September 25, 1984.
224 "I trust . . . 'Yes' ": Richard Potts, telephone interview, November 7, 1994.
224 "It was . . . sure": Mary Leakey, interview, Olduvai Gorge, September 25, 1984.
224 "They were . . . today": ibid.
224 "I then . . . job": ibid.
224 "So then . . . fire!": ibid.

Chapter 16. The Human with Ability

225 "a very primitive *Homo*": L. S. B. Leakey to Phillip V. Tobias, August 9, 1962 (NMK).
225 "mental Rubicon": Sir Arthur Keith, *A New Theory of Human Evolution,* p. 207.
225 "primitive . . . Australia": ibid., p. 205.
226 "distinctive human trait": W. E. Le Gros Clark, "Bones of Contention," *Journal of the Royal Anthropological Institute,* Vol. 89, 1959, p. 139.
226 "No structure found. . . . a quantitative one": Grafton Elliot Smith, quoted by Sir Arthur Keith in his lecture at the British Association meeting in Leeds; *The Times* (London), August 1, 1928, p. 15.
226 "It was . . . such a skull": Phillip V. Tobias, interview, Johannesburg, August 3, 1984.
226 "Father of Kenyan Archaeology," "Father": Jean Brown, letter to author, May 17, 1988.
226 "I have just . . . disappointed": L. S. B. Leakey to Phillip V. Tobias, October 12, 1961 (NMK).
226 "I agree with. . . . to Tobias": W. E. Le Gros Clark to Kenneth Oakley, December 6, 1962 (BMNH).
227 "This is disappointing . . . our hands": L. S. B. Leakey to Phillip V. Tobias, October 12, 1961 (NMK).
227 "[T]he child . . . *Zinjanthropus,*" to "These are . . . moment": Phillip V. Tobias to L. S. B. Leakey, May 1, 1962 (NMK).
227 "I do want. . . . point of view": L. S. B. Leakey to Phillip V. Tobias, December 28, 1962 (NMK).
227 "So far, I. . . . premature 'psychosclerosis'!!!": Phillip V. Tobias to L. S. B. Leakey, January 28, 1963.
227 "I do *NOT*. . . . tell me *why?*": L. S. B. Leakey to Phillip V. Tobias, February 11, 1963 (NMK).
227 "Face . . . sure," "We will . . . study": ibid.
228 "CRANIAL . . . AFFINITY": Phillip V. Tobias to L. S. B. Leakey, June 21, 1963 (NMK).
228 "does not seem. . . . Epi-Homo" to "present-day *Homo sapiens*": L. S. B. Leakey to Phillip V. Tobias, July 10, 1963 (NMK).

228 "short . . . nails" to "closely . . . living to-day": J. R. Napier, "Fossil Hand Bones from Olduvai Gorge," *Nature,* November 3, 1962, p. 411.
228 "a lot of plasticine," "That was. . . . delighted": Michael Day, interview, London, May 24, 1984.
229 "walking pattern . . . great": John Napier to L. S. B. Leakey, July 19, 1963 (NMK).
229 "by no means made up" to "the study continues": Phillip V. Tobias to L. S. B. Leakey, July 30, 1963 (courtesy Phillip V. Tobias).
230 "I regretted . . . associated": Mary Leakey, letter to author, April 21, 1989.
230 "Mary did not . . . a bit further": George Gaylord Simpson, *Concession to the Improbable: An Unconventional Autobiography,* p. 200.
230 "ulterior motive," "conservative": ibid., p. 199.
230 "a very good field man": Leo Laporte, telephone interview, May 26, 1989.
230 "determine . . . evidence": *Concession to the Improbable,* p. 199.
231 "There cannot . . . on earth": ibid., p. 201.
231 "[Their] field . . . factual": ibid., p. 200.
231 "It still seems . . . Royal Society": ibid.
231 "Eagerly . . . inside" to "running full out": L. S. B. Leakey, "Adventures in the Search for Man," *National Geographic,* January 1963, p. 134.
231 "George! . . . I've got it": *Concession to the Improbable,* p. 202.
231 "[The jaw] . . . look to it": ibid., p. 134.
231 "an enormous . . . development": "Adventures in the Search for Man," op. cit., p. 134.
231 "bit of. . . . highly important" to "sound horrible . . . field": *Concession to the Improbable,* p. 203.
232 "Last week. . . . quiet about it": L. S. B. Leakey to Melville Grosvenor, June 28, 1961 (NMK).
232 "very hesitant . . . primates": L. S. B. Leakey to George Gaylord Simpson, December 14, 1961 (NMK).
232 "revolutionizing . . . Africa": inscription on Louis and Mary Leakey's 1962 National Geographic Society Hubbard Medal.
232 "a descendant . . . stock," "new . . . towards man": L. S. B. Leakey, unpublished lecture at Constitution Hall, March 23, 1962 (NMK).
232 "These two . . . Oldoway" to "real man": ibid.
233 "It was . . . premolar": Mary Leakey, *Olduvai Gorge: My Search for Early Man,* p. 78.
233 "The mandible . . . immediately," "whilst . . . washed": Mary Leakey, *Olduvai Gorge,* Vol. 3, *Excavations in Beds I and II, 1960–1963,* p. 231.
233 "Confidentially . . . excavating": L. S. B. Leakey to Phillip V. Tobias, October 25, 1963 (NMK).
233 "being . . . bitch," "boyfriend": Mary Leakey to L. S. B. Leakey, November 25, 1963 (NMK).
234 "1,500 fragments . . . grain of rice": *Olduvai Gorge,* Vol. 3, p. 232.
234 "I think our . . . closer to Man": L. S. B. Leakey to Phillip V. Tobias, November 22, 1963 (NMK).
234 Taking into account. . . . correct interpretation: Phillip V. Tobias to L. S. B. Leakey, December 19, 1963 (NMK).
234 "It was that. . . . it wasn't Louis: Phillip V. Tobias, interview, Johannesburg, August 3, 1984.
235 "He was a . . . what he was": ibid.
235 "[M]y eyes . . . this is *Homo'* ": P. V. Tobias, *Olduvai Gorge,* Vol. 4, *The Skulls, Endocasts and Teeth of* Homo habilis, p. 25.
235 "[I]t is . . . living site": L. S. B. Leakey, P. V. Tobias, and J. R. Napier, "New Species of the Genus *Homo," Nature,* April 4, 1964, p. 9.
236 "not only . . . stone-tool making man": L. S. B. Leakey, *Adam's Ancestors* (1953), p. x.
236 "Many people. . . . been changed": Michael Day, interview, London, May 24, 1984.
236 "to . . . pattern" to "[I]n consequence. . . . United Nations": L. S. B. Leakey, press conference, Washington, D.C., April 4, 1964.

Chapter 17. Chimpanzees and Other Loves

237 "Louis was. . . . orangutanville": Mary Smith, interview, Washington, D.C., February 13, 1985.
237 "It was an. . . . 'I certainly will' ": ibid.
238 "hoping to get involved with animals": Jane Goodall, interview, San Francisco, April 13, 1985.
238 "So I rang up. . . . want?' ": Jane Goodall in *The Legacy of L. S. B. Leakey,* National Geographic Society film, January 9, 1978.
238 "the main . . . man himself": L. S. B. Leakey, *Adam's Ancestors* (1953), p. 185.
238 "with . . . rivers": L. S. B. Leakey, "The Fossil Apes of Lake Victoria: Have Man's Ancestors Been Found?," *The Listener,* September 21, 1951, p. 143.

239 "who . . . nameless," "failed utterly": L. S. B. Leakey to Elspeth Huxley, November 17, 1971 (NMK).
239 "I only . . . myself": L. S. B. Leakey to Walter Baumgartel, March 9, 1956 (NMK).
239 "There's a young . . . background": Cathryn Hosea Hilker, telephone interview, October 5, 1994. "Jane, I hate. . . . hate myself, really": Jane Goodall, interview, San Francisco, April 13, 1985.
239 "He asked me. . . . serious" to "I thought. . . . sores for Mary": ibid.
240 "the way . . . doors for me": to "Louis was. . . . disgusting": Jane Goodall, interview, Gombe Stream Research Center, Tanzania, October 11, 1987.
240 "intoxicating" to "that's what. . . . mistake": Jane Goodall, interview, San Francisco, April 13, 1985.
240 "I have found . . . training also," "shed some light . . . *Proconsul*": L. S. B. Leakey to Sherwood Washburn, November 19, 1957 (NMK).
241 "getting to know. . . . East African plains": Alan Gentry, interview, London, May 23, 1984.
241 "He thought there. . . . won't hurt you' ": Leighton Wilkie, telephone interview, September 29, 1989.
241 "extremely tolerant. . . . habituation": Irven DeVore, interview, Cambridge, Mass., February 20, 1985.
241 "He wanted someone . . . knowledge": Jane Goodall, *In the Shadow of Man,* p. 6.
241 "Sure I was. . . . since, say, 1900": Irven DeVore, interview, Cambridge, Mass., February 20, 1985.
242 "I met Jane. . . . vulnerable to me": Betty Howell, interview, Berkeley, February 4, 1986.
242 "brave pioneers": Leighton Wilkie, telephone interview, September 29, 1989.
242 "horrified," "Louis was. . . . hurt, I think": Jane Goodall, interview, Gombe Stream Research Center, October 11, 1987.
243 What I was. . . . way with him": ibid.
243 "Louis was . . . men's wives": Elwyn Simons, interview, New York City, April 10, 1984.
243 "I remember. . . . women as well": Irven DeVore, interview, Cambridge, Mass., February 20, 1985.
244 "the third Mrs. Leakey": Richard Leakey, interview, Kothiai River Camp, Kenya, September 3, 1987.
244 "Supposedly . . . Valley": Karl Butzer, telephone interview, December 6, 1986.
244 "full of energy. . . . Louis": Jill Donisthorpe, interview, Nairobi, September 1, 1989.
244 "That was Mother's. . . . until then": Priscilla Davies, interview, Harrow, England, June 5, 1984.
244 "looking up Father's . . . library": ibid.
244 "It was no. . . . in Louis's life": Frida Leakey, interview, Cambridge, May 27, 1984.
244 "very easy . . . minutes": Priscilla Davies, interview, Harrow, England, June 5, 1984.
244 "Priscilla arranged. . . . 'be my son,' " "You could. . . . identical": Colin Leakey, interview, Cambridge, November 1, 1987.
245 "I remember. . . . duplicity": ibid.
245 "Of course. . . . would be" to "such . . . women": Mary Leakey, interview, Nairobi, August 20, 1989.
245 "I think Mary. . . . solace from Rosalie": Richard Leakey, interview, San Francisco, February 17, 1989.
245 "We had rows. . . . quite young": Mary Leakey, interview, Nairobi, August 20, 1989.
245 "I used to. . . . Don't go, Daddy": Richard Leakey, interview, San Francisco, February 17, 1989.
245 "I saw his. . . . did some good": ibid.
246 "Louis asked. . . . I went": Rosalie Osborn, interview, Nairobi, August 28, 1989.
246 "Two girls. . . . believe": unpublished summary of New York Zoological Society meeting on field research on primates, December 4–5, 1959 (KNA).
247 "Louis bought. . . . pieces": Jane Goodall, interview, San Francisco, April 25, 1986.
247 "I am anxious. . . . should be there": David Anstey to L. S. B. Leakey, May 3, 1960.
248 "I could think . . . chimpanzees": *In the Shadow of Man,* p. 9.
248 "You will be. . . . at once": L. S. B. Leakey to Leighton Wilkie, July 26, 1960 (courtesy of Leighton Wilkie).
248 "I was really. . . . worse I felt": Jane Goodall, interview, San Francisco, April 13, 1985.
248 "carefully. . . . a tool": *In the Shadow of Man,* p. 35.
248 "major [scientific] periodical" "I am glad . . . sexes": Jane Goodall, *The Chimpanzees of Gombe: Patterns of Behavior,* p. 60.
249 "the first . . . toolmaking": *In the Shadow of Man,* p. 37.

249 "He said . . . stone tools' ": Jane Goodall, interview, San Francisco, April 13, 1985.
249 "Miss Goodall. . . . with the work": L. S. B. Leakey to Harold J. Coolidge, November 15, 1960 (NMK).
249 "Good morning": *In the Shadow of Man,* p. 50.
250 "That was. . . . Child' ": Jane Goodall, interview, Gombe Stream Research Center, October 11, 1987.
250 "Vanne is. . . . relaxed with her": Meave Leakey, interview, Nairobi, October 8, 1984.
250 "absolutely incredible. . . . How funny": Jane Goodall, interview, San Francisco, April 25, 1986.
250 "a fool": Mary Leakey, interview, Nairobi, September 1, 1989.
250 "best of friends": Jane Goodall, interview, Gombe Stream Research Center, October 11, 1987.
251 "make . . . world": Richard Leakey, interview, en route to Kothiai River Camp, Kenya, September 23, 1987.
251 "a blonde bimbo": Michael McRae, "Dilemma at Gombe," in *Continental Drifter,* p. 68.
251 "very arrogant," "mean and nasty": Jane Goodall, interview, Gombe Stream Research Center, October 11, 1987.
251 "appallingly rude": Richard Leakey, interview, Portland, Oregon, February 26, 1994.
251 "If only you. . . . hurts, you know": Jane Goodall to L. S. B. Leakey, July 30, 1963 (NMK).
251 "We couldn't have. . . . it wasn't": Jane Goodall, interview, San Francisco, April 13, 1985.
251 "But I had . . . young man": Richard Leakey, interview, Nairobi, October 16, 1987.

Chapter 18. Richard Makes His Move

252 "There were few. . . . archeology": Betty Howell, interview, Berkeley, February 4, 1986.
252 "Mary-mine," "did not . . . much": Alan Gentry, interview, London, May 23, 1984.
252 "some of the happiest times": Richard Leakey, interview, San Francisco, February 17, 1989.
253 "They were. . . . Ngong Hills": Christine McRae, interview, Nairobi, October 26, 1987.
253 "something I was good at," "an outdoor life in Africa": Richard E. Leakey, *One Life,* p. 40.
253 "They all three. . . . in the fire": Mary Leakey, interview, Nairobi, July 9, 1984.
253 "The Leakeys' sons. . . . 150 years later": Betty Howell, interview, Berkeley, February 4, 1986.
253 "If one wanted. . . . one's keep": Richard Leakey, interview, San Francisco, February 17, 1989.
253 "When I. . . . for granted," "early men and early women": Maxine Kleindienst, interview, San Francisco, April 21, 1986.
254 "at which point . . . tackle": John Pfeiffer, "Man Through Time's Mists," *Saturday Evening Post,* December 3, 1966, p. 50.
254 "We chased . . . which they did": Richard Leakey, interview, Nairobi, July 4, 1984.
254 "experimented. . . . a little zebra": Boyce Rensberger, "The Face of Evolution," *New York Times Magazine,* March 3, 1974, p. 52.
254 "long Leakey legs": Phillip V. Tobias, interview, Johannesburg, August 3, 1984.
254 "It was . . . children were": Alan Gentry, London, May 23, 1984.
254 "a sadness of any youth": Richard Leakey, interview on the BBC Castaway Radio Program, Desert Island Discs, June 13, 1981.
254 "lover of niggers": *One Life,* p. 35.
254 "Before I knew. . . . escape" to "He had . . . school": ibid.
255 "marvellous degree of independence": ibid., p. 42.
255 "develop. . . . in early oceans": Richard Leakey, interview on the BBC Castaway Radio Program, op. cit.
255 "it was . . . begin": *One Life,* p. 48.
255 "very prickly about being": Richard Leakey, interview on the BBC Castaway Radio Program, op. cit.
255 "they were too alike in temperament": Mary Leakey, interview, Nairobi, July 9, 1984.
255 "as fond of . . . less indulgence": Mary Leakey, *Disclosing the Past,* p. 141.
256 "I shall always . . . in my life": *One Life,* p. 54.
256 "Richard and Kamoya. . . . filled with bones" to "I told him. . . . the books": Garniss Curtis, interview, Berkeley, March 28, 1985.
256 "sedate, white-collar community"; "After Years of Terror, a Comeback for Kenya," *U.S. News & World Report,* January 23, 1959, p. 105.
257 "Louis had. . . . bush somewhere": Mary Leakey, interview, Nairobi, September 7, 1987.
257 "road to Inefficiency. . . . by Europeans": L. S. B. Leakey, broadcast on the Voice of Kenya, 1960 (NMK).

258 "Kenya today.... with complacency": L. S. B. Leakey, "Time for a Firm Hand in Kenya," *Sunday Telegraph,* May 28, 1961.
258 "In fact... Kikuyu": Charles Njonjo, interview, Nairobi, September 14, 1987.
258 "Kenya was.... my skin": Richard Leakey, interview, Nairobi, October 10, 1984.
258 "They were appalled.... were off": Richard Leakey, interview on the BBC Castaway Radio Program, op. cit.
259 "I knew.... was curious": Richard Leakey, interview, Nairobi, February 1983.
259 "new and marvelous places": Richard Leakey, lecture, San Francisco, February 16, 1989.
259 "Richard was radioing.... more experience,'" "Louis was looking.... broke down?": Glynn Isaac, interview, Cambridge, Mass., April 13, 1984.
259 "Anything that fell... been there" to "with a sense.... finding them": ibid.
260 "strong social conscience... solidarity": *One Life,* p. 71.
260 "Naturally.... in the lead!": ibid.
260 "I had enjoyed.... to Nairobi": ibid., p. 72.
260 "a very white.... to stand": Kamoya Kimeu, "Kamoya Kimeu's Stories," as transcribed by Alan Walker, 1986.
261 "None of us... excitement" to "It was... sticking out": Glynn Isaac, interview, Cambridge, Mass., April 13, 1984.
261 "The new lower.... species of *Homo":* L. S. B. Leakey, press conference, Washington, D.C., April 3, 1964 (NMK).
262 "Richard saw.... point for him": Glynn Isaac, interview, Cambridge, Mass., April 13, 1984.
262 "On that expedition... please me very much": Richard Leakey, interview, San Francisco, February 17, 1989.
262 "I started to have.... go to university": Richard Leakey, interview on the BBC Castaway Radio Program, op. cit.
262 "In many ways... Parks": *One Life,* p. 76.
262 "was going to do next": ibid., p. 75.
262 "[F]or once... Margaret": ibid., p. 76.

Chapter 19. A Girl for the Gorillas

263 "could... family": L. S. B. Leakey to Sir Lawrence Bragg, FRS, April 3, 1963 (Rice University archives, Texas).
263 "I have.... future": Brian Lapping, *End of Empire,* p. 442.
263 "I was.... as friends": Charles Njonjo, interview, Nairobi, September 14, 1987.
264 "Yes.... Chief Justice": *End of Empire,* p. 444.
264 "It was.... considered": Mary Leakey, *Disclosing the Past,* p. 140.
264 "honey bee... another": P. G. H. Boswell, unpublished autobiography, p. 234.
264 "Louis.... meat": Lita Osmundsen, interview, New York City, April 4, 1984.
264 "know... world": Mary Smith, interview, Washington, D.C., February 13, 1985.
265 "I certainly.... desperate": *Disclosing the Past,* p. 145.
265 "All.... in America": Dee Simpson, interview, Calico Hills Early Man Site, California, April 20, 1985.
265 "if... was it!": L. S. B. Leakey, lecture at the International Conference on the Calico Mountains Excavation, Bloomington, California, October 22, 1970.
265 "You know, Dee... along those lines": Dee Simpson, interview, Calico Hills Early Man Site, April 20, 1985.
266 "circumstantial evidence" to "the few.... must be so!": L. S. B. Leakey, lecture at the International Conference on the Calico Mountains Excavation.
266 "bawdiness and 'macho' ": Frances Burton, letter to author, June 17, 1986.
266 "[Tigoni] was.... wasting his time": Mary Leakey, interview, Nairobi, September 7, 1984.
267 "It couldn't have.... pretty devastating": Lita Osmundsen, telephone interview, March 16, 1984.
267 "Dee, dig here": Dee Simpson, interview, Calico Hills Early Man Site, April 20, 1985.
267 "I tried to... I couldn't": Mary Leakey, interview, Nairobi, July 9, 1984.
267 He might say.... classification: Phillip V. Tobias, interview, Johannesburg, August 3, 1984.
267 "We knew Leakey.... listening to him": Kamoya Kimeu, interview, Mill Valley, California, November 13, 1985.
268 "He loved to... he was": Richard Leakey, interview, San Francisco, November 17, 1987.
268 "I think Louis.... running out": Mary Smith, interview, Washington, D.C., February 13, 1985.

268 "[Louis]. . . . succeeds": Mary Leakey to Phillip V. Tobias, March 8, 1965 (courtesy of Phillip V. Tobias).
268 "I don't think . . . behaved": Mary Leakey, interview, San Francisco, May 5, 1987.
268 "I don't want . . . studies": Jane Goodall to L. S. B. Leakey, August 15, 1963 (NMK).
269 "I don't think. . . . wait a minute": Hugo van Lawick, telephone interview, November 20, 1987. "I didn't . . . did I?": Jane Goodall, interview, San Francisco, April 13, 1985.
269 "treasured fossil": Dian Fossey, *Gorillas in the Mist,* p. 2.
269 "Surely, God, these are my kin": Farley Mowat, *Woman in the Mists,* p. 14.
269 "Miss Fossey. . . . all these people": Dian Fossey's diary, quoted in *Woman in the Mists,* p. 21.
270 "That is where I belong": L. S. B. Leakey's notes from interview with Dian Fossey, March (n.d.) 1966 (NMK).
270 "Dr. Leakey . . . talking": Dian Fossey's diary, quoted in *Woman in the Mists,* p. 22.
270 "Mother! Mother! . . . to do it!": Kitty Price, telephone interview, January 20, 1986.
270 "I have. . . . its proper end": L. S. B. Leakey to Robert Hinde, October 28, 1966 (NMK).
270 "winner": L. S. B. Leakey to Dr. Thorpe, Cambridge University, January 3, 1967 (NMK).
270 "indubitably suitable candidate": L. S. B. Leakey to Leighton Wilkie, November 3, 1966 (NMK).
270 "determined . . . gorillas": L. S. B. Leakey to Dr. Thorpe, January 3, 1967 (NMK).
270 "launched . . . as soon as possible": L. S. B. Leakey to Leighton Wilkie, November 3, 1966 (NMK).
270 "So, you're the . . . are you," "It was . . . question": Dian Fossey, letter to author, October 15, 1984.
271 "intimidating thought": Dian Fossey's diary, quoted in *Woman in the Mists,* p. 28.
271 "[I] found him. . . . where they've been' ": *Gorillas in the Mist,* pp. 6–7.
271 "tent pole . . . after him": ibid., p. 7.
271 "To date . . . gorillas" to "peel . . . leaves": Dian Fossey to L. S. B. Leakey, January 31, 1967 (NMK).
272 "nosedive[s]": Dian Fossey's diary, quoted in *Woman in the Mists,* p. 33.
272 "only by . . . dear life": *Gorillas in the Mist,* p. 56.
272 "This is . . . congratulate you" to "[D]o not . . . a little": L. S. B. Leakey to Dian Fossey, February 17, 1967 (NMK).
272 "We'd watch. . . . us": Helen O'Brien, interview, Newport Beach, California, April 18, 1985.
272 "had . . . fatigue": L. S. B. Leakey, in interview for Channel 28, San Francisco, 1968 (n.d.).
273 "I'd told Louis. . . . find the artifacts": Vance Haynes, telephone interview, July 27, 1989.

Chapter 20. To the Omo

274 "This was. . . . found *Zinj*": Yves Coppens, interview, Paris, June 6, 1984.
274 "The Leakey family. . . . permission": Basil Cooke, interview, New York City, April 9, 1984.
275 "Louis. . . . I didn't": Clark Howell, interview, Berkeley, March 21, 1985.
275 "Basically, Louis. . . . to let go": Merrick Posnansky, telephone interview, March 5, 1990.
275 "He cheated. . . . paralyze his expedition": Ernst Mayr, telephone interview, May 10, 1988.
275 "as the dictator. . . . in Kenya": ibid.
276 "In the course. . . . I'll arrange it' ": L. S. B. Leakey, lecture at Constitution Hall for the National Geographic Society, January 1969.
276 "I am . . . importance": L. S. B. Leakey to Leonard Carmichael, April 29, 1966 (NMK).
276 "Young Frank Brown": L. S. B. Leakey to Clark Howell, September 19, 1966 (NMK).
277 "Not at those . . . same idea, yes" to "He was definitely . . . have admitted it": Margaret Avery, interview, Capetown, South Africa, August 6, 1984.
277 "I had lots. . . . in a hurry": Richard Leakey, interview, Nairobi, February 9, 1983.
277 "academic guidance": Richard E. Leakey, *One Life,* p. 78.
277 "I thought, 'Ah. . . . didn't go back": Richard Leakey, interview on BBC Castaway Radio Program, Desert Island Discs, June 13, 1981.
277 "Richard didn't see. . . . do it anyway,' " "he was . . . father": Margaret Avery, interview, Capetown, South Africa, August 6, 1984.
278 "Because I was Kenyan. . . . out of their house": Philip Leakey, interview, Nairobi, October 24, 1984.
278 "the ornery . . . rages": Mary Leakey, interview, San Francisco, May 30, 1987.
278 "with predictable results": Mary Leakey, *Disclosing the Past,* p. 108.
278 "always been . . . own man": Clark Howell, interview, Berkeley, March 21, 1985.
279 "The first. . . . space": Richard Leakey, interview, San Francisco, November 11, 1986.
279 "go and listen, but not speak" to "good faith": ibid.
279 "tremendous row": *One Life,* p. 104.

279 "Richard has. . . . altogether": L. S. B. Leakey to Melvin Payne, April 11, 1967 (NMK).
279 "Father. . . . [between us]": Richard Leakey, interview, San Francisco, November 11, 1986.
280 "implied criticism . . . work": Richard Leakey, interview, Nairobi, July 5, 1984.
280 "There was. . . . supported me": Richard Leakey, interview, San Francisco, November 11, 1986.
280 "ammunition . . . management," "It was a *tiny*. . . . involvement' ": *One Life,* p. 82.
280 "[Their] reputation. . . . reactions prevail": G. H. H. Brown to Richard Leakey, May 29, 1967 (NMK).
281 "If ever . . . for real": *One Life,* p. 88.
281 "We . . . spoon": Yves Coppens, interview, Paris, June 6, 1984.
281 "[A]ll preparations . . . not mine": L. S. B. Leakey to Camille Arambourg, May 3, 1967 (NMK).
281 "My father. . . . various areas," "I always . . . chaperone": Richard Leakey, interview en route to Kothiai River Camp, Kenya, September 23, 1987.
281 "I was the undisputed leader": *One Life,* p. 86.
282 "The number . . . staggering": Frank Brown to L. S. B. Leakey, September 21, 1966 (NMK).
282 "Nobody . . . success": Yves Coppens, interview, Paris, June 6, 1986.
282 "Louis would. . . . see that": Clark Howell, interview, Berkeley, March 31, 1985.
282 "Richard had. . . . East Africa": Paul Abell, interview, Kingston, Rhode Island, February 21, 1985.
282 "effete airs," "the French. . . . whatsoever": ibid.
283 "The atmosphere. . . . monkeys": *One Life,* p. 87.
283 "They were enormous brutes": Paul Abell, interview, Kingston, Rhode Island, February 21, 1985.
283 "counted 590 . . . 20 feet": L. S. B. Leakey to Melvin Payne, July 31, 1967 (NMK).
284 "They abandoned . . . saw us": Bob Campbell, interview, Nairobi, June 28, 1984.
284 "The colobus. . . . our surroundings": *One Life,* p. 89.
284 " 'Is that. . . . Richard": Kamoya Kimeu, interview, Nairobi, February 10, 1983.
284 "We were . . . bank": Paul Abell, interview, Kingston, Rhode Island, February 21, 1985.
284 "Even then . . . couldn't talk": Kamoya Kimeu, interview, Nairobi, February 10, 1983.
285 "The country. . . . situation," "and . . . French": Richard Leakey to Mary and L. S. B. Leakey, June 20, 1967 (NMK).
285 "With this. . . . content": *One Life,* p. 91.
285 "We were spread. . . . together": Paul Abell, telephone interview, March 7, 1990.
286 "We were . . . celebrated": ibid.
286 "RICHARD . . . JOSEPH": L. S. B. Leakey to Melvin Payne, June 29, 1967 (NMK).
286 "Skull . . . *Homo sapiens*": L. S. B. Leakey, handwritten notes, July 8, 1967 (NMK).
286 "I found. . . . delighted": L. S. B. Leakey to Melvin Payne, July 8, 1967 (NMK).
286 "not significantly . . . today": Christopher Stringer and Clive Bamble, *In Search of the Neanderthals,* p. 129.
287 "JOSEPH'S RELATIVE": L. S. B. Leakey to Melvin Payne, August (n.d.) 1967 (NMK).
287 "Clark didn't think. . . . produce that," "Richard. . . . the others": Paul Abell, telephone interview, March 7, 1990.
287 "I was . . . college": Clark Howell, interview, Berkeley, March 21, 1985.
287 "chafing . . . suggestions": Karl Butzer, telephone interview, December 6, 1986.
287 "the somewhat weather-beaten. . . . a bit ragged," "Richard's party . . . located": L. S. B. Leakey to Melvin Payne, July 31, 1967 (NMK).
287 "acid comments . . . Americans": Paul Abell, interview, Kingston, Rhode Island, February 21, 1985.
288 "I remember. . . . happened": Richard Leakey, interview, Nairobi, October 19, 1987.
288 "these professors": Glynn Isaac quoting Richard Leakey, interview, Cambridge, Mass., April 13, 1984.
288 "I was terribly. . . . 'tent-boy' ": Richard Leakey, interview, San Francisco, February 17, 1989.
288 "I already. . . . own show": "Puzzling Out Man's Ascent," *Time,* November 7, 1977, p. 53.
288 "This took . . . seen before": *One Life,* p. 92.
288 "wildly excited": L. S. B. Leakey to Melvin Payne, August 14, 1967 (NMK).
288 "I jumped out. . . . fossils": *One Life,* p. 93.

Chapter 21. Breaking Away

289 "I doubt . . . conceal it": Mary Leakey, *Disclosing the Past,* p. 142.
289 "Before Louis. . . . we are stars": Yves Coppens, interview, Paris, June 6, 1984.

290 "We always . . . freeze": Gilbert Grosvenor, interview, Washington, D.C., February 12, 1985.
290 "Louis always. . . . way": Peter Andrews, interview, London, November 3, 1987.
290 "Abominable Showman": *The Legacy of L. S. B. Leakey,* National Geographic Society film, January 9, 1978.
290 "great argument": L. S. B. Leakey, *By the Evidence,* p. 230.
290 "transitional stage," "root stock . . . near-men": L. S. B. Leakey, "Rich Fossil Beds of Fort Ternan," *The Times Supplement on Kenya, The Times,* December 12, 1963, p. ix.
291 "He'd say, 'Ron. . . . 'ROT!' ": Ron Clarke, interview, Johannesburg, South Africa, August 1, 1984.
291 "I was still. . . . should shut-up" to "One eminent, now-dead. . . . No hard feelings": David Pilbeam, interview, Cambridge, Mass., April 12, 1984.
291 "It is annotated. . . . through the pages": Yves Coppens, interview, Paris, June 6, 1984.
292 "a very early . . . himself": *By the Evidence,* p. 230.
292 "That's part of. . . . side branches": Elwyn Simons, telephone interview, April 22, 1988.
292 "[U]ntil . . . fossil": Walter Sullivan, "Bone Found in Kenya Indicates Man Is 2.5 Million Years Old," *New York Times,* January 14, 1967, p. 1.
292 "Dr. Leakey's claims. . . . unquestionably human": "Discovery Claimed to Put Man's Ancestry Back 6m. Years," *The Times* (London), January 16, 1967, p. 8.
292 "the white-haired . . . on earth": Barbara Tuff, "The Ascent of Man, Out of the Past: Dr. Leakey Looks to the Future," *Science News,* February 25, 1967, p. 188.
293 "a most . . . *africanus":* L. S. B. Leakey to Melvin Payne, August 8, 1967 (NMK).
293 "It was a piece. . . . every time' ": John Van Couvering, interview, New York City, April 5, 1984.
293 "had been . . . blunt instrument": L. S. B. Leakey, "The Gathering of Man: From Savannah to City and Beyond," lecture given at Teilhard de Chardin Conference, San Francisco, May 1, 1971.
293 "peculiar lump . . . edges": L. S. B. Leakey, "Upper Miocene Primates from Kenya," *Nature,* May 11, 1968, p. 529.
293 "to break . . . marrow": ibid., p. 530.
293 "He asked me. . . . nothing happened," "Too strong. . . . softer": Yves Coppens, interview, Paris, June 6, 1984.
294 "outlandish": Elwyn Simons, telephone interview, April 22, 1988.
294 "most unconvincing. . . . stone tool": Mary Leakey, interview, Nairobi, July 9, 1984.
294 "He came up. . . . kinds of things": Elwyn Simons, telephone interview, April 22, 1988.
294 "extensively . . . as well": David Pilbeam, "Notes on *Ramapithecus,* the Earliest Known Hominid, and *Dryopithecus," American Journal of Physical Anthropology,* Vol. 25, 1966, p. 2.
294 "still teaching. . . . such a site": Yves Coppens, interview, Paris, June 6, 1984.
295 "perhaps. . . . 'under' people": Richard E. Leakey, *One Life,* p. 94.
295 "He said something. . . . lots of fossils": Richard Leakey, interview, San Francisco, March 19, 1985.
295 "Father. . . . in the Omo": ibid.
295 "You can have . . . our door again": Melville Grosvenor, quoted in Sonia Cole, *Leakey's Luck,* p. 297.
296 "Father supported. . . . in public": Richard Leakey, interview, San Francisco, March 19, 1985.
296 "For me. . . . wrong reasons": *One Life,* p. 94.
296 "an interdisciplinary . . . Institution": Richard Leakey, interview, San Francisco, February 17, 1989.
297 "felt that . . . hands": Richard Leakey, interview, San Francisco, November 11, 1986.
297 "Either the Museum. . . . improved": *One Life,* p. 102.
297 "Sir Ferdinand was. . . . wanted the job": Richard Leakey, interview, en route to Kothiai River Camp, Kenya, September 23, 1987.
297 "I was then. . . . persisted": Perez Olindo, interview, Nairobi, August 22, 1989.
298 "This was no. . . . completely in charge": *One Life,* p. 103.
298 "supported . . . foreigner!": ibid., p. 102.

Chapter 22. Richard Strikes Oil

299 "great. . . . to find it": Eric Robins, "Anthropologist Richard Leakey Tracks the Grandfather of Man in an African Boneyard," *People,* January 8, 1979, p. 30.
300 "My father asked. . . . as his boss": Richard Leakey, interview en route to Kothiai River Camp, Kenya, September 23, 1987.
300 "Richard had never . . . do come along' ": Bernard Wood, interview, Nairobi, February 9, 1982.

300 "There is one. . . . life or limb": Richard Leakey to Bernard Wood, March 24, 1968 (NMK).
302 "blow[s] . . . time": Margaret Leakey (now Avery), East Rudolf Expedition Field Diary, June 2, 1968 (NMK).
302 "He was very . . . troops": Margaret Avery, interview, Capetown, South Africa, August 6, 1984.
302 "authoritative posture": Paul Abell, interview, Kingston, Rhode Island, February 21, 1985.
303 "Richard needed. . . . pretty hard": ibid.
303 "I had been . . . still with me": Richard E. Leakey, *One Life,* p. 99.
303 "So far. . . . ruined otherwise": Margaret Leakey (now Avery), East Rudolf Expedition Field Diary, June 7, 1968 (NMK).
303 "These were all . . . in the area": Paul Abell, interview, Kingston, Rhode Island, February 21, 1985.
303 "There are literally. . . . near future": Richard Leakey to L. S. B. Leakey, June 13, 1968 (NMK).
303 "struck oil": Mary Leakey, quoted in letter from Richard Hay to Mary Leakey, June 16, 1968 (NMK).
303 "was absolutely. . . . let up a bit": Paul Abell, telephone interview, June 26, 1990.
304 "Richard drove. . . . time": Paul Abell, interview, Kingston, Rhode Island, February 21, 1985.
304 "wind blowing like blue blazes," "The place had. . . . giraffe": ibid.
304 "The afternoons. . . . stifling hot," "We preferred. . . . get lost": Paul Abell, telephone interview, June 26, 1990.
305 "English air": John Harris, interview, Los Angeles, April 11, 1984.
305 "Dinner . . . table": Richard Leakey, interview, Kothiai River Camp, September 9, 1987.
305 "Richard was. . . . morally shoot": Paul Abell, interview, Kingston, Rhode Island, February 21, 1985.
305 "white as the snow," "an edge": Bernard Wood, interview, London, May 23, 1984.
305 "We felt . . . troops!": *One Life,* p. 100.
306 "crammed with fossils," "At Ileret. . . . every day": Paul Abell, telephone interview, June 26, 1990.
306 "[Father] told me. . . . entirely my site": *One Life,* p. 101.
306 "over modest . . . finds," "[Richard]. . . . Olduvai": L. S. B. Leakey to Melvin Payne, September 13, 1968 (NMK).
306 "a native collector . . . Omo": L. S. B. Leakey to J. K. R. Thorpe, March 16, 1943 (KNA).
307 "The East Rudolf. . . . and Central Africa": Richard Leakey to Leonard Carmichael, September 20, 1968 (NMK).
307 "in that gap. . . . in human evolution": L. S. B. Leakey to Melvin Payne, September 13, 1968 (NMK).
308 "One did not . . . work and live": Mary Leakey, *Disclosing the Past,* p. 148.

Chapter 23. Mining Hominids at Olduvai

309 " 'spare time,' " "did . . . results": Mary Leakey to Phillip V. Tobias, January 22, 1968 (courtesy of Phillip V. Tobias).
310 "more primitive . . . tool-making": Mary Leakey, *Olduvai Gorge,* Vol. 3, p. 1.
310 "[C]hoppers were useless. . . . in the pelvis": Mary Leakey, *Olduvai Gorge: My Search for Early Man,* p. 89.
311 "odd or funny": Pat Shipman, quoting Mary Leakey, interview, New York City, April 9, 1984.
311 "make decisions . . . one's finger on": Mary Leakey, in *The Legacy of L. S. B. Leakey,* National Geographic Society film, January 9, 1978.
312 "an important . . . requirements": *Olduvai Gorge,* Vol. 3, p. 267.
312 "represent . . . possible": ibid., p. 266.
312 "You can see. . . . localities": Richard Potts, telephone interview, November 7, 1994.
313 "[T]he existence . . . Bed II times": *Olduvai Gorge,* Vol. 3, p. 272.
313 "the N.G.S. . . . Bed IV," "a pistol . . . head": Mary Leakey to Phillip V. Tobias, February 20, 1968 (Mary Leakey's files).
314 "You really. . . . turn up": J. Desmond Clark to Mary Leakey, February 7, 1969 (Mary Leakey's files).
314 "mining hominids at Olduvai": Ofer Bar-Yosef to Mary Leakey, February 9, 1969 (Mary Leakey's files).
314 "an explorer. . . . progenitors": Sonia Cole, *Leakey's Luck,* p. 303.
315 "I must be independent": L. S. B. Leakey to R. Ascher, September 11, 1969 (NMK).
315 "When you see. . . . fast grounder": L. S. B. Leakey, quoted in John Pfeiffer, "Man Through Time's Mists," *Saturday Evening Post,* December 3, 1966, pp. 50–52.

315 "One caller. . . . when I get back' ": "Working Scientist Leakey No 'Head in Clouds' Scholar," *Riverside Press-Enterprise,* May 23, 1963.
315 "everything is under control": Richard Leakey to L. S. B. Leakey, November 20, 1968 (NMK).
315 "Your final. . . . on this point": L. S. B. Leakey to Richard Leakey, November 25, 1968 (NMK).
316 "The teeth . . . otherwise": Mary Leakey to Desmond Clark, February 7, 1969 (Mary Leakey's files).
316 "We still talked . . . friction": Mary Leakey, interview, Nairobi, July 9, 1984.
316 "the meticulous excavations," "I do not . . . ruled out": Mary Leakey to Vance Haynes, March 30, 1967 (Mary Leakey's files).
316 "Mary said . . . stop him' ": Marie Wormington, interview, Boulder, Colorado, March 25, 1990.
316 "Mary was trying. . . . enough evidence: Mary Smith, interview, Washington, D.C., February 13, 1985.
317 "He was being. . . . foundation": Helen O'Brien, interview, Newport Beach, California, April 18, 1985.
317 "It is a . . . distorted," "not in any . . . racialism": Mary Leakey to Phillip V. Tobias, December 15, 1967.
317 "the effect . . . us at Olduvai": L. S. B. Leakey to Phillip V. Tobias, February 24, 1968 (NMK).
318 "her originality . . . fieldwork," "Probably . . . missed": Phillip V. Tobias, speech about Mary Leakey at the University of Witwatersrand, March 30, 1958.
318 "When I first . . . about that' ": Glynn Isaac, interview, Cambridge, Mass., April 13, 1984.
318 "treated. . . . respects": Clark Howell, interview, Berkeley, March 21, 1985.
318 "Clearly now . . . result": Glynn Isaac, interview, Cambridge, Mass., April 13, 1984.
318 "To put it . . . longer": Phillip V. Tobias, interview, Johannesburg, South Africa, August 3, 1984.
318 "There was never. . . . distressing": Mary Leakey, interview, Nairobi, July 9, 1984.
319 "destructive behavior": Mary Leakey, quoted by Mary Smith, interview, Washington, D.C., February 13, 1985.
319 "Had Louis needed. . . . And I did": Mary Leakey, interview, Nairobi, July 9, 1984.

Chapter 24. Dearest Dian

320 "You, the young. . . . Omega": L. S. B. Leakey, "The Gathering of Man: From Savannah to City and Beyond," lecture given at Teilhard de Chardin Conference, San Francisco, May 1, 1971.
320 "Louis was tremendously. . . . not at them": Richard Leakey, interview, San Francisco, November 17, 1987.
321 "I remember him. . . . awesome": Bob Drewes, interview, San Francisco, April 18, 1983.
321 I was *so* moved. . . . like a religion: Betty Howell, interview, Berkeley, February 4, 1986.
321 "We all tended. . . . out in everyone": Rosemary Ritter, interview, Calico Hills Early Man Site, California, April 20, 1985.
321 "He was the center. . . . towards the end": Sonia Cole, in *The Legacy of L. S. B. Leakey,* National Geographic Society film, January 9, 1978.
322 "His charisma . . . spellbinding": Helen O'Brien, interview, Newport Beach, California, April 18, 1985.
322 "Louis was really. . . . would help me," "There was a crowd. . . . orangutans": Biruté Galdikas, telephone interview, February 23, 1985.
322 "There were all kinds. . . . to do it": ibid.
323 "known . . . study," "interviewed . . . people": L. S. B. Leakey, "Research Programme on Orang Outans," unpublished, undated report (NMK).
323 "He said he. . . . happy about that," "I think, for. . . . essentially done": Biruté Galdikas, telephone interview, February 23, 1985.
323 "was lame . . . used to be": L. S. B. Leakey to William E. Scheele, January 2, 1969 (NMK).
324 "I had always. . . . my apartment": Frances Burton, letter to author, June 17, 1986.
324 "working. . . . under you": Frances Burton to L. S. B. Leakey, November 21, 1965 (NMK).
324 "clear thinking": L. S. B. Leakey to Frances Burton, November 30, 1965 (NMK).
324 "a state of dreams": Frances Burton to L. S. B. Leakey, November 21, 1965 (NMK).
324 "I was overwhelmed. . . . question in": Frances Burton, letter to author, June 17, 1986.
324 "I'd smashed. . . . readily": Juliet Ament, letter to author, May 18, 1985.
324 "encouraged . . . worthwhile": ibid.
324 "Louis. . . . everybody": Juliet Ament, letter to author, March 31, 1985.
325 "[They] had been. . . . before long": Juliet Ament, letter to author, May 18, 1985.
325 "I came. . . . Louis," "He gave . . . twenty years": Peter Andrews, interview, London, May 22, 1984.
325 "It wasn't . . . going on," "Louis would. . . . to the next": ibid.

325 "always dig at him a bit": Meave Leakey, interview, Nairobi, October 8, 1984.
326 "We'd ask Louis. . . . away from him": Hugo van Lawick, telephone interview, November 20, 1987.
326 "a lone woman": Dian Fossey to L. S. B. Leakey, July 27, 1967 (NMK).
326 "Frankly, I . . . today": L. S. B. Leakey to Dian Fossey, July 8, 1967 (NMK).
327 "earmarked": Dian Fossey, *Gorillas in the Mist,* p. 15.
327 "She. . . . so much' ": Biruté Galdikas quoted in Harold T. P. Hayes, *The Dark Romance of Dian Fossey,* pp. 163–64; " 'You . . . rape me!' ": Anita McClellan, quoted in same book, p. 163.
327 "She has had. . . . gorillas] again": L. S. B. Leakey to Melvin Payne, August 1, 1967 (NMK).
327 "kindred . . . ways": L. S. B. Leakey to Dian Fossey, December 18, 1970.
327 "bellowing," "magnificent work": L. S. B. Leakey to H. Fairfield Osborne, August 1, 1967 (NMK).
328 "What new. . . . than men?": Fairfield Osborne to L. S. B. Leakey, July 31, 1967 (NMK).
328 "You will. . . . beyond description.": Dian Fossey to L. S. B. Leakey, September 28, 1967 (NMK).
328 "delighted" to "[A]s . . . at all": L. S. B. Leakey to Dian Fossey, October 13, 1967 (NMK).
328 "MEMBERS OF . . . CONTACT": Dian Fossey to L. S. B. Leakey, January 13, 1969 (NMK).
329 "I shall be. . . . as I know": L. S. B. Leakey to Dian Fossey, August 20, 1969 (NMK).
329 "care for me . . . real understanding": L. S. B. Leakey to Judy (Strong) Castel, January 30, 1969 (courtesy of Judy Castel); "Sometimes he would. . . . teeth": Judy Castel, telephone interview, November 12, 1990; "emphasized . . . meant to": L. S. B. Leakey to Judy (Strong) Castel, n.d. (courtesy of Judy Castel); "The man. . . . affection"; Judy Castel, telephone interview, November 12, 1990.
329 "Dian my dearest. . . . yourself. Louis": L. S. B. Leakey to Dian Fossey, October 17, 1969 (Dian Fossey archives).
330 "My dearest love. . . . myself to you. Louis": ibid.
330 "Dearest love . . . you so. Louis": ibid.
330 "I have such. . . . hear from you": L. S. B. Leakey to Dian Fossey, October 23, 1969 (Dian Fossey archives).
330 "I want you. . . . caring": L. S. B. Leakey to Dian Fossey, December 18, 1969 (Dian Fossey archives).
330 ". . .I've had no. . . . soon, soon, soon": L. S. B. Leakey to Dian Fossey, November 11, 1969 (Dian Fossey archives).
330 "Dearest Dian. . . . Louis": L. S. B. Leakey to Dian Fossey, December 18, 1969 (Dian Fossey archives).
331 "Just as I. . . . and will be": ibid.
331 "Don't know. . . . what a mess": Dian Fossey's diary, quoted in Farley Mowat, *Woman in the Mists,* p. 82.
331 "At last I. . . . or 6th," "quiet talk": L. S. B. Leakey to Dian Fossey, January 30, 1970 (Dian Fossey archives).
331 "simply had . . . things done": L. S. B. Leakey to staff at Tigoni, n.d., 1970 (NMK).

Chapter 25. Father and Son

333 "agitation, anxiety and anger": Michael Day to Mary Leakey, February 18, 1970 (NMK).
333 "[A]t present, I. . . . dealt with": L. S. B. Leakey to Tynka Robertson, February 13, 1970 (NMK).
334 "*not* alone," "surrounded . . . your load": Joan Travis to L. S. B. Leakey, February 11, 1970 (NMK).
334 "the many demands. . . . ten years]": Leonard Carmichael to L. S. B. Leakey, February 16, 1970 (NMK).
334 "Forgive me. . . . absolutely clear": L. S. B. Leakey to Leonard Carmichael, February 27, 1970 (NMK).
334 "exceedingly . . . from Olduvai": ibid.
334 "this action . . . finances": Leonard Carmichael to L. S. B. Leakey, March 12, 1970 (NMK).
334 "You explained . . . illness": Richard Leakey to L. S. B. Leakey, March 17, 1970 (Richard Leakey's files).
335 "seem harsh," "not . . . this way": ibid.
335 "He looked. . . . actions": Richard Leakey, interview, Longview, Washington, February 8, 1991.
335 "desperate . . . defending myself": ibid.
335 "He was very. . . . loyal to him": Richard Leakey, interview, San Francisco, November 17, 1987.
335 "I knew . . . in charge": Richard E. Leakey, *One Life,* p. 103.
335 "I was . . . Director": ibid., p. 197.
336 "I remember. . . . deference": Jean Brown, letter to author, May 17, 1988.

336 "the Centre. . . . trustees": John Karmali, interview, Nairobi, September 19, 1987.
336 "Father always. . . . place": Richard Leakey, interview, San Francisco, November 11, 1986.
336 "Centre a more national flavour": Richard Leakey to L. S. B. Leakey, February 27, 1969 (Richard Leakey's files).
337 "a preposterously . . . comparisons": *One Life,* p. 130.
337 "I had wanted. . . . Kenya and me": Richard Leakey, interview, New York City, October 22, 1990.
337 "He was furious. . . . Kenya": *One Life,* p. 105.
337 "It was simply. . . . stability": Richard Leakey, interview, San Francisco, November 11, 1986.
337 "stomp on his father": M. E. Morbeck, interview, Nairobi, August 16, 1984.
337 "Most of us. . . . cross-fire," "Louis would yell. . . . window": ibid.
338 "The first time . . . Kookyland": Alan Gentry, interview, London, May 23, 1984.
338 "It was very. . . . each other," "In that period. . . . toward that": Thomas Odhiambo, interview, Nairobi, August 22, 1989.
339 "I think that. . . . recuperate in": Dian Fossey to L. S. B. Leakey, 1970 (n.d.) (NMK).
339 "[Louis] really. . . . easily": Michael Day to Mary Leakey, March 20, 1970 (NMK).
339 "Interfering in gorilla matters": Richard Leakey, quoting his father in a letter to L. S. B. Leakey, April 1, 1970 (Richard Leakey's files).
339 "[These] have . . . Doctor's orders": L. S. B. Leakey to Richard Leakey, March 25, 1970 (Richard Leakey's files).
339 "On the gorilla. . . . was done": Richard Leakey to L. S. B. Leakey, April 25, 1970 (NMK).
339 "I cannot. . . . have done so": L. S. B. Leakey to Rochelle Porter, March 31, 1970 (NMK).
339 "money is so. . . . responsibility' ": Juliet Ament to L. S. B. Leakey, March 26, and April 28, 1970 (NMK).
340 "Since writing. . . . something heavy": L. S. B. Leakey to Richard Leakey, April 3, 1970 (NMK).
340 "You must surely. . . . stressing this": ibid.
340 "degree of . . . possibility": Dr. L. J. Grant to Richard Leakey, April 2, 1970 (NMK).
340 "I have received . . . " to ". . . will surely come": Richard Leakey to Dr. L. J. Grant, April 6, 1970 (NMK).
341 "kept in the dark": L. S. B. Leakey to Rochelle Porter, March n.d., 1970 (NMK).
341 "I do not . . . job himself": Mary Leakey, letter to author, November 19, 1990.
341 "One of his. . . . *obey him*": L. S. B. Leakey to Leonard Carmichael, April 6, 1970 (NMK).
341 "Louis was. . . . eclipse him": Jean Brown, letter to author, May 17, 1988.

Chapter 26. Jackpot at Koobi Fora

342 "I was always. . . . Richard run?' ": Bill Richards, telephone interview, November 9, 1990.
343 "The job. . . . 1 October": Richard E. Leakey, *One Life,* p. 106.
343 "This time. . . . frightened": ibid.
343 "entirely . . . possible": ibid., p. 107.
343 "Boy, I was nervous": Richard Leakey, interview, New York City, October 22, 1990.
343 "He is . . . seen": Leonard Carmichael, interview, Washington, D.C., March 1969.
344 "I had hopes . . . palaeontologist": *One Life,* p. 113.
344 "Even. . . . 'Horseshit!' ": Paul Abell, interview, Kingston, Rhode Island, February 21, 1985.
345 "Of course Richard. . . . position": Margaret Avery, interview, Capetown, South Africa, August 6, 1984.
345 "We were often. . . . believe in me": Richard Leakey, interview, San Francisco, February 17, 1989.
345 "She was . . . easily," "casually . . . trunk]": Frances Burton, letter to author, June 17, 1986.
345 "I'd just finished. . . . in London," "Louis was. . . . bother me": Meave Leakey, interview, Nairobi, October 8, 1984.
346 "Everyone had told. . . . wrong way": ibid.
346 "Louis's accountant. . . . very much": Richard Leakey, interview, Nairobi, October 19, 1987.
346 "long lecture," "He wasn't. . . . pursue it": Meave Leakey, interview, Nairobi, October 8, 1984.
346 "I had very. . . . without a Ph.D": Richard Leakey, interview, en route to Kothiai River Camp, Kenya, September 23, 1987.
347 "Initially. . . . right track": Richard Leakey, interview, San Francisco, February 17, 1989.
347 "They say that. . . . stuff": Margaret Avery, interview, Capetown, South Africa, August 6, 1984.
347 "Margaret. . . . done' ": Richard Leakey, interview, San Francisco, February 17, 1989.
348 "There were. . . . a lot": Richard Leakey, interview, San Francisco, March 19, 1985.
348 "ensure . . . conducted": Richard Leakey to Bryan Patterson, May 17, 1968 (NMK).
348 "There have. . . . Geologists": Richard Leakey to Bryan Patterson, April 29, 1968 (NMK).

348 "From what.... there' ": Richard Leakey, interview, San Francisco, March 19, 1985.
349 "Patterson felt.... fossils": Anna K. Behrensmeyer, interview, Washington, D.C., February 16, 1985.
349 "When Richard.... for myself: Anna K. Behrensmeyer, interview, Washington, D.C., February 14, 1984.
349 "Koobi Fora.... Olduvai": Richard Hay, telephone interview, January 18, 1991.
349 "He said.... commit myself": ibid.
350 "I felt.... teenagers": Richard Leakey, interview, en route to Kothiai River Camp, September 23, 1987.
350 "Many of these.... Zero": ibid.
350 "It's a funny... men": ibid.
350 "the opportunity of a lifetime": Vince Maglio, quoted by Anna K. Behrensmeyer, interview, Washington, D.C., February 14, 1985.
350 "We drove all.... outdoors," "professional... student": ibid.
350 "I was used.... any trouble," "fantastic... work": ibid.
351 "in strict confidence" to "I wanted.... stories": Richard Leakey to Kebbede Mikael, February 12, 1969 (Richard Leakey's files).
351 "It is... assessment": Kebbede Mikael to Richard Leakey, April 16, 1969 (Richard Leakey's files).
351 "competent Palaeontologist," "I see... in 1968": Richard Leakey to Kebbede Mikael, April 25, 1969 (Richard Leakey's files).
352 "RECOMMEND... LETTER": Richard Leakey to Kebbede Mikael, May 29, 1969 (Richard Leakey's files).
352 "And Arambourg.... *last gentleman*": Clark Howell, quoted by Paul Abell, interview, Kingston, Rhode Island, February 21, 1985.
352 "To this day... joy": *One Life,* p. 124.
353 "You were right.... on the site!": Richard E. Leakey, "In Search of Man's Past," *National Geographic,* May 1970, p. 725.
353 "I confess.... had no interest": *One Life,* p. 116.
353 "It is quite.... at Kay's site!": ibid., p. 115.
353 "as a matter of some urgency": Richard Leakey to Jack Miller, June 16, 1969 (Richard Leakey's files).
354 "The easiest... before hand": *One Life,* p. 117.
354 "explode... rifles," "ruefully... ammunition": "In Search of Man's Past," op. cit., pp. 727–28.
355 "a camel-riding... desert": ibid.
355 "We traveled.... antelope meat": ibid., p. 725.
355 "Meave!... australopithecine!": Richard Leakey, *Bones of Contention,* Survival Anglia, Ltd., film, 1978.
356 "peering... breathless": *One Life,* p. 122.
356 "That looks like... on earth!": Paul Abell, quoted in Koobi Fora Field Journal, August 7, 1969 (NMK).
356 "It's Richard": Mary Leakey, ibid.
356 "It's beautiful.... magnificent!": "In Search of Man's Past," op. cit., p. 726.
356 "You could just.... as good": Paul Abell, interview, Kingston, Rhode Island, February 21, 1985.
357 "I felt.... genus *Homo*": "In Search of Man's Past," op. cit., p. 731.
357 "But... Fora": Clark Howell, telephone interview, November 28, 1994.
357 "I have just... jackpot' ": Richard Leakey to Allen O'Brien, September 5, 1969 (Richard Leakey's files).

Chapter 27. Misadventure at Calico

358 "practically... found!," "Compared... Omo": L. S. B. Leakey to Melvin Payne, August 11, 1969 (NMK).
359 "I have allowed... in the bud": Sir Wilfrid Le Gros Clark to Kenneth Oakley, June 7, 1964 (BMNH).
359 "I think... ourselves": Jeanne Reinert, "The Man Dr. Leakey Dug Up," *Science Digest,* November 1966, p. 76.
360 "ordinary mammals": L. S. B. Leakey, quoted by Charles A. Betts, "Man's Tree—a New Root?," *Science News Letter,* May 29, 1965, p. 346.
360 "we will... australopithecine stage": L. S. B. Leakey to Le Gros Clark, January 6, 1964 (NMK).

360 "The time . . . development," "This is only the beginning": L. S. B. Leakey, quoted by Charles A. Betts, "Man Evolved Like Animals," *Science News Letter,* April 17, 1965, p. 243.
360 "with . . . deck": "Man's Tree—a New Root?," op. cit., p. 346.
360 "Man developed . . . emerged": L. S. B. Leakey, quoted by Charles A. Betts, "Man Evolved Like Animals," op. cit., p. 243.
360 "Why should. . . . special creation": L. S. B. Leakey, quoted by Jeanne Reinert, "The Man Dr. Leakey Dug Up," op. cit., pp. 74–75.
360 "review . . . known": L. S. B. Leakey, press release, April 1965 (NMK).
360 "man, not a 'near man' ": L. S. B. Leakey, lecture at L. S. B. Leakey Foundation, Pasadena, California, October 25, 1969 (NMK).
361 "I thought. . . . arguments": Richard Leakey, interview, Longview, Washington, February 8, 1991.
361 "Why did you ask them?," "My father . . . outsiders in": ibid.
361 "working . . . participation": Richard E. Leakey, *One Life,* p. 129.
362 "They always made. . . . other again": Peter Andrews, interview, London, March 22, 1984.
362 "At the moment. . . . plans": L. S. B. Leakey to Joan Travis, August 5, 1970 (NMK).
362 "I am already. . . . progress": ibid.
363 "I am counting . . . U.S.A. prehistory": L. S. B. Leakey to All Members of the Calico Dig, August 20, 1966 (NMK).
363 "You could hardly. . . . around things": Dee Simpson, interview, Calico Hills Early Man Site, April 20, 1985.
363 "Louis knew. . . . of the team": ibid.
363 "Together. . . . helped' ": L. S. B. Leakey to All Members of the Calico Dig, August 20, 1966 (NMK).
363 "Dr. Leakey. . . . work for him," "There were three. . . . opinion": Rosemary Ritter, interview, Calico Hills Early Man Site, April 20, 1985.
363 "feature": Dee Simpson, interview, Calico Hills Early Man Site, April 20, 1985.
363 "holding our breath": Rosemary Ritter, interview, Calico Hills Early Man Site, April 20, 1985.
363 "Louis used to. . . . round behind": Dee Simpson, interview, Calico Hills Early Man Site, April 20, 1985.
363 "a very . . . structure": L. S. B. Leakey to Richard Leakey, February 5, 1969 (NMK).
364 "major . . . prehistory," "[A]lthough . . . place": L. S. B. Leakey to Members of the Calico Crew, July 23, 1968 (NMK).
364 "I was never. . . . possibility": Clark Howell, telephone interview, March 15, 1991.
364 "Leakey is right. . . . America": François Bordes, quoted in letter from L. S. B. Leakey to the L. S. B. Leakey Foundation, October 25, 1969 (NMK).
364 "If Louis had. . . . cavalier": Emil Haury, telephone interview, March 19, 1991 (NMK).
364 "A lot of. . . . we will": Dee Simpson, interview, Calico Hills Early Man Site, April 20, 1985.
365 "like a mother hen," "He wanted everything. . . . were asked": Joan Travis, telephone interview, March 21, 1991.
365 "Louis thought. . . . decide": ibid.
365 "He had been. . . . America": L. S. B. Leakey addressing the International Calico Conference, October 22, 1970; quoted in Walter C. Schuiling, ed., *Pleistocene Man at Calico,* p. 99.
366 "It was like. . . . their Messiah": Vance Haynes, telephone interview, July 27, 1989.
366 "Why he's . . . people!": Mary Leakey, quoted by Glynn Isaac, interview, Cambridge, Mass., April 13, 1984.
366 "pretty damned old": Richard Hay, telephone interview, March 20, 1991.
366 "Kenneth asked. . . . answer that": ibid.
366 "It's not that. . . . make one," "One of the. . . . the cores,": J. Desmond Clark, telephone interview, March 21, 1991.
366 "They were . . . manufacture": ibid.
366 "He wasn't. . . . processes," "all . . . producing": Vance Haynes, telephone interview, July 27, 1989.
367 "Clements'. . . . Louis's face": Walter Schuiling, telephone interview, March 20, 1991.
367 "In the final . . . fire": Rainer Berger, "An Isotopic and Magnetic Study of the Calico Sites," in *Pleistocene Man at Calico,* p. 34.
367 "His grin . . . hall": Dee Simpson, telephone interview, March 20, 1991.
367 "We all kept. . . . remarks": J. Desmond Clark, telephone interview, March 21, 1991.
367 "People didn't. . . . sight to see": Karl Butzer, telephone interview, December 6, 1986.

368 "The great. . . . friends": Clark Howell, telephone interview, March 15, 1991.
368 "It's . . . evidence": Mary Leakey, *Disclosing the Past,* p. 144.
368 "That's the way. . . . again": Richard Hay, telephone interview, March 20, 1991.
368 "Mary knew. . . . bury it": ibid.
368 "I . . . be here," "silent . . . right!' ": Vanne Goodall, "The Legacy of L. S. B. Leakey," *TV Guide,* January 7, 1978, p. 22.
368 "We knew. . . . Yeti": Clark Howell, telephone interview, March 15, 1991.
368 "Louis was . . . deferentially": Richard Michael Gramly, telephone interview, September 2, 1986.
369 "Everything. . . . great mess": L. S. B. Leakey to Dian Fossey, January 12, 1971 (NMK).
369 "By the end . . . world": L. S. B. Leakey, taped letter to Vanne Goodall, January (n.d.) 1971 (courtesy of Richard Leakey).
369 "only an idiot": Andrew Hill, interview, Nairobi, October 19, 1984.
369 "But . . . as well!": ibid.
369 "generally lazy," "Then. . . . mobbed me": L. S. B. Leakey, taped letter to Vanne Goodall, op. cit.
370 "Drive quickly. . . . collapsing": ibid.
370 "My brain . . . die" to "I was much. . . . for the best": ibid.
370 "after about . . . at all" to "I can't . . . less": ibid.
371 "because . . . badly," "Quite. . . . tough": ibid.

Chapter 28. An Unstoppable Man

372 "Louis used to. . . . take its toll": Joan Travis, quoted in *The Legacy of L. S. B. Leakey,* National Geographic Society film, January 9, 1978.
373 "Everyone knew. . . . into the unknown": Alan Walker, interview, New York City, April 7, 1984.
373 "There was always. . . . with it": Karl Butzer, telephone interview, December 6, 1986.
373 "If you became. . . . off-scale": Elizabeth Meyerhoff, interview, Lake Baringo, Kenya, November 4, 1984.
374 "It put me. . . . enjoy the scene," "I came over . . . important' ": ibid.
374 "We said things. . . . primate behaviorist": Alan Walker, interview, New York City, April 7, 1984.
375 "Louis was. . . . over a rose" to "intelligent women": Elizabeth Meyerhoff, interview, Lake Baringo, Kenya, November 4, 1984.
375 "Louis had this. . . . in his hands": Derek Roe, interview, Oxford, England, November 27, 1984.
375 "understanding," "I don't want. . . . worthwhile": L. S. B. Leakey to Judy (Strong) Castel, February 9, 1969 (courtesy of Judy Castel).
376 "I remember. . . . on my face": Gabrielle Dolphin, journal from Kenya, January 20, 1971 (courtesy of Gabrielle Dolphin).
376 "secret dig" to "lessons in observation": ibid., February 24, 1971.
376 "It's becoming a second home": ibid., March 26, 1971.
376 "Louis has . . . animal": ibid., March 12, 1971.
377 "Gradually Louis fell. . . . of disgust": Melvin M. Payne, "Family in Search of Prehistoric Man," *National Geographic,* February 1965, p. 210.
377 "Louis's childlike. . . . possibilities": Gabrielle Dolphin, telephone interview, September 19, 1990.
377 "Louis was crippled. . . . set-up" to "He'd want. . . . was after": Gabrielle Dolphin, telephone interview, March 20, 1991.
377 "I wanted to believe. . . . guard": Gabrielle Dolphin, telephone interview, September 19, 1990.
377 "that he was . . . again": Gabrielle Dolphin, telephone interview, March 20, 1991.
377 "Louis was. . . . respectful of me": Toni Kay Jackman, telephone interview, January 4, 1991.
378 "We went to. . . . not mine": Cara Phelips to her parents, April 27, 1971; quoted in Sonia Cole, *Leakey's Luck,* p. 376.
378 "the audience rose. . . . applause": Phelips, ibid., May 1, 1971; quoted in *Leakey's Luck,* p. 377.
378 "When Louis fell . . . surprised": Joan Travis, quoted in *The Legacy of L. S. B. Leakey,* National Geographic Society film, January 9, 1978.
379 "I was foolish. . . . shoulder": L. S. B. Leakey to Rosemary Knocker, May 14, 1971 (NMK).
379 "Louis has. . . . letters": Cara Phelips to her parents, n.d.; quoted in *Leakey's Luck,* p. 378.
379 "Some of the trustees. . . . kindly": Joan Travis, telephone interview, April 24, 1991.
379 "Louis was terribly. . . . didn't make it": Dee Simpson, interview, Calico Hills Early Man Site, April 20, 1985.
380 "Louis couldn't walk. . . . next-of-kin?," "Louis was so. . . . recovery room": Joan Travis, telephone interview, April 24, 1991.

380 "Louis's ailments. . . . at his heels" to "anger and frustration": ibid.
380 "Louis was gaga. . . . too much": Jane Goodall, interview, Gombe Stream Research Center, Tanzania, October 11, 1987.
381 "Louis had gone. . . . Mary was aghast": Ron Clarke, interview, Johannesburg, South Africa, August 1, 1984.
381 "That behavior was. . . . discussing him": Richard Leakey, interview, Longview, Washington, February 8, 1991.
381 "With gratitude. . . . had to win": Joan Travis, telephone interview, April 24, 1991.
381 "Everybody . . . I do," "The man . . . unstoppable": Dee Simpson, interview, Calico Hills Early Man Site, April 20, 1985.
382 "We were. . . . Olduvai," "African . . . for you": Biruté Galdikas, telephone interview, February 23, 1985.
382 "woman-to-woman" to "A daughter": ibid.
383 "Oh, it's. . . . medals": Mary Leakey, quoted by Andrew Hill, interview, Nairobi, October 19, 1984.
383 "Louis said. . . . well": ibid.
383 "Our separation. . . . females coming in": Mary Leakey, interview, Nairobi, August 20, 1989.
383 "deplorably low standard": Mary Leakey, *Disclosing the Past,* p. 158.
383 "She looked me . . . own taste' ": Penelope Caldwell quoted in Sy Montgomery, *Walking with the Great Apes,* pp. 81–82.
383 "anonymous admirer" to "Why does Louis. . . . all of them": Biruté Galdikas, telephone interview, February 23, 1985.
384 "Louis. . . . women": *Disclosing the Past,* pp. 158–59.

Chapter 29. Roar of the Old Lion

385 "whoop like a hyena": Koobi Fora Field Journal, June 5, 1971 (NMK).
385 "a whole galaxy of hominid[s]. . . . the skull]": Mary Leakey to Phillip V. Tobias, August 11, 1970 (Mary Leakey's files).
386 "Your discoveries . . . paleontology!": Clark Howell to Richard Leakey, August 18, 1970 (NMK).
386 "remarkable . . . fossils": Sarah Bunney, "Fossilmanship in East Africa," *Nature,* September 3, 1971, p. 20.
386 "Louis sometimes. . . . encouraged them!," "Look, here. . . . own company": Alvin Gittins, transcribed interview, December 1975 (Leakey Foundation archives; courtesy of Joan Travis).
387 "However, in those. . . . unsuccessfully": L. S. B. Leakey, "Notes to the National Geographic Society on the Results of Richard's North-East Rudolf Expedition," 1971 (NMK).
387 "a sick . . . his career," "found. . . . adequately": Richard Leakey, quoted in "Puzzling Out Man's Ascent," *Time,* November 7, 1977, p. 77.
387 "looked like. . . . child": Gayle Gittins, transcribed interview, December 1975 (Leakey Foundation archives; courtesy of Joan Travis).
387 "Fossils. . . . specimens": Richard Leakey, interview, en route to Kothiai River Camp, September 23, 1987.
387 "It was. . . . lot smarter: Richard Leakey, interview, San Francisco, March 19, 1985.
388 "So he. . . . poorly trained": Alan Walker, interview, New York City, April 7, 1984.
388 "I used to. . . . going on": Andrew Hill, interview, Nairobi, October 19, 1984.
388 "I had shown. . . . just carried on": Alan Walker, interview, New York City, April 7, 1984.
388 "hardly walk . . . seeing": Richard E. Leakey, *One Life,* p. 140.
388 "the perennial . . . evolution": Richard Leakey to Gilbert Grosvenor, September 15, 1970 (NMK).
389 "The hominid room. . . . incredible": M. E. Morbeck, interview, Nairobi, August 16, 1984.
389 "It was . . . like this": Richard Leakey, interview, San Francisco, November 11, 1986.
389 "often complained . . . appalling conditions": Richard Leakey, interview, Longview, Washington, February 8, 1991.
389 "I told Louis . . . research rooms": Richard Leakey, interview, San Francisco, November 11, 1986.
389 "I planned . . . could do": Richard Leakey, interview, Longview, Washington, February 8, 1991.
389 "Louis told me. . . . run Tigoni": Richard Leakey, interview, San Francisco, November 11, 1986.
389 "in honour . . . this field": Richard Leakey to Gary W. Phillips, April 23, 1971 (NMK).
389 "Richard suddenly . . . after me," "to take. . . . just yet": L. S. B. Leakey to Joan Travis, April 23, 1971 (NMK).
390 "I must be . . . my son": L. S. B. Leakey to Gary W. Phillips, April 23, 1971 (NMK).
390 "I'm sure. . . . Louis furious": Andrew Hill, interview, Nairobi, October 19, 1984.

390 "I used to. . . . headaches," "very tense. . . . expects": Richard Leakey, interview, Longview, Washington, February 8, 1991.
390 "He wanted. . . . very uptight": Meave Leakey, interview, Nairobi, October 8, 1984.
391 "The idea was. . . . whole of Africa," "Tigoni was Louis's. . . . when you die?'": Thomas Odhiambo, interview, Nairobi, August 22, 1989.
391 "I didn't. . . . wouldn't see it": Richard Leakey, interview, Nairobi, September 17, 1989.
391 "If any major. . . . TIGONI": L. S. B. Leakey to Centre for Prehistory and Palaeontology staff, September (n.d.) 1970 (NMK).
391 "The last time. . . . tell everybody": Mary Smith, interview, Washington, D.C., February 13, 1985.
392 "I am very . . . two years": L. S. B. Leakey to Dian Fossey, September 19, 1971 (NMK).
392 "going . . . again": L. S. B. Leakey to Phillip V. Tobias, September 1, 1971 (NMK).
392 "in the not too distant future": L. S. B. Leakey to Dian Fossey, October 28, 1971 (NMK).
392 "With the enthusiasm. . . . ago" to "bring back . . . ancestors": "Once More into the Past," *Science News,* October 16, 1971, p. 259.
393 "Well, we were. . . . Do you?'": Robert Ardrey, transcribed interview, n.d. (Leakey Foundation archives).
393 "went around . . . pocket": Joan Travis, telephone interview, March 21, 1991.
393 "It was agonizing" to "over . . . days": L. S. B. Leakey to Mary Leakey, March 23, 1972 (courtesy of Joan Travis).
393 "to slow down. . . . bossy women'": Jean Brown, letter to author, May 17, 1988.
393 "My only problem . . . camp": L. S. B. Leakey to Dian Fossey, May 10, 1972 (NMK).
393 "My wife and. . . . Society": Basil Cooke, interview, New York City, April 9, 1984.
394 "He was so elated. . . . last hurrah": Joan Travis, telephone interview, June 24, 1991.
394 "is liable. . . . in England": L. S. B. Leakey to Mr. Jameson-Smith, January 8, 1971 (NMK).
394 "practical joke . . . wrong": L. S. B. Leakey to Glyn Daniel, June 6, 1972 (NMK).
394 "Perhaps you know. . . . *responsible'":* L. S. B. Leakey to Glyn Daniel, May 24, 1972 (NMK).
394 "polishing": Sonia Cole, *Leakey's Luck,* p. 399.
394 "never got further . . . rough draft," "as pure fantasy . . . cashing in on this": Mary Leakey to author, July 29, 1985.
394 "May I remind. . . . plus lectures": Joan Travis to L. S. B. Leakey, March 31, 1972 (NMK).
394 "Forgive . . . time": L. S. B. Leakey to Mr. and Mrs. Irving Levy, March 29, 1972 (courtesy of Joan Travis).
395 "Louis was an Alpha. . . . unpleasant": Peter Andrews, interview, London, May 22, 1984.
395 "I remember saying. . . . come true": Alan Gentry, interview, London, May 23, 1984.
395 "the charisma . . . leader": Anna K. Behrensmeyer, interview, Washington, D.C., February 16, 1985.
395 "He ran. . . . stock in": Anna K. Behrensmeyer, interview, Washington, D.C., February 14, 1985.
395 "very special keepsake" to "the balance . . . mind": Dian Fossey, letter to author, October 15, 1984.
396 "We'd gone. . . . team": Glynn Isaac, interview, Cambridge, Mass., April 13, 1984.
396 "It was like. . . . Koobi Fora": Betty Goerke, interview, Mill Valley, California, April 27, 1987.
396 "a stupid . . . good friends": *One Life,* pp. 140–41.
396 "I remember. . . . fossil beds": Anna K. Behrensmeyer, interview, Washington, D.C., February 16, 1985.
396 "He was like . . . his own": M. E. Morbeck, interview, Nairobi, August 16, 1984.
396 "Richard had. . . . 'expert' did": Paul Abell, interview, Kingston, Rhode Island, February 21, 1985.
397 "He would . . . for him": Anna K. Behrensmeyer, interview, Washington, D.C., February 16, 1985.
397 "We were finding. . . . place to be": Glynn Isaac, interview, Cambridge, Mass., April 13, 1984.
397 "made it . . . undisturbed": *One Life,* p. 172.
397 "I was amazed. . . . position!": Meave Leakey's diary, quoted in *One Life,* p. 139.
398 "some hominids": Meave Leakey, in the Koobi Fora Field Journal, June 17, 1972.
398 "I was not. . . . human beings": Bernard Ngeneo, interview, Kothiai River Camp, Kenya, September 1, 1987 (Kamoya Kimeu, translator).
398 "So we called . . . check": Kamoya Kimeu, interview, Kothiai River Camp, Kenya, September 1, 1987.
398 "What . . . hominid!": Richard Leakey, in *Bones of Contention,* Survival Anglia, Ltd., film, 1978.
398 "None . . . cranium": *One Life,* p. 148.
398 "1470. . . . had it all": Meave Leakey, interview, Nairobi, September 11, 1987.
399 "In no time. . . . 1970": *One Life,* p. 149.

399 "We have . . . edge pieces": Meave Leakey, quoted in Richard E. Leakey, "Skull 1470," *National Geographic,* June 1973, p. 820.
399 "It was larger . . . the brain?": *One Life,* p. 149.
399 "Historic moment . . . memorable": Koobi Fora Field Journal, September 9, 1972.
399 "The skull . . . like that": Meave Leakey, interview, Nairobi, September 11, 1987.
399 "I knew. . . . out of life": Richard Leakey, interview, San Francisco, November 17, 1987.
400 "The feeling. . . . 100 per cent": Pat Barrett to Joan Travis, October 30, 1972 (courtesy of Joan Travis).
400 "[It] was . . . discussion": *One Life,* p. 149.
400 "Louis was excited. . . . old times": Mary Leakey, *Disclosing the Past,* p. 159.
400 "In many ways. . . . time to talk": *One Life,* p. 150.
400 "I hadn't . . . would see him": Richard Leakey, interview, San Francisco, November 17, 1987.
400 "We'd finished. . . . your father?": ibid.

Chapter 30. An End and a Beginning

401 "the man . . . mind," "thrown . . . ancestry": *South African Rand Daily Mail,* October 3, 1972.
401 "Another heart. . . . Dead": Mary Leakey, *Disclosing the Past,* p. 159.
402 "I spoke to. . . . of us all": Priscilla Davies to Mary Leakey, October 19, 1972 (Mary Leakey's files).
402 "I was very upset . . . thought" to "Louis was. . . . been said": Richard Leakey, interview, en route to Kothiai River Camp, Kenya, September 23, 1987.
402 "ghastly waiting," "while crate . . . reached": *Disclosing the Past,* p. 160.
403 "somehow . . . other people": ibid.
403 "nobody . . . expeditiously": Richard Leakey, interview, San Francisco, February 17, 1989.
403 "to pay . . . Kenya," "for himself . . . human species": Charles Njonjo, eulogy for L. S. B. Leakey, All Saints Cathedral, Nairobi, October 6, 1972.
403 "left more . . . ourselves out": *Disclosing the Past,* p. 161.
404 "I still find . . . gone": Mary Leakey to Joan Travis, October 30, 1972 (NMK).
404 "Mary didn't. . . . was done": Richard Leakey, interview, San Francisco, February 17, 1989.
404 "I told Jonathan. . . . Ever Always": Richard Leakey, interview, en route to Kothiai River Camp, Kenya, September 23, 1987.
404 "the great love": Elizabeth Meyerhoff, interview, Lake Baringo, Kenya, November 4, 1984.
405 "nothing had been done": Rosalie Osborn, interview, Nairobi, August 28, 1989.
405 "I told Jonathan. . . . the nerve!": Richard Leakey, interview, en route to Kothiai River Camp, Kenya, September 23, 1987.
405 "it sits. . . . my grave": ibid.
405 "I love your. . . . be buried there": ibid.
405 "Actually, Father's. . . . dealt with": Richard Leakey, interview, San Francisco, November 11, 1986.
405 "He could. . . . integrity": Richard Leakey, interview, en route to Kothiai River Camp, Kenya, September 23, 1987.
405 "few hundred pounds": Richard Leakey, interview, New York City, October 22, 1990.
405 "Richard didn't seem. . . . on our own": Toni Kay Jackman, telephone interview, January 4, 1991.
406 "some California woman," "We don't know. . . . came of it": Richard Leakey, interview, San Francisco, November 11, 1986.
406 "The National Geographic . . . by Richard": Dian Fossey, letter to author, October 15, 1984.
406 "Mary. . . . outcast": Elizabeth Meyerhoff, interview, Lake Baringo, November 4, 1984.
406 "There was never. . . . for money": Richard Leakey, interview, Denver, August 30, 1991.
406 "Mary thought. . . . crowd": Helen O'Brien, interview, Newport Beach, California, April 18, 1985.
407 "It was a. . . . Rubbish!": Richard Leakey, interview, Denver, August 30, 1991.
407 "Life is being. . . . give me peace": L. S. B. Leakey to Dian Fossey, January 1972 (Dian Fossey archives).
407 "We go out. . . . his sake": Dian Fossey's journal, October 6, 1972.
407 "For me. . . . loss to me": Richard Leakey, interview, Denver, August 30, 1991.
407 "[I]t seemed . . . prehistorian": Richard E. Leakey, *One Life,* p. 155.
407 "Many of the men. . . . fountainhead has gone": Pat Barrett to Susan Dietrich, November 7, 1972 (NMK).
408 "What a curious. . . . unwise," "I know . . . critics": Richard Leakey to Bernard Wood, October 19, 1972 (NMK).
408 "One wants to. . . . fossil" to "prolonged . . . nomenclature": Bernard Wood to Richard Leakey, October 15, 1972 (NMK).

408 "wanted the limelight": Richard Leakey, interview, Nairobi, February 10, 1983.
409 "In spite of.... have done": L. S. B. Leakey to Joan Travis, December 4, 1971 (NMK).
409 "amateur scientist": Professor Lord Zuckerman, ed., *The Concepts of Human Evolution,* p. 64.
409 "I am.... what it is": Richard Leakey, quoted by Phillip V. Tobias, interview, Johannesburg, August 3, 1984.
410 "have astounded... of modern man": "Skull Pushes Back Man's Origins One Million Years," *New York Times,* November 10, 1972, p. 1.
410 "It would seem... ancestry": "The Oldest Man?," *Newsweek,* November 20, 1972, p. 137.
410 "Mr. Chairman.... specialists can work" to "I am quite... I have": *The Concepts of Human Evolution,* p. 64.
410 "went on... blushed": Phillip V. Tobias, interview, Johannesburg, August 3, 1984.
410 "I was as... furious": *One Life,* p. 152.
410 "against [his] will," "I can only... over the world": ibid., p. 153.
411 "might lead.... what it is": "Briton Finds 'Missing Link' Skull," *Evening Standard,* November 9, 1972.

Chapter 31. The Best Bones

412 "fantastically overcrowded": Glyn Daniel to Mary Leakey, January 30, 1973 (Mary Leakey's files).
412 "in recognition... science": Leonard Carmichael to Richard Leakey, January 12, 1973 (NMK).
412 "an innate flair... audience": Gilbert Grosvenor, interview, Washington, D.C., February 12, 1985.
413 "Father and Richard... needed": Philip Leakey, interview, Nairobi, October 20, 1987.
413 "Richard is... he wears": Lita Osmundsen, telephone interview, March 16, 1984.
413 "The Richard.... give money": Paul Abell, telephone interview, November 27, 1991.
413 "Everyone... expectations": ibid.
414 "really worked his crew," "I scratched... hominid": Paul Abell, telephone interview, January 11, 1992.
414 "in contrast... Rudolf," "KNM ER-1805... enigmatic": R. E. F. Leakey, "Further Evidence of Lower Pleistocene Hominids from East Rudolf, North Kenya, 1973," *Nature,* April 19, 1974, p. 655.
415 "The interesting thing.... raw liver": Richard Leakey, interview, Koobi Fora, Kenya, July 5, 1984.
415 "You couldn't go.... find one" to "And there... find them": Anna K. Behrensmeyer, interview, Washington, D.C., February 16, 1985.
415 "It had always.... comparative animals": Richard Leakey, interview, San Francisco, March 19, 1985.
416 "He was always... fossils": Paul Abell, telephone interview, January 11, 1992.
416 "Christ, you.... descents": Ralph Holloway, interview, New York City, February 8, 1985.
416 "It takes.... had found": Richard Leakey, interview, Kothiai River Camp, Kenya, September 4, 1987.
417 "The size.... comparable": "Further Evidence of Lower Pleistocene Hominids from East Rudolf, North Kenya, 1973," op. cit., p. 655.
417 "a remnant... Pleistocene," "may... boundary": "Further Evidence of Lower Pleistocene Hominids from East Rudolf, North Kenya, 1973," op. cit., p. 656.
417 "If one accepts.... population": "Should Fossil Hominids Be Reclassified?," *Nature,* April 19, 1974, p. 635.
418 "It's all part... side branches": Elwyn Simons, telephone interview, April 22, 1988.
418 "Obviously.... *Homo,* begin?" Richard Leakey, interview, en route to Kothiai River Camp, Kenya, September 23, 1987.
418 "caused... experts": "The Leakeys' Telltale Skulls," *Newsweek,* July 15, 1974, p. 72.
418 "I had the best.... money": Richard Leakey, interview, en route to Kothiai River Camp, Kenya, September 23, 1987.
418 "Paleontology's... Man": Andrew Jaffe, *International Wildlife,* May 1975, pp. 4–11.
418 "There was.... treasure," "The older guys.... many of them": M. E. Morbeck, interview, Nairobi, February 10, 1983.
419 "I was working.... dating tool": Basil Cooke, interview, New York City, April 9, 1984.
419 "golden oldie": Frank Fitch to Glynn Isaac, May 15, 1973 (NMK).
420 "I felt that.... trust him": Richard Leakey, interview, en route to Kothiai River Camp, Kenya, September 23, 1987.

420 "People ask . . . against them' ": Basil Cooke, interview, New York City, April 9, 1984.
420 "We would . . . from here?": Richard Leakey, interview, en route to Kothiai River Camp, Kenya, September 23, 1987.
420 "They are not congruent": Clark Howell, at "Stratigraphy, Paleoecology and Evolution in Lake Rudolf Basin," a symposium, Nairobi, September 1973.
420 "We already. . . . public": Anna K. Behrensmeyer, telephone interview, December 22, 1991.
420 "is considerable . . . ignored": Basil Cooke, "Suidae from Plio-Pleistocene Strata of the Rudolf Basin," in *Earliest Man and Environments in the Lake Rudolf Basin,* ed. Yves Coppens, et al., p. 261.
421 "There'd always . . . match," "We had. . . . get it right": Glynn Isaac, interview, Cambridge, Mass., April 13, 1984.
421 "20 more . . . skulls," Richard Leakey to Phillip V. Tobias, November 9, 1973 (Richard Leakey's files).
421 "I didn't feel. . . . winning": Richard Leakey, interview, en route to Kothiai River Camp, Kenya, September 23, 1987.
421 "Interpretation. . . . of their kind": Karl Butzer, "Dawn in the Rudolf Basin," *South African Journal of Science,* October 19, 1973, pp. 292–93.
422 "Don really. . . . never got": M. E. Morbeck, interview, Nairobi, August 16, 1984.
422 "I had. . . . insignificant" to "Done": Donald C. Johanson and Maitland Edey, *Lucy: The Beginnings of Humankind,* p. 135.

Chapter 32. The Gladiators' Clash

423 "unfairly knocked about": Richard Leakey, interview, en route to Kothiai River Camp, Kenya, September 23, 1987.
423 "Richard took me. . . . territorial" to "As a geologist. . . . noise": Richard Hay, telephone interview, December 23, 1991.
424 "We had. . . . data unsatisfactory": Glynn Isaac to Richard Leakey, December 20, 1973 (Richard Leakey's files).
424 "These two . . . the opposite": Brent Dalrymple to Glynn Isaac and F. C. Howell, December 2, 1974 (Richard Leakey's files).
424 "I said . . . long time": Richard Leakey, interview, en route to Kothiai River Camp, Kenya, September 23, 1987.
424 "Every Leakey . . . uninvited": Mary Leakey, quoted by Derek Roe, interview, Oxford, England, November 27, 1984.
425 "incompetent geologists": Mary Leakey, quoted by Garniss Curtis, interview, Berkeley, March 28, 1985.
425 "There's no. . . . doing": Richard Leakey, interview, San Francisco, November 11, 1986.
425 "It's embarrassing . . . night' " to "In public. . . . very angry": Richard Leakey, interview, en route to Kothiai River Camp, Kenya, September 23, 1987.
425 "At the Omo. . . . believed them": John Harris, telephone interview, December 19, 1991.
425 "adamant" "the only appropriate method": Frank Fitch and Jack Miller, quoted in "Minutes of the East Rudolf Research Project," February 3, 1974, Kingston, Rhode Island (NMK).
425 "If Fitch. . . . were right": Richard Leakey, interview, en route to Kothiai River Camp, Kenya, September 23, 1987.
426 "a typical Glynn compromise": ibid.
426 "It has given. . . . be okay": Paul Abell, telephone interview, March 3, 1992.
427 "And there. . . . weight": ibid.
427 "but he didn't . . . was" to "He wrote. . . . ordeal": ibid.
428 "Richard was . . . have done": ibid.
428 "That whole. . . . search," "gung-ho . . . back to work": Anna K. Behrensmeyer, interview, Washington, D.C., February 14, 1985.
428 "For a while . . . to eat": Anna K. Behrensmeyer, telephone interview, November 28, 1991.
428 "There is no way to tell": Glynn L. Isaac and John W. K. Harris, "Archaeology," *Koobi Fora Research Project,* Vol. 1, ed. Meave G. Leakey and Richard E. Leakey, p. 71.
428 "the prevailing . . . for all": Richard Leakey to Bill Garland, November 8, 1974 (Richard Leakey's files).
429 "his henchmen" Richard Leakey to Glynn Isaac, October 4, 1974 (Richard Leakey's files).
429 "There is every. . . . 100% effective! ": Richard Leakey to Jack Miller, November 26, 1974 (Richard Leakey's files).

429 "I am in. . . . hysterical": Jack Miller to Richard Leakey, December 6, 1974 (Richard Leakey's files).
429 "They continued . . . with it," "irrelevant": John Harris, telephone interview, December 19, 1991.
429 "gladiators' clash": Glynn Isaac to Richard Leakey, January 14, 1975 (Richard Leakey's files).
429 "powerful blow": Richard Leakey to Glynn Isaac, October 4, 1974 (Richard Leakey's files).
429 "It was a . . . around": Richard Hay, telephone interview, December 23, 1991.
430 "I thought . . . they thought," "they failed . . . praise": Garniss Curtis to Richard Leakey, August 30, 1975 (Richard Leakey's files).
430 "I'd heard. . . . right there": Richard Hay, telephone interview, December 23, 1991.
430 *"extremely*. . . . dating": Garniss Curtis, interview, Berkeley, March 28, 1985.
430 "I felt . . . meeting," Richard E. Leakey, *One Life,* p. 168.
430 "I was . . . pigs": Basil Cooke, telephone interview, December 20, 1991.
430 "There was. . . . this," "My wife. . . . Perfectly": Basil Cooke, interview, New York City, April 9, 1984.
431 "Well, it's . . . Bed IV" to "I just. . . . tuff": Roger Lewin, *Bones of Contention,* p. 208.
431 "It didn't take . . . fracas": Clark Howell, interview, Berkeley, March 21, 1985.
431 "I must confess. . . . father! ": Mary Leakey to Garniss Curtis, May (n.d.) 1975 (Mary Leakey's files).
431 "My intention. . . . methods": Richard Leakey to Glynn Isaac, June 13, 1975 (Richard Leakey's files).
431 "stand firm. . . . years": Richard Leakey to Glynn Isaac, June 16, 1975 (Richard Leakey's files).
432 "Glynn . . . ship": Richard Leakey to Bernard Wood, June 16, 1975 (Richard Leakey's files).

Chapter 33. On the Trail of *Homo erectus*

433 "Olduvai . . . to live": Mary Leakey, *Disclosing the Past,* p. 161.
434 "She can. . . . to come' ": Micky Day, interview, London, November 26, 1984.
434 "One easily . . . favor": Richard Hay, interview, Urbana, Illinois, February 4, 1985.
434 "She was allergic . . . around them": Philip Leakey, interview, Nairobi, August 22, 1987.
434 "I've always loved. . . . was fun": Helen O'Brien, interview, Newport Beach, California, April 18, 1985.
434 "She's basically. . . . prickly": Richard Hay, interview, February 4, 1985.
434 "Whatever her duties. . . . softies": Lita Osmundsen, interview, New York City, April 4, 1984.
435 "Not as though. . . . very busy": Mary Leakey, quoted by Elizabeth Vrba, interview, Pretoria, South Africa, August 2, 1984.
435 "Yet it is. . . . her character": Elizabeth Vrba, interview, Pretoria, South Africa, August 2, 1984.
435 "[A]bove all . . . at all": *Disclosing the Past,* p. 150.
435 "She always. . . . her fur": Richard Hay, interview, Urbana, Illinois, February 4, 1985.
435 "Mary belongs. . . . her camp": Glynn Isaac, interview, Cambridge, Mass., April 13, 1984.
435 "Mary did. . . . lunch": Mwongela Muoka, interview, Nariokotome River Camp, Kenya, September 17, 1987 (Joseph Mutaba, translator).
435 "clicking . . . eyes": *Disclosing the Past,* p. 150.
436 "It was . . . thrown in": Richard Hay, telephone interview, May 26, 1992.
436 "Well, if. . . . fair": Mary Leakey, interview, Olduvai Gorge, September 26, 1984.
436 "The dogs. . . . her children": Judith Shackleton, letter to author, March 18, 1988.
436 "They are . . . save them": Mary Leakey, interview, Olduvai Gorge, September 26, 1984.
436 "apparently. . . . to attack": *Disclosing the Past,* p. 134.
437 "treated animals better . . . people do": Mary Leakey, quoting an exchange between Tim White and Thure Cerling, interview, Olduvai Gorge, September 26, 1984.
437 "Once when. . . . second": Micky Day, interview, London, November 26, 1984.
437 "No, no, Smudge" to "They'd say. . . . eat her": Peter Jones, interview, Hadza, Tanzania, September 28, 1984.
437 "She could. . . . your work": Mwongela Muoka, interview, Nariokotome River Camp, September 17, 1987 (Joseph Mutaba, translator).
437 "dim-witted fellow": Mary Leakey, interview, Nairobi, September 1, 1989.
438 "The potential . . . opinion": Mary Leakey, *Olduvai Gorge: My Search for Early Man,* pp. 167–168, 178.
439 "fine Acheulian . . . years old": *Disclosing the Past,* p. 152.
439 "exceptionally sharp condition": *Olduvai Gorge: My Search for Early Man,* p. 106.
439 "[I]t may well. . . . thrilled!!!": Phillip V. Tobias to Mary Leakey, May 5, 1979 (NMK).

439 "It is not . . . seems likely": M. D. Leakey, "Discovery of Postcranial Remains of *Homo erectus* and Associated Artefacts in Bed IV at Olduvai Gorge, Tanzania," *Nature,* August 6, 1971, p. 383.
440 "hot . . . *Homo erectus!":* Mary Leakey to Derek Roe, June 6, 1971 (Mary Leakey's files).
441 "quite against expectation": *Disclosing the Past,* p. 153.
441 "at variance . . . WK industry": "Discovery of Postcranial Remains of *Homo erectus* . . . ," op. cit., p. 383.
441 "[B]ut the Bed IV. . . . sources": Mary Leakey to Desmond Clark, August 2, 1970 (NMK).
441 "If you plot. . . . supply is different": Richard Hay, interview, Urbana, Illinois, February 4, 1985.
441 "I would dearly. . . . certainty": *Disclosing the Past,* p. 153.
441 "[a]t sites . . . result": Kathy D. Schick and Nicholas Toth, *Making Silent Stones Speak,* p. 237.
442 "We were digging. . . . pits with care": Mary Leakey, interview, Olduvai Gorge, September 25, 1984.
442 "[I]t's a bit . . . *Homo erectus!":* Mary Leakey to Michael Day, May 15, 1972 (NMK).
442 "You couldn't. . . . ditches": Richard Hay, telephone interview, May 26, 1992.
442 "evidence . . . present": *Olduvai Gorge: My Search for Early Man,* p. 65.
442 "but others . . . a stick": Mary Leakey to Melvin Payne, March 28, 1972 (NMK).
442 "beautifully preserved . . . sides": Mary Leakey to Gordon Hanes, January 25, 1972 (NMK).
442 "Some of the small. . . . dried soil": *Olduvai Gorge: My Search for Early Man,* pp. 64–65.
443 "a nice . . . woman," "just . . . produced": Mary Leakey to Glynn Isaac, April 15, 1972 (NMK).
443 "You know . . . trouble," "It was me he meant": *Disclosing the Past,* p. 154.
443 "Virtually . . . ridges," "walls . . . thick": *Olduvai Gorge: My Search for Early Man,* p. 178.
444 "made by . . . extent": Mary Leakey, "Cultural Patterns in the Olduvai Sequence," *After the Australopithecines,* ed. Karl W. Butzer and Glynn L. Isaac, p. 492.
444 "All of those. . . . about them": Mary Leakey, interview, Nairobi, September 1, 1989.
444 "It seemed . . . washed out": *Disclosing the Past,* p. 164.

Chapter 34. Mother and Son

446 "This . . . followed": Mary Leakey, *Disclosing the Past,* p. 164.
447 "We plodded. . . . field trip": Alan Walker, interview, New York City, April 7, 1984.
447 "We got . . . bed": Richard Leakey, interview, San Francisco, November 11, 1986.
447 "Richard. . . . the picture": Peter Jones, interview, Nairobi, September 13, 1984.
447 "most arduous": *Disclosing the Past,* p. 168.
447 "sound wicket": Mary Leakey to Garniss Curtis, October 8, 1975.
448 "I played second . . . attention": Mary Leakey, interview, Olduvai Gorge, September 23, 1984.
448 "They were. . . . new light": Boyce Rensberger, "Man Traced 3.75 Million Years by Fossils Found in Tanzania," *New York Times,* October 31, 1975, p. 1.
448 "watertight": "Hominid Bones: Old and Firm at 3.75 Million," *Science News,* November 8, 1975, p. 292.
448 "I think . . . who died out": "Man Traced 3.75 Million Years by Fossils Found in Tanzania," op. cit.
448 "not unlike ourselves," "We can't . . . findings": "The Oldest Man," *Time,* November 10, 1985, p. 93.
448 "The winning team": Richard Leakey, interview, en route to Kothiai River Camp, Kenya, September 23, 1987.
449 "personal ambition . . . accuracy": Mary Leakey to Garniss Curtis, September 14, 1975 (Mary Leakey's files).
449 "I think Mary. . . . prickly": Richard Leakey, interview, San Francisco, November 11, 1986.
450 "She asked. . . . to help": Alan Walker, interview, New York City, April 7, 1984.
450 "I probably. . . . short life": Hilary Ng'weno, "Conversations with Philip and Richard Leakey," *The Nairobi Times,* March 2, 1980.
450 "Richard. . . . captivating": Bill Richards, telephone interview, November 9, 1990.
450 "Richard wanted. . . . anymore": Joan Travis, telephone interview, April 24, 1991.
450 "I wanted. . . . approach": Richard Leakey, interview, November 11, 1986.
451 "You could. . . . not working' ": ibid.
451 "I'm not. . . . them": Joan Travis, telephone interview, April 24, 1991.
451 "was a failure," "strong-arm tactics": L. S. B. Leakey Foundation Minutes, March 20, 1973 (Leakey Foundation archives).
451 "I don't think. . . . end of that": Ned Munger, interview, Pasadena, California, April 18, 1985.
452 "He can tell . . . get it": anonymous, Nairobi, August 16, 1987.

452 "Richard. . . . like this": Nancy Gonzales, telephone interview, December 19, 1991.
453 "increasingly despondent," "I seemed. . . . fossils": Richard E. Leakey, *One Life,* p. 177.
453 "a pause . . . essential": Richard Leakey and Glynn Isaac to East Rudolf Research Project Team Members, November (n.d.) 1974 (Richard Leakey's files).
453 "That was. . . . wind down," "My first season. . . . the unknown: Glynn Isaac, interview, Cambridge, Mass., April 13, 1984.
453 "Tim was. . . . 'Sure' " to "Tim had . . . reason": Milford Wolpoff, telephone interview, August 22, 1988.
454 "Once the plane. . . . commands": Boyce Rensberger, "The Face of Evolution," *New York Times Magazine,* March 3, 1974, p. 54.
454 "Richard . . . school" to "No . . . table": Bernard Wood, interview, London, May 23, 1984.
454 "strikingly similar": R. E. F. Leakey, "New Hominid Fossils from the Koobi Fora Formation in Northern Kenya," *Nature,* June 17, 1976, p. 575.
454 "I think . . . years ago": "Evolution Revolution," *Science News,* March 13, 1976, p. 164.
455 "Our curiosity. . . . complete skull?": Meave Leakey's field diary, August 1, 1975; quoted in *One Life,* p. 170.
455 "how he . . . imagine": Meave Leakey's field diary, August 2, 1975; *One Life,* pp. 170–71.
455 "It would . . . on it now!": Meave Leakey's field diary, August 2, 1975; *One Life,* p. 171.
455 "What a find. . . . million years old": *One Life,* p. 172.
456 "doing a jigsaw . . . guide": Peter Gwynne, "The Oldest Man," *Newsweek,* March 22, 1976, p. 59.
456 "a glorious . . . seeing": "Evolution Revolution," *Science News,* March 13, 1976, p. 164.
457 "Richard and Alan. . . . like a fool: Milford Wolpoff, telephone interview, August 22, 1988.
457 "competitors. . . . human evolution": "Evolution Revolution," op. cit., p. 164.
457 "dead-ringer": Peter Gwynne, "The Oldest Man," op. cit., p. 59.
458 "He [Johanson] . . . things": "Evolution Revolution," op. cit., p. 165.
458 "Our preliminary . . . hands today": ibid., p. 164.
458 "No *Australopithecus* . . . like that" "obviously . . . site": ibid., p. 165.

Chapter 35. A New Contender

459 "I liked. . . . all buddies": Richard Leakey, interview, San Francisco, March 19, 1985.
459 "sipping. . . . share": Andrew Jaffe, "Paleontology's Daring Young Man," *International Wildlife,* May 1975, pp. 4–11.
460 "Richard. . . . Leakey": Anna K. Behrensmeyer, interview, Washington, D.C., February 14, 1985.
460 "paleoanthropology's . . . supernova": Donald C. Johanson and Maitland Edey, *Lucy: The Beginnings of Humankind,* p. 185.
460 "imagination . . . inflamed," "was proof . . . fossils": ibid., p. 98.
460 "Louis not only . . . imagination": Donald Johanson, lecture in San Francisco, February 25, 1987.
460 "a glamorous . . . press," "another thrill. . . . famous": *Lucy,* p. 124.
461 "People were . . . dissertation": Maurice Taieb, interview, Berkeley, April 15, 1986.
461 "exceptional and fantastic," Maurice Taieb, *Sur la terre des premiers hommes,* p. 27.
461 "And the. . . . found hominids": Maurice Taieb, interview, Berkeley, April 15, 1986.
461 "The allure . . . tremendous," "Richard. . . . I?": *Lucy,* p. 131.
462 "much better": ibid., p. 132.
462 "all of them. . . . dreams": ibid., p. 133.
462 "We knew. . . . leader," "He wanted. . . . position": Jon Kalb, telephone interview, April 3, 1989.
462 "Louis was. . . . funds": Maurice Taieb, interview, Berkeley, April 15, 1986.
463 "Johanson was. . . . over his": Jon Kalb, telephone interview, April 3, 1989.
463 "it dawned . . . fossil": *Lucy,* p. 155.
463 "We could . . . for that!": Jon Kalb, telephone interview, April 3, 1989.
463 "I now had . . . my own": *Lucy,* p. 165.
463 "I was there. . . . before," "definitively competitive": M. E. Morbeck, interview, Nairobi, August 16, 1984.
464 "I see no. . . . too glad": Richard Leakey to Donald Johanson, May 29, 1974 (Richard Leakey's files).
464 "Dr. Johanson. . . . period": Mary Leakey, To Whom It May Concern, June 12, 1974 (Mary Leakey's files).

464 "Strange. . . . baffling": *Lucy,* p. 175.
464 "an unparalleled. . . . million years": D. B. Ottoway, "3-Million-Year-Old Human Fossils Found," *International Herald Tribune,* October 28, 1974.
464 "All previous. . . . years" to "revolutionary postulate": "Fossils in Ethiopia Said to Show Man as Million Years Older Than Believed," *New York Times,* October 26, 1974, p. 1.
465 "a trifle strong": Richard Leakey to Donald Johanson, November 4, 1974 (Richard Leakey's files).
465 "[T]here has been . . . reasonable," "These are minor. . . . importance": ibid.
465 "It was very. . . . starting": Maurice Taieb, interview, Berkeley, April 15, 1986.
465 "Don seemed. . . . going on": John Harris, telephone interview, April 9, 1985.
466 "I'm going . . . Leakeys": Donald Johanson, quoted by Maurice Taieb, interview, Berkeley, April 15, 1986.
466 "My pulse. . . . hominid," "the realization. . . . midday sun": Donald Johanson, "Ethiopia Yields First Family," *National Geographic,* December 1976, p. 793.
466 "an unknown . . . worker," "dazzling . . . Leakey": *Lucy,* p. 185.
466 "public . . . awhile": Glynn Isaac, quoting Richard Leakey in letter to R. Leakey, November 8, 1974 (Richard Leakey's files).
467 "Window dressing. . . . Lucy": Paul Abell, interview, Kingston, Rhode Island, February 21, 1985.
467 "Some people. . . . that sort": Frank Fitch to Richard Leakey, December 4, 1974 (Richard Leakey's files).
467 "Donald. . . . family name": Maurice Taieb, interview, Berkeley, April 15, 1986.
467 "I had the. . . . pick them up": *Lucy,* p. 213.
468 "Hey, Richard. . . . got you now": an eyewitness source, Nariokotome River Camp, Kenya, August 31, 1984.
468 "revolutionary . . . human evolution": Peter Gwynne, "The Oldest Man," *Newsweek,* March 22, 1976, p. 59.
468 "Johanson talked . . . Leakeys": Maurice Taieb, interview, Berkeley, April 15, 1986.
469 "I think . . . be the same": *Lucy,* p. 218.

Chapter 36. The Name Game

471 "have the small . . . species": R. E. F. Leakey, "New Hominid Fossils from the Koobi Fora Formation in Northern Kenya," *Nature,* June 17, 1976, p. 576.
471 "For myself . . . years ago": Mary Leakey to Charles Oxnard, May 6, 1976 (Mary Leakey's files).
471 "that the australopithecines . . . on a limb": Mary Leakey to Garniss Curtis, October 8, 1975 (Mary Leakey's files).
472 "just for the sake of devilry": Mary Leakey to Phillip V. Tobias, June 21, 1976 (Mary Leakey's files).
472 "This proved. . . . more came," "simpler to remove . . . off singly": Mary Leakey, *Disclosing the Past,* p. 172.
473 "made a considerable . . . people": ibid., p. 168.
473 "emotional . . . kind": ibid., p. 169.
473 "There was a . . . Mary": Peter Jones, interview, Nairobi, September 13, 1984.
473 "Philip took us. . . . at them" to "It was very exciting": Anna K. Behrensmeyer, telephone interview, December 4, 1992.
474 "fantastic site for footprints": Anna K. Behrensmeyer's field diary, July 25, 1974 (courtesy of Anna K. Behrensmeyer).
474 "Jesus. . . . a new eye": Richard Hay, telephone interview, December 29, 1992.
474 "These were some . . . my life" to "You can. . . . first showers": ibid.
475 "Mary had the . . . there, too?" to "She didn't like. . . . many of them": Peter Jones, interview, Cambridge, Mass., April 25, 1985.
475 "genuine": Mary Leakey to Bob Savage, November 27, 1977 (Mary Leakey's files).
475 "conquered [her] phobia": Mary Leakey to Marion Sterling, September 20, 1974 (Mary Leakey's files).
475 "On the basis . . . referred to *Homo":* D. C. Johanson and M. Taieb, "Plio-Pleistocene Hominid Discoveries in Hadar, Ethiopia," *Nature,* March 25, 1976, p. 296.
476 "I kidded her. . . . You're lucky": Donald C. Johanson and Maitland Edey, *Lucy: The Beginnings of Humankind,* p. 251.
476 "Instead of giving. . . . to contribute": Richard Leakey, interview, en route to Kothiai River Camp, Kenya, September 23, 1987.

476 "He is a . . . the *AJPA*": Mary Leakey to Lita Osmundsen, March (n.d.) 1976 (Mary Leakey's files).
477 "I'm not a . . . without shaking," "Tim knows. . . . wrong": Milford Wolpoff, telephone interview, August 22, 1988.
477 "He was bright . . . team," "unspeakably . . . to others": Richard Leakey, interview, San Francisco, March 19, 1985.
477 "grumbling" to "How could . . . like a child?": Kamoya Kimeu, interview, Nairobi, October 30, 1984.
478 "Their paper. . . . comment" to "He started shouting. . . . slammed the door": Richard Leakey, interview, en route to Kothiai River Camp, Kenya, September 23, 1987.
478 "to take a grip on yourself" to "consider it. . . . further work": Richard Leakey to Tim White, September 28, 1976 (Richard Leakey's files).
478 "The Leakeys. . . . own individuality": Anna K. Behrensmeyer, interview, Washington, D.C., February 14, 1985.
479 "As far as. . . . to Mary": Richard Leakey, interview, San Francisco, March 19, 1985.
479 "I think Tim. . . . defeated Richard": Milford Wolpoff, telephone interview, August 22, 1988.
479 "the L/H . . . enlargement," "What about. . . . adaptations": Donald Johanson to Mary Leakey, December 23, 1977 (Mary Leakey's files; also quoted in Roger Lewin, *Bones of Contention,* p. 249).
480 "doubtful . . . this stage": Mary Leakey to Donald Johanson, November 27, 1977 (Mary Leakey's files).
480 "A number. . . . agree" to "If you can . . . must stand down": Mary Leakey to Donald Johanson, January 9, 1979 (Mary Leakey's files).
480 "in a straightforward way": Don Johanson to Mary Leakey, February 4, 1978 (Mary Leakey's files).
480 "I never thought. . . . all along": Mary Leakey, interview, San Francisco, May 5, 1987.
481 "I think you'd. . . . from Laetoli," "Don devoted. . . . hard for her": Richard Leakey, interview, en route to Kothiai River Camp, Kenya, September 23, 1987.
481 "All the things. . . . was so impertinent": Mary Leakey, interview, San Francisco, May 5, 1987.

Chapter 37. Footprints for the Mantelpiece

482 "Mary showed me. . . . water on it": Michael Day, interview, London, November 26, 1984.
482 "75% certain" to "It probably ate. . . . would have too": "Leakey's Finds: Tracks of an Ancient Ancestor," *Time,* March 6, 1978, p. 53.
483 "Louise was Mary's new footprint expert," "in his cowboy. . . . anatomist that year": Michael Day, interview, London, November 26, 1984.
483 "She had. . . . on the dot,' " "fourth son": Peter Jones, interview, Nairobi, September 13, 1984.
483 "worked very closely. . . . right-hand man": Louise Robbins, telephone interview, March 24, 1985.
484 "I'd finished my. . . . it's a footprint": Paul Abell, telephone interview, December 8, 1992.
484 "And it *was* a footprint": Richard Hay, telephone interview, December 31, 1992.
484 "We all said. . . . exaggerating": Peter Jones, interview, Cambridge, Mass., April 25, 1985.
485 "We were all. . . . previous season": Louise Robbins, telephone interview, March 24, 1985.
485 "There was unqualified. . . . found it' ": Richard Hay, telephone interview, December 31, 1992.
485 "blurred . . . his feet," "loose . . . prints": Mary D. Leakey, "Footprints in the Ashes of Time," *National Geographic,* April 1979, p. 453.
485 "Now this . . . mantelpiece": John Reader, *Missing Links,* p. 15.
486 "a well-thought-out. . . . in one summer' ": Louise Robbins, telephone interview, March 24, 1985.
486 "We are having. . . . and all": Mary Leakey to Allen O'Brien, August 3, 1978 (Mary Leakey's files).
486 "We have found . . . developed": Mary Leakey to Dick Hoojier, September 12, 1978 (Mary Leakey's files).
486 "The fully bipedal. . . . am sure": Dick Hoojier to Mary Leakey, September 20, 1987 (Mary Leakey's files).
486 "It looks. . . . human aspect": Boyce Rensberger, "Prehistoric Footprints of Man-Like Creatures Found," *New York Times,* March 22, 1979, p. 16.
486 "experienced . . . doubt": "Footprints in the Ashes of Time," op. cit., p. 453.
487 "Tim showed me . . . my objection": Mary Leakey to Clark Howell, September 13, 1978 (Mary Leakey's files).
487 "That name rankled. . . . Afar": Paul Abell, telephone interview, December 8, 1992.
487 "It was kind. . . . my beasts' ": Richard Hay, telephone interview, December 30, 1992.

488 "It was a horrible . . . on end": Ernst Mayr, telephone interview, January 17, 1990.
488 "She felt . . . hominids": Richard Hay, telephone interview, December 31, 1992.
488 "White was talking. . . . times": Paul Abell, telephone interview, December 30, 1992.
488 "Needless . . . White": Mary Leakey, letter to author, July 29, 1985.
488 "He started . . . how to dig": Mary Leakey, interview, Laetoli, Tanzania, September 24, 1984.
488 "I could . . . expertise": Mary Leakey, *Disclosing the Past,* p. 177.
489 "It was. . . . print" to "their career strategy": Mary Leakey, interview, San Francisco, May 5, 1987.
489 "PLEASE. . . . MARY": Mary Leakey to Donald Johanson, August 22, 1978 (Mary Leakey's files).
489 "I trust . . . *fail":* Mary Leakey to Donald Johanson, September 12, 1978 (Mary Leakey's files).
489 "I would be . . . disturbed by it": Mary Leakey to Clark Howell, September 13, 1978 (Mary Leakey's files).
489 "difficult": Mary Leakey, interview, San Francisco, May 5, 1987.
489 "That's a Maasai. . . . avoid him": Ron Clarke, interview, Johannesburg, South Africa, August 1, 1984.
490 "state of euphoria": Mary Leakey to Clark Howell, September 13, 1978 (Mary Leakey's files).
490 "Call it *Hylobates . . . Australopithecus":* Mary Leakey, quoted in Roger Lewin, *Bones of Contention,* p. 259.

Chapter 38. Battling Over Bones

491 "We are in. . . . next July": Mary Leakey to Don Johanson, October 11, 1978 (Mary Leakey's files).
491 "I asked him. . . . one of them": Richard Leakey, interview, en route to Kothiai River Camp, Kenya, September 23, 1987.
492 "It is a. . . . on p. 5!": Mary Leakey to Phillip V. Tobias, November 3, 1978 (Mary Leakey's files).
492 "there is no . . . [in] 1939" to "gladly . . . name": Mary Leakey to Michael Day, February 18, 1979 (Mary Leakey's files).
493 "I've often thought. . . . point for her": Richard Hay, telephone interview, December 31, 1992.
494 "It seems . . . shootout," "I guess we are": Donald C. Johanson and Maitland Edey, *Lucy: The Beginnings of Humankind,* p. 291.
494 "where we . . . private": Anne Thurston, "Interview with Richard Leakey," *The Humanist,* January/February 1979, p. 26.
494 "found . . . anthropologist," "upset . . . new scenario": "Puzzling Out Man's Ascent," *Time,* November 7, 1977, p. 48.
494 "about 5 per cent": "Interview with Richard Leakey," op. cit., p. 24.
495 "He was . . . field": Glynn Isaac, interview, Cambridge, Mass., April 13, 1984.
495 "the development . . . team" to "if a . . . dies": "Interview with Richard Leakey," op. cit., pp. 27, 28.
496 "I even said. . . . platform": Mary Smith, interview, Washington, D.C., February 13, 1985.
496 "I am perfectly . . . found": "Finding Eve's Cousin," *Newsweek,* January 29, 1979, p. 81.
497 "Wondering . . . going to end": Richard Leakey, interview, en route to Kothiai River Camp, Kenya, September 23, 1987.
497 "I think Don. . . . *Australopithecus":* Boyce Rensberger, "Rival Anthropologists Divide on 'Pre-Human' Find," *New York Times,* February 18, 1979, p. 1.
497 "The material. . . . opinion": ibid.
497 "dead ringers": Roger Lewin, *Bones of Contention,* p. 170.
497 "I thought there'd . . . natural death": Richard Leakey, interview, en route to Kothiai River Camp, Kenya, September 23, 1987.
498 "But when pressed. . . . *Afarensis,*" "not very scientific": "The Leakey Footprints: An Uncertain Path," *Science News,* March 31, 1979, p. 196.
498 "I winced. . . . kiss of death": Mary Smith, interview, Washington, D.C., February 13, 1985.
498 "really . . . all about," "The footprints . . . *Australopithecus afarensis":* "The Leakey Footprints: An Uncertain Path," op. cit., p. 196.
498 "They were direct . . . man" to "We'll leave. . . . problem": ibid., p. 197.
499 "a wonderful . . . noses" to "enjoying it immensely": Paul Galloway, "The Evolution Revolution," *Chicago Sun-Times,* August 26, 1979, pp. 16–21.
499 "briefly . . . Cronkite news!": Mary Leakey to Paul Abell, April 3, 1979 (Mary Leakey's files).
499 "The footprints. . . . ever made": Paul Abell to Mary Leakey, March 31, 1979 (Mary Leakey's files).
499 "It is . . . single word!": Mary Leakey to Mike Mehlman, January 29, 1979 (Mary Leakey's files).
500 "As regards yourself . . . do not agree": Mary Leakey to Tim White, April 26, 1979 (Mary Leakey's files).

500 "She was . . . arrival" to "White. . . . funny": Ron Clarke, interview, Johannesburg, South Africa, August 1, 1984.
500 "to crazy . . . camp": Peter Jones, interview, Nairobi, September 13, 1984.
501 "clearly absurdly long," "You will see. . . . 'Lover's Lane!' ": Mary Leakey to Phillip V. Tobias, August 31, 1979 (Mary Leakey's files).
501 "was being. . . . forwards": Mary Leakey to Kenneth Oakley, August 22, 1979 (Mary Leakey's files).
501 "We talked. . . . each other": Paul Abell, telephone interview, December 8, 1992.
501 "Boys . . . really shocking!": Gertrude Caton-Thompson to Mary Leakey, September 30, 1979 (Mary Leakey's files).
501 "follow-the-leader," "holding on . . . in front": Mary Leakey to Melvin Payne, September 8, 1979 (Mary Leakey's files).
501 "If one . . . juvenile": Mary D. Leakey, "Tracks and Tools," The Emergence of Man, *Proceedings of the Royal Society and the British Academy,* p. 100.
502 "all chimps . . . frightened": Mary Leakey to Anna K. Behrensmeyer, October 30, 1979 (Mary Leakey's files).
502 "It is now. . . . prints": "Tracks and Tools," op. cit, p. 99.
502 "a wonderful season": Mary Leakey to Jean Jacques, October 24, 1979 (Mary Leakey's files).
502 "end-stage renal failure": Richard E. Leakey, *One Life,* p. 189.

Chapter 39. Richard Reborn

503 "found the days . . . long": Richard E. Leakey, *One Life,* p. 182.
504 "The institute really. . . . an idea": Andrew Hill, interview, Nairobi, October 19, 1984.
504 "Abuje . . . one day": John Harris, interview, Los Angeles, April 11, 1984.
505 "over-excess . . . trusted," "It quickly . . . archeology": ibid.
505 "step aside" "unable . . . is rampant": Richard Leakey to Professor T. R. Odhiambo, March 26, 1979 (Richard Leakey's files).
505 "One does not . . . these problems' ": Richard Leakey, interview, en route to Kothiai River Camp, September 23, 1987.
506 "All of this . . . kidney failure": *One Life,* p. 196.
506 "Richard had amazing. . . . he was in": Joan Karmali, interview, Nairobi, September 19, 1987.
506 "terrible blow": *One Life,* p. 185.
507 "I confess . . . goings on": ibid., p. 184.
507 "half-hearted attempt": ibid., p. 185.
507 "Richard. . . . Fine' " to "He just would. . . . getting transfusions": Kathy Eldon, interview, Nairobi, November 14, 1984.
507 "I would try. . . . the end": Meave Leakey, interview, Nairobi, October 8, 1984.
508 "[T]o my utter. . . . a rest," "kidney complaint": *One Life,* p. 185.
508 "Tears were quietly. . . . friends": ibid., p. 188.
509 "I think that. . . . might be short": Mary Smith, interview, Washington, D.C., February 13, 1985.
509 "It was an. . . . just once more!": *One Life,* p. 189.
509 "We never thought. . . . temperament for it": Meave Leakey, interview, Nairobi, October 8, 1984.
510 "I can remember. . . . relationship": ibid.
510 "to experiment," "I rushed around. . . . my life": Philip Leakey, interview, Nairobi, October 24, 1984.
510 "That one is. . . . just like that": Teresia Ng'anga, interview, Nairobi, October 29, 1987.
510 "But I decided . . . too small": Philip Leakey, interview, Nairobi, October 24, 1984.
510 "unprincipled and irresponsible": Richard Leakey, interview, en route to Kothiai River Camp, Kenya, September 23, 1987.
510 "[F]or more . . . my house" to "I had my pride": *One Life,* p. 192.
511 "in the cold light of day": ibid., p. 193.
511 "He is my. . . . same for me": Hilary Ng'weno, "Conversations with Philip and Richard Leakey," *Nairobi Times,* March 2, 1980, p. 5.
511 "They. . . . ever imagined," "I think. . . . No way!": Meave Leakey, interview, Nairobi, October 8, 1984.
511 "major undertaking": *One Life,* p. 193.
511 "I can remember. . . . shivering": Meave Leakey, interview, Nairobi, October 8, 1984.
512 "Meave found. . . . he was": Alan Walker, interview, New York City, April 7, 1984.
512 "I spent. . . . my body": *One Life,* p. 194.

512 "It . . . feet": Richard Leakey to Mary Leakey, September 4, 1979 (Mary Leakey's files).
512 "I am told . . . family": Mary Leakey to Allen O'Brien, August 27, 1979 (Mary Leakey's files).
512 "it was. . . . affair": "Conversations with Philip and Richard Leakey," op. cit., p. 5.
513 "It was. . . . tense": Meave Leakey, interview, Nairobi, October 8, 1984.
513 "I think . . . stiff": *One Life,* p. 200.
513 "a number . . . friends," "short . . . unprintable": ibid., p. 201.
513 "It was hard. . . . in his leg": Meave Leakey, interview, Nairobi, October 8, 1984.
513 "God, I must . . . this": ibid.
513 "my blood . . . urine," "Although. . . . been reborn": *One Life,* p. 7.
514 "I began to . . . transplants," "The shock was unbelievable": ibid., p. 202.
514 "You know. . . . anymore,' " "He'd been . . . weaker": Meave Leakey, interview, Nairobi, October 8, 1984.
515 "People were . . . that," "so . . . Richard": Anthony Marshall, interview, New York City, February 8, 1985.
515 "I'd never. . . . his critics": ibid.
515 "After Tony left. . . . a fuss" to "That was the worst. . . . really was": Meave Leakey, interview, Nairobi, October 8, 1984.
516 "Pleurisy. . . . agony" to "Breathe. . . . Breathe! ": Richard Leakey, interview, San Francisco, November 11, 1986.
516 "magnificent effort. . . . lived": *One Life,* p. 202.

Chapter 40. "How Very Human"

517 "[I]t . . . good! ": Richard Leakey to Anna K. Behrensmeyer, April 8, 1980 (Richard Leakey's files).
517 "probably liv[e] . . . pace": Hilary Ng'weno, "Conversations with Philip and Richard Leakey," *The Nairobi Times,* March 2, 1980, p. 5.
517 "There was. . . . empire": John Karmali, interview, Nairobi, September 19, 1987.
518 "It seemed as. . . . having it all' ": Andrew Hill, interview, Nairobi, October 19, 1984.
518 "Richard's move. . . . goodwill": David Pilbeam, interview, Cambridge, Mass., February 19, 1985.
518 "All of these. . . . true friends were": Richard Leakey, interview, Kothiai River Camp, Kenya, September 5, 1987.
518 "He had a. . . . needed support": Alan Walker, interview, Nariokotome River Camp, August 31, 1984.
519 "He was. . . . swear word": Richard Leakey, interview, en route to Kothiai River Camp, Kenya, September 23, 1987.
519 "Tim has. . . . that bad": Mary Smith, interview, Washington, D.C., February 13, 1985.
519 "That's when. . . . Mary and me": Richard Leakey, interview, en route to Kothiai River Camp, Kenya, September 23, 1987.
520 "I am fine . . . dreadful I am": Richard Leakey to Laurie Benz, March 6, 1981 (Richard Leakey's files).
520 "We're not talking. . . . important": "Bones and Prima Donnas," *Newsweek,* February 16, 1981, p. 77.
520 "the one thing . . . one thought": "Conversations with Philip and Richard Leakey," op. cit., p. 5.
520 "Don had been. . . . to go on" to "I asked him . . . of the creationists' ": Richard Leakey, interview, Nariokotome River Camp, Kenya, August 31, 1984.
521 "I don't know. . . . up to Cronkite": Donald C. Johanson and James Shreeve, *Lucy's Child,* p. 119.
521 "I kept thinking . . . just leave": Richard Leakey, interview, Nariokotome River Camp, Kenya, August 31, 1984.
521 "Our announcement. . . . family tree": Donald C. Johanson and Maitland Edey, *Lucy: The Beginnings of Humankind,* p. 303.
521 "to review . . . feature," "disappointing": ibid., pp. 299–300.
521 "After all. . . . to others": Alan Walker, interview, Kothiai River Camp, August 31, 1987.
521 "much too . . . my taste": Richard Leakey, *Origins Reconsidered,* p. 115.
521 "[We] can. . . . disagreement": Richard Leakey, interview, Nairobi, February 16, 1983.
521 "I greatly regret. . . . due course": *Lucy,* p. 304.
522 "Richard had. . . . record": *Lucy's Child,* p. 119.
522 "We brought . . . ancestry": Walter Cronkite, "Cronkite's Universe," broadcast in May 1981 by CBS Television Network.

522 "I realized. . . . like this": Richard Leakey, interview, Nariokotome River Camp, Kenya, August 31, 1984.
522 "There has been . . . tree": Donald Johanson, "Cronkite's Universe."
522 "I've heard it. . . . or wrong, "But . . . wrong": Richard Leakey, "Cronkite's Universe."
522 "I've brought along . . . a spot for": Donald Johanson, "Cronkite's Universe."
522 "For me! . . . not an artist": Richard Leakey, "Cronkite's Universe."
523 "I felt such . . . the trap": Richard Leakey, interview, Nariokotome River Camp, Kenya, August 31, 1984.
523 "I think . . . do that": Richard Leakey, "Cronkite's Universe."
523 "aback": *Lucy's Child,* p. 120.
523 "And what . . . in its place?": Donald Johanson, "Cronkite's Universe."
523 "A question mark! " "I would love. . . , wrong": Richard Leakey, "Cronkite's Universe."
523 "Richard was. . . . then it was": Milford Wolpoff, telephone interview, August 22, 1988.
523 "no good explanations": Richard Leakey to Glynn Isaac, May 21, 1981 (Richard Leakey's files).
523 "He had. . . . out of joint": Mary Smith, interview, Washington, D.C., February 13, 1985.
523 "It was after. . . . business," "If I'd. . . . at the top": Richard Leakey, interview, en route to Nariokotome River Camp, Kenya, August 28, 1984.
524 "within . . . arriving": Ron Watkins, interview, London, November 20, 1984.
524 "The Miocene. . . . were his": Richard Leakey, interview, en route to Nariokotome River Camp, Kenya, August 28, 1984.
524 "[P]erhaps. . . . being tested! ": Richard Leakey to Alan Walker, May 13, 1981 (Richard Leakey's files).
525 "I wanted . . . put the camp" to "It was a . . . got on well": Richard Leakey, interview, Portland, Oregon, February 26, 1994.
525 "It was terribly. . . . been to before": Richard Leakey, interview, Koobi Fora, Kenya, July 5, 1984.
526 "[W]e . . . derived": R. E. F. Leakey and A. Walker, "New Higher Primates from the Early Miocene of Buluk, Kenya," *Nature,* November 14, 1985, p. 175.
526 "had relit [his] fire": Richard Leakey, interview, en route to Nariokotome River Camp, Kenya, August 28, 1984.
526 "Once I started. . . . going to do": ibid.
526 "across . . . Turkana," "wondering . . . secrets": *Origins Reconsidered,* p. xix.
526 "I'd asked . . . Fora": Richard Leakey, interview, Portland, Oregon, February 26, 1994.
526 "That's where . . . bipedalism": Richard Leakey, interview, Nairobi, February 16, 1983.
527 "You might . . . them": *Origins Reconsidered,* p. 81.
527 "We have. . . . Skeletons?" ibid., p. 33.
527 "a small . . . good condition": Alan Walker's field diary, West Turkana Expedition, August 23, 1984, (also quoted in *Origins Reconsidered,* p. 26).
527 "How he . . . know": *Origins Reconsidered,* p. 26.
527 "Seldom . . . hopeful": ibid., p. 34.
528 "It is . . . humans": Alan Walker, telephone interview, November 28, 1994.
528 "We've found. . . . skull! ": *Origins Reconsidered,* p. 40.
528 "to where . . . dry earth": ibid.
528 "beautifully . . . *Homo erectus":* Alan Walker's field diary, West Turkana Expedition, August 24, 1984 (also quoted in *Origins Reconsidered,* p. 40).
528 "[R]ight beneath . . . lifetime," "breathtaking moment": *Origins Reconsidered,* pp. 44–45.
528 "[A]s a result . . . human eyes": ibid., p. 46.
528 "This is the. . . . taking pictures! ": Alan Walker's field diary, West Turkana Expedition, September 12, 1984 (also quoted in *Origins Reconsidered,* p. 46).
528 "You'd have . . . as complete": *Origins Reconsidered,* p. 46.
528 "I wonder what. . . . bountiful land": Mary Leakey, interview, Nariokotome River Camp, September 5, 1984.
529 "There is nothing . . . hominid bone": Richard Leakey, interview, Turkana Boy excavation, August 29, 1984.
529 "Such nonsense! . . . decisions! ": Richard Leakey, interview, en route to Kothiai River Camp, Kenya, September 23, 1987.
529 "I consider. . . . distance myself": Richard Leakey to Donald Johanson, March 28, 1983 (Richard Leakey's files).

529 "The reason . . . scoundrel": Richard Leakey to Donald Johanson, May 18, 1983 (Richard Leakey's files).
530 "a strapping youth": Richard Leakey, press conference, Nairobi, October 18, 1984.
530 "Everyone. . . . eluded us," " 'Remarkable.' . . . he looked": Richard Leakey's field diary, September 20, 1984; *Origins Reconsidered,* p. 63.

Chapter 41. Grande Dame of Archeology

532 "It is the Leakeys . . . science" to "We've had. . . . astonishing": John Noble Wilford, "The Leakeys: Rocking the 'Cradle of Mankind,' " *New York Times,* October 30, 1984, sec. III, p. 1.
532 "medium-sized": Donald C. Johanson and Maitland Edey, *Lucy: The Beginnings of Humankind,* p. 37.
532 "deep-rooted": Richard Leakey, interview, Nairobi, February 10, 1983.
533 "well-organized," "a [religious] . . . legacy": Mary Leakey, lecture, American Museum of Natural History, New York, April 10, 1984.
534 "The extent. . . . anti-Leakey sentiment": Frank Brown to Richard Leakey, April 9, 1984 (Richard Leakey's files).
535 "This was . . . reason" to "pep talks": Mary Leakey, *Disclosing the Past,* p. 209.
535 "[I've] almost. . . . Nairobi traffic": Mary Leakey to Ethel Payne, February 14, 1983 (Mary Leakey's files).
535 "Well. . . . hit a one! ": Andrew Hill quoting Mary Leakey, interview, Cambridge, Mass., February 19, 1985.
536 "It will. . . . used to be": Mary Leakey to Richard Hay, May 23, 1983 (Mary Leakey's files).
536 "We worried. . . . in demand": Valerie Leakey, interview, Nairobi, September 29, 1987.
536 "see. . . . grandchildren": Mary Leakey to Mollie Leakey, February 10, 1984 (Mary Leakey's files).
536 "soldier-straight," "fairly . . . aisle": Alan Walker, Kothiai River Camp, Kenya, August 31, 1987.
537 "such lies" to "Because Donald . . . the first": Maurice Taieb, interview, Berkeley, April 15, 1986.
537 "terribly upsetting," "It was . . . violation": Richard Leakey, interview, San Francisco, November 11, 1986.
537 "It's an unwritten. . . . important": Anna K. Behrensmeyer, telephone interview, June 3, 1985.
537 "sad . . . circumstances": Richard Hay to Mary Leakey, July 19, 1985 (Mary Leakey's files).
538 "I saw . . . hopes," "since . . . contact Mary": Donald C. Johanson and James Shreeve, *Lucy's Child,* p. 36.
538 "just enough. . . . exhausted": ibid., pp. 38–39.
538 "She just . . . happened": Richard Leakey, interview, San Francisco, November 11, 1986.
538 "what had . . . hut": *Lucy's Child,* p. 133.
538 "enemy forces": Mary Leakey to Bob Harm, October 16, 1985 (Mary Leakey's files).

Chapter 42. A New Challenge

539 "Instead. . . . references": Richard Leakey to Karla Jennings, April 10, 1985 (Richard Leakey's files).
539 "edg[ing] . . . Hall": Mary Leakey to Ethel Payne, October 10, 1985 (Mary Leakey's files).
540 "the two problems": Mary Leakey to Mary Smith, October 10, 1985 (Mary Leakey's files).
540 "I can confirm. . . . my idea": Richard Leakey to Charles Jaffin, January 7, 1986 (Richard Leakey's files).
540 "Tim White's . . . ignored": Charles Jaffin to Richard Leakey, December 11, 1985 (Richard Leakey's files).
540 "If you . . . action": Richard Leakey to Charles Jaffin, January 7, 1986 (Richard Leakey's files).
540 "My information . . . letter": Richard Leakey to Michael H. Malone, July 22, 1986 (Richard Leakey's files).
541 "frustratingly barren": Richard Leakey, *Origins Reconsidered,* p. 122.
541 "I got. . . . quite a fuss": ibid., p. 125.
542 "Then I. . . . hominid!" to "Look at . . . tooth roots": ibid., p. 124.
542 "would . . . bushier?" to "I am even. . . . for some": ibid., p. 126.
543 "Naming . . . do that": Alan Walker, interview, Kothiai River Camp, Kenya, August 29, 1987.
544 "We said, 'Those . . . sometime,' " "It was. . . . new to science": Richard Leakey, interview, Portland, Oregon, February 26, 1994.
545 "I was . . . that," "we never . . . feet": *Origins Reconsidered,* p. 134.
546 "[It] is . . . since Lucy": "Redrawing the Family Tree," *Time,* August 18, 1986, p. 64.

546 "Whatever the final . . . stated": A. Walker, R. E. Leakey, J. M. Harris, and J. F. Brown, "2.5-Myr *Australopithecus boisei* from west of Lake Turkana, Kenya," *Nature,* August 7, 1986, p. 522.
546 "This throws . . . others," "implies . . . just one": "Skull May Rattle Theories of Evolution," *San Francisco Chronicle,* August 7, 1986.
546 "There was . . . *afarensis*": Bruce Bower, "Family Feud: Enter the 'Black Skull,' " *Science News,* January 24, 1987, p. 58.
546 "nobody . . . human tree": Donald C. Johanson and James Shreeve, *Lucy's Child,* p. 131.
546 "Not since . . . uncovered," "in a state . . . ferment": Charles Petit, "Fossil Hunter's Find May Shake Man's Family Tree," *San Francisco Chronicle,* November 13, 1986, p. 4.
546 "Leakey's. . . . scratched the surface," "The old man . . . the '60s": David Pilbeam, telephone interview, September 22, 1988.
547 "This could. . . . a bit": "Fossil Hunter's Find May Shake Man's Family Tree," op. cit., p. 4.
547 "The Leakeys. . . . three days": *Lucy's Child,* p. 181.
547 "Whoa! . . . next to it": Institute of Human Origins, press release, May 20, 1987.
547 "The twists . . . at times": Michael Day to Mary Leakey, October 10, 1986 (Mary Leakey's files).
548 "My congratulations. . . . O.H. 62": Mary Leakey to Donald Johanson, September 24, 1986 (Mary Leakey's files).
548 "competitive instincts": *Origins Reconsidered,* p. 111.
548 "partial skeleton," "*really* scrappy": Alan Walker, interview, September 11, 1987.
548 "We sort of. . . . sudden": " 'Exciting' Find on Human Ancestor," *San Francisco Chronicle,* May 21, 1987, p. 3.
549 "That's part of. . . . tossed into it": Richard Leakey, interview, San Francisco, October 22, 1994.
549 "highly improbable": D. E. Lieberman, D. R. Pilbeam, and B. A. Wood, "A Probablistic Approach to the Problem of Sexual Dimorphism in *Homo habilis,*" *Journal of Human Evolution,* Vol. 17, 1988, pp. 503–12.
550 "He's beginning . . . like Louis! ": Mary Leakey, interview, September 7, 1987.
550 "The man. . . . interesting": M. E. Morbeck, interview, Nairobi, August 16, 1984.
550 "His ego . . . paleontology": Harry Merrick, interview, Nairobi, February 14, 1983.
550 "I think . . . their forties": Richard Leakey, interview, en route to Nariokotome River Camp, Kenya, August 29, 1984.
550 "the incredible nastiness," "These fights . . . in the field": "Fossil Hunter's Find May Shake Man's Family Tree," op. cit., p. 4.
551 "Richard has. . . . especially Richard": Mary Leakey to George Lindsay, December 3, 1987 (Mary Leakey's files).
551 "I need a new challenge": Richard Leakey, interview, San Francisco, February 29, 1988.
551 "the new Director . . . immediate effect": *The Standard* (Nairobi), April 21, 1989, p. 1.
551 "I plan to succeed": *Origins Reconsidered,* p. xiv.

Epilogue

552 "It occurred . . . tree": Eugene Linden, "Richard the Lionhearted," *Time,* July 19, 1993, p. 51.
553 "the only . . . alive": Richard Leakey, telephone interview, September 1, 1993.
553 "I always . . . quit": Richard Leakey, telephone interview, March 25, 1994.
553 "I take off . . . time": Richard Leakey, interview, San Francisco, October 22, 1994.
554 "Louise is . . . banner": Richard Leakey, telephone interview, July 16, 1993.

Bibliography

Articles and Books by Louis, Mary, Meave, Richard, and Wilfrida Leakey

(Note: This is not a comprehensive bibliography of publications by the Leakeys, but a list of those books and articles that I consulted.)

Leakey, L. S. B. *Adam's Ancestors.* 1934. Revised and reprinted. New York: Harper, 1953.

———. "Adventures in the Search for Man." *National Geographic,* January 1963, pp. 132–52.

———. "African Ancestries, Interpretation of Fossil Finds in Kenya." *The Times,* August 23, 1946, p. 5.

———. *Animals of East Africa.* Washington, D.C.: National Geographic Society, 1969.

———. *By the Evidence: Memoirs, 1932–1951.* New York: Harcourt Brace Jovanovich, 1974.

———. "Calico and Early Man." In *Pleistocene Man at Calico,* edited by Walter C. Schuiling. Redlands, Calif.: San Bernardino County Museum Association, 1979.

———. *Defeating Mau Mau.* London: Methuen, 1954.

———. "Earliest Man, Discoveries in Tanganyika." *The Times,* March 9, 1935, p. 11.

———. "Earliest Man in East Africa." *East African Standard,* October 10, 1931, p. 1.

———. "Early Man in Kenya: Acheulean Camp Sites." *The Times,* October 4, 1946, p. 3.

———. "The Evolution of Man." Letter to the editor, *Discovery,* August 1964, pp. 48–49.

———. "Exploring 1,750,000 Years into Man's Past." *National Geographic,* October 1961, pp. 564–589.

———. Field Journals of the British-Kenya Miocene Expedition, 1947–49. NMK.

———. Field Journals from Losodok, 1951; Rusinga, 1954–55; and Mfwangano, 1955. NMK.

———. Field Journal, Olduvai Gorge, 1959. NMK.

———. "Finding the World's Earliest Man," *National Geographic,* September 1960, pp. 421–35.

———. "The Food of *Proconsul:* Fossilised Fruits and Seeds Which Are Expected to Throw Light on the Life of the Prehistoric Primates in Kenya." *Illustrated London News,* August 26, 1950, pp. 334–35.

———. "The Fossil Apes of Lake Victoria: Have Man's Ancestors Been Found?" *The Listener,* September 21, 1950, pp. 143–48.

———. "Fossil Human Remains from Kanam and Kanjera, Kenya Colony." *Nature,* October 10, 1936, p. 643.

———. "The Giant Animals of Prehistoric Tanganyika." *Illustrated London News,* June 19, 1954, pp. 1047–51.

———. "A Giant Child Among the Giant Animals of Olduvai?: A Huge Fossil Milk Molar Which Suggests That Chellean Man in Tanganyika May Have Been Gigantic." *Illustrated London News,* June 28, 1958, pp. 1103–1105.

———. "*Homo habilis, Homo erectus* and the Australopithecines." *Nature,* March 26, 1966, pp. 1279–81.

———. "The Juvenile Mandible from Olduvai." *Nature,* July 22, 1961, pp. 417–18.

———. *Kenya: Contrasts and Problems.* London: Methuen, 1936.

———. "Lower Dentition of *Kenyapithecus africanus.*" *Nature,* March 2, 1968, pp. 827–30.

———. "Lower Miocene Invertebrates from Kenya." *Nature,* April 12, 1952, pp. 624–25.

———. "Man's African Origin." *Annals of the New York Academy of Sciences,* vol. 96 (1962), pp. 495–503.

———. *Mau Mau and the Kikuyu.* London: Methuen, 1952.

———. "A Miocene Anthropoid Mandible from Rusinga, Kenya." *Nature,* September 18, 1943, pp. 319–20.

———. "New Finds at Olduvai Gorge." *Nature,* February 25, 1961, pp. 649–51.

———. "A New Fossil Skull from Olduvai." *Nature,* August 15, 1959, pp. 491–93.

———. " 'Newly' Recognized Mandible of *Ramapithecus.*" *Nature,* January 10, 1970, pp. 199–200.

———. "The Oldoway Human Skeleton." *Nature,* May 14, 1932, p. 578.

———. *Olduvai Gorge: A Report on the Evolution of the Hand-Axe Culture in Beds I–IV.* Cambridge: Cambridge University Press, 1951.

———. *Olduvai Gorge, 1951–1961: A Preliminary Report on the Geology and Fauna.* Cambridge: Cambridge University Press, 1965.

———. "A Pre-Historian's Paradise in Africa: Early Stone Age Sites at Olorgesailie." *Illustrated London News,* October 5, 1946, pp. 382–85.

———. *Proceedings of the First Pan-African Congress on Prehistory.* Oxford: Blackwell, 1952.

———. "Recent Discoveries at Olduvai Gorge, Tanganyika." *Nature,* April 19, 1958, pp. 1099–1103.

———. "Remains of Man with Oldowan Culture at Olduvai." In *Actes du IVe Congrès Panafricain de Préhistoire.* Tervuren, Belgium: Royal Museum of Central Africa, 1962.

———. Reports of the East African Archaeological Expedition, 1928–29. BMNH.

———. Reports of the East African Archaeological Expedition, 1931–32. NMK.

———. Reports of the East African Archaeological Expedition, Fourth Season, 1934–35. NMK.

———."Rich Fossil Beds of Fort Ternan." *The Times,* December 12, 1963, p. ix.

———. *The Southern Kikuyu.* 3 vols. London: Academic Press, 1977.

———. *Stone Age Africa: An Outline of Prehistory in Africa.* London: Oxford University Press, 1936.

———. *The Stone Age Cultures of Kenya Colony.* Cambridge: Cambridge University Press, 1931.

———. *The Stone Age Races of Kenya.* London: Oxford University Press, 1935.

———. "A 'Stupendous Discovery': The Fossil Skull from Olduvai. *Illustrated London News,* September 12, 1959, pp. 217–19.

———. "Time for a Firm Hand in Kenya." *Sunday Telegraph,* May 28, 1961, p. 14.

———. "Upper Miocene Primates from Kenya." *Nature,* May 11, 1968, p. 529, pp. 527–28.

———. "Was Kenya the Centre of Human Evolution?" *Illustrated London News,* August 24, 1946, pp. 198–201.

———. *White African.* London: Hodder & Stoughton, 1937.

———, ed. *Adam or Ape: A Sourcebook of Discoveries About Early Man.* Cambridge, Mass.: Schenkman, 1970.

Leakey, L. S. B.; J. F. Evernden; and G. H. Curtis. "Age of Bed I, Olduvai Gorge, Tanganyika." *Nature,* July 29, 1961, pp. 478–79.

Leakey, L. S. B., and Vanne Morris Goodall. *Unveiling Man's Origins: Ten Decades of Thought About Human Evolution.* Cambridge, Mass.: Schenkman, 1969.

Leakey, L. S. B., and M[ary]. D. Leakey. *Excavations at the Njoro River Cave: Stone Age Cremated Burials in Kenya Colony.* Oxford: Clarendon Press, 1950.

———. "Recent Discoveries of Fossil Hominids in Tanganyika: At Olduvai and Near Lake Natron," *Nature,* April 4, 1964, pp. 5–7.

Leakey, L. S. B., and W. E. Le Gros Clark. "British-Kenya Miocene Expeditions: Interim Report." *Nature,* February 5, 1955, p. 234.

———. "The Miocene Hominoidea of East Africa." *Fossil Mammals of Africa,* no. 1. London: British Museum of Natural History, 1951.

Leakey, L. S. B.; Ruth De Ette Simpson; and Thomas Clements. "Archaeological Excavations in the Calico Mountains, California: Preliminary Report." *Science,* March 1, 1968, pp. 1022–23.

Leakey, L. S. B.; P. V. Tobias; and J. R. Napier. "New Species of the Genus *Homo.*" *Nature,* April 4, 1964, pp. 7–9.

Leakey, Mary [D.]. *Africa's Vanishing Art: The Rock Paintings of Tanzania.* New York: Doubleday, 1983.

Leakey, M[ary]. D. "Cultural Patterns in the Olduvai Sequence." In *After the Australopithecines,* edited by Karl W. Butzer and Glynn L. Isaac. The Hague: Mouton, 1975.

———. *Disclosing the Past.* London: Weidenfeld and Nicolson, 1984.

———. "Discovery of Postcranial Remains of *Homo erectus* and Associated Artefacts in Bed IV at Olduvai Gorge, Tanzania." *Nature,* August 6, 1971, pp. 380–83.

———. "Excavation of Burial Mounds in Ngorongoro Crater." *Tanzania Notes and Records,* no. 66 (December 1966), pp. 123–36.

Leakey, Mary D. "Footprints in the Ashes of Time." *National Geographic,* April 1979, pp. 446–57.

———. *Olduvai Gorge: My Search for Early Man.* London: Collins, 1979.

———. *Olduvai Gorge.* Vol. 3, *Excavations in Beds I and II, 1960–1963.* Cambridge: Cambridge University Press, 1971.

———. *Olduvai Gorge.* Vol. 5, *Excavations in Beds III, IV & the Masek Beds.* Cambridge: Cambridge University Press, 1994.

———. "Olduvai Gorge, 1911–1975: A History of the Investigations." In *Geological Background to Fossil Man,* edited by W. W. Bishop. Edinburgh: Scottish Academic Press, 1978.

———. "Recent Discoveries of Hominid Remains at Olduvai Gorge, Tanzania." *Nature,* August 16, 1969, p. 756.

———. "A Review of the Oldowan Culture from Olduvai Gorge, Tanzania." *Nature,* April 30, 1966, pp. 462–66.

———. "Tracks and Tools." In *The Emergence of Man: Proceedings of the Royal Society and the British Academy.* Cambridge: Cambridge University Press, 1981.

Leakey, M[ary]. D.; R. J. Clarke; and L. S. B. Leakey. "New Hominid Skull from Bed I, Olduvai Gorge, Tanzania." *Nature,* July 30, 1971, pp. 308–12.

Leakey, M[ary]. D., and R. L. Hay. "Pliocene Footprints in the Laetolil Beds at Laetoli, Northern Tanzania." *Nature,* March 22, 1979, pp. 317–23.

Leakey, M[ary]. D.; R. L. Hay; G. H. Curtis; R. E. Drake; M. K. Jackes; and T. D. White. "Fossil Hominids from the Laetolil Beds." *Nature,* August 5, 1976, pp. 460–66.

Leakey Mary [D.], and Louis Leakey. *Some String Figures from North East Angola.* Pasadena: L. S. B. Leakey Foundation, 1981.

Leakey, Mary [D.], and Derek Roe. *Olduvai Gorge.* Vol. 5, *Excavations in Beds III, IV & the Masek Beds.* New York: Cambridge University Press, 1994.

Leakey, Meave G., and Richard E. *Koobi Fora Research Project,* Vol. 1, *The Fossil Hominids and an Introduction to Their Context, 1968–1974.* Oxford: Oxford University Press, 1978.

Leakey, Richard E. "In Search of Man's Past." *National Geographic,* May 1970, pp. 712–34.

———. *The Making of Mankind.* New York: E. P. Dutton, 1981.

———. *One Life: An Autobiography.* London: Michael Joseph, 1983.

———. "Skull 1470." *National Geographic,* June 1973, pp. 819–29.

Leakey, R[ichard]. E. F. "Evidence for an Advanced Plio-Pleistocene Hominid from East Rudolf, Kenya." *Nature,* April 13, 1973, pp. 447–50.

———. "Further Evidence of Lower Pleistocene Hominids from East Rudolf, North Kenya, 1971." *Nature,* June 2, 1972, pp. 264–69.

———. "Further Evidence of Lower Pleistocene Hominids from East Rudolf, North Kenya, 1972." *Nature,* March 16, 1973, pp. 170–73.

———. "Further Evidence of Lower Pleistocene Hominids from East Rudolf, North Kenya, 1973." *Nature,* April 19, 1974, pp. 653–56.

———. "New Hominid Fossils from the Koobi Fora Formation in Northern Kenya." *Nature,* June 17, 1976, pp. 574–76.

———. "New Hominid Remains and Early Artefacts from Northern Kenya." *Nature,* April 18, 1970, pp. 223–24.

Leakey, R[ichard]. E. [F.], and M[eave]. G. Leakey. "A New Miocene Hominoid from Kenya." *Nature,* November 13, 1986, pp. 143–46.

———. "A Second New Miocene Hominoid from Kenya." *Nature,* November 13, 1986, pp. 146–48.

Leakey, Richard E., and Roger Lewin. *Origins: The Emergence and Evolution of Our Species and Its Possible Future.* New York: E. P. Dutton, 1977.
———. *Origins Reconsidered.* New York: Doubleday, 1992.
———. *People of the Lake: Mankind and Its Beginnings.* New York: Doubleday, 1978.
Leakey, R[ichard] E. [F.], and A. C. Walker. "*Australopithecus, Homo erectus* and the Single Species Hypothesis." *Nature,* June 17, 1976, pp. 572–74.
———. "Further Hominids from the Plio-Pleistocene of Koobi Fora, Kenya." *American Journal of Physical Anthropology,* June 1985, pp. 135–63.
———. "The Hominids of East Turkana." *Scientific American,* August 1978, pp. 54–66.
———. "*Homo Erectus* Unearthed." *National Geographic,* November 1985, pp. 624–29.
———. "New Higher Primates from the Early Miocene of Buluk, Kenya." *Nature,* November 14, 1985, pp. 173–75.
———. "On the Status of *Australopithecus afarensis.*" *Science,* March 7, 1980, p. 1103.
———, eds. *The Nariokotome* Homo erectus *Skeleton.* Cambridge, Mass.: Harvard University Press, 1993.
Leakey, Wilfrida. "Priscilla in Darkest Africa." In *The Boat Train,* edited by M. A. Hamilton. London: George Allan and Unwin, 1934.

Other Sources

"After Years of Terror, a Comeback for Kenya." *U.S. News & World Report,* January 23, 1959, p. 105.
Baxter, E. J. "Mau Mau—The Terror That Has Come to Kenya and What It Means." Nairobi: East African News Review, Ltd., n.d.
Behrensmeyer, A. K. "Preliminary Geological Interpretation of a New Hominid Site in the Lake Rudolf Basin." *Nature,* April 18, 1970, pp. 225–26.
Berger, Rainer. "An Isotopic and Magnetic Study of the Calico Sites." In *Pleistocene Man at Calico,* edited by Walter C. Schuiling. Redlands, Calif.: San Bernardino County Museum Association, 1979, pp. 31–34.
Betts, Charles. "Man Evolved Like Animals." *Science News Letter,* April 17, 1965, p. 243.
———. "Man's Tree—A New Root?" *Science News Letter,* May 29, 1965, p. 346.
Boswell, P. G. H. "Human Remains from Kanam and Kanjera, Kenya Colony." *Nature,* March 9, 1935, p. 371.
———. "The Search for Man's Ancestry." *Nature,* November 18, 1950, p. 839.
———. Unpublished autobiography. Manuscript at the archives of the University of Liverpool and Imperial College of Science and Technology.
Bower, Bruce. "Family Feud: Enter the 'Black Skull.'" *Science News,* January 24, 1987, pp. 58–59.
"The British Association/Leeds Meeting Opened." *The Times,* August 1, 1928, p. 15.
Brockway, Fenner. *African Journeys.* London: Gollancz, 1955.
Broom, Robert. *Finding the Missing Link.* London: Watts, 1950.
———. "A New Fossil Anthropoid Skull from South Africa." *Nature,* September 19, 1936, pp. 476–77.
Brown, F.; J. Harris; R. Leakey; and A. Walker. "Early *Homo erectus* Skeleton from West Lake Turkana, Kenya." *Nature,* August 29, 1985, pp. 788–92.
Bunney, Sarah. "Fossilmanship in East Africa." *Nature,* September 3, 1971, p. 20.
Burkitt, M. C. *South Africa's Past in Stone and Paint.* Cambridge: Cambridge University Press, 1928.
Butzer, Karl. "Dawn in the Rudolf Basin." *South African Journal of Science,* October 19, 1973, pp. 292–93.
Butzer, Karl, and Glynn Isaac, eds. *After the Australopithecines.* The Hague: Mouton, 1975.
Campbell, Bernard. "Just Another 'Man-Ape'?" *Discovery,* June 1964, pp. 37–38.
Clark, J. Desmond, "Louis Seymour Bazett Leakey." *Proceedings of the British Academy, 1973,* p. 456.
———. "A Personal Memoir." In *A History of African Archaeology,* edited by Peter Robertshaw. London: James Currey, 1990.
Clark, Ronald W. *The Life and Work of J. B. S. Haldane.* New York: Coward-McCann, 1968.
Clarke, Alan. "Pritt Protest Halts Mau Mau Trail." *Daily Herald,* January 8, 1953, p. 2.
Cole, Sonia. *Leakey's Luck: The Life of Louis Seymour Bazett Leakey, 1903–1972.* London: Collins, 1975.
Cooke, Basil. "Suidae from Plio-Pleistocene Strata of the Rudolf Basin." In *Earliest Man and Environments in the Lake Rudolf Basin,* edited by Yves Coppens, et al. Chicago: University of Chicago Press, 1976.

Curtis, G. H.; T. Cerling; Drake; and Hampel. "Age of KBS Tuff in Koobi Fora Formation, East Rudolf, Kenya." *Nature,* December 4, 1975, pp. 395–98.
Cutler, W. E. Field diary from British Museum Dinosaur Expedition to Tendagaru, Tanganyika, 1924–25. BMNH.
Daniel, Glyn. "Dr. Louis Leakey." *The Times,* October 12, 1972, p. 18.
Dart, Raymond. *Adventures with the Missing Link.* London: Hamish Hamilton, 1959.
Davis, P. R. "Hominid Fossils from Bed I, Olduvai Gorge, Tanganyika: A Tibia and Fibula." *Nature,* March 7, 1964, pp. 967–68.
Day, Michael [H.]. *Guide to Fossil Man.* Chicago: University of Chicago Press, 1986.
Day, M[ichael]. H.; M[ary]. D. Leakey; and C. Magori. "A New Hominid Fossil Skull (L.H. 18) from the Ngaloba Beds, Laetoli, Northern Tanzania." *Nature,* March 6, 1980, pp. 55–56.
Day, M[ichael]. H.; Mary D. Leakey; and Todd R. Olson. "On the Status of *Australopithecus afarensis.*" *Science,* March 7, 1980, pp. 1102–1103.
Day, M[ichael]. H., and J. R. Napier. "Hominid Fossils from Bed I, Olduvai Gorge, Tanganyika: Fossil Foot Bones." *Nature,* March 7, 1964, pp. 669–70.
"Discovery Claimed to Put Man's Ancestry Back 6m. Years." *The Times,* January 16, 1967, p. 8.
Dolphin, Gabrielle. Journal from Kenya. Courtesy of Gabrielle Dolphin.
"Early Human Remains in East Africa: Report of a Conference at Cambridge Convened by the Royal Anthropological Institute." *Man,* no. 65 (1933), p. 210.
"Early Man in East Africa: Further Investigation." *Nature,* November 10, 1934, p. 730.
East Rudolf Expedition Field Diaries, 1968. NMK.
Eliot, Charles. *The East African Protectorate.* London: Edward Arnold, 1905.
"Evolution Revolution." *Science News,* March 13, 1976, pp. 164–65.
Ferrell, J. E. "Bone Wars." *Image* (Sunday magazine of the *San Francisco Examiner*), August 23, 1987, pp. 14–35.
"Finder Says Fossil Links Ape and Man." *New York Times,* September 4, 1959, p. 1.
Findlay, George. *Dr. Robert Broom, F.R.S.* Cape Town: A. A. Balkema, 1972.
Fitch, F. J., and J. A. Miller. "Radioisotopic Age Discriminations of Lake Rudolf Artefact Site." *Nature,* April 18, 1970, pp. 226–28.
Fossey, Dian. *Gorillas in the Mist.* Boston: Houghton Mifflin, 1983.
"Fossil Apes in Kenya/Finds Near Lake Victoria." *The Times,* December 30, 1947, p. 3.
"Fossil Finds Diversify Ancient Apes." *Science News,* November 22, 1986, p. 324.
"Fossils in Ethiopia Said to Show Man as Million Years Older Than Believed." *New York Times,* October 26, 1974, p. 1.
Fuchs, Vivian. Field diary from Louis Leakey's 1931 expedition. Courtesy of Sir Vivian Fuchs.
Galdikas, Biruté M. F. *Reflections of Eden.* Boston: Little, Brown, 1995.
Galloway, Paul. "The Evolution Revolution." *Chicago Sun-Times,* August 26, 1979, pp. 16–19.
Geminus. "It Seems to Me." *New Scientist,* March 2, 1961, p. 552.
Goodall, Jane. *The Chimpanzees of Gombe: Patterns of Behavior.* Boston: Harvard University Press, 1986.
———. *In the Shadow of Man.* Boston: Houghton Mifflin, 1971.
Goodall, Vanne. "The Legacy of L. S. B. Leakey." *TV Guide,* January 7, 1978, p. 22.
"The Great Iam." *Time,* September 30, 1966, pp. 110–12.
Gwynne, Peter. "Finding Eve's Cousin." *Newsweek,* January 29, 1979, p. 81.
———. "The Oldest Man." *Newsweek,* March 22, 1976, p. 59.
Halberstam, David. "Bones of Earliest Human, a Child, Reported Dug Up in Tanganyika." *New York Times,* February 25, 1961, p. 1.
Hall, Francis George. Letters, 1892–1895. 3 vols. Rhodes House, Oxford, England.
Hawkes, Jacquetta. *Mortimer Wheeler: Adventurer in Archaeology,* London: Weidenfeld & Nicolson, 1982.
Hay, Richard L. *Geology of the Olduvai Gorge: A Study of Sedimentation in the Semiarid Basin.* Berkeley: University of California Press, 1976.
———. "The KBS Tuff Controversy May Be Ended." *Nature,* April 3, 1980, p. 401.
———. "Olduvai Gorge: A Case History in the Interpretation of Hominid Paleoenvironments in East Africa." In *East Africa: Establishment of a Geologic Framework for Paleoanthropology,* edited by L. F. Laporte. Boulder, Colo.: Geological Society of America, 1990.
Hayes, Harold T. P. *The Dark Romance of Dian Fossey.* New York: Simon & Schuster, 1990.
Hillaby, John. "Fossil Gives Clue to the 'First Man.' " *New York Times,* October 8, 1959, p. 44.
Hobley, C. W. "W. E. Cutler" (obituary). *Nature,* September 12, 1925, p. 406.

"Hominid Bones: Old and Firm at 3.75 Million." *Science News,* November 8, 1975, p. 292.

Hopwood, A. Tindell. "Miocene Primates from Kenya." *Journal of the Linnaean Society of London (Zoology),* 260 (1933), pp. 437–64.

Howell, F. C. "Pliocene/Pleistocene Hominidae in Eastern Africa: Absolute and Relative Ages." In *Calibration of Hominoid Evolution,* edited by W. W. Bishop and J. A. Miller. New York: Scottish Academic Press for the Wenner-Gren Foundation for Anthropological Research, 1972.

Huxley, Elspeth. *Nellie: Letters from Kenya.* London: Weidenfeld and Nicolson, 1980.

———. *Out in the Midday Sun: My Kenya.* New York: Viking, 1987.

———. *White Man's Country.* 2 vols. London: Chatto and Windus, 1956.

Isaac, Glynn L. *Olorgesailie: Archeological Studies of a Middle Pleistocene Lake Basin in Kenya.* Chicago: University of Chicago Press, 1977.

Isaac G[lynn]. L.; R. E. F. Leakey; and A. K. Behrensmeyer. "Archaeological Traces of Early Hominid Activities, East of Lake Rudolf, Kenya. *Science,* September 17, 1971, pp. 1129–34.

Jaffe, Andrew. "Paleontology's Daring Young Man." *International Wildlife,* May 1975, pp. 4–11.

Jeffries, Michael. "Briton Finds 'Missing Link' Skull." *Evening Standard,* November 9, 1972, p. 1.

Johanson, Donald [C.]. "Ethiopia Yields First Family." *National Geographic,* December 1976, pp. 790–811.

Johanson, Donald C., and Maitland Edey. *Lucy: The Beginnings of Humankind.* New York: Simon & Schuster, 1981.

Johanson, Donald C., and James Shreeve. *Lucy's Child.* New York: Morrow, 1989.

Johanson, D[onald]. C., and M. Taieb. "Plio-Pleistocene Hominid Discoveries in Hadar, Ethiopia." *Nature,* March 25, 1976, pp. 293–97.

Johanson, D[onald]. C., and T. D. White. "On the Status of *Australopithecus afarensis." Science,* March 7, 1980, pp. 1104–1105.

———. "A Systematic Assessment of Early African Hominids." *Science,* January 26, 1979, pp. 321–30.

Johanson, Donald C.; Tim D. White; and Yves Coppens. "A New Species of the Genus *Australopithecus* (Primates; *Hominidae*) from the Pliocene of Eastern Africa." *Kirtlandia,* no. 28 (1978), pp. 2–14.

Johanson, Donald C., et al. "New Partial Skeleton of *Homo habilis* from Olduvai Gorge, Tanzania." *Nature,* May 21, 1987, pp. 205–9.

"Kattwinkel's Heirs." *Time,* March 10, 1961, p. 44.

Keith, Arthur. *The Antiquity of Man.* London: Williams & Norgate, 1915.

———. "Australopithecinae or Dartians." *Nature,* March 15, 1947, p. 377.

———. *An Autobiography.* New York: Philosophical Library, 1950.

———. "Early East Africans." *Nature,* February 2, 1935, pp. 163–64.

———. *New Discoveries Relating to the Antiquity of Man.* London: Williams & Norgate, 1931.

———. *A New Theory of Human Evolution.* London: Watts & Co., 1948.

"Keith, Sir Arthur." In *Lives of the Fellows of the Royal College of Surgeons of England 1952–1964,* edited by H. O. B. Robinson and W. R. LeFanu. Edinburgh: E. & S. Livingstone, 1970.

Kelley, Jay, and David Pilbeam. "The Dryopithecines: Taxonomy, Comparative Anatomy, and Phylogeny of Miocene Large Hominoids." In *Comparative Primate Biology.* Vol. 1, *Systematics, Evolution and Anatomy,* edited by J. Erwin. New York: Alan R. Liss, 1986.

Kenyatta, Jomo. *Facing Mount Kenya: The Tribal Life of the Gikuyu.* London: Secker and Warburg, 1953.

Klein, Richard. *The Human Career.* Chicago: University of Chicago Press, 1989.

Koobi Fora Field Journals, 1969–1972. NMK.

Lapping, Brian. *End of Empire.* New York: St. Martin's Press, 1985.

"The Leakey Footprints: An Uncertain Path." *Science News,* March 31, 1979, pp. 196–97.

"Leakey: The Man with the Million Year Mind." *Rand Daily Mail,* October 3, 1972.

"Leakey's Finds: Tracks of an Ancient Ancestor." *Time,* March 6, 1978, p. 53.

"The Leakeys' Telltale Skulls." *Newsweek,* July 15, 1974, p. 72.

Le Gros Clark, Wilfrid [E.] *Chant of Pleasant Exploration.* London: E. & S. Livingstone, 1968.

———. "Fossil Hominoids from Kenya." *Proceedings of the Zoological Society of London,* vol. 122 (August 1952), pp. 273–86.

———. "The Importance of the Fossil Australopithecinae in the Study of Human Evolution." *Science Progress,* vol. 35 (1947), pp. 377–95.

———. *Man-apes or Ape-men: The Story of Discoveries in Africa.* New York: Holt, Rinehart and Winston, 1967.

———. "New Palaeontological Evidence Bearing on the Evolution of the Hominoidea." *Quarterly Journal of the Geological Society,* vol. 105 (1950), pp. 225–64.

———. "Observations on the Anatomy of the Fossil Australopithecinae." *Journal of Anatomy,* vol. 81 (1947), pp. 300–32.

Le Gros Clark, W[ilfred]. E., and L. S. B. Leakey. *Fossil Mammals of Africa,* no. 1. London: British Museum of Natural History, 1951.

Lewin, Roger. *Bones of Contention.* New York: Simon & Schuster, 1987.

Lieberman, D. E.; D. R. Pilbeam; and B. A. Wood. "A Probablistic Approach to the Problem of Sexual Dimorphism in *Homo habilis." Journal of Human Evolution,* vol. 17 (1988), pp. 503–12.

"The Life and Times of Dear Boy." *The Sunday Post,* May 22, 1960.

"Likely to Be Long Hearing Says Magistrate." *East African Standard,* December 5, 1952.

Linden, Eugene. "Richard the Lionhearted." *Time,* July 19, 1993, p. 51.

Loveridge, Arthur. *Many Happy Days I've Squandered.* New York: Harper & Brothers, 1932.

"Man and His Ancestors." *The Times,* October 21, 1933, p. 12.

"Mau Mau at Close Quarters." *The Times,* November 25, 1954, p. 7.

Mayr, Ernst. "Taxonomic Categories in Fossil Hominids." In *Cold Spring Harbor Symposia on Quantitative Biology.* Vol. XV, *Origin and Evolution of Man.* Cold Spring Harbor, N.Y.: The Biological Laboratory, 1950.

McHenry, H. M. "How Big Were the Early Hominids?" *Evolutionary Anthropology* 1 (1992), pp. 15–20.

———. "Tempo and Mode in Human Evolution." *Proceedings of the National Academy of Sciences,* vol. 91 (1994), pp. 6780–86.

McRae, Michael. *Continental Drifter.* New York: Lyons & Burford, 1993.

Millar, Ronald William. *The Piltdown Men.* New York: St. Martin's Press, 1972.

"Miocene Ape's Skull at Oxford, Mrs. Leakey Hands Over Her Find." *Evening News,* November 1, 1948.

" 'Missing Link' Found in Africa." *New York Times,* January 23, 1947, p. 25.

"Missing Link Proof Goes on Show." *News Chronicle,* October 8, 1959.

"Modern Man Not So Old/African Discovery Questioned/Dr. Leakey and a Geologist/Piltdown Skull Still Holds Its Place." *Morning Post,* March 7, 1935.

Montgomery, Sy. *Walking with the Great Apes.* Boston: Houghton Mifflin, 1991.

"More Secrets from the Past: Oboyoboi Gorge." *Punch,* March 8, 1961.

Mowat, Farley. *Woman in the Mists.* New York: Warner Books, 1987.

" 'Murder Verdict' on Hominid." *The Times,* February 26, 1961, p. 9.

Murray-Brown, Jeremy. *Kenyatta.* New York: E. P. Dutton, 1973.

Napier, J. R. "Fossil Hand Bones from Olduvai Gorge." *Nature,* November 3, 1962, pp. 409–11.

"New Species of Man: Ancestors from 'Afar.' " *Science News,* January 20, 1979, p. 86.

Ng'weno, Hilary. "Conversations with Philip and Richard Leakey." *The Nairobi Times,* March 2, 1980.

Oakley, Kenneth. "The Earliest Tool-makers." *Antiquity,* March 1956, p. 7.

"Oldest Fragment of Man Disputed/British Professor Reports He Found in Africa No Support for Leakey Discovery Claim." *New York Times,* March 8, 1935, p. 23.

"The Oldest Man." *Time,* November 10, 1985, p. 93.

"The Oldest Man?" *Newsweek,* November 20, 1972, p. 137.

"Once More into the Past." *Science News,* October 16, 1971, p. 259.

Ottoway, D. B. "3-Million-Year-Old Human Fossils Found." *International Herald Tribune,* October 28, 1974.

"The Pan-African Congress on Prehistory." *Nature,* February 15, 1947, pp. 216–18.

Patterson, Tom. "Working Scientist Leakey No 'Head in Clouds' Scholar." *Riverside Press-Enterprise,* May 23, 1963.

Payne, Melvin M. "Family in Search of Prehistoric Man." *National Geographic,* February 1965, pp. 194–231.

Petit, Charles. " 'Exciting' Find on Human Ancestor." *San Francisco Chronicle,* May 21, 1987, p. 3.

———. "Fossil Hunter's Find May Shake Man's Family Tree." *San Francisco Chronicle,* November 13, 1986, p. 4.

Pfeiffer, John. "Man Through Time's Mists." *Saturday Evening Post,* December 3, 1966, p. 50.

Pilbeam, David. "The Earliest Hominids." *Nature,* September 28, 1968, pp. 1335–38.
———. "Notes on *Ramapithecus,* the Earliest Known Hominid, and *Dryopithecus." American Journal of Physical Anthropology,* vol. 25 (1966), pp. 1–5.
Plummer, Thomas, and Richard Potts. "Hominid Fossil Sample from Kanjera, Kenya: Description, Provenance, and Implications of New and Earlier Discoveries." *American Journal of Anthropology,* vol. 96 (January 1995), pp. 7–23.
Potts, Richard. *Early Hominid Activities at Olduvai.* New York: Aldine de Gruyter, 1988.
"Prehistoric Man in East Africa/Opal Mining in 4,000 B.C." *The Times,* May 22, 1938, p. 13.
"Puzzling Out Man's Ascent." *Time,* November 7, 1977, p. 53.
Quiggin, A. Hingston. *Haddon, the Headhunter.* Cambridge: Cambridge University Press, 1942.
Reader, John. *Missing Links.* Boston: Little, Brown, 1981.
Reed, Charles A. "A Short History of the Discovery and Early Study of the Australopithecines: The First Find to the Death of Robert Broom (1924–1951)." In *Hominid Origins: Inquiries Past and Present,* edited by Kathleen J. Reichs. New York: University Press of America, 1978.
Reinert, Jeanne. "The Man Dr. Leakey Dug Up." *Science Digest,* November 1966, p. 76.
Rensberger, Boyce. "The Face of Evolution: The Leakeys' Findings Show How Our Tools Have Shaped Our Brains." *New York Times Magazine,* March 3, 1974, p. 13.
———. "Man Traced 3.75 Million Years by Fossils Found in Tanzania." *New York Times,* October 31, 1975, p. 1.
———. "Prehistoric Footprints of Man-Like Creatures Found." *New York Times,* March 22, 1979, p. A16.
———. "Rival Anthropologists Divide on 'Pre-Human' Find." *New York Times,* February 18, 1979, p. 1.
"Rethinking Human Evolution." *Nature,* December 9, 1976, pp. 507–508.
Robins, Eric. "Anthropologist Richard Leakey Tracks the Grandfather of Man in an African Boneyard." *People,* January 8, 1979, p. 30.
Rosberg, Carl G., Jr., and John Nottingham. *The Myth of "Mau Mau": Nationalism in Kenya.* Stanford: Hoover Institution Press, 1966.
Russell, Loris S. *Dinosaur Hunting in Western Canada.* Toronto: Royal Ontario Museum, University of Toronto, 1966.
Schick, Kathy D., and Nicholas Toth. *Making Silent Stones Speak.* New York: Simon & Schuster, 1993.
Schuiling, Walter C., ed. *Pleistocene Man at Calico.* Redlands, Calif.: San Bernardino County Museum Association, 1979.
"Science in East Africa: Mr. Leakey's Work." *The Times,* April 30, 1937, p. 21.
"Should Fossil Hominids Be Reclassified?" *Nature,* April 19, 1974, p. 635.
Simpson, George Gaylord. *Concession to the Improbable: An Unconventional Autobiography.* New Haven: Yale University Press, 1978.
"A Skull at Least 600,000 Years Old." *The Guardian,* October 8, 1959.
"Skull May Rattle Theories of Evolution." *San Francisco Chronicle,* August 7, 1986, p. 5.
Slater, Montagu. *The Trial of Jomo Kenyatta.* London: Secker & Warburg, 1955.
Stringer, Christopher, and Clive Bamble. *In Search of the Neanderthals.* New York: Thames and Hudson, 1993.
Sullivan, Walter. "Bone Found in Kenya Indicates Man Is 2.5 Million Years Old." *New York Times,* January 14, 1967, p. 1.
———. "Skull Pushes Back Man's Origins One Million Years." *New York Times,* November 10, 1972, p. 1.
Taieb, Maurice. *Sur la terre des premiers hommes.* Paris: Rober Laffont, 1985.
Taieb, Maurice, et al. "Geological and Palaeontological Background of Hadar Hominid Site, Afar, Ethiopia." *Nature,* March 25, 1976, pp. 289–97.
"That Skull." *Daily Herald,* November 19, 1948.
Thurston, Anne. "Interview with Richard Leakey." *The Humanist,* January/February 1979, pp. 24–26.
Tildesley, M. L. "The Status of the Kanam Mandible and the Kanjera Skulls." *Man,* vol. 33 (December 1933), pp. 200–211.
Tobias, Phillip V. "The Kanam Jaw." *Nature,* March 26, 1960, p. 946.
———. "Memorable Moments with Louis Leakey." *The L. S. B. Leakey News,* no. 12, Fall 1978.
———. "The Olduvai Bed I Hominine with Special Reference to Its Cranial Capacity." *Nature,* April 4, 1964, pp. 3–4.

———. *Olduvai Gorge.* Vol. 4, *The Skulls, Endocasts and Teeth of* Homo habilis. Cambridge: Cambridge University Press, 1991.

———. "A Re-Examination of the Kanam Mandible." *Proceedings of the 4th Pan-African Congress of Prehistory.* Tervuren, Belgium: Royal Museum of Central Africa, 1962.

Tuff, Barbara. "The Ascent of Man, Out of the Past: Dr. Leakey Looks to the Future." *Science News,* February 25, 1967, p. 188.

Tuttle, Russell H. "Paleoanthropology Without Inhibitions." *Science,* May 15, 1981, p. 798.

Visitors' Guide to Kariandusi Prehistoric Site. Nairobi: National Museums of Kenya, n.d.

Visitors' Guide to the Hyrax Hill Site. Nairobi: National Museums of Kenya, n.d.

Walker, A.; R. E. Leakey; J. M. Harris; and J. F. Brown. "2.5-Myr *Australopithecus boisei* from West of Lake Turkana, Kenya." *Nature,* August 7, 1986, pp. 517–22.

Washburn, Sherwood. "Analysis of Primate Evolution." In *Cold Spring Harbor Symposia on Quantitative Biology.* Vol. XV, *Origin and Evolution of Man.* Cold Spring Harbor, N.Y.: The Biological Laboratory, 1950.

Wayland, E. J. The Age of the Oldoway Human Skeleton. NMK.

White, Tim D. "New Fossil Hominids from Laetolil, Tanzania." *American Journal of Physical Anthropology,* vol. 46 (March 1977), pp. 197–229.

White, T. D., and J. M. Harris. "Suid Evolution and Correlation of African Hominid Localities." *Science,* October 7, 1977, p. 13–21.

"The White Kikuyu." *The Observer,* July 18, 1954, p. 3.

Wilford, John Noble. "The Leakeys: Rocking the 'Cradle of Mankind.'" *New York Times,* October 30, 1984, III, p. 1.

———. *The Riddle of the Dinosaur.* New York: Knopf, 1985.

Wood, B. A. "Early Homo: How Many Species?" In *Species, Species Concepts and Primate Evolution,* edited by W. H. Kimbel and L. B. Martin. New York: Plenum Press, 1993, pp. 485–522.

———. *Koobi Fora Research Project.* Vol. 4, *Hominid Cranial Remains.* Oxford: Clarendon Press, 1991.

Woodward, A. S. "Missing Links Among Extinct Animals." *Report of the British Association.* Birmingham: British Association, 1913.

Wymer, Norman. *The Man from the Cape.* London: Evans Brothers, 1959.

Zuckerman, Professor Lord, ed. *The Concepts of Human Evolution.* London: Academic Press, 1977.

Index